PROPERTY OF NAPP SYSTEMS INC.
TSG DEPT.

Plastics Extrusion Technology

Edited by Friedhelm Hensen

Plastics Extrusion Technology

Edited by Friedhelm Hensen
In collaboration with W. Knappe and H. Potente

With contributions from
H. Bongaerts, S. Braun, Prof. Dr. H. Breuer, Prof. Dr. H.-G. Fritz, H.-J. Gohlisch, Prof. Dr. F. Hensen,
H. Herrmann, R. Hessenbruch, P. John, Dr. K. Kircher, Dr. P. Klenk, K.-D. Kolossow, Dr. W. May,
Dr. M. Mayer, W. Mücke, Prof. Dr. W. Predöhl, Dr. F. Ramm, P. Reitemeyer, K.-F. Roesch,
W. Rüger, H. P. Schneider, E. Schöllkopf, P. Stamprech, H. Tenner, G. Winkler

Translated and revised
by D. Sandiford

With 708 Figures and 84 Tables

Hanser Publishers, Munich Vienna New York

Distributed in the United States of America by
Oxford University Press, New York
and in Canada by
Oxford University Press, Canada

Distributed in USA by
Oxford University Press
200 Madison Avenue, New York, N.Y. 10016

Distributed in Canada by
Oxford University Press, Canada
70 Wynford Drive, Don Mills, Ontario M3C IJ9

Distributed in all other countries by
Carl Hanser Verlag
Kolbergerstraße 22
D-8000 München 80

The use of general descriptive names, trademarks, etc., in this publication, even if the former are not especially identified, is not to be taken as a sign that such names, as understood by the Trade Marks and Merchandise Marks Act, may accordingly be used freely by anyone.

While the advice and information in this book are believed to be true and accurate at the date of going to press, neither the authors nor the publishers can accept any legal responsibility for any errors or omissions that may be made. The publisher makes no warranty, express or implied, with respect to the material contained herein.

Library of Congress Cataloging-in-Publication Data

Plastics extrusion technology / edited by F. Hensen : with
  contributions from H. Bongaerts . . . [et al.].
    p.   cm.
  Bibliography: p.
  Includes index.
  ISBN 0-19-620760-2
  1. Plastics—Extrusion.   I. Hensen, F. (Friedhelm)
  II. Bongaerts, H.
  TP1175.E9P56 1988
  668.4′13——dc19   88-16266
                                                                 CIP

CIP-Titelaufnahme der Deutschen Bibliothek

Plastics extrusion technology / ed. by Friedhelm Hensen in
collab. with W. Knappe and H. Potente. With contributions
from H. Bongaerts . . . — Munich ; Vienna ; New York : Hanser ;
New York : Oxford University Press, 1988
  Dt. Ausg. u.d.T.: Handbuch der Kunststoff-Extrusionstechnik. Bd. 2
  ISBN 3-446-14589-3 (Hanser)
  ISBN 0-19-520760-2 (Oxford Univ. Press)
  NE: Hensen, Friedhelm [Hrsg.]; Bongaerts, Horst [Mitverf.]

ISBN 3-446-14589-3 Carl Hanser Verlag, Munich, Vienna, New York
ISBN 0-19-520760-2 Oxford University Press, New York
Library of Congress Catalog Card Number 87-060832

All rights reserved. No part of this book may be reproduced or transmitted in any form or by any means, electronic or mechanical, including photocopying or by any information storage and retrieval system, without permission from the publisher.

Copyright © Carl Hanser Verlag, Munich, Vienna, New York 1988
Printed in the Federal Republic of Germany by Konrad Triltsch,
Druck- und Verlagsanstalt Würzburg GmbH, Würzburg

List of editors and authors

Dipl.-Ing. H. Bongaerts, c/o Barmag Barmer Maschinenfabrik AG, D-5630 Remscheid 11
S. Braun, c/o Barmag Barmer Maschinenfabrik AG, D-5630 Remscheid 11
Prof. Dr.-Ing. H. Breuer, c/o Fachhochschule Niederrhein, D-4150 Krefeld 1
Prof. Dr.-Ing. H.-G. Fritz, c/o Institut für Kunststofftechnologie, Universität Stuttgart,
    D-7000 Stuttgart 1
Dipl.-Ing. H.-J. Gohlisch, c/o Paul Troester Maschinenfabrik, D-3000 Hannover 89
Prof. Dr.-Ing. F. Hensen, c/o Barmag Barmer Maschinenfabrik AG, D-5630 Remscheid 11
Dipl.-Ing. H. Herrmann, c/o Werner & Pfleiderer, 7000 Stuttgart 30
Dipl.-Ing. R. Hessenbruch, c/o Barmag Barmer Maschinenfabrik AG, D-5630 Remscheid 11
Ing. grad P. John, c/o Hoechst Aktiengesellschaft, D-6230 Frankfurt/M. 80
Dr.-Ing. K. Kircher, c/o Bayer AG, D-5090 Leverkusen, Bayerwerk
Dr.-Ing. P. Klenk, c/o P. K. H. T. S., Plastik Know-How u. Technologie Service,
    D-8000 München 90
Prof. Dr. W. Knappe, c/o Institut für Kunststoffverarbeitung, Montanuniversität,
    A-8700 Leoben
K.-D. Kolossow, c/o Hermann Berstorff Maschinenfabrik GmbH, D-3000 Hannover 1
Dr.-Ing. W. May, c/o Paul Troester Maschinenfabrik, D-3000 Hannover 89
Dr.-Ing. M. Mayer, c/o Barmag Barmer Maschinenfabrik AG, D-5630 Remscheid 11
Dipl.-Ing. W. Mücke, c/o Filterwerk Mann & Hummel GmbH, D-7140 Ludwigsburg
Prof. Dr.-Ing. H. Potente, Universität Paderborn, Fachbereich 10, D-4790 Paderborn
Prof. Dr.-Ing. W. Predöhl, c/o Reifenhäuser GmbH & Co, D-5210 Troisdorf 15
Dr.-Ing. F. Ramm, c/o Paul Troester Maschinenfabrik, D-3000 Hannover 89
P. Reitemeyer, c/o Reifenhäuser GmbH & Co, D-5210 Troisdorf 15
Dipl.-Ing. K.-F. Roesch, c/o Polytype Maschinenfabrik AG, CH-1701 Fribourg
Dipl.-Ing. W. Rüger, c/o Paul Troester Maschinenfabrik, D-3000 Hannover 89
Dipl.-Ing. H. P. Schneider, c/o Krauss-Maffei AG, D-8000 München 50
Dipl.-Ing. E. Schöllkopf, Polytype Maschinenfabrik AG, CH-1701 Fribourg
Ing. P. Stamprech, c/o Hoechst Aktiengesellschaft, Werk Gendorf, D-8269 Burgkirchen
Ing. H. Tenner, c/o Paul Leistritz Maschinenfabrik GmbH, D-8500 Nürnberg
Dipl.-Ing. G. Winkler, c/o Windmöller & Hölscher Maschinenfabrik, D-4540 Lengerich

Translation editor

D. Sandiford, M. A., Plastic Communication International, CH-1007 Lausanne

Preface

In this volume, "Plastics Extrusion Technology", we present the state of the art in the technical design and operation of plastics extrusion lines.
The chapters of this book have been written by well-known experts in the various fields of plastics extrusion technology. As the aim was to treat the subject in as practice-oriented a way as possible, the choice of authors was not limited to polymer scientists: professionals with many years of experience and extensive knowledge in their own special fields, some of whom have contributed to the development of the systems dealt with in this book, were also invited to contribute.
Contents and presentation aim at making a helpful tool available to technologists and scientists, and providing an authoritative learning aid and reference book for the student. Clarity and terminology also make this book understandable for non-experts, and worth reading for those with a general technical interest.
As one of the co-authors I consider it a pleasure to thank all those involved for their cooperation in the preparation of this book.

Friedhelm Hensen   Remscheid, Summer 1988

Contents

1 **Introduction** 25

2 **Compounding lines** (H. Herrmann) 26

   2.1 Purpose of plastics compounding and the unit operations 26
   2.2 Development and economic importance of compounding technology 26
   2.3 Compounding lines 27
   2.4 Compounding of polyolefins 29
      2.4.1 Plant for melt feed 29
      2.4.2 Systems for concentration of melt solutions 30
      2.4.3 Compounding systems for polyolefin powder 32
      2.4.4 Screen changer and pelletizing systems 35
   2.5 Compounding of PVC and other temperature-sensitive polymers 37
   2.6 Compounding of polystyrene and styrene copolymers 41
   2.7 Compounding of engineering plastics 41
      2.7.1 Systems for thermoplastic engineering plastics 42
      2.7.2 Systems for production of filled thermoplastics 42
      2.7.3 Systems for the production of reinforced thermoplastics 45
      2.7.4 Pelletizing systems for thermoplastic engineering plastics 48
      2.7.5 Systems for thermosets 48
   2.8 Compounding of elastomers 49
   2.9 Direction of development 50
   References for Chapter 2 54

3 **Extrusion of pipes, profiles and cables** (W. Predöhl, P. Reitemeyer) 56

   3.1 Manufacture of pipes 56
      3.1.1 Market significance and products 56
      3.1.2 Extrusion of pipes 57
         3.1.2.1 Lines for the manufacture of pipes from rigid PVC 57
         3.1.2.1.1 Extruders for producing pipes from rigid PVC 57
         3.1.2.1.2 Pipe die heads 59
         3.1.2.2 Lines for producing pipes from polyolefins 61
         3.1.2.2.1 Extruder for polyolefin pipes 61
         3.1.2.2.2 Pipe die heads 62
         3.1.2.3 Calibration systems for smooth-wall pipes 65
         3.1.2.3.1 Calibrating inserts in vacuum tank-calibration units 66
         3.1.2.3.2 Internal air pressure calibration 67
         3.1.2.3.3 Design of pipe die heads and sizing dies 68
         3.1.2.4 Cooling baths 68
         3.1.2.5 Pipe haul-off units 72
         3.1.2.6 Cutting units 73
         3.1.2.7 Winding, discharge, socketing 74
         3.1.2.8 Measurement and control 76
         3.1.2.9 Corrugated pipe lines 78
         3.1.2.9.1 Products 78
         3.1.2.9.2 Manufacturing Processes 78
   3.2 Extrusion of profiles 79
      3.2.1 Market significance 79

	3.2.2	Products	80
		3.2.2.1 Rigid PVC profiles for the construction- and furniture industries	81
		3.2.2.2 Plasticized PVC profiles for the automobile industry	81
		3.2.2.3 Profiles from other materials	81
	3.2.3	Processes for the manufacture of profiles of differing shapes and dimensions from various raw materials	83
		3.2.3.1 Profile extrusion die	84
		3.2.3.2 Calibration	85
		3.2.3.2.1 Vacuum calibration	85
		3.2.3.2.2 Drawing calibration	86
		3.2.3.2.3 Short vacuum calibration	87
		3.2.3.2.4 Segment- or template calibration	87
		3.2.3.2.5 Vacuum block calibration	88
		3.2.3.2.6 Vacuum tank calibration	89
		3.2.3.2.7 Cooling die processes	89
		3.2.3.3 Measurement and control	90
3.3	Extrusion lines for the cable industry		90
	3.3.1	Extrusion line for the insulation of optical fibers	90
	3.3.2	Extrusion line for the insulation of fine wires up to 0.5 mm² cross-section	91
	3.3.3	Extrusion line for the insulation of single wires and stranded wires	92
	3.3.4	Extrusion line for the production of cables and wires from crosslinked polyethylene	92
	3.3.5	Extrusion line for the insulation of telephone wires	92
	3.3.6	Extrusion line for the sheathing of conductors and fine cables	93
	3.3.7	Extrusion line for the sheathing of heavy cables in a tandem process	93
References for Chapter 3			94

4 Extrusion of blown films (G. Winkler) 95

4.1	Introduction		95
4.2	Raw materials		95
4.3	Criteria for characterization of films		97
4.4	LDPE blown film lines		99
	4.4.1	Extruders	99
	4.4.2	Dies	101
	4.4.3	Cooling and calibration	102
		4.4.3.1 External cooling	102
		4.4.3.2 Internal cooling	104
		4.4.3.3 Calibration	104
	4.4.4	Haul-off unit	105
	4.4.5	Winders	106
		4.4.5.1 Contact winders	106
		4.4.5.2 Central winders	107
		4.4.5.3 Auxiliary equipment	108
	4.4.6	Thickness uniformity	109
4.5	HDPE blown film lines		110
4.6	LDPE blown film lines		113
	4.6.1	LDPE lines in general	113
		4.6.1.1 Thin film lines	114
		4.6.1.2 Heavy-duty film lines	115
4.7	LLDPE (linear low-density PE) lines		117

4.8	Coextrusion	118
4.9	Special processes	118
4.10	Automation	119
	References for Chapter 4	123

5 Blown film coextrusion (R. Hessenbruch) 125

- 5.1 Introduction 125
- 5.2 Blown film coextrusion 125
- 5.3 Polymers, combinations and properties 128
  - 5.3.1 Polymers 128
    - 5.3.1.1 Support materials 128
    - 5.3.1.2 Bonding agents 129
    - 5.3.1.3 Barrier materials 129
- 5.4 Advantages of coextrusion 130
  - 5.4.1 Conversion 130
    - 5.4.1.1 Improved weldability 130
    - 5.4.1.2 Improvement in conversion properties 131
    - 5.4.1.3 Handling of packages 131
  - 5.4.2 Cost reduction 131
    - 5.4.2.1 Production costs 131
    - 5.4.2.2 Raw material costs 131
- 5.5 Quality improvement 134
- 5.6 Fields of application 135
  - 5.6.1 Polyolefin combinations 135
  - 5.6.2 Combinations with barrier properties 137
- 5.7 Line layout 137
  - 5.7.1 Extruders 137
  - 5.7.2 Filters 138
  - 5.7.3 Rotation device 138
  - 5.7.4 Blown film dies 139
    - 5.7.4.1 Two-layer blown film dies 139
    - 5.7.4.2 Two-layer blown film dies with additional bonding layer . . . 139
    - 5.7.4.3 Three-layer blown film dies 139
    - 5.7.4.4 Five-layer blown film die 139
  - 5.7.5 Bubble cooling 141
  - 5.7.6 Downstream equipment 141
  - 5.7.7 Special equipment 141
    - 5.7.7.1 Collapsing frame 141
    - 5.7.7.2 Slitting device 141
    - 5.7.7.3 Oscillating device 141
- 5.8 Automation 141
  - 5.8.1 Pressure measurement 141
  - 5.8.2 Width measurement 142
  - 5.8.3 Thickness control 142
  - 5.8.4 Additional equipment 142
- References for Chapter 5 142

6 Flat film extrusion using chill-roll casting (H. Bongaerts) 143

- 6.1 Introduction 143
- 6.2 General principle of flat film extrusion 143
  - 6.2.1 Extrusion 144

6.2.2	Molding	145
6.2.3	Cooling	148
6.2.4	Winding	153

6.3 Raw materials, film types, properties and applications 156
    6.3.1 Raw materials and product dimensions 156
    6.3.2 Properties and applications of flat film and thermoformable sheet ... 156
    6.3.3 Properties and fields of application of coextruded flat films and thermoformable sheet 160
6.4 Process technology and performance of flat film extrusion lines 163
6.5 Flat film production lines 168
    6.5.1 Line concepts 168
        6.5.1.1 Chill roll system 168
        6.5.1.2 Lines for thermoformable film and sheet 172
    6.5.2 Line layout and individual units 172
        6.5.2.1 Raw material feed and scrap recycling 172
        6.5.2.2 Extruders 174
        6.5.2.3 Filtration equipment 175
        6.5.2.4 Melt metering pumps 175
        6.5.2.5 Slot dies 176
        6.5.2.6 Coextrusion systems 178
        6.5.2.7 Chill roll take-off units 181
        6.5.2.8 Take-off lines for thermoformable film and sheet 186
        6.5.2.9 Wind-up equipment 188
    6.5.3 Measurement and control technology, and automation 191
    6.5.4 Special designs of flat film lines 194
References for Chapter 6 198

7 Production of films and sheets by the roll-stack process (H. Breuer) 203

7.1 Introduction 203
7.2 Roll-stack arrangements 203
    7.2.1 Polishing-stack operations 204
    7.2.2 Calendering mechanisms 207
    7.2.3 Differences between polishing stacks and calenders 208
7.3 Quality of films and sheets 216
    7.3.1 Polished films and sheets 216
    7.3.2 Calendered films 216
    7.3.3 Raw material content of films and sheets 217
    7.3.4 Special quality aspects of films and sheets 218
    7.3.5 Applications of film and sheet 221
7.4 Production lines for films and sheets 223
    7.4.1 Raw material compounding 223
    7.4.2 Construction of sheet extrusion lines 224
        7.4.2.1 Extruders and kneaders 225
            7.4.2.1.1 Extruders 225
            7.4.2.1.2 Mixers and kneaders 227
            7.4.2.1.3 Combined extruders 227
        7.4.2.2 Screen changers 228
        7.4.2.3 Slot dies 229
        7.4.2.4 Polishing stacks and calenders 231
            7.4.2.4.1 Deformations of the polishing stack 231
            7.4.2.4.2 Polishing-stack drives 235

	7.4.2.4.3 Special roll arrangements	235
	7.4.2.4.4 Thermal processes on polishing stacks	236
	7.4.2.5 Embossing, laminating, and coating	237
	7.4.2.6 Cooling sections	237
	7.4.2.7 Take-off units with cutting equipment	239
	7.4.2.8 Sheet off-loaders	240
	7.4.2.9 Film winders	240
7.4.3	Factors affecting production and quality	241
	7.4.3.1 Scale of production	241
	7.4.3.2 Energy usage	241
	7.4.3.3 Production limitations	244
	7.4.3.4 Influences on quality	245
7.4.4	Special designs of line	245
	7.4.4.1 Coextrusion lines	245
	7.4.4.2 Coating lines	247
	7.4.4.3 Ribbed hollow-profile lines	247
7.4.5	Auxiliary equipment for sheet extrusion lines	248
7.4.6	Monitoring systems and operator comfort	249
References for Chapter 7		251

8 Manufacture of oriented films (F. Hensen) ... 257

8.1 Introduction ... 257
8.2 Development of oriented films and their commercial importance ... 257
8.3 Theory of the orienting process ... 257
    8.3.1 Processes for the manufacture of oriented films ... 263
    8.3.2 Lines for the production of oriented films ... 263
    8.3.3 Extrusion dies and cooling unit for the production of oriented film ... 272
        8.3.3.1 Temperature control of the films ... 274
    8.3.4 Stretching units ... 274
References for Chapter 8 ... 282

9 Extrusion of film tapes (F. Hensen) ... 284

9.1 Introduction ... 284
9.2 Development and market significance of film tape extrusion ... 284
9.3 Methods of manufacture of film tapes ... 285
9.4 Characteristics of film tapes ... 286
9.5 Film tape production methods ... 290
9.6 Description of tape lines ... 292
    9.6.1 Extruding ... 293
    9.6.2 Filtering ... 295
    9.6.3 Film formation ... 295
    9.6.4 Slitting ... 296
    9.6.5 Feeding ... 297
    9.6.6 Heating ... 298
    9.6.7 Stretching ... 298
    9.6.8 Fibrillating ... 299
    9.6.9 Wind-up ... 300
9.7 Design variations of tape lines ... 301
9.8 Automation of tape lines ... 305
References for Chapter 9 ... 305

10 Monofilament extrusion (S. Braun) 307

- 10.1 Introduction 307
- 10.2 Theory 307
- 10.3 End product 308
- 10.4 Process 309
- 10.5 Lines 311
  - 10.5.1 Line concepts 311
    - 10.5.1.1 Monofilament line with two-stage drawing 311
      - − pre-drawing in a water bath
      - − secondary drawing in a hot-air oven
    - 10.5.1.2 Monofilament line with two-stage drawing 312
      - − pre-drawing in a hot-air oven
      - − secondary drawing in a hot-air oven
    - 10.5.1.3 Monofilament line with single-stage drawing 315
      - − drawing in a hot-air oven
    - 10.5.1.4 Monofilament line with single-stage drawing 316
      - − drawing in a water bath
  - 10.5.2 Line components 317
    - 10.5.2.1 Polymer feed 317
    - 10.5.2.2 Extruder 317
    - 10.5.2.3 Monofilament die 318
    - 10.5.2.4 Spinning pumps 320
    - 10.5.2.5 Quenching bath 322
    - 10.5.2.6 Monofilament drying 322
    - 10.5.2.7 Draw units 323
    - 10.5.2.8 Drawing bath 325
    - 10.5.2.9 Hot-air oven 326
    - 10.5.2.10 Winding 326
    - 10.5.2.11 Aspirator unit 327
    - 10.5.2.12 Process control systems 327
- 10.6 Layout of the line 328
- References for Chapter 10 330

11 Extrusion coating and laminating (E. Schöllkopf, K. Roesch) 331

- 11.1 Paper, film and foil converting 331
- 11.2 Coating and laminating 331
- 11.3 Material combinations 332
- 11.4 Adhesion 332
  - 11.4.1 Adhesion values 332
  - 11.4.2 Improving the adhesion 333
    - 11.4.2.1 Flame pretreatment 333
    - 11.4.2.2 Corona discharge 333
    - 11.4.2.3 Ozone showers 334
    - 11.4.2.4 Primers 334
  - 11.4.3 Adhesion-related coating- and laminating parameters 335
    - 11.4.3.1 The temperature of the extruded PE-film 335
    - 11.4.3.2 Surface conditions of the carrier web and the laminating web 336
    - 11.4.3.3 The chill roll temperature 336

		11.4.3.4	Nip pressure between chill roll and impression cylinder	336

              11.4.3.5 The thickness of the extruded film 336
     11.4.4 Coextrusion as adhesion promoter 336
11.5 Machine design considerations 337
     11.5.1 Web widths 337
     11.5.2 Web speeds 337
     11.5.3 Coating weights of extruded films 337
     11.5.4 Extruder size 338
11.6 Machine layout criteria 338
     11.6.1 The coating principle 338
     11.6.2 Extrusion coating and laminating machinery 339
     11.6.3 Coating machine with primer coating station 341
     11.6.4 Coating and laminating machine with primer station for the laminating substrate 342
     11.6.5 Extrusion coating and laminating machine with primer coating and overlacquering station 343
     11.6.6 Tandem extrusion coating and laminating machine 343
     11.6.7 Multipurpose extrusion coating and laminating machine 347
     11.6.8 Tandem lacquering and laminating machine with extrusion coating and laminating facility 348
11.7 Machine components 350
     11.7.1 Unwinds 350
     11.7.2 Rewinds 352
     11.7.3 Web-tension controls 354
              11.7.3.1 Dynamic control system 354
              11.7.3.2 Static control system 355
     11.7.4 Coating unit for primers, lacquers, glues and dispersions 356
     11.7.5 Drying tunnels 356
     11.7.6 Drying and dry laminating 358
     11.7.7 Web turner-bars 359
     11.7.8 Edge-guiding equipment 359
     11.7.9 Drive and control systems 360
     11.7.10 Process reference parameters 361
References for Chapter 11 362

12 Extrusion blow molding (H.-G. Fritz) 363

12.1 Introduction 363
     12.1.1 Techniques of blow molding 363
     12.1.2 The historical development of extrusion blow molding 363
     12.1.3 The economic importance of extrusion blow molding 366
12.2 Polymers for extrusion blow molding 367
12.3 Classification of extrusion blow molding processes 369
     12.3.1 Continuous tube extrusion blow molding 370
     12.3.2 Extrusion blow molding with intermittent tube extrusion 370
12.4 Extrusion blow molding lines 371
     12.4.1 Description of individual units 371
              12.4.1.1   Extruder systems 371
              12.4.1.1.1 Conventional single-screw extruders 374

	12.4.1.1.2	Extruders with a forced-feed section (grooved extruders)	380
	12.4.1.2	Dies	388
	12.4.1.2.1	Tube dies for continuous production of parisons	389
	12.4.1.2.2	Dies for continuous multilayer tube extrusion	412
	12.4.1.2.3	Dies for intermittent extrusion of tube	414
	12.4.1.2.4	Wall-thickness control	419
	12.4.1.2.5	Mold clamping units	423
	12.4.1.2.6	Tube cutting devices	424
	12.4.1.2.7	Calibration and blowing station, stretching and pre-weld devices	424
12.4.2	Complete blow molding lines		425
	12.4.2.1	Automated blow molding machines	425
	12.4.2.2	Accumulator-head blow molding machines	427
References for Chapter 12			428

13 Extrusion of foamed intermediate products with single-screw extruders
(K.-D. Kolossow) 430

13.1 Introduction 430
13.2 Phases of the manufacturing process 430
    13.2.1 Raw materials and recipes 431
    13.2.2 Nucleating agents and additives 433
    13.2.3 Blowing agents 434
    13.2.4 Feeding of raw materials 435
    13.2.5 Extrusion 436
    13.2.6 Foaming 437
    13.2.7 Post-expansion 441
    13.2.8 Lamination 441
    13.2.9 Scrap recovery 441
13.3 Foam Products 443
13.4 Processes 444
    13.4.1 Single-stage process 444
    13.4.2 Two-stage process 447
    13.4.3 Multistage process 447
13.5 Single-screw extrusion systems 448
    13.5.1 Single-screw lines 449
    13.5.2 Tandem single-screw lines 455
    13.5.3 Single-screw scrap recovery lines 459
13.6 Plant components 460
    13.6.1 Storage, conveying, and mixing equipment for raw materials . . . 460
    13.6.2 Extruder components 461
    13.6.3 Screen changer 463
    13.6.4 Dies and calibrating devices 463
    13.6.5 Storage, conveying, and dosing systems for the blowing agent . . . 465
    13.6.6 Take-off equipment 466
    13.6.7 Winders 467
    13.6.8 Laminators 468
    13.6.9 Cutting and stacking devices 469
References for Chapter 13 470

14 Extrusion of foamed intermediate products with twin-screw extruders
(P. Klenk, H. P. Schneider) 471

- 14.1 Introduction 471
- 14.2 Structure of formulations 472
  - 14.2.1 PVC types 472
  - 14.2.2 Stabilizers 472
  - 14.2.3 Lubricants 473
  - 14.2.4 Fillers 473
  - 14.2.5 Pigments 473
  - 14.2.6 Blowing agents 473
  - 14.2.7 Foaming aids 475
  - 14.2.8 Mixing technique 475
  - 14.2.9 Formulations 476
- 14.3 Theory 477
  - 14.3.1 Foaming operation 477
  - 14.3.2 Foaming methods 478
- 14.4 Extrusion foaming methods 479
  - 14.4.1 Methods using chemical blowing agents 479
    - 14.4.1.1 Coextrusion of full profiles and sheets from two different material streams (Reifenhäuser) 479
    - 14.4.1.2 Extrusion of foamed full profiles (Ugine Kuhlmann) . . 479
    - 14.4.1.2.1 Full profiles from one material stream 479
    - 14.4.1.2.2 Full, part-skinned profiles from one material stream . . 480
    - 14.4.1.2.3 Hollow profiles from one material stream 480
    - 14.4.1.2.4 Coextrusion of full and hollow profiles from two different material streams 481
    - 14.4.1.3 Extrusion of woodlike sheets from one material stream (Sekisui Kaseihin Kogyo) 481
    - 14.4.1.4 Extrusion of full profiles from a split material stream (Scherer & Trier) 482
    - 14.4.1.5 Extrusion of pipes with foamed walls (Société Armosic) 482
  - 14.4.2 Processes using physical blowing agents 483
- 14.5 Plant for foamed semifinished products 483
  - 14.5.1 Extruder 483
  - 14.5.2 Screw design 484
  - 14.5.3 Die head 485
  - 14.5.4 Calibration unit, cooling unit 486
  - 14.5.5 Take-off 486
  - 14.5.6 Cutting to length, punching and stacking 487
  - 14.5.7 Embossing, printing 487
- 14.6 Selection criteria 487
- References for Chapter 14 487

15 Crosslinking of plastics after extrusion (K. Kircher) 489

- 15.1 Introduction 489
- 15.2 Modification of the properties of polyethylene by crosslinking . . . 489
- 15.3 Uses for crosslinked polyethylene 490
- 15.4 Crosslinking processes 490
  - 15.4.1 Chemistry of radical crosslinking 491
  - 15.4.2 Silane crosslinking processes 495

15.5 Checking the degree of crosslinking 496
15.6 Implementation of crosslinking 497
    15.6.1 Heat-activated peroxide-radical crosslinking 497
        15.6.1.1 Steam crosslinking processes 498
        15.6.1.2 Gas crosslinking lines (radiation methods) 501
        15.6.1.3 Other gas crosslinking lines 504
        15.6.1.4 Comparison of the economics of steam- and radiation
                crosslinking 505
        15.6.1.5 Crosslinking in liquid baths 505
        15.6.1.6 Special processes 506
    15.6.2 Crosslinking by UV light 507
    15.6.3 Electron-beam crosslinking 507
    15.6.4 Implementation of silane crosslinking 509
    15.6.5 Crosslinking of polyethylene foam 510
15.7 Criteria for choosing different crosslinking processes 510
References for Chapter 15 510

16 Extrusion of Elastomers (H.-J. Gohlisch, W. May, F. Ramm, W. Rüger) 513
16.1 Extruders 513
    16.1.1 The task of the elastomer extruder 513
        16.1.1.1   Hot-feeding 513
        16.1.1.2   Cold-feeding 513
        16.1.1.2.1 Systems for plastication of elastomers 514
        16.1.1.3   Vacuum extruders 519
    16.1.2 Extruder design 520
        16.1.2.1   Drive units 520
        16.1.2.2   Drive transmission 520
        16.1.2.3   Gearboxes 522
        16.1.2.4   Thrust bearings 523
        16.1.2.5   Hopper units 523
        16.1.2.6   Barrel liners 523
        16.1.2.7   Extruder barrels 524
        16.1.2.8   Extruder screws 525
        16.1.2.9   Temperature control of extruders 525
    16.1.3 Extrusion heads 526
        16.1.3.1   Pork-chop heads 526
        16.1.3.2   Pelletizer heads 526
        16.1.3.3   Strainer heads 527
        16.1.3.4   Profile heads 528
        16.1.3.5   Tube heads 528
        16.1.3.6   Slit-tube heads 529
        16.1.3.7   Crossheads 529
        16.1.3.7.1 Crosshead dies for sheathing 530
        16.1.3.7.2 Crosshead dies for two- and three-layer sheathing . . 530
        16.1.3.7.3 Sheathing of non-circular electrical conductors
                or of metal profiles 531
        16.1.3.7.4 Covering flat profiles 531
        16.1.3.7.5 Double- or multiple crosshead dies 531
        16.1.3.8   Special crosshead dies for filament and wire coating . 531
        16.1.3.9   Slot dies 531
        16.1.3.10   Tread heads 532
        16.1.3.11   Piggyback extrusion heads 532

		16.1.3.12	Single-roll roller heads	534
		16.1.3.13	Two-roll roller heads	534
		16.1.3.13.1	Roller-head machines	534
		16.1.3.13.2	Roller-die units	535
	16.1.4	Auxiliary equipment		535
		16.1.4.1	Screen changers	535
		16.1.4.2	Die changers	536
		16.1.4.3	Head changers	536
		16.1.4.4	Feeding devices	536
		16.1.4.4.1	Granule- or pellet-feed equipment	536
		16.1.4.4.2	Strip- or slab-feed equipment	537
		16.1.4.4.3	Slab-cutting and feeding devices	537
16.2	Continuous vulcanization lines			537
	16.2.1	General		537
	16.2.2	CV pipe		539
	16.2.3	Hot-air vulcanization		539
	16.2.4	Fluid-bed vulcanization		540
	16.2.5	Salt-bath vulcanization		541
	16.2.6	Infrared lines		543
	16.2.7	Helicure vulcanization		543
	16.2.8	UHF vulcanization		544
	16.2.9	Smear-head vulcanization		545
	16.2.10	Rotation vulcanization		547
16.3	Cooling			547
	16.3.1	General		547
	16.3.2	Cooling processes		548
		16.3.2.1	Immersion cooling	548
		16.3.2.2	Spray cooling	548
		16.3.2.3	Contact cooling	548
		16.3.2.4	Air-cooling lines	549
	16.3.3	Cooling lines		549
		16.3.3.1	Slab-cooling lines	550
		16.3.3.2	Strip-cooling lines	550
		16.3.3.2.1	Lines for the simultaneous cooling of several strips	550
		16.3.3.2.2	Capstan lines for cooling strips of raw compound	551
		16.3.3.2.3	Mesh drums for cooling strips of raw compound	551
		16.3.3.3	Pellet-cooling lines	551
		16.3.3.4	Cooling lines for treadstrip	551
		16.3.3.5	Side-strip-cooling lines	552
		16.3.3.6	Sheet-cooling lines	552
		16.3.3.7	Profile-cooling lines	552
16.4	Measurement and control equipment			552
	16.4.1	General		552
	16.4.2	Extruder controls		553
		16.4.2.1	Feed devices	553
		16.4.2.2	Temperature control	553
		16.4.2.3	Pressure/screw-speed control	553
		16.4.2.4	Melt-temperature/screw-speed control	554
	16.4.3	Haul-off-speed control		554
	16.4.4	Measurement and control of product dimensions		554
		16.4.4.1	Profile measurement	555
		16.4.4.2	Thickness measurement	555

		16.4.4.3	Width measurement	556
		16.4.4.4	Weight measurement	556
		16.4.4.5	Control of measured quantities	556
	16.4.5	Control of set values		557
References for Chapter 16				557

17 Fiber extrusion (M. Mayer) ... 561

- 17.1 Introduction ... 561
    - 17.1.1 Areas of application for man-made fibers ... 562
    - 17.1.2 The principle of fiber extrusion ... 562
- 17.2 Extrusion section (section I) ... 563
    - 17.2.1 Extruders ... 563
        - 17.2.1.1 History of development ... 563
        - 17.2.1.2 Design and function of a spinning extruder ... 563
        - 17.2.1.3 Screw design ... 565
        - 17.2.1.4 Mixing of additives ... 567
    - 17.2.2 Manifold ... 568
    - 17.2.3 Spinning head ... 570
        - 17.2.3.1 Gear pumps for spinning ... 571
        - 17.2.3.2 Spin pack ... 574
    - 17.2.4 Quenching ... 577
        - 17.2.4.1 Spinneret blanketing ... 578
        - 17.2.4.2 Monomer exhaust ... 578
    - 17.2.5 Heating system ... 579
    - 17.2.6 Filters ... 579
    - 17.2.7 Examples from the extrusion section (section I) ... 583
- 17.3 Filament treatment section (section II) ... 587
    - 17.3.1 Machines for take-up or deposit of freshly spun fibers ... 589
    - 17.3.2 Machines for drawing freshly spun filaments ... 595
    - 17.3.3 Machines for drawing and further treatment of freshly spun filaments ... 600
- 17.4 Important components of the filament treatment section (section II) ... 601
    - 17.4.1 Take-up heads ... 601
    - 17.4.2 Tow piddler ... 606
    - 17.4.3 Draw unit ... 606
    - 17.4.4 Crimper ... 606
    - 17.4.5 Cutter ... 608
    - 17.4.6 Godets ... 608
    - 17.4.7 Spin finish application ... 610
    - 17.4.8 Texturing jet ... 611
    - 17.4.9 Automation ... 612
- References for Chapter 17 ... 613

18 Ram Extrusion of PTFE and UHMWPE (P. Stamprech) ... 614

- 18.1 Description of the process ... 614
- 18.2 Processing machines ... 615
- 18.3 The extrusion die ... 616
    - 18.3.1 The extrusion pipe ... 617
        - 18.3.1.1 Choice of material ... 618

	18.3.2	The mandrel	619
	18.3.3	The ram	620
18.4	Processing problems		620
	18.4.1	Dimensioning the die	621
	18.4.2	Surface quality of the die	622
	18.4.3	Extrusion speed	622
	18.4.4	Temperature control	623
18.5	Materials for ram extrusion		623
18.6	Extrusion of solid profiles		625
18.7	Extrusion of pipes and hollow profiles		628
	18.7.1	Powder densification rate	630
	18.7.2	Ram stroke depth	630
	18.7.3	The movable mandrel	630
References for Chapter 18			632

19 Extrusion welding (P. John) 633

19.1	Introduction	633
19.2	Extrusion welding	633
19.3	Fields of application	634
19.4	Extrusion welding equipment	634
	19.4.1 Extruder unit with movable welding head	634
	19.4.2 Mobile welding apparatus	635
	19.4.3 Welding apparatus for manual transfer of the filler rod from the extruder to the welding gap	636
	19.4.4 Welding apparatus with extruder-like plasticating chamber	637
	19.4.5 Manual welding extruders	637
19.5	Welding shoes	638
19.6	Preparation of the seam	638
19.7	Start-up of welding apparatus	639
19.8	Welding (or filler) material	639
19.9	Preheating of the base material	639
19.10	Welded seams	640
19.11	Finishing the weld	640
19.12	Practical examples	640
19.13	Economic efficiency	641
19.14	Welding factor	641
	19.14.1 Short-term welding factor	641
	19.14.2 Long-term welding factor	642
19.15	Chemical resistance	642
19.16	Quality assurance	643
19.17	Testing	643
19.18	Prospects	644
References for Chapter 19		644

20 Feeding of extruders (W. Mücke) 646

20.1	Introduction	646
20.2	Screw conveyors	646

20.3	Pneumatic conveying systems		647
	20.3.1 Single-machine conveying systems		648
		20.3.1.1 Suction/pressure conveying units	648
		20.3.1.2 Suction conveying units	648
	20.3.2 Multiple-machine conveying systems		649
		20.3.2.1 Pressure conveying	649
		20.3.2.1.1 Pressure conveying systems incorporating pipe-switching junctions	650
		20.3.2.1.2 Systems with ring lines	650
		20.3.2.2 Suction conveying	651
	20.3.3 Filter systems		652
		20.3.3.1 Cleaning filters by pressure scavenging	653
		20.3.3.2 Cleaning filters by suction scavenging	653
		20.3.3.3 Cleaning filters by scavenging with recirculated air	653
20.4	Auxiliary equipment		654
	20.4.1 Metering systems		654
		20.4.1.1 Metering by redirecting granulate flow	655
		20.4.1.2 Metering without a main component	655
		20.4.1.3 Metering of all components	655
	20.4.2 Drying systems		656
		20.4.2.1 Drying with ambient air	657
		20.4.2.2 Drying with dried air	657
		20.4.2.2.1 Pellet dryers using air drying	660
References for Chapter 20			661

21 Extrusion recycling of plastics waste (H. Tenner) 662

21.1	Introduction	662
21.2	Recycling via the melt – stages of the process	663
	21.2.1 The individual steps in the process and their purposes	663
21.3	Applications, products, special properties	672
	21.3.1 Standard recycling of production waste	673
	21.3.2 Washing and recycling of contaminated waste	675
	21.3.3 Integrated recycling and compounding, including filling and alloying	676
	21.3.4 Reclaiming of engineering plastics waste	676
	21.3.5 Direct recycling from film scrap to blown film	678
21.4	Typical line design concepts and line components	680
	21.4.1 Classification of reclaim lines according to their uses	681
	21.4.2 Classification of reclaim extruders with respect to design criteria	683
	21.4.3 Wear and wear prevention in reclaim extruders	689
	21.4.4 Size reduction: cutting mills	690
	21.4.5 Metal detection and metal separation	692
	21.4.6 Bulk storage and homogenization	693
	21.4.7 Extruder feeding – the crammer feeder	694
	21.4.8 Screen changer – melt filter	694
	21.4.9 Pelletizing	699
21.5	Energy balance – specific energy consumption	702
	21.5.1 Energy consumption in size reduction	703
	21.5.2 Energy consumption in extrusion	704
References for Chapter 21		706

1 Introduction

F. Hensen

Plastics are superior to many conventional materials, in both their physical properties and the variety of ways in which they can be processed.

This fact became recognized particularly during the period of great economic growth after 1950, and has led, during the rapid development since then, to the present world production of over 50 million tons annually. Despite this large amount, sufficient quantities of raw materials are available to meet foreseeable future demand for plastics.

The particularly useful chemical and physical properties of plastics make possible an extraordinarily large number and variety of applications. Molding during processing owes its simplicity to the basic characteristic of thermoplastic resins: that they change reversibly from the solid to the molten state within a relatively low temperature range, over which the plastic materials are highly viscous. This behavior pattern is related to their macromolecular structure. It is the reversibility of the change of state, in particular, that permits the molding operation to be shifted from the raw material producer to the processor.

Machine manufacturers have the task of developing extrusion lines for the processor's special requirements. They offer him a range of solutions to production problems in the form of complete extrusion lines, utilizing their own developments and the scientific findings of research institutes, often in cooperation with the applications engineers of raw material producers.

Extrusion lines consist of a logical in-line arrangement of the machines, auxiliary equipment, and measuring and control intruments that are required for the production of semi-finished and finished products from thermoplastics. Almost all extrusion lines are based on the principle of transforming a polymer by means of an extruder from the solid to the easily moldable plastic state, then discharging it through an extrusion die in a predetermined cross-sectional shape, solidifying it by cooling, possibly drawing it or subjecting it to further treatment, and stacking it as semi-finished product or winding or packing it as finished product.

The rapid growth of the industry produced many suppliers of extrusion lines, and led to strong competition, involving a wealth of inventions and innovations. The competition in technology that developed produced a large variety of such lines, which were brought to a high level of technical sophistication.

When one studies trade magazines and visits the large plastics exhibitions, it becomes clear that the era of rapid development of new processes and basic concepts for extrusion lines is over, and that development now only takes place in small steps, on existing lines, in the design of individual units and their automation. Now is therefore the right time to record the level of knowledge achieved, and to compile it in the form of a reference book.

The book contains 21 chapters, and, starting with feeding sytems and compounding lines, deals with extrusion lines for the production of semi-finished goods in all industrially useful shapes and sizes, including specialties; it also covers fiber and filament extrusion and, finally, waste recycling.

2 Compounding Lines

H. Herrmann

2.1 Purpose of plastics compounding and the unit operations

Compounding is required to turn plastics raw materials into processable compounds for specific applications, and involves the use of various unit operations. In compounding we consider thermoplastics, thermosets and elastomers. In terms of unit operations, compounding consists of combining such processes as:

- the mixing of polymer with additives (stabilizers, lubricants, plasticizers, colorants, fillers, flame retardants, cross-linking agents, blowing agents, etc.) and thoroughly dispersing these ingredients [1],
- reinforcing the polymers with glass- or carbon fibers,
- alloying of various polymers with each other,
- the homogenizing of simple polymer melts, or achieving desired flow behavior by using controlled shear conditions.

Compounding also deals with separation processes such as:

- removal of volatiles (residual monomer or solvents, water) [2],
- concentration of polymer solutions,
- filtering of polymer melts to separate particulate impurities [3], and also
- pelletizing, the step needed to arrive at an easily handled and properly processable particulate form of the compound [4].

In general, all these unit operations assume melting of the polymer, and therefore take place in a high-viscosity phase. Exceptions are premixing operations of free-flowing feedstocks (powder, pellets, chips, crumbs, etc.). Other unit operations in polymer compounding may include size reduction of the raw polymer, reject and trim scrap, as well as the fine grinding of certain polymers. However, this article will not deal with these particular operations.

2.2 Development and economic importance of compounding technology

After the introduction of major polymers in the 1950's, large-scale continuous compounding operations began in the 1960's. References to the concept of compounding were found in the plastics industry after 1933. The term "compounding" was first coined in the July issue of the German magazine "Kunststoffe", in an article by *K. Brandenburger* on the "Compounding of modern high speed resins" [5]. The origins of polymer compounding technology, however, reach back into the 1870's. We can easily distinguish four time periods [6]:

- 1870–1910: Origins
  Invention and development of discontinuous mixers, kneaders and evaporators. Introduction of discontinuous processing.
- 1910–1935: Realization of discontinuous compounding for rubber, cellulose, celluloid, shellac, casein and similar resins.
- 1935–1960: Invention and development of continuous dewatering presses, screw kneaders and screw evaporators. Introduction of continuous compounding concepts for thermoplastics and thermosets of all types. Continued use of discontinuous operations for rubber- based products.

– 1960–1985: Large-scale realization of continuous compounding methods for thermoplastics and thermosets of all types and continued use of discontinuous methods for rubber.

According to one estimate [7], about 56% of all polymer production in the western world was compounded on screw machines in 1977. This figure is perhaps on the low side for world plastics production today. Assuming the level of world production of plastics (in 1983) to be about 53 million tons/y, at least 30 million tons of this goes through screw-type compounding machines. In addition, approximately 8 million tons/a of PVC is compounded in high-speed mixers to produce dry blend, or in vessels to produce plastisol. One may also assume that at least 70% of all polymers is compounded on some form of machinery prior to final conversion into an end product.

2.3 Compounding lines

Compounding lines for plastics consist of the basic compounding machinery and peripheral equipment, appropriately arranged to do a specific job. The main compounding machine is generally a continuously operating kneader, a screw evaporator, or a special extruder [8].

The design of continuous screw kneaders must be adaptable to a number of unit operations including plastication, mixing and homogenizing, dispersing, removal of volatiles, filtering of the melt and pelletizing, or extrusion with open discharge for calender feed or for thermosets. In some cases, and in general for smaller weights up to two tons per hour, it is also desirable to form the melt directly into a final extruded shape.

Screw kneaders are commonly characterized by screws of special design for specific end products.

The aim is to develop a controllable range of shear intensity at specific locations, and to achieve intensive distributive mixing of the polymer melt by means of repeated subdivision of the stream, change of direction of the stream and interchange of the material layers in the screw channels.

Another characteristic is a relatively high specific energy (main drive power in relation to machine free-volume [1]), in order to achieve energy conversion by shear heating much beyond what could be derived from external heating.

Single- and twin-screw extruders are specifically designed for plastication and extrusion. For improved plastication and distributive mixing, these screw extruders are equipped with special mixing and shear elements. If one continues the modification of the normal, relatively undisturbed material flow in a standard screw by introducing more controlled shear, then one moves from a basic screw extruder in the direction of a screw kneader. The transition between the two types of unit is continuous.

In specific cases, the following special units are used:
– discontinuous intensive mixers for compounding ABS, or for production of polymer pigment concentrates with frequent color changes and small lot sizes,
– discontinuous double-arm kneaders for unsaturated polyester/glass-fiber mixes,
– stirred vessels for dissolving of polyurethanes, certain rubbers and PVC plastisols,
– dispersing molds for plastisol and liquid colorants,
– roll mills for forming polymer melts into strips in line with an intensive mixer in calendering, and for compounding of certain special thermosets.

In the context of this review of extrusion technology, we shall not discuss these types of compounding machines [9].

In general, some auxiliary equipment is needed and, depending on the particular task, may include the following items:

- silos for storage and handling of feedstocks and for the finished compounded material, mostly in pellet form,
- discontinuous or continuous premixer,
- feeding units for free-flowing materials, for melts and liquids, including feed screws, gravimetric belt feeders, weight loss feeders, feed hoppers with crammers or stirrers, rotary valves, gear pumps, melt feed valves, piston pumps, and injection valves,
- units for protection against foreign matter, generally using mechanical, magnetic, or inductive principles,
- screening and screen-changing devices for filtering particulate impurities and dirt from the melt,
- pelletizing units with the appropriate pneumatic or water-based pellet transport system,
- pellet cooler or pellet dryer,
- bagging units,
- temperature control systems for the compounding machines using electrical heating, oil heating, water cooling or pressurized water units,
- vacuum pumps for the removal of volatiles from polymers, especially water ring pumps, mechanical pumps, or steam jets with their necessary separators and condensers, and
- control cabinets and control centers with appropriate control, display and recording systems.

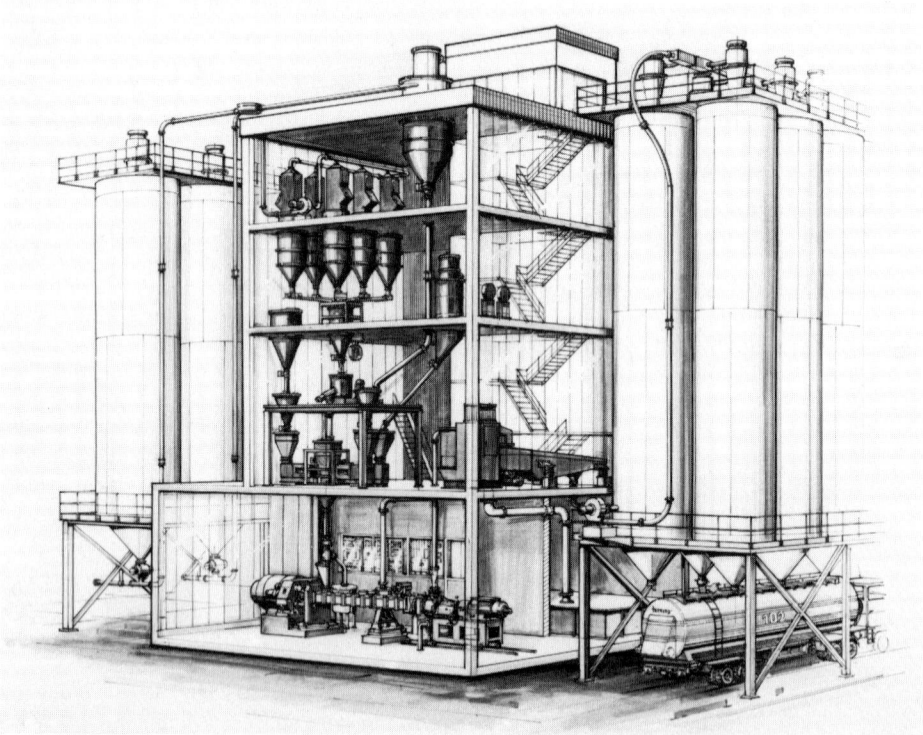

Figure 1. Model of a polyolefin compounding plant

Figure 1 shows a model of a typical polyolefin compounding system with silos, day bins, premixers, pellet driers and plastifying units, with the main screw kneader placed on the ground floor.

Compounding is generally required for practically all plastics; however, the degree of sophistication of compounding depends very much on:

- the particular task,
- the type of polymer,
- its origin, and
- the end use of the material.

Accordingly, the compounding techniques have become highly differentiated with many different and valid technical solutions. From a practical point of view, we can distinguish five main areas of plastics compounding:

- polyolefins [10],
- PVC and other temperature-sensitive plastics [11],
- polystyrene and styrene copolymers,
- engineering plastics,
- elastomers.

For each of these five areas, a clearly identifiable and typical state-of-the-art technology and machine design has evolved worldwide.

2.4 Compounding of polyolefins

Polyolefins are produced in melt form, as melt solutions or as powder.

2.4.1 Plant for melt feed

Single-screw extruders, with a length of 8 to 12 D and mixing devices toward the end of the screw, are used for pelletizing LDPE. Simultaneous dispersive mixing of additives is carried out by pumping them into the melt as liquids via side extruders. Rates of 20 and 29 tons per hour have been achieved with screw diameters of 500 and 600 mm, screw speeds of 90 and 80 rpm and main drive capacities of 1900 and 2300 kW [12], respectively.

In many cases, LDPE requires additional homogenizing. Good homogenization can show up as better stretchability of the melt in film production and in improved optical properties.

The original two-stage homogenization of LDPE (discharge from the reactor, incorporation of stabilizers, and pelletizing, followed by plastication, homogenization and final pelletizing) has been replaced more and more by a single-step operation (direct feed of the melt into a screw kneader, homogenization and pelletizing).

For single-step melt homogenization, extruders with a length of about 24 D and Maillefer screw geometry (Figure 2), or intermeshing, co-rotating twin-screws with a length of about 21 D, are used with kneading blocks and/or shear elements. Table 1 gives machine and rate data for melt-homogenizing extruders [12 to 14].

Depending on the pressure in the low-pressure separator, the LDPE melt has a residual ethylene concentration of 0.1 to 0.2%. With rear venting under atmospheric pressure, the ethylene concentration in the first-stage discharge or in the homogenizing extruder can be reduced to 500 to 800 ppm. The rest remains in the pellet and must be removed as it diffuses to the surface by passing air through the silo.

With increased environmental concern in recent years, there has been a demand for reduction of residual ethylene to less than 50 ppm. Once this residual concentration is reached, the expense of ethylene removal in silos can be eliminated. Residual ethylene concentrations of 50 ppm can be achieved only by use of vacuum gassing at 30 to 50 mbar. Best results are achieved with simultaneous rear and forward venting [15 to 17].

For processing of LDPE melts, screw machines with rates of up to 25 t/h are used. Further development toward larger extruder sizes or other machine types now appears unlikely (see

Table 1), as the demand for LDPE has been reduced by the arrival of LLDPE, which is obtained as a melt solution or as a powder.

Table 1. Throughput rates in homogenizing extruders for LDPE melts

Machine type	Screw diam. D mm	Max. screw speed n min^{-1}	Max. drive power N kW	Rate G t/h
Single-screw homogenizing extruder KE	300 400 500 600	115 85 70 60	1000 1800 3000 4000	4.2 to 5.5 7.5 to 9.5 11.8 to 15 17 to 22.3
Intermeshing co-rotating twin-screw extruder ZSK	160 220 280	250 180 140	600 1200 2000	3 to 5 6.5 to 10 10 to 15

Figure 2. Single-screw homogenizing extruder for LDPE melts, with screen changer and underwater granulation: Type KE 600 24D. (Photo: Berstorff, Hannover, West Germany)

2.4.2 Systems for concentration of melt solutions

In the solution polymerization of polyolefins, it is advantageous not to pursue the classic method of steam stripping, since drying the wet PE involves an expensive separation of water from the solvent, but to go instead to direct concentration of the polymer solution. Above a polymer concentration of 85%, the use of screw machines is indicated because of the high viscosity of the product.

HDPE injection molding resins can be concentrated from 15% down to 0.05% solvent in single-screw units operating in cascade, with degassing during transfer between the two screw machines. With screw diameters of 300 mm, rates of 7 t/h have been achieved. For higher-viscosity products, such low residual solvent concentration cannot be achieved, because the solvent bubbles do not break through the surface fast enough owing to the high melt strength of the polymer [14, 17].

For such high-viscosity PE melts, multiscrew machines with both forward and rear degassing have been used successfully, including counter-rotating and non-intermeshing twin-screws with lengths up to 60 D, and co-rotating intermeshing twin-screws of length of about 30 D. Tables 2 and 3 show machine data and rates. At this stage, degassing by stripping is

2.4 Compounding of polyolefins

possible and practiced. Even high-viscosity melts are extended to thin layers by the action of the screws, and the solvent-vapor bubbles are broken by this action [16, 18].

Table 2. Machine data and rates of non-intermeshing, counter-rotating twin-screw machines for degassing

Screw diam. D mm	Screw speed n min^{-1}	Drive power N kW	Rate G t/h
150	100 to 180	120 to 250	0.67 to 1.25
200	87 to 150	220 to 460	1.2 to 2.3
250	78 to 135	370 to 750	2.0 to 3.75
305	70 to 120	600 to 1,200	3.3 to 6.0
380	63 to 110	970 to 2,000	5.4 to 10
460	58 to 100	1,500 to 3,000	8.3 to 15

Table 3. Machine data and rates of intermeshing, co-rotating twin-screw extruders (ZSK) for the concentration of melts

Screw diam. D mm	Screw speed n min^{-1}	Drive power N kW	Rate G t/h
130	180 to 300	150 to 240	0.85 to 1.5
170	150 to 250	300 to 490	1.75 to 3
240	140 to 230	680 to 1,100	4 to 7
300	110 to 180	1,200 to 2,000	7 to 12

Screw evaporators of type ZSK with 300 mm screw diameter and rates of 8 to 10 t/h are proven technology for the concentration of LLDPE melt from 15% solvent to 500 ppm residual solvent (Figure 3).

In certain PP production systems, extraction clarification of the isotactic PP yields large quantities of atactic PP solutions of 2 to 5% concentration. Film evaporators are best suited for recovery of the solvent (recovery of the polymer is less significant). In single-stage film evaporator systems, residual concentrations of 1 to 5% can be achieved; in two-stage units 0.2 to 0.5% [19].

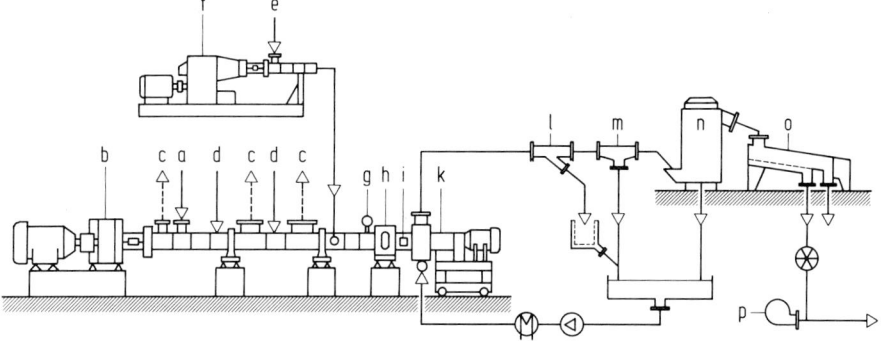

Figure 3. Screw-evaporator Type ZSK 300 for concentrating LLDPE solutions, with pressurizing gear pump, screen changing equipment, and underwater granulation
a LLDPE solution, *b* twin-screw kneader ZSK, *c* degassing, *d* drag agent, *e* additive/masterbatch, *f* ZSK side machine, *g* start-up valve, *h* gear pump, *i* screen change equipment, *k* underwater granulation UG, *l* start-up diverter, *m* water trap, *n* granule drier, *o* classifier, *p* pneumatic conveyor

2.4.3 Compounding systems for polyolefin powder

The compounding of powder polyolefins has remained basically the same over the years, involving:
- plastication of polyolefin powder,
- dispersing and mixing of additives,
- removal, if required, of small quantities of volatiles,
- filtering of the melt,
- pelletizing.

Throughput rates have increased substantially together with reactor output, with rates of 10 to 22 t/h being the norm. In certain cases, lines with higher rates are being planned. As we are dealing exclusively with very high rate compounding units here, only machinery which can handle this type of production rate safely and uniformly is of value. Single-screw units have a practical rate limit of about 5 t/h. Intermeshing, counter-rotating-screw machines present mechanical problems related to their screw action which limit screw diameters to a maximum of 170 mm, and are thus not capable of achieving the required rates [20, 21].

Twin-screw systems have proven successfull that allow the use of large screw diameters, locally fixed placement of the plasticating zone, and the use of special kneading and mixing elements; these include:

- non-intermeshing, counter-rotating twin-screw systems in two-stage design with downstream single-screw melt extruder or gear pump (Figure 4),
- intermeshing, co-rotating twin-screws as a single-stage (Figure 5) or a two-stage system with subsequent gear pump for pressure generation (Figure 6).

Table 4 contains the sizes and rates of twin-screw compounding machines available on the market [22 to 24].

In the development of specialized polyolefin materials (pipe, tape, blow molding grades, injection molding, etc.) special demands arose which made it necessary to further refine the zones where the melt is subjected to a particular level of shear and temperature. Considering broad product mixes, it is therefore necessary to be able to vary the amount of energy input without changing screw geometry (which is impractical for big machines). For this purpose, control valves were developed to influence the effectiveness of plastication and of energy input. These are available nowadays on practically all large-scale compounding machines (Table 4).

For reasons of economy, relatively high screw speeds are needed (see Table 4). For co-rotating machines, circumferential speeds of up to 3.1 m/sec are common, and for two-stage non-intermeshing, counter-rotating-screw systems, circumferential speeds of 5 to 8 m/s have been reached.

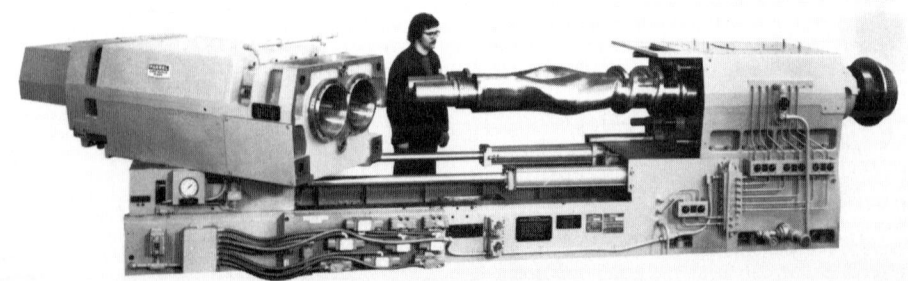

Figure 4. Twin-screw, non-intermeshing counter-rotating screw kneader Type 12 FCM. (Photo: Farrel Machinery Group, Ansonia, USA)

2.4 Compounding of polyolefins

Figure 5. Co-rotating double-screw kneader Type ZSK 300, screw housings and screws. (Photo: Werner and Pfleiderer, Stuttgart, West Germany)

Figure 6. Co-rotating double-screw kneader Type ZSK 300, with pressurizing gear pump and underwater pelletizing. (Photo: Werner and Pfleiderer, Stuttgart, West Germany)

The maximum possible screw speed is determined by mechanical considerations and by the maximum operating temperature for the product. The latter depends on the type of material, and lies between 230 and 260 °C, although in some cases even lower limits are set. In order to achieve these temperatures with fast-rotating large-scale production units, the following prerequisites must be fulfilled:

– short screw length,
– high efficiency plastication elements to avoid temperature peaks,
– sufficiently high torques to allow a high degree of fill,
– energy-efficient pressure build-up zone.

Accordingly, non-intermeshing counter-rotating twin screws with downstream melt extrusion are normally designed with screw lengths of 5 to 9 D, and single-stage intermeshing, co-rotating twin screws with lengths.

Effective plastication elements must produce uniform shear as well as good distributive mixing in order to achieve intensive mixing of melt and unplasticated solids.

The discharge parts of the extrusion system (screen changer, pelletizing, etc.) generate pressures between 150 and 300 bar. This high pressure leads to lengthy backing up of material in the screw machine and involves additional energy consumption, which in turn leads to high temperatures on the fast-rotating screw.

Table 4. Machine data and rates of large machines for compounding powdery polyolefins (HDPE, PP, LLDPE)

Machine type	Screw diam. D mm	Max. screw speed n min^{-1}	Max. drive power N kW	Rate G t/h	Plastication unit	Control mechanism	Discharge mechanism
Non-intermeshing, counter-rotating screws, two-stage Type FCM (Farrel)	229 (9") 305 (12") 381 (15")	375 400 320	1,102 2,200 2,940	5.5 to 6.8 9.1 to 11 14.5 to 18	kneading rotors	movable throttle valve, changeable material discharge cross-section	single discharge screw or gear pump
Non-intermeshing, counter-rotating screws, two-stage Type CIM-S (Japan Steel Works)	160 200 250 320	630 590 545 490	750 1,260 2,100 3,750	3.5 to 5.5 5 to 8 8.5 to 14 15 to 25	kneading rotors	axially adjustable conical annular clearance between screw shaft and housing	single discharge screw
Intermeshing, co-rotating twin-screws, single-stage, Type ZSK (Werner & Pfleiderer)	170 240 300	300 250 198	1,550 3,200 5,200	4 to 7 7 to 12 12 to 18	kneading disks	turnable throttle valve, changeable discharge cross-section	– – –
Intermeshing, co-rotating twin-screws, two-stage (gear pump) Type ZSK + 'Polyrex' (Werner & Pfleiderer)	170 240 300	300 250 198	1,550 3,220 5,200	4 to 8 7 to 13 12 to 22	kneading disks	turnable throttle valve, changeable discharge cross-section	'Polyrex' 280/180 'Polyrex' 320/224 'Polyrex' 400/280

Figure 7. Housing of a pressurizing gear pump Type Polyrex 320 for polyolefins. (Photo: Maag Zahnräder AG, Zurich, Switzerland)

Non-intermeshing, counter-rotating screws have relatively poor pressure build-up capability. In order to avoid an unacceptable rise in temperature resulting from back-up length, and in order to keep the machine short, the pressure build-up is carried out by separate short, slowly rotating, single-screw melt extruders.

Intermeshing, co-rotating twin screws provide a better basis for pressure build-up in a relatively short screw length. Therefore, in many cases the increased energy consumption (and thus the material temperature increase, caused by material back-up) remains within acceptable limits. This is assuming, of course, that screw geometry in the pressure-development zone is optimized. Therefore, in many cases, it is possible to operate with a single-stage concept.

2.4 Compounding of polyolefins

Gear pumps offer the best possible efficiency for pressure generation. Where low material temperatures are a special requirement, a combination of screw kneader with downstream gear pump is often used to generate pressure with a minimum of temperature rise.

In the past few years, large-scale gear pumps for melt rates up to 22 t/h at high viscosities and with back pressures of 300 bar have become available. They were developed by Maag in Zurich, in a joint effort with Werner & Pfleiderer, and sold under the tradename "Polyrex" (Figure 7). "Polyrex" gear pumps with rates of 20 t/h are a proven state-of-the-art [25, 26].

Machines that are used for polyolefin powders are also used for compounding polyolefins available in pellet form.

2.4.4 Screen changer and pelletizing systems

To deal with the high rates of compounding polyolefins, suitable screen-changer and pelletizing systems were developed; these are also used for polystyrene. In order to accommodate large filter surfaces in as small a space as possible, SZW screen changers use filter cartridges instead of flat filter plates (Figure 8). With these cylindrical filter cartridges, the filtering surface area can be up to 3.5 times that available with flat filter plates for a given cross-section of melt channel. It is economically very important not to have degraded product entering the melt stream during screen changes. When this happens, the dirty material has to be flushed through as off-spec material for as much as half an hour, wasting substantial quantities of product at the very high operating rates of polyolefin compounding units.

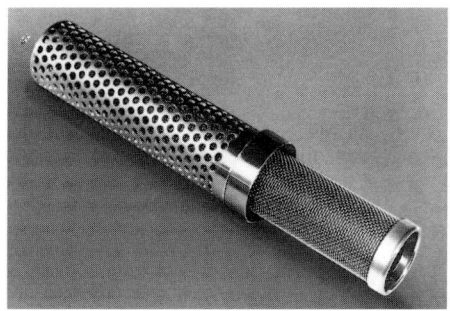

Figure 8. Filter candle with support basket for screen change equipment Type SWZ. (Photo: Werner and Pfleiderer, Stuttgart, West Germany)

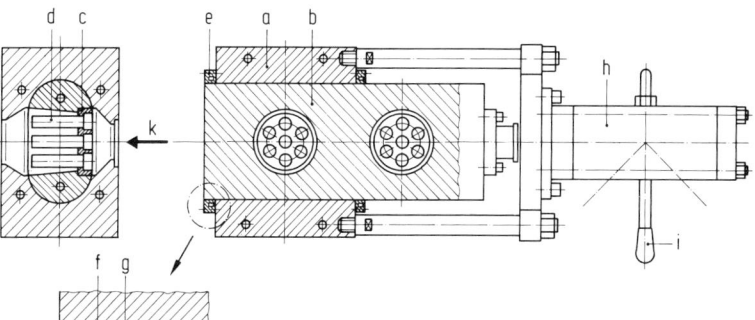

Figure 9. Screen change unit Type SWZ with filter basket and thermal sealing for contamination-free operation
a housing, *b* slide plate, *c* support plate, *d* filter element, *e* cooling plate, *f* cooling plate, *g* heat barrier plate, *h* hydraulic cylinder, *i* hand-control slider, *k* melt feed

Figure 10. Underwater granulator Type UG 500, with screen changer: granulator hood open. Output range for polyolefins up to 25 t/h. (Photo: Werner & Pfleiderer, Stuttgart, West Germany)

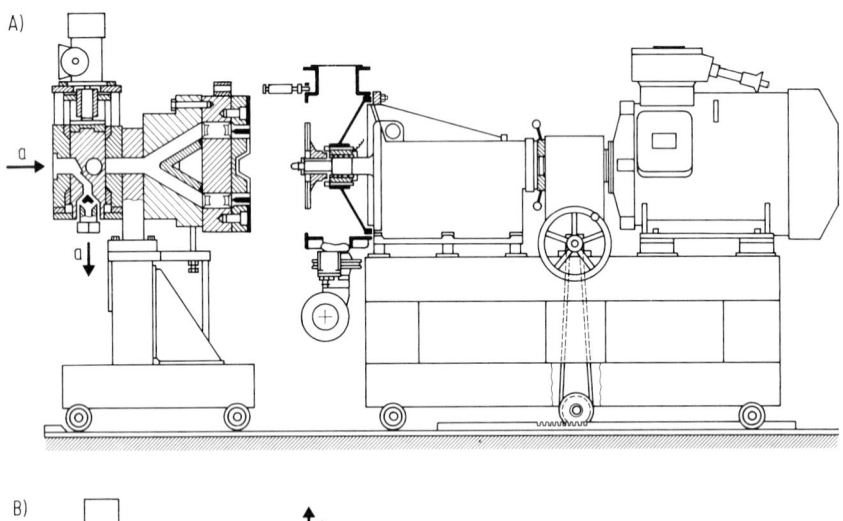

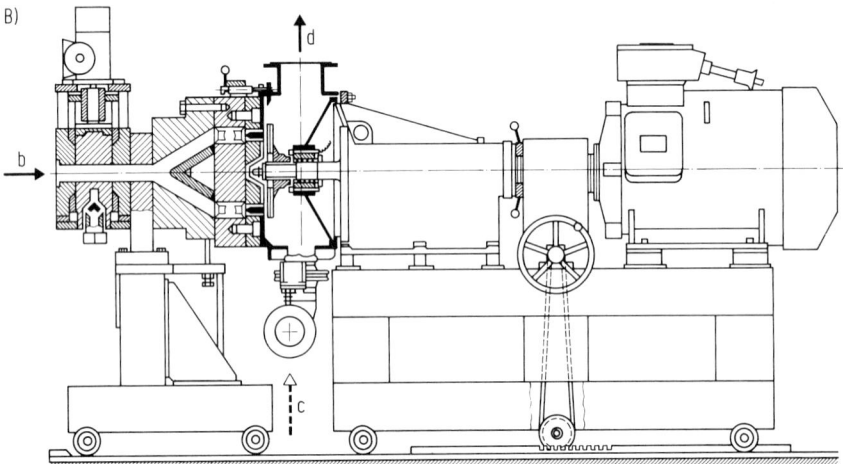

Figure 11. Underwater granulation with start-up valve
A) Granulator hood open, start-up valve in dump position, B) start-up valve in operating position
a start-up product, b material inflow, c water inflow, d granules in water stream

Therefore, in large-scale production, it is important that no degraded material particles enter the product stream during the movement of the screen changer plate. The screen changer shown in Figure 9 fulfills this requirement. Here, the screen changer plate has an oval cross-section which fits tightly into the screen changer housing. Sealing of the melt is achieved by freezing the product along the entire circumference of the screen changer plate. In this design, dead spots, corners and unswept areas are avoided [3].

For pelletizing polyolefins on big machines, underwater pelletizers with rates up to 25 t/h are used. For HDPE and LLDPE with a melt index of $2.16/190\,°C \leq 10$ (g/10 min), it is possible to use water-ring pelletizers up to rates of 15 t/h. Figure 10 shows an underwater pelletizing unit (type UG 500) for rates up to 25 t/h. This picture clearly shows that screen changers and pelletizing units for such large production machines are more than ancillary pieces of equipment. We are talking about highly developed machines embodying a significant amount of know-how. Start-up valves can be successfully used in combination with underwater pelletizers for easier start-up and to avoid false starts. Figure 11 shows underwater pelletizers and start-up valves in start position and in operating position. With the use of such start-up valves, it is possible to avoid freezing the melt in some of the bores of the pelletizing die during start-up. After the start-up valve is placed in operating position, each bore of the plate sees full rate immediately. The introduction of water can be accurately coordinated in time sequence with the arrival of the product at the bore exits. The activation of the start-up valve and the admission of water into the underwater pelletizer are achieved by pressing one switch [4, 26].

2.5 Compounding of PVC and other temperature-sensitive polymers

The premixing of PVC powders with additives is generally carried out in high-speed mixers. This state-of-the-art technology is well known and remains unchanged [27, 11].

In many cases – for example in pipe extrusion of unplasticized PVC – it is preferred to feed PVC premix in powder form straight into the extrusion machine. In other cases, pelletized PVC compounds are preferred, especially plasticized PVC for cable and shoe production, and in the case of rigid PVC for blown bottles or containers.

Screw machines with greatly differing designs are used for compounding the premix, using the following unit operations [11]:

– plastication,
– distributive and dispersive mixing of additives,
– degassing (if appropriate) of small amounts of volatiles,
– filtering of the melt,
– pelletizing or feeding into a calender.

In comparison with polyolefins, the production rates are much smaller:

– rigid PVC; 1 t/h, in rare cases up to 3 t/h
– plasticized PVC: 2 t/h, in rare cases up to 6 t/h

It is of critical importance in production planning with extremely shear- and temperature-sensitive PVC resins, to maintain a uniform and controllable product temperature and dwell time. Accordingly, the following machine systems are used:

– intermeshing, counter-rotating twin-screw machines [11],
– plasticators, as specialized machines for plasticized PVC [11, 28],
– Ko-Kneaders with single-screw discharge units [11, 29],
– planetary extruders with single-screw discharge units [11],
– intermeshing, co-rotating twin-screw machines with single-screw discharge (Kombiplast) [11, 28].

When feeding a calender, the downstream single-screw pelletizing unit is eliminated. Intermeshing, counter-rotating twin screws are widely used for the direct extrusion of PVC premixes. Figure 12 shows these units used with a maximum screw diameter of 170 mm at rates of 1.5 t/h for rigid and up to 3 t/h for plasticized PVC.

Intermeshing, counter-rotating twin-screw machines are limited to a screw diameter of 170 mm and circumferential speeds of as low as 0.2 to 0.3 m/s. This arises from the characteristics of PVC and the operating principle of this type of machine, which cause significant wear problems.

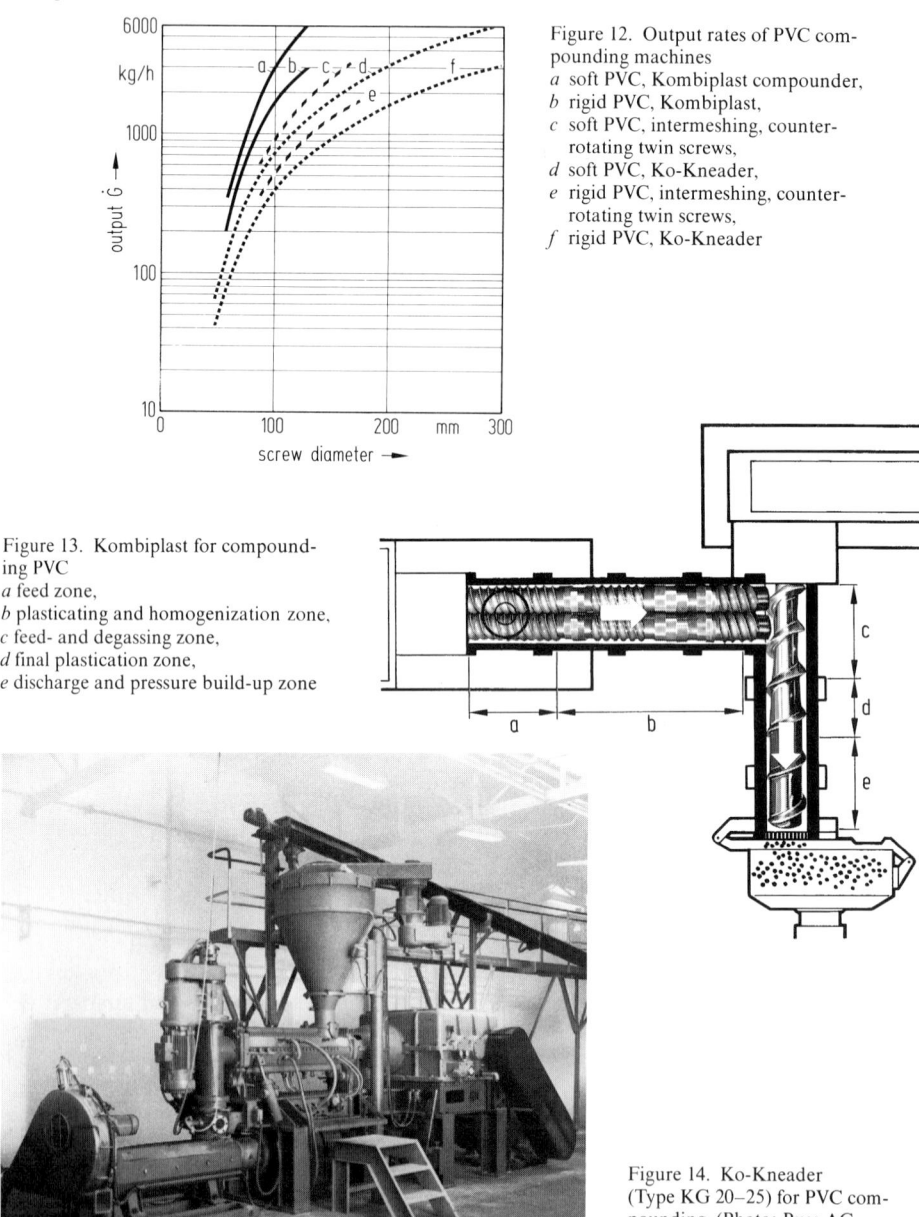

Figure 12. Output rates of PVC compounding machines
a soft PVC, Kombiplast compounder,
b rigid PVC, Kombiplast,
c soft PVC, intermeshing, counter-rotating twin screws,
d soft PVC, Ko-Kneader,
e rigid PVC, intermeshing, counter-rotating twin screws,
f rigid PVC, Ko-Kneader

Figure 13. Kombiplast for compounding PVC
a feed zone,
b plasticating and homogenization zone,
c feed- and degassing zone,
d final plastication zone,
e discharge and pressure build-up zone

Figure 14. Ko-Kneader (Type KG 20–25) for PVC compounding. (Photo: Buss AG, Basel, Switzerland)

2.5 Compounding of PVC and other temperature-sensitive polymers

Ko-Kneaders and co-rotating twin-screw machines can be used for higher rates and, because of their action, significantly higher circumferential speeds than with the intermeshing, counter-rotating machines are permissible.

However, because of the temperature sensitivity of PVC, it is possible to use high screw speeds only in combination with open machine discharge. To generate pressure for pelletizing, it is necessary to use separate short, slowly rotating single-screw pumps for the Ko-Kneader, in addition to using the high-speed co-rotating twin-screw machines (Figure 13).

Plasticating action must be accurately and reproducibly adjusted, and this need has led to the development of special control valves. Figure 14 shows a Ko-Kneader with an additional vertical control screw in the adaptor between Ko-Kneader and single screw. By changing the screw speed of the first screw, the back-up of PVC material in the Ko-Kneader is regulated and thus the amount of energy input is affected. Figure 15 shows the throttle valve in the Kombiplast (co-rotating twin screw with downstream slowly rotating single screw). The throttle valve is at the end of the twin screw, and by lengthwise displacement of the screw shafts relative to the barrel assembly, the amount of exposure of the polymer to the plasticating elements can be changed, and thus the amount of energy input can be changed continuously. Figure 16 shows the relationship between energy input and product temperature for different PVC formulations as a function of the throttle-valve position.

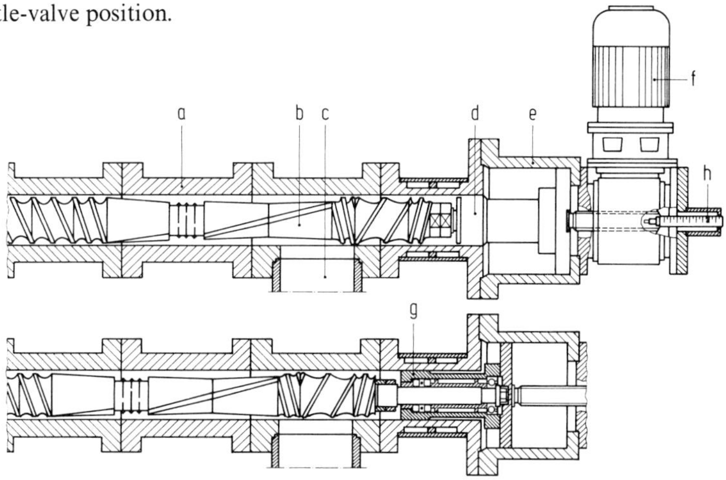

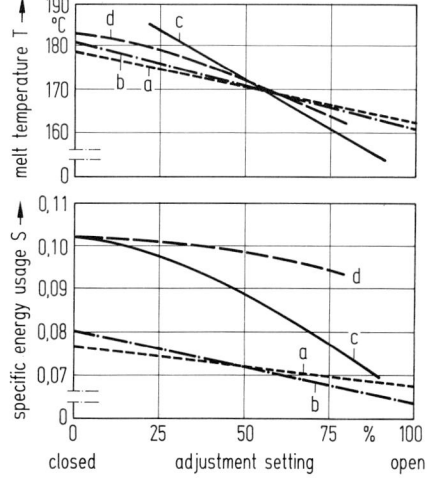

Figure 15. Kombiplast for PVC compounding: control unit for setting the degree of gelling. above: set for minimum gelling, below: for maximum gelling.
a housing,
b kneading elements,
c transfer pipe to discharge screw,
d shaft bearings,
e bearing guide housing,
f drive motor,
g slidable bearing bush,
h adjustment scale

Figure 16. Material temperature (top) and specific energy consumption (bottom) as a function of the throttle setting on a Kombiplast 800 with various PVC recipes
a soft PVC tube material,
b soft PVC cable material,
c rigid PVC extrusion material,
d rigid PVC injection molding material

Figure 17. Eccentric air die-face pelletizer for PVC. (Photo: Werner & Pfleiderer, Stuttgart, West Germany)

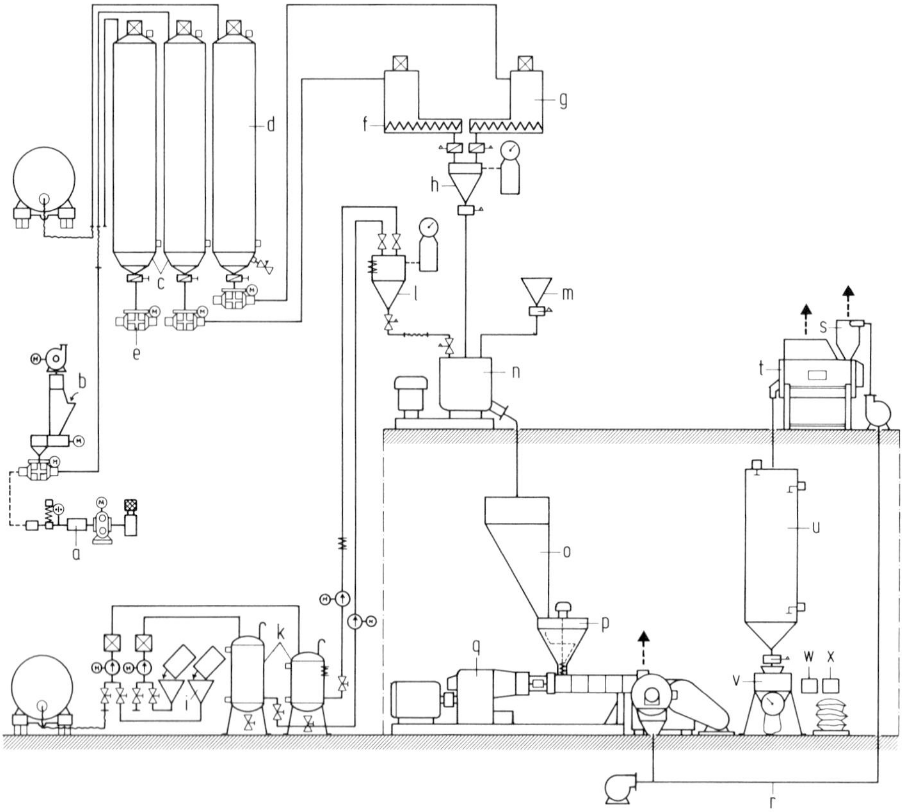

Figure 18. Flow diagram of a PVC compounding plant. (Photo: Werner & Pfleiderer, Stuttgart, West Germany) *a* blower station, *b* sack feed, *c* PVC silos, *d* chalk silos, *e* blower station, *f* day-tank and discharge screw for PVC, *g* day-tank and discharge screw for chalk, *h* tank scales for PVC and chalk, *j* plasticizer feed from drums, *k* plasticizer tanks, *l* plasticizer tank scales, *m* feed point for additives, *n* high-speed mixer, *o* intermediate tank, *p* stuffer screw, *q* Kombiplast KP, *r* pellet transport, *s* cyclone, *t* pellet cooler, *u* intermediate silo, *v* bagging scales, *w* bag sealing unit, *x* bag sewing unit

PVC is pelletized in general with hot die-face cutting in air, usually with eccentric knife assemblies (Figure 17).
Figure 18 shows a flow sheet for a PVC compounding system.
Other temperature-sensitive thermoplastics such as crosslinkable PE, foamable PE, or thermoplastic rubbers are compounded in systems similar to those for PVC.

2.6 Compounding of polystyrene and styrene copolymers

Most polymerization systems for polystyrene yield polystyrene as a melt. For pelletizing such melts, it is sufficient to use gear pumps in association with pelletizing units (strand pelletizers up to 5 t/h or underwater pelletizers).
The main task with polystyrene is to remove monomer down to a residual content of 500 ppm (food quality level). Starting concentrations of monomer, depending on the polymerization system used, vary from 15 to 20% to 0.2 to 0.5% [30].
For degassing, the following units are used:
- flash kettle with discharge pumps (mostly internally developed by polystyrene manufacturers),
- thin film evaporators [19],
- single-screw extruders,
- intermeshing, co-rotating twin-screw machines,
- combinations of flash kettles with screw machines.

Screw machines are used primarily when, together with the degassing of monomers, it is necessary to incorporate additives such as plasticizers, lubricants or colorants (color concentrate introduced via side-extruders or in the form of liquid color). Single-screw extruders with a screw diameter of 200 mm and screw length of 36 D can handle 3.5 t/h of high-impact polystyrene in reducing styrene concentration from 0.2 to 0.05% [17].
A co-rotating, twin-screw machine of type ZSK 170, used with feedstock containing 15% styrene, can bring the residual styrene concentration down to 300 ppm at a rate of 3.5 t/h. In this case, the addition of 2% water at two locations as stripping medium results in foam degassing (increased surface for mass transfer).
For SAN and ABS, residual acrylonitrile of 5 ppm is required and can be achieved in co-rotating twin-screws of type ZSK. The starting level of acrylonitrile is between 20 and 500 ppm.

2.7 Compounding of engineering plastics

Engineering plastics include a large number of thermoplastics, all thermosets and all reinforced and filled plastics. In typical compounding tasks, the following unit operations are required:
- plastication.
- distributive and dispersive mixing of additives (stabilizers, colorants, flame retardants),
- removal of small amounts of volatiles,
- filling if indicated,
- reinforcing, if indicated,
- alloying with other polymers, if indicated,
- pelletizing.

Engineering plastics are produced in very small quantities compared with standard bulk polymers. Because of the differentiated and very technical end-uses, a large number of formulations and special compounding conditions are required. Typical rates are therefore in the range of 100 to 1500 kg/h. In special cases, rates of up to 5 t/h have been achieved.

2.7.1 Systems for thermoplastic engineering plastics

Materials include polyamide (PA), polyester (PET, PBT), polycarbonates (PC), acrylonitrile-butadiene-styrene copolymer (ABS), polyacetal (POM), cellulose acetate (CA), polyurethane (PUR), polymethyl-methacrylate (PMMA), modified polyphenyloxide (PPO), polyphenylenesulfide (PPS), polysulfone (PSO) and polytetrafluorethylene (PTFE).
Many different machine systems are used:

- single-screw machines up to 150 mm screw diameter and screw length of 25 to 30 D [1],
- Ko-Kneaders up to 200 mm screw diameter and rates up to 1.5 t/h [1, 31],
- intermeshing, co-rotating twin-screw machines with screw diameters up to 133 mm and rates up to about 3000 kg/h (in special cases up to 170 mm screw diameter and rates up to 5 t/h) [1, 32, 33, 34, 35, 36].

Table 5 shows machine data for different types of compounding machines for engineering plastics.

Table 5. Data for engineering plastics compounding machines

Machine type	Screw diam. D mm	Max. screw speed n min^{-1}	Max. drive power N kW
Intermeshing, co-rotating twin-screw machine, Type ZSK (Werner & Pfleiderer)	40	500	34
	58	450	90
	70	400	144
	92	350	275
	133	300	714
Intermeshing, co-rotating screws, Type ZE/ZE-A (Bestorff)	40/43	450	30
	60/64	400	95
	75/80	350	160
	90/96	350	350
	130/140	300	800
Ko-kneader (Buss)	46	240	17
	70	240	34
	100	240	136
	140	240	360
	200	240	590

2.7.2 Systems for production of filled thermoplastics

The following operating conditions are important for achieving desired product quality with filled thermoplastics:

- uniform and dependable feeding of the polymer and fillers into the compounding unit
- effective degassing in the melt zone in order to assure complete wetting of the filler material by the polymer melt,
- good dispersive mixing of fillers into the melt while reducing to a minimum the heat- and shear history of the polymer matrix, to avoid polymer degradation.

At the beginning of their development, engineering plastics used a low level of filler concentration compounded in standard single- or twin-screw extruders. In response to special requirements in the feeding of fillers and to achieve precise product treatment under efficient conditions, it became necessary to use specialized screw kneaders. These have now almost entirely replaced standard extruders. In most cases, co-rotating twin-screw machines are used in single- or two-stage versions, as well as specialized single-screw compounding machines. Thus, controlled product movement through each stage of the process is assured, even with high filler concentrations. Also, the high shear forces necessary for proper dispersion of the fillers can be achieved together with homogeneous distribution of the filler in specially designed mixing zones.

Fillers used depend on the application and include minerals such as calcium carbonate, talc, mica, asbestos, and so on, as well as organic products such as wood flour, cellulose and cotton. Obviously, the various fillers are very different in terms of their physical structure. Superficially, all of them appear to be more or less free-flowing powders. A microscopic check reveals, however, varying structures (prisms, cubes, flakes, fibers or spherical structures as small as 10 µm in size) with resulting significant differences in flow characteristics.

This leads to the conclusion that, with such widely differing raw material properties, it is impossible to design single, standardized, optimal feed systems, especially when filler concentrations ranging from 20 to 80% by weight have to be accommodated.

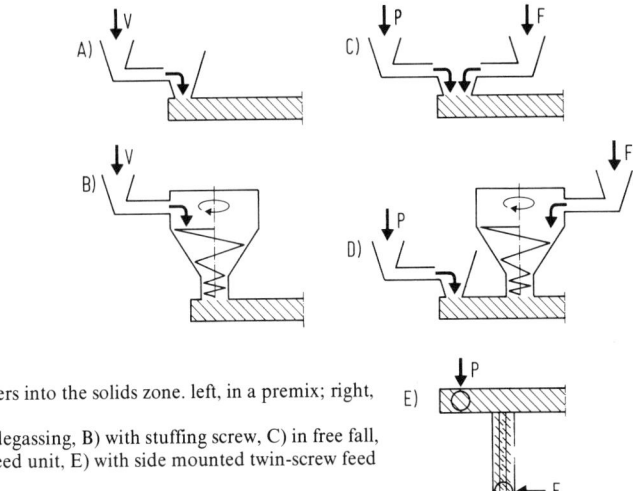

Figure 19. Feed of fillers into the solids zone. left, in a premix; right, by separate dosing
A) in free fall, partial degassing, B) with stuffing screw, C) in free fall, D) with single-screw feed unit, E) with side mounted twin-screw feed unit
F filler, P polymer, V premix of polymer and filler

In order to cope with this great variety of requirements, several feed systems were developed to provide economically and technically optimized feed into the compounder. Figures 19 and 20 show how fillers can be fed together with a polymer in the solids-conveying section, or into the melt. Feeding into the solids section is used with large particles or difficult-to-disperse and untreated fillers, which need high shear forces for adequate dispersion. In the case of powder polymers, a premix with a filler is generally prepared, and fed by gravity or with crammer screws. For pelletized polymers, it is advisable to feed the polymer separately into the feed throat of the compounder, or to downstream feed ports by gravity or with vertical or horizontal single- or twin-screw feeders.

Fillers are added to the melt when they are of a fine particle size requiring distributive mixing only, with very little particle size reduction. Addition of fillers to the melt greatly reduces wear. The use of specially designed feed barrels or flanged feed screws is common. Finally, it is possible to combine both approaches, especially when very large filler concentrations are required.

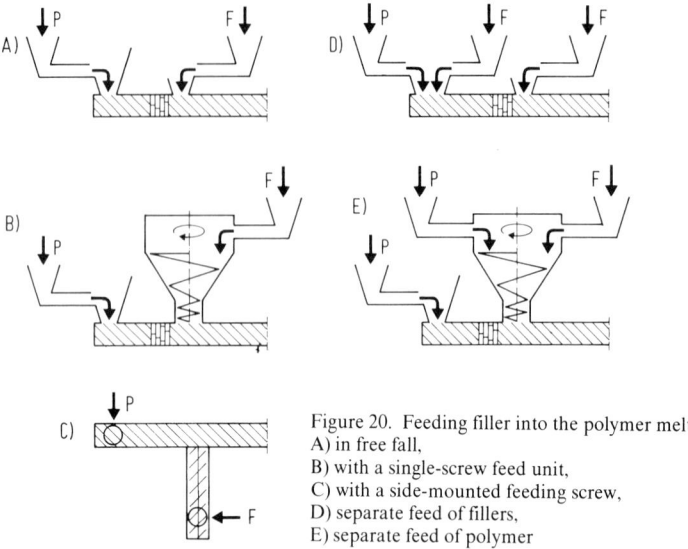

Figure 20. Feeding filler into the polymer melt
A) in free fall,
B) with a single-screw feed unit,
C) with a side-mounted feeding screw,
D) separate feed of fillers,
E) separate feed of polymer

The proper choice of feed system is essential to attain optimal use (maximum output) from a production system. Figures 21 and 22 show two typical examples based on co-rotating twin-screw compounders used in a single-stage set up.

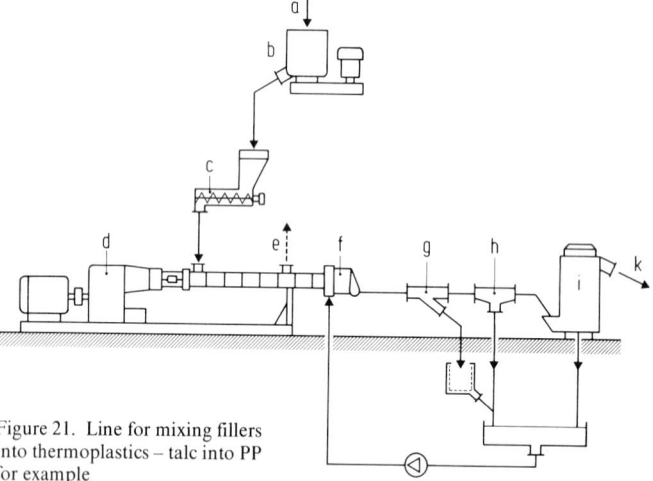

Figure 21. Line for mixing fillers into thermoplastics – talc into PP for example

a feed of PP, talc, and additives, *b* high-speed mixer, *c* dosing screw, *d* twin-screw kneader ZSK, *e* vacuum degassing, *f* water-cooled die-face granulator, *g* start-up diverter, *h* prelim. dewatering chute, *i* pellet drier, *k* to bagging

All fillers contain finely divided air and moisture inclusions. Both have to be removed from the polymer and the additives during the compounding process, together with other volatiles. Otherwise the wetting of the filler will not be adequate, and porous products will result. No mechanical deficiency necessarily results from such porous pellets; however, in thin-walled injection-molded parts made from such pellets, there may be surface defects which can be optically detrimental, and which may detract from the usefulness of the product. Results from well degassed and poorly degassed products are shown in Figures 23 and 24.

2.7 Compounding of engineering plastics

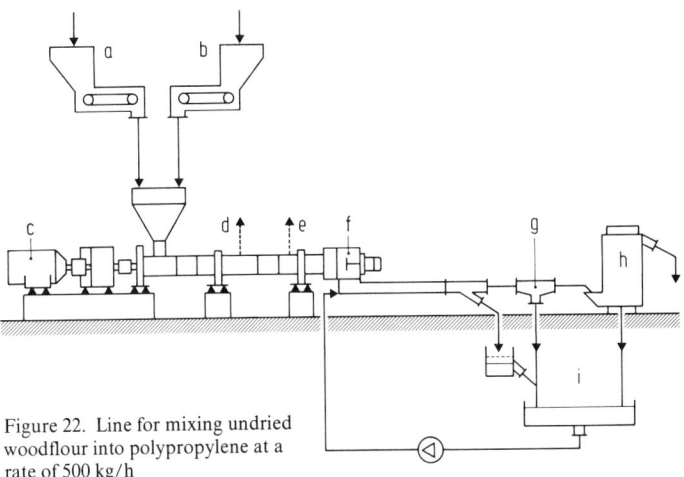

Figure 22. Line for mixing undried woodflour into polypropylene at a rate of 500 kg/h
a PP metering unit, *b* woodflour dosing unit, *c* co-rotating double-screw kneader ZSK, $D = 90$ mm, *d* degassing through free opening, *e* degassing under vacuum, *f* pelletizer, *g* pre-dewatering chute, *h* drier, *i* water circulation unit

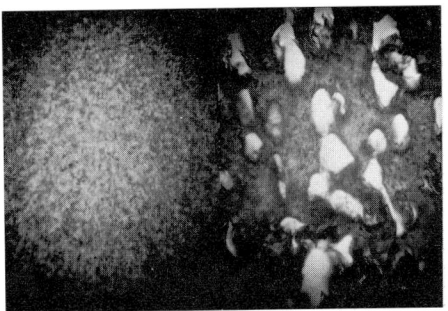

Figure 23. Photo-micrograph of well degassed (left) and poorly degassed, porous pellets (right). Magnification ×35. (Photo: Werner & Pfleiderer, Stuttgart, West Germany)

Figure 24. Moldings made from well degassed (left) and poorly degassed (right) pellets

Compounding systems for filled polymers use at least one degassing zone in the melt section downstream of the filler addition. In most cases this zone is operated under vacuum. It is generally not sufficient to use feed-hopper venting [31 to 36].

2.7.3 Systems for the production of reinforced thermoplastics

Typical reinforcing materials are glass fibers, carbon fibers, steel fibers, cellulose fibers and other organic fibers. The most widely used reinforcing agent is glass fiber. Glass fiber is added to the polymer mostly in the form of free-flowing, chopped textile glass (fiber length between 3 and 12 mm) up to a concentration of 50% by weight. Feeding is carried out almost exclusively through a second feed port into the polymer melt, thus reducing machine wear as well as attrition of the glass fiber, which would occur if feeding took place upstream of the plastication zone. Fibers are mixed homogeneously with the polymer melt, and volatiles are removed from the product by vacuum degassing, to achieve optimum wetting of the fibers by the thermoplastic matrix. Pelletizing is generally done in stranding or hot die-face cutting units [33, 37].

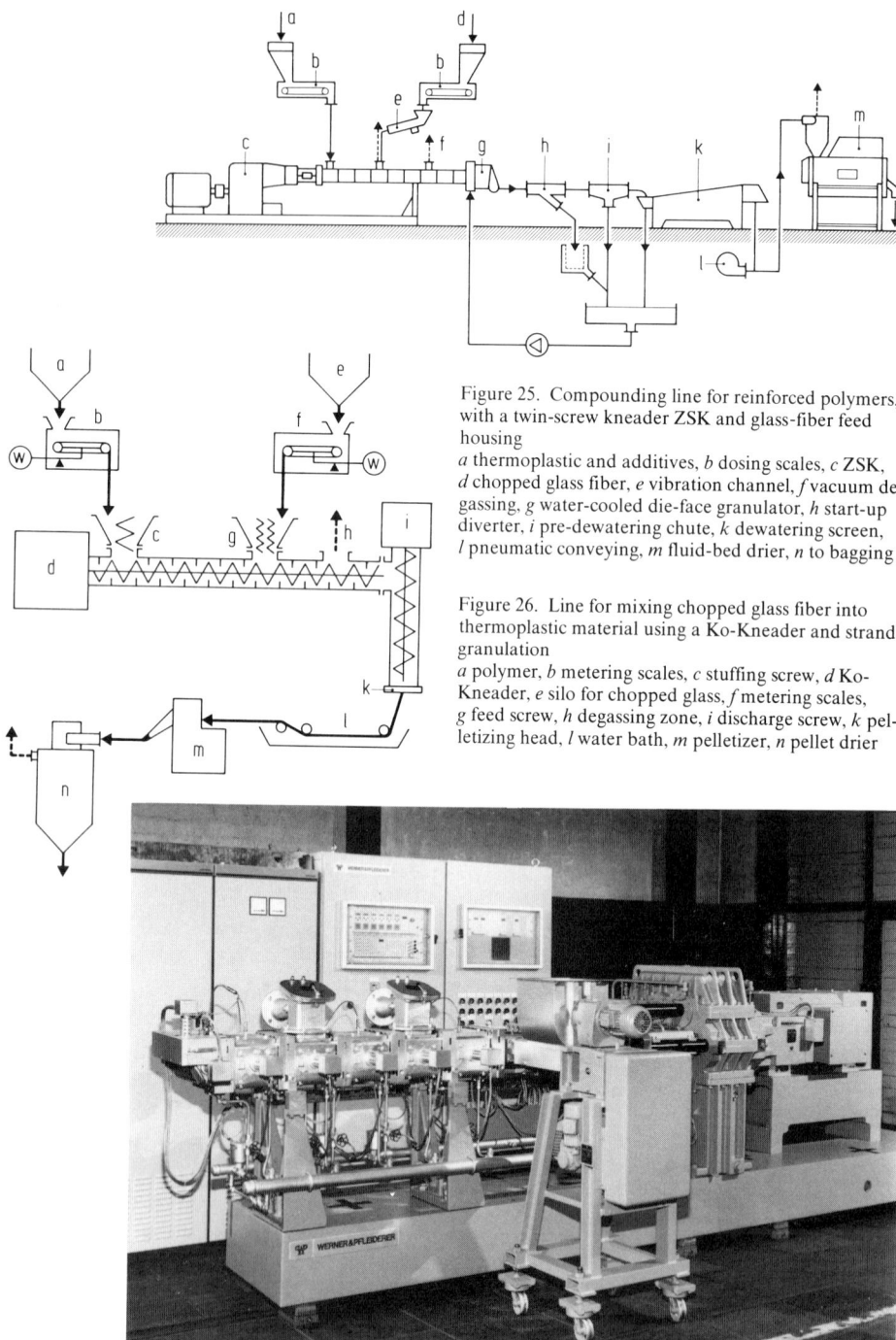

Figure 25. Compounding line for reinforced polymers, with a twin-screw kneader ZSK and glass-fiber feed housing
a thermoplastic and additives, *b* dosing scales, *c* ZSK, *d* chopped glass fiber, *e* vibration channel, *f* vacuum degassing, *g* water-cooled die-face granulator, *h* start-up diverter, *i* pre-dewatering chute, *k* dewatering screen, *l* pneumatic conveying, *m* fluid-bed drier, *n* to bagging

Figure 26. Line for mixing chopped glass fiber into thermoplastic material using a Ko-Kneader and strand granulation
a polymer, *b* metering scales, *c* stuffing screw, *d* Ko-Kneader, *e* silo for chopped glass, *f* metering scales, *g* feed screw, *h* degassing zone, *i* discharge screw, *k* pelletizing head, *l* water bath, *m* pelletizer, *n* pellet drier

Figure 27. Co-rotating twin-screw unit Type ZSK 70 with coupled horizontal twin-screw feed unit Type ZS-B 70 for glass fiber or fillers (Photo: Werner & Pfleiderer, Stuttgart, West Germany)

2.7 Compounding of engineering plastics

Figure 25 shows the flow sheet of a typical unit using chopped fiber.
The polymer or polymer premix is fed into the first feed port of the screw kneader of type ZSK. Melting takes place in barrel sections 2 and 3. In barrel section 4, the chopped glass is added to the polymer melt. In the downstream barrel section, the glass is wetted by the melt, homogenized, and properly mixed in. Volatiles are removed in barrel section 7. Barrel section 8 serves to generate pressure to overcome the pressure drop of the die.
Figure 26 shows a similar system based on a Ko-Kneader in a two-stage arrangement [31].
The various manufacturers of compounding machines have developed different methods of feeding the fiber into the melt stream. Single- or twin-screw, horizontal- or vertical-feed screws are used.
It is advantageous to use gravity feed through special feed ports without auxiliary screws, or twin-screw side-feeders flanged onto the main compounder, as shown in Figure 27.
With co-rotating twin-screws, fibers can also be fed in the form of endless rovings, which are automatically pulled at a steady rate into the processing section by the screws (Figure 28). Rovings are used where chopped fibers of a free-flowing nature are not available, or where rovings are available at a substantial price advantage.

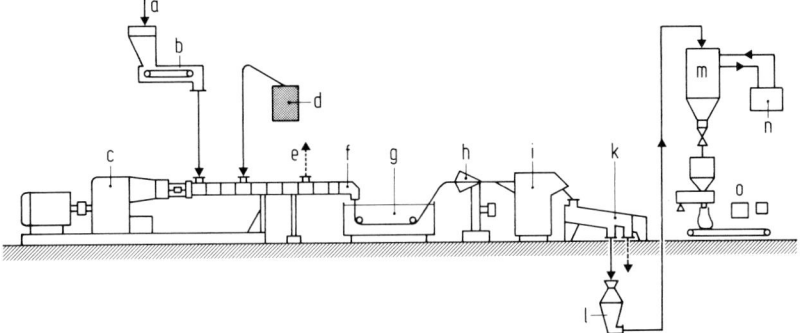

Figure 28. Mixing of glass rovings and strand granulation
a Thermopastic + additive, *b* dosing scales, *c* twin-screw kneader ZSK, *d* glass rovings, *e* vacuum degassing, *f* pelletizing head, *g* water bath, *h* blow-off unit, *i* pelletizer, *k* classifier, *l* pneumatic conveyor, *m* dried-material holder, *n* drier, *o* bagging station

The screw geometry for the production of reinforced engineering plastics has to have a great variety of different screw zones in order to cope with the multitude of unit operations (e.g. infeed of rovings, plastication, degassing, mixing of the fibers, generation of pressure, degassing under vacuum, generation of pressure for pelletizing, etc.). Such variety of screw zones of different geometry is best achieved in machines designed on a modular principle, using screw elements that are assembled on a screw shaft. In almost all co-rotating twin-screws, the Ko-Kneader and some counter-rotating twin-screw machines, such modular principles are used.
The modular principle has additional advantages, considering the wear that is inevitable when reinforcing polymers with mineral-based reinforcing materials.
Sections of high wear can be economically replaced without changing the rest of the screw, resulting in significantly reduced costs. With the proper use of wear-resistant materials, a screw-lifetime of at least 5000 hours in the glass-fiber mixing zone, and of at least 12,000 hours in the barrel section, can be achieved with 30% glass-filled nylon 6.
For co-rotating twin-screw machines, massive oval wear inserts of special steel (Figure 29) are used to protect the barrel sections. The screw elements themselves are made from fully hardened steel, protected by layers of sintered hard-wearing metal carbides.
Compound elements with a hardened steel outer layer and a ductile core have now been developed.

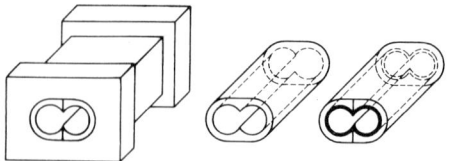

Figure 29. Wear-protected ZSK housings with oval sleeves

2.7.4 Pelletizing systems for thermoplastic engineering plastics

For pelletizing, lace extrusion with or without automatic lace guidance, and hot die-face cutting systems are used. Air/water pelletizing units with horizontal water screen or water vortex have been in use for quite some time. Water-ring pelletizers are used mainly for POM and filled polyolefins at rates of over 1 t/h.

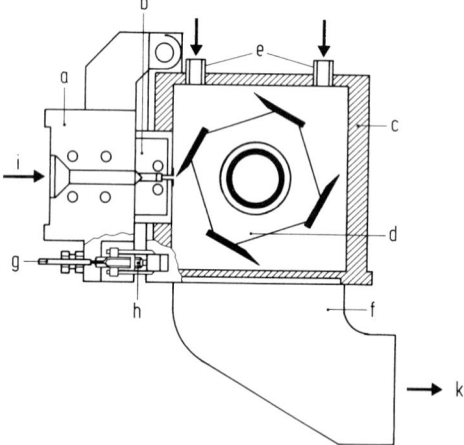

Figure 30. Rotating knife granulator for throughputs of up to 1 t/h
a pelletizer head, *b* perforated strip, *c* hood, *d* cutting mill, *e* water feed nozzle, *f* pellet-catching tray, *g* high-speed chuck, *h* adjustable stop, *i* material input, *k* pellet in water stream

In a newly developed die-face pelletizing system (Figure 30) for rates up to 1000 kg/h, the melt is cut directly at the die by a rotating knife assembly. Water used to cool the pellets is mostly in the form of an air/water mixture, with a water spray added. The freshly cut pellets are cooled on the flight path from the die plate to the exit port from the pelletizer, and carried out with the cooling water along a special cooling track into the pellet drier [38].

2.7.5 Systems for thermosets

Originally, thermosets (phenol-formaldehyde, amino resins, unsaturated polyesters, epoxy resins) were compounded on roll mills or, depending on product type, on other discontinuous compounding systems. Roll mills have now been largely replaced by continuous kneaders such as the Ko-Kneader, planetary extruder, intermeshing, co-rotating twin-screw extruder, and intermeshing, counter-rotating twin-screw machines. At the end of the compounding process, the material is discharged at ambient pressure in chip form and then ground to small particle size.

In order to improve the feed behavior on injection molding machines, certain thermosets are now supplied in pellet form. Figure 31 shows a compounding line for producing cylindrical pellets of thermosetting materials based on a Ko-Kneader followed by a special pelletizing screw [39].

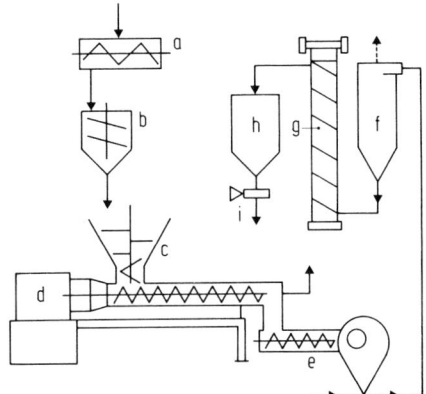

Figure 31. Flow diagram of a line for pelletizing thermosetting molding compounds
a premixing, *b* intermediate store, *c* buffering and feeding (feed hopper), *d* melting, homogenization, impregnation (Ko-Kneader), *e* pelletizing (pelletizing screw), *f* separation of pellets, *g* cooling, *h* silo storage, *j* bagging

2.8 Compounding of elastomers

Compounding of elastomers is conditioned by:
- the form in which the rubber is supplied (mainly bales, and very rarely as pellets, crumb or powder),
- the large number of components in a typical formulation,
- an unusual variety of formulations, especially in the area of technical rubber goods.

As a result, discontinuous, high-intensity mixers (Banburys) are widely used in the tire industry, as well as in formulations for technical rubber goods [40, 41].
Screw kneaders do find increasing use in all cases where the rubber is supplied in free-flowing form, and where enough of a particular formulation has to be produced to justify an economically feasible, continuous operation.
For such continuous operations, the same types of machines are used as in the compounding of thermoplastics (e.g. intermeshing, co-rotating twin-screw extruders of type ZSK, single screw Ko-Kneaders, planetary extruders and non-intermeshing, counter-rotating screw kneaders of type FCM). There are also special machines such as single-screw pin extruders, single-screw EVK kneaders with shear and distributive mixing elements, or integrated two-stage continuous kneaders MVX (M = mixing, V = venting, X = extruding). If more than 10 components are used in a formulation, a premix is indicated and is usually prepared on a high-speed mixer. If a smaller number of components are used, separate feeding is feasible [42 to 44].
Continuous compounding of elastomers is being practiced in certain applications in the cable, shoe and insulation industries (Figure 32). In most cases, we are dealing with a highly filled formulation based on EPDM, EVA or thermoplastic rubber. For the borderline case of elastomer-modified thermoplastics, the same continuous screw kneaders are used as in other areas of thermoplastic compounding.

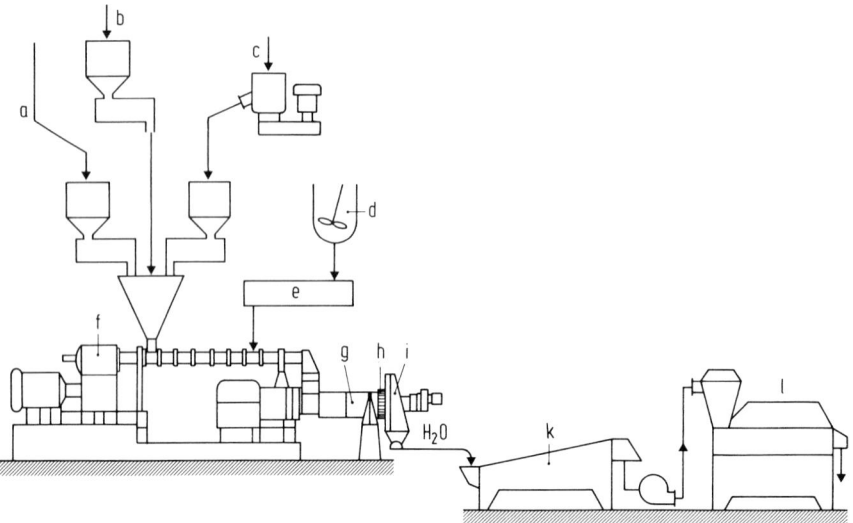

Figure 32. Production of EPDM cable material (medium-high voltage range)
a EPDM silo, b filler silo, c additives, d peroxide, e piston pump, f ZSK, g delivery screw ESA, h screen, i eccentric pelletizer, k swing screen, l pellet drier GFC

2.9 Direction of development

The machinery- and process-related developments in the last few years have been concerned largely with the continued development of known and proven machine systems.
With respect to their basic principle, compounding machines have remained about the same. Single-screw extruders, Ko-Kneaders, planetary extruders, plastificators, intermeshing co- and counter-rotating twin-screw machines, and non-intermeshing, counter-rotating screws, have been optimized and further developed so that each machine system offered today is in its third, fourth or fifth generation. Figure 33 demonstrates the kind of development that has taken place, by comparing the ZSK 83 of the first ZSK generation with a ZSK 70 of the fifth generation.

High on-stream factor

When designing machines for large throughputs of 20 t/h or more, special consideration must be given to ensuring a high on-stream factor. This is especially true because modern polymerization systems are built in such a way that the entire reactor output is compounded on a single compounding machine, and because economics dictate the elimination of large buffer silos between reactor and compounding section.
Today, in this type of circumstance, an entire compounding line for polyolefins, including main drive and peripherals, must have an overall on-stream factor of 95 to 97% on a five-year average. Figure 34 shows a typical on-stream diagram over seven years in large production units of the ZSK type including main drive, screen changer and underwater pelletizing system. In such a system, there are a large number of components that must function properly all at the same time in order to achieve such high on-stream factors. This example should give an idea of the very high level of reliability needed for this multitude of line components (motors, couplings, bearings, gearboxes, pumps, oil systems, barrel sections, screws, die plates, knives, valves, heat exchangers, recording and measurement instruments,

2.9 Direction of development

Figure 33. Upper: ZSK 83 of 1957. Lower: ZSK 70 of 1984. (Photo: Werner & Pfleiderer, Stuttgart, West Germany)

relays, switches, etc.). A high on-stream factor in high-rate production machinery can be more significant than the initial purchase cost of the equipment when sophisticated economic calculations for such a plant are made.

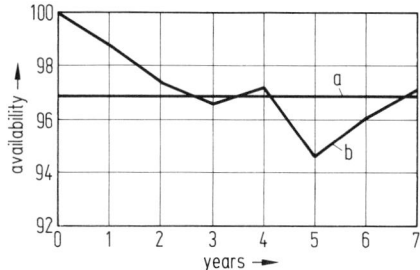

Figure 34. Typical availability diagram of large polyolefin compounding machines (Type ZSK) a average value over five years, b general: overhaul at 5 years

High screw speeds

The throughput rate of compounding machinery is directly proportional to the screw speed, and thus a continued demand for high rates and low price/rate relationships has led to higher and higher screw speeds. Limits of screw speed are set by machine wear and stress on the product. For co-rotating twin-screw machines circumferential speeds of 3 m/s are quite usual.

High torque capabilities

The throughput rate of a compounding machine is also directly proportional to applied torque. High torque capabilities are also necessary in order to operate screw machines with a high degree of fill. Partially filled screws running at high speeds because of lack of torque, generate high product temperatures which are generally not desirable for quality reasons. Table 6 shows the torque capabilities of a ZSK 83 of the first ZSK generation compared with a ZSK 92 of the fifth generation. If we use the torque divided by the third power of the screw diameter as a basis, the torque capabilities have been increased by 61%. Such a substantial increase was possible only after the development of very highly rated speed reducers and distribution gears, and high-torque screw shafts. In those cases where a modular principle is used, with a screw shaft having slide-on screw elements, the coupling between screw element and screw shaft becomes very important.

Table 6. ZSK Screw torque from 1957 to 1984

Machine generation	Machine type	Screw diam. D mm	Torque M N·m
First generation (1957)	ZSK 83/700	83	1,750
Fifth generation (1984)	ZSK 92/v	92	3,850

Low specific energy input

The throughput rate of a compounding machine is inversely proportional to the specific energy demanded (kWh/kg). In order to maximize throughput rate and at the same time reduce energy cost and product stress, it is desirable, at each stage of the process, to arrive at the lowest possible specific energy consumption. This can be achieved by developing optimum screw geometries with high-efficiency plasticating elements and energy-efficient mixing elements, and by reducing machine length to a minimum consistent with the process requirements. All machine producers have made efforts to reduce specific energy input, and it has thus also been possible to reduce product temperatures; this is a desirable result, usually, for product quality reasons.

Today there is a method available for any compounding requirement. Further developments are aimed at three basic targets:

- improvement of product,
- reduction of price/rate relationship,
- reduction of operating costs.

Improving product quality in most cases requires:

- less stress on the product at the lowest possible product temperature,
- uniform and reproducible conditions being maintained in the compounding process to hold narrow tolerances on product properties.

Figure 35. Controller Type EPC 11 for automatic data gathering and fault analyses, automatic start-up and shut-down on continuous compounding lines. (Photo: Werner & Pfleiderer, Stuttgart, West Germany)

2.9 Direction of development

In order to achieve these aims, the following measures are the focus of further effort:
- development of dependable, robust and reliable feeding- and operational techniques,
- automation of the entire compounding system by use of microcomputer technology. This measure ensures that previously proven operating conditions can be automatically reproduced and the process continuously monitored, with automatic data recording and failure analysis; in addition the start-up and shut-down sequences can be automated. Figure 35 shows a control system for a continuous compounding system (type EPC 11) based on a microprocessor and a programmable controller [45].
- ensuring constancy of product quality by on-line measurement of essential product properties, with automatic feedback, to eliminate deviations from standards;
- further development of compounding machinery in the direction of relatively large screw depth, high available torque, shorter machine length, mixing elements with minimum shear input, uniformly operating melting elements and throttling valves which allow variation of pressure build-up in specific areas of the screw;
- development of compounding processes where several heat histories are combined, to avoid multiple melting and intermediate pelletizing, with the aim of reducing energy consumption. Such a combination of different conversion steps may be used to build a line for compounding and final conversion into an end product, where it can be technically and economically justified. A typical example of this is the production of highly filled polymer sheet [46].

Improvements in economics require:

- low operating costs
- low ratio of investment to output rate

Reduction of operating costs implies achieving:

- lower energy costs by reducing specific energy requirements,
- lower personnel costs by automating the compounding process,
- lower repair costs by using machines with a high on-stream factor and good wear and corrosion protection,
- high on-stream factor (low down-times and short turn-around) by using machines with high on-stream factors and high reliability, and self-diagnosing systems for early identification of potential problems,
- low reject rate and off-spec production by improved feeding technology, process automation, and on-line measurement of product properties with feedback for quality control.

The following equation applies to price/rate relationships:

$$S = \frac{N}{\dot{G}} \approx \frac{M \cdot n}{\dot{G}}$$

$$\frac{P}{\dot{G}} \approx \frac{P \cdot S}{M \cdot n}$$

G rate (kg/h)
N available drive power (kw)
M available torque (mkp)
n screw speed (rpm)
S specific energy input (kwh/kg)
P investment cost ($)

It follows from these relationships that to reduce the price/rate relationship, the specific energy input has to be reduced and the available torque has to be increased. The chosen screw speed should be as high as possible; machine wear and product quality will determine the limits here.

In addition, there is an obvious requirement to reduce the machine price by intelligent design and efficient manufacturing methods.

The technical and economic targets identify the need for automation, on-line measurement of product properties with closed-loop feedback, improved feed systems, high available torque and the lowest possible energy consumption for the given process. These are the targets for almost all of the development work being done in the field of compounding.

References for Chapter 2

For the convenience of the reader the English titles of all publications in languages other than English are shown in parentheses.

[1] VDI-K (Ed.): Mischen von Kunststoffen (Mixing of Plastics). VDI-Verlag, Düsseldorf, 1983.
[2] VDI-K (Ed.): Entgasen von Kunststoffen (Degassing of Plastics). VDI-Verlag, Düsseldorf, 1980.
[3] VDI-K (Ed.): Filtrieren von Kunststoffschmelzen (Filtration of Polymer Melts). VDI-Verlag, Düsseldorf, 1981.
[4] VDI-K (Ed.): Granulieren von thermoplastischen Kunststoffen (Pelletizing of Thermoplastic Materials). VDI-Verlag, Düsseldorf, 1974.
[5] *Brandenburger, K.:* Kunststoffe 23 (1933), pp. 149/153.
[6] *Herrmann, H.:* Kunststoffe 75 (1985), pp. V/XVI.
[7] *Schneider, W., Zettler, H. D., Jeckel, G.:* VT-Verfahrenstechnik 12 (1978), pp. 477/485.
[8] *Herrmann, H.:* Schneckenmaschinen in der Verfahrenstechnik (Screw Machines in Processing Technology). Springer Verlag, Berlin, Heidelberg, New York, 1972.
[9] *Herrmann, H.:* Mischer, Kneter, Granuliervorrichtungen (Mixing, Kneading and Pelletizing Equipment), in *Johannaber, F., Stoeckhert, K. (Ed.):* Kunststoff-Maschinen-Führer (Plastics Machinery Guide). Carl Hanser Verlag, München, Wien, 1984.
[10] VDI-K (Ed.): Aufbereiten von Polyolefinen (Compounding of Polyolefins). VDI-Verlag, Düsseldorf, 1983.
[11] VDI-K (Ed.): Aufbereiten von PVC (Compounding of PVC). VDI-Verlag, Düsseldorf, 1979.
[12] Extruder für die Verarbeitung von Kunststoffen (Extruders for Processing Plastics). Company brochure Berstorff, Hannover, West Germany.
[13] *Anders, D.:* Gummi, Asbest, Kunststoffe 33 (1980), pp. 614/621.
[14] *Anders, D.:* Kunststoffe 66 (1976), pp. 250/257.
[15] *Werner, H.:* Kunststoffe 71 (1981), pp. 18/26.
[16] *Werner, H.:* Mehrwellensysteme (Multiscrew Systems). In: Entgasen von Kunststoffen (Degassing of Plastics). VDI-Verlag, Düsseldorf, 1980.
[17] *Anders, D.:* Entgasen von Einschneckenextrudern für die Aufbereitung von Thermoplasten (Degassing of Thermoplastics in Single-screw Extruders for the Compounding of Thermoplastics). In: Entgasen von Kunststoffen (Degassing of Plastics). VDI-Verlag, Düsseldorf, 1980.
[18] Melt Feed Twin Screw Vent Extruder. Company brochure JSW. The Japan Steel Works Ltd., Tokyo, Japan.
[19] *Heimgartner, E.:* Entgasen in Dünnschichtverdampfern (Degassing in Thin-layer Evaporators) In: VDI-K (Ed.): Entgasen von Kunststoffen (Degassing of Plastics). VDI-Verlag, Düsseldorf, 1980.
[20] *Herrmann, H., Burkhardt, U.:* Kunststoffe 68 (1978), pp. 19/26.
[21] *Herrmann, H., Burkhardt, U.:* Kunststoffe 68 (1978), pp. 753/758.
[22] FCM-Continuous Mixer. Company brochure Farrel Machinery Group. USM Corp., Ansonia, USA.
[23] CIM-S. Continuous Intensive Mixer Super Series. Company brochure JSW. The Japan Steel Works Ltd., Tokyo, Japan.
[24] ZSK, Aufbereitungsmaschinen für Kunststoffe (ZSK Compounding Machines for Plastics). Company brochure Werner & Pfleiderer, Stuttgart, West Germany.
[25] *Kapfer, K., Eise, W., Herrmann, H.:* Integrated Gear Pump Optimizes Pressure Build-up and Discharge in Extruder Operations. 41st Antec, Chicago, 1983.

[26] *Mauch, K.:* Aufbereiten von Polyolefinen auf zweiwelligen, gleichläufigen Schneckenknetern (Compounding of Polyolefins on Co-rotating Double-screw Kneaders). In: VDI-K (Ed.): Aufbereiten von Polyolefinen (Compounding of Polyolefins). VDI-Verlag, Düsseldorf, 1984.
[27] *Schiffers, H. (Ed.):* Grundlagen der Aufbereitung von Polyvinylchlorid (Fundamentals of PVC Compounding). VDI-Verlag, Düsseldorf, 1978.
[28] Aufbereitungsanlagen für PVC- und VPE-Granulat (Compounding Lines for PVC and PE Granules). Company brochure Werner & Pfleiderer, Stuttgart, West Germany.
[29] Kontinuierliche Aufbereitung und Granulierung von Hart- und Weich-PVC auf Buss-Ko-Kneter-Anlagen. Baureihe KG und UKG (Continuous Compounding and Pelletization of Rigid and Soft PVC on Buss Ko-Kneader Lines. The MDK/E range). Company brochure Buss AG, Basel, Switzerland.
[30] *Hess, K. M.:* Kunststoffe 69 (1979), pp. 199/203.
[31] Kontinuierliche Aufbereitung von gefüllten und verstärkten Thermoplasten auf Buss-Ko-Kneter-Anlagen. Baureihe MDK/E (Continuous Compounding of Filled and Reinforced Thermoplastics on Buss Ko-Kneader Lines. The MDK/E range). Company brochure Buss AG, Basel, Switzerland.
[32] *Mauch, K.:* Kunststoffe 71 (1981), pp. 266/271.
[33] Der neue ZSK-Hochleistungs-Compounder (The New ZSK High-capacity Compounder). Company brochure Werner & Pfleiderer, Stuttgart, West Germany.
[34] *Hess, K. M.:* Kunststoffe 73 (1983), pp. 282/286.
[35] *Dienst, M.:* Kontinuierliche Aufbereitung im Zweischnecken-Extruder (Continuous Compounding on Twin-screw Extruders). Kunststoff-J. (1984), pp. 14/15.
[36] *Uhland, E.:* Einarbeiten von Füllstoffen und flüssigen Additiven in Thermoplaste (Mixing of Fillers and Liquid Additives into Plastics). In: VDI-K (Ed.): Mischen von Kunststoffen (Mixing of Plastics). VDI-Verlag, Düsseldorf, 1983.
[37] *Hess, K. M.:* Verstärken von Thermoplasten mit Fasern (Reinforcement of Thermoplastics with Fibers). In: VDI-K (Ed.): Mischen von Kunststoffen (Mixing of Plastics). VDI-Verlag, Düsseldorf, 1983.
[38] Messerwalzen-Granuliervorrichtung MWG (Knife-roll Granulation Unit MWG). Company brochure Werner & Pfleiderer, Stuttgart, West Germany.
[39] Kontinuierliche Aufbereitung von härtbaren Formmassen auf Buss-Ko-Kneter Anlagen. Baureihe PRD und DUG. (Continuous Compounding of Thermosetting Compounds on Buss Ko-Kneader lines. The PRD and DUG Ranges). Company brochure Buss AG, Basel, West Germany.
[40] VDI-K (Ed.): Der Mischbetrieb in der Gummi-Industrie (The Mixing Operation in the Rubber Industry). VDI-Verlag, Düsseldorf, 1984.
[41] *Lehnen, J. P.:* Kunststoffe 74 (1984), pp. 19/21.
[42] *Capelle, G., Meier, G.:* Kunststoffe 72 (1982), pp. 392/394.
[43] *Harms, E. G.:* Kunststoffe 74 (1984), pp. 33/35.
[44] *Wälty, O.:* Gummi, Asbest, Kunststoffe 31 (1978), pp. 414/418.
[45] Mikrorechnersteuerung für Extruder EPC 11 (Microprocessor Control for Extruder Type EPC 11). Company brochure Werner & Pfleiderer, Stuttgart, West Germany.
[46] *Cormont, J. J. M., u. a.:* Kunststoffe 73 (1983), pp. 599/602.

3 Extrusion of pipes, profiles and cables

W. Predöhl, P. Reitemeyer

3.1 Manufacture of pipes

3.1.1 Market significance and products

Pipes hold second place to film with regard to quantity in the extrusion of thermoplastics. More than 1.2×10^9 DM worth of pipes and shaped parts is manufactured from rigid PVC and polyolefins annually in the Federal Republic of Germany. The historical growth of this output is shown in Figure 1. The proportion of shaped parts is some 12 to 15%.

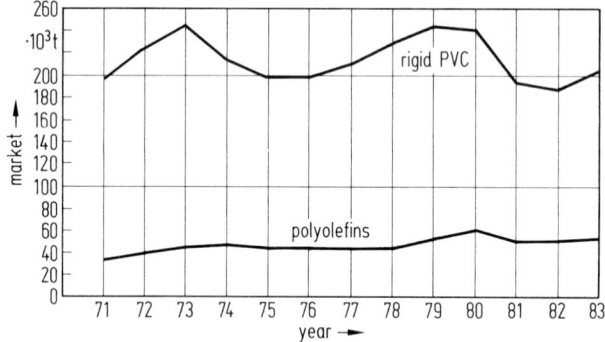

Figure 1. Pipes and formed parts from rigid PVC and polyolefins, market in the Federal Republic of Germany

This production can be broken down into the following products:
- conduit pipes (PVC, HDPE)
- pressure pipes (PVC, HDPE)
- cable- and electro conduit pipes (PVC, HDPE, LDPE)
- drainage pipes (PVC, HDPE)
- industrial pipes (PVC, HDPE, PP)
- house drainage pipes (PVC, PP)
- gutter- and drainage pipes (PVC)
- gas pipes (PVC, HDPE)
- pipes for domestic drinking water installation (HDPE)

The range of products is completed by semi-finished products for industrial processing, such as pipes made from other thermoplastic materials, e.g. polyamide pipes for the automobile industry, pipes and hoses from plasticized PVC for the medical sector, with polyamide- or polyester webbing for agriculture, biaxially stretched tubes for shotgun cartridges etc.

Most pipes and formed parts are destined for use in the construction sector, and the consumption is therefore dependent on the current level of activity in the building sector, which is also reflected in the annual GNP.

Some typical products with their most important features are listed in Table 1.

3.1 Manufacture of pipes

Table 1. Pipe products

Rigid PVC Pipes

Application	Standard	PVC K-value	Stabilizer
Drinking water pipes	DIN 19532	67 to 70	Pb, Sn, CaZn
Sewage pipes	DIN 19534	67 to 70	Pb, Sn
Drainage pipes	DIN 1187	64 to 67	Pb
Corrugated insulating pipes	DIN 49018	64 to 67	Pb
Cable conduit pipes	DIN 8062	64 to 67	Pb
Gutter down-pipes	DIN 8062	64 to 67	Pb, Sn, BaCd

Polyolefins and Polyamide pipes

Raw Material	MFI-value		Density	Crystalline melt range	Remarks
	g/10 min	Method	g/cm^3	°C	
HDPE	0.4 to 0.7	190/5	0.953 to 0.958	127 to 131	for pressure water pipes and drainpipes colored black with carbon black
LDPE	0.3 to 0.6	190/2.16	0.928 to 0.938	105 to 110	for pressure water pipes and drainpipes colored black with carbon black
PP	1.0 to 3.0	230/5	0.905 to 0.915	158 to 164	for pipe production colored grey
PP	1.2 to 2.0	230/5	0.935 to 9.945	158 to 164	for house drainpipes colored grey
PA 6			1.12 to 1.15	218 to 222	brake and fuel pipes for the automobile industry
PA 6.6			1.12 to 1.16	250 to 255	
PA 11/12			1.03 to 1.05	170 to 185	

3.1.2 Extrusion of pipes

Pipes made from thermoplastics are continuously extruded. Only wound pipes that are used in the conduit sector are produced in fixed lengths.
Twin-screw extruders are used for PVC, single-screw extruders for polyolefins and other thermoplastic materials. They plasticate the resin and convey it through a pipe die head. A haul-off draws the pipe through the calibration- and cooling section. Rigid pipes are cut to length and flexible pipes are wound.

3.1.2.1 Lines for the manufacture of pipes from rigid PVC

A line suitable for processing PVC is shown in Figure 2.

3.1.2.1.1 Extruders for producing pipes from rigid PVC

At present, twin-screw extruders with counter-rotating screws are used. As pipe formulations are less sensitive to shear in the melt than profile formulations, the screws can rotate faster. The large throughputs that are thereby achieved make long barrels necessary, so that the required amount of heat can be introduced into the PVC. Table 2 gives a comparison between typical machine data for pipe- and profile machines.

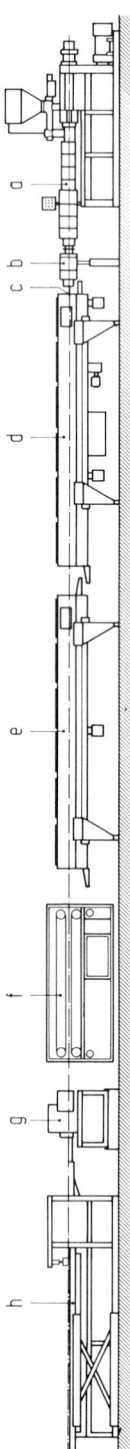

Figure 2. Plant for manufacturing pipes from rigid PVC
a twin-screw extruder, *b* pipe die head with die insert, *c* calibrating die, *d* vacuum tank calibration unit, *e* cooling unit, *f* haul-off, *g* cutting unit, *l* socketing machine

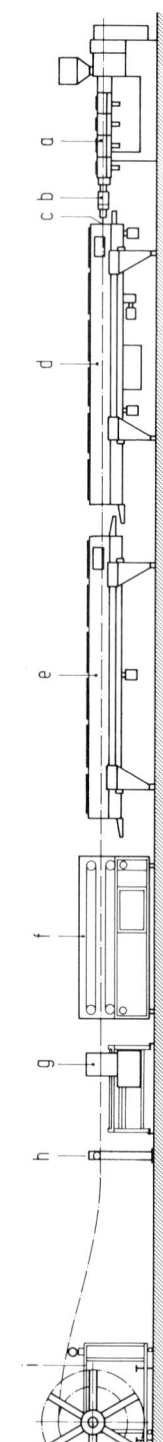

Figure 6. Production line for polyolefin pipes
a single-screw extruder, *b* pipe die head with die insert, *c* sizing die, *d* vacuum tank calibration unit, *e* cooling unit, *f* haul-off, *g* cutting unit, *h* pressure pad, *i* pipe winder

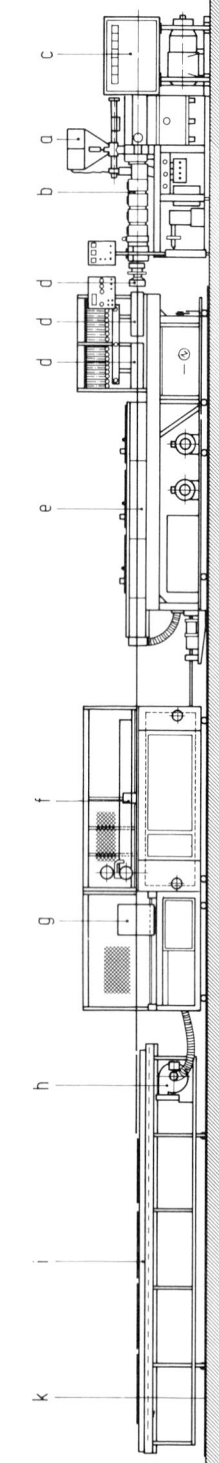

Figure 29. Extrusion plant for window main profiles
a dosing unit, *b* twin-screw extruder, *c* temperature and drive control cabinet, *d* profile die (die and calibration), *e* calibration base with vacuum tank, *f* haul-off, *g* cutting unit, *h* swarf exhaust unit, *i* discharge unit, *k* rail track

3.1 Manufacture of pipes

Table 2. Characteristic data for twin-screw extruders

Screw diameter mm	Screw length L/D	Rotation speed range min^{-1}	Drive output kW	Specific energy kWh/kg
Twin-screw extruders for pipes				
60 to 70	18 to 22	35 to 50	15 to 25	0.10 to 0.14
80 to 90	18 to 22	30 to 40	28 to 40	0.10 to 0.14
100 to 110	18 to 22	25 to 38	58 to 70	0.10 to 0.14
120 to 140	18 to 22	20 to 34	65 to 100	0.10 to 0.14
Twin-screw extruders for profiles				
45 to 55	16 to 18	35 to 50	10 to 16	0.12 to 0.16
60 to 70	16 to 18	30 to 40	16 to 20	0.12 to 0.16
80 to 90	16 to 18	20 to 30	20 to 28	0.12 to 0.16

3.1.2.1.2 Pipe die heads

Processing of PVC, which is thermally unstable, demands perfect flow channels: this need in essence is met by dies with a spider mandrel retainer (Figure 3).

The PVC melt comes from the extruder through a narrow inlet channel. The melt stream then fills out the relatively large volume of the die head and flows around the displacement mandrel. This is fixed by means of the spider mandrel retainer in the melt stream. The spider dimensions must be such that they can absorb the pressure on the mandrel. They have streamline profiles, to avoid stagnation of melt on the inflow and outflow edges.

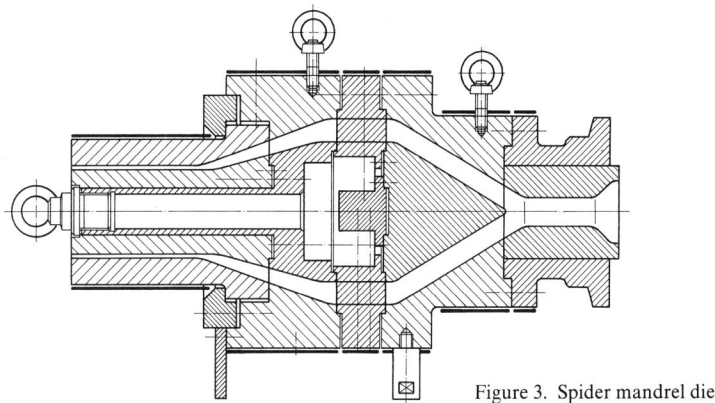

Figure 3. Spider mandrel die

In the extrusion of PVC pipes, the pipe die head, in addition to dividing the melt, must also fulfill the task of reducing thermal and mechanical inhomogeneities in the melt. The flow channels must therefore be dimensioned in relation to the throughput, so that during the time in the die head, thermal inhomogeneities can be eliminated and the melt that issues from the twin screws in a strained state can relax. Furthermore, it is necessary that the diameter and cross-section of the flow channel in the mandrel retainer area be made larger than the die diameter and exit cross-section. This enables the melt that has been divided into ring segments by the spider to be fused together again. The die inserts are flange-connected to the die housing. They can be off-centered in relation to the mandrel with the aid of centering screws to even out melt inhomogeneities. It is recommended, in the case of larger pipe dimensions, to heat the die mandrel with heater bands or heater cartridges. The requisite heat sensor and current connections are inserted through the hollow spider legs.

Each pipe die head can be used for a particular working range. By changing the die insert and mandrel, pipes of varying wall thickness and diameter can be produced. The ratio of the flow channel cross-sections in the mandrel-retainer area to that in the die-insert area is used to characterize the arrangement.

Figure 4 gives the general correlation. The upper limit results from the increasing flow resistance of the pipe die head, the lower limit from the increasingly poor welding together of the flow streams.

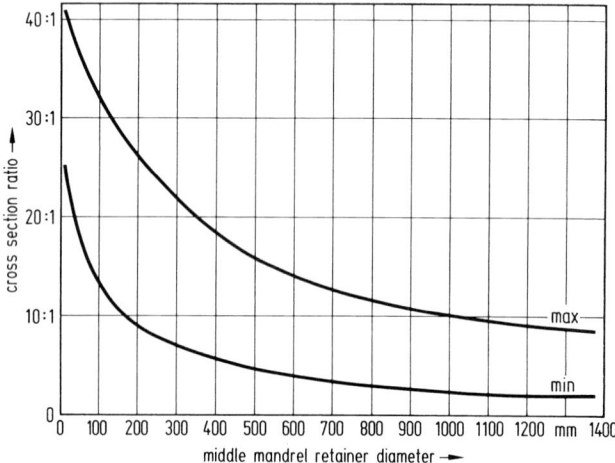

Figure 4. Bimodal curve for cross-section ratios in rigid PVC pipe die heads

Flow-channel volume, allocation of die insert to the pipe die heads, flow resistance and permissible dwell times of the melt in the die, determine throughput capacity ranges (Figure 5). If one takes into account all the given criteria, one obtains the working ranges of the pipe die heads listed in Table 3, in relation to the DIN pressure levels. The machine manufacturer or pipe manufacturer can select the appropriate dies corresponding to the required pipe dimensions.

In designing die inserts, the swelling of the extrudate must be taken into account. Normally, the following dimensions are calculated:

– outer diameter of the die (D_2) = outer diameter of the pipe (D)
– inner diameter of the die (D_1):

$$D_1 = \sqrt{D_2^2 - (D^2 - (D - 2 S_R)^2) \cdot \frac{100}{100 + q}}$$

– S_R thickness of the pipe wall.
– q die swell %.

The die swell coefficient is about 10 to 35%, depending on formulation and extrusion conditions.

3.1 Manufacture of pipes

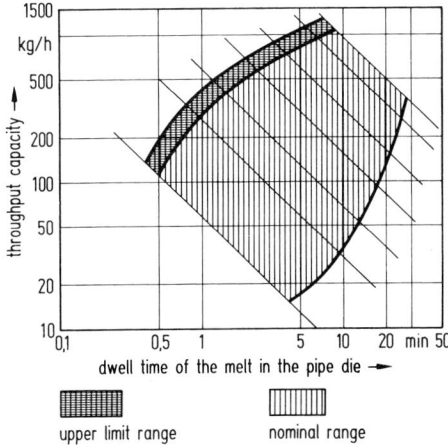

Figure 5. Throughput capacity range of rigid PVC pipe die heads

(Legend: upper limit range / nominal range)

Table 3. Working ranges of PVC pipe die heads. Where areas intersect both die sizes can be used

pipe outer diameter [mm]	pipe die head type	wall thickness according to DIN 8062 [mm] DIN-series				
		2	3	4	5	6
12	1					1,4
16	1				1,2	1,8
20	1				1,5	2,3
25				1,5	1,9	2,8
32	2			1,8	2,4	3,6
40			1,8	1,9	3,0	4,5
50			1,8	2,4	3,7	5,6
63			1,9	3,0	4,7	7,0
75	3	1,8	2,2	3,6	5,6	8,4
90		1,8	2,7	4,3	6,7	10,0
110		2,2	3,2	5,3	8,2	12,3
125		2,5	3,7	6,0	9,3	13,9
140	4	2,8	4,1	6,7	10,4	15,6
160		3,2	4,7	7,7	11,9	17,8
180		3,6	5,3	8,6	13,4	20,0
200	5	4,0	5,9	9,6	14,9	22,3
225		4,5	6,6	10,8	16,7	25,0
250		4,9	7,3	11,9	18,6	27,8
280	6	5,5	8,2	13,4	20,8	
315		6,2	9,2	15,0	23,4	
355		7,0	10,4	16,9	26,3	
400	7	7,9	11,7	19,1	29,7	
450		8,9	13,2	21,5		
500		9,8	14,6	23,9		
560	8	11,0	16,4	26,7		
630		12,4	18,4	30,0		
710		14,0	20,7			

3.1.2.2 Lines for producing pipes from polyolefins

A line for manufacturing pipes from polyolefins (PO) is shown in Figure 6 (see page 58). Pipes up to a diameter of approx. 160 mm can be wound up as long as the walls are not too thin. Thin walls lead to flattening of the pipes and to the danger of breaking. Larger pipes are cut into fixed lengths, as with PVC pipes, and stacked. The manufacture of PO pipes is comprehensively dealt with in [1, 2, 3].

3.1.2.2.1 Extruders for polyolefin pipes
(Single-screw extruders)

Over the years, the throughput of extruders has been considerably increased in parallel with the uprating and strengthening of drives, gearboxes, bearing assemblies and an increase in barrel length. Thus for an extruder with a screw diameter of 90 mm, and HDPE we have:

1961   90 kg/h,
1971   180 kg/h,
1981   280 kg/h.

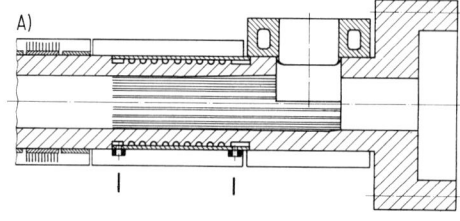
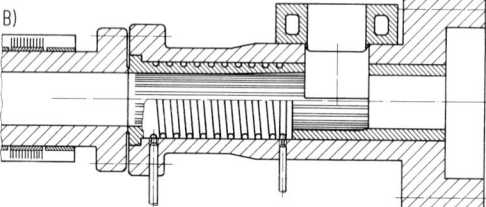

Figure 7. Assisted-conveying feed zone systems
A) standard, B) wet liners

After 1981/82, even higher throughput capacity became possible by equipping the extruder (30 D) with grooved feed bushes, and direct water cooling. Figure 7 shows the difference between the systems in the area of material feed, and Table 4 gives the throughput and torque requirements of the two systems. This maximum-performance extruder has not yet displaced the standard system.

Table 4. Comparison between two types of extruder for PO pipes

Design of cooled, grooved feed zone	Screw diameter mm	Screw length L/D	Drive output kW	Gearbox torque Nm	max. throughput	
					HDPE kg/h	PP kg/h
Standard	90	25	90	8 000	250	220
	120	25	145	18 000	450	350
Intensively cooled	90	30	132	11 000	450	420
	120	30	240	24 000	700	600

3.1.2.2.2 Pipe die heads

At first, the classical die head with spider mandrel retainer was used in processing polyolefins. With increasing throughputs and increasing pipe diameters and wall thicknesses, orientation in the material and uneven flow of the melt increased. The consequences were uneven wall thicknesses, reduction in strength, and shrinkage at elevated temperatures.

So the conception of the *mandrel retainer followed by a perforated ring* as retarder disk developed for the lower diameter range up to approx. 400 mm. This perforated ring serves to normalize the melt flow and reduces the effect of the flow disturbance through the spider (Figures 8 and 9).

With larger pipe dies, especially for pipes > 400 mm diameter, only a *perforated ring* is used by some die manufacturers. Here the mandrel retainer can also be omitted, because the perforated ring takes up the large axial forces.

In order to satisfy demands for high pipe quality, other types of die were developed. Here the spiral mandrel die and filter basket die head systems deserve particular mention.

3.1 Manufacture of pipes

Figure 8. Perforated disk mandrel-retainer. (Photo: BASF, Ludwigshafen, West Germany)

Figure 9. Flow configuration with spider mandrel die with additional perforated disk. (Photo: BASF, Ludwigshafen, West Germany)

The centrally fed *spiral mandrel die* (Figure 10) guides the melt through radially arranged distributor bores to the spiral channels. The melt leaves the channels through the overflow gap in an axial direction, until it again flows axially. In this way, the melt stream is evenly distributed and conveyed at once through an annular gap to the die exit [4, 5].

Nowadays the optimal geometry of the spiral distributor is calculated by computer.

The calculation assumes that the pressure loss in the gap and in the spiral channel is the same:

$$\Delta p_s = \Delta p_w$$

With $$\Delta p = \frac{12\, \eta \cdot \dot{V} \cdot L}{B \cdot H^3 \cdot f_p}$$

this gives

$$\left(\frac{12 \cdot \eta \cdot \dot{V} \cdot L}{B \cdot H^3 \cdot f_p} \right)_s = \left(\frac{12 \cdot \eta \cdot \dot{V} \cdot L}{B \cdot H^3 \cdot f_p} \right)_w$$

Δp pressure loss
η viscosity
f_p coefficient of flow
$\dot{V}$ volume flow
L length
B width
H height

Subscripts:
s gap, w spiral

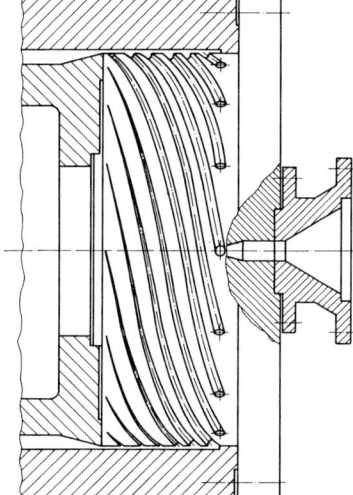

Figure 10. Spiral mandrel retainer system

When the viscosity curve has been calculated using the appropriate mathematical formula, the volume flows in the gap $\dot{V}_s$ and in the spiral $\dot{V}_w$ can be calculated from the relationships $\eta = f(\dot{\gamma})$ and $\dot{\gamma} = f(v)$; this is an iterative procedure. ($\dot{\gamma}$ = shear rate).

The geometry is then altered until an optimum melt distribution quality and pressure drop is attained.

Spiral distributor dies have the advantage over spider mandrel retainer dies that the spiral distributor need not be larger in diameter than the die insert. The average pipe diameter is close to the diameter of the spiral distributor, see Figure 3.11. So these dies, in addition to the other advantages, can also be made smaller and lighter.

Pipe die heads with spiral distributors are today used for pipes from 10 to 1600 mm outer diameter, and are classified according to DIN pressure levels for pipes, see Table 5.

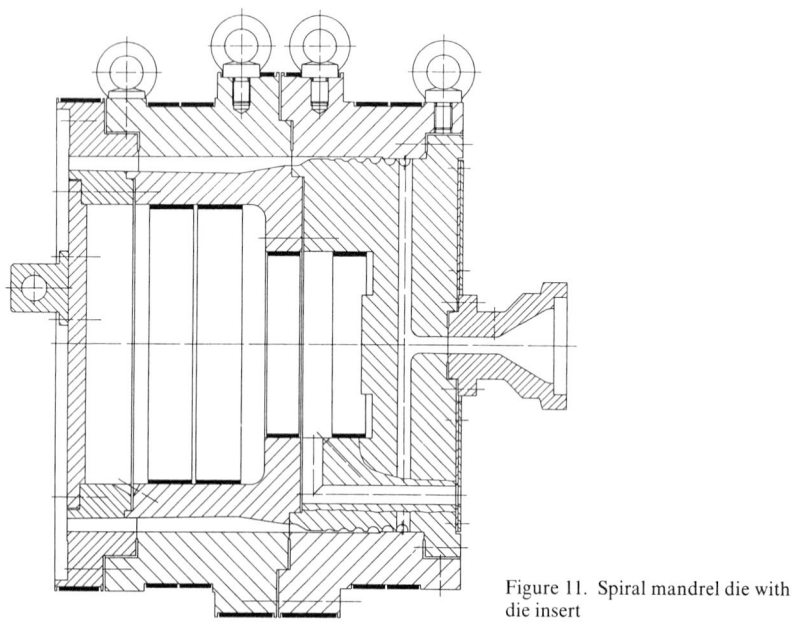

Figure 11. Spiral mandrel die with die insert

Table 5. Pipe dimensions (PO) available from a series of die heads

Type	Pipe range – outer diameter mm	Emerging cross-section cm²
A	6 to 32	0.1 to 3.8
B	16 to 63	1.1 to 17.3
C	12 to 90	0.6 to 21
D	50 to 200	3 to 104
E	110 to 280	10 to 205
F	225 to 450	38 to 527
G	355 to 630	95 to 1150
H	450 to 800	300 to 1070
I	710 to 1250	480 to 1670
K	1000 to 1600	945 to 2340

With the *filter basket die*, Figure 3.12, the perforated disk is so arranged in relation to the mandrel retainer that the melt stream flows radially outward. The plastic melt is diverted 90° twice in the filter basket. A disadvantage of this construction is that the re-feed borings cannot be included without also dividing the melt stream. It is helpful if a disk with webs or large bores is inserted as an intermediate flange. Here the bores for compressed air and electrical power can be made radially. The filter basket then conceals the spider marks.

The reason for development of this design was the desire to manufacture pipes with the smallest possible die head.

3.1 Manufacture of pipes

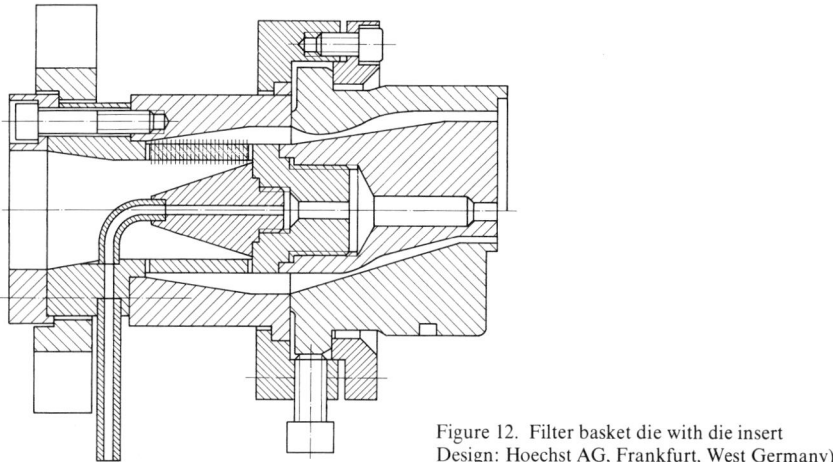

Figure 12. Filter basket die with die insert
Design: Hoechst AG, Frankfurt, West Germany)

Design of the die inserts

The design of die inserts (which determine the pipe dimensions) is carried out independently of the die head type used. The following guidelines are normally followed in the design of dies when polyolefins are processed:

1. Die diameter D = pipe diameter × 1.05 (with small and thin-walled pipes, the factor 1.05 becomes greater).
2. Die gap S = nominal wall thickness.
3. Length of the exit gap: 8 to 15 × S. The length becomes shorter with increasing pipe diameter and larger wall thickness.

Although the melt leaving the die swells up, the exit cross-section is made larger than the cross-section of the pipe. In this way the volume contraction resulting from crystallization is offset. Tension is produced by the extension of the melt between die and calibration units, and the transport of the melt in the calibration unit, up to hardening of the outer layer, is thereby stabilized.

3.1.2.3 Calibration systems for smooth-wall pipes

The tube of thermoplastic melt leaving the die head must be cooled, and calibrated by shaping. Great significance is attached to the calibration, both for exact dimensions and for stresses in the pipe wall. One differentiates basically between outer and inner calibration. Although inner calibration leads to a more favorable distribution of stress within the pipes, its application is limited to certain exceptions. It is complicated and difficult to handle and so pipes are usually sized from outside. Combination of the two is not possible.
Two systems have proved themselves satisfactory for external calibration: vacuum tank calibration and internal air pressure calibration.
Vacuum tank calibration (Figure 13) has generally predominated in the last few years. The size range at the present time lies between 4 and 1000 mm pipe external diameter. The molten plastic tube is drawn through a calibrator mounted in the vacuum tank, shortly after leaving the die. The calibrator is either a sleeve- or a disk type, depending on the particular thermoplastic.
The pressure against the calibrator is provided by the difference between that inside the pipe and the lower pressure in the vacuum tank.

Various pre-cooling devices are used to suit the viscosity and adhesive properties of the different thermoplastic materials.

The vacuum-tank calibration unit consists of a vacuum-tight cooling tank with centrally arranged spray nozzles. The tank can also be filled with water. As a rule, vacuum-tank calibration units are divided into two chambers, where the length of the first chamber is 1/6 to 1/7 of the total length. The chambers have their own control circuits, so that it is possible to work in graduated pressure stages.

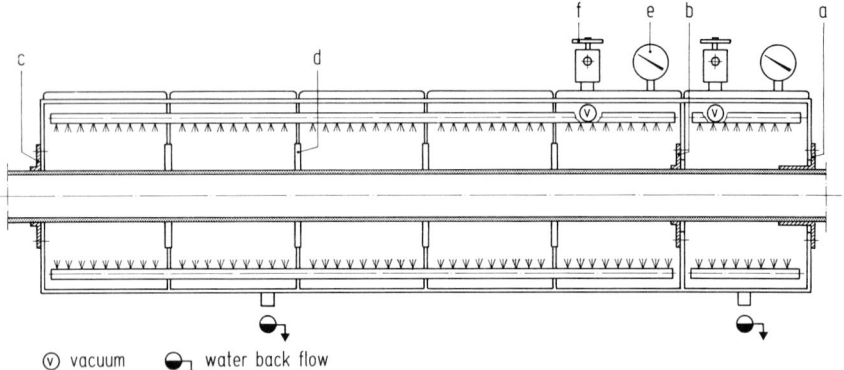

Figure 13. Vacuum-tank calibration with spray cooling
a calibrating die, *b* intermediate wall sealing, *c* exit sealing, *d* sizing ring, *e* vacuum meter, *f* air inlet valve

3.1.2.3.1 Calibrating inserts in vacuum-tank calibration units

These units exist in a wide variety of designs and in widely differing raw materials. Their design depends on the properties of the melt, e.g. viscosity, tendency to sticking, behaviour when cooling, contraction, and diameter of the melt tube, as well as wall thickness and haul-off speed.

Materials primarily used are those which possess good thermal conductivity, e.g. red bronze, brass and aluminum.

A one-piece perforated sleeve, Figure 14, is sufficient for low extrusion speeds and thermoplastic materials which cool down quickly and without shape distortion (rigid PVC).

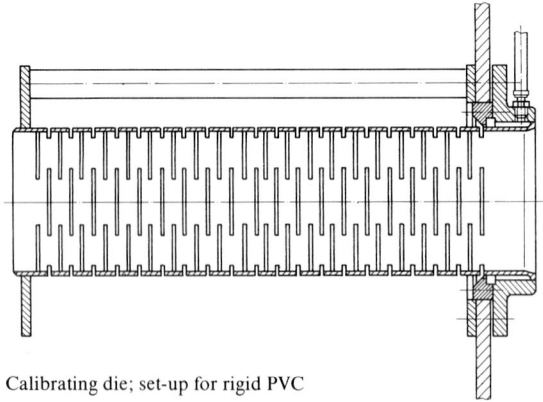

Figure 14. Calibrating die; set-up for rigid PVC

A sleeve with fixed pre-cooling is to be recommended with polyolefins for medium and slow haul-off speeds. Higher speeds, and all polyolefins, demand adjustable pre-cooling, see Figure 15.

3.1 Manufacture of pipes

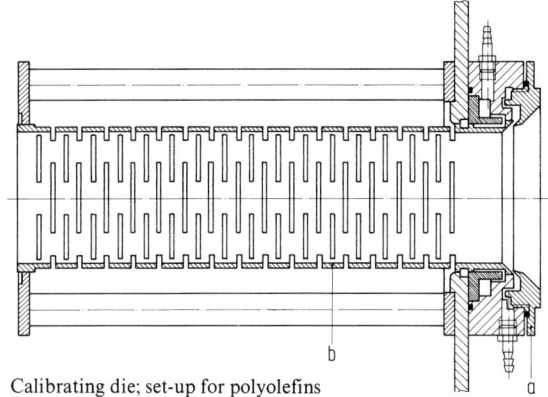

Figure 15. Calibrating die; set-up for polyolefins
a pre-cooling, *b* sleeve

The combination of sleeve/disk calibration (with adjustable pre-cooling according to requirements) is preferred for small pipes that are extruded at high speeds, and for thermoplastics that have a strong tendency to sticking.

The arrangement shown in Figure 16 allows the still-plastic tube to be exactly calibrated in a relatively large cooling bath, so that the required diametral accuracy can be obtained.

Greater flexibility is possible if the arrangement of individual disks and their dimensions are adapted to the extruded product.

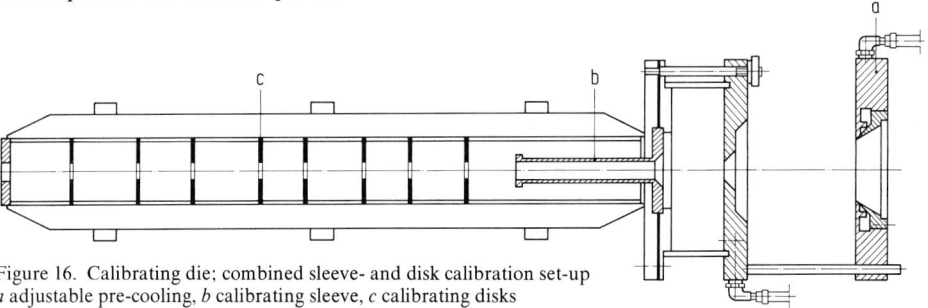

Figure 16. Calibrating die; combined sleeve- and disk calibration set-up
a adjustable pre-cooling, *b* calibrating sleeve, *c* calibrating disks

3.1.2.3.2 Internal air pressure calibration (Figure 17)

Calibration by use of internal air pressure is still a widely used process for larger PVC and PO pipes. A sizing sleeve is placed centrally and at a small distance behind the pipe die head, and a sealing plug with flexible sealing lips is attached to the pipe die head at a fixed

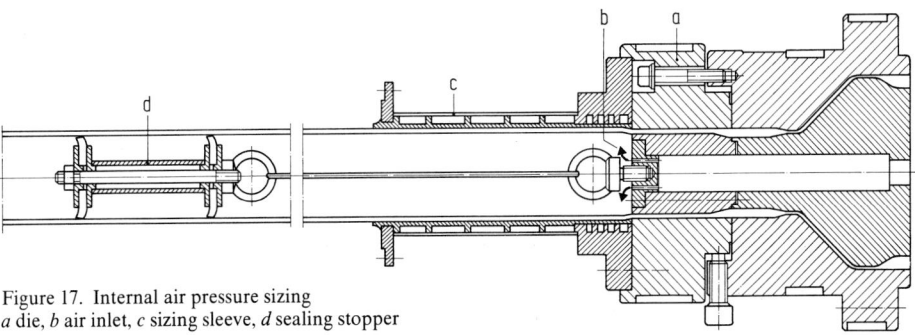

Figure 17. Internal air pressure sizing
a die, *b* air inlet, *c* sizing sleeve, *d* sealing stopper

distance inside the pipe. An excess pressure is produced in the space between die and plug by compressed air, which serves to press the pipe outer wall against the inner side of the water-cooled sizing sleeve.

3.1.2.3.3 Design of pipe die heads and sizing dies

In order to obtain the required pipe dimensions and to keep the extrusion process as stable as possible, the key dimensions shown in Figure 18 must be adapted as exactly as possible to the extruded material when designing the die insert and the sizing die, see Table 6.

Melts of high viscosity swell after leaving the die, while melts of low viscosity must sometimes be considerably extended between die and calibration unit, to keep them under sufficient tension.

Table 6. Die dimensions and positioning of calibration units based on empirical results

Pipe raw material	Outer diameter of pipe mm	A mm	B mm	S in % of nominal wall thickness	a at pressure 6 to 10 mm	b mm	Haul-off speed m/min
PVC	20	20	20.16	64	30	100 to 150	20 to 35
	160	160	161.3	79	150	500 to 600	2.0 to 3.5
PE	20	21	21	100	20	150	25 to 30
	160	168	167.2	100	40 to 175[3])	640	1.2 to 2.2
PP	20	21	21	100	20	150	25 to 30
	160	168	167.2	100	40 to 175[3])	640	1.0 to 2.0
				S mm			
PA 12[1])	8	14.2	8.6	2.0	25	130	55 to 60
PA 12[2])	20	28.3	20.85	3.3	25 to 30[3])	130	12 to 15
PA 12[2])	22	30.0	23.0	3.3	35 to 50[3])	130	10 to 12

[1]) wall thickness 1 mm
[2]) wall thickness 2 mm
[3]) dependent on wall thickness

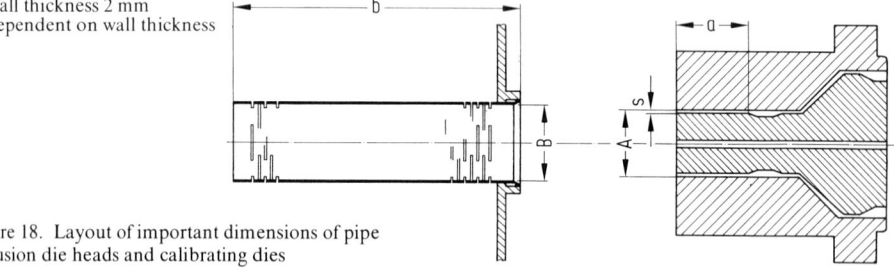

Figure 18. Layout of important dimensions of pipe extrusion die heads and calibrating dies

3.1.2.4 Cooling baths

The pipes must be cooled sufficiently in the calibrating tank and in the bath which follows to retain enough form stability to withstand the stresses in the haul-off, in the winder or in the cutting unit.

Cooling is carried out in the water bath or by spraying with water; both methods give high heat transfer values and uniform cooling. When there is no central water cooler for the extrusion plant, the heat is removed from the cooling water by a heat exchanger in the line itself. The following methods are used for *assessing the required length of cooling bath:*

3.1 Manufacture of pipes

The simplest way of calculating the length of cooling bath required is by use of the Fourier number Fo, which is a dimensionless constant that one can interprete as a time ratio:

$$Fo = \frac{a \cdot t}{X^2} \tag{1}$$

a temperature conductivity $\dfrac{mm^2}{s}$
t cooling time s
X pipe wall thickness mm

By equating the Fo figures from one or more trials with the plant equipment on changing temperature conductivity (change of material) and/or thickness (change in dimension), one obtains the new cooling time t and, from the haul-off speed V_{ab}, the new cooling length l required:

With $\quad l = V_{ab} \cdot t \tag{2}$

and $\quad Fo = $ const. we obtain,

$$\frac{a_1 \cdot l_1}{V_{ab} \cdot X_1^2} = \frac{a_2 \cdot l_2}{V_{ab2} \cdot X_2^2} \tag{3}$$

subscript 1: known case
subscript 2: case to be designed

Here all boundary conditions, such as cooling water temperature T_u, coefficient of heat transfer α, melt temperature and final temperature of the pipe, T_m and T_e, must be kept constant.

This procedure is only permitted with very large Fo-coefficients, i.e. very small pipe wall thicknesses or very long cooling times, so that the pipe is cooled down almost to ambient temperature; or with sufficiently large thicknesses, when the ratio of thermal resistance X/λ to the external heat transfer resistance $1/\alpha$ is no longer influential. This ratio is called the Biot number and is valid for $Bi > 100$:

$$Bi = \frac{\alpha \cdot X}{\lambda} \tag{4}$$

α thermal transmission coefficient $\dfrac{W}{m^2 K}$

λ thermal conductivity $\dfrac{W}{m K}$

X pipe wall thickness m

A simple example will clarify this. The cooling behaviour of an HDPE pipe, diameter 110 mm, wall thickness 10 mm with $V_{ab} = 2.2$ m/min and a cooling length of 22 m, is well known. The question is what cooling length is required for a pipe with half the wall thickness at the same melt throughput, i.e. twice the haul-off speed.

Equation (3) transformed gives

$$l_2 = l_1 \cdot \frac{a_1}{a_2} \cdot \frac{V_{ab2}}{V_{ab1}} \cdot \frac{X_2^2}{X_1^2}$$

with $\quad \dfrac{a_1}{a_2} = 1, \quad \dfrac{V_{ab2}}{V_{ab1}} = 2, \quad \dfrac{X_2}{X_1} = 1/2$

one obtains

$$l_2 = 22 \text{ m} \cdot 1 \cdot 2 \cdot \tfrac{1}{4} = 11 \text{ m}.$$

A more exact calculation is shown in [6]. There the pipe wall is treated as a flat sheet, which is valid so long as the wall thickness is considerably smaller than the inner diameter.
The fundamental equation is an infinite series of which, for $Fo > 0.5$, the first term is normally sufficient.

$$\Theta = C(m) \cdot f(m) \cdot \exp(-m^2 \cdot Fo). \tag{5}$$

Θ is a non-dimensional (normalized) temperature.

$$\Theta = \frac{T - T_u}{T_m - T_u} \quad \text{(at any point in time } t\text{)} \tag{6a}$$

$$\Theta_o = \frac{T_o - T_u}{T_m - T_u} \quad \text{(on the pipe outer wall at time } t\text{)} \tag{6b}$$

$$\Theta_i = \frac{T_i - T_u}{T_m - T_u} \quad \text{(on the pipe inner wall at time } t\text{)} \tag{6c}$$

$$\bar{\Theta} = \frac{\bar{T} - T_u}{T_m - T_u} \quad \text{(normalized c mean temperature at time } t\text{)} \tag{6d}$$

T_u cooling-medium temperature
T_m melt temperature
T temperature at position X at time t
T_o temperature on pipe outer wall at time t
T_i temperature on pipe inner wall at time t
$\bar{T}$ caloric mean temperature at time t

$$C(m) = \frac{\sin m}{m + \sin m \cdot \cos m} \tag{7}$$

$$f(m) = \cos\left(m \cdot \frac{x}{X}\right) \quad \text{for } \Theta, \Theta_o, \Theta_m \tag{8a}$$

$$f(m) = \frac{\sin m}{m} \quad \text{for } \bar{\Theta} \tag{8b}$$

m is a transcendental function of the Bi number

$$m = Bi \cdot \frac{\cos m}{\sin m}. \tag{9}$$

x running length in thicknesses (X) direction.

$\bar{T}$ represents the temperature which an ideally insulated piece of pipe with a radial temperature profile would finally reach. Θ is a direct measure of the proportion of the original enthalpy difference between molten pipe and the pipe at cooling medium temperature that exists at time t in the pipe. One can extend the data of the above example with the following details, which can be taken from pipe extrusion (HDPE):

$$\alpha = 300 \frac{W}{m^2 K}, \quad \lambda \approx 0.3 \frac{W}{m K} \quad \text{and} \quad a \approx 0.15 \frac{mm^2}{s}$$

3.1 Manufacture of pipes

so one obtains

$$Bi_1 = \frac{300 \cdot 0.01}{0.3} = 10 \quad \text{and} \quad Fo_1 = \frac{0.15 \cdot 22}{\frac{2.2}{60} \cdot 10^2} = 0.9.$$

This gives the following results, by calculation with equation (5) for the original example (diameter 110 mm, wall thickness 10 mm, haul-off speed 2.2 m/min):

$$\Theta_o = 0.029, \quad \Theta_i = 0.201 \quad \text{and} \quad \bar{\Theta} = 0.140$$

and with $T_m = 210\,°C$ and $T_u = 15\,°C$ by equation (6)

$$T = \Theta\,(T_m - T_u) + T_u$$

$$T_o = 21\,°C, \quad T_i = 54\,°C \quad \text{and} \quad \bar{T} = 42\,°C.$$

One can also use the diagrams in [6] instead of the calculation.
Now, if one maintains the Fo value constant as above,

$$Fo_2 = Fo_1$$

and assumes that the thermal transmission coefficient does not change, then if $Bi_2 = 5$ for the example to be calculated:

$$\Theta_o = 0.067, \quad \Theta_i = 0.262 \quad \text{and} \quad \bar{\Theta} = 0.193$$

or

$$T_o = 28\,°C, \quad T_i = 66\,°C \quad \text{and} \quad \bar{T} = 53\,°C.$$

One can see that with the same Fo value the thinner pipe has not cooled down so much as the thicker pipe. If the point where the same degree of cooling is attained is to be determined, the non-dimensional temperatures must be equated:

$$\bar{\Theta}_2 = \bar{\Theta}_1.$$

This gives $Fo_2 = 1.09$ and the required cooling length

$$l_2 = V_{ab2} \cdot Fo_2 \cdot \frac{X_2^2}{a_2} = 4.4 \frac{m}{min} \cdot 1.09 \cdot \frac{5^2\,mm^2}{0.15\,\frac{mm^2}{s}} = 13.3\,m.$$

By this calculation process using the example, one obtains a cooling section that is 21% longer than by simply observing the Fo formula. In order to simplify the calculation, approximate forms of equations (7) to (9) in [7] and [8] were used.

According to [9], air-cooling sections can be added between individual water-cooling sections, so long as they are smaller than 10% of the total length.

Although the thermal transmission coefficient with air only amounts to 4 to 12 W/m² K and therefore no heat dissipation worthy of mention takes place by comparison with water cooling, temperature equalization occurs in the pipe wall, and provides a larger heat flow in the next water section, as a result of larger differences in temperature between the pipe outer wall and the cooling medium.

The accuracy of both processes depends upon conditions remaining constant, which they do not:

1. *Material properties.* The temperature dependence of material properties can be found in [10, 11]. The temperature conductivity of HDPE falls slightly as it cools from 0.12 mm²/s at 250 °C, rises rapidly as the melt crystallizes and reaches the value 0.195 mm²/s at 20 °C.
The mean value to adopt is a question of experience.

2. *Coefficient of thermal transmission.* The coefficient of thermal transmission changes during cooling. Regularities are largely unknown. Experiments to calculate the value from the Nu value (compare [6, 12]), have been carried out either empirically on particular lines, or only under such ideal conditions that they are not relevant to practical extrusion lines. The following mean values apply in pipe extrusion:
Water cooling: $\alpha = 300$ to 600 W/m^2 K
Spray cooling: $\alpha = 700$ to 1000 W/m^2 K
3. The temperature at the beginning of the cooling section is not constant.

The numerical differentiation process does not suffer from these variations. Its graphical version is known as the Binder-Schmidt process. The amount of calculation is much greater than with the other processes. The process calculates temperatures at discrete points along a radius and at equal time intervals, so the starting temperature profile and material values need no longer be prescribed constant. Thus the coefficient of thermal transmission can vary, and the simplification of assuming the sheet to have infinite radius and treating it as a flat sheet is no longer necessary. It is then possible to determine the inherent coefficients of thermal transmission in an iterative way from the temperatures measured, which can then serve as the basis for the method of approximation.

An uneven distribution of internal stress in the pipe results from cooling the pipes from outside during manufacture, with compressive stresses in the outer area, and tensile stresses on the inside [13]. Partially crystalline PO pipes show particularly large stress differences. Methods of determining internal stresses are given in [14]. An interrruption of the cooling effect between cooling baths can reduce internal stresses, and internal cooling would improve matters.

3.1.2.5 Pipe haul-off units

Pipe haul-off units have the task of pulling the pipe away from the die without jerking and at constant speed through the calibration unit and cooling section. The available haul-off power must be greater than the friction forces arising in the calibration unit and the seals of the cooling section, and, with internal air calibration, the braking force of the sealing plug must also be overcome.

Haul-off units of varying designs are used, differing with the diameter of the pipe and the raw material. When selecting haul-off units it is usual to choose those which can work with the lowest specific force; a high coefficient of friction between pipe and haul-off belt and a long contact length keep the gripping forces low.

The extremes of caterpillar design are the simple two caterpillar units and those with twelve caterpillars, the latter being used for thin-walled large pipes. There are haul-off units with endless belts or multipad caterpillar belts. Belt systems run more evenly and smoothly, especially at high haul-off speeds. However, when damage occurs at one place the whole belt must be changed instead of a single pad.

The following factors must be taken into account for optimum choice and decision on the suitability of a haul-off unit:

a) *Pipe diameter.* The maximum opening width, or the free passage in the haul-off unit, should be about 50 mm larger than the largest pipe outer-diameter to be hauled off.
b) *Wall thickness and inherent rigidity of pipes.* Pipes with low inherent rigidity and thin walls require haul-off units with several long caterpillars. For haul-off units with two caterpillars rubber pads or rubber belts of low Shore hardness are used.
c) *Haul-off speed and haul-off power.* Haul-off speed and haul-off power should be so specified that there is a reserve of approx. 20 to 25%.

The arrangement of the profile chains or belt supports is very important with pipe haul-off units:

3.1 Manufacture of pipes

- With several caterpillar units the supports for the lower ones should be centrally adjustable and the upper supports individually adjustable.
- The lower caterpillars should have fixed bearings and the upper caterpillars moving bearings.
- There are various systems for power transmission from the drive to the haul-off unit. Figure 19 shows a modern bearing and drive system with cardan shafts.

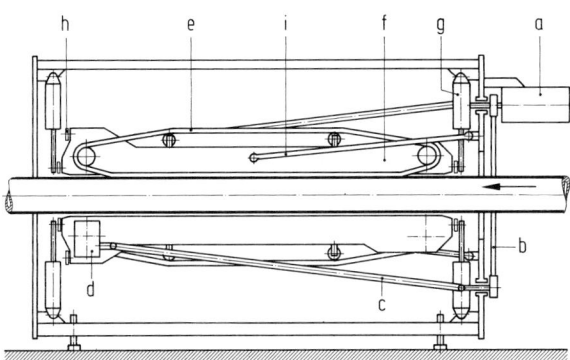

Figure 19. Haul-off unit, Reifenhäuser system
a drive motor,
b belt drive,
c cardan shaft,
d screw gear,
e haul-off chain,
f caterpillar support,
g pressure device,
h transverse link,
i connecting rod

Normally a DC drive motor is used, for the following reasons:
- synchronization with the extruder or downstream units,
- large control range.

Table 7 gives performance data on a wide range of haul-off units.

Table 7. Haul-off units

Working range pipe outer mm diameter	Number of haul-off caterpillars	Drive power kW	Haul-off force N	Haul-off speed m/min	
				min	max
0 to 63	2	1.1	2 000	0.8	20
	2	2×0.75	3 000	2.5	50
	2	2×0.75	1 500	5.0	100
20 to 125	2	3.5	8 000	0.4	10.0
20 to 200	4 or 6	4.2/5.2	7 500/12 000	0.4	15.7/10
40 to 250	4 or 6	5.0/5.2	15 000	0.4	15.7/10
63 to 400	4 or 6	5.2	20 000	0.1/0.2	2.5/5.0
110 to 630	4 or 6	5.2	20 000/30 000	0.1	2.5
400 to 800	4 or 8	5.5	22 000/36 000	0.07	3.3
450 to 1 220	8	5.7	45 000	0.07	1.7

Pipes with thickened wall sections are produced for *forming pressure-pipe sockets*, and to do this, special control of the pipe haul-off units is required. The haul-off speed is reduced for a short time without changing the extruder speed so that the wall becomes thicker over a short distance. The overall pipe length and the dimensions of the thick section are adjustable.

3.1.2.6 Cutting units

Selection of an appropriate cutting unit depends on the following factors:

a) the cut form and the quality of cut required,
b) pipe diameter and wall thickness,

c) type of raw material,
d) cut-off length,
e) form of swarf.

Various cutting systems to take care of these criteria have been developed, see Figure 20. *Simple fly cutters* are often used with small pipe diameters and soft raw materials.

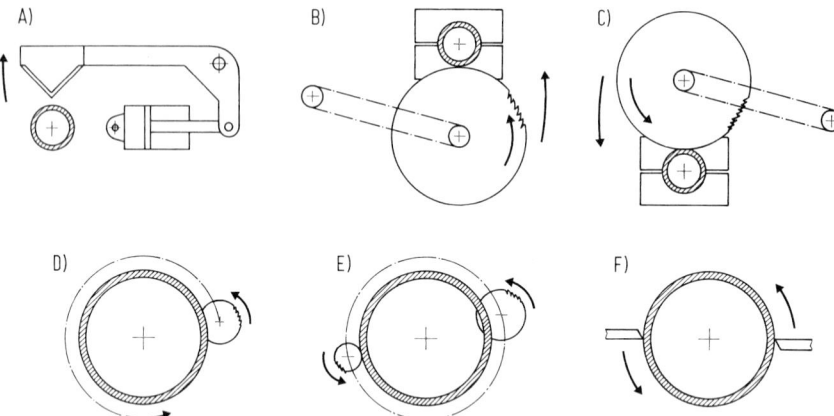

Figure 20. Cutting systems
A) guillotine, B) radial saw (saw blade from below), C) radial saw (saw blade from above), D) planetary saw, E) planetary saw with milling cutter, F) parting device with chamfering attachment

The *radial saw* is normally used for pipes up to about 200 mm diameter. Care must be taken that no swarf sticks to the inside or outside of the pipe as a result of electrostatic charge.
The pipe is guided by jaws that clamp the pipe at the beginning of the sawing process. The transport carriage, on which the radial saw is mounted, is synchronized to production speed. Larger pipes must be cut with a *planetary saw*. If required, an attachment rotating with the saw carriage can be fitted for chamfering one or both sides of the pipe.
Automatic parting-off machines are always used where swarf is to be avoided. These devices have the additional advantage of shorter cycle times with smaller pipe diameters and higher extrusion speeds. One or more parting tools are rotated at high speed around the pipe and inserted into it at such a rate that the swarf produced is of fixed length and heavy enough not to remain hanging on the pipe, and is not so long that it wraps itself around the pipe or becomes entangled in the area of the cutting blade.

3.1.2.7 Winding, discharge, socketing

Particular attention must be given to the further treatment of pipes after cutting, because in this area of the plant the operations are very labor-intensive, and automation here can lead to the greatest possible savings.

Winding

Only those plastic pipes which undergo no permanent deformation through bending should be wound. Polyolefin pipes, PA hoses, plasticized PVC pipes or pipes from other soft materials may be wound up.
The following points must be taken into consideration when choosing a winder:

1. Core diameter of the swift: the core diameter of the swift should be no smaller than 20 to 25 times the outer diameter of the pipe which is to be wound.

2. Outer diameter of the swift: the outer diameter of the swift should be about 100 to 200 mm larger than the largest outer diameter of the wound pipe coils. Winding diameters depend on the pipes and pipe lengths. Loading- and transport regulations must be observed with coil diameters larger than 2500 mm.
3. Extrusion speed of pipes to be wound: At extrusion speeds of less than 2 m/min., one can work with a single-station winder, but above 2 m/min., a dual winder should be used. At speeds of greater than 20 m/min., semi- or fully automatic winding stations with automatic tying-off units are to be recommended, particularly with short lengths of wound pipe.

Pipe winders that are employed in extrusion lines with high degrees of automation must be included in the line control systems.

Discharge

Simple units are designed as troughs, which tip the cut pipe onto pallets or platform wagons (on one or both sides).
For heavy pipes of large diameter the tipping units are designed for roller operation, and the rollers are individually driven.
Conveyor belts or swiveling devices are used as needed for further processing or intermediate storage.

Socketing units

The only pipes that can be socketed are those made from materials that are form-stable after molding. Rigid PVC is particularly suitable for this process, but PP, HDPE, LDPE and ABS can also be socketed.
In general, three types of socket form are used, see Figure 21.

Solvent socket

The solvent socket is only used with rigid PVC or ABS pipes. Its application lies in the field of pressure pipes, drainage pipes, electro- and cable-protection conduits. The disadvantage of this socket is that the pipe connection cannot be taken apart once it has been stuck together, and it does not allow longitudinal movement.

Drainage socket

There are two socket forms for drainage pipes:

a) with a drop-in O-ring or lip,
b) with a fixed, integrated lip ring.

The O-ring design is used mainly with rigid PVC or ABS pipes and the lip ring with PP and LDPE. The fixed, integrated lip ring is locked in by an additional process directly after socketing, by upsetting the pipe end.

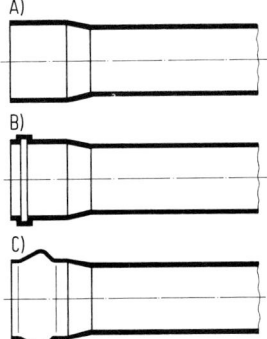

Figure 21. Socketing

Pressure-pipe socket

The pressure-pipe socket with a drop-in lip ring is used mainly with rigid PVC or ABS pipes. To prevent weakening of the pipe in the area of the socket, the pipes are socketed at a position where the wall has been thickened during extrusion. There are also processes in which upsetting to produce wall thickening during socketing is carried out.

In another process short pieces of pipe are pushed over the pipe in the socket area to strengthen it before socketing, and these pieces of pipe are then also shaped.

For polyolefin pipes the socket is produced as an injection molded part that is joined to the pipe by the hot-plate welding- or friction welding process.

The following points must be taken into consideration when choosing a socketing machine:

1. The diameter range to be processed on the machine.
2. The pipe lengths to be socketed on the machine.
   For pipe lengths of 4000 to 6000 mm or longer, it is sufficient in most cases if the machines are equipped with a heating station (hot air or two-section contact device) and a socketing unit.
   For pipe lengths of 1000 to 5000 mm, machines for socketing both short and long pipes can be used.
   For pipe lengths under 1000 mm, machines with a number of heating- and socketing stations are used.
3. The types of socket required, including:
   a) adhesive sockets,
   b) beading sockets for drainage pipes,
   c) beading sockets for pressure pipes.
   There are socketing machines for an individual type of socket; also universal machines.
4. The sealing ring to be used:
   When the type of sealing ring has been chosen, it can be decided how the socketing unit should be designed.
   With beading sockets for drainage pipes:
   – two-part or three-part external forming shells with blowing mandrel or hard rubber mandrel
   – two-part or three-part external forming shells with mechanically expanding mandrel
   With pressure pipes:
   – two-part or three-part external forming rings with expanding mandrel
   – expanding mandrel with vacuum
   – water-pressure chamber with expanding mandrel
5. Multiple extrusion:
   In this case, the machine must be enlarged and the number of units increased to handle socketing of two- or three-fold extruded piping, at the same time as taking points 1 to 4 into account.

The socketing machines in general use can handle pipe diameters of 10 to 710 mm.

3.1.2.8 Measurement and control

The manufacture of pipes, in principle, is a very stable process. For this reason, efforts to equip pipe extrusion plants with more measurement and control devices have not yet progressed very far.

However, large pipe lengths and the increasing proportion of raw material costs in pipe production costs led very early on to the use of devices for measuring pipe wall thickness in extrusion lines.

In particular, two measuring processes have become common [15, 16]:

– Ultrasonic on-line measurement, suitable for wall thickness control from about 2 mm and pipe diameters from 15 mm upward.
– The induction measurement process, for smaller pipes with thicknesses up to 6 mm max. and diameters from 6 to 63 mm.

In both cases, the measuring device should be positioned as close as possible to the calibration unit, in the interest of effective control.

3.1 Manufacture of pipes

The ultrasonic measuring head (Figure 22) runs reversibly over 360 degrees on a ring that surrounds the pipe. A film of water couples the sonic head to the pipe wall to transmit sonic pulses in the megahertz range and receive the reflections. In this way, the full pipe circumference is checked during one cycle. The thickness target value and maximum permitted tolerances are fed into a microprocessor-based comparator unit, with the measured value to provide:
- video display of the pipe thickness profile, presented as the measured value at 45 degree positions around the circumference. The minimum and maximum deviations for one cycle of measurement are stored.
- target value print-out for control of haul-off speed to correct the lengthwise wall thickness profile.

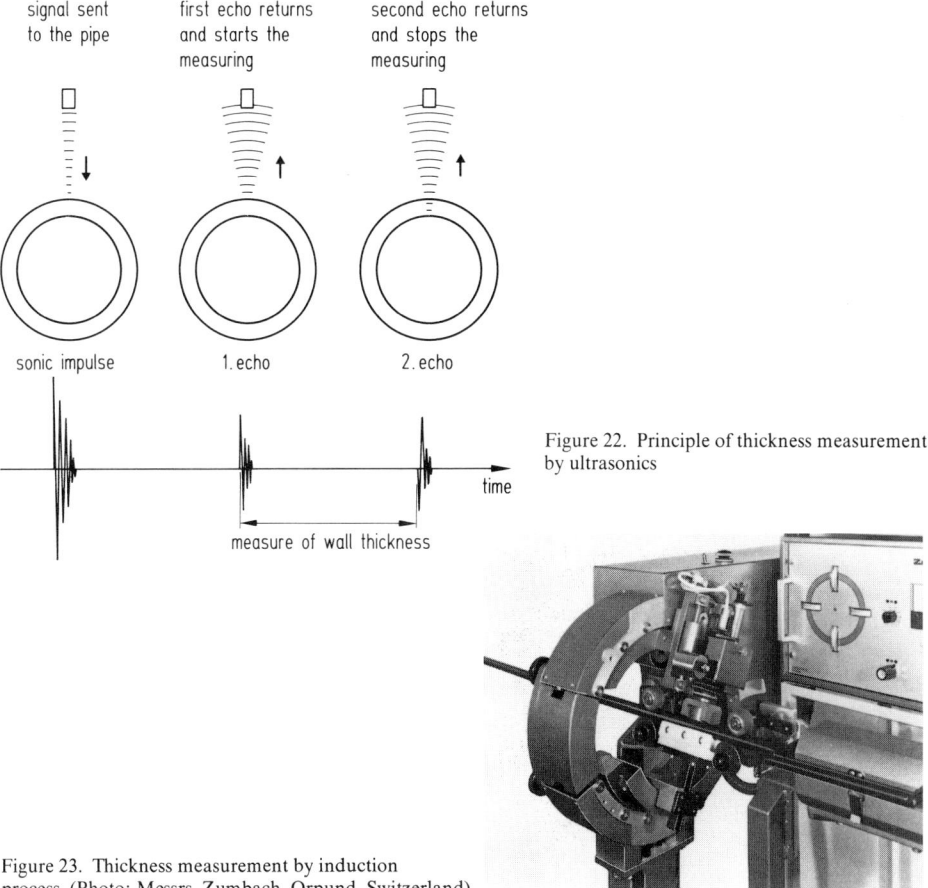

Figure 22. Principle of thickness measurement by ultrasonics

Figure 23. Thickness measurement by induction process. (Photo: Messrs. Zumbach, Orpund, Switzerland)

In the *induction measuring process* (Figure 23), an iron core is inserted into the pipe to serve as a magnetic 'reflector' for the measuring head. The measuring head, which is fixed to a ring surrounding the pipe, moves reversibly over an arc of almost 360°, carrying a horseshoe magnet; this pulls the 'reflector' radially against the inner wall of the pipe. As the measuring head turns, the reflector moves around the inner circumference of the pipe separated from the head by the pipe wall. The thickness determines the inductance of a magnetic circuit, and an electrical signal proportional to this is used for display and control purposes. To prevent axial movement, the 'reflector' is attached to an extension of the core mandrel, and rotates around it.

In addition to simply displaying the pipe wall thickness and providing closed-loop control of the haul-off speed, it is possible to center the pipe die automatically [17].

Additional efforts include the closed-loop control of the vacuum in the calibrating tank, to maintain the desired diameter in the production of smaller, high-precision pipes.

3.1.2.9 Corrugated pipe lines

3.1.2.9.1 Products

Corrugated pipes have a wide range of applications. These pipes have either parallel ring grooves or a continuous helical groove.

The advantage of these pipes lies in their great flexibility and high back-fill load-bearing capacity. The first PVC corrugated pipes were manufactured in 1958/1959, and were used as electrical conduit in above-ground construction, displacing the familiar Bergmann pipes of the time. Development led to the manufacture of corrugated pipes of larger diameters, which are now used as drainage pipes. In 1960/1961, corrugated pipes accounted for only 3 to 4% of the drainage pipes laid in the Federal Republic of Germany; the proportion today is 99%.

In 1983, the following quantities of pipes were produced in the Federal Republic of Germany:

Drainage pipes:	some 80 to 100 × 10^6 m, of which
	a) 40% for agriculture
	b) 60% for above-ground construction
Drainage in underground and foundation construction:	approx. 20 × 10^6 m
Electrical conduit:	approx. 80 to 90 × 10^6 m
Technical pipes:	approx. 10 to 12 × 10^6 m

In the last few years, new and interesting areas of application have emerged for corrugated pipes, which required process development for other thermoplastics, such as polyolefins, polyamide, fluoropolymers [18].

3.1.2.9.2 Manufacturing Processes

The manufacture of corrugated pipes is, in principle, the continuous blow molding of pipe blanks.

The cylindrical part of the pipe die head extends into the closed area of a revolving mold-block chain. The plastic tube is pressed against the profiled, revolving mold-block halves by internal air pressure, Figure 24. As it passes through the forming machine it is cooled by contact with the mold-blocks, and by the time it reaches the end of the chain the tube must be sufficiently cooled to leave the rotating mold blocks in a stable form.

Today, extrusion speeds up to 30 m/min. can be handled, using mold-block cooling (air or water) irrespective of raw material and pipe diameter.

The demand for savings in raw material led to the development of double-walled pipe. These pipes consist of two layers, corrugated on the outside and smooth on the inside. They are mainly used as sewer pipes and cable-protection conduit, where a smooth inner surface is essential. The advantage is a saving in material of some 25 to 30% compared with solid pipe [19].

Double-walled pipes are produced by the coextrusion process (see Figure 25).

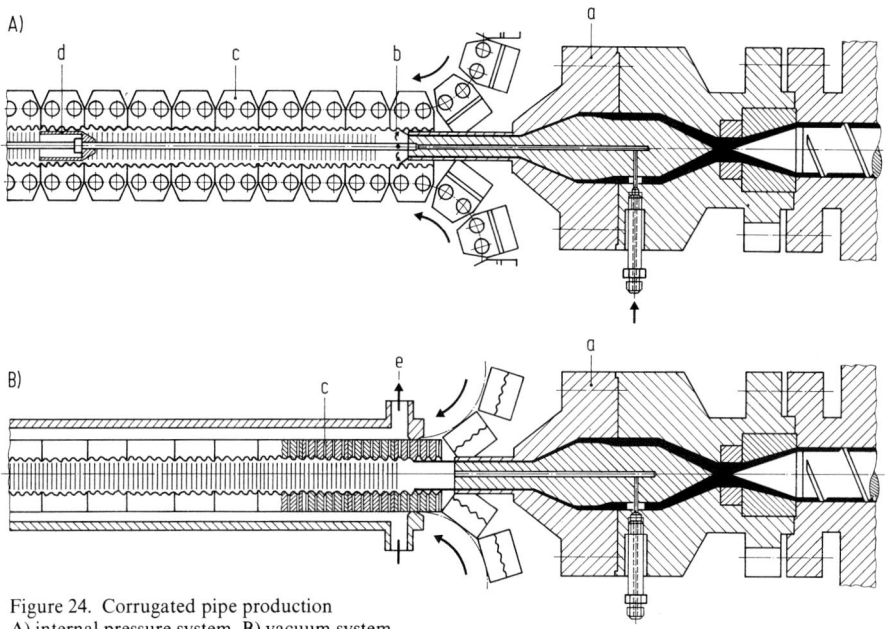

Figure 24. Corrugated pipe production
A) internal pressure system, B) vacuum system
a pipe die head with die insert, *b* compressed air inlet, *c* shaping die, *d* sealing stopper, *e* vacuum connection

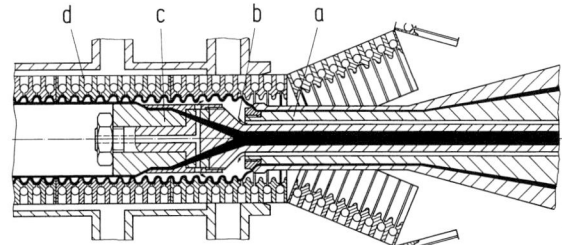

Figure 25. Double-walled corrugated pipe production
a die for coextrusion,
b throughflow guide,
c mandrel extension for inner layer,
d shaping die

The basic principle of manufacture is the same as with standard corrugated pipes: but when the first tube has been formed, the second tube is laid smoothly onto the inner surface of the still-plastic corrugations and welded to the first with the aid of a sizing mandrel.

Corrugated pipes can now be produced with diameters ranging from 6 to 650 mm.

3.2 Extrusion of profiles

3.2.1 Market significance

Profiles of all kinds are increasing their market share, and their absence from certain areas is now difficult to imagine, since they have completely replaced some traditional raw materials. Furthermore, there are almost no boundaries to creativity regarding the shape and design of profiles, see Figure 26.

Applications in the construction industry hold first place, followed by furniture and vehicle manufacturing. However, plastic profiles are gaining more and more importance in the area of electronics, and for use in areas of special packaging and communication technology.

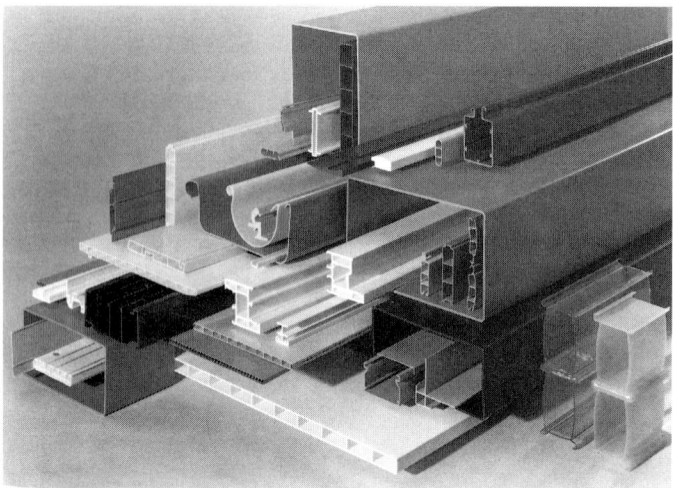

Figure 26. Extruded profiles

3.2.2 Products

It is extremely difficult to classify profiles. However, with a simple profile, Figure 27, it is relatively easy to show the different influences the profile type has on die design, on calibrating- and cooling methods to be used, as well as on other processing parameters involved in the production of such combinations of cross-section and raw material.

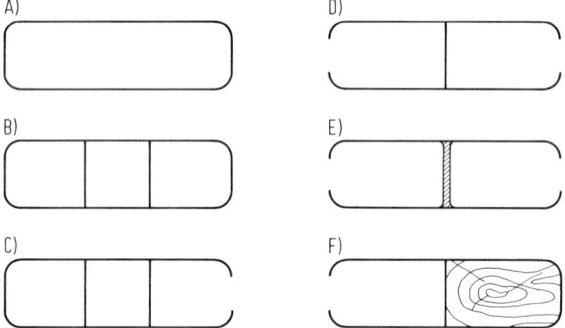

Figure 27. Continuous profile forms
A) profile pipe, B) hollow profile, C) chamber profile, D) solid profile, E) composite profile, F) core profile

Explanations of the types of profile shown in Figures 26 and 27:

a) Profile pipes usually have even wall thicknesses with corners that can be round- or sharp edged, according to whether a pipe- or profile die is used.
b) Hollow profiles are closed hollow forms, i.e. the outer jacket encloses only space. Different wall thicknesses are possible, also supporting ridges and sharp edges.
c) Chamber profiles are profiles with closed and open spaces, possibly with intermediate ridges. Different wall thicknesses are also possible.
d) Full profiles have no hollow chambers at all and all conceivable shapes and cross-sections are possible.

e) Compound profiles, also called sector- and multi-component profiles, embrace all the shapes described; however, they can combine individual sections of the same material in another color, of the same material with a different hardness, of different materials that can be homogeneously compounded, or they may combine several of these possibilities.
f) Core profiles are hollow- or full profiles with a core from a different type of material, e.g. steel or wood. The basis of the manufacturing process is full or partial sheathing.

3.2.2.1 Rigid PVC profiles for the construction- and furniture industries

A distinction must be made between profiles for external use and profiles for internal use. Profiles which are exposed to weathering must contain special stabilizers for protection against UV-radiation and weathering.
The following are some of the most widely used profiles in the construction industry:

External application: Window- and door profiles, window sills, roll shutters, roof guttering; fascia-, fence- and balcony profiles; light-wall profiles, double- or triple-walled sheets for greenhouses, terrace- and pergola coverings.

Internal application: Skirting boards, wall- and ceiling-covering profiles, installation profiles, conduit profiles, door frames, curtain rails and roller box profiles.
In the furniture industry, there is a basic distinction between load bearing and decorative profiles.

Load-bearing profiles: Main body profiles, drawer profiles, cappings, handle profiles, stopper rails, shelf bases, connection profiles, roll joints.

Decorative profiles: Decorative strips, handle profiles with decorative inserts, edge veneers and bandings.

3.2.2.2 Plasticized PVC profiles for the automobile industry

Plasticized PVC profiles are being used more and more today in the automobile industry, mostly in association with backing materials such as stainless steel or aluminum, to achieve further weight reduction as well as to increase functionality and value.
The profiles that are most widely used nowadays in automobile construction are:
Door frame profiles with gridband insert, glazing profiles with flocking, sealing profiles, weatherstrip, profiles for upholstery, tread strip (mostly a combination of rigid and plasticized PVC), shock absorber- and ram protection profiles, side protection strips, and trim strips with a film insert or backing material.

3.2.2.3 Profiles from other materials

About 80 to 85% of all profiles are still produced from PVC; however, an increasing trend to higher value materials can be observed, particularly when special mechanical or chemical properties must be attained.

– Foamed profiles, mostly of PVC, have already obtained a large market share in the construction and furniture industries, as skirting boards, roller blind box elements, shelf bases and decorative strips in furniture, picture frame strips, and so on.
– Profiles made from polystyrene (PS), styrenebutadiene (SB) and acrylonitrile butadiene styrene (ABS) are mainly used for furniture profiles, e.g. as main body parts and corner connection pieces. PS profiles are used for fluorescent light-tube covers.

- Profiles from PMMA have shown enormous rates of growth because of their very good light stability and resistance to ageing. Light covers, with or without embossing, have particularly proven themselves, as have double- and triple-walled sheets for greenhouse coverings, wind protection on terraces and pergola covers; or as solar collector elements, where, particularly in the greenhouse area, the outstanding translucency and insulating property of the multishell building elements offer enormous advantages.
- The range of applications for PC profiles is similar to that of PMMA, though PC possesses an inherently higher flexural strength; it is, however, subject to ageing after long exposure to sunlight, and it yellows slightly. Recently, trials have been undertaken to improve the surface of PC profiles with a coextruded or lacquered PMMA layer and thereby combine the advantages of both raw materials.
- The main areas of application for PP profiles today are in the construction of solar collectors, as cardboard substitute where moisture is a problem, as a thermal break for aluminum windows and in the construction of ventilation shafts.
- Although polyethylene (PE) profiles are very difficult to calibrate, certain mass products have made an impact for special applications in the last few years, e.g. heat shrinkable collars made of cross-linked PE for joining or repair of telephone cables, or multi-application profiled pipes as cable-protection conduits for the new generation broadband communication systems.
- Cellulose acetobutyrate (CAB) has established itself for trim strips, because of its outstanding ageing resistance and good mechanical properties.
- As many profiles must fulfill various functions at the same time, the market is increasingly demanding profiles which can assume another function, e.g. simultaneously an optical and a mechanical function, in combination with other raw materials. A window profile that illustrates all the possibilities of a multicomponent profile is shown in Figure 28.

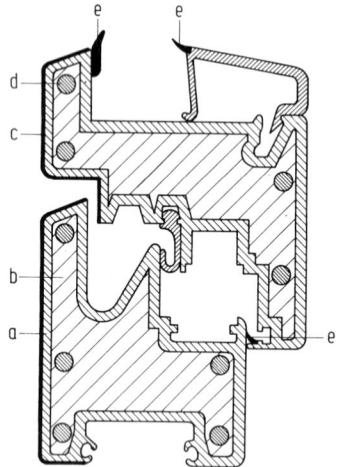

Figure 28. Multicomponent profiles
a rigid PVC casing layer,
b filler material from thermosetting resin with silica-glass hollow spheres,
c acrylic covering layer (PMMA) – coextruded,
d glass fiber strands,
e flexi-PVC sealing lips – coextruded
Illustration: "SCHOCK thermo-massiv" system

- Solid rods from all common thermoplastics (e.g. PA, POM, PE, PC, PMMA etc.).
Solid rods are extruded in the same way as profiles – through sizing sleeves. Quality requirements regarding homogeneity and freedom from blowholes are more stringent than with normal extruded profiles, and place special demands on the die and the downstream equipment during production. Most commonly called for are PA and POM solid rods, which are used as machining stock for gear-wheels, bearing bushes and so on.

3.2.3 Processes for the manufacture of profiles of differing shapes and dimensions from various raw materials

Critical factors for the quality of the end product and the efficiency of the production process are the correct layout and optimum combination of extruder, shaping and sizing die, calibrating and cooling unit and haul-off and cutting devices, right up to the discharge station. The appropriate processing technique and equipment, and the die design, are decided upon with the following points in mind:

- the profile shape,
- the weight of the profile,
- the outer and inner profile dimensions and contours,
- the grade of resin and form of feedstock.

The profile shape indicates directly the shaping and sizing die design to be used; the grade of the resin determines the extruder type (single- or twin-screw extruder) to be used and its design features.

The basic design of a profile extrusion line is illustrated (Figure 29, see page 58) by equipment for the manufacture of thick-walled window profiles.

The following selection criteria for individual components of the line, which are dependent on profile and resin, must be observed:

Extruder: The extruder is matched to the required production capacity. It is particularly important in profile extrusion to attain the lowest possible melt temperature with the smallest possible screw speed. This leads to a high melt viscosity, which has an advantageous effect on the calibrating process. Table 2 summarizes the characteristics of some standard twin-screw extruders (see Chapter 3.1.2.1.1). PVC mixtures for profile extrusion melt faster than pipe mixtures, and therefore extruders for PVC profiles work at lower screw speeds. For the same reason the specific power consumption is higher.

Profile die: The task of the shaping die and die insert is to distribute the melt over the total profile cross-section and transform it into the desired profile. The flow channel must be so designed that the melt flow rate at all points of the cross-section is virtually the same. The dwell time of the melt in the die should be short and vary little around the profile. A sufficiently long parallel flow before exit from the die aids steady flow and promotes good surface qualities. The calibration system then takes over the task of stabilizing the shaped melt stream and the final shaping. Calibration is carried out with the help of vacuum and cooling.

Calibration unit and cooling unit: A solid base is required for the construction of the calibrating and cooling units. Auxiliary units such as vacuum pumps, circulating pumps and cooling blowers, should be vibration-free. The base must be adjustable both vertically and sideways and movable lengthwise for matching with the various processing techniques and profile contours. Cooling channels, and spray or water baths, in addition to vacuum tanks with spray or flooding possibilities, are used. In special cases, e.g. with PMMA and PC processing, additional cooling is installed.

Haul-off unit: A variety of haul-off systems are available, and particular choice depends on the shape, calibrating process and surface nature of the profiles to be produced:

- roller haul-offs: for small profiles, flexible profiles and where demand for tractive power and pressure is small.
- belt haul-offs: for profiles with surfaces that are sensitive to pressure and scratching, and flexible profiles, where tractive forces less than 5000 N are required.
- pad-chain haul-offs: for profiles which require high tractive forces, and profiles which are unstable and tend to twist, and therefore must be guided without torsion by matching pads whose pressure can be adjusted.

For all three systems the profile must be evenly transported i.e. at constant speed and under constant pressure; and the upper caterpillar must be adjustable to compensate for its weight especially with profiles that are sensitive to pressure.

Cutting unit: Various cutting systems are used; choice depends on the material to be cut, the extrusion speed, the cut length, and the cutting quality demanded. Standard traveling cutting units, synchronized with extrusion speed, use radial saws which are equipped with hard metal blades or diamond cutting disks, depending on the material to be cut. Short-cycle cutting units with circular or fly-cutting blades are available for special situations.

Discharge unit: Depending on length and take-up capacity, and the product properties, simple tipping units with collecting braces or sliding tables with rollers for packing, or automated stacking and packing units with suction heads and conveyor belts, are used. Winding units are used mostly with flexible profiles.

Basic plant construction has now been dealt with. In the case of special products, slitting devices like those needed for light slits in roller blind profiles or for inserting openings in wiring ducts, can be built into the line.

Marking- and printing units, e.g. for company codes and for surface embossing, and laminating devices for decorative, protective and adhesive purposes, can also be integrated into the line.

In describing the various processes used in profile extrusion, it is useful to notice how the die is designed.

As a general rule, one differentiates between "hot" and "cold" parts of the die. Only the die insert or shaping die, also called the profile extrusion die, is "hot".

The "cold" part of the die is formed by the calibration unit and the dimensional and shape-dependent stabilizing elements in the downstream equipment, see Figure 30.

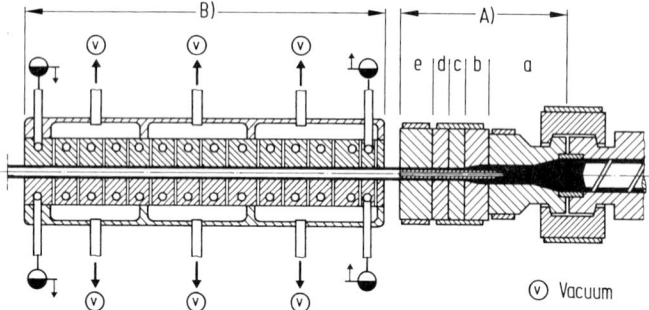

Figure 30. Profile die – basic design
A) cold die part (vacuum tank calibration unit), B) hot die part (profile extrusion die)
a connection piece with laminar flow zone, *b* Intermediate- or transition plate, *c* Core- or mandrel retainer plate with distributor tip, *d* Regulating plate, *e* Mouthpiece disk with parallel guide

3.2.3.1 Profile extrusion die

The construction of a profile die is determined by the size and shape of the profile and by the raw material.

The following basic layout applies to profiled pipes, hollow and chamber profiles (see Figure 31): adaptor with flow zone, intermediate- or transition plate, mold- or mandrel retainer-plate with distributor tip, regulating plate and mouthpiece disk with parallel guide.

For solid, cored and composite profiles, the design is similar; however, there is no mold or mandrel retainer plate, so the transition plate is extended correspondingly.

3.2 Extrusion of profiles

For flexible profiles, e.g. flexible PVC profiles, so-called deflector mandrel dies are used, which have inside them a displacement torpedo, tapered in the forward direction. A short distance ahead of this, a mouthpiece plate with the desired profile shape is positioned. The advantage of this idea is that, by merely changing the mouthpiece plate, other profiles of a similar cross-section can be made. This design is not suitable for raw materials that are thermally unstable.

Figure 31. Profile extrusion die components

For wide flat profiles – double wall sheet and carton profiles for example – sheet dies with a coat hanger channel, restrictor bar regulation, and adjustable die lips are used. This type of design is very expensive, but allows material flow and wall-thickness distribution to be influenced during production; in the case of standard dies, changes are possible only by mechanical refinishing in the flow channel area. Depending on die size and form of construction, heating is carried out by resistance heating elements or by heating cartridges, in zones that can be individually controlled.

3.2.3.2 Calibration

The calibration unit takes care of shape and dimensional stabilization for rigid profiles. Their inner surface, which contacts and shapes the product, is adapted in stages to the outer contour of the extruded profile to take account of throughput and shrinkage. The type and construction of the calibration unit is strongly influenced by the profile type and the resin used.

3.2.3.2.1 Vacuum calibration

The most widely used calibration method, vacuum calibration, is based on a patented process going back to 1958, and is used today with a very wide variety of shapes and types of profile.
The melt is pulled into the calibration unit and sucked by a partial vacuum against the cooled walls. In this way the outer dimensions are determined and the surface is formed. The arrangement of the cooling and vacuum channels is largely based on practical experience, since with most complex profile forms a calculation would be too difficult and costly.

Calibrator blocks, the most widely used form of calibrator, employ brass, steel or aluminum for the shaping section and aluminum for the upper and lower parts, see Figure 32. The vacuum suction channels are arranged in the form of slits or borings alternately with the cooling borings, crosswise to the extrusion direction, and divided into individually adjustable sections.

Figure 32. Vacuum calibration unit (block calibrator).
(Photo: Reifenhäuser, Troisdorf, West Germany)

The intensive effect of the vacuum and cooling in the entry section is reduced with increasing solidification of the profile contour, and maintained only at a level sufficient to counteract contraction. The length of the effective calibrating section is determined by the required throughput and the thermal conductivity of the molding material used, as well as by the profile cross-section and profile type. Hollow and chamber profiles can be cooled only from the outside. Thus, the webs inside are deprived of direct cooling and the profile must, therefore, be held to the sides of the calibrator by air pressure long enough for the webs to reach full strength.

The temperature gradient in the calibrating die must be set so that the surface of the running profile is sufficiently cooled and stabilized to overcome the friction against the walls of the calibrating die. If the layer fixed by cooling is not thick and strong enough to overcome the frictional resistance, the profile will elongate and can tear.

With materials that can stick to the walls, such as polyolefins, sizing aids in the form of spray rings and blower nozzles located ahead of the calibrator are required, in order to prevent sticking.

3.2.3.2.2 Drawing calibration

Drawing calibration is the simplest calibrating procedure.

The continuous profile length leaving the die head is held in shape by passing through holes in disks arranged one behind another in a water bath. The disks are arranged to taper in the direction of haul-off so that the profile is subjected to draw-down by tractive force. This method is suitable only for simple profile cross-sections and profiles with low surface-quality requirements, as only very little surface polishing is possible and the smallest excess feed into the calibration unit, e.g. through conveying fluctuations in the extruder, can lead to tearing.

3.2 Extrusion of profiles

3.2.3.2.3 Short vacuum calibration

A derivative of the preceding method, a short vacuum calibrator, known as the TECHNO-FORM profile precision drawing process, has been developed and automated in conjunction with a sensor control system (see Figure 33).

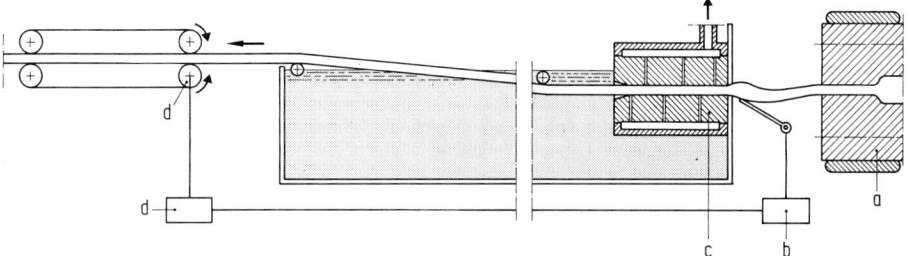

Figure 33. Short vacuum calibration unit with regulating system
a die, *b* sag sensor, *c* short vacuum calibration unit, *d* haul-off with speed regulator

An inductance sensor continually scans the sag between die and calibration unit and sets the haul-off speed by means of an on-line closed-loop control unit, to ensure that a slight excess of feed at the calibrator inlet can be maintained, and that the optimum operating point remains set. Additional straightening elements are arranged behind the drawing calibrator itself, to guarantee the straightness of the free side of the profile. This leads to high dimensional accuracy. Direct water cooling leads to an additional increase in throughput.

3.2.3.2.4 Segment- or template calibration

A further variant of drawing calibration is the so-called segment- or template calibration, which, as with short calibration, also uses partial vacuum, but only with indirect cooling or temperature control (Figure 34). In order to assist the cooling process air is blown through distributor pipes adapted to the profile contour, giving gentle cooling. This method is used particularly with materials which are sensitive to stress cracking, like PMMA, for the manufacture of light covers and similar products subjected to thermal loads in use, where the lowest possible after-shrinkage and freedom from stress are required. Rigid PVC guttering and polycarbonate light panels are also produced in this way, using a cooled vacuum tank whose design is dependent above all on the resin used. The individual segments are arranged as half templates staggered with respect to one another, so that guidance over a greater length of product is attained. In the case of roof guttering manufacture, a shaping element is inserted to aid transition from the round emerging cross-section to the profile contour.

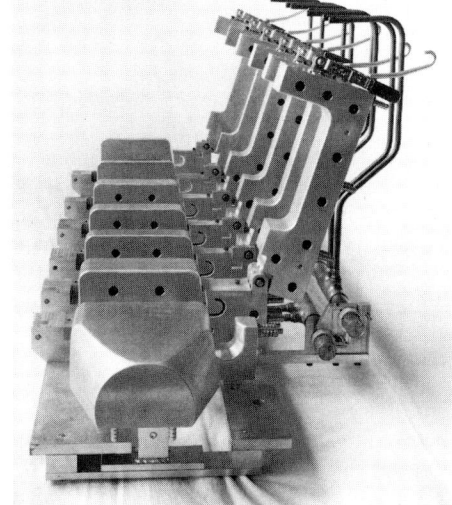

Figure 34. Segment- or template calibration

3.2.3.2.5 Vacuum-block calibration

The calibrating concept that is most frequently used nowadays in profile extrusion, is indirectly cooled, vacuum-block calibration, which has already been briefly described. This form of construction is used predominantly with hollow- and chamber profiles. Depending on wall thickness, extrusion speed and material being processed, block lengths of between 200 and 800 mm are arranged one behind another, followed by an appropriately designed post-cooling section.

For thin-walled profiles, such as roller blind- and wall-cladding profiles, one to three block calibrators of 600–700 mm length including an air-cooling channel, are used. By means of one or more extractors this channel provides all-round airflow and gentle cooling to the profile length, so that the remaining heat in the inner ribs is slowly removed and distortion of the profile avoided.

For thick-walled profiles, such as window profiles or foamed profiles, this arrangement would not be adequate with today's required throughput capacities, and for this reason several other methods, as shown in Figure 35, are employed.

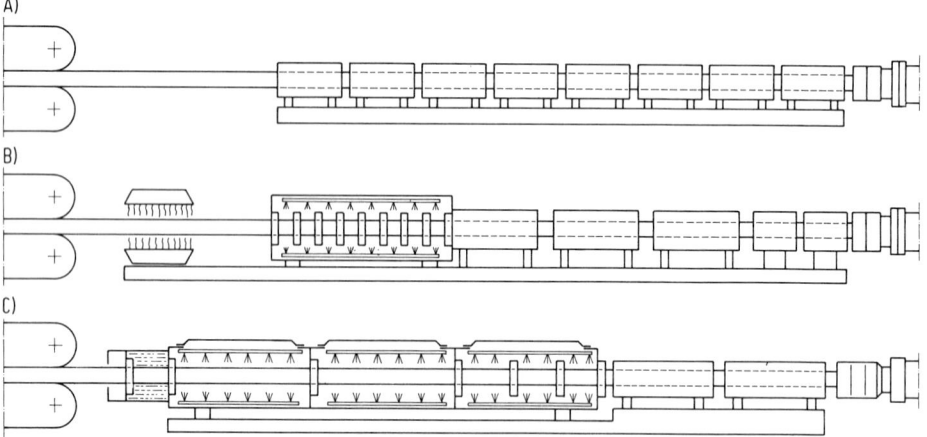

Figure 35. Calibrating systems for thick-walled profiles
A) calibrator blocks of equal length in series, B) calibrator blocks of differing length with spray tank and radiant heater unit, C) calibrator blocks of the same length with vacuum spray tank

Version A shows the arrangement of eight calibrators, which are aligned on a clamping plate.

Version B also shows the arrangement of several calibrators one behind the other, but in this case with alternate short and longer blocks, and an in-line spray tank with stabilizing elements in the form of support disks. The radiant heater section placed at a distance after the tank is used to remove stress from the profiles.

Version C is a less expensive modern variant. Here only two block calibrators 600 to 700 m in length are used for the same throughput as in A and B. A vacuum tank 3 m long with a few contoured support disks connected in series is used to take care of the final cooling and stabilization of the profile by means of water spray and vacuum.

Further post-cooling units can be added as required to the arrangement described under A and B, e.g. spray tunnels or water baths; intermediate water baths are frequently provided between individual block calibrators.

In general, a downward trend is evident in calibration costs, in spite of increasing throughput quantities. This is being achieved by intensification of the calibrating sections as in de-

3.2 Extrusion of profiles

sign C, and by minimizing energy consumption. If more calibrators are employed, more vacuum pumps must be used, with the result that additional stoppages can occur.

3.2.3.2.6 Vacuum-tank calibration

With profile manufacture using vacuum-tank calibration, pre-calibration in a short block of traditional design has proved effective. By this means it is also possible to shape complicated cross-sections with hollow chambers precisely. In the downstream, enclosed vacuum calibration section, the running profile receives its final cooling and shape stabilization by means of disk elements arranged one behind the other, with inner contours corresponding to the external contours of the profile, see Figure 36.

Figure 36. Vacuum tank with short vacuum calibration unit for profile- and pipe production. (Photo: Reifenhäuser, Troisdorf, West Germany)

3.2.3.2.7 Cooling die processes

The production of solid rods and thick sheets for use as technical semi-finished products is carried out by a process that is very similar to traditional profile extrusion, see Figure 37. The dies are separated into heated and cooled areas between which heat flow is interrupted by insulating materials, or the cross-sectional area is reduced to a minimum between the hot and cold parts of the die.

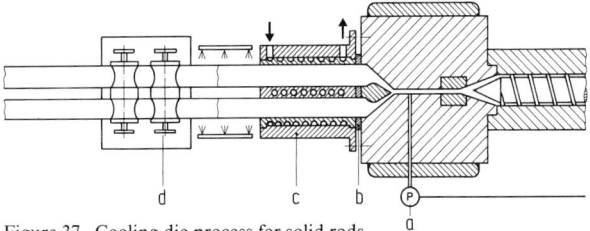

Figure 37. Cooling die process for solid rods
a pressure measurement for control of screw speed, *b* heat barrier, *c* cooling die, *d* brake haul-off

By observing the variation of significant dimensions one can monitor the dimensional condition of the profile before entry into the calibration unit and, by closed-loop control of haul-off speed, influence dimensional accuracy.

In the heated part, the profile cross-section is completely formed, and on contact with the cooling die, a shape-stable jacket is gradually formed, which encloses a still-plastic core of melt. The contraction caused by the volume shrinkage on cooling produces a virtual extra conveying pressure on the melt. Conveying rate must be sustained constant by closed loop melt-pressure control.

The haul-off unit is designed as a braking and transporting unit. Because cooling of solid rods is a very slow process they are usually produced with multistrand dies, in order to make economic production possible, particularly with cross-sections under 100 mm diameter.

3.2.3.3 Measurement and control

In addition to the normal measurement and control tasks involved in extrusion, the closed-loop control of the most important profile dimensions is particularly important in profile extrusion.

Another possibility is to control the feed to the calibration unit with the aid of a sensor on the profile before entry into the calibration unit. A well defined, constant melt stream entering the calibration unit makes for good dimensional accuracy and optimum cooling conditions, as contact with the walls of the calibration unit will vary little (Figure 33).

The higher the friction in the calibration unit, the more the profile produced will be extended and the more it will shrink when heated. For this reason, traction force measurement between haul-off and calibration unit can be provided to generate a signal by which the haul-off speed, material throughput rate, and partial vacuum in the calibration unit can be affected.

A further aid to monitoring profile extrusion lines is the measurement of the raw material feed-rate before entry into the machine, to provide a basis for keeping the feed-rate constant.

3.3 Extrusion lines for the cable industry

Because a wide variety of products are used by the cable industry (differing cross-sections and differing insulation materials), there are also a wide variety of lines available.

The total range of plants will not be dealt with in detail here; the discussion will be limited to the most usual types of line:

- extrusion line for the insulation of optical fibers,
- extrusion line for the insulation of fine wires (mini-wire),
- extrusion line for the insulation of single wires or stranded wires,
- extrusion line for the production of high-current cable, insulated with crosslinked polyethylene,
- extrusion line for the insulation of telephone wires,
- extrusion line for the sheathing of conductors and lightweight cables,
- extrusion line for the sheathing of heavy cables.

The construction of these lines is described below.

3.3.1 Extrusion line for the insulation of optical fibers (Figure 38)

A loose or solid insulation must be provided to protect the coated, primary fibers, without changing the attenuation value for light transmission in the fiber. Thus, particular emphasis is laid on the handling of the fiber in this line.

Construction of a line for the insulation of a single fiber:

3.3 Extrusion lines for the cable industry

- driven pay-off with traverse to prevent wandering of the fiber,
- capstan haul-off with tension measuring and regulating device to maintain tension constant on the line (100 N),
- extruder with 25, 30 or 35 mm barrel diameter. Design with a special die for loose insulation, which permits the introduction of gel fillings,
- a cooling section, divided into various sections, that makes possible stage-by-stage cooling with temperature-controlled water,
- capstan haul-off and traversing winder (two winding positions are often used),
- diameter-monitoring device with the ability to measure in two planes.

In this way, the quality of the wire can be checked even with loose insulation.

Figure 38. Partial view of an extrusion line for the insulation of optical fibers. (Photo: Reifenhäuser, Troisdorf, West Germany)

3.3.2 Extrusion line for the insulation of fine wires up to 0.5 mm² cross-section

As a rule, this type of line is designed as a universal line for processing all the usual thermoplastics, and fluoropolymers.
Basically, the line consists of:

- driven pay-off, matched to the spools used,
- wire preheating device,
- extruder with 25, 30 or 35 mm screw diameter. The screws are matched to the particular raw material. In the case of fluoropolymer processing, the barrel is given a special fluoro-resistant armor-plating.
  The screws for fluoropolymers are produced from Hastelloy C. Sheathing dies have fixed, centered die inserts (from Hastelloy C for fluoropolymers).

- cooling section with telescopic part, if necessary, with various zones for stage-by-stage cooling,
- a diameter-monitoring device permits the control of diameter by variation of haul-off speed,
- high-voltage spark tester for checking insulation quality,
- capstan haul-off with downstream winder. Size of the winder is dependent on spools,
- central line control, in which all control devices and operating elements of the line are combined. Synchronization of all line components in order to guarantee a controlled start-up and shutdown of the line.

3.3.3 Extrusion line for the insulation of single wires and stranded wires

As a rule is designed for two cross-section ranges (0.5 to 16 or 1.5 to 35 mm^2) and is suitable for the processing of all usual raw materials.

The lines are normally designed for continuous operation, i.e. the change from full to empty spool is carried out without interrupting production. Production speeds lie between 500 and 1500 m/min., depending on product. Basic line components:

- dual overhead pay-off, designed for the particular spools. Equipped with a welding device for joining wire ends and a wire-straightening device.
- wire preheating device,
- extruder with fixed, centered sheathing die. According to the required throughput, extruders are used in the size range of 50 to 120 mm, and often also 150 mm,
- cooling sections with several channels,
- diameter-monitoring device,
- high-voltage spark tester,
- capstan haul-off with dancer control,
- semi- or fully automatic dual winder.
  Dual winders are available for spools with diameters from 500 to 1250 (1300) mm. The transfer of the wire from full to empty spool is carried out automatically when the desired wound length has been reached. With fully automatic winders, the full spool is also changed automatically. Recently, winding systems have been developed for which spools are not required. The wires are wound on a separate bracket, and then tied off and shrunk into a film pack. For many applications, this is a solution that saves both space and transport costs.
- concentricity-measuring device,
- central line control.

Recently, lines have been equipped with freely programmable control, partially also with microprocessor control. These contribute greatly to rationalization and quality improvement in cable production.

3.3.4 Extrusion line for the production of cables and wires from crosslinked polyethylene

These lines are equipped with completely different components and technology, and are not dealt with here.

3.3.5 Extrusion line for the insulation of telephone wires

New processes had to be found in order to keep up with the rapid developments in the field of telephone wire production. High speeds (up to 2500 m/min.) were only made possible by the integration of the wire annealing machine into the extrusion line. New measuring and control devices became necessary to guarantee the quality required of the wires.

Line construction:

- drawing machine with downstream resistance annealer,
- wire preheating device of inductive type, temperature-controlled by microprocessor,
- extruder with 70 or 80 mm diameter barrel for the insulation layer of PE or foamed PE, equipped with a fixed, centered two-layer die, to which an
- extruder with 35 to 45 mm diameter barrel is connected. The main skin is applied by this extruder,
- cooling trough with several channels, and a motorized movable telescopic part,
- dual capstan haul-off, as a rule built into the cooling trough, with downstream dancer control,
- automatic dual winder for spools of 500 or 630 mm diameter,
- measuring and monitoring devices:
  diameter-monitoring device for non-contact measurement, capacity-measuring device for monitoring degree of foaming. The measured values from both devices are fed back to a microprocessor. This sets the position of the telescopic cooling trough, the haul-off speed and the temperature of the final barrel zone in relation to the diameter and degree of foaming.

3.3.6   Extrusion line for the sheathing of conductors and fine cables

Very simple construction in comparison with the lines hitherto described.

- pay-offs matched to the spools used,
- extruder with screw diameters between 90 and 150 mm,
- cooling trough,
- dual capstan- or belt haul-offs, according to cable dimension,
- wind-up.

Depending on the cable construction, it may be necessary to introduce a filler layer under the sheath; if so, a second extruder is used.

3.3.7   Extrusion line for the sheathing of heavy cables in tandem process

Similar in construction to the line previously described.

- traversing pay-offs,
- belt haul-off as pushing device,
- extruder for inner sheath, sheathing die head,
- extruder 150 mm diameter for outer sheath, sheathing die head,
- cooling trough,
- belt haul-off,
- wind-up of traversing type,
- line control system with all operating and indicating elements.

A trend towards higher operating reliability with better product quality can be observed on all lines. Great attention is paid to the economical and scrap-free application of cable and insulating material. This leads increasingly to processor-controlled lines, which support operating personnel in improving efficiency.

References for Chapter 3

For the convenience of the reader the English titles of all publications in languages other than English are shown in parentheses.

[1] *Gebler, H., et al.:* Kunststoffe 70 (1980) 4, pp. 186/192, 5, pp. 246/253, and 7, pp. 390/395.
[2] *Schiedrum, H.-O.:* Kunststoffe 73 (1983) 1, pp. 2/8.
[3] *Gebler, H.:* Kunststoffe 73 (1983) 2, pp. 73/76.
[4] *Michaeli, W.:* Extrusionswerkzeuge für Kunststoffe (Extrusion Dies for Plastics), Carl Hanser Verlag, Munich, Vienna, 1984.
[5] *Wortberg, H., Schmitz, K. P.:* Unpublished work at IKV, Technical Faculty Aachen (1980).
[6] VDI-Wärmeatlas, VDI Verlag, Düsseldorf, 1977.
[7] *Menges, G., et al.:* Kunststoffe 72 (1980) 5, pp. 257/263.
[8] *Haberstroh, E.:* Dissertation RWTH Aachen, 1981.
[9] *Kamp, W., Kurz, H. D.:* Kunststoffe 70 (1980) 5, pp. 257/263.
[10] VDMA: Kenndaten für die Verarbeitung thermoplastischer Kunststoffe. Teil 1: Thermodynamische Kenndaten (Characteristic Data for the Processing of Thermoplastics. Part I: Thermodynamic Characteristic Data). Carl Hanser Verlag, Munich Vienna, 1979.
[11] *Dierkes, A., et al.:* Tagungshandbuch zum 9. Kunststofftechnischen Kolloquium des IKV (Proceedings of the 9th IKV Plastics Technical Colloquium), Aachen, 1978, pp. 25/29.
[12] *Kleindienst, U.:* Dissertation, TH Stuttgart, 1976.
[13] *Potente, H., Reinke, M.:* Maschinenmarkt 87 (1981) 34, pp. 689/692.
[14] *Gebler, H., Racké, H. H.:* Kunststoffe 72 (1982), pp. 33/38.
[15] *Fröhlich, S.:* Kunststoffberater (1984) 4, pp. 27/30, and 5, pp. 35/39.
[16] *Andersen, K. V.:* Kunststoffberater (1982) 12, pp. 33/35.
[17] Kunststoffe 73 (1983) 1, p. 8.
[18] *Ehnert, M.:* Kunststoffe 73 (1983) 1, pp. 13/16.
[19] *Dalhoff, W.:* Kunststoffe 74 (1984) 1, pp. 54/56.
[20] *Pütz, D.:* Fertigungsverfahren und -anlagen für PE- und VPE-isolierte Energiekabel und Leitungen (Production Processes and Manufacturing Lines for PE- and XPE-insulated Energy Cables and Conductors), in: VDI-K (Ed.): Kabel und isolierte Leitungen (Cables and Insulators), VDI-Verlag, Düsseldorf, 1984.

4 Extrusion of blown films

G. Winkler

4.1 Introduction

Film blowing is the most important method for producing polyethylene films, and will therefore figure prominently in the following discussion. It is estimated that some 90% of all PE films are produced on blowing lines. And it can be seen from the overall usage statistics for LDPE that the extrusion blowing process is in a dominant position. West European usage of LDPE in 1976 was about 3 mio. t, of which 1.9 mio. t was turned into film [1]. This process offers optimum efficiency and other advantages, like variability in the width and thickness dimensions, and the outstanding mechanical properties obtainable by biaxial orientation.

Machinery development went through a build-up phase — as did other areas of plastics processing techniques — under pressure from the efforts to raise the quantity and quality of the end product in order to satisfy increasing demands. Thus, by comparison with 1955, it has been possible to raise the output of medium-sized extrusion units by about a factor of two or three. In spite of the large increase in machine costs that also took place over this period, it was possible to reduce the price/output ratio of a blown film line, and so investment- and operational unit costs are more favorable with modern, high-performance lines than with older ones. Using more precise market intelligence it has been possible to establish that the demand some years ago for quantitative improvement in performance yielded an increasing quality consciousness. This can be recognized in the fact that use of coextrusion to manufacture superior multilayer films shows a disproportionately high growth rate, and test equipment for quality control is being increasingly used. The move towards high-quality film is also based on the demand to be able to produce the ideal film for a given purpose. Thus the usual range of film properties, like dimensional precision, strength, extension at break, shrinkage, weldability, conversion efficiency ("machinability"), among others, broadened to include additional quality characteristics relevant to the end use, such as gas barrier, moisture barrier, chemical resistance and so on.

4.2 Raw materials

The range of ethylene polymers and copolymers available for the manufacture of blown films is very extensive. The film producer thus has the possibility of meeting end product requirements by the appropriate choice of raw material. In practice, polyethylene grades are distinguished by the following criteria:

– manufacturing process,
– MFI value (melt-flow index according to DIN 53 735),
– molecular weight distribution,
– density (according to DIN 53 479).

LPDE (Low-density polyethylene)
    Production process: high pressure

For the following PE melt-flow index ranges the indicated core applications have become solidly established:

MFI (190 °C/21.6 N) in (g/10 min)
MFI 0.3   sack- and heavy-duty film, shrink film
MFI 0.7 to 1 carrier bags, general packaging film, refuse bag film
MFI ~ 2   thin films, laminating films, thin shrink films
MFI ~ 4   thin films.

HDPE (High-density polyethylene)
    Production process: low pressure
MFI (190 °C/50 N) = 0.2 to 0.6 g/10 min, packaging films, bag films.

MFI value

The MFI value is an international indicator of melt behavior. It gives information about the viscosity, the drawability and mechanical properties of the polymer:

High MFI value:
− large melt drawability,
− low melt viscosity and therefore low heat generation in the extruder.

Low MFI value:
− small melt drawability
− high melt viscosity and therefore large heat generation in the extruder,
− high tensile strengths can usually be obtained.

Figure 1 shows the dependence of impact strength on density and melt index.

Molecular weight distribution

The breadth of the molecular weight distribution has effects upon the melt strength (bubble stability) and on the strength properties.

Broad molecular weight distribution:
− high melt strength, good bubble stability,
− general reduction of strength characteristics.

Narrow molecular weight distribution:
− low melt strength and therefore poor bubble stability,
− general increase of strength characteristics.

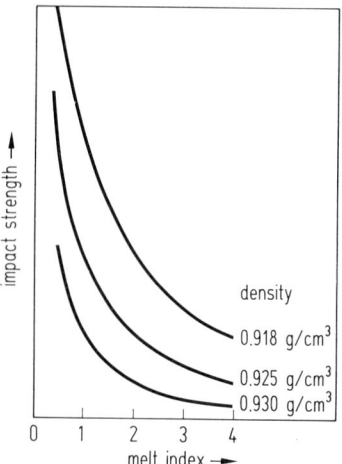

Figure 1. Dependence of impact strength on melt index and density (general relationships)

Density

The density range for LDPE is 0.91 to 0.94 g/cm³. The density of the material gives the processor information about the mechanical properties to be expected, the drawability, and the gas-barrier properties.
In Table 1 values for an LDPE and an HDPE film are given for comparison.
As with choice by MFI value, so certain density ranges have become established for particular uses:

− sack- and heavy-duty film   0.92 to 0.93 g/cm³
− carrier bags   0.92 to 0.925 g/cm³
− thin films   0.925 g/cm³.

The density of HDPE ranges from 0.94 to 0.96 g/cm³.

4.3 Criteria for characterization of films 97

Table 1. Comparison of LDPE and HDPE film properties [2]

Material		LDPE	HDPE along/across
Density	g/cm³	0.91 to 0.94	0.94 to 0.97
Tensile strength DIN 53455	N/cm²	1 500 to 2 000	4 500 to 3 500
Extension at break DIN 53455	%	600	650 to 450
Tear strength DIN 53455	N/100 mm	8	32 to 27
Tensile impact toughness DIN 53448	N/cm²	20 000	20 000 to 18 000
Water vapor permeability DIN 53122	g/m²/24 h	3	1 to 1.5
Max. use temperature	°C	≈ 80	110 to 115

Special grades

In addition to the LDPE and HDPE materials used for the majority of applications, extensive use is also made of other types with a similar set of properties:

- EVA (Ethylene-vinylacetate copolymers)
  MFI (190 °C/21.6 N) = 0.3 to 4 g/10 min
  Density 0.915 to 0.92 g/cm³
  Proportion of VA: 2 to 10% (max. 18%)
  EVA films are finding acceptance in preference to LDPE where the following properties are specially required:
  - large extension at break
  - high impact strength
  - improved gas-barrier properties
  - improved heat sealability.
- MM.HDPE (Medium-molecular-weight HDPE)
  MFI (190 °C/50 N) = 1 to 5 g/10 min.
  Uses: small bags, pouches.
- Ionomers
  Density: 0.94 to 0.96 g/cm³
  Ionomer films exhibit very high strength and provide high transparency. They also have very good heat-seal characteristics.
- LLDPE (Linear LDPE)
  MFI (190 °C/21.6 N) = 0.5 to 2 g/10 min.
  The importance of LLDPE is growing steadily and it is being used like LDPE.

4.3 Criteria for characterization of films

The characterization of a film or a film composite must take conversion operations and end uses into account. There are many criteria and test methods used for quality characterization of blown films.
The basis of these are the relevant DIN and ASTM standards, the information leaflets and guidelines issued by VDMA (Association of German Machinery & Production Line Con-

structors), GKV (Association of the Plastics Processing Industry), and EUROMAP (Committee of European Plastics and Rubber Machinery Producers).
The following criteria have been adopted in practice:

Dimension control

Film thickness. The thickness variation along (MD) and across (TD) the machine running direction is a particularly important quality criterion which strongly influences secondary processability and the way the material can be used.
The variations in longitudinal thickness should be less than ± 2% from the average. In the transverse direction, depending on film thickness, blow-up ratio, material, and film speed, the variation can be ± 3% to ± 15%.
Film width. Maintenance of narrow width tolerances is achieved by the use of calibration and control systems. The width tolerance achieved with a modern high-performance line lies between ± 1 and ± 2 mm.

Machine running

Machine running characteristics serve as a measure of the convertibility of films. They are determined by dimensional accuracy, stiffness, frictional and slip behavior, tendency to sticking (blocking flatness) and other properties. Assessment of machinability is carried out almost exclusively by experiment, or is based on experience. Machine running characteristics can be influenced by appropriate adjustment of the operational parameters or by modification of the raw material employed.
The stiffness of the film is very important in relation to transfer and conveying devices on packaging equipment, sack- and bag-machines.
By blocking is meant the situation when layers of film stick to one another. It is caused by the film being too hot when it is squeezed by the haul-off unit and at the wind-up station, or because of deficiencies in the material itself. During extrusion, low-molecular-weight olefinic compounds migrate to the surface of the film, where they can lead to blocking. By use of an internal air exchange system these compounds are extracted from inside the bubble, and the tendency to blocking is reduced.

Optical quality criteria

Air as a cooling medium really does offer a cost-effective means of energy removal, but has the disadvantage that it cools the film down slowly and the surface becomes roughened by air turbulence. The kind of clarity that is possible with the chill-roll process cannot be obtained with blown film.
For characterization of the optical properties, several methods of measurement are being used [3]; the most important of these are:
− gloss measurement (ASTM D 2457),
− measurement of the reflection and scattering effect at the film surface,
− haze measurement (ASTM D 1003),
− passing a beam of white light through the film,
− measurement of "see-through" clarity (ASTM D 1746),
− measurement of the sharpness of an image viewed through the film.

For optical and quality characterization of a film, a count of gel particles and impurities is also used [8]. Counting devices are being developed, but have yet to be used in production operations. In practice it is primarily the experience of the specialist that is called on in judging the optical quality of a film.

4.4 LDPE blown film lines

Strength, extension, shrinkage

The following well-known test methods are used:
- tensile test (stress at break, extension at break)   DIN 53 455
                                                         ISO R/1184
                                                         ASTM 882-67
- tear propagation resistance
  on trapezoidal, and on   DIN 53 363
  Graves' angle test specimens   DIN 53 515
- edge tear resistance   DIN 40 634
- puncture test (energy, force to damage)   DIN 53 373
- dart drop test   ASTM D 1709-62 T

There are no standard tests for measuring shrinkage and shrinkage forces. But one widely adopted method for determining the shrinkage is to observe dimensional changes in a film specimen 15 mm wide, 100 mm long, after 20 s immersion in a glycerine/water mixture at 120 °C [5].
The determination of shrinkage requires a much greater effort and is therefore usually only carried out in larger laboratories.

4.4 LDPE blown film lines

Blown film lines consist essentially of five elements:
- extruder,
- die unit,
- cooling and calibration unit,
- haul-off unit,
- wind-up.

4.4.1 Extruders

Extruders used for LDPE blown film production are predominantly slow-running single-screw units, with the following range of dimensions:
- screw diameter, D: 40 to 200 mm,
- screw length: 20 D to 30 D.

The speed of these extruder screws is limited to a maximum surface velocity of 0.8 to 1.2 m/s.
Table 2 shows the maximum throughput for a given drive power achievable on blown film lines. The values relate to modern high-performance lines with 30 D screws. High-performance extruders for blown film have very efficient, sleeved feed zones, which have the following features (Figure 2):
- lengthwise grooves which taper conically in the transport direction,
- intensive cooling of the grooved sleeve,
- good heat barrier between the heated barrel and the grooved sleeve,
- relatively small screw-flight depth.

The operating principle, in brief, is as follows:
the lengthwise grooves and the shallow screw channels produce stable granule bridges which receive a strong component of shear in the transport direction from the screw flights, since the grooves inhibit rotational motion of the granulate. Intensive cooling and heat insulation is needed to suppress premature plastication.

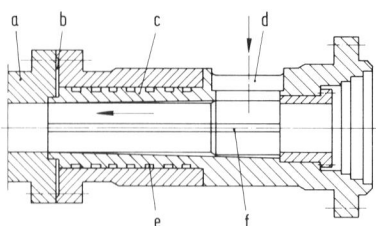

Figure 2. High efficiency feed zone
a extruder barrel, *b* thermal break,
c grooved sleeve, *d* feed opening,
e cooling spiral, *f* grooves

Table 2. Output data for LDPE high-performance blown film lines

Screw diameter D mm	Maximum throughput kg/h	Drive power extruder kW
60	200	55
90	400	110
120	550	170
150	900	300

Table 3. Maximum throughputs of HDPE-performance blown film lines

Screw diameter D mm	Maximum throughput kg/h
35	40
50	80
65	140
75	200
90	260

This extrusion principle makes it possible to operate with low extrusion temperatures and, in addition, the transport rate is independent of the back pressure over a wide range [6 to 8]. The extruder screw is subdivided into the following functional areas:
– feed and compression,
– plastication with the aid of shear elements,
– homogenization zone with mixing elements.

The screw geometry must be designed so that the melt is processed and mechanically homogenized at the lowest possible temperature [13].

Materials used for screws are principally chrome steels and nitrided steels with nitride-hardened surfaces. The screw flights are usually protected with a wear-resistant alloy, to provide a longer service life, particularly for the processing of highly pigmented color concentrates.

The drive is usually a shunt-wound DC motor, steplessly adjustable over the range 1:10. Shunt-wound AC motors are the rule only for small lines with outputs of less than 200 kg/h. The speed of these drives can be varied over the range 1:3 to 1:6.

Power transfer is achieved either by means of a direct-coupled gearbox, or by use of a belt drive. So that the cooling zone for the film shall be as long as possible, a low-level method of construction has been adopted for extruders. Figure 3 shows a low-level high-performance 30 D extruder with a direct-coupled drive. The extruder barrel is made of bimetal or of nitrided steel. Extruder heating is normally carried out by ceramic heating bands, which are divided into three to five controllable heating/cooling zones, corresponding to the different working zones of the extruder. Air is used to cool the barrel.

A melt filter is located between the extruder and the blowing head to remove impurities and foreign bodies. These filters can work continuously or discontinuously [8, 9]. In practice, discontinuous filters that use plates or cassettes are preferred. The use of the cartridge principle is also frequently met with. In this type of unit the filter cloth is made to form a tube. This arrangement is preferred for large-surface-area filters [10].

With screws that are not optimal, static mixers can be used in front of the blowing die to improve melt homogeneity. Disadvantages of these devices are the additional loss of pressure they cause and the danger of contamination (dead spots) because of their complicated form.

4.4 LDPE blown film lines

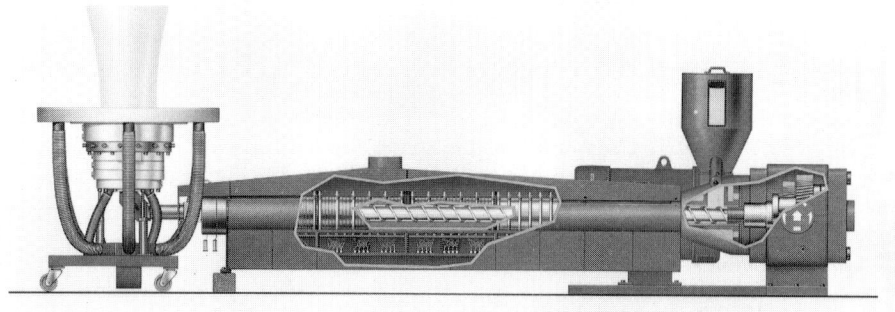

Figure 3. High-performance blown film extruder with die.
(Photo: Windmöller und Hölscher, Lengerich, West Germany)

4.4.2 Dies

The film blowing head, the die on a blown film line, shapes the melt in a narrow annular gap. This shaping process must be carried out free of blemishes and at the lowest acceptable temperature.

Three kinds of blowing head, in particular, are used industrially [7, 8, 9, 11, 12]:

- side-fed dies,
- spider-type dies,
- spiral mandrel dies.

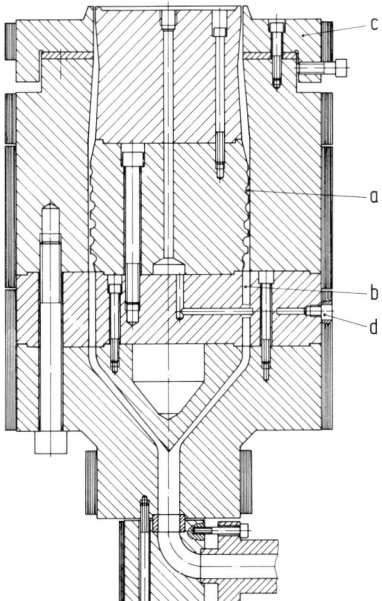

Figure 4. Spider-type die
a smear device,
b spider ring,
c centering,
d entry for internal air

Side-fed dies are hardly used at all now. The melt stream is distributed around the mandrel by heart-shaped grooves on the outer surface. Where the two melt streams meet, weld marks appear which are visible in the blown film; moreover the die head has to be recentered whenever the melt throughput rate is changed, because of the change in forces on the mandrel from side feeding.

The spider-type die (Figure 4) is much used in industry. The term "spider" derives from the support ring with radial elements (spiders' legs) that connect the mandrel, the inner component of the die, solidly to the die body. This spider, or breaker plate, which divides the melt for a short time, causes weld lines to appear. These must be eliminated by use of a smearing device which causes the main axial stream to be enveloped by diverted melt flowing circumferentially. The advantage of the spider die derives from axial feeding, which favors uniform distribution of the melt at the die exit. Such dies are used in particular when tight thickness tolerances have to be met. Because of the axial forces, and the associated loading on the mandrel support, die size is limited.

The spiral mandrel die is the form of construction most favored at present for high output operations. The melt is fed axially into the die and then through radial channels to the spiral mandrel. The dimensions of the spiral are chosen to give a melt distribution along its length that is as uniform as possible. Optimization of the design can be simplified with the help of computer programs [14, 15], Figure 5. Figure 6 shows a centrally fed spiral mandrel die with internal cooling.

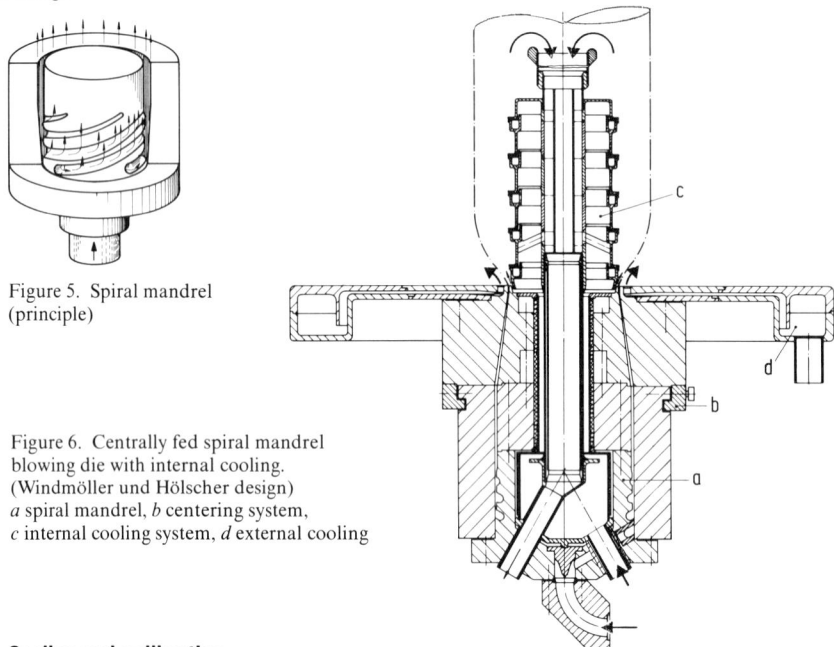

Figure 5. Spiral mandrel (principle)

Figure 6. Centrally fed spiral mandrel blowing die with internal cooling. (Windmöller und Hölscher design)
a spiral mandrel, *b* centering system, *c* internal cooling system, *d* external cooling

4.4.3 Cooling and calibration

4.4.3.1 External cooling

The melt emerging from the die gap is blown in the thermoplastic state and drawn down to final dimensions. The deformation process stops at the freeze line, at which the changeover from the plastic to the solid state occurs.

Bubble cooling is carried out by air emerging from a cooling ring mounted directly on the die outlet. Air volume, air speed, and the direction of the air stream, as well as air temperature, determine the effectiveness of cooling. This area has been extensively studied using model calculations [8] and this, on the basis of approximation equations, has resulted in relevant influences being better taken into account in design planning. Optimal design in the region of the die outlet and the cooling ring, however, can only be arrived at by experiment. Cooling rings used nowadays almost without exception employ the labyrinth system of design with single-stage or two-stage pressure equalization chambers, Figure 7.

The cooling ring is judged by three criteria:
- cooling capacity,
- bubble stability,
- uniformity of the airstream.

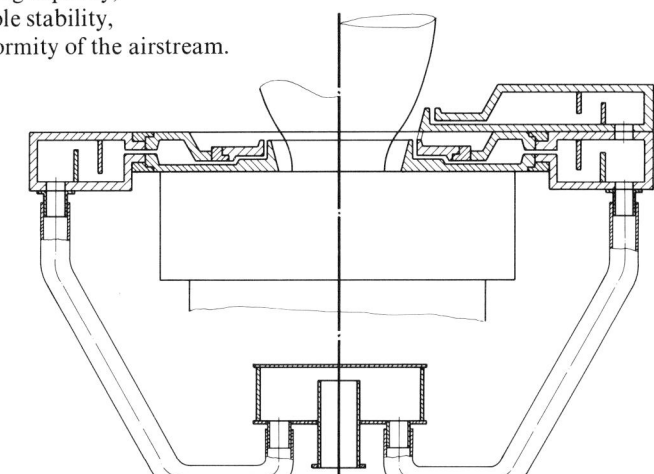

Figure 7. External cooling rings of single- and two-stage types

The shape of the cooling-ring lips, the angle of air impingement and the ratio of air volume to air speed, have a critical influence on optimizing the design of the cooling ring, and for a well-balanced compromise on cooling performance.

The production of sack film is a particular exception. In order to obtain good slip resistance one needs a rough surface; therefore a more powerful airstream of greater cooling intensity directed at the bubble, is required.

Cooling rings for production lines are generally operated with blowers of 15 to 75 m^3/min capacity at pressures of 400 to 800 mm water gauge, depending on the size of the blowing head and the throughput. The use of cooled air produces an improvement in performance, and leads to better stability of the cooling process, since the influence of temperature variation (day/night) is excluded; this contributes to constancy of film properties.

A combination often met in practice is the use of an iris diaphragm above the cooling ring [9], Figure 8. The cold airstream produces a partial vacuum in the chamber formed by the

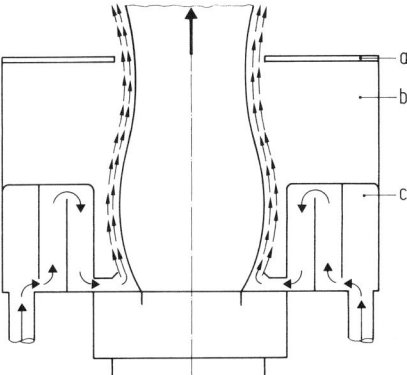

Figure 8. Cooling ring with vacuum chamber and iris diaphragm (principles)
a iris diaphragm, *b* vacuum chamber, *c* air cooling ring

diaphragm and the cooling ring, which causes the bubble to inflate. If the bubble diameter increases, the gap between the bubble and the diaphragm narrows, and the vacuum effect is reduced, so the bubble diameter is reduced again. In other words, the bubble diameter is automatically stabilized by the cooling air.

4.4.3.2 Internal cooling

A significant improvement in performance is achieved with internal cooling by internal air exchange, and this has now become established as a normal feature of high-performance film blowing technology. In Figures 9 and 10 the influence of cooling air temperature on lines with and without internal cooling is shown schematically. The 30 to 50% improvement in performance which is achieved comes about for the following reasons: the cooling surface area is doubled, and higher air velocities (with correspondingly higher heat transfer coefficients) can be used on the outside of the bubble, because of the stabilizing effect of the calibration arrangement with its higher internal pressure. In addition, fresh air is fed into the bubble and directed against it by a stack of distributor disks, Figure 6. The cooling system is equipped with a pressure blower for bringing in fresh air, and a suction blower for pulling air into the bubble through a tube mounted axially inside it. A very useful side effect of internal air exchange is that it removes volatiles from inside the bubble.

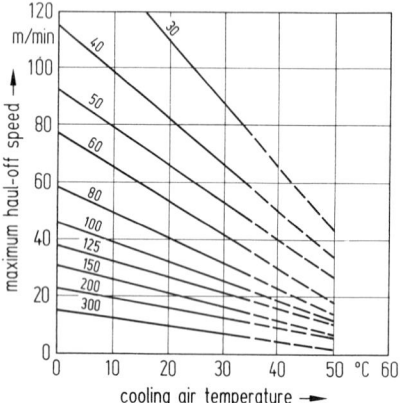
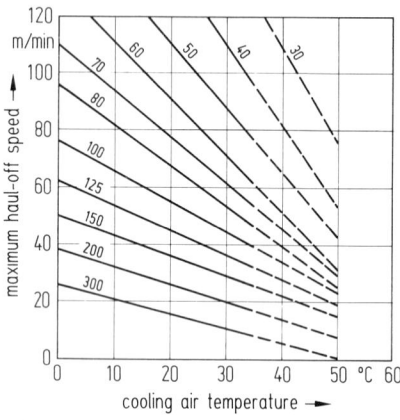

Figure 9. Maximum haul-off speed as a function of cooling air temperature and film thickness, without internal cooling (experimental values)

Figure 10. Maximum haul-off speed as a function of cooling air temperature and film thickness, with internal cooling (experimental values)

In [7, 16, 22] an internal air exchange system is described in which the function of an internal air stabilizer, and the aerodynamic self-balancing of a special external cooling ring, are of particular importance. The self-balancing action is based on the principle that two cooling airstreams are formed that produce a carrying and supporting effect related to the flow resistance, to which the form of the bubble itself gives rise.

4.4.3.3 Calibration

Calibration equipment is needed on lines with internal cooling. This is a basket-like construction with individual free-running rollers on curved support arms, which can be adjusted in height and also by about 1:2 in diameter, Figure 11. The bubble diameter is scanned by a sensitive contact device and the internal pressure held constant by valves. By use of large amounts of air, diameter variations are very quickly corrected, Figure 12.

4.4 LDPE blown film lines

Figure 11. Blowing head with internal air exchange, calibration and external rotating IR thickness measurement system (transmission principle). (Photo: Windmöller und Hölscher, Lengerich, West Germany)

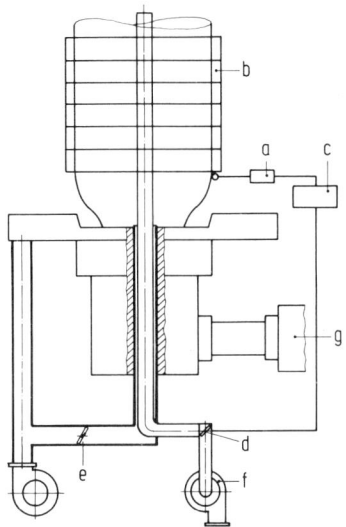

Figure 12. Schematic diagram of control system for bubble diameter and degree of basket filling
a value monitor,
b calibrating basket,
c regulator,
d throttle valve,
e throttle valve for external air,
f vacuum blower,
g extruder

The dimensional constancy achieved by the calibration and control system lies between about ± 1 and ± 2 mm. Diameter control is achieved by corrections to the bubble pressure, by increasing or reducing the internal air volume.
The calibration basket is set at such a height that the frost line always lies below the first set of rolls. Film faults will occur if the basket is set higher or lower than this.
Calibration baskets are also used on film lines that do not have internal air exchange.

4.4.4 Haul-off unit

While the components so far mentioned determine the performance of the line and the film quality, those to be described now have a particularly strong effect upon roll quality.
Haul-off units include collapsing frames and haul-off or squeeze rolls. Edge guides, also, are extensively used just before the collapsing frames, in order to ensure the stability of the bubble and that the film is fed accurately into the frame.
The collapsing frame is constructed from two or more wedge-shaped, angularly adjustable surfaces made of either wood or free-running rollers, Figure 13.
The geometrical relationships in collapsing a film bubble are indicated in Figure 14. To prevent folds occurring during collapse, one must keep the deformation forces as small as possible, by working with very low friction and the smallest possible opening angle [9].
The two haul-off rolls are either rubberized steel, or a combination of rubber and steel rolls. The steel roll is usually cooled. The pressure of the haul-off or squeeze roll is adjustable so as to avoid inversion breaks in the side fold region. Additional cooling grids or, as illustrat-

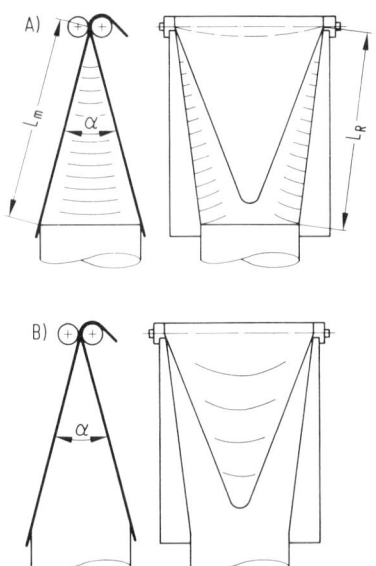

Figure 13. Haul-off unit consisting of collapsing rolls, cooling rolls, haul-off rolls and reversing turning bar system. (Photo: Windmöller und Hölscher, Lengerich, West Germany)

Figure 14. Creasing during bubble collapse because of:
A) length difference, B) frictional resistance

ed in Figure 13, cooling rolls are used to achieve accurate thermal control of the film web and to avoid blocking of the two film layers, which might occur if they are at too high a temperature as they pass through the squeeze rolls.

4.4.5 Winders

Film winders are divided into two categories:
− contact winders
− center-driven winders

4.4.5.1 Contact winders

The contact winder (also called circumferential-, surface-, or drive-roll winder) is the type most used on blown film lines [8, 17].
The rubberized or chromed roll is driven, and the winding shaft or the film-roll is pressed against this roll. Application of roll pressure is achieved mechanically or pneumatically. To obtain good package build-up it is important that the axis of the winding shaft lie parallel to that of the contact roll; this is normally ensured by using two coupled support arms along which the film roll is displaced as the diameter increases, with the result that the drive-roll pressure should remain nearly constant, Figure 15.
For the manufacture of heavy-duty film, contact winders of simple and robust construction are used. For more demanding types of film, thin films for example, contact winders are also used; but they are extended by the use of a pre-haul-off and a sensitive web tension con-

troller, Figure 16. The pre-haul-off ensures that the actual winding point is relieved of the high web tension that is usually necessary to transport the film from the wind-up to the winder. The web-tension controller, in the form of a dancer roll, permits a definite winding tension to be set. The storage effect of the dancer roll makes it possible to achieve reliable, troublefree roll changes.

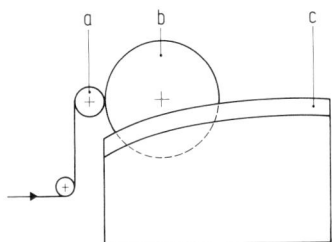
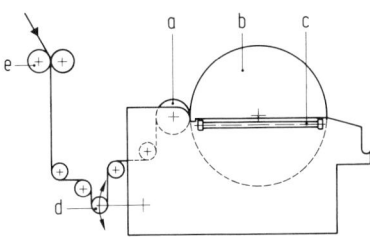

Figure 15. Contact winder (schematic)
a contact roll, b film roll, c winding curve

Figure 16. Contact winder (schematic)
a contact roll, b film roll, c pneumatic cylinder, d dancer roll, e tensioning rolls

Figure 17. Contact winder with two winding stations (double winder).
(Photo: Windmöller und Hölscher, Lengerich, West Germany)

Figure 17 shows a double-winder design of contact winder. The film bubble is separated at the winder and taken to two wind-up stations.

The advantages of contact winders lie in their simplicity of operation and stable construction method. Disadvantages are that very soft, low-tension wound rolls cannot be produced, and faults at a particular position on the roll (high-spots in the film) lead to interruptions in conversion operations. And very short, fat rolls can only be produced with great difficulty. Maximum roll diameter is about 1500 mm, and maximum width some 3200 mm.

4.4.5.2 Central winders

With central winders (also known as axial or direct winders) the winding shaft is direct driven. The drives (DC motors) are designed so that as the film roll diameter increases –

and the motor speed decreases — the torque increases to keep the web tension constant. Dancer-roll control, Figure 18, and computer-controlled drives carry out these tasks [8, 17]. Central winders can produce soft rolls of wound film at very low web tension. This process is aided by air that is drawn in between layers during wind-up, and the buffer effect, which compensate film thickness differences. Such rolls are less sensitive to shrinkage after wind-up.

Central winders are technically expensive, and are used in film blowing for demanding wind-up tasks. It is also advantageous to operate with a combination of the two wind-up principles, that is with a driven contact roll and a separately driven winding shaft, with the film roll running under pressure from the contact roll, or under separate control that maintains a fixed gap between the two.

The use of a central winder or even a combination of contact and central winding, is recommended when low-slip films, for example EVA products, or very soft rolls of film have to be produced.

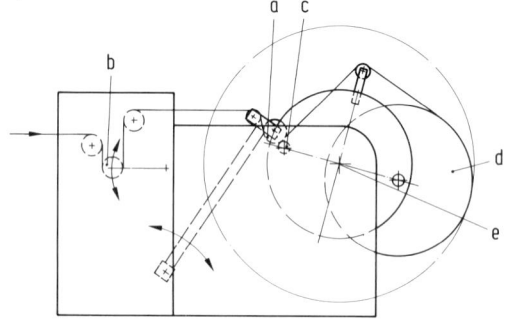

Figure 18. Central winder (schematic diagram)
a swiveling pressure roll,
b dancer roll,
c winding shaft in start position,
d finished roll of film,
e capstan

4.4.5.3 Auxiliary equipment

Winders are usually provided with a range of auxiliary devices:

- guillotines,
- longitudinal slitters,
- edge or center control,
- pretreatment,
- edge trim evacuation.

Guillotines

The guillotine cuts the film web between the completed roll and the new one during roll change. Two cutting principles are employed:

- flying knife,
- chopping knife or saw-tooth knife.

The flying knife is driven across the film by a pneumatic cylinder, cutting as it moves. The chopping knife teeth strike the film web simultaneously across its full width. Both systems have been accepted industrially.

Longitudinal slitters

Lengthwise slitters cut the double layer of film running into the winder into two webs, and — if required — cut each web into several smaller strips (multiple reels). There are two main slitting systems in use:

- blade cutters,
- rotary knife cutters.

Blade cutters are used for film thicknesses up to 250 μm. Precise blade guidance and constant web tension are essential. Normal razor blades, or somewhat stiffer industrial blades, are employed for this work.

In order to withstand the wear encountered when cutting highly pigmented films, special blades are often used — those coated with tungsten carbide for example; and the blades are also carried in a traverse carriage to avoid rapid blunting.

The rotary knife cutter is predominantly used for very thick and hard films. The diskshaped knife runs in a groove, so that the cutting process is effected either by the edges of the groove and the rotary knife, or by pressure against a hardened roll surface.

Edge or center guidance

These systems ensure precision entry of the film web into the winder. Edge control guarantees that the film goes into the winder with one side accurately located. Center control guides the centerline of the film so that variations in the film width are halved at the two edges of the roll.

Pretreatment

Films have to be surface treated if printing, sticking, or laminating are to be undertaken in a later conversion process [18]. The pretreatment station, usually a corona discharge device, is installed between the haul-off unit and the winder.

Edge-strip evacuation

By trimming the film bubble it is possible to produce two flat film webs with absolutely precise edges. The edge trim is removed from the winder by blowers and is either returned directly to the extruder or regenerated. Return feed to the extruder can be carried out in the following ways:

– regranulation and mixing the granulate with fresh material,
– production of shreds in a cutting mill and mixing in the granule hopper,
– compounding in a second, dedicated smaller extruder, which feeds direct to the main extruder (by-pass principle).

4.4.6 Thickness uniformity

To obtain a cylindrical film roll, the thickness variations occurring in the web must be spread evenly across the full width of the roll, so as to guarantee troublefree conversion. The tubular film process offers various possibilities:

– rotating haul-off/winder combination,
– rotating extruder and die,
– rotating die,
– rotating or reversing haul-off elements.

Rotating haul-off/winder combination

The haul-off/winder combination is mounted on a rotating framework, and rotates about the axis of the film tube. The effect of this is that all faults in the film due to extrusion or environmental conditions that exist at the wind-up are spread out, so that high-quality rolls can be produced. This system is found quite often on small lines, especially for HDPE (Figure 19); but it is not suitable for wide, high-output LDPE lines, since the cooling section is too short with the vertical-downward layout, for one reason, and also because the design complexity and the expense of a rotating LDPE wind-up is too great.

Figure 19. Rotating haul-off combination.
(Photo: Windmöller und Hölscher, Lengerich, West Germany)

Rotating extruder with die

This technique, also, is limited for mechanical design reasons to smaller extruder units, and is therefore very largely applied to the manufacture of HDPE films.

Rotating die

The rotating blowing head distributes the faults originating in the die and at the cooling ring, and thus essentially eliminates the thickness errors put into the film.
The sealing problems within the die have been fully solved, with the result that this variant is one of the most used in LDPE film blowing technology.

Reversible haul-off

The reversible haul-off unit not only eliminates thickness errors originating in the die, but also those that can arise because of environmental influences, calibration, and so on. With this correction system very high quality rolls of film can be produced, Figure 20. The equipment consists of roll sets and rotating cylinders that are turned about a vertical axis by gearbox drive, in such a way that the length of film between the haul-off and the winder remains constant. A full description is given in [19]; in the same reference a description of a quite different kind of reversing haul-off is also given: the 45° turn-bar system, Figure 21. This system is used in particular with narrow film widths, especially for heavy-duty film.

4.5 HDPE blown film lines

HDPE blown film, also described as high-molecular (HM) film, is much less important than LDPE in the European packaging sector. This is not so in Japan. Yet because of its mechanical properties and special paper-like characteristics it has come to occupy an established place in the film conversion industry [21, 22]. A comparison of LDPE and HDPE films is

4.5 HDPE blown film lines

Figure 20. Reversing haul-off unit with horizontal turning bar. (Photo: Windmöller und Hölscher, Lengerich, West Germany)

Figure 21. Reversing haul-off unit with 45° reversing beam. (Photo: Windmöller und Hölscher, Lengerich, West Germany)

given in Table 1. The property improvements over LDPE, which are briefly described below, result from the particular molecular structure of HDPE:

– high viscosity over the complete range of shear rates,
– tendency to strong flow orientation because of the linear molecular structure,
– molecular breakdown at high shear rates,
– crosslinking (gel particles) at temperature peaks,
– high melt temperatures,
– post-shrinkage of the film,
– tendency to creasing.

Extruders

HDPE film can equally well be produced on small lines with extruders having short screws of 16 to 25 D, and maximum screw surface speeds of 1.4 m/s, as on lines with conventional extruders with special screws. The extruders are equipped with feed zones (grooved sleeves) of high transport efficiency, and screws with shear and mixing zones.

The grooved sleeve operates as a stable solids-conveying pump, and makes possible a high enough specific throughput (kg/h per rev/min) for the extruder temperature to be kept below 240 °C and for thermal breakdown of the material to be avoided.

The high viscosity and surface hardness can lead to increased screw and barrel wear: for this reason hardened steels and hard metal alloys are preferred nowadays for screws and grooved sleeves.

Table 3 reviews the maximum output levels attainable on modern HDPE lines.

Dies

Design calculations on dies have to take pressures of 400 to 800 bar into account. The spiral mandrel die is the best suited to accept high working pressures, because of the design principle followed. The rheological design of the melt channels and spirals has to be carried out in such a way that excessive shear rates are avoided. The use of hard-chromed surfaces is recommended to suppress melt fracture and to avoid build-up of surface deposits (plate-out).

The die diameter generally lies between 30 and 200 mm.

Cooling and calibration

The shape of an HDPE bubble is distinctly different from that of LDPE, Figure 22. Blow-up of the tube occurs, at the earliest, 5 to 8 diameters from the die. The blow-up ratio usually lies between 4:1 and 6:1.

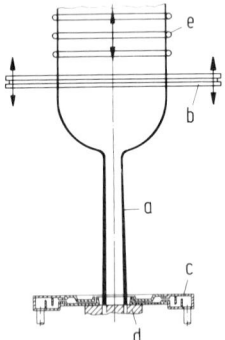

Figure 22. Cooling and calibration of HDPE blown film (schematic diagram)
a bubble neck,
b iris diaphragm with height adjustment,
c cooling ring,
d blowing head,
e calibrating basket with height adjustment

The cooling ring and the ring-lips are shaped to stabilize the bubble form, as well as to provide intensive cooling. Iris diaphragms with height adjustment, mounted just above the freeze line, are used for calibration.

Haul-off unit

The haul-off unit is fitted with a wood-slat collapsing frame and a large number of side elements to guide the relatively stiff film, and to make it possible to work with a very small angle of opening during collapse of the bubble. To ensure that the film temperature in the collapse region is correct, the haul-off is often adjustable in height.

Winder

In HDPE blown film technology contact winders are used almost exclusively. It is advisable to use tensioning rolls to isolate the wind-up from the high tension of the film web. The auxiliary equipment described in 4.5.3 is also installed on HDPE winders. On compact winders, edge control of the web can be omitted.

Thickness distribution

There are three preferred systems in normal use for spreading out unavoidable thickness variations in the film (Figure 23):

– rotating extruder with die,
– rotating haul-off/winder combination,
– rotating die/cooling ring unit.

Lines with internal air exchange can also be used for producing HDPE film, and although no particular improvement in performance is to be expected, there are several advantages worth noting:

4.6 LDPE blown film lines

- improved bubble stability,
- improved cooling,
- evacuation of volatiles.

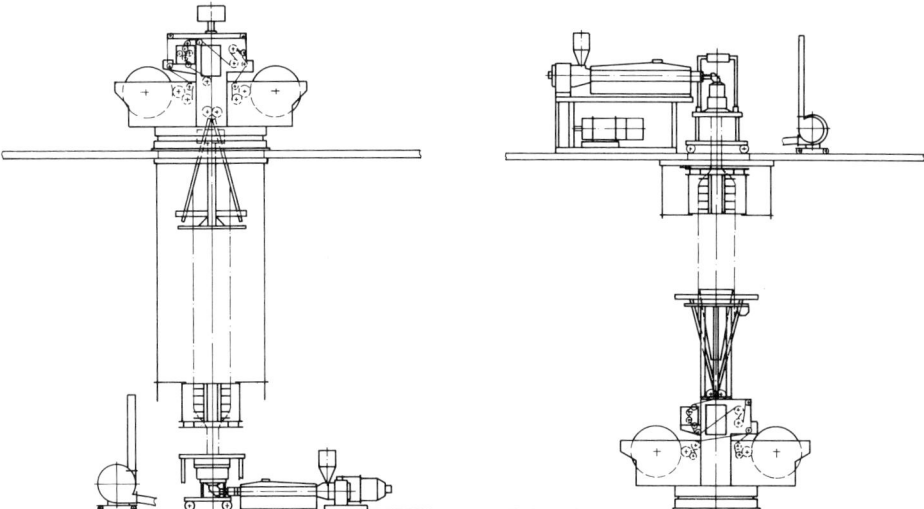

Figure 23. Vertical arrangement of HDPE blown film line. (Windmöller und Hölscher design)

4.6 LDPE blown film lines

4.6.1 LDPE lines in general

LDPE films are divided into groups related to their particular areas of application. The rough summary given below lists some characteristic parameters and data:

heavy-duty sack film: film thickness 150 to 200 mm
                            blow ratio 1:1.3 to 1:1.6
                            die diameter 200 to 250 mm
                            MFI (190 °C/21.6 N) = 0.3 g/10 min

carrier-bag film: film thickness 40 to 70 mm
                            blow ratio 1:1.7 to 1:2.3
                            die diameter 250 to 400 mm
                            MFI (190 °C/21.6 N) = 0.7 to 1 g/10 min

heavy-duty shrink film: film thickness 100 to 250 mm
                            blow ratio 1:3 to 1:4.5
                            die diameter 350 to 650 mm
                            MFI (190 °C/21.6 N) = 0.3 g/10 min

thin shrink film: film thickness 30 to 100 mm
                            blow ratio 1:3 to 1:4
                            die diameter 150 to 350 mm

wide film: film thickness 100 to 250 mm
                            width 2500 to 10,000 mm (double collapsed bubble width)

laminating film: film thickness 20 to 100 mm
                            film thickness 15 to 30 mm (LDPE copolymers, LLDPE)

It has certainly been possible to expand still further the spectrum of films produced by the tubular process, by use of some of the ideas frequently used industrially e.g. to make form-and-fill films, pouch films, and so on.

Since optimally designed, dedicated lines for making the products listed above are seldom required by the processor, in practice flexible lines are required that give the film producer the ability to manufacture a number of products.

Two kinds of line, in particular, have arisen as a result of market demands for the manufacture of two different ranges of films:

- thin film lines,
- heavy-duty film lines.

4.6.1.1 Thin film lines

The concept of the thin film line is not clearly defined, but one generally understands the following types of film to be included among fine films [23, 24]:

- laminating films,
- thin shrink films,
- form-and-fill films,
- bag films,
- general packaging films.

a extruder,
b die with internal air-exchange system,
c calibrator,
d bubble collapse,
e reversing haul-off device,
f centerline control,
g double contact winder,
h tension rolls,
e dancer rolls

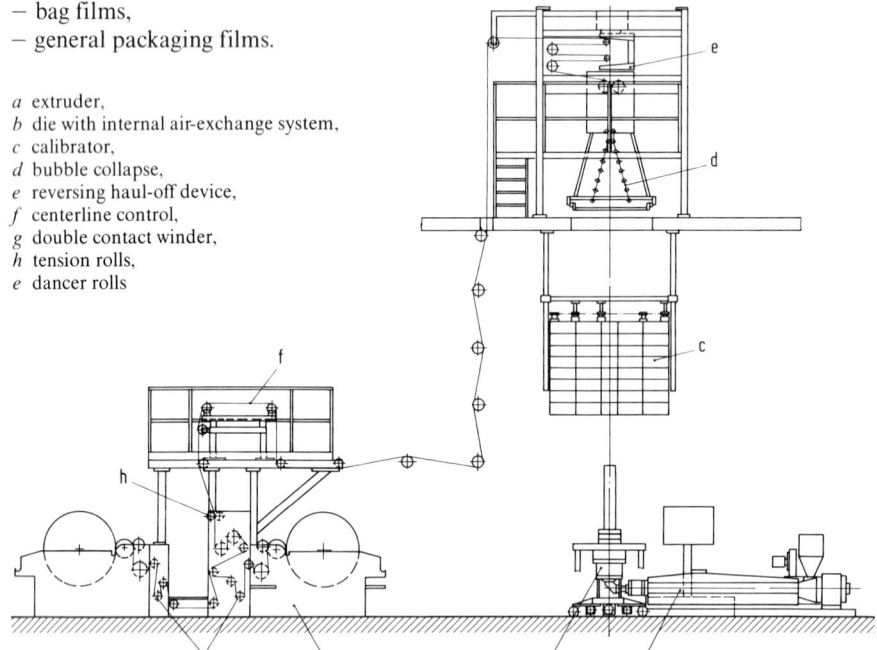

Figure 24. Thin film line. (Windmöller und Hölscher design)

The designation "thin" applies to films thinner than 100 µm, and the above-mentioned products are, in general, thinner than this.

The essential components of a thin film line are those described in Section 4, having particular regard to line flexibility and film quality. Criteria to be noted are:

- blowing heads that can be assembled with different die inserts; this provides big variability in blow ratio and film width,

4.6 LDPE blown film lines

- calibration basket with the largest possible working range,
- high specification on the roll quality, and good film flatness,
- thin films require very sensitive web-tension control, since quite small forces can produce large extensions. The contact type of winder is most widely used.

If requirements are particularly severe (for example as with laminating films) or with special products like LDPE copolymers, central winders or combined central/contact winders are preferred. The winder normally comprises a pretreatment station, web-edge control, and two wind-up positions:

- a collapsing unit with a very low frictional resistance is needed. The roll device has proved suitable for thin flexible films.
- film cooling, internal and external, must be optimized with regard to the following two criteria:
  a) intensive cooling to achieve high transparency,
  b) lowest possible air turbulence at the film surface to avoid roughening the surface and losing gloss and transparency.

Compliance with tight tolerances is of economic importance for reasons connected with the conversion of the film, since some 75% of the production costs are raw material costs.

With laminating films, quality criteria like thickness tolerance, flatness, roll quality, and optical properties are of special importance.

The thin film lines in use today are normally equipped with 60 to 90 mm extruders, and occasionally with 120 mm extruders.

Working area:

- 60 mm extruder,
  die diameter 150 to 350 mm,
  width maximum 1500 mm, double layflat width;
- 90 mm extruder,
  die diameter 250 to 500 mm,
  width maximum 2000 mm, double layflat width.

Typical thin film lines are shown in Figures 24 and 25.

Figure 25. Thin film line with rotating die, internal air exchange system, double winder with roll core magazine and with automatic open- and closed-loop controls.
(Photo: Reifenhäuser, Troisdorf, West Germany)

4.6.1.2 Heavy-duty film lines

Whilst thin film lines are designed to be as flexible as possible, for heavy-duty film there are five basic designs of line which are more or less set up for a given product, and which, therefore, can be considered to be single product lines, e.g. for:

- heavy-duty sack films,

- refuse-sack film,
- heavy-duty shrink films,
- carrier-bag film,
- wide film.

Example 1:

Figure 26 is a schematic diagram of a typical sack-film line. Figure 27, a twin-die version with two blowing heads, is an economically efficient arrangement.

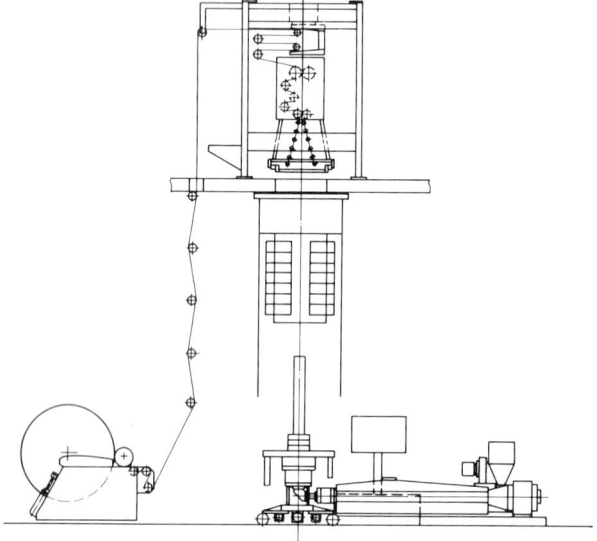

Figure 26. Heavy-duty sack-film line. (Windmöller und Hölscher design)

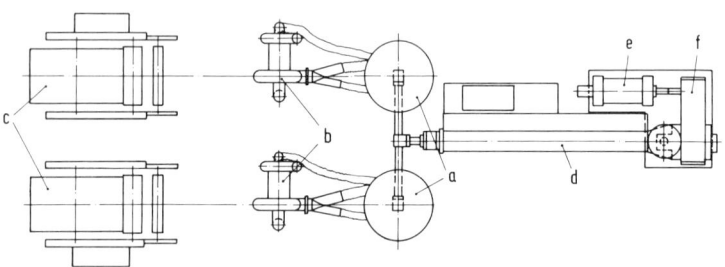

Figure 27. Heavy-duty sack-film line with two blowing heads. (Windmöller und Hölscher design)
a blowing heads, *b* air blowers for external air and internal air exchange, *c* contact winder, *d* extruder, *e* drive, *f* gearbox

Example 2:

Figure 28 is a schematic diagram of a shrink-film line with a rotating blowing head and gusseting device. Extruder sizes lie between 120 and 150 mm. The winder is of the contact type, with pre-tension- and dancer rolls for web-tension control. The width of shrink-film lines lies between 1800 and 3500 mm.

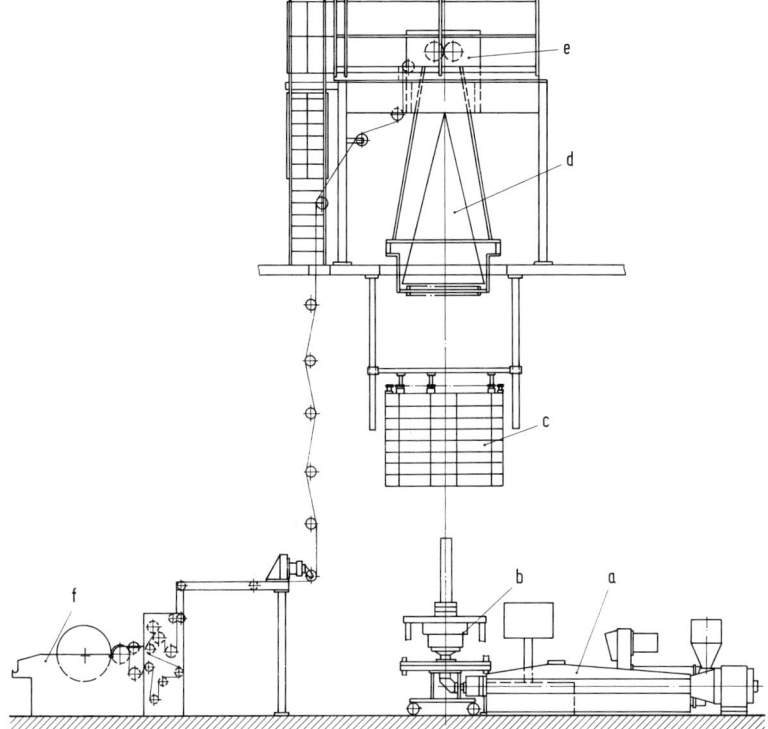

Figure 28. Heavy-duty shrink-film line with rotating blowing head and gusseting device. (Windmöller und Hölscher design)
a extruder, b rotating blow-head with internal air exchange system, c calibration basket, d bubble collapse with gusseting device, e haul-off, f contact winder with tension rolls and dancer rolls

4.7   LLDPE (linear low-density PE) lines

This raw material has displaced LDPE in some applications because of its special mechanical properties and the low cost of production (low pressure process). A comparison with LDPE brings out the special features of this material:

– for the same strength a 25 to 50% reduction in film thickness is possible,
– extension at break is about 200% greater.

In the USA about 30% of all LDPE has been replaced by LLDPE, but in Europe LLDPE is only used for stretch films, in refuse sacks, as a second component in LDPE and HDPE, and in coextrusion applications.

The machinery producer has to make modifications to take account of the particular characteristics of this material, which relate especially to the molecular structure and the rheological behavior:

– Viscosity, Figure 29:
  The higher viscosity relative to LDPE requires a higher screw torque or drive torque, and modification to the screw geometry, in order to produce the lowest possible melt temperature.
– Narrow molecular-weight distribution:

This reduces the strength in the plastic state and lowers bubble stability, and calls for a special cooling technique to be used. The external cooling system has the additional task of supporting and stabilizing the bubble. For this reason two-stage cooling rings like that illustrated in Figure 7, or the by-pass version, are quite often used to produce a venturi effect (partial vacuum).

– Melt fracture:

To counteract melt fracture, which is a coarse deformation of the surface of the extrudate as it emerges from the die, the die gap must be increased (Figure 30).

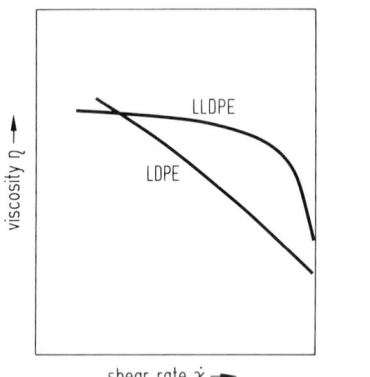

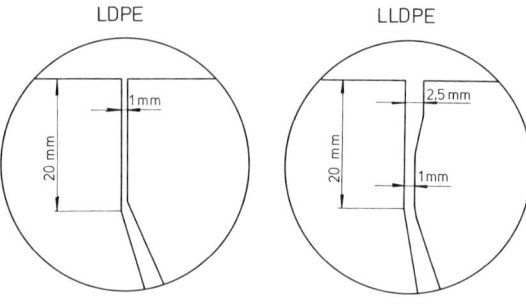

Figure 30. Die contours for LDPE and LLDPE with experimental values

Figure 29. Viscosity as a function of shear rate for LDPE and LLDPE (general principles)

4.8 Coextrusion

Because of the importance of coextrusion, it is dealt with separately in Chapter 5.

4.9 Special processes

The blowing process extensively used with PE is also used with a number of other thermoplastic materials, which purely in terms of quantity play very much a subsidiary role. The most important of these, and the special processes used, will be dealt with in note form.

Polystyrene (PS)

The tubular process is used almost exclusively for producing foamed film. There are two versions of the foam process [31]:

– use of polystyrene containing blowing agent,
– direct gassing by mixing a blowing agent in the extruder.

Since the foamed film is relatively thick, it is drawn over a cooling mandrel, and then slit open. Foamed PS lines usually operate horizontally.

Polypropylene (PP)

One of the most important properties of PP is transparency, which can only be attained by chilling the melt immediately after extrusion. The process used for this is slot-die chill-roll casting. This method has shown itself to be far superior to the tubular process in quality and performance.

For solving the cooling problem with blown film, a variety of special water-cooling devices, described in [8, 9, 32], are available:
- water-fed cooling sleeves (indirect water cooling),
- production of a water film between the cooling sleeve and the film (direct water cooling),
- spraying the tube with water,
- extrusion into a water bath.

The use of this process is limited to special areas.

Polyamide (PA)

PA is often used as a coextrudate in blown film manufacture; however extrusion of simple films is classified as a specialized process [8, 32].
As with PP, the preferred production method for PA films is slot-die extrusion, although the mechanical properties transverse to the machine direction can be improved by biaxial orientation in the bubble blowing process.

Polyvinylchloride (PVC)

The predominant technique for manufacture of PVC films is calendering. But the high investment costs involved are a continuing incentive to use the blowing process [8, 33, 34]. Single- and double-screw extruders are used industrially, the latter making possible the direct processing of powder.
Special demands are made on the extrusion process, in order that the product suffer as gentle thermal treatment as possible in a short time without development of high pressures as it passes through the die. Because of the thermal instability of the material, no satisfactory solution has yet been found which can remove blowing from the "special process" category.

4.10 Automation

The first-line task of measurement and control devices in the extrusion process is to assure the quality of the end product. Modern blown film lines are therefore equipped with a range of devices that can be divided into two groups:
- control of the thermodynamic quantities
- control of the geometry (dimensional control) of the finished or semi-finished product.

Thermodynamic quantities

Open- and closed-loop control, or even simple control of the thermodynamic quantities melt pressure and temperature, are extensively discussed in specialist journals and textbooks [35 to 37].
The state of the art is that the melt pressure and temperature produced by the extruder before the polymer enters the die are indicated and monitored. This is necessary in order to avoid so-called "cold starts", and to be able to monitor the operational state of the extruder during production.

Dimensional control

Dimensional control relates to film width and film thickness. Because of higher raw-material prices, and increased cost of labor over the last few years, increasing use has been made of width and thickness measurement systems for quality control. A further development step possible with these measurement systems is to build up control systems, in order to move

from the simple monitoring situation towards closed-loop control and automation. With the help of microelectronics this system has led to complex closed-loop control systems on individual blown film lines, and even to overall data-gathering and management systems, with the possibility of running several lines from a single control point [38, 39].
Often the way into automation has occurred through installing individual control systems on existing blown film lines, in order to eliminate specific weak spots, and to prepare the management for the introduction of modern techniques of control and data processing.
The following control systems are being used:
- Width control:
  Control of film width by correcting the calibrating basket diameter or the volume of air inside the bubble,
- Thickness control (mean value):
  a) Contactless thickness measurement:
     By integrating all the thickness values measured across the width of the film or around the bubble circumference, a mean value is arrived at. Comparison with the set (desired) value of thickness yields the appropriate correction to the haul-off speed.
  b) Feedstock weighing system:
     The mean film thickness can be indirectly determined by the use of a continuously operating feedstock weighing system. If the film width is known, the current value of thickness can be calculated using simple mathematical relationships, on which closed-loop control of the production speed can be based.
- Basket height control:
  The distance from the freeze line to the calibration basket is controlled by scanning the film temperature, Figure 31, with the effect that the set distance is kept constant. In this way virtual constancy of bubble shape during production can be ensured.
- Control of film thickness tolerances:
  A precondition for the control of thickness (overall tolerance) is a thickness measurement system that covers every point on the film web. A system equipped with microprocessors that provides fully automatic, closed-loop control of thickness is now on the market. The principle of operation depends on the fact that if there is a slight change of melt temperature in the flow channels of the die, the film thickness will be influenced in a specific way. This effect can be brought about, as shown in Figure 32, by means of cooling channels that are fed with air.

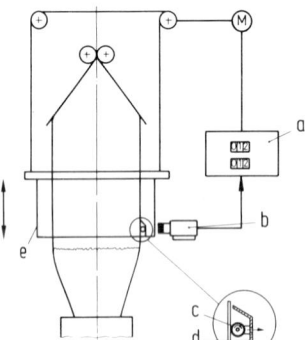
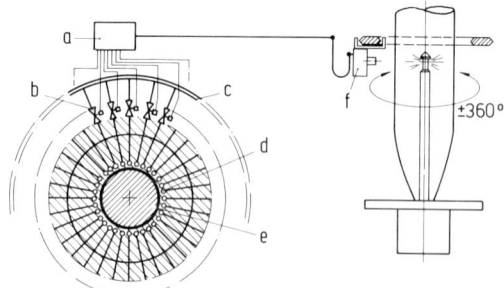

Figure 31. Basket height control.
(Windmöller und Hölscher design)
a regulator, b contactless temperature measurement, c probe roll, d bubble, e calibration basket

Figure 32. Control of film thickness tolerance.
(Windmöller und Hölscher design)
a microprocessor, b valve, c compressed air, d cooling channel, e die gap, f IR thickness measuring equipment (transmission method)

4.10 Automation

Connected systems; process control systems

Coupling and interlinking several control systems on a single line has become possible in the last few years; the operating principle is as follows:
The control elements on a line are adjusted physically, after start-up, to the values wanted for production. This process is initiated by a single-function system like a punched-tape reader, for example, or by calling up stored data with the keyboard. After the setting operation is completed the control system takes over supervision of the line, as described above. Figures 33 and 34 show the control panel of this system.

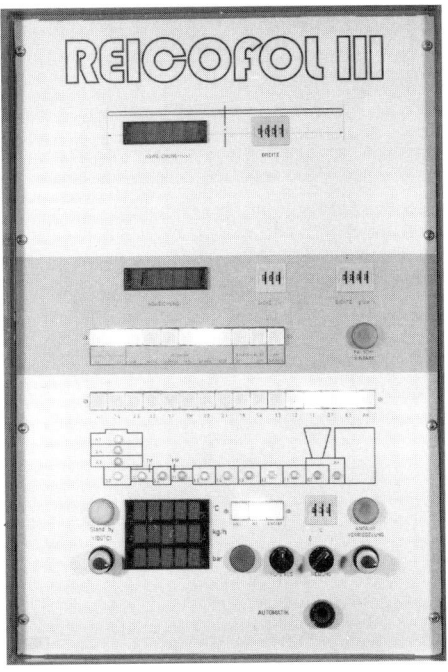

Figure 33. Control panel of an automated blown film line. (Photo: Reifenhäuser, Troisdorf, West Germany)

Figure 34. Control panel of an automated blown film line. (Photo: Windmöller und Hölscher, Lengerich, West Germany)

A further step towards data gathering, and open- and closed-loop control of several lines, has already been put into practice. Thus Figure 35 shows the VDU and keyboard of a central control computer, from where production can be monitored and controlled automatically. These systems already bear some resemblance to purely commercially directed computers, and therefore also make possible the execution of orders, scheduling of production, and so on. Film manufacture can be made fully automatic with these systems. However, it should not be overlooked that film products place extreme demands on machine and process technology, because of the enormous variety of dimensions, and demands for particular mechanical or optical properties, and also because of the very different raw materials that are used. So it can be assumed for the future that although the specialist can be backed up by the computer, he cannot be entirely replaced by it.
Figure 36 shows a microprocessor-controlled simulation device on which the operation of an automated blown film line can be demonstrated or, more precisely, can be simulated. This equipment is of primary importance for personnel training, and for internal operational studies.

 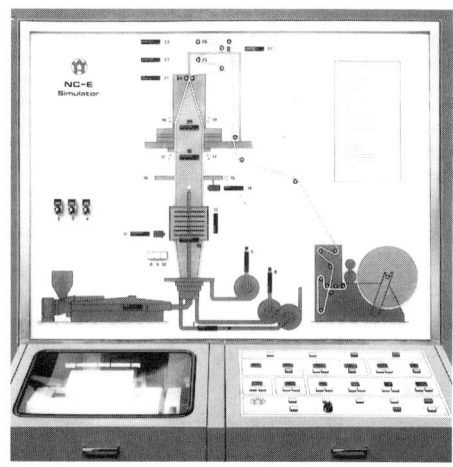

Figure 35. Control panel of a command computer. (Photo: Windmöller und Hölscher, Lengerich, West Germany)

Figure 36. Simulator for an automated blown film line. (Photo: Windmöller und Hölscher, Lengerich, West Germany)

The final illustration of a blown film operation, Figure 37, should provide the reader with a lasting impression of the state of modern LDPE blown film technology.

Figure 37. High-output line for LDPE blown film. (Photo: Windmöller und Hölscher, Lengerich, West Germany)

References for Chapter 4

For the convenience of the reader the English titles of all publications in languages other than English are shown in parentheses.

[1] *Keller, R., Kress, G., Pleßke, P.:* Kunststoffe 67 (1977) 10, pp. 583/585.
[2] *Krause, M., Verse, N., Ploens, J.* in VDI-K (Ed.): Extrudieren von Schlauchfolien (Extrusion of Tubular Film), VDI-Verlag, Düsseldorf, 1973.
[3] *Fritz, H. G.:* Kunststoffe 59 (1969) 8, pp. 393/397, pp. 478/512.
[4] *Hess, K. M.:* Kunststoffe 59 (1969) 8, pp. 512/514.
[5] *Müller, J.* in VDI-K (Ed.): Extrudieren von Schlauchfolien (Extrusion of Tubular Film), VDI-Verlag, Düsseldorf, 1973.
[6] *Boes, D.:* Industrie-Anzeiger 95 (1973) 35, pp. 729/732. Industrie-Anzeiger 95 (1973) 52, pp. 1118/1121.
[7] *Fischer, P.:* Plastverarbeiter 30 (1979) 3, pp. 117/125.
[8] *Predöhl, W.:* Technologie extrudierter Kunststoffolien (The Technology of Plastics Film Extrusion), VDI-Verlag, Düsseldorf, 1979.
[9] *Kress, G.* in VDI-K (Ed.): Extrudieren von Schlauchfolien (Extrusion of Tubular Film), VDI-Verlag, Düsseldorf, 1973.
[10] *Hensen, F.:* Plastverarbeiter 33 (1982) 12, pp. 1447/1454.
[11] *Ast, W.:* Kunststoffe 66 (1976) 4, pp. 186/192.
[12] *Kleindienst, U.:* Kunststoffe 63 (1973) 7, pp. 423/427.
[13] *Krämer, A., Winter, H. H., Martin, G.* in VDI-K (Ed.): Extruder als Plastifiziereinheit (The Extruder as a Plastication Unit), VDI-Verlag, Düsseldorf, 1977.
[14] *Rao, N.:* Fortran Programme für Kunststofftechniker (Fortran Programs for Plastics Technologists), Carl Hanser Verlag, München Wien, 1979.
[15] *Procter, B.:* SPE Techn. Pap. XVII, 29 (1971), pp. 211/218.
[16] *Ast, W.:* Kunststoffe 68 (1978) 6, pp. 343/347.
[17] *Kähler, F. J.* in VDI-K (Ed.): Extrudieren von Schlauchfolien (Extrusion of Tubular Film), VDI-Verlag, Düsseldorf, 1973.
[18] *Schmelding, H.* in VDI-K (Ed.): Extrudieren von Schlauchfolien (Extrusion of Tubular Film), VDI-Verlag, Düsseldorf, 1973.
[19] *Hartung, A.* in VDI-K (Ed.): Extrudieren von Schlaufolien (Extrusion of Tubular Film), VDI-Verlag, Düsseldorf, 1973.
[20] *Krause, M., Köhler, F. J.:* Kunststoffe 61 (1971) 10, pp. 731/737.
[21] *Kahl, T.:* Plastverarbeiter 24 (1973) 4, pp. 211/216.
[22] *Ast, W.* in VDI-K (Ed.): Kühlen von Extrudaten (Cooling of Extrudates), VDI-Verlag, Düsseldorf, 1978.
[23] *Hershey, G.* in VDI-K (Ed.): Extrudieren von Schlauchfolien (Extrusion of Tubular Film), VDI-Verlag, Düsseldorf, 1973.
[24] *Predöhl, W.* in VDI-K (Ed.): Extrudierte Feinfolien und Verbundfolien (Extruded Thin Films and Film Laminates), VDI-Verlag, Düsseldorf, 1976.
[25] *Land, W., Schulte, H.:* Verpackungs-Rundschau 9 (1971), pp. 1152/1158.
[26] *Wortberg, J.:* Plastverarbeiter 28 (1977) 4, pp. 178/184.
[27] *Hessenbruch, R.:* Papier- und Kunststoffverarbeiter 3 (1982), pp. 17/20.
[28] *Fischer, P., Wortberg, J.:* Kunststoffe 74 (1984) 1, pp. 28/32.
[29] *Hensen, F., Hessenbruch, R., Bongaerts, H.:* Kunststoffe 71 (1981) 9, pp. 530/538.
[30] *Kurtz, S. J., Scarola, L. S., Miller, J. C.:* Plastics Engineering, June 1982, pp. 45/48.
[31] *Rapp, B.* in VDI-K (Ed.): Extrudieren von Schlauchfolien (Extrusion of Tubular Film), VDI-Verlag, Düsseldorf, 1973.
[32] *Dreyer, W.* in VDI-K (Ed.): Extrudieren von Schlauchfolien (Extrusion of Tubular Film), VDI-Verlag, Düsseldorf, 1973.
[33] *Pirot, E.* in VDI-K (Ed.): Extrudieren von Schlauchfolien (Extrusion of Tubular Film), VDI-Verlag, Düsseldorf, 1973.
[34] *Pirot, E.:* Kunststoffe 61 (1971) 4, pp. 222/225.

[35] *Wiegand, G.:* Kunststoffe 69 (1979) 6, pp. 306/316.
[36] *Görmar, H., Pütz, H., Schwan, W., Allerdisse, W., Wiegand, G., Berndtsen, N., Schiedrum, H. O.* in VDI-K (Ed.): Messen an Extrusionsanlagen (Measurement on Extrusion Lines), VDI-Verlag, Düsseldorf, 1978.
[37] *Görmar, H., Recker, H., Pütz, H., Klinge, G.* in VDI-K (Ed.): Messen und Regeln beim Extrudieren (Measurement and Control in Extrusion), VDI-Verlag, Düsseldorf, 1981.
[38] *Wortberg, J., Scholl, K., Upmeier, H., Lang, F.* in VDI-K (Ed.): Rechnergesteuerte Extrusionsanlagen (Computer-controlled Extrusion Lines), VDI-Verlag, Düsseldorf, 1981.
[39] *Ast, W., Halter, H.:* Kunststoffe 68 (1978) 6, pp. 343/347.

5 Blown film coextrusion

R. Hessenbruch

5.1 Introduction

The sixties saw the development of processes for making multilayer blown films in one operation and their application in the production of LDPE double-layer films. The main field of application was for two-colored milk-pouch film which was dyed white/black or white/brown. At the same time heavy-duty bag films with improved properties were introduced into the market in the USA. As many fields of application, particularly in the food sector, set moisture- and gas-barrier requirements which could no longer be fulfilled by one raw material, the seventies saw the further development of coextrusion of incompatible raw materials such as PA and LDPE as well; the adhesion required was achieved by inserting an adhesive layer by extrusion or by applying gas treatment [1].

The perfection of this technology and the related economic advantages have led to high growth rates in blown film coextrusion, particularly as the introduction of the five- and seven-layer technology opened up further fields of application and provided additional economic benefits. Table 1 shows some of the combinations possible on a modern blown film line. With the same basic equipment it is possible to achieve the layer combinations A to D by exchanging the die and adding extruders.

5.2 Blown film coextrusion

In contrast to cast film extrusion, where the individual layers are usually combined in a multilayer adaptor and distributed across the width of a cast film die, blown film extrusion operates with separate melt channels (Figures 1 and 2).

The advantage of the separate melt channels lies in the fact that raw materials of different viscosities are easier to combine, with the tolerances of the individual layers being mostly determined by the design of the melt channels. The layer thickness ratios are easily adjustable by way of the extruder speeds, without parts of the blown film die having to be exchanged. The individual layers are then joined in the outlet area. The main melt streams are guided separately until they leave the die, and a common die gap can be created by fixing additional die lips. The common die gap protects the individual gaps against damage and oxidation.

The dimensions of the individual gaps for the main layers are, in the case of universal blown film dies, the same for all gaps. The gap widths range from 0.7 to 1 mm, and the common die gap from 1.2 to 1.5 mm. This die design permits layer thickness ratios from 1:2 to 1:3 when the same or similar materials are used. If greater thickness differences are required, materials with lower viscosity must be used for the thinner layers. If that is not possible, the die gap geometry must be adapted to the layer-thickness relationships. In the case of coextrusion dies, which can also extrude thin adhesive layers, this measure is already accounted for by reduced gap widths. The great market significance of blown film extrusion is due both to the continuously improved technology and to the development of new raw materials. Furthermore, there are the following process-related advantages which are a feature of the blown film process [2]:

– The blown film process is well known to film producers, handling is simple and conversions can be carried out quickly.

Table 1. Film combinations possible with the blown film coextrusion system

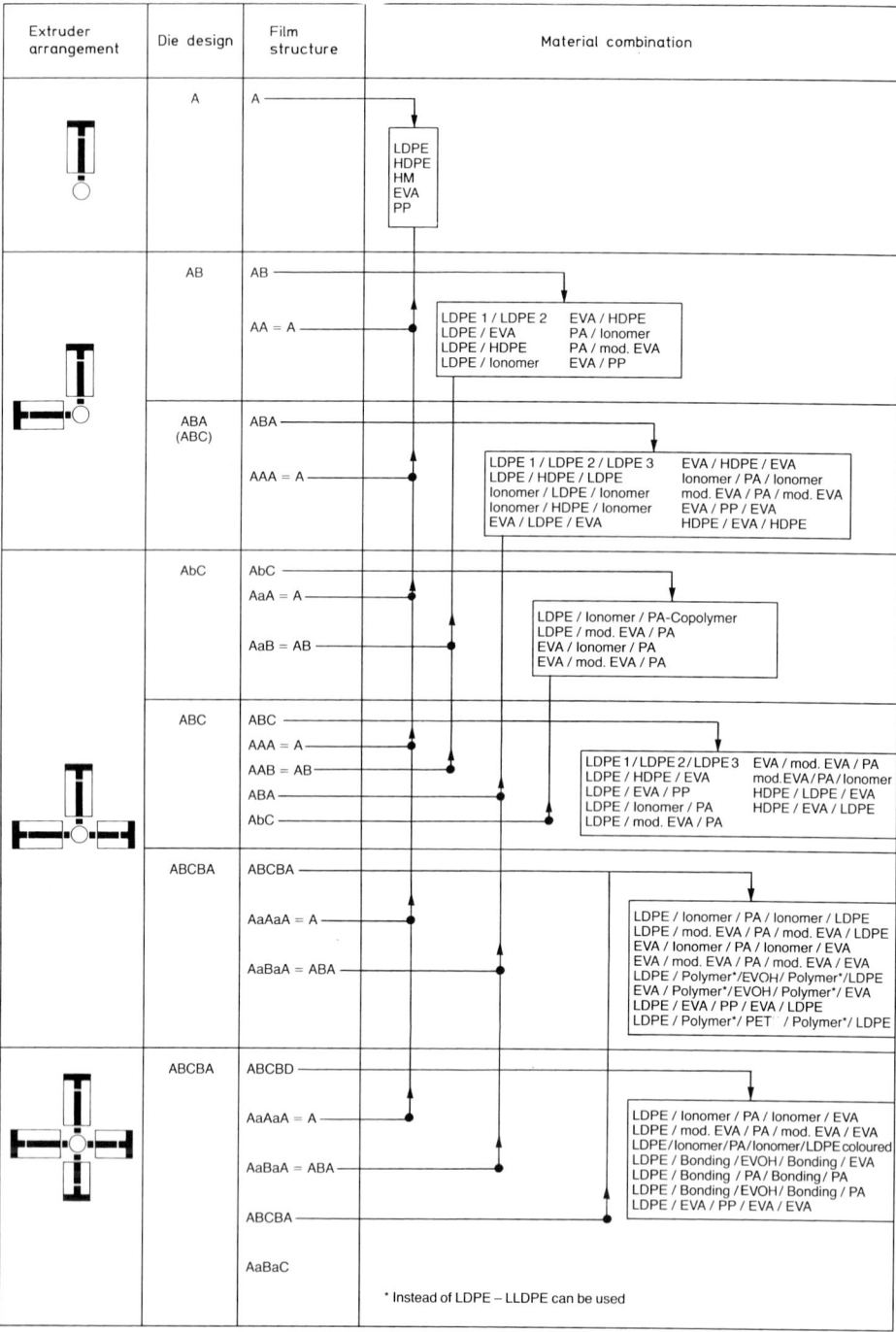

* Instead of LDPE – LLDPE can be used

- It is possible to produce very thin layers within a range from 2 to 5 µm.
- The lines are very flexible, and different film widths are obtained by selecting higher or lower blow-up ratios for the film tube. The film thickness is controlled by way of the take-off speed or the extruder output.
- By rotating the die and cooling ring, the nip roll station or the winding station, it is possible to produce cylindrical packages waste-free.
- The film can be wound as tube, half-tube, laterally gusseted- or flat film.
- By varying blow-up ratios it is possible to improve the strength in the transverse direction, and the shrinkage properties.
- By slitting the tube it is also possible to operate waste-free when winding it as flat film. This is of particular advantage in the case of multicomponent films, as recycling treatment and possible applications of reground material are limited.

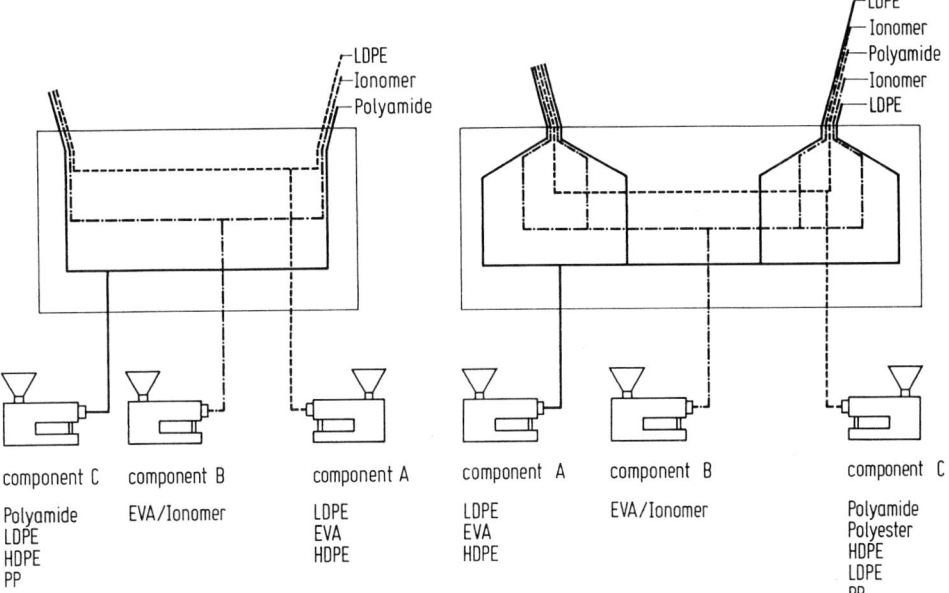

Figure 1. Diagram of three-layer coextrusion   Figure 2. Diagram of five-layer coextrusion

In coextrusion materials with very different processing temperatures are used. Homogenizing and metering take place in the extruder under optimum conditions for the specific material. Screw geometry and temperature level are designed to suit the materials to be processed. The separate temperatures of the individual components cannot be maintained in the part of the die through which all melts are guided jointly. In the die the temperature is set for the material with the highest processing temperature. Processing temperatures for the various materials are shown in Table 2.

On account of the higher temperature in the die, the heat content of the melt may increase. This reduces the cooling capacity, which determines the output. As an indicator, one may assume that with the combinations containing EVOH or polyamide, approximately 70 to 80% of the output of an LDPE film can be achieved.

As the bubble stability decreases with increased melt temperature, it becomes necessary to operate with a smaller amount of cooling air. But this can be compensated by cooling the air.

Table 2. Processing temperatures of various polymers

Polymers	Processing temperature °C	Polymers	Processing temperature °C
LDPE MFI = 0.3 to 1 g/10 min	180 to 200	Ionomers	180 to 220
LDPE MFI = 1.5 to 4 g/10 min	150 to 180	PC	260 to 290
EVA	150 to 200	Copolyamides	200 to 250
MDPE	190 to 210	Polyamide 6	240 to 260
HDPE	210 to 240	Polyester	260 to 280
LLDPE	190 to 230	EVOH	200 to 220

5.3 Polymers, combinations and properties

5.3.1 Polymers

Raw materials are classified as support, adhesion, or barrier materials, depending on their function.

5.3.1.1 Support materials (Table 3)

The task of the carrier layers in coextrusion lies in achieving good mechanical strength, weldability, moisture barrier, transparency or colorability, and printability, as well as in improving conversion behavior. These requirements are fulfilled to a high degree by polyolefins. In addition, there are a large number of different polyolefin grades available, which differ in rigidity, transparency and slip agent content.

Table 3. Support materials, properties and special aspects of coextrusion

Polymers	Properties	Special aspects of coextrusion
Support materials general	reasonably priced, good processability, very good availability and variety of types, good weldability, good printability	moisture barrier, increased strength, good transparency, good colorability, improved converting properties
LDPE	good processability, good transparency, good weldability, large variety of types, good availability, good water resistance, rel. low price	application in the MFI range from 0.3 to 2 g/10 min higher density = rigidity
EVA	with low VA content, same as LDPE improved weldability improved puncture resistance with VA contents above 8 to 10%: very good puncture resistance, good adhesion to PP (VA content > 18%), tendency to film blocking	like LDPE, preferably used on layers to be welded temperature-sensitive, must be purged with LDPE before extruder is shut down. As outer layer it can only be collapsed without creases with a roller collapsing frame

5.3 Polymers, combinations and properties

Table 3. (continued)

Polymers	Properties	Special aspects of coextrusion
MDPE	higher rigidity than LDPE, higher strength, higher melting point, transparent, with narrow molecular-weight distribution more difficult to process	preferably as outer layer in order to prevent it from sticking to the welding beam; better converting properties
HDPE	high rigidity, good strength, low transparency	high strength is achieved in combination with LDPE when operating with a long neck
LLDPE	good transparency, good strength, relatively cheap, good weldability	no special blown-film head design necessary; can also be processed blended with LDPE
Ionomers	very good weldability, adhesion to PA, good puncture resistance, special types with increased shrinkage	because of aggressiveness special steels are required for melt channels; preferably as inner layer; ionomers must be purged out of the extruder before it is shut down
PC	very good transparency, good temperature resistance, high surface gloss, good strength	high processing temperature necessary; application limited to special cases

5.3.1.2 Bonding agents (Table 4)

Are thermoplastic, extrudable polymers which adhere to different types of polymers [3]. There are bonding agents available for use as 'tie layers' (TL) between all common combinations (Table 6). Depending on application, peel-strengths between 500 g/15 mm strip width and "inseparable" are required.

Table 4. Bonding agents, properties and special aspects of coextrusion

Polymers	Properties	Special aspects of coextrusion
In general	good availability for all standard types, large MFI range	normal thicknesses: 3 to 5 μm; must be purged out before stopping the plant
Ionomers	good adhesion to LDPE and PA, EVA and LLDPE	larger primer layers improve mechanical properties of the film
Modified EVA	large number of types adapted for composite structures with LDPE, LLDPE, EVA, PA, EVOH, PC, PET	good processing; depending on composite, the adhesion can change up to two weeks after extrusion

5.3.1.3 Barrier materials [4–7]

Are thermoplastic polymers with improved gas-barrier properties. A list of the common types is compiled in Table 5.

Table 5. Barrier materials, properties and special aspects of coextrusion

Polymers	Properties	Special aspects of coextrusion
Barrier material general	polymer with good gas-barrier properties, with foodstuffs: also aroma protection	required O_2-permeability approx. $1 \frac{cm^3}{m^2 \cdot 24\,h \cdot bar}$ for very sensitive foodstuffs such as wine, fruit juices; for meat, sausage, etc. from 20 to 80 $\frac{cm^3}{m^2 \cdot 24\,h \cdot bar}$
Copolyamides	specially for co-extrusion good transparency, good strength, trouble-free extrusion, good gas barrier, good availability	processing at 220 to 250 °C; application preferably with three-layer films, prevention of curling tendency by water-bath treatment; collapse with rollers necessary
Polyamide 6	better gas barrier than copolyamides, good transparency with special types, lower price than copolyamide, good availability	because of strong curling tendency with three-layer films it should preferably be used with five-layer films; processing temperature 240 to 260 °C; limited blow-up ratio
Polyester	good transparency, good gas barrier, as copolyester processable at low temperatures	processing temperature from 260 to 280 °C; small importance in blown film coextrusion
EVOH	very good gas-barrier (30 to 50 times better than PA), good transparency, high rigidity, increased brittleness, hygroscopic	temperature-sensitive, crease-break danger with thicknesses above 15 μm, preferably used with five-layer films

5.4 Advantages of coextrusion

Coextrusion is employed for three reasons:
- improvement of conversion behavior,
- cost reduction,
- quality improvement.

Depending on the application, these criteria can be achieved individually or in combination.

5.4.1 Conversion

5.4.1.1 Improved weldability

An increase of speed on bag-making machines as well as an improvement of the welding seam quality is achieved if the layers to be welded are made of EVA or LLDPE. In specific cases, e.g. if bags are filled with greasy goods and the sealing area may have contact with grease or oil, it is useful to have an inner layer of an ionomer.

In order to prevent the outer layer from sticking to the welding beam, it should be made of a material with a higher melting point, such as higher-density LDPE, MDPE, HDPE or PA.

5.4.1.2 Improvement in conversion properties

In addition to the general requirements such as good tolerances, flatness and reel quality, it is important for automatic welding machines and packing machines that the films can be transported and deposited well.
This requirement can be fulfilled by using MDPE or HDPE as the center or outer layer, as these materials have higher rigidity and run better on the machine.
For form–fill–and–seal machines it is important to have, in addition to a precise width, a film that lies flat and has little tendency to curl. As differences in physical properties cause, for example, an LDPE/HDPE film to have a strong curling tendency, it is necessary to use films with symmetrical structures (LDPE/HDPE/LDPE), Figure 3.
This also applies to barrier films, where five-layer films have important advantages over three-layer films.

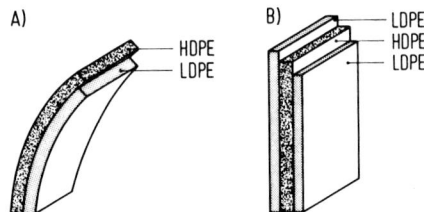

Figure 3. Comparison of two-layer (A) and three-layer film (B)

5.4.1.3 Handling of packages

Coextrusion offers the possibility of producing tailor-made film combinations for packaging for specific purposes. Sacks can be made suitable for stacking by means of an EVA outer layer or a rough, foamed outer layer, while for bags to run trouble-free in secondary operations a smooth HDPE layer or an LDPE layer with a high slip agent content is required.

5.4.2 Cost reduction

5.4.2.1 Production costs

The coextrusion process, being a single-step process, cuts out coating or laminating, and brings process-related cost benefits. In addition, indirect costs associated with the older processes, storage of semi-finished goods and multiple edge trimming, are avoided.

5.4.2.2 Raw material costs

Production costs can be reduced most by savings in raw material costs, which can be achieved in two ways:
− by reducing film thickness
− by using polymers with a lower price or better properties.

One must also remember that the individual layers in a coextruded film can be drawn thinner than a single layer of film can.

Example:

In the case of heavy-duty sacks materials with an MFI of 0.2 g/10 min are used, which have a minimum draw-down thickness of approximately 40 µm. In combination with high-molecular-weight (HM)-HDPE, this material can be drawn down to approximately 10 μm without loss of properties.

This means that a 50 μm carrier-bag film made of LDPE can be replaced by a three-layer film (LDPE/HM-HDPE/LDPE) of total thickness 30 μm and approximately the same properties. The saving in raw material costs is calculated as follows:

Subscript 1: LDPE, Subscript 2: HDPE, Subscript S: Sandwich

$$P_1 = \text{price} \left[\frac{\text{DM}}{\text{kg}}\right], \quad t_1 = \text{film thickness [μm]}, \quad \varrho_1 = \text{density [g/cm}^3\text{]}$$

Yield
$$A_1 = \frac{1000}{(t_1 \cdot \varrho_1 \cdot P_1)} \left[\frac{\text{m}^2}{\text{DM}}\right] \tag{1}$$

Similarly for P_2, t_2, ϱ_2 and A_2.
Then for the multilayer sandwich film we have:

Yield
$$A_S = \frac{1000}{(t_S \cdot \varrho_S \cdot P_S)} \left[\frac{\text{m}^2}{\text{DM}}\right] \tag{2}$$

with
$$t_S = (t_1 + t_2 + \ldots) \quad [\text{μm}] \tag{3}$$

$$\varrho_S = (t_1\varrho_1 + t_2\varrho_2 + \ldots) \cdot \frac{1}{t_S} \quad [\text{g/cm}^3] \tag{4}$$

$$P_S = (t_1\varrho_1 P_1 + t_2\varrho_2 P_2 + \ldots) \cdot \frac{1}{(t_S \cdot \varrho_S)} \left[\frac{\text{DM}}{\text{kg}}\right] \tag{5}$$

Consequent yield:

$$A_S = \frac{1000}{(t_1\varrho_1 P_1 + t_2\varrho_2 P_2 + \ldots)} \left[\frac{\text{m}^2}{\text{DM}}\right]. \tag{6}$$

$$A_S = y \left[\frac{\text{m}^2}{\text{DM}}\right] \text{ or as reciprocal value}$$

Film price $P_F = \dfrac{1}{y} \left[\dfrac{\text{DM}}{\text{m}^2}\right]$ (7)

If the relevant raw material prices are inserted into these formulae, the significant advantage of coextruded film becomes very apparent.

With barrier films large differences in raw material costs can also be found for the same or similar film properties.

The polymers used for these films are basically copolyamides, polyamides and EVOH. Their O_2 permeability can be calculated from the formula:

$$K = P \cdot d \, \frac{\text{cm}^3}{\text{m}^2 \cdot 24\,\text{h} \cdot \text{bar}} \cdot \text{μm} \tag{8}$$

K permeability constant
P permeability
d film thickness

The permeability constant has the following values:

5.4 Advantages of coextrusion

Copolyamide in three-layer film: $K = 1500 \dfrac{\text{cm}^3}{\text{m}^2 \cdot 24\,\text{h} \cdot \text{bar}} \cdot \mu\text{m}$

Copolyamide in five-layer film: $K = 1000 \dfrac{\text{cm}^3}{\text{m}^2 \cdot 24\,\text{h} \cdot \text{bar}} \cdot \mu\text{m}$

Polyamide 6 in five-layer film: $K = 700 \dfrac{\text{cm}^3}{\text{m}^2 \cdot 24\,\text{h} \cdot \text{bar}} \cdot \mu\text{m}$

EVAL in five-layer film: $K = 15 \dfrac{\text{cm}^3}{\text{m}^2 \cdot 24\,\text{h} \cdot \text{bar}} \cdot \mu\text{m}$

Using these values the thickness for a specific permeability can be calculated.

Example: film with O_2-permeability of $10 \dfrac{\text{cm}^3}{\text{m}^2 \cdot \text{bar} \cdot 24\,\text{h}}$

For PA 6: $d = \dfrac{K}{P} = \dfrac{700}{10} = 70\,\mu\text{m}$

For EVAL: $d = \dfrac{15}{10} = 1.5\,\mu\text{m}$

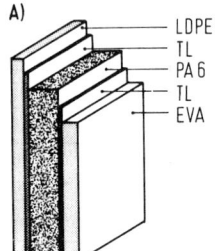
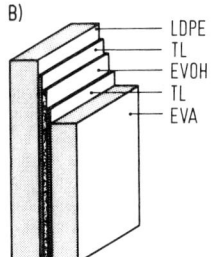

Figure 4. Comparison of barrier film with PA6 (A) and EVOH (B)
TL tie layer

As the theoretical thickness of 1.5 μm is not feasible in practice, layer thicknesses of 4 to 5 μm should be chosen to maintain the required minimum values for mass throughput and flow speed.

A comparison of the two films can be based on the following film combinations with the better mechanical values of PA not being taken into consideration.

PA combination (Figure 4A) (TL = tie layer)
EVA/TL/PA6/TL/LDPE
35 μm 5 μm 70 μm 5 μm 35 μm

EVOH combination (Figure 4B)
EVA/TL/EVOH/TL/LDPE
65 μm 5 μm 5 μm 5 μm 65 μm

Total thickness 150 μm,
of which   EVA   35 μm   ϱ = 0.92 g/cm³
           TL   10 μm   ϱ = 0.94 g/cm³
           PA6   70 μm   ϱ = 1.13 g/cm³
           LDPE 35 μm   ϱ = 0.92 g/cm³

Total thickness 145 μm,
of which   EVA   65 μm   ϱ = 0.92 g/cm³
           TL   10 μm   ϱ = 0.94 g/cm³
           EVOH   5 μm   ϱ = 1.19 g/cm³
           LDPE 65 μm   ϱ = 0.92 g/cm³

For the PA combination the following formula applies:

$$A = \dfrac{1000}{d_{\text{EVA}} \cdot \varrho_{\text{EVA}} \cdot P_{\text{EVA}} + d_{\text{TL}} \cdot \varrho_{\text{TL}} \cdot P_{\text{TL}} + d_{\text{PA}} \cdot \varrho_{\text{PA}} \cdot P_{\text{PA}} + d_{\text{LDPE}} \cdot \varrho_{\text{LDPE}} \cdot P_{\text{LDPE}}} \left[\dfrac{\text{m}^2}{\text{DM}} \right]$$

If the same calculation is done for the EVOH combination, one will find that this film has an approximately 80% higher yield, despite EVOH being about three times as expensive as PA 6.

Similar calculations are also possible for comparing three-layer with five-layer films, or copolyamides with polyamide 6 combinations.

5.5 Quality improvement

The main reason for coextrusion is to combine the desirable properties of the individual components in one film in order to increase the quality of that film. The selection criteria are gas- and water permeability, strength, weldability, transparency and temperature resistance.

Table 6. Properties of various polymers

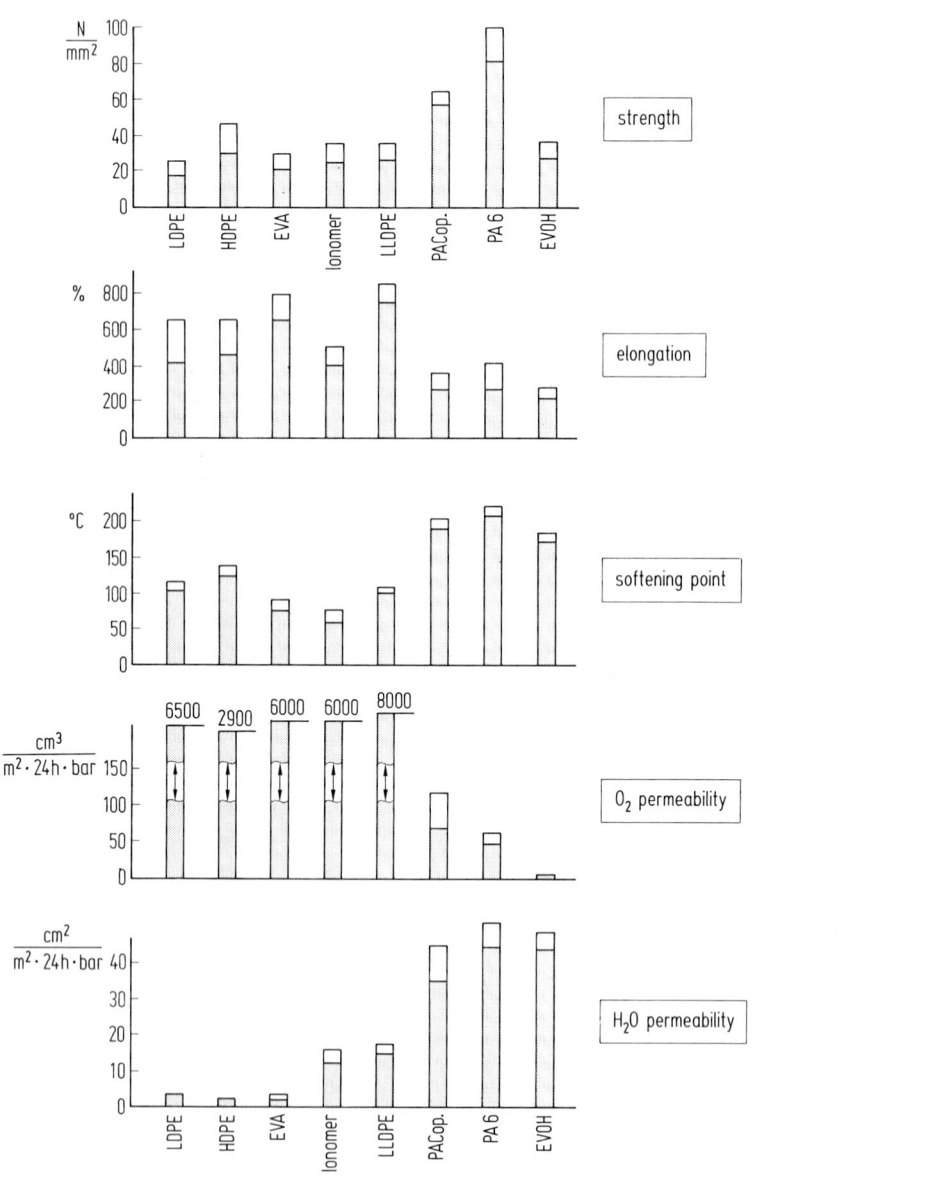

The relevant properties of the various polymers have been compiled in Table 6. The values given are indicative only, as there may be considerable variations for a given type of polymer.

Table 6 shows clearly that there is no one thermoplastic which possesses all the desired properties. The decision on which materials to use to form multilayer films will therefore depend largely on the properties required, but also on the price achievable.

For an optimum combination of polymers one must also take into consideration whether the chosen materials will adhere to each other or if bonding agents must be used. Table 7 shows the interrelationships for the polymers commonly used in the blown film process.

Table 7. Mutual bonding of various polymers

Polymer	LDPE	HDPE	EVA	Ionomer	LLDPE	PP	PACop.	PA 6	PET	EVOH
LDPE	+	+	+	o	+	-	-	-	-	-
HDPE	+	+	+	o	+	-	-	-	-	-
EVA	+	+	+	o	+	o	-	-	-	-
Ionomer	*	*	*	+	o	-	o	o	-	-
LLDPE	+	+	+	o	+	-	-	-	-	-
PP	*	*	*	*	*	+	-	-	-	-
PACop.	*	*	*	*	*	*	+	+	-	+
PA 6	*	*	*	*	*	*	+	+	-	+
PET	*	*	*	*	*	*	*	*	+	-
EVOH	*	*	*	*	*	*	+	+	*	+

← with tie layer → / without tie layer

- + good adhesion without tie layer
- − no adhesion
- o adhesion dependent on type of polymer
- * good adhesion with tie layer, suitable bonding agents available

Within the limits that coextrusion offers one can develop a film structure on the basis of the adhesion- and properties tables. Whether such a combination will meet expectations can only be decided by tests or field experience. In practice, optimum multilayer films have been found for many fields of application by selection within the various polymer groups.

5.6 Fields of application (Table 8)

5.6.1 Polyolefin combinations

Multilayer polyolefin films are replacing pure LDPE films, with increased strength at smaller thicknesses, improved weldability and better converting properties being the main criteria.

Table 8. Material combinations, special properties and important applications of coextruded blown films

Material combinations	Special properties	Most important fields of application
Double-layer film		
1. LDPE/LDPE	pinhole-free (multicolored)	milk film, carrier bags, general packaging
2. LDPE/EVA	good weldability, sterilizable	heavy-duty bags, stretch packaging, medical articles
3. HDPE/EVA	sterilizable	blood plasma, bakery goods, foodstuffs
4. HDPE/LDPE	good strength	bakery goods, foodstuffs, tomato concentrate
5. LDPE/ionomer	good weldability, puncture-resistant	dairy products, foodstuffs, medical instruments, general packaging
6. LLDPE/LDPE LLDPE/EVA	high elasticity good surface adhesion	stretch film
7. Ionomers/EVA	grease-proof	coconut, biscuits
8. Ionomers/PA	gas- and aroma-tight	meat, sausage, ham, fish, foodstuffs, cheese
Three-layer film, symmetrical		
9. LDPE/HDPE/LDPE	weldable on both sides, reduced curling tendency	like 4, pet food, cornflakes
10. EVA/PP/EVA	like 9	like 9
11. EVA/HDPE/EVA	like 3	like 9, cornflakes
Two-layer film with tie-layer (TL)		
12. LDPE/TL/PA	gas-, water- and aroma-tight	foamed PS granulate, meat, sausage, cheese, ham, fish, ready-made meals, hops
13. EVA/TL/PA	like 12 in hot-air channel good hot tack properties	like 12
14. Mod. EVA/TL/PA		like 12, vacuum packing for ham (shrinkable)
Three-layer film		
15. LDPE/HDPE/EVA	good weldability, good rigidity	bakery goods, foodstuffs
16. LDPE/EVA/PP	like 15	like 15
Five-layer film		
17. LDPE/TL/PA/TL/ LDPE or LLDPE/TL/ PA/TL/LLDPE	no curling tendency, improved barrier properties, as PA protected against moisture absorption, improved layer adhesion, weldable on both sides	like 12
18. EVA/TL/PA/TL/EVA	like 17	like 17
19. LDPE/TL/EVAL/ TL/LDPE	like 17	like 17, fish meal, wine packaging, milk powder (casein)
20. EVA/TL/EVAL/ TL/EVA	like 19	like 19

Note: LDPE can also be replaced by LLDPE (linear low density PE)

5.6.2 Combinations with barrier properties

These films contain polyolefins and a polymer with a low gas permeability such as polyamide, PET or EVOH. The proportion of barrier material depends on the package and the properties and shelf life required.

5.7 Line layout

The layout of coextrusion lines differs from that of conventional lines principally in the extrusion area (Figure 5); and the large variety of film combinations also has to be taken into consideration downstream, because of bubble collapse and wind-up differences. Since some of the polymers are very expensive it is also important to have control of the individual extruder outputs.

Figure 5. Extrusion unit of a coextrusion blown film line.
(Photo: Barmag, Remscheid, West Germany)

5.7.1 Extruders

As can be seen from Table 1, the number of extruders depends on the film combination and therefore on the die design. In order to realize the set-up for three-layer polyolefin combinations and for four-component, five-layer barrier films, the extruders are equipped with different screws and feed sections.

Extruder A: inner layer
Feed section: grooved
Screw: polyolefin screw with high specific output, equipped with shearing and mixing section
To work with: polyolefins

Extruder B: tie layer
Feed section: smooth
Screw: multi-purpose screw with shearing and mixing section
To work with: polyolefins, polyamides

Extruder C: barrier layer
Feed section: smooth
Screw: special screw with reduced shearing
To work with: polyamides, EVOH, polyolefins with low outputs

Extruder D: outer layer
Feed section: smooth
Screw: multi-purpose screw with shearing and mixing section
To work with: polyolefins, polyamides

With this arrangement the line remains flexible because, if a three-layer blown film head is fitted, assignment of the extruders to the individual layers can be achieved by assembling the film head on the rotation device in different positions.

The extruders used are between 30 and 90 mm in diameter, with the individual sizes being determined by the size of the blown film die, the layer thickness ratios and each extruder's share of total output.

5.7.2 Filters

As the total output of a coextrusion line is distributed among several extruders and the output of an individual component may be very small, the design must be suitably adapted with regard to filter size and melt channel cross-sections. For operational reasons the components should be easily exchangeable.

As some materials are sensitive to temperature, good and uniform heating is also very important.

5.7.3 Rotation device (Figure 6)

This unit [8] causes the film die and cooling ring to rotate in order to distribute film thickness differences across the wound reel. As the melt channels on rotating parts must operate

Figure 6. Four-component rotating device with filter adaptors, designed to accommodate various coextrusion film dies. (Photo: Barmag, Remscheid, West Germany)

under high pressure and high temperatures, without leaking into each other or to the outside, conventional seals cannot be used. The sealing problem has been solved satisfactorily by means of special sealing rings to which melt pressure is applied from the inside.

The various film dimensions result in different circumferential bubble speeds which can be compensated by smooth adjustment of rotation speed. If at all possible, the rotation device should be a sub-unit separate from the film die as this simplifies cleaning and maintenance work. Furthermore, by fitting the film head in different positions on the rotation device this approach makes it possible to assign extruders to different individual layers. A further advantage is that blown film dies of different design can be fitted on the same rotation device.

5.7.4 Blown film dies (Figure 7)

Film dies have considerable influence on the film quality. And they also determine the flexibility of the entire line, as the design must be such that many film combinations with varying layer thicknesses and materials can be processed.

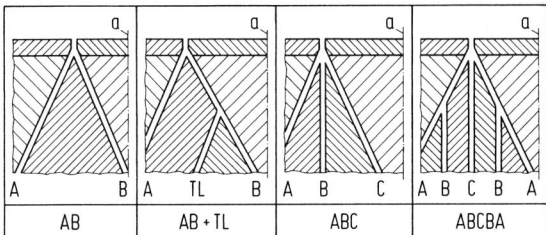

Figure 7. Schematic diagram of multilayer blown film dies

5.7.4.1 Two-layer blown film dies

These are primarily used for polyolefin combinations. Special designs are also used for LDPE/PA combinations, when an adhesion-promoting gas is blown between the two layers (Alkor process).

5.7.4.2 Two-layer blown film dies with additional bonding layer

These dies were developed for LDPE/PA combinations. A thermoplastic bonding agent is extruded between the two main layers by a third extruder. These tie layers are approximately 5 μm thick.

5.7.4.3 Three-layer blown film dies (Figure 8)

These dies are normally designed as multipurpose dies. In conjunction with two extruders they permit symmetrical combinations with an ABA structure. With the appropriate material supply, and with three extruders it is possible to extrude the combination ABC, but also AAB = AB, ABA, AAA or AB + TL.

5.7.4.4 Five-layer blown film die

This type of unit offers the possibility of producing symmetrical combinations from polymers which will not normally stick together without a bonding agent. Thus the curling tendency can be avoided, and handling and converting become easier. These blown film dies permit particularly advantageous extrusion of combinations with barrier materials, since the barrier layer lies in the center and is protected against moisture absorption, which would cause increased gas permeability. The flow channel dimensions for the individual layers are

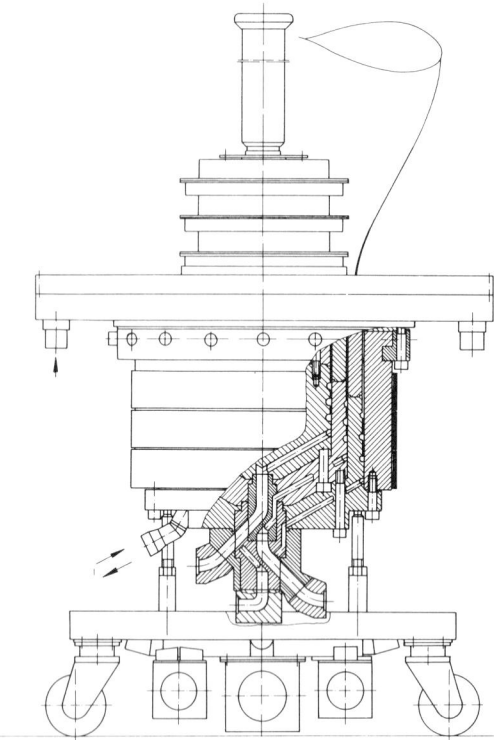
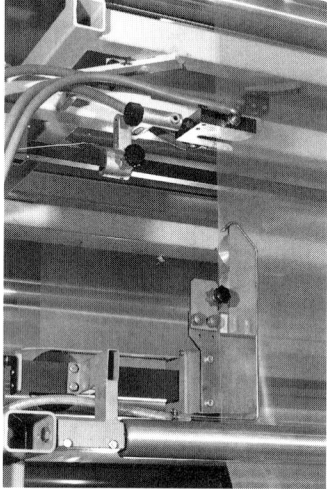

Figure 8. Three-layer blown film die with internal bubble cooling, system Windmöller & Hölscher

Figure 9. Film slitting device in conjunction with a web guiding system. (Photo: Barmag, Remscheid, West Germany)

adapted to the proportion of each layer in the total film to achieve favorable melt residence times and flow speeds.

The five-layer blown film die is suitable for both three- and four-component combinations with an ABCBA or ABCBD structure (also Table 1). The advantages achieved in comparison with three-layer film are:

- saving in raw material costs due to thinner barrier layers,
- no curling tendency,
- sealable on both sides,
- improved flex-crack resistance,
- hygroscopic barrier materials are protected against moisture absorption, as they are positioned in the center.

Additional advantages with four components:

- improved weldability of inner layer,
- improved converting properties,
- coloration of only the outer layer possible,
- combination with EVOH and PACop possible.

The increasing popularity of the five-layer film is based on these advantages.

To avoid frequent cleaning of the dies the lines must be thoroughly purged before they are shut down. The outlet area must be protected against oxidation by being covered with silicone grease and polymer.

5.7.5 Bubble cooling

Basically the same cooling principle as with single-layer lines is used, i.e. purely external- or a combination of external and internal cooling (Figure 8).
But with an increasing number of layers internal cooling is more difficult to realize, since dies become very large with provision for internal air exchange, and a temperature gradient within the blown film head, created by the inlet and outlet pipes for the internal air, will lead to melt distribution and transparency problems with some polymers. For this reason internal cooling is rarely used with five-layer films.

5.7.6 Downstream equipment

The haul-off and wind-up units, and other downstream equipment, differ only in detail from units used on conventional lines.

5.7.7 Special equipment

5.7.7.1 Collapsing frame

As very smooth and very rough films have to be handled, it is advisable to use collapsing frames equipped with both rollers and wooden slats.

5.7.7.2 Slitting device

To avoid generating edge trim, slitting units that operate in conjunction with web guiding control systems are used (Figure 9).
And to allow even slightly blocked, sticky film to be slit, the slitting blade holders carry a compressed-air jet so that the collapsed film is slightly inflated before it is slit, to enable the film to slide easily.

5.7.7.3 Oscillating device

Normal wind-up of thick film results in rolls with edge build-up; to avoid this effect the cutting position can be varied by oscillating the sensor of the web-guiding system over a few millimeters. For edge trimming, the cutting device is oscillated.

5.8 Automation

Because very expensive polymers are used and extruded as very thin layers, output control on the individual extruders is important.

5.8.1 Pressure measurement

Pressure indicators with maximum and minimum level contacts prevent excess pressures by switching off extruder drives if a maximum level is reached; and in effect monitor the minimum output rates and provide a signal if extrusion pressure drops below an adjustable minimum value, to prevent production of films in which individual polymers are not present in the specified quantities.

5.8.2 Width measurement

Commercially available units can be used.

5.8.3 Thickness control

Continuous thickness measurement on individual layers is not yet possible. However, a device is used which measures the output of the individual extruders continuously. The feed hoppers are supported on load cells which measure the weight off-take continuously and transmit the measured values to a microprocessor. The latter also receives information on film width, specific weights and production speed. After the desired values have been fed in, the computer calculates the required outputs for the individual extruders and provides closed-loop control of screw speeds. The individual extruders are controlled and readjusted independently.

With this equipment polymer quantities can be kept to specification even if the production speed is changed.

5.8.4 Additional equipment

Use of automatic winders, and control of the film web by means of a dancer roll, or scanning systems and the equipment described earlier, make it possible to operate coextrusion lines fully automatically, thus contributing to increased profitability.

References for Chapter 5

For the convenience of the reader the English titles of all publications in languages other than English are shown in parentheses.

[1] VDI-K (Ed.): Extrudieren von Schlauchfolien (Extrusion of Tubular Film), VDI-Verlag, Düsseldorf, 1973.
[2] *Hensen, F., Hessenbruch, R., Bongaerts, H.:* Kunststoffe 9 (1981).
[3] *Nagano, R.:* 8th Adhesives and Film Conversion Seminary, Munich, 1983.
[4] Company brochure, Bayer, Leverkusen.
[5] Company brochure, Emser Werke, Zurich.
[6] Company brochure, BASF, Ludwigshafen.
[7] Company brochure, Kuraray, Japan.
[8] DE-PS 2.855.607.C2 (1978), Barmag.
[9] *Hensen. F.:* Kunststoffe 59 (1969), pp. 3/8.

6 Flat film extrusion using chill-roll casting

H. Bongaerts

6.1 Introduction

The production of film by roll casting is one of the classic extrusion processes. During the past few years the requirements of the film market have led to intensive developments in flat film technology and created a mature production method for the fabrication of precision films from widely differing materials.
When processing partially crystalline thermoplastics in particular, this process enables the best possible use to be made of material properties, while holding optimum dimensional tolerances. Films produced on flat film lines are mainly used in the packaging industry. Packaging of both foodstuffs and technical products has undergone intense development towards more complex, often specialized, types of film. Most thermoplastics can be processed on flat film lines. The severe requirements that film packaging materials have to meet demand high-quality precision production, with the properties of individual materials often having to be combined, e.g. by coating, laminating or coextrusion. Cast films feature excellent optical and dimensional properties; chill-roll cooling permits high take-off speeds and also influences the morphological structure of the film very favorably.
Thermoformable single- and multilayer rigid films for packaging of groceries, dairy produce and technical products have achieved considerable importance in the packaging market.
The film producer now has high-speed lines at his disposal which provide efficient production of high-quality products.

6.2 General principle of flat film extrusion

Flat film extrusion is based on the principle of shaping a melt that has been plasticated and homogenized in the extruder, into a planar structure, cooling and stabilizing this structure by means of roll contact and then winding it up as a trimmed working width. Depending on film thickness and application, a distinction is made particularly between fine film (10 to 50 µm), thicker cast film (100 to 400 µm) and thermoformable sheet (0.2 to 2.5 mm). The first two types mentioned are produced on chill-roll lines; thermoformable sheet with a thickness of about 0.3 mm upwards is produced on special lines on which contact with the cooling roll is produced by a roll with a defined polishing gap.
All flat film types are either wound after being trimmed to finished dimensions, or undergo direct conversion in-line in a process like uniaxial or biaxial drawing, or thermoforming. Construction and operation of a chill-roll casting line are illustrated in Figure 1.
By combining several extruders, it is possible to coextrude multilayer films, and to combine similar or even different polymers. By comparison with blown films, the flat films have better transparency, gloss, crystallinity, rigidity, and thickness tolerance, while blown films (these are mostly made from polyethylene) offer advantages with regard to tensile strength in machine (MD) and transverse (TD) directions; and simple alteration of film width by changing the blow-up ratio [1].
The chill-roll process offers greater flexibility with regard to polymer choice and permits high throughput rates with great efficiency. Some films – polypropylene, polyamide and polyester for example – are primarily or even exclusively produced on chill-roll lines.

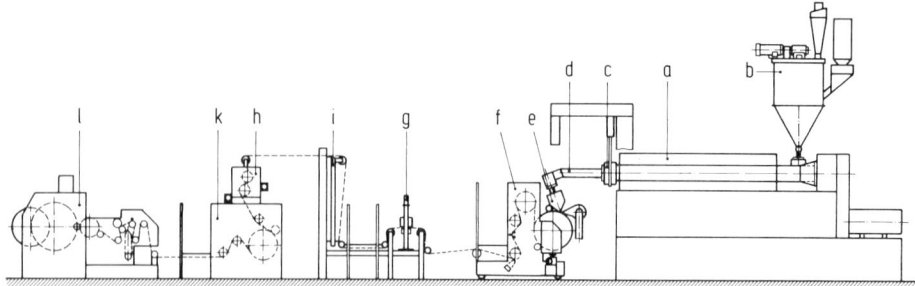

Figure 1. Flat film line – chill-roll principle
a extruder E4/33D, *b* mixing and metering unit, *c* filter, *d* adaptor system, *e* slot die, *f* chill-roll take-off unit, *g* thickness scanning system, *h* surface treatment unit, *i* film oscillating unit, *k* intermediate take-off unit, *l* winder

For the extrusion of thermoformable products either compact lines are used, or polishing stack systems such as those employed for the production of sheet.
Figure 2 shows one of the most common arrangements for thermoformable film and sheet [2]; in this case elements of chill-roll lines are combined with sheet take-off systems. Flat films and sheet for thermoforming are mainly made from SB, ABS, PP and HDPE resins, and also from PET and rigid PVC using special line designs.

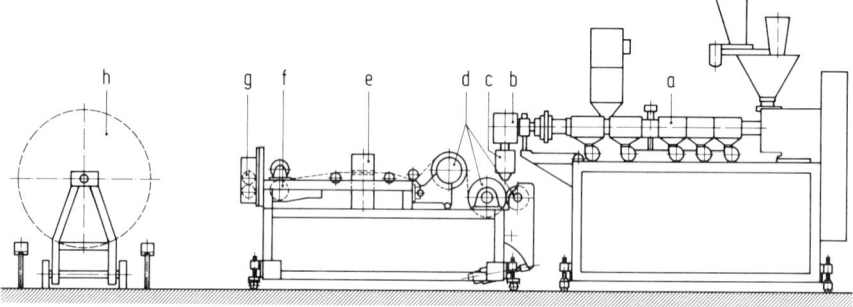

Figure 2. Line for the production of single- and multilayer thermoformable films [2]
a extruder, *b* adaptor, *c* die, *d* cooling rolls, *e* thickness scanner, *f* take-off unit, *g* edge trimming, *h* winding station

6.2.1 Extrusion

For the production of flat film, single-screw extruders are normally used. The severe requirements the melt must meet with regard to homogeneity, temperature uniformity and pressure constancy demand a well-balanced extruder design [3, 4].
The extruders used have a barrel length of 27 D to 33 D and provide uniform output at high plastication rates.
Adaptation of the feed zone for different resins and the use of powerful drives permit optimum extrusion conditions to be achieved. For processing polyolefins in particular, the extruder can be equipped with a water-cooled, grooved sleeve in the feed zone [5, 6], for efficient material transport.
In the choice of screw geometry for flat film extrusion, special consideration must be given to reworking edge trim or film scrap generated on the line. The multisectional screws used

6.2 General principle of flat film extrusion

are equipped with shearing and mixing elements, which ensure that the material fed back is completely molten and thermally well homogenized [7].

To ensure good stability of the manufacturing process and uniform product quality, it is necessary to employ precise control of the most important process variables. The relation between actual and target (or 'desired') values within the extruder control section is shown in Figure 3. In addition to keeping screw speed constant, it is necessary to achieve constant actual values for the barrel temperatures in the equilibrium operating condition. For melt control purposes, the temperature and pressure of the melt are measured in the flow channel after the screw. If the sensor is properly constructed, the melt temperature can serve as the target value for a reference controller in a "cascade" temperature control system for the barrel heating/cooling zones [8].

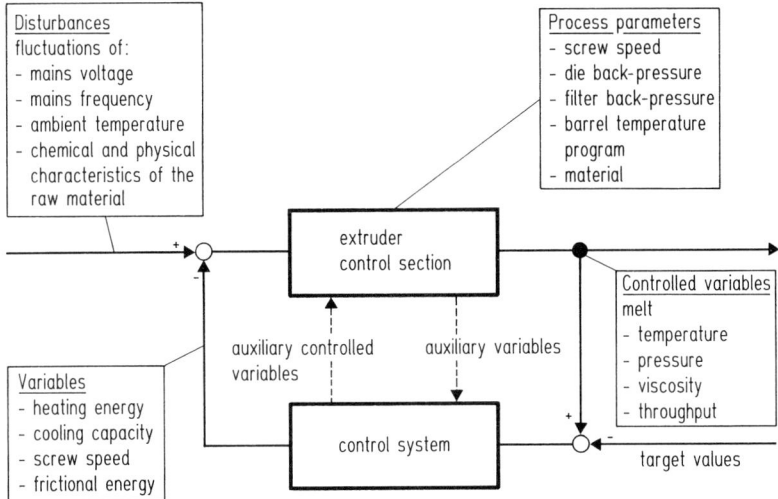

Figure 3. Extruder control system − schematic [8]

Since a long melt channel is often needed between extruder and slot die on flat film lines, it is of advantage to have static mixing elements in the adaptor ahead of the die to homogenize the melt.

For processing hygroscopic materials such as SB and ABS, extruders are provided with a degassing facility.

PVC powder is processed on twin-screw extruders or special single-screw extruders of cascade design. With cascade extruders the steps of feeding, conveying and plastication, and then of homogenizing are carried out under optimum conditions in two extruders equipped with separate D.C. drives. During the transfer of the agglomerated PVC mass to the second extruder, intensive degassing takes place [9].

6.2.2 Molding

In flat film extrusion the extruded melt is transformed into a rectangular flat shape by means of a slot die, whose cross-section is shown in Figure 4.

In order to achieve a good thickness tolerance across the film width it is particularly important to have the correct flow conditions, i.e. the correct geometric balance between manifold and dam (restrictor bar) zone (Figure 5) in addition to sensitive die-lip adjustment and even temperature distribution. Correct dimensioning permits maintenance of

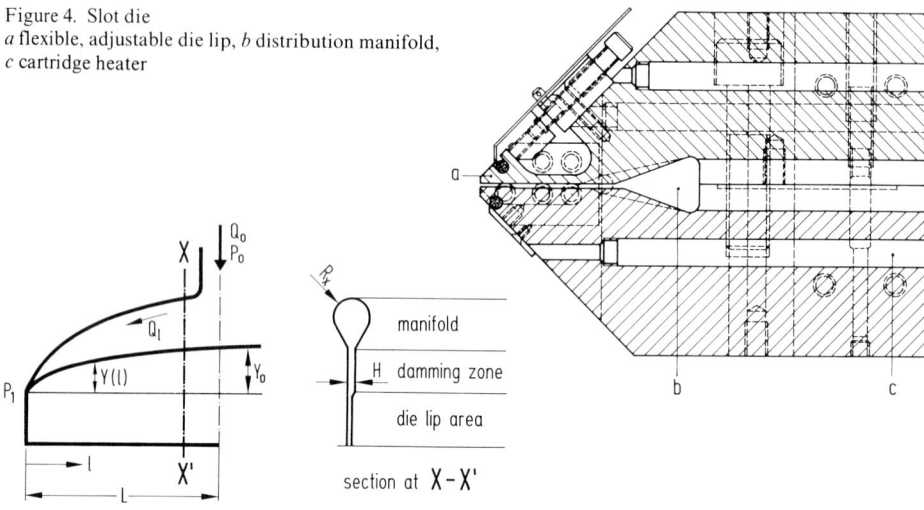

Figure 4. Slot die
a flexible, adjustable die lip, *b* distribution manifold, *c* cartridge heater

Figure 5. Diagram of the melt distributor system (manifold)
Q_0 throughput at die inlet, P_0 pressure at die inlet, Y_0 max. island length, $Y_{(l)}$ island length as function of width coordinates, l width coordinate, L half die width, H gap, Q_1 throughput as function of width coordinate, P_1 pressure in die lip area

constant throughput across the die width for all flow paths, with a uniform pressure drop from the die inlet to the die outlet.

Different materials often differ considerably in their rheological behavior. So-called "flow curves" of individual polymers determined in rheometers supply information about viscosity behavior at different processing temperatures (Figure 6).

The apparent viscosity, η_a, decreases with increasing shear rate in the case of structurally viscous melts. According to [10, 11] the following relationship applies:

$$\tau = \eta_a \cdot \dot{\gamma} \tag{1}$$

τ shear strain
$\dot{\gamma}$ shear rate
η_a apparent viscosity

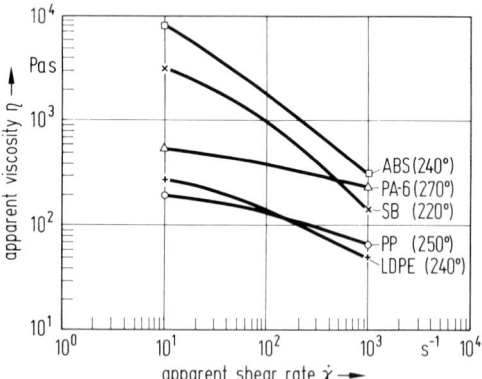

Figure 6. Flow curves of various thermoplastics at their respective processing temperatures

Viscosity also depends strongly on temperature. Calculation methods for the design of slot dies are mostly based on the *Prandtl-Eyring* flow law [12, 13]:

$$\dot{\gamma} = C \sinh \frac{\tau}{A} \tag{2}$$

(C and A are material parameters dependent on temperature and shear rate.)

6.2 General principle of flat film extrusion

In [14, 15, 16] calculation methods are discussed by means of which manifold geometries can be calculated. These calculations are based on the assumption that the melt is at the same temperature everywhere. For reasons of symmetry, only one die-half is examined; the gap height H in the "dam" or "island" area is assumed to be constant across the die width. The profile of the nearly circular manifold cross-section is calculated with the following formula:

$$R(l) = R_0 \left(\frac{2l}{B}\right)^{1/3} \tag{3}$$

$R(l)$ radius as function of the width co-ordinate
R_0 max. manifold radius
l width co-ordinate
B width

$$Y(l) = Y_0 \left(\frac{2l}{B}\right)^{2/3} \tag{4}$$

$Y(l)$ island length as function of the width co-ordinate
Y_0 max. island length

Values R_0 and Y_0 are related to each other through the laminar flow representative viscosities $\bar{\eta}$ for pipe and slot, by the equation:

$$Y_0 = \frac{\bar{\eta}_R \, H^3 \, B^2}{\bar{\eta}_s \, 4 \pi \, R_0^4} \tag{5}$$

$\bar{\eta}_R$ representative viscosity in the pipe
$\bar{\eta}_s$ representative viscosity in the gap
H gap height

Equations (3) to (5) form the basis for dimensioning the manifold system for slot dies. But the dimensions of the manifold are subject to special boundary conditions such as the limiting design pressure loading and the feasibility for manufacture; they must therefore be optimized for a practical design.
A special boundary condition, in particular for the extrusion of thermally sensitive thermoplastics, is the requirement for a uniform, short residence time of the melt within the die. If the condition for constant shear rate and viscosity is achieved throughout the melt distribution system, the die layout becomes independent of changes of operating point.
According to [14], the following formula is obtained, in this case, for values R_0 and Y_0:

$$R_0 = \left(\frac{B \cdot H^2}{\sqrt{6 \pi}}\right)^{1/3} \quad \text{and} \tag{6}$$

$$Y_0 = \left(\frac{9 \pi B^2 H}{16}\right)^{1/3} \tag{7}$$

The mean residence time t along any flow path is therefore:

$$t = \frac{B \cdot H \cdot Y_0}{\dot{V}} \tag{8}$$

t time
$\dot{V}$ volume throughput

This results in approximately the same thermal stress for each melt component across the outlet width.

For the design of slot dies, electronic computer programs [17, 18] have been developed over the past few years.

Two different processes are used for coextrusion of flat films. With the adaptor method, the melt streams to be combined are already united in an adaptor, or feed block, before entering the die and are then extruded together through a conventional flat film die. In the case of multi-manifold dies, the melts are guided separately by manifolds which are independent of each other, and the individual melt layers, which are already distributed across the die width, are usually brought together within the die prior to leaving it.

In the design of distribution systems for multilayer extrusion the same basic criteria used for single-layer extrusion apply. With the adaptor principle for multilayer extrusion, the melt distribution depends on the method of unification in the adaptor located upstream of the die manifold. Different melt viscosities and different proportions of the individual components cause a different pattern of shear rate and flow speed with multilayer flow. For a stable combined flow, it is advantageous to have the low-viscosity melt next to the wall to produce a sliding-film effect.

6.2.3 Cooling

The film cross-section profile produced in the slot die must be solidified by cooling. After leaving the die, the melt passes across a short gap, where a fairly high degree of draw down in the extrusion direction takes place; it then makes contact with the casting roll where cooling to below freezing temperature or to the roll temperature takes place.

The rate of heat transfer between cast film and roll surface is critical. When producing film from partially crystalline materials, the cooling rate over the crystallization range is of particular importance as this has considerable influence on product properties. The quantity of heat to be extracted for a specific polymer can be determined from the relationship between temperature and specific enthalpy Δh shown in Figure 7. For a particular throughput $\dot{m}_K$ (kg/h) the amount of heat to be extracted is:

$$Q = \dot{m}_K \cdot \Delta h \tag{9}$$

Q heat quantity
$\dot{m}_K$ specific throughput
h enthalpy

In flat film extrusion, most of the cooling occurs on the casting roll and the rest on additional cooling rolls (Figure 8).
The total cooling path length, x_k, is:

$$x_k = R_1 \cdot \varphi_1 + R_2 \cdot \varphi_2 \tag{10}$$

x_k cooled length
R roll radius
φ angle of wrap

For heat exchange between film and roll the following conditions apply:

- heat flow within the film, through the film thickness in the direction of the roll surface;
- heat flow from the roll surface in the direction of the inner wall of the roll jacket;
- heat flow from the inner roll wall to the cooling medium.

The most important quantity for the effectiveness of the cooling system is the heat transfer coefficient $\alpha_{contact}$ [19, 20, 21, 22].

6.2 General principle of flat film extrusion

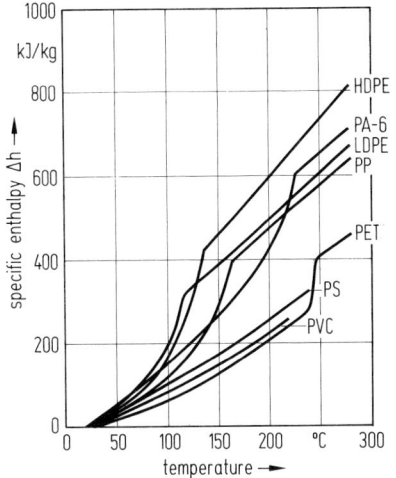

Figure 7. Specific enthalpy of various thermoplastics

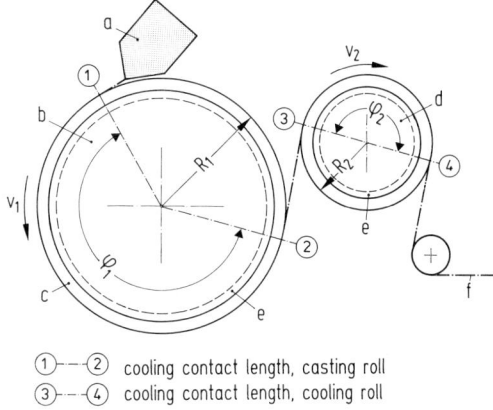

①--② cooling contact length, casting roll
③--④ cooling contact length, cooling roll

Figure 8. Cooling contact length and arrangements of casting and cooling rolls on a chill-roll line
a slot die, *b* casting roll, *c* roll jacket, *d* cooling roll, *e* cooling agent, *f* film, *v* speed, *R* roll radius, φ angle of wrap

Heat extraction by convection and radiation hardly needs to be taken into consideration with chill-roll lines, unless additional cooling systems such as air showers are used. Auxiliary devices of special importance for achieving good contact between film and roll include air knives, suction chambers and electrostatic pinning devices.

Unless they are used, a thin layer of air may be drawn in between film and roll, or the heat transfer may be adversely affected by additives diffusing out of the polymer and being deposited on the roll surface. The heat transfer at the contact surface is determined by the heat conduction equation:

$$Q = \lambda \cdot A \, (T_F - T_w)/s \tag{11}$$

A contact area
λ thermal conductivity
T_F film temperature
T_w roll temperature
s film thickness

The contact temperature, θ_k, depends on the heat penetration numbers:

$$b_i = \sqrt{\varrho_i \cdot C_{p_i} \cdot \lambda_i}$$

b heat penetration number
ϱ density
C_p specific heat,

whence:

$$\theta_k = \frac{b_F \cdot \theta_F + b_w \cdot \theta_w}{b_F + b_w} \tag{12}$$

θ temperature
F film
w roll

When extruding thin flat film, cooling to roll temperature occurs completely on the casting roll; the cooling behavior as a function of contact time is shown diagrammatically for the roll side and the air side of the film, in Figure 9.

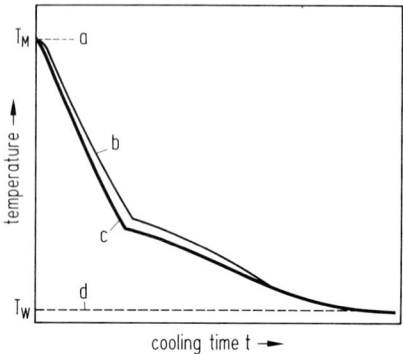

Figure 9. Cooling curve of a flat film
a melt temperature, *b* free surface, *c* roll contact surface, *d* roll temperature

The dependence of properties on cooling rate must be taken into consideration, particularly when processing partially crystalline polymers. Calculations and properties are given in [23, 24, 25]. A typical curve of cooling rate for a PP flat film is shown in Figure 10. The heat flow from the roll inner wall to the heat transfer medium (TM) takes place according to the following formula:

$$Q = \alpha_{TM} \cdot A (T_W - T_{TM}) \qquad (13)$$

α_{TM} heat transfer coefficient of transfer medium
T_W roll temperature
T_{TM} temperature of heat transfer medium
A cooling surface area

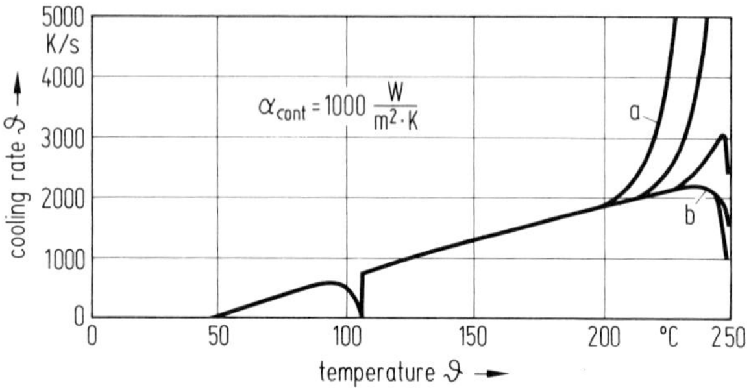

Figure 10. Cooling speeds in a PP film of 40 µm thickness, cooled on one side [21]
a roll contact surface, *b* free surface

The flow rate, $\dot{m}_{TM}$, of the heat transfer medium required for a heat extraction rate Q, is given by equation (9):

$$\dot{m}_{TM} = \frac{Q}{C_{TM} \cdot \Delta T_{TM}} = \frac{\dot{m}_k \cdot \Delta h}{C_{TM} \cdot \Delta T_{TM}} \qquad (14)$$

$\dot{m}_{TM}$ flow rate of transfer medium
ΔT temperature difference

The permissible temperature difference between incoming and outgoing TM should be $< 2\,°K$. A schematic arrangement of a chill roll with a heat exchange unit is shown in Figure 11.
A relatively simple method for calculating the length of chill-roll contact required on a line is given in [25].

6.2 General principle of flat film extrusion

Calculations are based on the *Fourier* number F_0, the continuity equation and examples taken from cast film extrusion.
The *Fourier* number describes the degree of cooling:

$$F_0 = \frac{a \cdot t_k}{s^2} \tag{15}$$

with $a = \dfrac{\lambda}{\varrho \cdot c}$, the temperature conductivity

λ heat conductivity
ϱ density
c specific heat
t_k cooling time
s film thickness

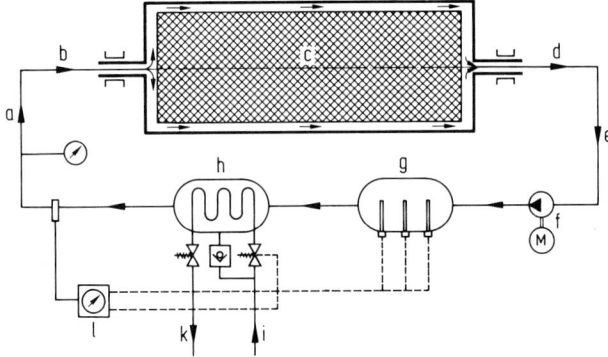

Figure 11. Principle of a chill roll with temperature control unit
a flow feed, *b* inlet, *c* roll core, *d* outlet, *e* flow return, *f* pump, *g* heating unit, *h* heat exchanger, *i* inlet cooling water, *k* outlet cooling water, *l* three term controller

The cooling time is obtained from:

$$t_k = \frac{x_k}{v} \tag{16}$$

with x_k = contact length and v = take-off speed. This gives us the *Fourier* number:

$$F_0 = \frac{\lambda}{\varrho \cdot c} \cdot \frac{x_k}{v \cdot s^2} \tag{17}$$

With the same degree of cooling and the same boundary conditions the following applies:

$$F_{0\,\text{unknown}} \cong F_{0\,\text{known}} \tag{18}$$

By using the continuity equation, which gives the throughput, $\dot{m}_K$, as a function of film width B, film thickness s, take-off speed v, and density ϱ, the following result is obtained:

$$F_0 = \left(\frac{\lambda}{c} \cdot \frac{B}{s} \cdot \frac{x_k}{\dot{m}_K}\right) \tag{19}$$

or, for the cooling contact length:

$$(x_k)_{\text{unknown}} = \frac{c}{\lambda} \cdot \frac{s}{B} \cdot \dot{m}_K \cdot (F_0)_{\text{known}} \tag{20}$$

More specific investigations and calculation methods, some with the help of computers, can be found in the literature [21, 26].
When thicker film or thermoformable sheet are produced, the cast web passes through a pre-set gap between the cooling roll and a smoothing roll. This ensures good heat transfer to the chill roll. For cooling partially crystalline thermoplastics especially, the contact time

on the first chill roll must be such that the transfer of the film or sheet to the second chill roll takes place before the crystallization temperature range is reached on the air side of the film.

Figure 12 shows the cooling behavior of thermoformable PP sheet for two different periods of contact on the first chill roll [27].

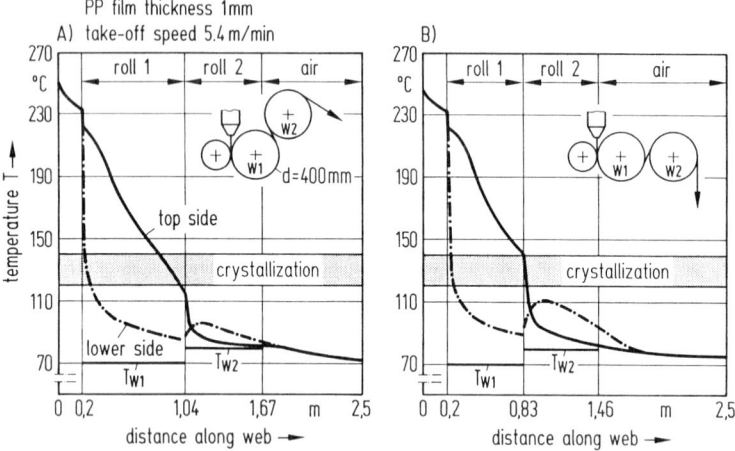

Figure 12. Cooling of a film web [27]

Morphology favorable for the thermoforming process is achieved in the film by proper choice of chill-roll temperature and cooling rate. With the correct roll-contact length an even, fine crystalline structure can be obtained. Figure 13 compares the different spherulitic structures (degrees of crystallization) obtained with fast and slow cooling [27].

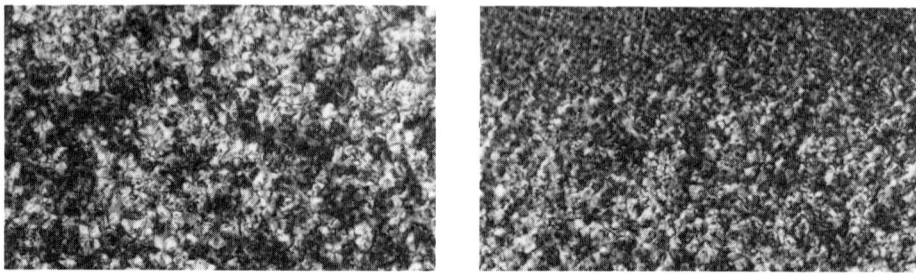

Figure 13. Effects of slow (left) and fast cooling (right) on crystalline structure of polypropylene films [27]

Material P 5200; sheet thickness 1.5 mm, magnification 250 times; $T_{melt} = 230\,°C$

$T_{chill\ roll} = 100\,°C$
$dT/dt = 1\,K/s$
density at $20\,°C = 0.905\,g/cm^2$
degree of crystallization $\varkappa = 50\%$

$T_{chill\ roll} = 40\,°C$
$dT/dt = 10\,K/s$
density at $20\,°C = 0.891\,g/cm^2$
degree of crystallization $\varkappa = 37\%$

$E_l > E_S$ $\sigma_{B_L} > \sigma_{B_S}$ $\varepsilon_{B_L} < \varepsilon_{B_S}$

Cooling of amorphous thermoplastics, such as ABS and PS/SB, differs considerably from that in PP processing and is primarily determined by cooling capacity.

6.2 General principle of flat film extrusion

6.2.4 Winding

Winding makes it possible to store the film compactly, but it must be done in such a manner that the films can be processed and utilized without problems. Basically, there are two winding methods, using different types of drive: contact winding – also called surface winding – and center winding, in which the axis of the film roll is driven (Figure 14).

If the film is sensitive to tension, center winding or a combination of center and surface winding is used. The torque, Md, applied by the take-up drives is the critical factor, and the power, N, required is:

$$N = K \cdot Md \cdot n \qquad (21)$$

N power
Md torque
n roll speed
K constant

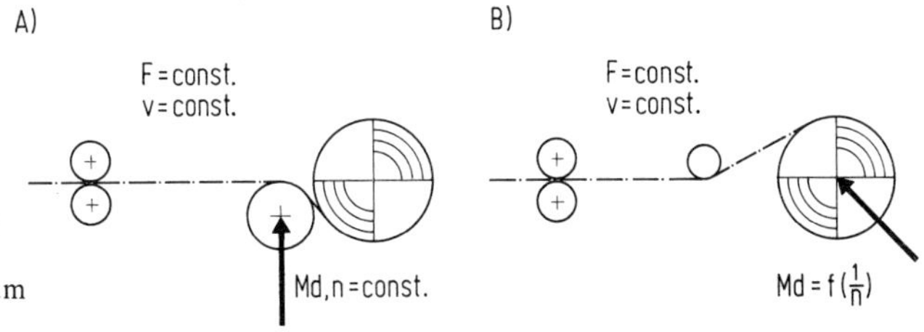

Figure 14. Diagrams of surface (A) and central winding (B) methods
F tension, v speed, Md torque, n r.p.m

The film speed, v, and the web tension F are important factors in the winding process. Using the package diameter, D, one can calculate the torque Md and the roll speed n:

$$Md = F \frac{D}{2} \qquad (22)$$

$$n = \frac{v}{\pi \cdot D} \qquad (23)$$

F tension
D diameter
v surface speed

With constant web speed and tension, the torque and the speed are proportional and inversely proportional to the package diameter, respectively:

$$Md_1 = \frac{F}{2} \cdot d_1; \quad n_1 = \frac{v}{\pi D_1} \qquad (24, 25)$$

$$Md_2 = \frac{F}{2} \cdot d_2; \quad n_2 = \frac{v}{\pi D_2} \qquad (26, 27)$$

During the winding process the torque, Md_1, follows a hyperbolic course:

$$Md = \frac{F \cdot v}{2} \cdot \frac{1}{n} \qquad (28)$$

If one looks at the storage capacity (wound length, L) of a package of film with a certain thickness s, core diameter, d, and final diameter, D, one obtains:

$$L = \frac{\pi}{s} \cdot \frac{(D^2 - d^2)}{4} \qquad (29)$$

L film length
s film thickness
d core diameter

The number of film layers is given by the number of turns, w:

$$w = \frac{R-r}{s} = \frac{D-d}{2s} \tag{30}$$

With center winding the ratio of final diameter to initial diameter has an important effect on package quality, and is called the winding factor, q:

$$q = \frac{D_{max}}{d_{min}} = \frac{R_{max}}{r_{min}} \tag{31}$$

If q is inserted into equation (29), one obtains the length of film on the roll as follows:

$$L = \frac{\pi}{s} \cdot R^2 \cdot \left(1 - \frac{1}{q^2}\right) \tag{32}$$

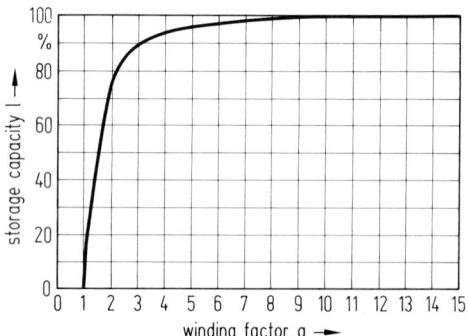

Figure 15. Storage capacity of a film package

The effective storage capacity of a package is reduced in proportion to the core diameter. As Figure 15 illustrates, however, a storage capacity of 96% can be achieved with a winding factor of only 5:1. A high q-value puts additional control demands on the winder drive system [28].

A winder must be able to provide a constant or hyperbolically decreasing web tension, depending on the type of film condition. A schematic illustration of the tension variation in relation to the package diameter is given in Figure 16 for various winding characteristics.

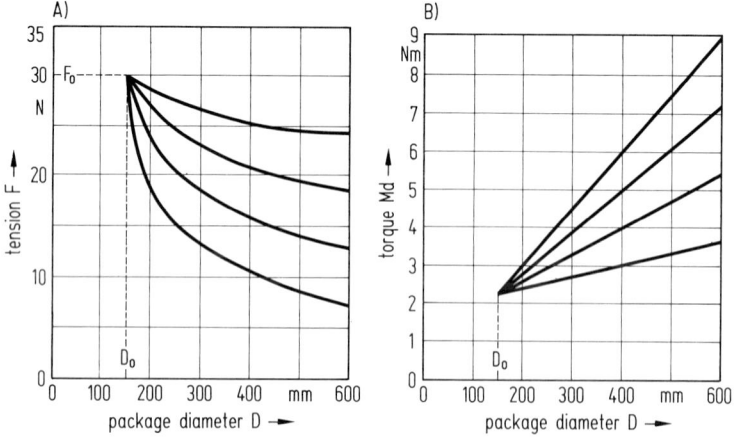

Figure 16. Winding characteristics with D.C. drives
A) tension force curve, B) torque reference time

6.2 General principle of flat film extrusion

The package drive must therefore be able to provide a constant torque, or one which rises linearly with package diameter (Figure 16). This task can be carried out by a thyristor-controlled D.C. motor. The simplest method is a control system which provides a torque proportional to the diameter of the package. The diameter can be scanned mechanically and an appropriate signal transmitted to a control unit, as shown in Figure 17A). Another method involves using a computer to obtain the diameter, to generate the relevant signal, and to transmit it to the control unit (Figure 17B).

A measure of the web tension is the motor torque, which is proportional to the current it draws; the armature current is a measure of the torque and the armature voltage a measure of the speed [28].

With these indirect systems of tension measurement, disturbing influences caused by friction losses in the mechanical transmission elements are not taken into consideration. When winding films which are very sensitive to tension, the winding power required is often lower than the friction losses in the system. In that case, the change of tension with package diameter must be controlled by direct measurement of web tension. Using a load cell tension controller, as shown in Figure 17C), or a dancer-roll system, web tension in front of the take-up position is recorded and transmitted to a control circuit [29, 30].

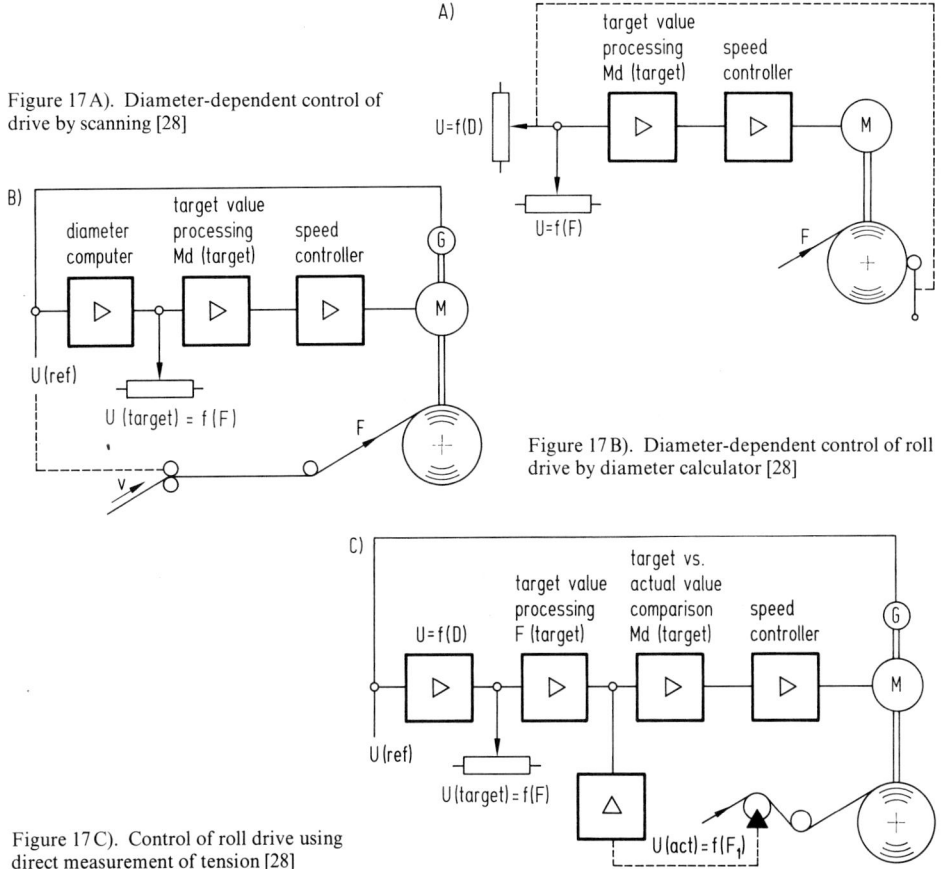

Figure 17A). Diameter-dependent control of drive by scanning [28]

Figure 17B). Diameter-dependent control of roll drive by diameter calculator [28]

Figure 17C). Control of roll drive using direct measurement of tension [28]

When winding flat films, it is usually necessary to have a specific winding characteristic, with the tension decreasing over the diameter. The required winding curve is controlled directly as a function of the signal transmitted by the diameter calculator and of the web tension measurement. Winding must be adjusted to a pre-set characteristic dependent on film properties, in order to produce a hard or soft (decreasing web tension) package and to provide for possible shrinkage of the film web − as may, for example, occur on account of additional crystallization − on the package. The large amount of effort that has gone into measurement and control technology results in constant film quality throughout the full package diameter, as well as reproducible winding results. In general, the rolls produced on flat film lines have a diameter between 300 and 600 mm. The high film speeds require the use of turret winders with automatic roll-change facilities.

6.3 Raw materials, film types, properties and applications

6.3.1 Raw materials and product dimensions

When looking at flat film extrusion as a whole, we find that a wide range of polymers is used and that there is a comprehensive thickness range from about 10 to 2500 µm. The main types of film and their dimensions, which depend on application and later conversion operations, are listed in Table 1:

Table 1. Main types of film and dimensions

Film type	Thickness range µm	Film widths mm
Flat films FF	10 to 150 100 to 400	1200 to 3200 800 to 2200
Thermoformable films TF	100 to 400 400 to 2500	450 to 1600 600 to 1400
Orientable films OF [1)] Uniaxial (UO) Biaxial (BO)	30 to 600 30 to 2500	600 to 1800 400 to 1500

1 Values apply to unoriented primary films

Flat cast films have thicknesses from about 10 to 150 µm, and even up to about 400 µm for special applications. These films are mainly used for packaging purposes, with PE, PP and PA-6 materials being produced in the greatest quantities.
Film and sheet for thermoforming are primarily made from SB, ABS, PP, HDPE, PET and rigid PVC.
For uniaxially or biaxially oriented films PP, HDPE, PET, PA-6, PC and PS are used; for these purposes flat film extrusion is directly combined with the stretching process.
The raw materials (Table 2) used in flat film extrusion must ensure good film quality with regard to purity, homogeneity, additives and stabilizers, and absence of gels. Unvarying melt viscosity permits constant, high-quality film production. Table 3 shows common viscosity ranges of some resins used for cast film and thermoformable sheet production.

6.3.2 Properties and applications of flat film and thermoformable sheet

Films are subject to widely varying requirements which depend on application and further processing [31, 32]. Control testing is done on the basis of DIN, ASTM, and ISO standards.

Table 2. Materials used for slot-die extrusion of films and sheet

Abbreviations	Designations	Density g/cm^3	Film type
Polyolefins:			
LDPE	Low-density polyethylene	0.918 to 0.035	FF
HDPE	High-density polyethylene	0.94 to 0.954	FF, TF, OF
LLDPE	Linear low-density polyethylene	0.917 to 0.94	FF, OF
EVA	Ethylene vinyl acetate	0.93 to 0.94	FF
IONOMER	Ethylene-methacylic acid-copolymer	0.94	FF
PP	Polypropylene	0.896 to 0.907	FF, TF, OF
PP-Cop.	PP block copolymer	0.9 to 0.91	FF, TF
PP-R	PP random copolymer	0.9	FF, OF
Hetero-polymers:			
PA 6	Polyamide	1.13 to 1.15	FF, OF
PA 6.6	Polyamide	1.12 to 1.15	FF, OF
PA Cop.	PA copolymers	1.10	FF
PET	Polyethylene terephthalate	1.33 to 1.36	FF, TF, OF
PBT	Polybutylene terephthalate	1.3	FF, TF
PETG	Copolyester	1.20 to 1.27	FF, TF
PC	Polycarbonate	1.2	FF, OF
CA	Cellulose acetate	1.3	FF, TF
CAB	Cellulose acetobutyrate	1.19	FF, TF
CP	Cellulose propionate	1.21	FF, TF
PPO	Polyphenyleneoxide	1.06	FF, TF
Halogens:			
PVC rigid	Polyvinyl chloride (rigid)	1.38 to 1.40	FF, TF, OF
PVC soft	Polyvinyl chloride (soft)	1.20 to 1.30	FF
PVDC	Polyvinylidene chloride	1.6	FF, TF
PVDF	Polyvinylidene fluoride	1.78	FF
PAN	Polyacrylonitrile	1.18 to 1.27	FF, TF
Styrene polymers:			
PS	Polystyrene, general purpose	1.05	TF, OF
SB	S-butadiene copol. (impact res.)	1.05	TF
SAN	Styrene-acrylonitrile copol.	1.08	TF
ABS	Acrylonitrile-butadiene-styrene copol.	1.04 to 1.06	TF
Elastomers:			
PUR	Polyurethane	1.21	FF

FF = flat films; TF = thermoformable films; OF = orientable films

Table 3. Viscosity data (MFI; η_{rel}) of some materials used in extrusion of flat film and thermoformable sheet

| Polymer type | Film type | Melt flow index MFI g/10 min | | | | Solution viscosity η_{rel} |
		190 °C/21.6 N	230 °C/21.6 N	200 °C/50 N	220 °C/100 N	
LDPE	FF	1.5 to 4.5	–	–	–	–
HDPE	FF, TF	1.5 to 8.0	–	–	–	–
LLDPE	FF	1.0 to 6.0	–	–	–	–
PP	FF	–	5.0 to 12.0	–	–	–
PP	TF	–	1.5 to 5.0	–	–	–
PA 6	FF	–	–	–	–	2.8 to 4.0
PET	FF, TF	–	–	–	–	1.6 to 2.0
SB	TF	–	–	3.0 to 5.0	–	–
ABS	TF	–	–	–	5.0 to 8.0	–

FF = flat film, thickness range: 10 – 400 µm
TF = thermoformable film; thickness range: 200 – 2500 µm

Table 4. Quality and performance characteristics of films

Mechanical properties	Physical properties	Product technical properties
Tensile strength (MD/TD) Elongation at break (MD/TD) Impact strength – unnotched – notched Tensile modulus Flexural modulus Hardness Tear propagation resistance Tear initiation resistance Toughness Stiffness	water absorption sealability weldability seam strength aroma barrier gas barrier water vapour barrier odor neutrality taste neutrality food contact purity antistatic behavior thermoformability friction coefficient interlaminar adhesion surface tension	width tolerance thickness tolerance weight per unit area flatness rigidity machinability crystallinity printability ink adhesion slip curling tendency elastic recovery stretchability static build-up blocking tendency
Thermal properties	Optical properties	Electrical properties
Melt temperature Melting range Heat resistance Shrinkage Cold resistance Thermal expansion Thermal conductivity Heat-stress endurance	transparency haze gloss gel count film structure homogeneity purity color tint surface uniformity	dielectric loss factor

Table 5. Comparison of mechanical and optical properties of flat films

Properties		Film type			
		PP	PA 6	LDPE	LLDPE
Film thickness		25 µm	25 µm	25 µm	25 µm
Tensile strength, MD DIN 53455	N/mm^2	40 to 50	45 to 50	20 to 30	20 to 30
Tensile strength, TD DIN 53455	N/mm^2	30 to 40	40 to 50	10 to 15	15 to 20
Elongation at break, MD, DIN 53455	%	450 to 600	350 to 400	200 to 350	500 to 800
Elongation at break, TD, DIN 53455	%	350 to 550	300 to 400	300 to 400	500 to 800
E-modulus DIN 53457	N/mm^2	500/900	550/750	350/450	200/300
Gloss (45 °C) ASTM D 2457 – 6 ST	%	80 to 90	85 to 95	60 to 80	85 to 95
Haze ASTM D 1003-61	%	1.5 to 2.5	2.5 to 3.5	6.0 to 10.0	1.0 to 2.5
Transparency/clarity ASTM D 1746	%	65 to 75	60 to 70	40 to 60	60 to 70

6.3 Raw materials, film types, properties and applications

In addition, further methods of comparison for monitoring quality and processability are employed during production and conversion. An outline of various test and quality characteristics is shown in Table 4. Apart from the physical values, optical properties are of special importance with cast films. Typical values for the most common cast film materials are listed in Table 5 [33, 34, 35]. The main field of application is the packaging industry. The films, which are usually transparent, are converted — with or without print — into bags, or used for lamination with other films and/or substrates. Table 6 gives an outline of the preferred fields of application and properties for the various film types [36].

Table 6. Applications and properties for various types of film

Film type	Application products	Polymers											Film thickness range μm
		PP	LDPE	LLDPE	HDPE	PA	PET	PVC soft	PVC rigid	PS/SB	ABS	PUR	
Flat films (thin films)	bag films	×	×		×	×		×					20 to 80
	lamination film	×	×	×	×	×	×	×		×	×	×	15 to 150
	cut blanks	×	×	×		×		×					20 to 40
	covering film		×	×								×	15 to 150
	stretch films			×				×					20 to 40
	twist films	×							×				20 to 50
	household films		×	×				×					10 to 25
	textile packaging	×	×										20 to 80
	flower packaging	×	×										20 to 40
	sweets	×	×	×									20 to 50
	foodstuffs	×	×	×	×	×		×	×				20 to 80
	bread	×	×										25 to 50
	cold cuts	×	×										20 to 30
	office folders	×	×					×	×		×		60 to 300
	clear filing envelopes	×	×					×					60 to 150
	protective covers	×	×		×			×	×				80 to 200
	sterilization film					×							25 to 50
	medical packaging					×	×		×				30 to 120
	separators	×	×				×						20 to 150
	decorative film	×				×	×		×		×		20 to 200
	diaper liners		×	×				×					20 to 30
	colored tapes				×								12 to 20
	adhesive tapes	×	×	×	×								80 to 400
	embossed films	×	×	×	×			×	×		×	×	20 to 400
Thermo-formable films	lids	×			×			×					200 to 400
	containers	×			×		×	×	×	×			400 to 2500
	beakers	×						×	×				600 to 1000
	trays	×					×	×	×	×			200 to 800
	skin packaging		×	×		×	×						30 to 400
	blister packaging	×					×		×				100 to 1500

Films and sheet for thermoforming feature good dimensional stability, excellent surface quality and a homogeneous structure. For the thermoforming process it is advantageous to have a uniform, low shrinkage. For a film thickness of 1 mm, longitudinal shrinkage with SB should be about 8 to 10% (DIN 16955) and up to 10 to 14% with ABS (DIN 16956). The thermoforming process is carried out in the thermoplastic state with SB and ABS, and in the thermoelastic region just below the crystallite melting point with the partially crystalline material PP, using the principles employed in the SPPF process (solid phase pressure forming) developed by Shell [37].

6.3.3 Properties and fields of application of coextruded flat films and thermoformable sheet

The various requirements for packaging films can often only be fulfilled by combining different materials. Multilayer combinations of different films whose nature depends on the packaging task, or products laminated or coated in combination with other products, such as paper or aluminum, are used. Alongside lamination, coextrusion is gaining increasing importance, as it permits the production of a packaging material which is optimized for the particular packaging task.

The film structure can be either symmetrical or asymmetrical (Figure 18); and thermoplastics of differing properties which adhere poorly to each other can be combined by using thin layers of adhesive between them. Special requirements for improved sealability or barrier properties can be produced very cost-effectively by coextruding such extremely thin layers.

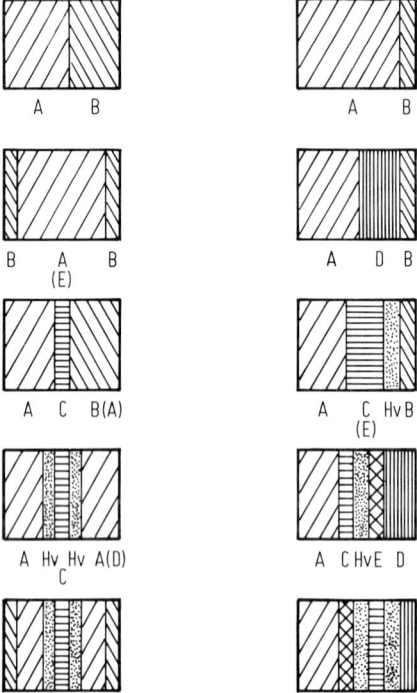

Figure 18. Structure of multilayer films
l. h. column: symmetrical layer structure
r. h. column: asymmetrical layer structure
A main layer, support layer; B outer or top layer (heat seal, gloss, antistatic, or colored); C barrier layer; D main layer, colored layer; E rework layer; Hv tie layer

6.3 Raw materials, film types, properties and applications

The properties and applications of coextruded cast films and thermoformable sheet made from various material combinations are compiled in Table 7 [38, 39, 40, 41]. Microtome cross-sections of coextruded films are shown in Figure 19 [38].

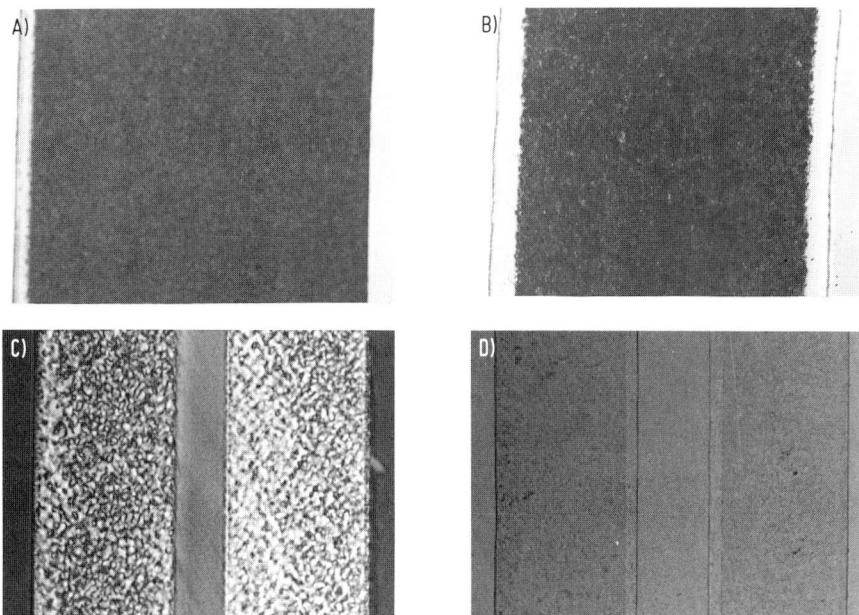

Figure 19. Microtome cross-sections of coextruded flat films. (Photo: Barmag, Remscheid, West Germany)
A) Double-layer film, top layer 10 µm (random copolymer), base layer 290 µm PP colored;
B) Three-layer film, outer layer 25 µm PP, base layer 300 µm chalk-filled PP, outer layer 25 µm PP;
C) Three-layer film, outer layer 90 µm PA 6, barrier layer 28 µm EVAL, outer layer 90 µm PA 6;
D) Five-layer film, outer layer 190 µm PP, tie layer 10 µm, barrier layer 90 µm EVAL, tie layer 10 µm, outer layer 190 µm PP

Table 7. Material combinations, properties and applications of coextruded flat films

Material combination	Special properties	Most important applications	Film type
Two-layer film			
LLDPE/EVA	stretchability,	stretch film,	FF
(LDPE/EVA)	sealability	packaging film, laminating film	
PP/EVA	transparency, rigidity, sealability	packaging film laminating film	FF
SB/PS	surface brilliance, antistatic	drinking cups, containers for dairy products, plates	TF
SB-pigmented/SB white	decorative surface,	containers for dairy products	TF
PP/PP random copolymer	good weldability, good printability	biaxially oriented films containers for bread, margarine, jam, dairy products	OF TF
PP/PP foam	smooth surface, good printability, good weldability	packaging films for consumer goods, packaging containers uniaxially oriented films	OF
PA 6/ionomer	good sealability	packaging film for meat, cheese	FF
PET/PETG		trays	TF

FF = flat films; TF = thermoformable films; OF = orientable films; TL = tie layer (adhesive)

Table 7. (Continued)

Material combination	Special properties	Most important applications	Film type
Two-layer film with tie layer (TL)			
PA6/TL/LDPE	good sealability	packaging films for cheese	FF
PETG/TL/SB	good barrier properties, good transparency	sausage thermoformed containers	TF
PE/TL/SB	good sealability grease resistance	dairy products margarine	TF
Three-layer film with symmetrical layer structure			
PP/PP + chalk/PP	less shrinkage, smooth outer layer, good printability, good thermoforming properties	menu, fast-food trays lids, containers, packaging film	TF FF
PP-cop./PP + rework/PP-cop.	low-temperature-resistant, low materials costs	freezer packaging	TF
PA6/EVAL/PA6	barrier properties, gas barrier sealabilty	laminating film	FF
PP-random, cop./PP-homop./PP-random cop.		technical film biaxially oriented film	FF OF
LDPE/EVAL/LDPE, PP/EVA/PP	high impact resistance, non-sticky outer layers, good transparency	bag production	
EVA/LLDPE/LDPE (EVA/LLDPE/EVA)	high stretchability, good adhesion	stretch film for pallet wrapping, adhesive packaging film	FF
Ionomer/PA/Ionomer	good sealability, good transparency, good thermoforming properties, reduced curl	vacuum packing	FF
Three-layer film with asymmetrical layer structure			
PS/SB colored/SB white	brilliant surface antistatic surface, decorative color layer	dairy products containers, trays	TF
Four-layer film			
SB white/SB rework/SB brown/PS	brilliant surface, UV barrier	dairy products	TF
PS/SB/TL/PETG	good barrier property, aroma tightness, good sealability, good thermoforming properties	vacuum packing	FF
PA6/EVAL/TL/LDPE			FF
Five-layer film			
PP/TL/EVAL/TL/PP	gas, water-vapor and aroma barrier can be sterilized	ready-made meals fruit juices	FF TF
SB/TL/EVAL/TL/SB	gas, water-vapor and aroma barrier	milk products, meat dishes	TF
LDPE/TL/PA6/TL/LDPE	good barrier properties protection of PA against moisture	vacuum packing, thermoformable sheet	
PP-random cop./TL/PA6/TL/PP	good sealability, good transparency, thermoforming properties	meat, sausage, cheese, ham, packaging	

FF = flat films;   TF = thermoformable films;   OF = orientable films; TL = tie layer (adhesive)

For improved sealing properties special polymer types such as ionomers, EVA, EAA, and random PP copolymers are used, and PA, PET, PAN, EVOH, PVDC are employed as barrier materials. Table 8 lists permeability values for various polymers [32, 42, 43].

Table 8. Permeability values of various polymers used for film manufacture (thickness $s = 25$ µm)

Material	Gas permeability $\dfrac{cm^3}{m^2 \cdot d \cdot bar}$		Water vapor permeability $\dfrac{g}{m^2 \cdot d}$
	O_2	CO_2	H_2O
LDPE	6000	3000	20
HDPE	2000 to 2800	9000	5
PP	2500 to 3600	9500	8 to 10
PS	6000	17000	100 to 120
ABS	2000	6000	70 to 100
PETG	400	2000	60
PET	155	465	93
PVC	120 to 200	240	20 to 40
PA 6	40 to 60	400	160
PAN	12 to 13	17	78
PVdC	1.6	4.7	1.6
EVOH	0.4	1.3	60

Oxygen permeability (measured according to DIN 53380) can be reduced greatly by using EVOH, PVDC and PAN; in laminates, good barrier effects against water vapor (measured according to DIN 53122) are provided by PVDC, HDPE, PP and PVC. Good protection against aroma loss and adverse effects on flavor is achieved with PET, PETG and PA [44, 45].

In the coextrusion process special attention must be paid to recycling and/or refeeding the edge trim. Depending on the film structure and material combination, the chopped edge trim can be either directly added to a main layer, or extruded as a separate layer in a multilayer combination.

6.4 Process technology and performance of flat film extrusion lines

The process steps required for flat film extrusion and the relevant influential factors are shown in Figure 20, set out in the direction of material flow [36]. Depending on the size of the line, extruders with screw diameters between 75 and 150 mm are used. The effective screw length is usually 30 or 33 D. A special design of feed zone – smooth or grooved cooled sleeve – permits effective and stable conveying with high throughput, optimized for the particular material. Table 9 summarizes extruder capacities for various film casting materials [2, 46].

Chill roll films

In order to obtain film of uniform high quality with good transparency, flatness and surface structure, cooling must be carried out under defined conditions specific to the particular film type and polymer.

Table 10 outlines the normal ranges of melt and roll temperatures for various materials. With partially crystalline thermoplastics like PP, PA, PET the cooling process determines

Table 9. Capacities of extruders used for production of flat films and thermoformable sheet

Cast film extrusion (film thickness range: 20 to 150 µm)

Screw diameter	Screw length	Throughput kg/h			
		PP	PA 6	LDPE	LLDPE
75	30 to 33 D	140 to 160	90 to 110	180 to 200	160 to 180
90	30 to 33 D	220 to 240	130 to 150	250 to 280	220 to 240
105	30 to 33 D	280 to 300	180 to 200	340 to 380	300 to 340
120	30 to 33 D	350 to 400	260 to 300	450 to 500	400 to 450
150	30 to 33 D	500 to 550	380 to 420	680 to 750	600 to 650

Thermoformable film extrusion (film thickness range: 0.4 to 2.0 mm)

Screw diameter	Screw length	Throughput kg/h			
		PP	SB	ABS	PET
75	30 to 36 D	180 to 200	300 to 320	220 to 250	120 to 140
90	30 to 36 D	260 to 290	450 to 500	360 to 400	180 to 220
105	30 to 36 D	320 to 350	600 to 650	450 to 480	240 to 280
120	30 to 36 D	480 to 550	750 to 850	600 to 650	320 to 360
150	30 to 36 D	650 to 750	1100 to 1200	850 to 900	480 to 540

Table 10. Ranges of melt and roll temperatures used in flat film extrusion casting

Material	Melt temperature range °C	Roll temperature range °C	
		casting roll	cooling roll
PP	230 to 260	15 to 40	15 to 30
PA 6	260 to 280	70 to 120.	90 to 140
LDPE	220 to 250	30 to 60	20 to 30

the crystalline structure, and this in turn influences the optical properties of films, among other things [35, 47, 48]. With increasing roll temperature the transparency and clarity achievable for a certain film thickness decrease, but these conditions favor producing film with better machinability, dimensional stability, slip and gloss. Figure 21 shows haze as a function of casting-roll temperature, for a 40 µm PP extrusion-cast flat film, with other process variables such as melt temperature, screw speed and take-off speed held constant. A slight reduction in haze can be achieved by increasing the melt temperature; but this may lead to increased volatilization of additives contained in the raw material and thus to condensation on the roll surface, which in turn has adverse effects on film cooling and optical properties. With an increase in line throughput — higher take-off speed with the same film thickness — unevenness in the film and its fine structure may increase; reduction of melt dwell time in the die may result in elastic deformations not being able to relax sufficiently. In addition to the process parameters, material-specific influences are also important for achieving good optical properties [49, 50].

Considerable changes of the spherulite structure in the production of PA-6 cast film are brought about by very slight variations in casting roll temperature. Low roll temperatures and high cooling speeds lead to fine-crystalline structure, which provides good optical properties and good thermoforming behavior.

6.4 Process technology and performance of flat film extrusion lines

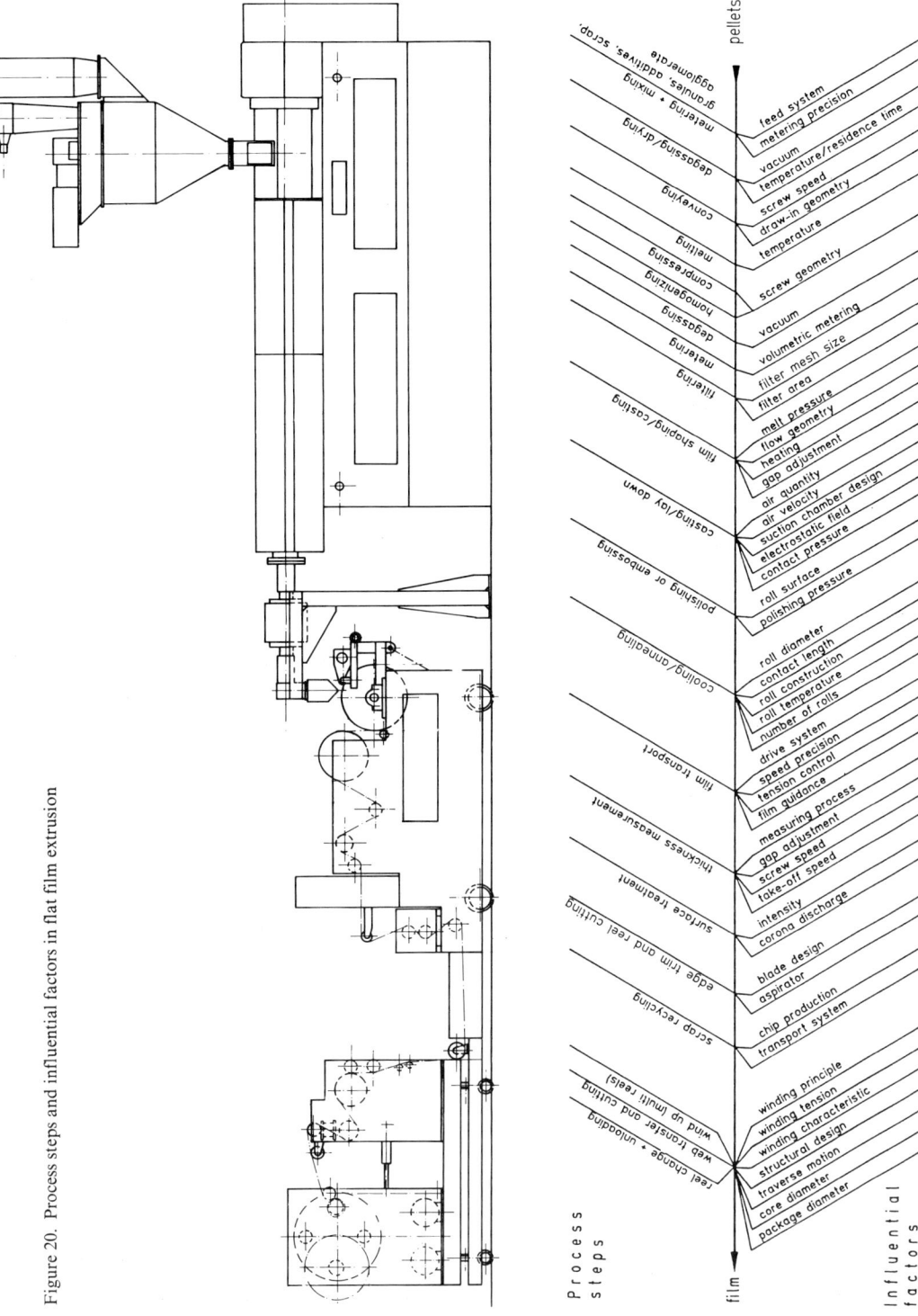

Figure 20. Process steps and influential factors in flat film extrusion

Particularly with polyamide films it is possible to raise crystallization by increasing the temperature (120 to 150 °C) of an annealing roll without significant spherulite formation; this produces film which runs well on conversion machinery with little adverse effect on optical properties.

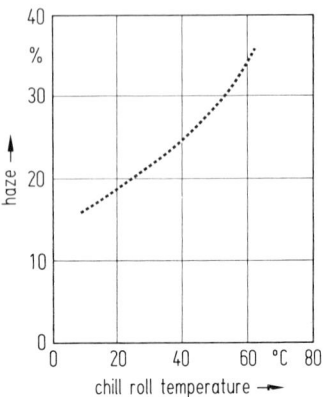

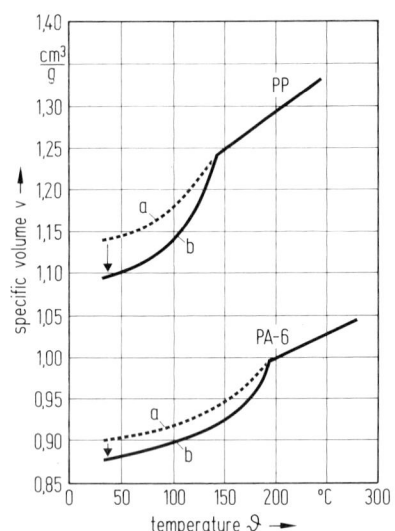

Figure 21. PP flat film: haze as a function of casting-roll temperature

Figure 22. Specific volume as a function of temperature and cooling rate for PP and PA 6 [35]
a rapid cooling, *b* slow cooling

Changes of film dimensions during storage are also highly dependent on the spherulite structure and the cooling conditions. If films have been cooled rapidly, crystallization on the roll causes additional shrinkage, particularly with PP films. These films must be wound with extreme care in order to prevent hardening of the package. The condition of the film can be characterized by means of the density or the specific volume. Figure 22 shows specific volume as a function of temperature for rapid and slow cooling of PP and PA melts [35, 47]. With high cooling rates the specific volume below the crystallization temperature is higher; this means, particularly with PP films, that there is a high potential for additional crystallization, and, consequently, for a change in optical properties and for dimensional shrinkage. Structure and properties of PA cast films, produced under different conditions, are shown in detail in [47]. In addition to the influence of cooling conditions on the structure and post-crystallization of PA films, water absorption during storage causes a change in film dimensions which can partially counteract the shrinkage caused by post-crystallization. With two-layer coextruded sandwich films, PE/PA for example, the additional crystallization and the water absorption in the PA layer cause warping of the film. By analogy with the bi-metal effect, the sandwich film curls towards the layer with the greater contraction. The curling tendency can be reduced by controlling cooling conditions and by conditioning with water vapor [34, 51]. The condition of the roll surface is of great importance for producing highly transparent, glossy films. This surface may be hard chrome-plated and highly polished, or have a dull finish. For high take-off speeds special matt finishes are used. Rougher roll surfaces permit more even contact of the film, but have adverse effects on the surface gloss. For specially structured films it is possible to emboss the casting roll surface. PP cast films with an "orange peel" surface, for example, are embossed directly on the casting roll; the film is also pressed against the casting roll by an air knife in this case.

Figure 23 shows the take-off speeds that are possible at present for various polymers and film thicknesses, based on a casting diameter of 800 mm in this case.

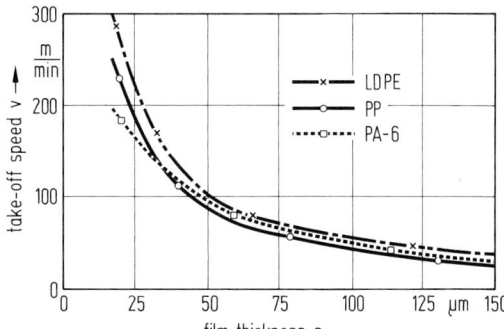

Figure 23. Maximum production speeds for flat films as a function of polymer and film thickness with an 800 mm diameter casting roll

Films and sheet for thermoforming

Thermoformable films require particularly good melt homogeneity, as mechanical and thermal inhomogeneity may lead to tearing during the thermoforming process. The extruder feed section must be able to accept a mixture of various raw materials, a color master batch and the addition of ground material from start-up waste, edge trimming and punching grid. When processing PS/SB and ABS, vented extruders are used. In the degassing zone, moisture and any monomers, or other low-boiling-point components present, are extracted. When extruding films for packaging foodstuffs, low melt temperatures should be aimed for, to avoid reduction in molecular weight and the formation of monomers.

With degassing screws, the second screw section determines the extruder throughput capacity. For processing PP homopolymers and copolymers, multisection screws with mixing and shearing components are used. In the extrusion of chalk- or talc-filled PP, degassing screws are also used. Figure 24 shows the design principle of a PP mixing screw and an SB degassing screw. In addition to good melt homogeneity it is necessary to achieve even and symmetrical cooling of the film web. When producing PP sheet for thermoforming it is important that the film surface can be polished when it makes contact with the second chill roll and that the temperature is still above the crystallization range. This is important for achieving a fine crystalline structure. Table 11 outlines the ranges of melt temperature and roll temperature used in the production of various thermoformable films [2].

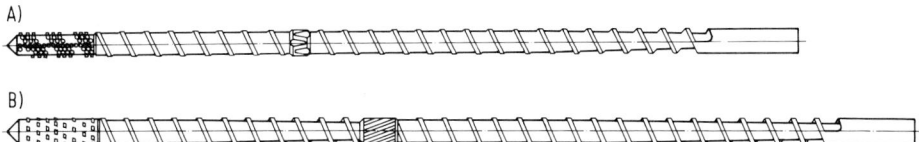

Figure 24. Screw design principles for PP and PS/SB processing
A) PP mixing and shearing screw, B) SB degassing screw

The thermoforming characteristics and product properties of partially crystalline thermoplastics like PP are influenced considerably more by processing conditions than is the case with amorphous PS/SB films. In order to achieve good shape stability in the thermoformed articles, no anisotropic shrinkages must occur during the forming process.

In the manufacture of thermoformable films less than 400 μm thick, the melt is pressed against the cooling roll by an air knife instead of by the smoothing roll. Figure 25 shows the variation of permissible take-off speed with raw material and film thickness, using two 400 mm diameter rolls.

Table 11. Ranges of melt and roll temperature ranges used in thermoformable film and sheet extrusion casting

Material	Melt temperature range °C	Chill roll temperature range °C
PP	230 to 260	15 to 60
SB	210 to 230	50 to 90
ABS	220 to 240	60 to 100
PET	280 to 285	15 to 60

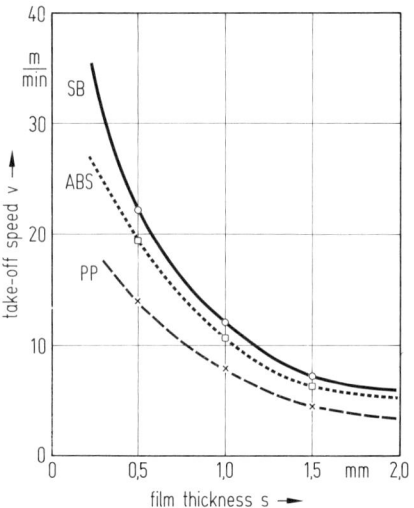

Figure 25. Maximum production speeds using two 400 mm dia. chill rolls for thermoformable films as a function of polymer and film thickness

6.5 Flat film production lines

6.5.1 Line concepts

6.5.1.1 Chill-roll system

The line design is determined by the range of polymers to be processed and by the throughput and speed range for the different film thicknesses. Figure 26 (see opposite) is a general flow chart of the flat film extrusion casting process.

Depending on the capacity range of the line, chill rolls from 400 to 1200 mm dia. are used in the arrangements shown in Figure 27. In the majority of cases a casting roll with a large wrap angle is combined with one or two subsequent chill or annealing rolls.

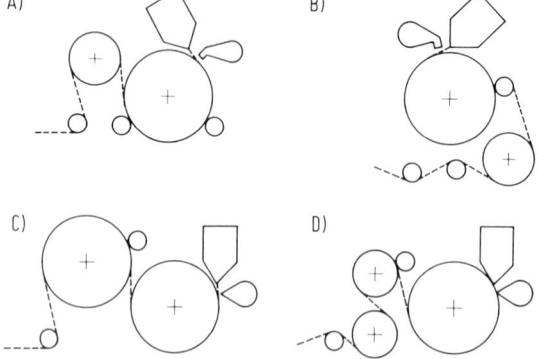

Figure 27. Roll arrangements on chill-roll lines

The slot die is usually positioned vertically above and tangentially to the casting roll. Film guidance depends on the operating and installation conditions of down-stream equipment such as thickness gauges, surface treatment units, web tension measuring devices, cutting devices, and so on.

In order to achieve good package formation with uniform hardness, the film web must be oscillated across the extrusion direction before final trimming.

6.5 Flat film production lines

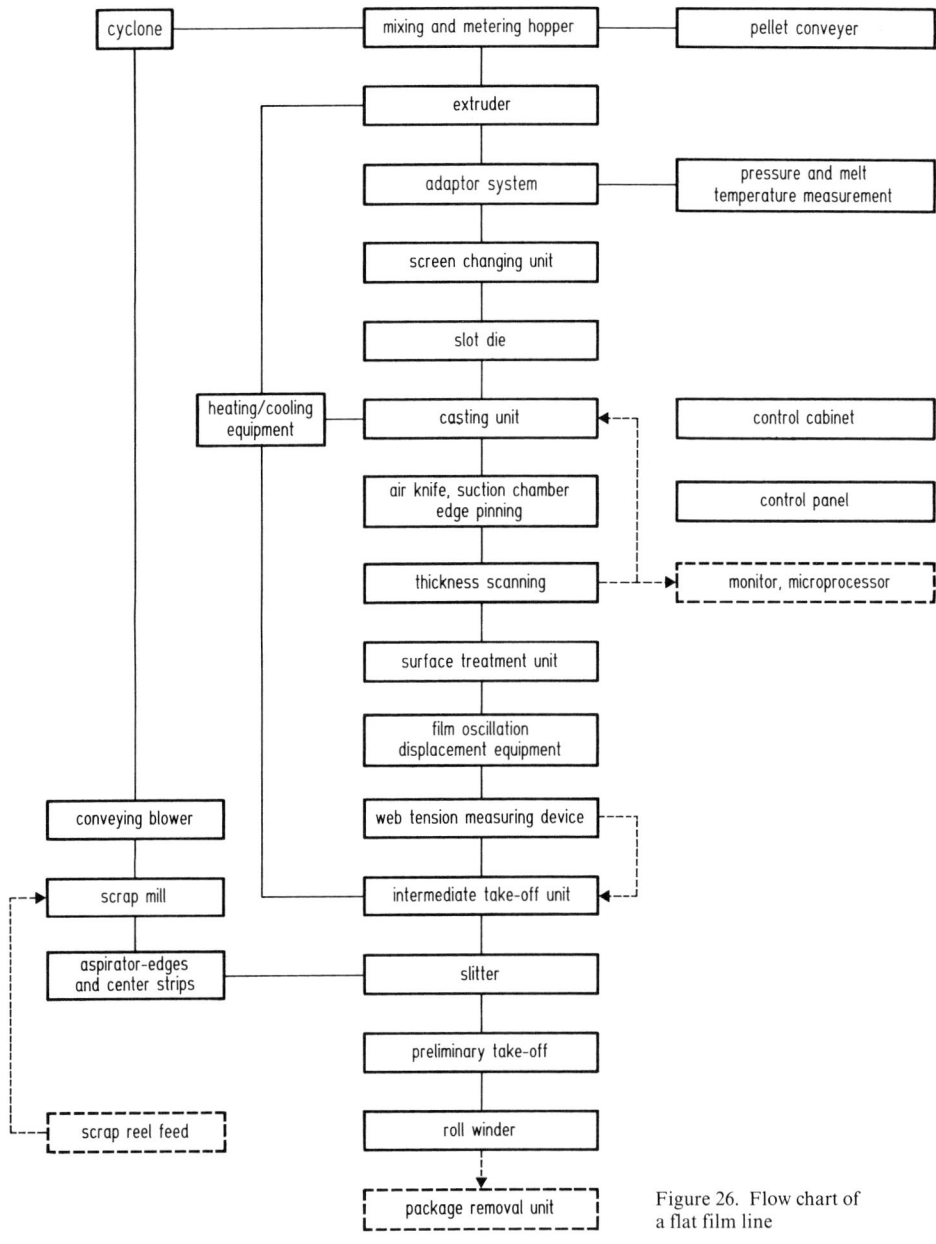

Figure 26. Flow chart of a flat film line

In practice, the various listed methods are employed, with web traverse between casting unit and take-up being used most frequently:

- traversing the extruder with the die,
- traversing the film web within the line by means of an oscillating device,
- traversing the take-up unit and cutting device together.

Figure 28 shows one version of a chill-roll line [52].

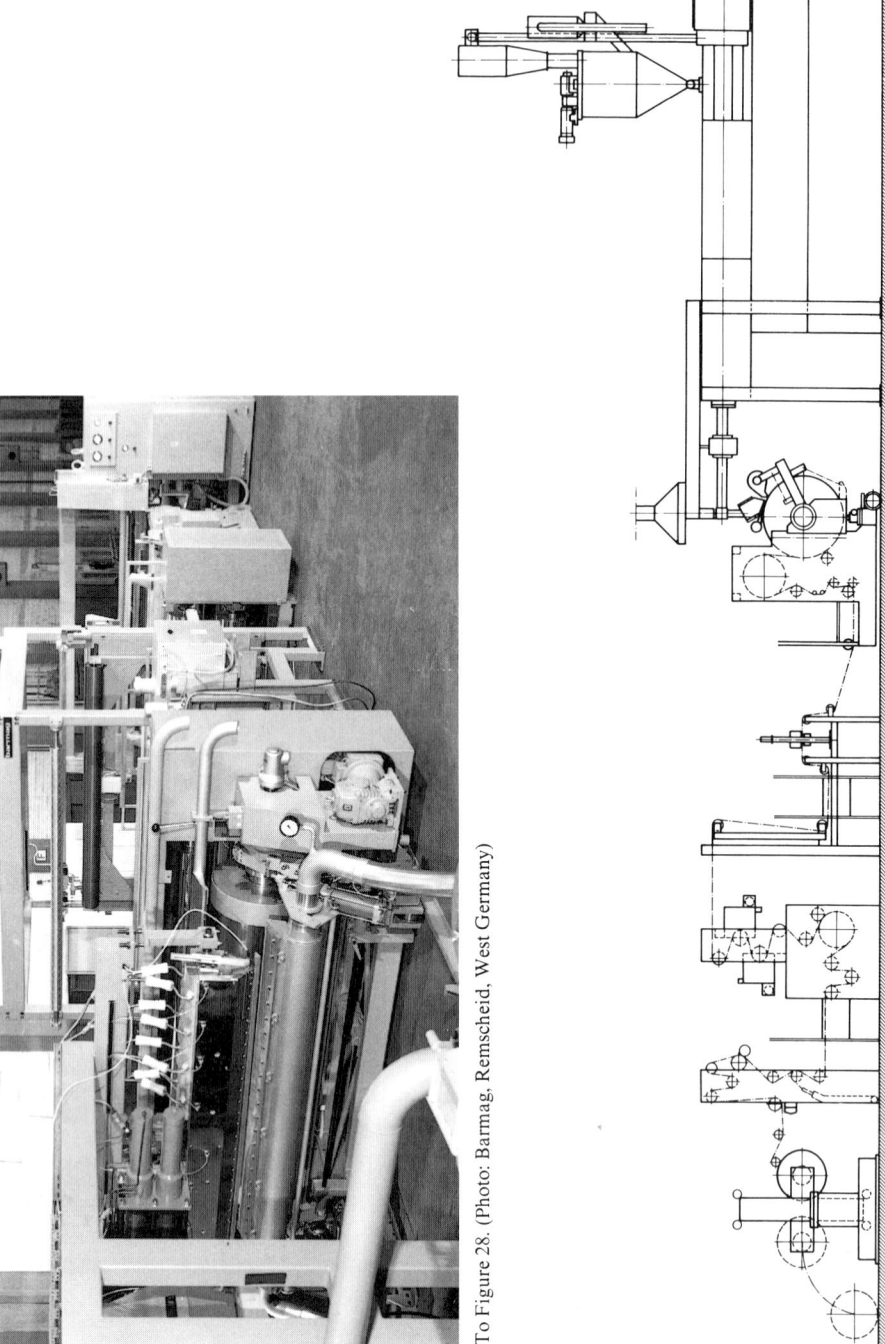

To Figure 28. (Photo: Barmag, Remscheid, West Germany)

To Figure 28

6.5 Flat film production lines

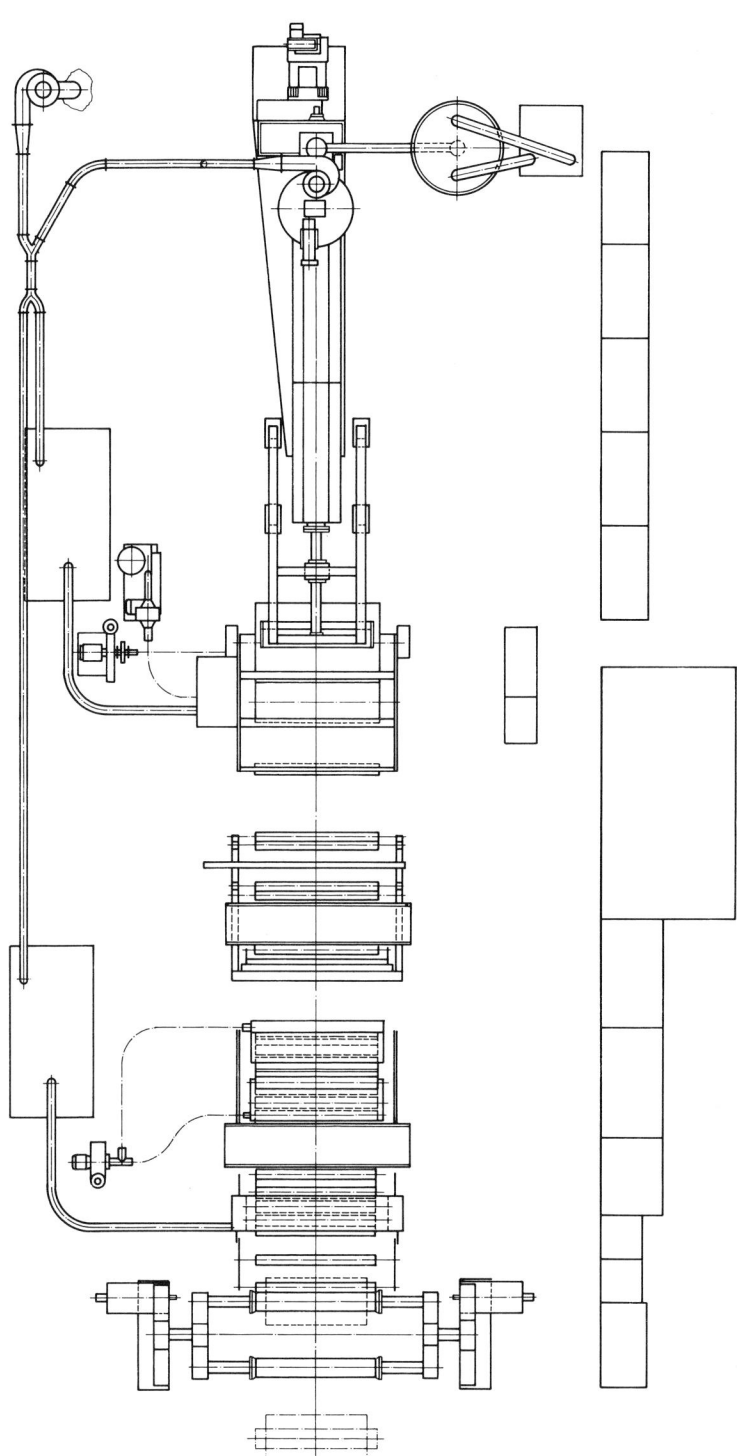

Figure 28. Plan and elevation drawings, and general view of a flat film line [52]

6.5.1.2 Lines for thermoformable film and sheet

For the production of polished thermoformable film or sheet, the line concepts shown in Figure 29 are used. The roll diameters lie within a range of 200 to 600 mm. The horizontal arrangement of the polishing and chill rolls is preferred, especially when low-viscosity melts are processed.

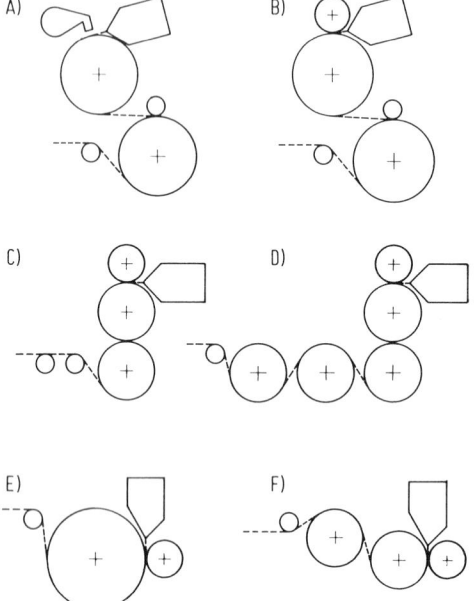

Figure 29. Roll arrangements on thermoformable film lines

With the in-line process – the direct combination of the extrusion line with the thermoforming machine – there are economic advantages from energy saving, as well as process technology advantages. By utilizing the remnant heat within the film, the energy needed for reheating is reduced [53].

6.5.2 Line layout and individual units

6.5.2.1 Raw material feed and scrap recycling

Resins for flat film extrusion are usually supplied in pellet form. In most cases they are conveyed to the extruder hopper by pneumatic devices, either direct from the silo or from containers positioned next to the extruder. Additives like color master-batch or concentrates of slip agents, anti-blocking agents, and antistatic agents can be mixed directly into the pellets using metering and mixing units above the extruder hopper, or added in separate mixing silos.

Hygroscopic polymers like PA and PET must be carefully dried before they are fed to the extruder, to prevent undue decrease in viscosity and preserve the superior properties of the product. Drying temperatures and moisture contents are listed in Table 12. To ensure that resin feed is oxygen-free when processing PA or PET, the feed zone can be operated under partial vacuum (approx. 15 to 25 mbar) or be blanketed with nitrogen, and the feed section

6.5 Flat film production lines

of the extruder and the screw bearings on the gearbox side be sealed by means of radial sealing rings. The pellets are then supplied through a double vacuum-hopper.

Table 12. Drying conditions for various thermoplastics

Material	Normal initial moisture % H$_2$O	Residual moisture content % H$_2$O	Drying temperature, max. °C
PA 6	1.0 to 2.0	< 0.08	75
PET	0.2 to 0.2	< 0.01	160
PS/SB	0.2 to 0.6	< 0.02	80
ABS	0.2 to 0.6	< 0.02	80

Drying may be carried out in hopper driers, supplied with dry air (dew point $\leq -30\,°C$) at a particular temperature. In order to ensure constant resin temperatures over a prolonged period of time and thus to stabilize the extrusion process, it is advisable to preheat the pellets, even if the resin is not hygroscopic.

Various methods are employed for recycling the edge trim or center strips generated in cast film production. The amount of waste can lie between 10 and 25%, depending on the line width and the product.

After size reduction the waste is either re-compounded separately and then mixed with the virgin material as recycled granulate or agglomerate, or is fed to the extruder in chopped

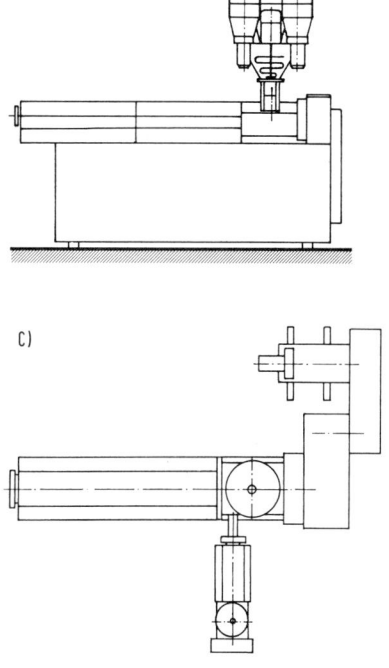
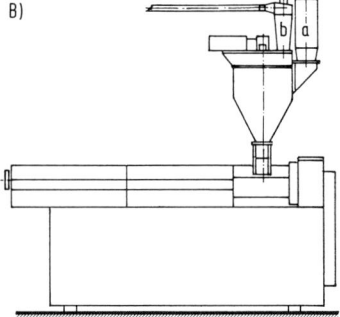
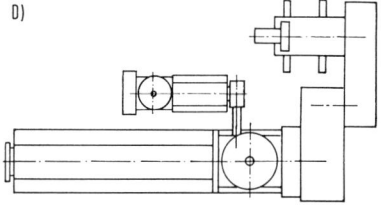

Figure 30. Various possibilities for edge trim chopping and recycling [54, 55, 56, 57]

A) mixing hopper
    a pellets, *b* agglomerate or regranulate
B) metering and mixing hopper with metering screw
    a pellets, *b* chopped film
C) chopped film recycling extruder
D) recycling extruder for continuous edge trim

form. The chopped material is either accepted by an auxiliary extruder, and compressed and fed into the main extruder, or mixed into the fresh feedstock by a metering and mixing screw in a twin-chamber hopper.

By varying the metering screw speed, which is synchronized with that of the extruder screw, the percentage of rework can be adjusted.

It is also possible to rework continuous edge strips (unchopped) using a separate extruder with a special feed section and to feed the plasticated rework into the main extruder near the feed zone.

Figure 30 shows the different schemes for edge-trim recycling [54, 55, 56]. In film and sheet production for thermoforming, the waste generated in start-up, off-standard rolls, in edge trim and in punching grid is ground up and mixed with fresh feedstock. The return rate lies between 20 and 50%, depending on production conditions. Because of the higher bulk density of regrind from rigid film and sheet (thickness range 0.4 to 2.5 mm), and its good free-flow ability, constant conveying and mixing rates can be achieved with volume metering and mixing devices. The metering station must be designed to accept different resins, reground material and, in the case of coloration processes, color master-batch.

6.5.2.2 Extruders

In order to meet the high quality requirements for the melt, single-screw extruders with an effective screw length of 30 or 33 D are used in flat film extrusion. The construction principle of extruders used for processing most thermoplastics is illustrated in Figure 31. Powerful, thyristor-controlled D.C. drives provide good adaptability to the various polymers. Thanks to the use of a belt drive between motor and gear unit, extremely smooth running is achieved. The feed zone is of a variable design and can be equipped with either a grooved sleeve for high conveying efficiency or a smooth sleeve. This design rationale permits optimum adaptation to different materials, particularly with coextrusion lines. The feed section is either cooled or temperature-controlled. The barrel is lined with a centrifugally cast, wear-resistant metal alloy to give a long service life. The extrusion screw has mixing and homogenizing sections, and may also have shearing elements of appropriate size to ensure optimum melt temperature and homogeneity.

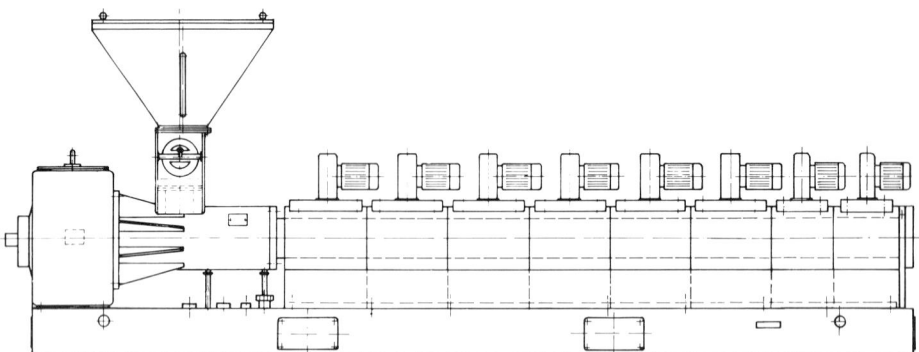

Figure 31. Single-screw extruder with air cooling

By correct choice and dimensioning of the feed section of the extruder and the screw, surge-free conveying is achieved even when edge trim from the line is being reworked.

The screws used in flat film extrusion are normally nitrided, but can also be armored with wear-resistant layers on the flights. The barrel is divided into several heating/cooling zones, with air cooling being the most frequently used system. To reduce heat losses by radiation, important with high-temperature processing polymers like PET, extruders are covered with heat insulation. Extruders with barrel venting are used for processing PS/SB and ABS. The moisture content, which normally lies between 0.1 and 0.2%, can be reduced to such an extent by means of a high capacity vacuum pump, that stable extrusion and bubble-free product quality are achieved.

6.5.2.3 Filtration equipment

Impurities may have strong adverse effects on the production process and the subsequent utility of flat films. Thus the homogenized melt must have solid particles removed before it is shaped in the die. Filtering is usually done by means of multilayer filter cloths supported by a perforated disk. The filtering effect is determined by the filter mesh sizes — normally up to approx. 10 000 mesh/cm^2.

In the course of film production the screen pack must be exchanged with a frequency depending on the degree of contamination and the permitted pressure drop. Screen changing devices can be continuous (no interruption of extrusion) or discontinuous; with the latter, production must be halted for the change to be carried out. Figure 32 shows diagrams of some designs of screen changing devices [57, 58, 59]. Screen carriers are usually moved hydraulically.

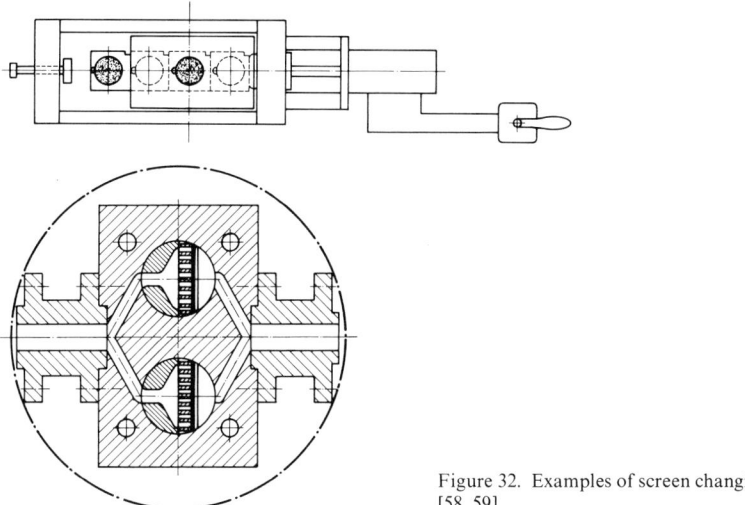

Figure 32. Examples of screen changing devices [58, 59]

The screen diameter depends on the screw diameter. In order to achieve ultra-fine filtration with mesh sizes less than 60 µm and little drop in pressure, large-area filters of long-life or nonstop design with switch-over are used. Such filters commonly have an effective area of between 0.1 to 0.4 m^2, depending on throughput and the mesh size [60].

6.5.2.4 Melt metering pumps

In order to improve thickness uniformity for high-quality films, a gear pump can be integrated into the extrusion process as a metering device. Melt metering pumps have been common in the extrusion spinning of low-viscosity materials for a long time; they have linear metering characteristics and provide steady, constant melt output [61, 62]. Throughput fluctuations, which can occur, for example, when the amount of chopped film rework is increased, can be offset by using a pump. A pump unit consists of the pump housing, the metering gear wheels, the melt supply- and discharge pipes, and a drive. Heating can be effected either electrically or with liquid agents. The drive, a thyristor-controlled D.C. shunt motor, gives precise control of the pump speed (Figure 33).

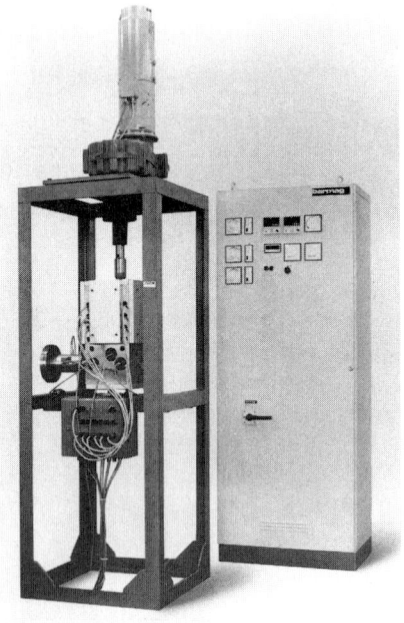

Figure 33. Metering pump unit with drive and control cabinet.
(Photo: Barmag, Remscheid, West Germany)

With the extruder/pump combination the pressure on the feed side of the pump can be set freely within a certain range, to act as the set value for controlling the extruder drive. The screw speed is controlled continuously with reference to the pump feed pressure.

The melt metering pump holds pressure constant to $< \pm 1\%$ measured at the entry to the slot die. The melting conditions in the extruder can be optimized by means of the adjustable, constant counter-pressure. Adaptation to the various extruder sizes and output ranges is achieved by choosing a pump size with the appropriate volume delivery per revolution. The pump sizes used for PA, PET and PP flat film extrusion deliver a volume of approx. 100 to 600 cm^3/rev. The speed of the pump gearwheels usually ranges from 10 to 40 r.p.m.

If metering pumps are used, films of improved quality and tighter tolerances can be produced and, consequently, raw material economies can be achieved. In addition, the linear metering characteristics provide for better process control [63, 64].

6.5.2.5 Slot dies

The slot die, in combination with the other units of the line, has the important task of producing a film which is dimensionally accurate and has very narrow tolerances. Die widths up to 3500 mm are used on cast film lines. The die outlet width must be larger than the effective width of the film to be produced, by the amount needed for lateral neck-in, film traverse, and edge trimming. This additional width may be as much as 250 to 300 mm, depending on production conditions.

In order to keep the distance short between the die exit and the casting roll, the die must be shaped to take the casting position in relation to the chill roll into consideration.

Flat film dies are used both with and without restrictor bars (Figure 34 [65]).
Restrictor bars are not required if the viscosity ranges of the raw materials to be used lie close together and the film thickness is up to about 200 µm.

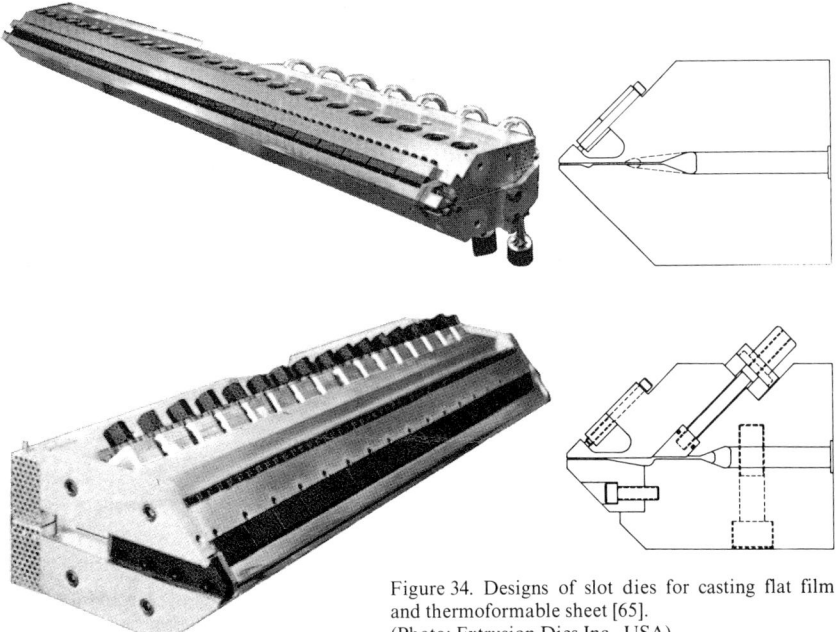

Figure 34. Designs of slot dies for casting flat film and thermoformable sheet [65].
(Photo: Extrusion Dies Inc., USA)

One die-half is designed with a flexible lip to allow the die gap to be sensitively adjusted by push screws or push/pull bolts with differential threads. The melt-contact surfaces are usually hard-chrome plated and the die-lip area is hardened. High surface quality ($R_t < 0.2$ µm) ensures a favorable flow free of die marks.

The slot die is divided across its width into several control sections; heating is effected electrically by cartridges (300 to 500 W per cartridge) or by flat heater plates. With cartridges, insulation fixed to the die body is commonly used to reduce heat radiation and to stabilize operating conditions.

The electric heating systems and the cable layout are designed to allow rapid dismantling and cleaning.

Tolerance optimization on manually adjustable dies depends on the training and experience of the operators. Figure 35 shows a so-called "automatic" die [65] whose exit gap can be adjusted by expansion of the die bolts ("expansion bolts"). Each die bolt has a heating cartridge parallel to it, over which air is blown onto the bolt. If the applied voltage changes, the temperature and length of the bolt change, and with them the die gap at that point. The voltage can be altered manually by a potentiometer, or by a microprocessor. The processor takes care of die-gap corrections by using signals proportional to the actual film thickness registered by a thickness gauge at the corresponding positions on the film. See also Section 6.5.3. To provide for the extrusion of a wide range of thicknesses of thermoformable sheet and film (say 0.5 and 2.5 mm), slot dies are equipped with an adjustable die lip or with an exchangeable lip, so that the best gap setting can be found for the thickness of film to be extruded.

With thermally insensitive materials, external deckling of the die gap is used to make it possible to produce various film widths with very little edge trimming.

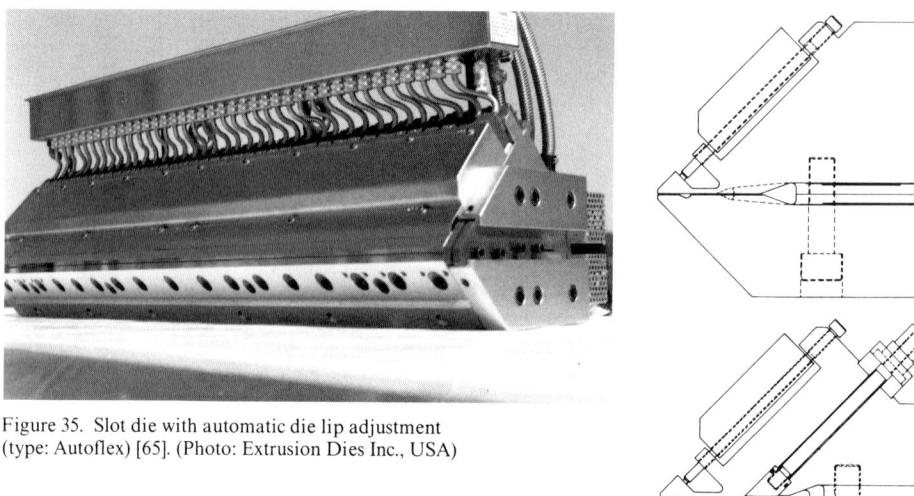

Figure 35. Slot die with automatic die lip adjustment (type: Autoflex) [65]. (Photo: Extrusion Dies Inc., USA)

6.5.2.6 Coextrusion systems

Two different coextrusion methods are used for the production of multilayer films and sheet: *the feed-block system and the multichannel die process*. In addition, it is possible to use a combination of the two for certain film constructions. Figure 36 summarizes the different variations that are possible.

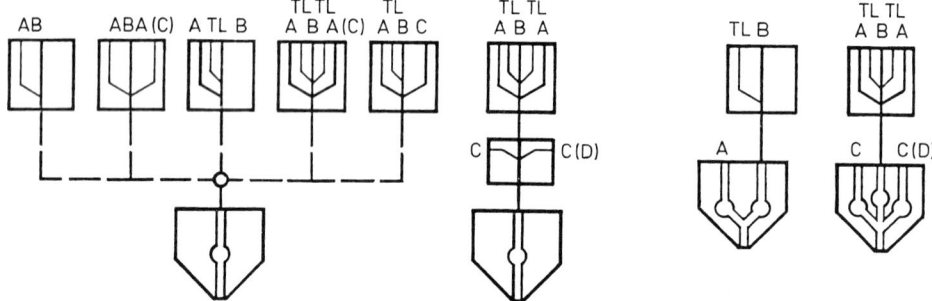

Figure 36. Possible variations (schematic) with adaptor coextrusion and combination with multichannel dies

With the adaptor principle (feed block) the individual melt components are layered by a flow plate or supply channel system so that they flow laminarly into a conventional slot die together, and are then shaped to film width.

Figure 37 shows some common layer arrangements at the die entrance used in feed block coextrusion [34].

In the coextrusion process using a multichannel die, the individual melt streams are distributed across the film width in separate channels, and are then usually brought together inside the die upstream of the lip area, and discharged through a common die gap. Figure 38 shows different versions of multichannel coextrusion dies [34, 36].

The advantages and disadvantages of the two coextrusion processes, and selection criteria, are compared in Table 13 [38, 66]. The adaptor method requires considerably lower investment than the multimanifold die. This is very obvious when there are a large number

6.5 Flat film production lines

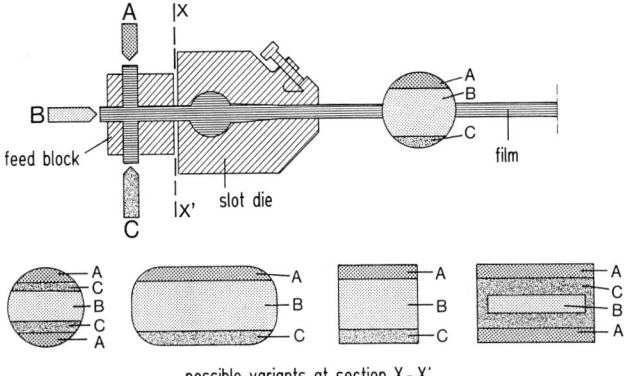

Figure 37. Illustration of feed-block coextrusion with various designs of the input geometry [39]

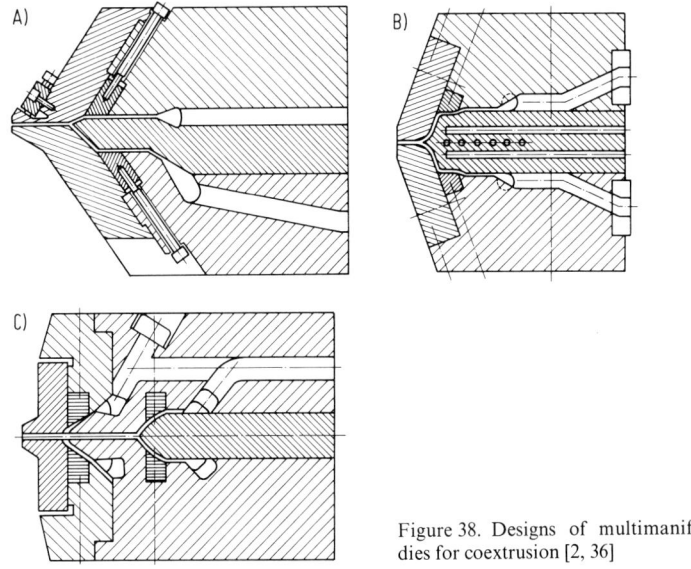

Figure 38. Designs of multimanifold slot dies for coextrusion [2, 36]

of layers. Laminar adaptor systems with exchangeable melt distribution elements are available; these make possible an optimum arrangement with regard to number and position of the layers, layer thicknesses and rheological conditions for any requirement. The adaptor method is frequently used when thin layers are required – either as a top layer (heat-seal layer) or a center layer (barrier layer).

As the flat film process always involves edge trimming, it is essential, from an economy point of view, that the edges of the film consist of a single material as far as possible, so that it can be recycled in the process. Figure 39 shows the structure of a five-layer coextrusion with a recyclable edge trim. The flow of the layers in the edge area can be optimized by means of adjustable devices in the feed-block system [67, 68].

It is possible to leave the film margins free of a covering layer; but throughput-independent operation is usually only achievable with multichannel die coextrusion; in which case the widths of the channels for the covering layers are reduced in proportion to the size of the required free area.

Table 13. Comparison of feed-block and multichannel die systems for the production of coextruded flat films

Selection criterion	Feed-block process	Multichannel die process
Investment costs	relatively low	relatively high; depends on number of layers
Number of layers	nearly unlimited, up to 9 layers already practiced	limited, usually 2, but 3 and 4 possible
Handling	relatively easy, no regulation of individual layers	more expensive, because individual layers have to be regulated
Thickness deviation of individual layers	± 10%	± 5%
Permissible viscosity difference in components	1 : 2 to 1 : 3	larger than 1 : 3
Outer layer extrusion (< 10%)	preferred with die width > 1 m	preferred with die width < 1 m
Extrusion of thermally sensitive materials	better in center layer (no contact with metal)	better in covering layers (special manifold design)
Flexibility	better, easy variation in number and position of layers by exchange of parts	low, number of layers pre-set

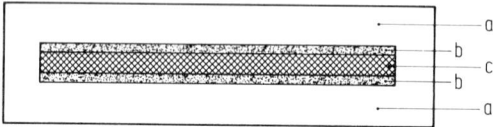

Figure 39. Film structure of a five-layer coextrusion from three components with recyclable edge area
a base layers, *b* tie layers, *c* barrier layer

Figure 40. Flat film coextrusion unit for the production of five-layer films from three components (Photo: Barmag, Remscheid, West Germany)

If materials with very different rheological behavior are to be coextruded, or if materials which differ greatly in their processing temperature are to be combined, the combination of a feed block with a multichannel die is commonly employed; for example with the laminar adaptor, connected to one flow channel of a two-channel die.

The coextrusion system that is best in a particular situation depends on film structure and material combination. Figure 40 shows a flat film coextrusion unit for manufacturing five-layer films from three different polymers.

6.5.2.7 Chill roll take-off units

For good quality film production by the chill-roll process, the dimensions of the unit must be matched to the throughput in order to ensure even cooling of the film. Design detail depends on the polymer, film dimensions and speed range. Table 14 lists the most important characteristics of chill-roll take-off units.

Table 14. Outline of design characteristics and technical data of chill-roll take-off lines

Roll surface width	mm	1200 to 3600
Casting roll dia.	mm	400 to 1000
Cooling roll dia.	mm	200 to 600
Roll surfaces		
– roughness R_t	μm	highly polished $R_t \leq 0.1$ to 0.3
		matt or coarse matt
– cylindricity	mm	± 0.01 to 0.02
– concentricity	mm	± 0.01 to 0.02
Temperature range	°C	Water: 15 to 90
		15 to 130 (150)
		Oil: 50 to 160
Temperature precision	°C	$\Delta t \pm 1$ and 1.5 resp.
Drive		thyristor-controlled D.C. motor
Control range		1:10 to 1:15
Max. take-off speed	m/min	100 to 400
(depends on roll diameter and film product)		

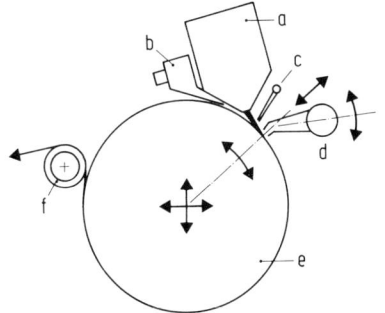

Figure 41. Casting position on a chill-roll take-off unit. (Photo: Barmag, Remscheid, West Germany)
a slot die, *b* suction chamber, *c* edge pinning (air jets), *d* air knife, *e* casting roll, *f* lift-off roll

The extruded film is cooled on the casting roll which may have a diameter of 400 to 1200 mm, depending on line size. Figure 41 shows the unit set up for casting, with casting roll, slot die, edge-pinning device, air knife, suction chamber and lift-off roll. The internal structure of the roll (Figure 42 [69]) is designed for stability and to provide the required cooling capacity; but, in addition, properties like concentricity, cylindricity, vibration-free running, and speed uniformity are of special importance. The roll surface (and this is to some extent determined by the film material) is usually highly polished (roughness $R_t <$ 0.3 μm), but for higher take-off speeds can also be provided with special matt chrome-plating. A special roll design with an exchangeable jacket permits dismantling for cleaning and removal of deposits. It is also possible to mount several roll jackets with different surface finishes on one basic roll body [69]. The casting roll is usually driven by a thyristor-controlled D.C. shunt motor with high-precision speed control.

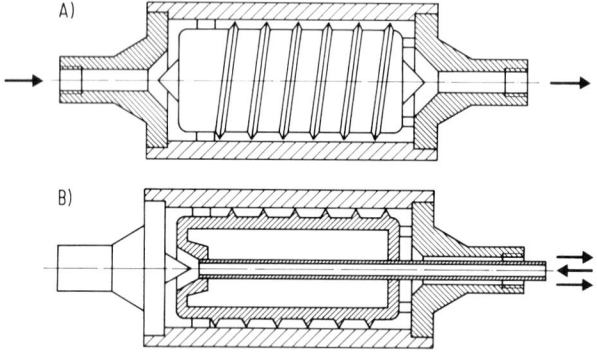

Figure 42. Designs of chill rolls [69]
A) chill roll with spiral distributor element and straight-through flow;
B) chill roll with spiral distributor element one-sided supply and return flow

After the film leaves the casting roll, it is led over one or two cooling or thermal conditioning rolls – depending on the film type – where final cooling and stabilizing take place. The cooling roll is separately driven by a D.C. shunt motor, or is designed for adjustment by a precision control gearbox in relation to the speed of the casting roll. In order to achieve good contact between the film and the surface of the casting roll, lay-on aids are used: such as air knives, suction chambers or, for polymers like PET, electrostatic pinning devices. The air knife (Figure 43) consists of a pipe, fed with air from both sides, having a precision-made slit whose size can be adjusted over a range from approx. 0.2 to 2.5 mm. The knife holder can be precisely and reproducibly positioned to obtain the best quenching point for given film conditions. The knife is fed with filtered air by a variable output blower. For certain production conditions the air can be cooled or heated. The knife can be heated to prevent condensation of volatiles – slip agent for example – on its surface. A suction chamber arranged between slot die and casting roll aids film contact and serves to remove volatiles leaving the melt; positional adjustment is like that for the air knife.

To minimize neck-in between the slot die and the contact line with the casting roll, edge pinning devices like air jets, or electrodes to create an electrostatic charge, are used. Along the circumference of the casting roll there are "cleaning" or lift-off rolls. These can be engaged pneumatically and their position can usually be altered to adapt to the various cooling and speed conditions with different film types. In this way condensation of polymer additives on the casting roll surface can be largely prevented. The lift-off roll ensures that film take-off across the width of the casting roll is uniform.

The contact point is optimized by lengthwise movement of the chill-roll take-off unit and by vertical adjustment of the casting roll or of the entire take-off unit; the movements can be carried out manually or by motor power. In order to achieve a precise cooling gradient along the film web, separate units are provided for controlling the inlet temperature of the heat-transfer agent. Depending on the film type, different processing temperatures are required: for PP films the temperatures lie between 15 and 40 °C and with PA-6 be-

6.5 Flat film production lines

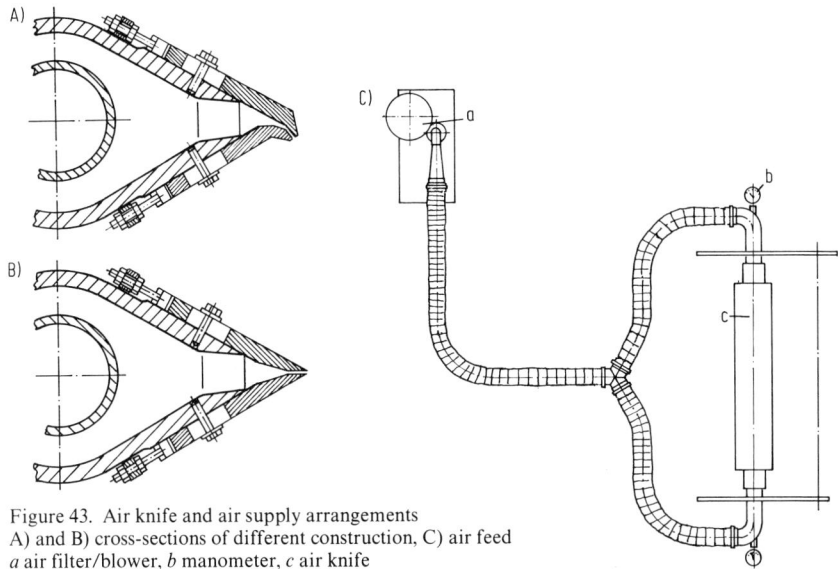

Figure 43. Air knife and air supply arrangements
A) and B) cross-sections of different construction, C) air feed
a air filter/blower, *b* manometer, *c* air knife

tween 80 and 140 °C, depending on film speed and crystallinity. A central cold water installation with condenser cooling can be used to supply the individual units.
The temperature variation of the heat transfer agent entering the chill rolls is $< 0.5\,°C$ with modern line designs. In order to obtain a uniform roll surface temperature (say $\Delta K < \pm 1\,°C$), high flow speeds are required in the guide channels of the double-jacketed roller.

Film guidance and downstream equipment

After leaving the chill-roll unit the film passes through several guide elements such as deflector idler rolls, spreader rolls, and a web tension measuring device before it reaches the wind-up station. The web transport line has the following in-line auxiliary equipment: thickness scanner, surface treatment unit, web oscillating unit, intermediate or cooling take-off, longitudinal cutting equipment, edge aspirators, and ion generators. The idler rolls are free-running cylinders that permit delicate films to be transported at low web tension. For certain products it may be necessary to use driven rolls, or to drive the roll-axle to offset bearing friction. In certain sections of the line, ahead of the longitudinal cutting equipment for example, driven spreader rolls are used to ensure crease-free running of the film web; in this case the spreading effect is adjustable. In order to achieve reproducible production and to avoid excessive tension during film transport, it is necessary to measure the web tension by using either dancer systems or measuring rolls (Figure 44) running in load-cell-bearings [70]. The tension controllers are connected to an electronic unit that provides control, display and limit-value signals. The load-cell units can be installed either horizontally or vertically, and the film web must enter and leave at constant entry and exit angles. Although very narrow thickness tolerances are achieved in the flat film process, it is necessary to oscillate the film web transverse to the extrusion direction before it is cut and wound, in order to achieve uniform package hardness. Slight thickness variations could otherwise lead to hard spots in the reel and to the formation of 'piston rings'. In addition to oscillating the extruder and die transversely, or the take-up unit and longitudinal cutter, it is also possible to use an oscillating unit

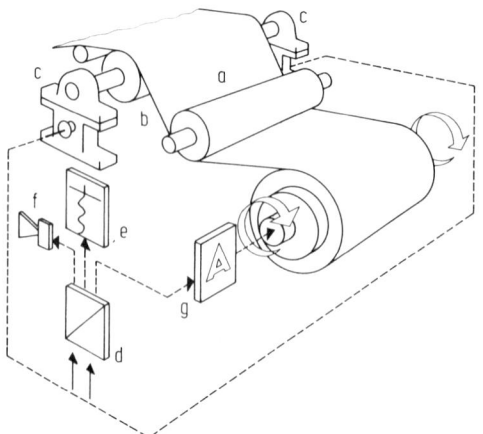

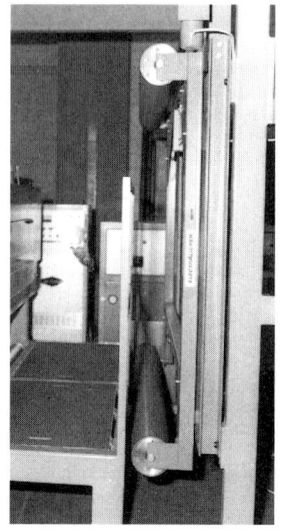

Figure 44. Functional diagram of a web tension measuring roll [70]
a film web, *b* measuring roll, *c* web tension load cell, *d* electronics unit, *e* display, print out, *f* alarm signal, *g* drive control

Figure 45. Film oscillating unit.
(Photo: Barmag, Remscheid, West Germany)

within the line to change the path of the web. The film edge follows an infrared sensor carried on a moving frame. Traverse path and oscillating speed are smoothly adjustable. The film path and the arrangement of an oscillating unit can be either horizontal or vertical (Figure 45). The oscillating unit is positioned after the thickness-measuring equipment so that there is a precise relationship on the running film web between the measuring point and the relevant position in the slot die, to optimize tolerances. The thickness-measuring sensors are non-contact devices that use radioactive isotope or infrared radiation in transmission, and traverse the running web in a carriage to provide a transverse profile of the film. An electronic evaluation system processes the signal supplied by the measuring head to provide target vs. actual value comparison, indication of deviations, minus or plus tolerances, and mean values. Measured values can be displayed continuously either on a chart recorder or on a visual display unit (VDU). Various operating programs for controlling the measuring head are available. Furthermore, it is possible, by the use of a microprocessor, to store product data and to control the take-off speed and/or the screw speed, e.g. as a function of the mean value of the measured profile. Correlation with the lip adjustment screws on the slot die is effected either by means of an appropriate scale on the traversing frame of the thickness-measuring unit or by a display on the monitor screen (Figure 46).

For many cast film applications, e.g. printing, coating or laminating, it is necessary to pretreat the film surface to facilitate good adhesion of print, colors or adhesives. The film surface tension can be modified on one or both sides by means of corona discharge [71, 72]. Special treatment stations are integrated into the line for this purpose. The corona discharge comes from electrodes connected to a high-voltage generator, usually located at several positions on the circumference of the roll around which the film is wrapped. Treatment intensity is variable in eight steps of 32 to 56 Nm/m and is checked on the film surface by means of test liquids [72]. The usual range is from around 38 to 44 Nm/m. The treatment intensity may be adversely affected by additives such as lubricants and anti-blocking agents.

By using segment electrodes the treatment can be limited to selected portions of the film width, if necessary: treatment is disadvantageous for the welding process. A double-sided treatment station is shown in Figure 47 [73]. Blowers serve to extract the ozone produced in the treatment station.

Corona treatment causes heating of the film, which requires intensive cooling before it can be wound. Wrap-round take-off systems comprising one or several driven cooling rolls, depending on line design, are used. The film web undergoes final trimming in the entry section of the take-up unit; the edge strips, and center strips from multireel slitting, are removed by an aspirator system.

Figure 46. Thickness scanning unit with VDU [88].
(Photo: Barmag, Remscheid, West Germany)

Figure 47. Corona treatment station [73].
(Photo: Barmag, Remscheid, West Germany)

The slitters used are of various designs and, in some cases, employed in conjunction with grooved rolls. Examples are shown in Figure 48. A special design of multiple-blade holder, or rotating holders, makes it possible to change blades without interrupting the slitting process. Between the longitudinal slitter and the take-up position it is necessary to provide a preliminary take-off unit to keep the tension required for film slitting and transport independent of the wind-up tension.

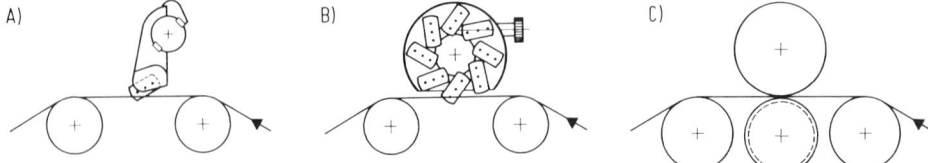

Figure 48. Designs of longitudinal slitting devices
A) industrial blade cut (single), B) industrial blade cut (revolver head), C) scissors cut

6.5.2.8 Take-off lines for thermoformable film and sheet

For the production of thermoforming film and sheet over a thickness range from 0.4 to 2.5 mm, dedicated lines of compact design, Figure 49 [74], are used as well as multipurpose lines which, in principle, can also be employed.

Figure 49. Take-off line of compact design for thermoformable film and sheet [74].
(Photo: Reifenhäuser, Troisdorf, West Germany)

Depending on the thermoplastic to be processed and the throughput range, the number of polishing and cooling rolls lies between two and five, with the first polishing roll usually having the smaller diameter (200 to 300 mm). The diameters of the cooling rolls range from 400 to 600 mm. With partially crystalline thermoplastics it may be necessary to choose the diameters of the first polishing and cooling rolls so as to achieve optimum crystallinity, but complete cooling of the film web has to take place at higher speeds on additional rolls. The roll surface normally has a highly polished, hard-chrome-plated finish. Widths range from 800 mm to 1600 mm.

After both sides of the film have been polished, the sheet usually passes through a thickness measuring station and is hauled off by a twin-roll take-off unit, the speed of which can be adjusted to the reference drive of the polishing/chilling unit. Edge trimming is done by means of circular blades which may be placed before or after the take-off unit. For temperature control of the polishing and cooling rolls, closed-circuit pressurized-water units, with a control range from 20 to 130 °C, are employed.

In some cases additional equipment for antistatic treatment and lamination of a glossy layer is used.

In the manufacture of thermoforming films less than 0.4 mm thick, the film is brought into contact with the chill roll by an air knife installed instead of the polishing roll (Figure 50) [75].

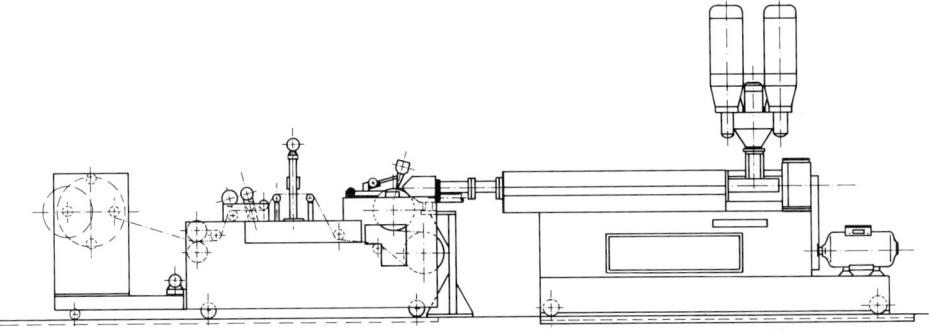

Figure 50. Line design for single- and multilayer thermoformable films < 400 μm; film pinning with air knife. (Photo: Barmag, Remscheid, West Germany)

Special requirements apply to extrusion lines for producing rigid PVC sheet. In addition to the specialized nature of the extrusion, designs of polishing roll-stack with high line-contact pressure and suitable roll construction are required. Figure 51 shows an extrusion line with cascade extruder and special roll stack for a film width of 1300 mm max. and a film thickness range of 0.2 to 1.5 mm. The roll stack is closed hydraulically in this case. If the roller conveyor and the stacker are adapted accordingly, this line can also be used for the production of thicker sheets from rigid PVC [75].

Figure 51. Polishing stack with cascade extruder for the production of films from rigid PVC. (Photo: Barmag, Remscheid, West Germany)

6.5.2.9 Wind-up equipment

Flat film lines employ various types of winders, differing in technical standard and degree of automation according to production speed, film material, film thickness range and package diameter. Films which are insensitive to web tension, usually those more than 200 µm thick, are wound at relatively low take-up speeds — usually lower than 30 m/min — on single- or multiposition centrally driven stationary winders. Film transfer on roll change is done by hand in this case. For thicker flat films and sheet for thermoforming, large roll winding systems (Figure 52) with hydraulic drives are also used, and package diameters of up to 2 m can be achieved [76].

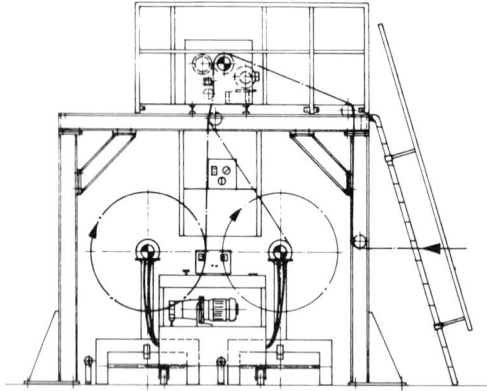

Figure 52. Large roll winding system with preliminary take-off and transverse cutter [76]

A more expensive design is the turret winder (Figure 53) which offers roll-change and operational advantages in comparison with stationary winders. Each winding position is D.C. motor driven with tension and winding-characteristic control. The changeover drive is usually a three-phase motor. The winding of tension-sensitive films at high production speeds requires higher-quality control systems for the winder drive, and precise film

6.5 Flat film production lines

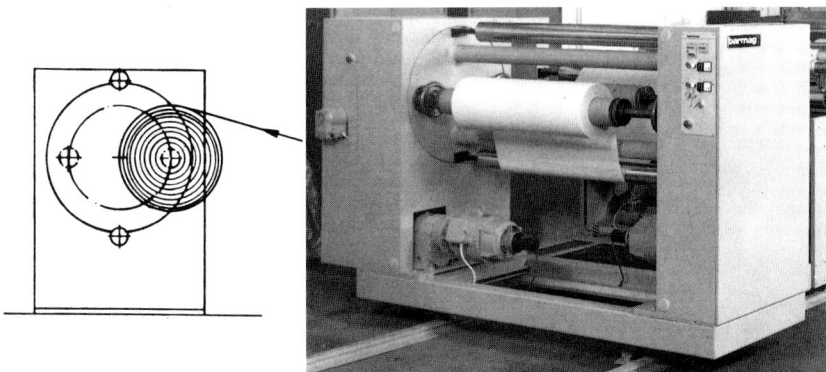

Figure 53. Twin-turret winder. (Photo: Barmag, Remscheid, West Germany)

guidance which is self-adjusting to the package diameter, ahead of the take-up position. Roll change, web changeover and cutting of the web, are fully automatic. In the case of the automatic twin-turret winder shown in Figure 54, which has an automatic roll-change system, a film guide is mounted in front of the take-up position, in a slide that is hydraulically withdrawn as the roll diameter increases. The guide roller either presses against the film package with a slight pneumatically adjustable contact pressure, or is held at a small distance from the surface of the film. This gap-winding procedure is preferred for pressure-sensitive films and is made possible by the use of a photocell which registers the package diameter and causes the guide slide to move away continuously. Instead of a guide slide, there are also designs where the lay-on or guide rolls are arranged in a hydro-pneumatically driven rocker arm [30].

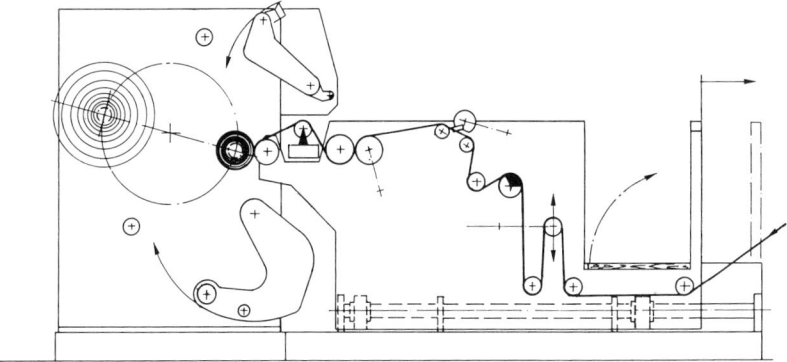

Figure 54. Turret winder with automatic roll-change system [77]

A short distance before the winding position is a D.C. driven preliminary take-off unit, which accepts the web movement- and cutting tension, and the tachogenerator, which transmits the web speed to the winder drive-control systems.

When the preset length of film – recorded by the length counter – has been reached, the new roll station is swung into position by a reversing drive and synchronized to winding speed. Then the web-transfer and cutting devices, Figure 55 [77], are engaged. If the film web is sufficiently wrapped round the new core, winding can be safely started after cutting with an impact knife even if the core surface has not been treated with an adhesive. If several ready-slit widths of film are to be wound on one core, precise, simultaneous transfer

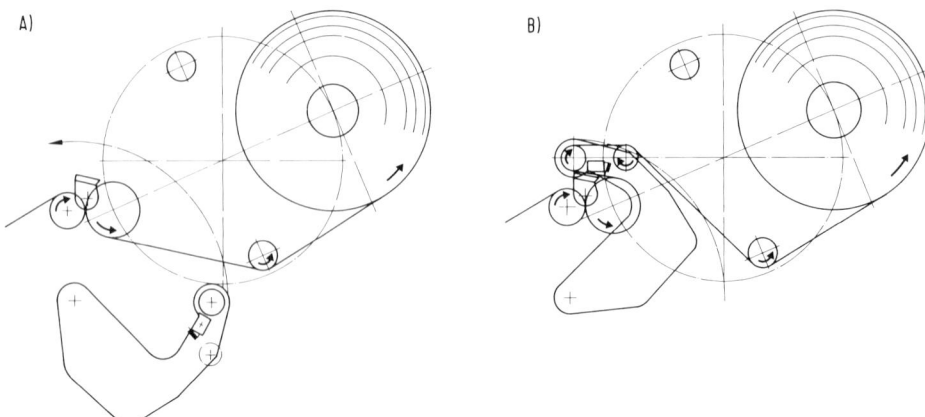

Figure 55. Roll-change system with automatic web transfer and cutting device [77]
A) basic position: full roll in removal position, empty core in waiting position, changeover system in rest position; B) change position: roll-change system engaged, serrated knife ready for cutting

Figure 56. Automatic turret winder [77]. (Photo: Stahlkontor Weser, Hameln, West Germany)

and cutting of the individual film webs is essential. After roll changeover has been completed, the finished film package can be taken from the rear position on a trolley or a lowering device. During the transfer procedure the guide roll is set to contact pressure. Definite, firm winding of the core layers is ensured by the central drive. Figure 56 shows one design of an automatic turret winder from the back, or operator side. The winding tension and winding characteristics are adjustable to suit the various film types. Information on web speed and package diameter is processed electronically, for the purpose of controlling the D.C. take-up drives and programming the winding characteristic. If there are strict precision requirements, direct web-tension control is used. Tension is monitored by a measuring roll or a dancer-roll system and the package taper-controlled to the programmed winding characteristic. The production of high-quality film packages requires cores with suitable concentricity and stability. The cores are held either by chucks at each end or on shafts which are usually capable of holding several cores if they are of an air-expandable design. Cores with an internal diameter of 70, 76 (3″), 152.4 (6″), 203.2 (8″) mm are available to suit the maximum package diameter and the package width to be wound. The winding shafts are supported in mechanically or pneumatically operated chucks. The package diameter usually reaches some 300 to 600 mm. With certain cast films, multireel winding can only be done effectively by using separate friction-drive elements on the winding shaft for each reel; even very small thickness variations will inevitably lead to variations in the diameters of the individual packages wound on a solid shaft, and this in turn will lead to different package hardnesses. An alternative to the technically elaborate winding of film reels by friction-drive elements is to put the finished full web width through a separate slitting process, using slitter-rewinder machines; but this does involve additional edge-trim waste and extra production time.

6.5.3 Measurement and control technology, and automation

In order to increase the productivity of extrusion lines, special efforts in measurement and control technology are required, as well as improvements in design. Attractive opportunities for automation are available from the use of microprocessors, which are very cost-effective [78, 79].

From an economic point of view the objectives of automation may include the following [80]:

- quality improvement,
- material savings,
- production increase,
- energy savings.

In film production the results are, in addition to tighter control of the processing parameters, simplification in operation and increased operational safety. The following paragraphs list the most important parameters of the flat film extrusion process and the measuring equipment used:

Melt temperature: measured by resistance thermometers or thermocouples in the adaptor system between screw and slot die [78].

Melt pressure: measured by pressure transducer [78], usually in two positions: in the adaptor between screw-tip and screen changer – the extruder is stopped by a cut-out switch upon reaching an adjustable maximum pressure – and in the adaptor before the slot die.

Temperatures: measured by resistance thermometer or thermocouples in the heating zones of the barrel, in the adaptor system, on the screen changer unit, on the slot die, at the entry and exit points of the heat-exchange fluid to casting and chill rolls, and in the air feed to the air knife.

Position of the air knife and pressure of the air, and

Roll gap and *contact pressure* in the case of polishing-nip lines.

Speeds and *power consumption* of the D.C. drives for extruder screw and metering units.

Speeds: measured by tachogenerators of analog or digital design or by pulse counters on the casting, cooling and take-off rolls.

Web tension: measured on rolls equipped with load cells, or by dancer rolls in the transport section and in the area ahead of the winder.

Liquid flow: measured in the cooling or heating agent at the casting and chill rolls, and in the liquid-controlled sections of the extruder.

Film thickness: continuous, non-contact measurement by traversing sensors using radiometric principles (infrared, β or γ radiation); display as longitudinal or transverse profile on a pen recorder or a monitor screen with printer attached.

Quality control: measurement of light transmission (transparency) on-line in the film transport section by Transmissiometer [80].

The measured values are registered on a multichannel recorder or, if a data logger is used, displayed on a monitor or printed for process monitoring purposes.

In addition to recording data, process control systems also provide closed-loop control of process values.

Control loops, for example for temperatures, pressures and speeds, are incorporated in the process computer as Direct Digital Control (DDC) and generally include a program for automatic heat-up, and for alterations of process data and controller parameters for start-up and shutdown operations, or production and/or format changes [81].

Alarm signals are given if preset limit values for the individual process parameters are exceeded in either direction, which ensures increased operational safety. For certain film products the operating data and line settings are stored on data loggers and can be called up as needed. This allows the starting time to be reduced because the line is automatically adjusted to optimum production conditions. Comparison with previously stored data simplifies the elimination of faults and optimization of process control.

The use of an adaptive control (AC) system utilizing statistical or physical process models permits optimization which is independent of the operating personnel, i.e. automatic adaptation of the control system to altered process conditions [80, 82, 83]. This requires that the quality-determining properties of the product (e.g. film thickness, transparency) be pre-calculated by means of a process model from process values which can be measured without delay, and be used as target values for control purposes. Subsequent measurement of these values permits checking and correction of the process models. Furthermore, the controller parameters can be optimized. In practical application, automatic control of film thickness in longitudinal and transverse directions has now been very widely accepted for use on flat film extrusion lines [78, 82].

Automatic film-thickness control

When an automatic die (Figure 57) is used, the film thickness can be optimized automatically if the system includes a thickness scanner and a microprocessor [84, 85, 86]. The scanner – which usually works on the radiometric principle – transmits the actual film-thickness transverse profile to the computer; this compares target and actual values and then corrects the die gap setting. The lip gap is adjusted by expansion bolts; the number of bolts depends on the die width, and they are marked in the thickness-profile display on the monitor screen. Each bolt is fixed in a block which is equipped with a blown-air channel and an electric cartridge-heater. Another design has the cartridge located along the axis of the adjustment screw [87].

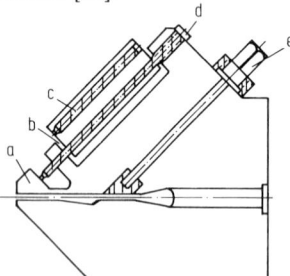

Figure 57. Arrangement of adjustment devices with an automatic die
a flexible lip, *b* cooling channel, *c* heating cartridge, *d* lip adjustment, *e* restrictor bar adjustment

Figure 58 shows the schematic layout of an automatic die-control system for the flat film extrusion process. The heat supplied to the individual expansion bolts is controlled by the computer in accordance with the control algorithm so that the deviation from target value (thickness tolerance) of the film across the width is minimized. The maximum adjustment range with automatic lip-gap adjustment is usually about ± 0.25 mm; greater adjustments of the die gap must be carried out manually.

Figure 58. Scheme for automatic thickness control for the flat film extrusion process
a thickness scanner, *b* take-off unit cooling roll, *c* die with expansion bolt, *d* extruder, *e* drive, *f* tachogenerator, *g* power adjuster for expansion bolt, *h* thickness measurement, *i* control unit and printer, *k* microprocessor, *l* monitor

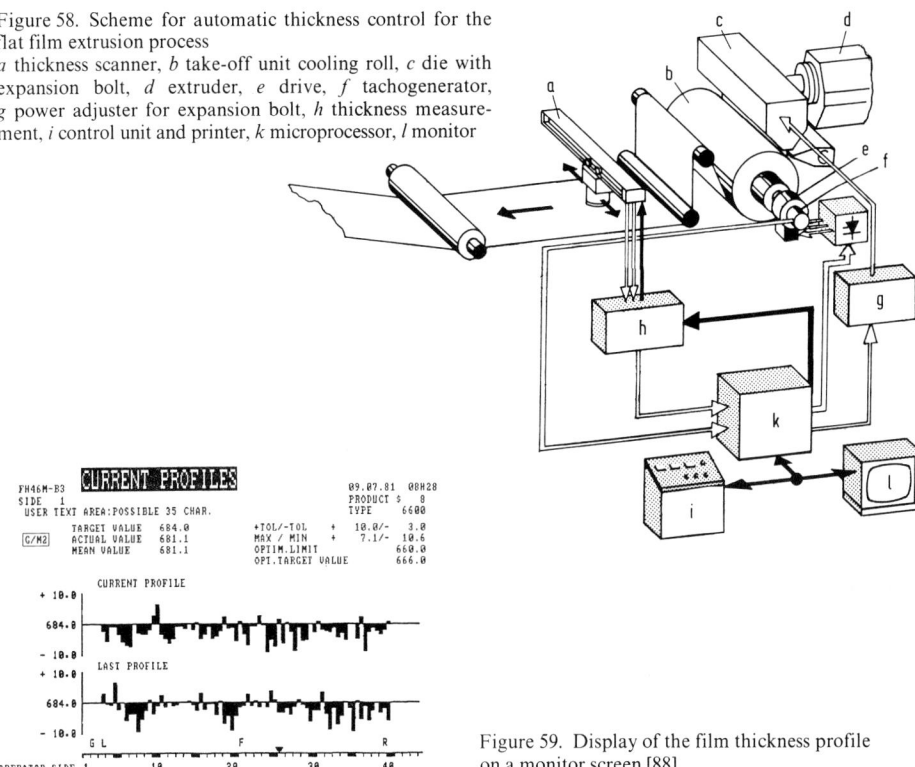

Figure 59. Display of the film thickness profile on a monitor screen [88]

Optimization is carried out by simultaneously adjusting expansion bolts (number depending on the control program) with the effect of their reciprocal influence and film neck-in being taken into account. Figure 59 shows the monitor screen display, with profile, profile memory and product parameters scanned by the measuring system [88]. The upper part of the picture shows the last transverse thickness profile measured, the lower part the profile from the previous traverse. The display also shows the relevant mean value of thickness.

Other information available on the VDU includes, for example: *Trend diagram* (Figure 60) which shows the thickness profile averaged over a certain production period, and *Roll record* (Figure 61) which provides an average of all profiles measured during the running period

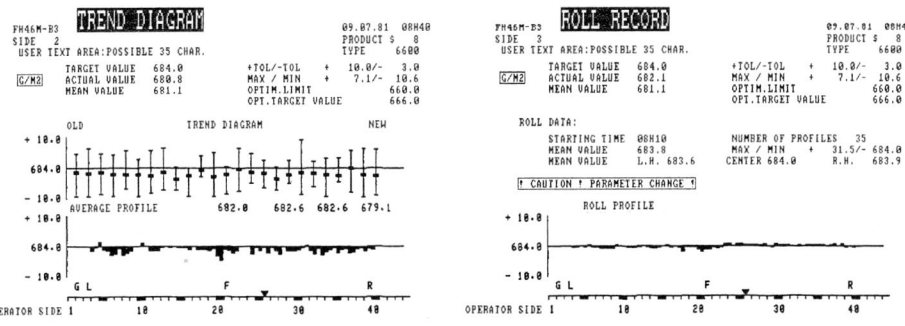

Figure 60. Trend diagram [88]

Figure 61. Roll record [88]

of the current roll [88]. To indicate the operating condition of the automatic die, the monitor shows the temperature and/or the relevant heating power of the expansion bolts as a bar chart.

Further optimization of the film thickness in conjunction with a computer-aided thickness measuring system can take place in the longitudinal direction (MD). The take-off speed, and in some cases the extruder screw speed as well, are used as adjustment elements for MD thickness deviations.

In the case of an automatic control system, e.g. for take-off speed as a function of film thickness, the scatter of actual thickness can be reduced, and consequently the target value can be brought closer to some specification minimum thickness; this can result in considerable savings in materials.

Selection of the equipment for automated process control depends greatly on the existing production conditions and should take the following aspects into consideration:
- good ratio between extra expenditure for MC technology and product improvement (quality and price),
- optimum price/efficiency ratio of the system,
- high degree of reliability of MC technology/self-monitoring/self-testing,
- high degree of efficiency of software/good software maintenance,
- reliable and fast repair service,
- clear layout of line/operator-friendliness/qualifications required of operators,
- compatibility of MC technology with overall design.

6.5.4 Special designs of flat film lines

Some more specialized line components are outlined below:

Take-off units for embossed flat films

Special films, like those made of polyethylene for the hygiene sector (baby pants, hospital accessories, etc.), are either embossed on separate lines or produced directly during extrusion on a casting roll having the appropriate embossing pattern. The film is pressed against the casting roll by a rubberized roll. The line otherwise corresponds to a normal chill roll line [89].

Figure 62. Single-roll casting of film for biaxial drawing. (Photo: Barmag, Remscheid, West Germany)

6.5 Flat film production lines

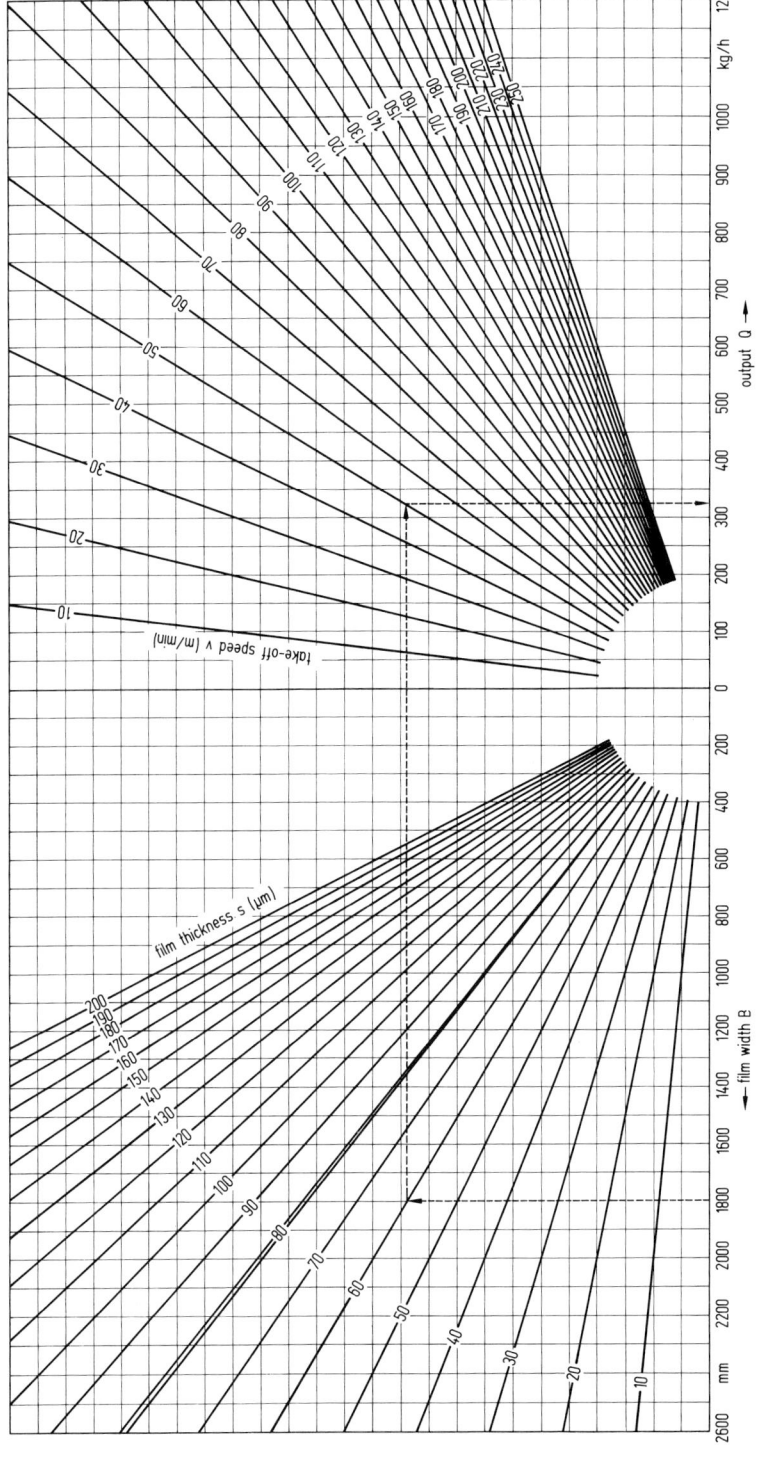

Figure 63. Nomogram: Correlation of output Q (kg/h), take-off speed v (m/min), film width B (mm) and film thickness s (μm) for cast flat films in the thickness range from 10 to 200 μm. Specific weight $\varrho = 1.0$ g/cm^3

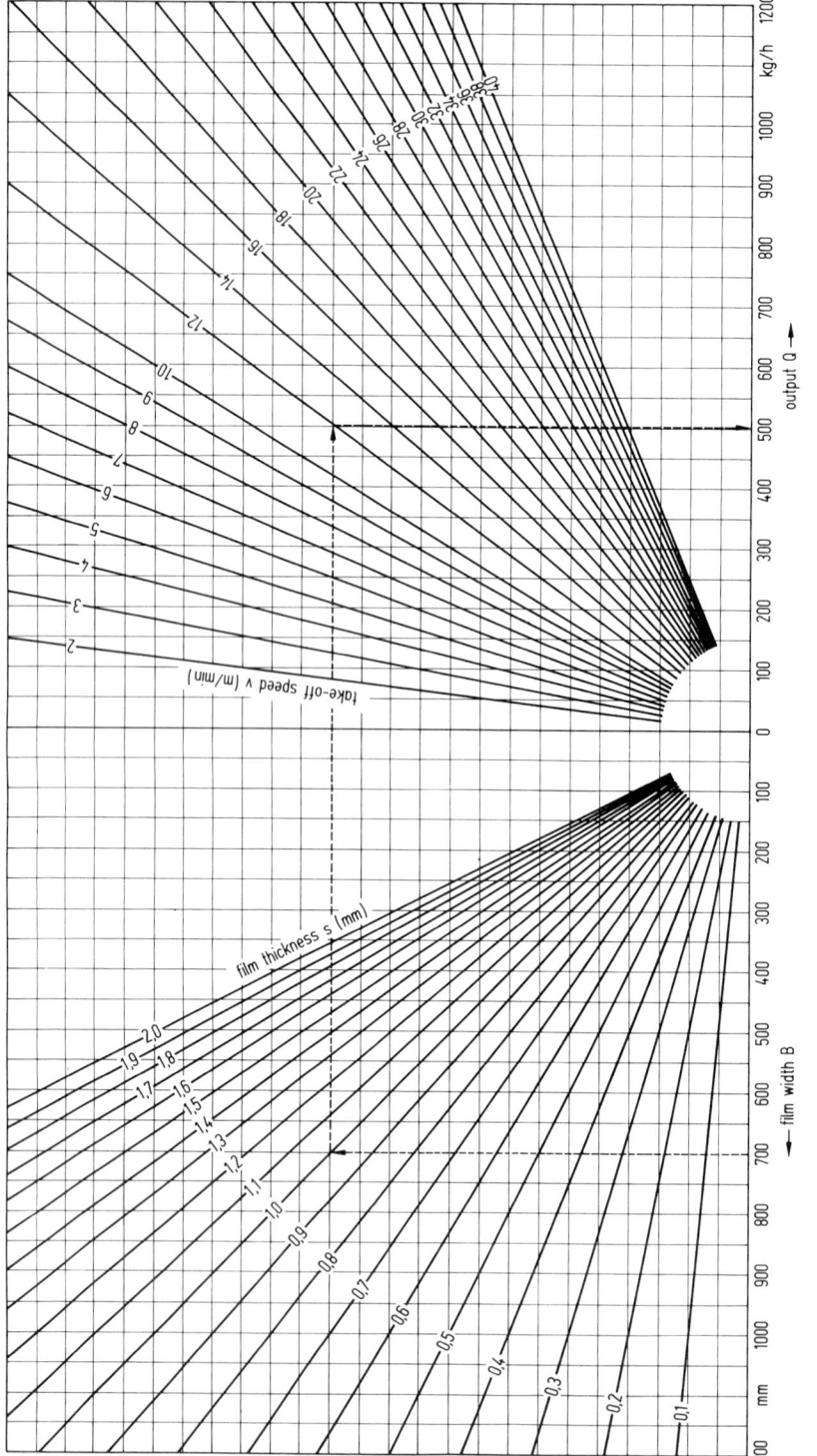

Figure 64. Nomogram: Correlation of output Q (kg/h), take-off speed v (m/min), film width B (mm) and film thickness s (mm) for flat films and sheet in the thickness range of 0.1 to 2.0 mm. Specific weight $\varrho = 1.0$ g/cm^3

6.5 Flat film production lines 197

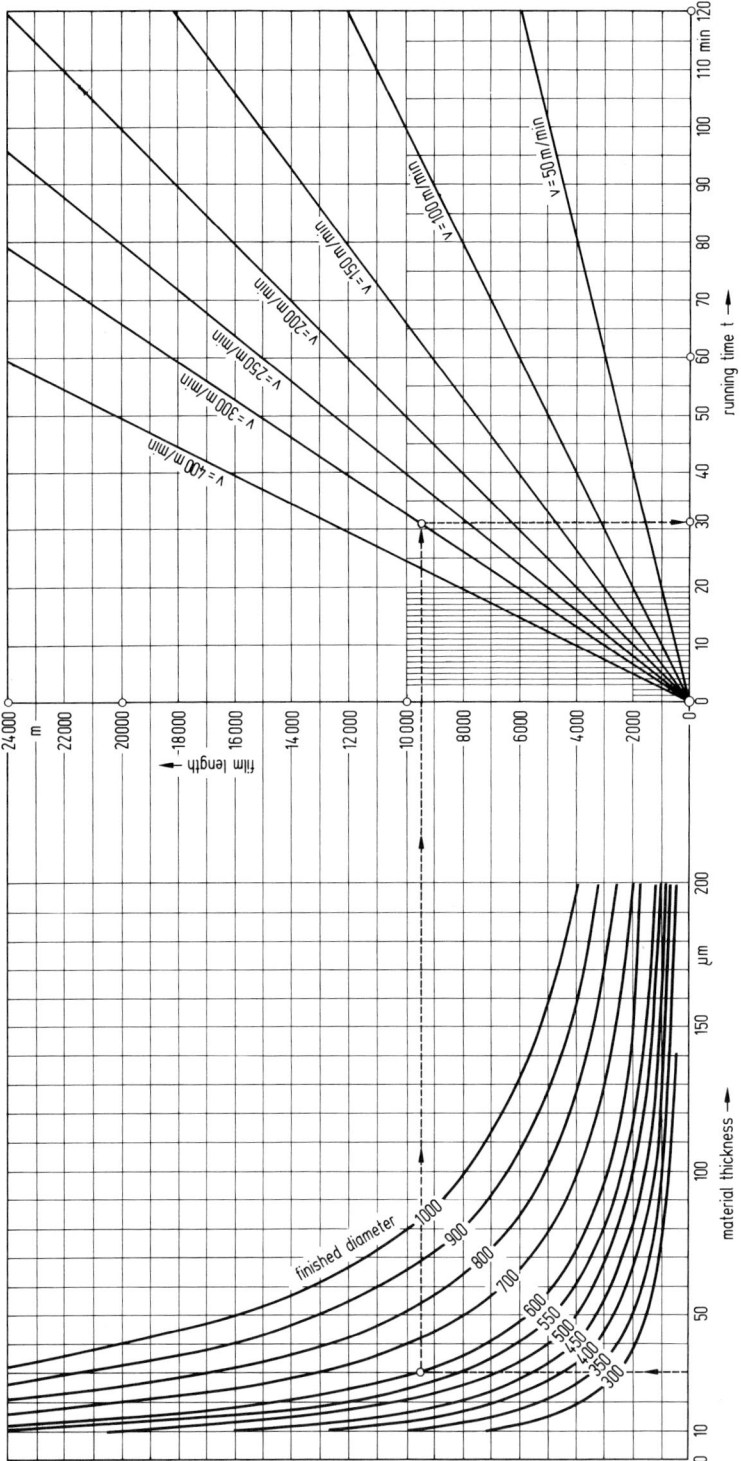

Figure 65. Nomogram: Correlation between material thickness, finished package diameter, film length, take-off speed and running time [90]

Water-bath cooling

Flat films can be cooled in a water bath for certain applications instead of by a chill roll. Cast PP or HDPE film for in-line drawn-tape lines, is made in a water bath to provide balanced cooling on the two sides. The water that is carried along must be thoroughly removed from the moving film.

Take-off lines for biaxially orientable films

Cast polypropylene or polyester films for biaxial orienting lines are produced on the chill roll principle. Figure 62 shows a single-roll take-off unit with adjustable lift-off roll. The diameter of the casting roll lies between 800 and 2400 mm, depending on the material to be processed, film thickness range and specific cooling capacity. Because of the different stretch ratios that can be applied later, line widths in the range between 800 and 1200 mm are used for PP and between 800 and 2000 mm with PET. For the manufacture of thicker feed-films (0.8 to 2.5 mm), additional cooling of the side of the film not in contact with the roll, by means of air showers or water baths is required. PP films are held against the casting roll by an air knife, and PET films by electrostatic pinning devices.

References for Chapter 6

For the convenience of the reader the English titles of all publications in languages other than English are shown in parentheses.

[1] *Rahlfs, H.:* Extrudieren von ein- und mehrschichtigen Flachfolien (Extrusion of Single- and Multilayer Flat Films). In: Extrudierte Feinfolien und Verbundfolien, Reihe: Ingenieurwissen (Extruded Fine Films and Composite Films in the Engineering Know-How Series), VDI-Verlag, Düsseldorf, 1973.

[2] *Predöhl, W.:* Folienextrusion mit Coextrusion (Coextrusion of Films), VDI-Verlag, Düsseldorf, 1980.

[3] *Ficher, P., Görner, H. P., Herner, M., Kosel, U.:* Maschinen- und verarbeitungstechnische Fortschritte bei Herstellung von Kunststoff-Tafeln und -Folien (Progress in Machinery and Processing Technology for Manufacturing Sheet and Film), Kunststoffe 61 (1971), pp. 342/355.

[4] *Menges, G., Hegele, R., Kosel, U.:* Gedanken zu einem neuen Extruderkonzept (Thoughts on a New Extruder Concept). Kunststoff-Rundschau 7 (1971).

[5] *Rautenbach, R., Pfeiffer, H.:* Modellrechnung zur Auslegung der förderwirksamen genuteten Einzugszone von Einschneckenextrudern (Model Design Calculations on the Conveying Efficiency of Grooved Feed Zones on Single-Screw Extruders). Kunststoffe 72 (1982), pp. 137/143.

[6] *Rautenbach, R., Pfeiffer, H.:* Durchsatz- und Drehmomentverhalten genuteter Einzugszonen von Einschneckenextrudern (Throughput and Torque Behavior of Grooved Feed Zones of Single-Screw Extruders). Kunststoffe 72 (1982), pp. 262/265.

[7] *Krämer, A.:* Optimieren der verfahrenstechnischen Auslegung von Einschneckenextrudern (Optimization of the Design of Single-Screw Extruders). Kunststoffe 68 (1978), pp. 12/19.

[8] *Meißner, M.:* Systeme zur Massetemperatur-Regelung am Extruder (Systems for Control of Melt Temperature on Extruder). Kunststoff-Berater 16 (1971), pp. 1149/1156.

[9] *Hensen, F., Gathmann, E.:* Fortschritte in der Extrusionstechnik durch Optimieren getrennter Verfahrensschritte (Progress in Extrusion Technology by Optimization of the Individual Steps in the Process). Kunststoffe 64 (1974), pp. 343/349.

[10] *Knappe, W., Schönewald, H.:* Anwendung der temperaturinvarianten Auftragung rheologischer Daten für die Auslegung von Düsen (Use of a Temperature-invariant Plot of Rheological Data in the Design of Dies). Kunststoffe 60 (1970), pp. 657/665.

[11] *Menges, G.:* Werkstoffkunde der Kunststoffe (Materials Data on Plastics). Carl Hanser Verlag, München, Wien, 1979.
[12] *Schenkel, G., Kühnle, H.:* Zur Bemessung der Bügellängenverhältnisse bei Mehrkanal-Extrudierwerkzeugen für Kunststoffe (On the Calculation of Die Land Dimensions in Multichannel Extrusion Dies for Plastics). Kunststoffe 73 (1983), pp. 17/22.
[13] *Görmar, H.:* Beitrag zur verarbeitungsgerechten Dimensionierung von Breitschlitzwerkzeugen für thermisch instabile Thermoplaste, insbesondere PVC-hart (On the Dimensioning of Slot Dies for Processing Thermally Unstable Thermoplastics, Especially Rigid PVC). Dissertation RWTH Aachen, 1968.
[14] *Wortberg, J.:* Werkzeugauslegung für Ein- und Mehrschichtextrusion (Die Design for Single- and Multilayer Extrusion). Dissertation RWTH Aachen, 1978.
[15] *Wortberg, J.:* Breitschlitzwerkzeuge für Ein- und Mehrschichtextrusion (Slot Dies for Single- and Multilayer Extrusion), in: Berechnen von Extrudierwerkzeugen, Reihe: Ingenieurwissen (Calculations for Extrusion Dies, in the Engineering Know-how Series). VDI-Verlag, Düsseldorf, 1978.
[16] *Masberg, U.:* Einsatz zur Methode der finiten Elemente zur Auslegung von Extrusionswerkzeugen (Use of the Finite Element Method for the Design of Extrusion Dies). Dissertation RWTH Aachen, 1981.
[17] *Wortberg, J.:* Werkzeugauslegung mit rheologischen Prüfdaten und Stoffwertefunktionen (Die Design Using Rheological Test Data and Materials Functions). Plastverarbeiter 34 (1983), pp. 1220/1224.
[18] *Wortberg, J., Tempeler, K.:* Breitschlitzwerkzeug mit großer Verarbeitungsbreite (Slot Dies with a Wide Operating Range). Kunststoffe 73 (1983), pp. 404/406.
[19] *Michaeli, W.:* Zur Analyse des Flachfolien- und Tafelextrusionsprozesses (An Analysis of the Process for Flat Film and Sheet Extrusion). Dissertation RWTH Aachen, 1975.
[20] *Hensen, F.:* Kühlen von Tafeln, Tiefziehfolien und Reckfolien (Cooling of Sheets, Thermoforming Film and Stretch Film). In: Kühlen von Extrudaten. Reihe: Ingenieurwissen (Cooling of Extrudates, in the Engineering Series). VDI-Verlag, Düsseldorf, 1978.
[21] *Haberstroh, E.:* Analyse von Kühlstrecken in Extrusionsanlagen (Analysis of Cooling Sections on Extrusion Lines). Dissertation RWTH Aachen, 1981.
[22] *Menges, C., Haberstroh, E., Jancke, W.:* Systematische Auslegung von Kühlstrecken in Folien-, Tafel- und Rohrextrusionsanlagen (Systematic Design of Cooling Sections for Film-Sheet-, and Pipe Extrusion Lines). Kunststoffe 72 (1982), pp. 332/336.
[23] *Dietz, W.:* Bestimmung der Wärme- und Temperaturleitfähigkeit von Kunststoffen bei hohen Drücken (Determination of the Heat- and Temperature Conductivity of Plastics at High Pressures). Kunststoffe 66 (1976), pp. 161/167.
[24] *Knappe, W.:* Die thermischen Eigenschaften von Kunststoffen und ihre Bedeutung für die Kunststoffverarbeitung (The Thermal Properties of Plastics and their Importance in Plastics Processing). Kunststoffe 66 (1976), pp. 297/304.
[25] *Ast, W.:* Einfache Methode zur Auslegung von Kühlstrecken bei der Kunststoff-Extrusion (Simple Design Methods for Cooling Sections on Plastics Extrusion Lines). Kunststoffe 69 (1979), pp. 186/193.
[26] *Michaeli, W., Junk, P. B., Wortberg, J., Dierkes, A., Predöhl, W.:* Produktbeeinflussung durch Werkzeug und Folgeaggregate bei der Extrusion (The Influence of the Die and Downstream Equipment on the Extruded Product). Plastverarbeiter 27 (1976), pp. 490/495.
[27] *Wortberg, J.:* Berechenbarkeit des Extrudier- und Tiefziehprozesses (The Calculability of the Extrusion and Thermoforming Processes). In: Extrudieren und Tiefziehen von Packmitteln. Reihe: Ingenieurwissen (Extrusion and Thermoforming of Packaging Material, in the Engineering Technology Series). VDI-Verlag, Düsseldorf, 1980.
[28] *Kähler, M.:* Herstellung einschichtiger Flachfolien auf Extrusionsanlagen (Manufacture of Single-layer Flat Films on Extrusion Lines). Company brochure, Maschinenfabrik Stahlkontor Weser Lenze GmbH. & Co., D-3250 Hameln 1 (1981).
[29] *Klein, H.:* Herstellung selbstklebender Materialien − wickeltechnische Probleme (Production of Self-adhesive Materials − Winding Problems). Plastverarbeiter 27 (1976), pp. 375/382.
[30] *Weiss, H. L.:* Control Systems for Web-fed Machinery. Converting Technology Company, Milwaukee, WI, USA, 1983.
[31] *Kühne, G.:* Verpacken mit Kunststoffen (Plastics Packaging). Carl Hanser Verlag, München, Wien, 1974.

[32] *Schricker, G.:* Anforderungen an Verpackungsfolien zum Schutze des Packgutes (Specifications for Films Used in Protective Packaging). In: Verpacken mit Kunststoff-Folien. Reihe: Kunststofftechnik (Packaging with Plastics Films, in the Plastics Technology Series). VDI-Verlag, Düsseldorf, 1982.
[33] *Predöhl, W.:* Herstellen extrudierter Folien (Manufacture of Extruded Films). In: Verpacken mit Kunststoff-Folien. Reihe: Kunststofftechnik (Packaging with Plastics Films, in the Plastics Technology Series). VDI-Verlag, Düsseldorf, 1982.
[34] *Predöhl, W.:* Technologie extrudierter Kunststoff-Folien (Technology of Extruded Plastics Films). VDI-Verlag, Düsseldorf, 1979.
[35] *Harnier, A. v.:* Kühlen von Flachfolien (Cooling of Flat Films). In: Kühlen von Extrudaten. Reihe: Ingenieurwissen (Cooling of Extrudates, in the Engineering Know-how Series). VDI-Verlag, Düsseldorf, 1978.
[36] *Hensen, F., Bongaerts, H.:* Entwicklungsstand bei der Breitschlitzfolienextrusion (The State of Development of Slot-die Film Extrusion). Plastverarbeiter 30 (1979), pp. 441/449.
[37] *Beijen, J. M., Oepkes, J.:* Ein neues Formverfahren für PP-Folien (A New Shaping Process for PP Films). Kunststoffe 65 (1975), pp. 666/669.
[38] *Hensen, F., Hessenbruch, R., Bongaerts, H.:* Entwicklungsstand bei der Koextrusion von Mehrschichtblasfolien und Mehrschichtbreitschlitzfolien (The State of Development in Coextrusion of Multilayer Blown Films and Multilayer Slot-die Films). Kunststoffe 71 (1981), pp. 530/538.
[39] Company brochure Dow Chemical, USA: Feedblock Coextrusion. 1976.
[40] *Sneller, J.:* New-breed Coextrusions are Low in Cost, High in Performance. Modern Plastics International (1981) 12, pp. 46/48.
[41] *Lanquetot, H.:* New Barrier Structures for Rigid Containers. In: COEX '83, Conference Handbook, Düsseldorf, Oct. 1983. Scotland Business Research, Inc., Princetown, N.J., U.S.A.
[42] *Roder, H.:* Kunststoff-Folien in der Verpackungstechnik (Plastics Films in Packaging Technology). In: Verpacken mit Kunststoff-Folien. Reihe: Kunststoff-Technik (Packaging with Plastics Films, in the Plastics Technology Series). VDI-Verlag, Düsseldorf, 1982.
[43] *Kautz, G.:* Das Eigenschaftsbild der extrudier- und tiefziehbaren Kunststoffe (Properties of Extruded, Thermoformable Plastics Films). In: Extrudieren und Tiefziehen von Packmitteln. Reihe: Ingenieurwissen (Extrusion and Thermoforming of Packaging Materials, in the Engineering Know-how Series). VDI-Verlag, Düsseldorf, 1980.
[44] *Ikari, K.:* Oxygen Barrier Properties and Applications of Kuraray EVAL Resins. In: COEX '82, Conference Handbook, Princetown, N.J., Nov. 1982, Scotland Business Research, Inc., Princetown, N.J., U.S.A.
[45] *Maack, H., Rogiers, E.:* High Barrier Coextrusions for Retortable Food Packaging. In: COEX '83. Conference Handbook, Düsseldorf, Oct. 1983. Scotland Business Research, Inc., Princetown, N.J., U.S.A.
[46] *Predöhl, W., Herres, N.:* Extrudieren von Flachfolien, Tiefziehfolien und Tafeln (Extrusion of Flat Films, Thermoformable Films, and Sheets). Kunststoffe 71 (1981), pp. 659/665.
[47] *Harnier, A. v.:* Einfluß des Kühlens auf Folieneigenschaften (Influence of Cooling on Film Properties). Kunststoffberater 4 (1980), pp. 32/36.
[48] *Wellenhofer, P.:* Monoaxiales und biaxiales Recken mit besonderer Berücksichtigung des biaxialen Flachfolienstreckens (Uniaxial and Biaxial Drawing with Particular Reference to Stretching of Flat Films). In: Folien, Gewebe, Vliesstoffe aus Polypropylen. Reihe: Ingenieurwissen (PP Films, Woven-, and Non-woven Fabrics, in the Engineering Know-how Series). VDI-Verlag, Düsseldorf, Düsseldorf, 1979.
[49] *Michaeli, W., Junk, P. B., Wortberg, J., Dierkes, A.:* Produktbeeinflussung durch Werkzeug und Folgeaggregate bei der Extrusion (The Susceptibility of the Product to the Die and Downstream Equipment on Extrusion Lines). Plastverarbeiter 27 (1976), pp. 529/538.
[50] *Kautz, G., Schumacher, F.:* Extrudieren von Flachfolien und Tafeln (Extrusion of Flat Films and Sheets). Kunststoffe 67 (1977), pp. 588/593.
[51] *Predöhl, W.:* Extrudieren von ein- und mehrschichtigen Schlauchfolien (Extrusion of Single- and Multilayer Tubular Films). In: Extrudierte Feinfolien und Verbundfolien. Reihe: Ingenieurwissen. (Extruded Fine Films and Multilayer Films, in the Engineering Know-how Series). VDI-Verlag, Düsseldorf, 1976.

[52] Company brochure Barmag, Barmer Maschinenfabrik AG., Remscheid, West Germany.
[53] Company brochure Adolf Illig, Maschinenbau (Reifenhäuser), Heilbronn, West Germany.
[54] Company brochure Colortronic Reinhard + Co., Friedrichsdorf/Ts., West Germany.
[55] Company brochure Process Control Corporation, Atlanta, U.S.A.
[56] Company brochure Reifenhäuser GmbH. & Co., Troisdorf, West Germany.
[57] Company brochure Plastik Maschinenbau GmbH, Kelberg, West Germany.
[58] Company brochure J. Kreyenberg & Co. GmbH, Münster-Kinderhaus, West Germany.
[59] Company brochure Bolton-Emerson, SA, Lausanne, Suisse.
[60] *Hensen, F., Siemetzki, S.:* Verbesserte Extrusionstechnologie durch Großflächen-Schmelzefiltration (Extrusion Technology Improved by Use of Large-area Melt Filtration). Kunststofe 70 (1980), pp. 753/758.
[61] *Schneider, K.:* Zahnradpumpen für Polymerschmelzen und -lösungen (Gear Pumps for Polymer Melts and Polymer Solutions). Kunststoffe 68 (1978), pp. 201/206.
[62] *McKelvey, J. M., Maire, U., Haupt, F.:* How Gear Pumps and Screw Pumps Perform in Polymer Processing Applications. Chemical Engineering 83 (1976) 27, pp. 94/101.
[63] Neue Präzisions-Zahnrad-Förderpumpen (New Precision Conveying Gear Pump). Barmag, Remscheid, West Germany.
[64] *Sneller, J. A.:* Extruder Add-ons and Design Variations Squeeze Out Big Productivity Bonuses. Modern Plastics International (1983) 12, pp. 26/28.
[65] Extrusion Dies Inc., Chippewa Falls, Wisconsin, USA.
[66] Coextrusion à flux laminaire (Coextrusion with Laminar Flow). Plastiques Flash (1980) No. 132, pp. 82/85.
[67] Coextrusion Takes a Giant Step Into the Future. Modern Plastics International (1983), Aug., pp. 14/18.
[68] *Coeren, P.:* Performance and Economics of Barrier Coextrusion Coatings. In: COEX '83, Conference Handbook Düsseldorf, Oct. 1983. Scotland Business Research, Inc., Princetown, N.J., USA.
[69] E. Derichs, Maschinen- und Apparatebau, Krefeld-Fischeln, West Germany.
[70] *Hohmann, H.:* Bahnspannungsmessung (Web Tension Measurement). Plastverarbeiter 32 (1981), pp. 989/992.
[71] *van der Linden, R.:* Die Corona-Behandlung von PE-Folien (Corona Treatment of PE Films). Kunststoffe 69 (1979), pp. 71/75.
[72] Prüfung des Vorbehandlungseffektes und der Druckfarbenhaftung bei Hohlkörpern aus Polyethylen (Testing of Surface Treatment Effects and Printing-Ink Adhesion on Polyethylene Blow Molded Containers). BASF-Kunststoffe, TI-TKS-06d, 82313, Dec. 1980.
[73] Company brochure Kalwar GmbH & Co. KG, Halle, West Germany.
[74] Company brochure Reifenhäuser GmbH & Co., Troisdorf, West Germany.
[75] Company brochure Barmag, Barmer Maschinenfabrik AG, Remscheid, West Germany.
[76] *Maier, E., Primeßing, H.:* Rationalisieren durch verbesserte Wickeltechnik bei der Produktion und Verarbeitung von Kunststoff-Folien (Rationalization by Improved Winding Techniques in the Production and Conversion of Plastics Films). Kunststoffe 72 (1982), pp. 693/695.
[77] Company brochure Stahlkontor Weser Lenze GmbH & Co., Hameln, West Germany.
[78] *Wiegand, H. G.:* Prozeßautomatisierung beim Extrudieren und Spritzgießen von Kunststoffen (Automation of the Extrusion and Injection Molding Processes for Plastics). Carl Hanser Verlag, München, Wien, 1979.
[79] *Schwab, E.:* Mikrorechnereinsatz zur Prozeßführung bei Extrusionsanlagen (Use of Microprocessors for Controlling Extrusion Lines). Plastverarbeiter 34 (1983), pp. 736, 739.
[80] *Predöhl, W.:* Optimierung des Betriebsverhaltens von Flachfolienanlagen mit Hilfe eines Rechners (Optimization of the Operational Behavior of Flat Film Lines with the Help of a Computer). In: Rechnergesteuerte Extrusionsanlagen. Reihe: Kunststofftechnik (Computer-controlled Extrusion Lines, in the Plastics Technology Series). VDI-Verlag, Düsseldorf, 1981.
[81] *Dormeier, S.:* Aufbau von Prozeßrechnerprogrammen zur Steuerung von Extrusionsanlagen (The Design of Computer Programs for Controlling Extrusion Lines). In: Rechnergesteuerte Extrusion? Reihe: Ingenieurwissen. (Computer-controlled Extrusion? in the Engineering Know-how Series). VDI-Verlag, Düsseldorf, 1976.

[82] *Bergweiler, E.:* Prozeßsteuerung bei der Flachfolien- und Tafelextrusion durch Mikrorechner (Control of the Flat Film- and Sheet Extrusion Processes by Computer). Dissertation RWTH Aachen, 1981.
[83] *Lemke, H.-J.:* Regelungstechnische Methoden bei der Automatisierung von Kunststoffverarbeitungsmaschinen (Methods of Control in the Automation of Plastics Processing Machinery). Kunststoffe 71 (1981), pp. 355/359.
[84] *Sneller, J.:* Getting More for the Money in Sheet and Film Gaging. Modern Plastics International (1983), March, pp. 36/39.
[85] Gauge Bands Eliminated Through Automatic Die Control. Plastics Technology (1981) Sept.
[86] *Galli, E.:* Inline Thickness Control. Part I: Sheet extrusion. Plastics Machinery & Equipment (1982) May, pp. 23/28.
[87] Company brochure Egan Machinery, Sommerville, N.J., U.S.A.
[88] Company brochure FAG Kugelfischer, Georg Schäfer & Co., Erlangen, West Germany.
[89] Company brochure Egan Machinery, John Brown Plastics Machinery, Sommerville, N.J., U.S.A.
[90] Bahnlängenspeicherung in Vorrats- und Fertigrollen. Der Einfluß der Hülsendurchmesser (Storage of Continuous Webs on Stock-rolls and Finished Rolls. The Influence of Core Diameter). Papier + Kunststoff-Verarbeiter 12 (1971), pp. 42/48.

7 Production of films and sheets by the roll-stack process

H. Breuer

7.1 Introduction

Films and sheets are produced as flat webs which are wound up as intermediate products, or cut into sheets or boards, as appropriate. In this context when we speak of films we mean flat films, in contrast to blown tubular film, but only so long as the web can be wound up. And this is usually possible, with some variation from plastic to plastic, up to about 2 mm thickness. For greater thicknesses one speaks of sheet. In the lower range of thicknesses from 0.02 to 0.2 mm, films are often called fine films or thin films [1].
Such unoriented films are produced on lines of various designs, as follows:

- cast film lines, also known as chill-roll lines, for film 0.02 to 0.5 mm thick,
- flat film lines for film from 0.2 to 2 mm thick,
- calender lines for thicknesses from 0.05 to about 0.8 or 1 mm,
- extrusion lines for film thicknesses from 0.5 to 2 mm, and sheets from 2 to about 25 mm thick.

Since the first two kinds of process are dealt with in Chapter 6, only the last two are discussed in what follows in this chapter.
A common feature of calendering and sheet extrusion lines is the roll stack; but its calendering function is distinct from the polishing function it carries out on extrusion lines.
The materials normally processed on sheet extrusion lines are: PS, ASA, SB, PVC, PMMA, PE, PP, PC, CAB, and PA. And on calender lines: soft, semi-rigid, and rigid PVC [2]. The intermediate products made from these materials are transformed into containers, packaging, beakers, and components for use in the building, automotive, and engineering industries; they are also made into furniture. Films and sheets can be produced with flat, glossy or matt surfaces on one or both sides, or with embossed surfaces [3]. The special mechanical and processing techniques required for efficient processing of the plastics materials mentioned, and the relevant features of the machines, are described below.

7.2 Roll-stack arrangements

The differences between roll stacks on calendering and on sheet extrusion lines, apart from their characteristic size, are found in the number and arrangement of the rolls, the length/diameter ratio of the rolls, the operating loads, the temperature control, the roll-bearings and roll-axis movements, and in usage efficiencies.
Calenders and polishing stacks on sheet extrusion lines create the plastic web from the melt, determine its size and polish it. However, there are differences in the way the roll stack is fed, and in the state of the product web as it is taken off the last roll. The different types of calenders are:

- melt roll-laminating calenders, and
- film-drawing calenders,

and the different types of polishing stack:

- cooling and polishing arrays,
- temperature-conditioning stacks.

It can be seen, therefore, that roll-stack treatment can be matched to the processing characteristics of each material.

7.2.1 Polishing-stack operations

The polishing stack, shown as a component of the sheet extrusion line in Figure 1, is fed with the viscoelastic melt from the slot die. To obtain perfect film or sheet quality, thorough plastication, dispersion and homogenization of the melt must be achieved in the extruder, and the product web must be formed steadily and pulsation-free in the slot die.

Figure 1. Sheet extrusion line. (Photo: Reifenhäuser, Troisdorf, West Germany)

This web, also called a sheet or curtain, is drawn into the polishing-roll gap where, as indicated in Figure 2, it comes under increasing pressure as the gap narrows. The pressure reaches its maximum before the gap is at its narrowest. Because of the non-Newtonian, structural viscosity of the melt, it relaxes only beyond the narrowest point (so-called memory effect). Studies of this behavior have shown that the pressure in the melt at the narrowest point is still half the maximum [4].

The consequence [5] is that the velocity distribution shown in Figure 2 results in an average value of velocity, v, given by the equation:

$$v = v_\mathrm{w} + (1/2\,\eta)\,(y^2 - h^2/4)\,\mathrm{d}p/\mathrm{d}z \quad \mathrm{m\,s^{-1}}. \tag{1}$$

In this representation it is assumed that the circumferential speeds, v_w, of the two rolls forming the gap are the same. The viscosity η is a power-law-corrected value [6], and h the local roll gap.

7.2 Roll-stack arrangements

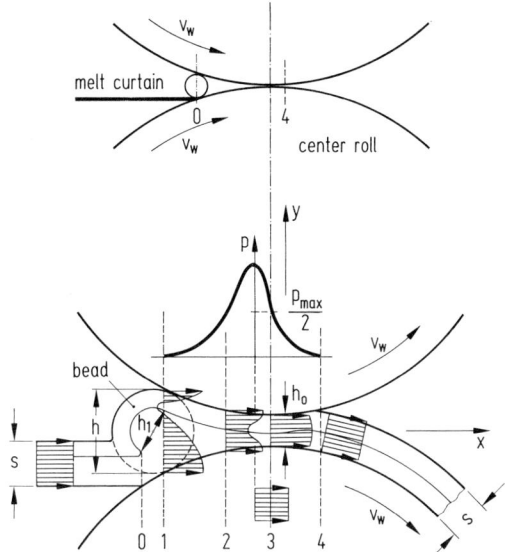

Figure 2. Relationship between flow and pressure in the polishing-roll gap

On the entry side of the gap a bead of melt is produced (Figure 2). This ensures that the roll gap is well filled for producing smooth, perfect surfaces on the films and sheets. The bead overhangs the in-running melt web by only a few millimeters [4].

The pressure of the melt filling the gap, operating across the roll width, gives rise to a separating load. Because of the way the polishing stack and the rolls are built, this gap loading usually has to be limited to 500 N/cm, or exceptionally 800 N/cm. At higher loads the rolls are so greatly bowed that the thickness difference across the roll width becomes technically and economically unacceptable. And even in the frame of the roll stack (Figure 1) the deformations can become so great that a reproducible absolute measurement of the roll gap is difficult. Rather, combined gap (i.e. bearing) loading/displacement measurement is recommended, which makes calibration of the relevant polishing stack necessary [6].

The various viscoelastic materials that are shaped by the polishing stack differ in their viscosity and thermal conductivity, and, because of their structure (amorphous or crystalline) require different bead thicknesses and roll temperatures. From this it follows that there are very many factors that affect the pressure in the roll gap. A first attempt at a rough determination of the gap force associated with the separating load is described in [7], using the equation:

$$F \approx \eta \cdot d_w \cdot l_w \cdot v_{w2} (1/h_0 - 1/H) \quad \text{N.} \tag{2}$$

Investigations based on this have led to the following generalization [8]:

$$p_{max} = d_w^2 \cdot v_{w2} \cdot \eta (x/((x^2 + r\,h)^2)) \quad \text{N mm}^{-2}. \tag{3}$$

Further details on the behavior of plastics on rolls can be found in [9].

One sees that the nip force (roll opening pressure) depends on many factors aside from those already mentioned. These include roll dimensions, the diameter d_w, the loaded width l_w, and the speed setting of the middle roll v_{w2} of the polishing stack.

The polishing stack affects the film and sheet quality not only in the nip, but also during roll wrap-around. During this phase, orientation and internal stresses are produced in the viscoelastic web to an extent determined by the roll temperature, the dwell time on the rolls, the thickness of the web, and whether the polymer is amorphous or crystalline. These effects have a strong influence on the film or sheet flatness that is attainable.

The unavoidable unsymmetrical temperature profiles that are created in the film or board during cooling — temperatures higher inside than out — produce strain differences across the thickness because freezing happens at different times in different places. These bring about internal stresses in the film or sheet without external loading [10]. As Figure 3 makes clear, this effect shows up particularly with thicker sheets on change from one wrap-round direction to the other. The temperature settings on the rolls therefore must be such that the unsymmetrical temperature profile in the plastic mass is equalized as it passes from the fluid to the solid state.

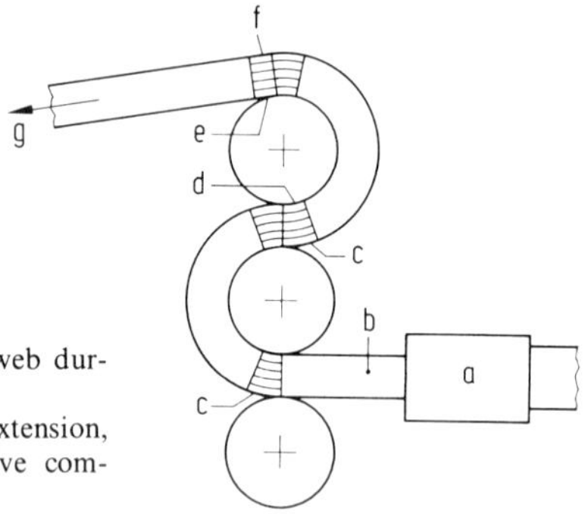

Figure 3. Stretching and compression in the web during wrap around the rollers
a die slot, *b* stretching in the melt web, *c* extension, *d* compression, *e* relative extension, *f* relative compression, *g* haul-off

This balance is more difficult to attain with crystalline materials than with amorphous ones, since the rolls must be very intensively cooled to prevent extrudate from sticking to the surface, and crystallization, which impairs transparency. In this case the equalization of internal stresses can be improved by downstream annealing, using infrared radiation or heat treatment on additional rolls.

With partially crystalline polymers this annealing treatment means that there is increased crystallization and, along with it, a considerable amount of shrinkage. Amorphous polymers are run at higher temperatures on the anneal-polish stack and achieve better stress equalization, as well as lower shrinkage.

The temperature gradient in the sheet arises from:

— Heat transfer by convection, $\dot{Q}_C$, where

$$\dot{Q}_C = \alpha (\theta_w - \theta_u). \tag{4}$$

θ_w, θ_u are roll and environment temperatures respectively. The heat transfer coefficient, α, is temperature-dependent.

— Heat transfer by radiation:

$$\dot{Q}_r = e \cdot C_S [(\theta_w/100)^4 - (\theta_u/100)^4]. \tag{5}$$

e is the emission coefficient for the particular material/film or sheet thickness combination, and C_S is the black-body radiation number with the value 5.77 W m^{-2} K^{-2}.

— Heat transfer by conduction:

$$\theta_C = (b_1 \cdot \theta_1 + b_2 \cdot \theta_2)/(b_1 + b_2). \tag{6}$$

θ_1 is the instantaneous surface temperature and $b_1 = \sqrt{\lambda_1 \varrho_1 C_{p1}}$ is the thermal diffusivity of the film or sheet, and θ_2 and $b_2 = \sqrt{\lambda_2 \varrho_2 C_{p2}}$ relate to the contact rolls. The moment a contact temperature is established, heat conduction occurs.

— Heat transfer by shearing:
the energy put in by the roll drive is transformed into an amount of heat ϕ, in the roll gap. To take account of this the heat transfer equation is expanded in [4] thus:

$$\mathrm{d}\theta = \frac{\lambda}{\varrho\,C_p} \cdot \frac{\partial(\partial\theta)}{\partial x^2} \cdot \mathrm{d}t + \frac{\phi}{\varrho\,C_p} \cdot \mathrm{d}t. \tag{7}$$

7.2.2 Calendering mechanisms

A calender, unlike a polishing stack, is not fed with fully plasticated melt. With a film-drawing calender the preplasticated and gelled, gas-free plastomer is fed into the first nip, usually in strand form. The first gap performs the task of distributing the mix across the width of the roll, while the next two rolls take care of thorough plastication and homogenization of the melt, and forming it to the required dimensions. A modern film-drawing calender is shown in Figure 4. Because of the greater toughness of the feed mass compared with extruder-fed melt, the build-up of the kneading stock required for homogenization, and the smaller gap, separation forces of up to 8 KN/cm arise in the gap — some ten times greater than on polishing stacks. These high loadings are reflected in the construction of the rolls of large diameter/width ratio, which have massive journals, bearing housings, and frames which are double-wall hollow castings of great flexural stiffness.

Figure 4. Film-drawing calender.
(Photo: Berstorff, Hannover, West Germany)

The flow pattern of the melt in the gap during forming is shown in Figure 5. The patterns shown are based on experimental results reported in [8] and [11 to 16]. With its typical rotational motion, the kneading stock on calenders is distinct from the hold-up bead on polishing rolls. These roll-circulating motions produce a homogeneous, fully plasticated melt.

The pressure distribution given in Figure 5 is similar to that in polishing nips. Equations (1) and (3) can be used to calculate the flow velocities and the highest pressure in the roll gap, taking into account the fact that the speeds of the two rolls are usually different.

The theoretical equation for the mean pressure in the roll gap is given in [15], and has been rewritten to embody the quantities used in equations (1) to (3) above, thus:

$$p_{\max} = K\,(v_w/h_0)^n \cdot (\sqrt{d_w/h_0}) \cdot f_1(n). \tag{8}$$

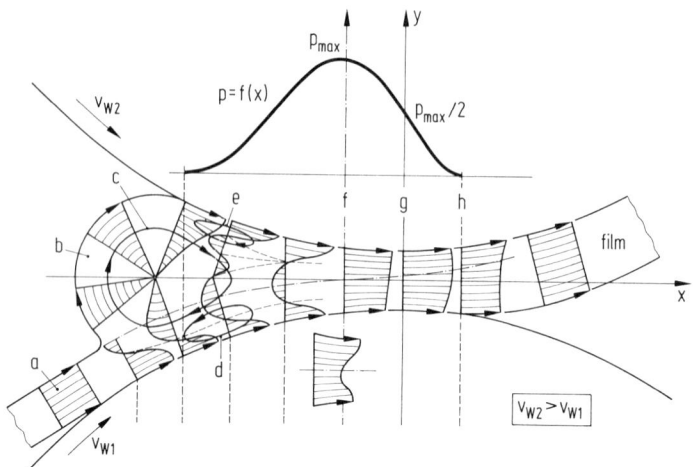

Figure 5. Flow and pressure relationships in the calender roll gap
a melt curtain, b knead,
c main vortex, d feed vortex,
e exit vortex, f flow-boundary,
g narrowest gap, h release point

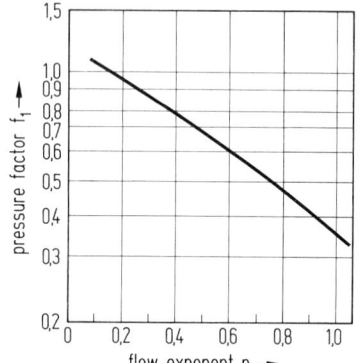

Figure 6. Pressure factor as a function of the flow exponent

In this equation K is the coefficient of the flow power law $\tau = K\dot{\gamma}^n$, with $K = K_0 \cdot e^{-\beta\vartheta}$, n the exponent and f_1 the pressure factor, which varies with n as shown in Figure 6. In [15] it is shown that empirical equation (3) agrees well with theoretical equation (5).

The high strength specification of film-drawing calenders also applies to melt roll-laminating calenders [15], which have three or four rolls. PUR as well as PVC is processed on these calenders. By extrusion coating is meant the covering of a solid carrier web with a thin layer of melt. These melts are produced by single-screw extruders with specially designed screws, or by planet-roll extruders, and cast from slot dies.

The heat transfer laws embodied in Eqs. (4) to (7) apply similarly to the calender.

7.2.3 Differences between polishing stacks and calenders

The different flow processes in the roll gaps of polishing stacks and calenders are the origin of the construction differences of the two systems.

In each case the rolls have to meet very severe demands: high bend stiffness, very precise roundness, uniform temperature profile, little temperature fall-off at the edges, smooth surface, corrosion resistance, reversible mounting, easy maintenance and handleability.

Because of the low loading levels on polishing rolls, the section moduli required are smaller. Thus for polishing stacks, rolls like those shown in Figure 7 are very often used. The advantage of these is that one can have one or more helical channels for the heat exchange fluid between the walls and close to the surface, and so achieve rapid-response temperature control.

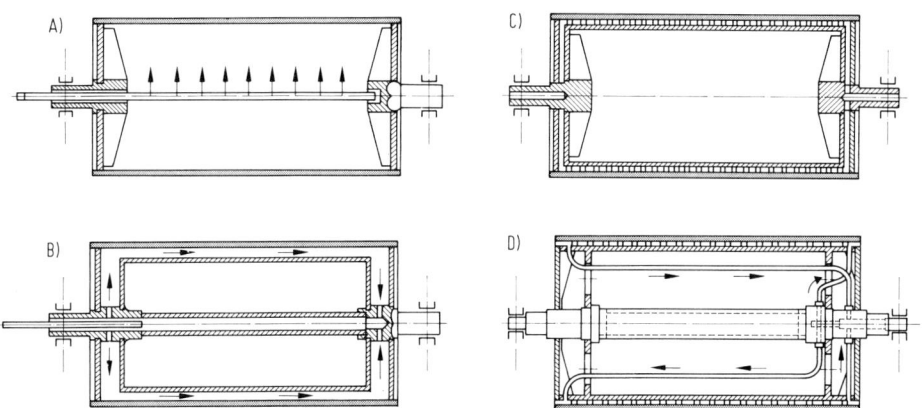

Figure 7. Double-wall rolls for polishing units
A) roll half-filled by water via the jet-pipe, B) double wall with a displacer body, C) double wall with forced circulation by helical channels, D) double-wall with forced circulation by two opposing helical channels

A particular form of double-wall construction is the quick-change sleeve roll. The pitch of the helix in these jacketed rolls decreases smoothly from the coolant feed end to the outlet, while the channel width narrows steadily and the flow velocity increases [101].

With calender rolls such a direct effect upon heat transfer is not possible, because the section modulus required to keep roll deflection low involves a greater wall thickness and a larger diameter/width ratio. One therefore finds that instead of the jacketing systems, one of three other systems is used, depending on the precision of temperature profile or speed of response required (Figure 8). The location, number, diameter, and arrangement of the channels in peripherally bored rolls affect the performance. It should be noted that such borings have a big effect on the deflection and strength of the roll. An exact calculation is difficult on account of the detail of their radial structure, but approximate calculations have been carried out [18].

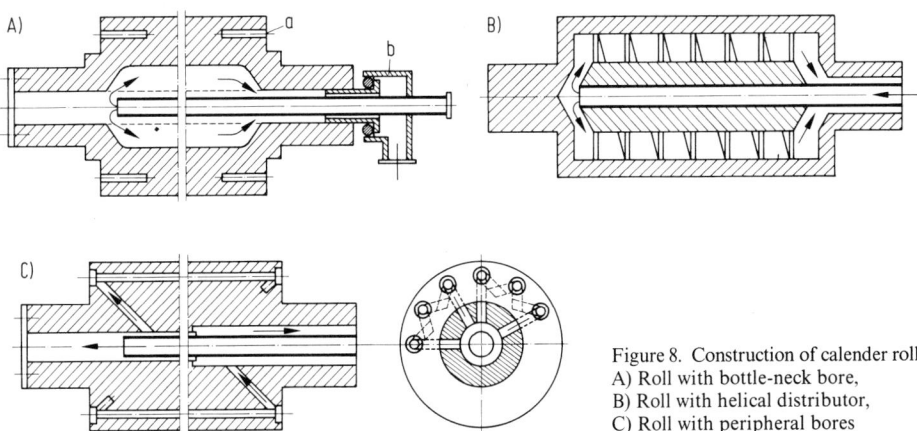

Figure 8. Construction of calender roll
A) Roll with bottle-neck bore,
B) Roll with helical distributor,
C) Roll with peripheral bores

Details of the optimal distribution of material in the roll cross-section for maximum section modulus, as a function of diameter and wall thickness, can be found in [17]. The theoretical calculation of deflection is also discussed. From this, and by derivation from [18], one obtains an equation for roll deflection that can be used in practical situations:

$$y_{max} = \frac{5 \cdot q \cdot l_q^4}{384 \cdot EL}(1+(4.8 \cdot l_a/l_q)). \tag{9}$$

The quantities used are indicated in Figure 9.

For a known roll loading and specified product dimensions, one can determine the deflection at the center of the roll roughly by using Eq. (9), or Figure 9 and fixed values of the quantities in Figure 10. Here $D = D_w \cdot K_1$ with D_w the outer diameter of the roll, and $K_1 = 1 - 1/(D_w/d)^4$, with d the inner diameter of the rolls [18, 19, 20]. One can see from this relationship that even at relatively small line-loadings in the roll gap the deflection can be considerable.

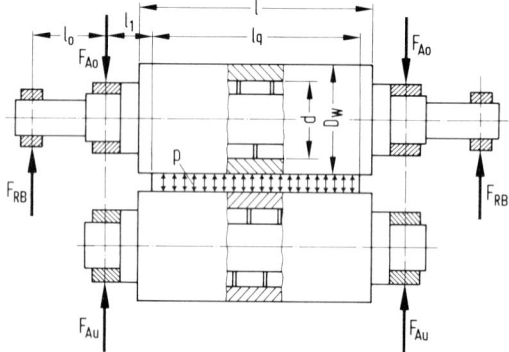

Figure 9. Reference dimensions on a roll with roll-bending

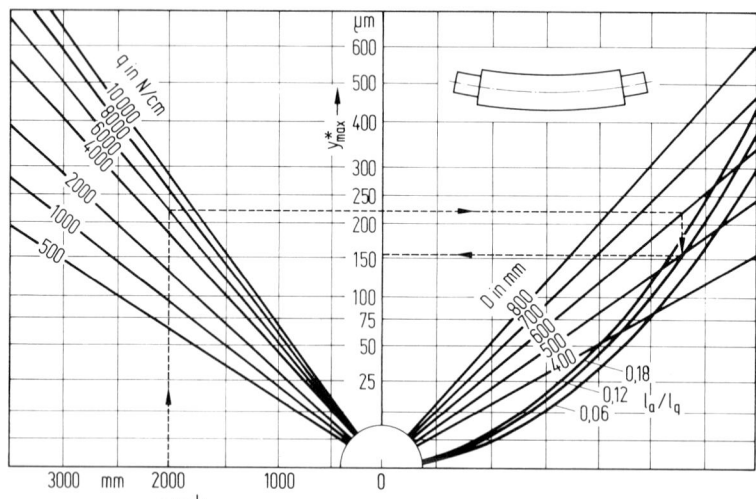

Figure 10. Determination of the deformation at the center of the roll

Polishing rolls and calenders are different from this point of view. The relatively small shape changes that occur under load on polishing rolls normally do not require compensation. But deflections on calender rolls, which cause unacceptable malformation of the film, must certainly be compensated by built-in design features.

7.2 Roll-stack arrangements

There are three possibilities:

- roll bombage or crowning, by grinding the roll profile parabolic with the greatest diameter at the middle,
- counter bending of the roll – often called roll bending,
- non-parallel setting of the roll axes, called roll crossing.

Crowning of rolls has not much to recommend it, since it applies only to one set of operating conditions and the range of uses for the calender is thus diminished. Better, because they are adjustable, are the other methods [11, 21]. They require design effort on calenders that is basically not needed on polishing stacks. The roll-bending force can be calculated with the following equation, relating the quantities appearing in Figure 10:

$$F_{RB} = \frac{5 \cdot q \cdot l_q^2 \cdot \left(1 + 4.8 \cdot \frac{l_a}{l_q}\right)}{48 \cdot l_0}. \tag{10}$$

The correlations from this relationship are summarized diagrammatically in [18]. The lighter construction of polishing stacks allows positional changes to be made, so with modern polishing stacks one can set up the rolls in the ways shown schematically in Figure 11 [22]. This is not possible with calenders. Rather, one must choose dedicated construction arrangements for different processing tasks, like those compiled in Figure 12 [11, 18, 23].

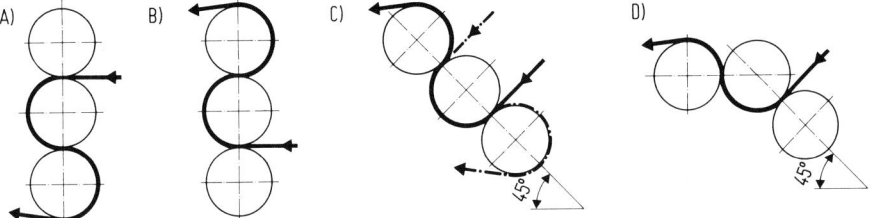

Figure 11. Roll arrangements in polishing units
A) wrap round of 360°, B) wrap round of 360°, C) wrap round of approx. 320°, D) wrap round of approx. 240°

Further differences between calenders and polishing stacks appear in the roll-bearing arrangements. Rolls on polishing stacks basically use anti-friction bearings, preferably self-aligning ball bearings or roller bearings, not only because of the compactness of construction, but also because of the ease with which rolls can be changed. With calenders, hydrostatic slide bearings are used in addition to roller bearings.

Polishing stacks in some cases have a central drive system that keeps all the rolls at the same speed; in others the rolls are driven individually. Which kind of drive is preferred for a particular process will be dealt with later. Calenders always have individual drives.

Individual drives used for polishing stacks are mostly electric motors with reduction gearboxes, and are small enough to be flange-mounted direct on the roll-bearing housing. With calenders, because of their large size and the need for special roll-journal movements (alignment, roll bending, roll crossing) the drives and gear housings are mounted separately on the side of the roll frame [24]. Drive transfer from there by means of universal couplings is illustrated in Figure 13.

A further point of difference between polishing stacks and calenders is in the system of roll adjustment. With polishing stacks the upper and lower rolls are moved with reference to the fixed middle roll in various ways:

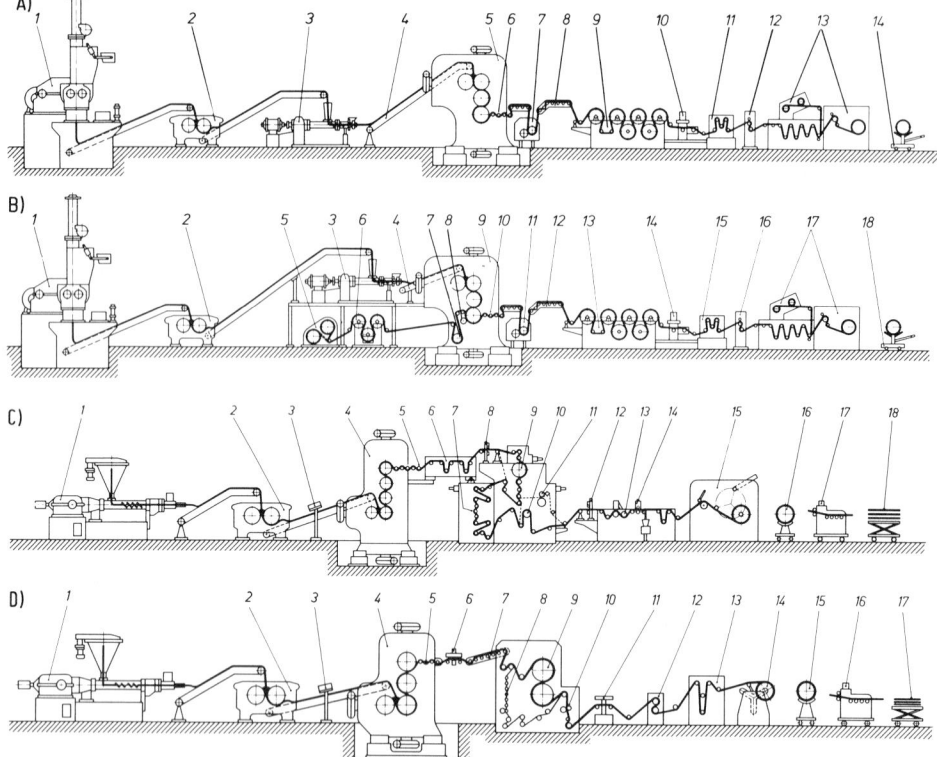

Figure 12. Construction of calenders. (Photo: Berstorff, Hannover, West Germany)
Line A): Manufacture of soft PVC films of various qualities, Line B): Manufacture of soft PVC films by doubling, Line C): Manufacture of rigid PVC films by the Luvitherm process, Line D): Manufacture of rigid PVC films by the high-temperature process

1 internal mixer for non-continuous plastication, and extruder for continuous plastication
2 roll unit for haul-off, homogenization and degassing
3 strainer-extruder for homogenization and distribution on line C); metal separator on line D)
4 swivel-mounted feed unit on lines A) and B); 4- or 5-roll- L- or F-calender on lines C) and D)
5 4-roll-F-, -Z-, -S- or -L-calender on line A); un-wind for films and fabric on line B); multi-roll take-off on lines C) and D)
6 multi-roll take-off on line A); preheating on line B); roller track on line C); thickness measurement unit on line D)
7 embossing unit on line A); preheating roller on line B); roll stretching section on line C); roll annealing unit on line D)
8 roll annealing unit on line A); doubling equipment on line B); thickness measuring equipment on line C); roll stretching section on line D)
9 cooling and annealing unit on lines A) and D); 4-roll -F-, -Z-, or -S-calender on line B); Luvitherm roll on line C)
10 thickness measurement equipment on line A); haul-off unit on line B); cooling and annealing unit on line C) longitudinal cutting unit on line D)
11 longitudinal cutting unit on line A); embossing unit on lines B) and C); thickness measurement unit on line D)
12 braking rolls for taking up the winder tension on lines A) and D); roll annealing unit on line B); thickness measurement unit on line C)
13 winder on line A); cooling and annealing rolls on line B); braking and pendulum rolls for taking up the winder tension on lines C) and D)
14 transport of finished rolls on line A); thickness measurement unit on line B); longitudinal cutting unit on line C); turret winder on line D)
15 longitudinal cutting unit on line B); winder on line C); transport of finished rolls on line D)
16 braking rolls to take up the winder tension on line B); transport of finished rolls on line C); transverse cutting unit on line D)
17 winder on line B); transverse cutting unit on line C); sheet stacker on line D)
18 transport of finished roll on line B); sheet stacker on line C)

7.2 Roll-stack arrangements

- by coaxial pneumatic or direct hydraulic cylinders,
- by means of levers pushed against return springs by pneumatic cylinders acting from one side,
- by means of pneumatic cylinders acting from both sides,
- coaxially by means of leadscrew gearboxes.

Figure 13. Drive unit of a calender. (Photo: Battenfeld, Bad Oeynhausen, West Germany)

Calender rolls, however, are always shifted together by leadscrew gearboxes ("screw-downs") because of the large forces involved; high gear is used first of all, and finally, for very small movements, low gear. This kind of two-speed adjustment is also being fitted to modern polishing stacks to an increasing extent [6].

Because of their relative lightness, and for technical reasons that will be explained later, polishing roll stacks are designed to be mobile and vertically adjustable. Both movements can be carried out by lead screws or electric motors. For precise adjustment of the roll gap a hardened wedge-ram is mounted direct on the bearing housing. Several types of product now have prestressed bearings to minimize play and misalignment.

Measurement of roll gap in practice is usually carried out by hand using a feeler gage on the run-out side of the rolls. Digital sensors are now available which signal the roll gap and take care electrically of the necessary adjustment by means of a leadscrew gearbox.

A final important difference between polishing stacks and calenders that must be mentioned is in the temperature control of the rolls. Roll temperatures on calenders usually lie between 160 and 200 °C, and on polishing stacks between 40 and 150 °C, depending on whether partially crystalline or amorphous or high-melting materials are being processed. Thus the following heat transfer media are used on polishing rolls:

- water circulating at 40 °C in a closed system, heated for start-up and cooled during production,
- water at 70 to 80 °C circulating in a closed system,
- pressurized water at 120 to 140 °C at the roll inlet, or
- oil at 80 to 160 °C.

On calenders one uses:

- superheated steam, or
- oil.

Which of these media is used on calenders is a question of local conditions and economics.

Influences on machinery	Influences on methodology

a) – heating and cooling capacity
– drive power and torque
– feeding aid (cooling medium)
– range of screw speed
– kind of temperature control system
– screw diameter
– screw length
– venting on the barrel
– shape and position of the venting zone
– screw geometry:
  three zones without venting
  or six zones with venting
  – flight depth
  – zone length
  – shear- and mixing-barrier
– screw temperature control
 	
b) – shape of screw-tip
– shape of adaptor
– valve installation
– screen changer with breaker plate
  – diameter of breaker plate
  – number of cavities
  – number of screens
  – mesh-number of screens
– channel diameter
– channel length
 	
c) – width of the slot die
– die lip gap
– length of die lips
– shape of melt flow channels
– flow distribution
– position and shape of restrictor bar
– temperature control
 	
d) – roll surface area
– width of roll
– roll diameter
– roll wall thickness for acceptable bending
– control of heat-exchange medium
– heating and cooling capacity
– drive power
– speed range of rolls
 	
e) – drive power
– surface hardness of take-off rolls
– speed difference between polishing unit and haul-off
 	
f) – length of air-cooling section
 	
g) film web tension as function of:
– roll diameter
– film width
– film thickness
– surface friction
	h) Delivery condition of the raw materials:
	– form and size of pellets
	– manner of coloring:
	  – colored granules
	  – master-batch
	  – color dry-mixed
	– drying method
	– drying temperature
	– quantity of waste added:
	  – origin
	  – compatibility with virgin material
	– kind of waste:
	  – chopped film
	  – ground material
	  – repelletized
	– uniformity of grain size
	– proportion of powder
	
	i) – heating time (extruder)
	– screw speed:
	  – at start-up
	  – rate of increase
	  – in production
	– temperature control:
	  – at start-up
	  – in production
	– degree of vacuum for venting
	– power consumption of extruder drive
	
	k) – temperature control of adaptor and screen-changer
	– temperature across die width
	– temperature on die lips
	– adjustment of restrictor bar
	– melt temperature on screw tip
	– melt pressure at screw tip
	
	l) – heating time (calender, etc.)
	– temperature of rolls
	– surface speed of rolls on introduction of the melt curtain
	– distance between slot die and roll gap
	– surface speed of rolls during manufacture
	– gap between the feed rolls
	– gap between the outlet rolls
	– line-load in the feed roll gap
	– relative speed between polishing unit and haul-off

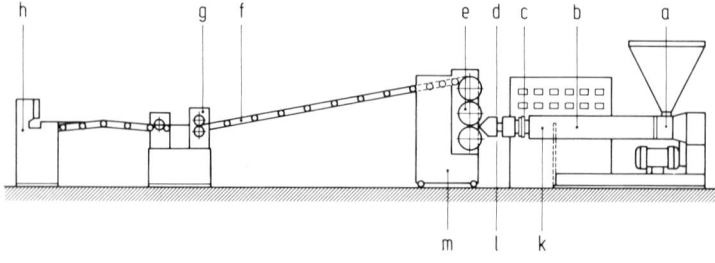

7.2 Roll-stack arrangements

Influences on machinery

a) – screw diameter
   – screw length
   – screw geometry:
     – flight depth
     – length of zones etc.
   – thermal control of screw
   – heating and cooling capacity
   – drive-power of extruder
   – screw speed range
   – temperature control method on extruder

b) – shape of screw tip
   – surface area of screw tip
   – shape of adaptor
   – surface area of adaptor
   – channel diameter
   – channel length

c) – width of slot die
   – die lip gap
   – length of die lips
   – shape of flow channels
   – flow distribution
   – heating capacity on the die
   – temperature control on the die

d) – roll surface area
   – roll width
   – roll diameter
   – roll wall thickness for acceptable bending
   – mechanical and thermal bombage
   – drive-power of rollers
   – speed of roll surface
   – heating and cooling capacity
   – crossing of the center roll

e) web tension as function of:
   – diameter of roll
   – film width
   – film thickness
   – surface friction

Influences of methodology

f) Compounding of PVC:
   – selection of raw PVC:
     M-, S- or E-PVC
     K-value
   – components of recipe
   – mixing temperature
   – mixing time
   – cooling temperature
   – timing of component addition

g) – heating time
   – screw speed:
     – at start-up
     – rate of increase
     – in production
   – temperature control
     – at start-up
     – in production
   – screw temperature control
     – temperature
     – intensity
   – temperature control of upper and lower vacuum hopper for degassing (alt. to barrel degassing)
   – level of vacuum in the lower hopper
   – extruder-drive current

h) adaptor temperature control
   – temperature distribution on die-width
   – melt temperature at the die gap

i) – heating time
   – roll temperature on introduction of the melt
   – surface speed of the rolls
   – time and method
     of adjusting the main roll
   – temperature of rolls during production
   – friction between the main rolls
   – roll gap
   – size, overhang and shape of the knead
   – feed angle of the web
   – temperature of the take-off roll
   – surface speed of the take-off roll
   – adjustment and pressure of the take-off roll
   – temperature of annealing rolls
   – surface speed of annealing roll
   – adjustment of annealing rolls

Waste addition:
   – origin
   – compatibility
   – quantity added
   – kind of waste:
     – chopped
     – repelletized
     – flakes

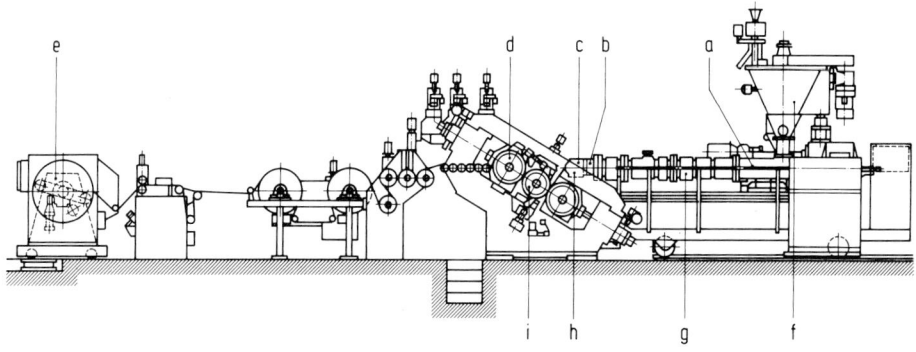

7.3 Quality of films and sheets

The general discussion of quality which follows deals with the processing conditions required to obtain films in various recognizable states, their internal morphological structure, the quality criteria, and the test methods employed.

7.3.1 Polished films and sheets

The machinery used for manufacturing extruded and roll-polished films and sheets must be equipped to be as fully flexible as possible. This means that the raw material must be chosen and compounded, and/or the processing parameters adjusted, for particular end-product uses [1].
The factors that are built into the sheet manufacturing system, and which are required by the processing technology, are summarized in the following review.
Because of the great number of such factors, it is obvious that running such a line calls for specialized knowledge, and that control systems extending to full automation of a sheet extrusion line require extensive electronic installations.
There is demand not just for natural, uncolored films and sheets, but also for highly transparent, opaque, colored transparent or color-coated products, and for glossy or matt films and sheets. Furthermore, it should be possible to produce various embossing effects, laminated products, surface coatings, and coextrusions, and to rework waste.
The line shown in Figure 1 can have facilities added to it to carry out these additional tasks.

7.3.2 Calendered films

A calender line, and particularly the calender roll-stack which is the core element, is built on a very large scale and is heavily dimensioned. And it is thus destined only for producing films of high quality.
Calenders are never used for manufacturing thick sheets, even though it is technically possible to do so with appropriate equipment. For sheet, particularly thick sheet, the line load needed in the roll gap for polishing is very small, so a calender is greatly over-dimensioned for that purpose.
A second point is that once the bank of viscoelastic material is formed in the roll gap (Figure 5), after previous dispersion on mill rolls or in a kneader (see Section 7.4.2.1.2), it is thoroughly plasticated, homogenized and relaxed. These functions can be carried out effectively on the melt in the roll gap only at very high pressures, approaching 800 bar.
It can be readily recognized, therefore, that it is not possible to produce good-quality sheets greater than 1 mm thick with this technique on a calender. The pressure required for the development of the flow pattern shown in Figure 5, which (see Eq. (3)) is directly proportional to the viscosity of the material in the gap, and inversely to the square of the gap width, cannot be developed.
The extrusion-calendering line has been chosen from the three possibles for use later in illustrating the machinery and processing variables that are relevant for sheet extrusion [25]. This calender unit is fed from a slot die or from a strand dispenser. The traditional film-drawing calender receives its raw material feed from a combined internal mixer and sheet roll mill (often fitted with a strainer) or direct from a Ko-Kneader or planet-roll extruder [21, 26, 27, 28].
From this outline it is evident that a high level of experience is demanded of the operating personnel, and that considerable expenditure on control technology and automation is required.

7.3.3 Raw material content of films and sheets

Calenders are used for producing films from single-component suspension and emulsion PVC in hard, semi-hard, and soft grades; from LDPE, HDPE, PP, and polyurethane; and from ABS/PVC blends.

Extrusion lines use many basic raw materials and compounds for sheet manufacture. In addition, multilayer composites can be produced, and waste can be reworked.

First we have the calendering materials:

- Suspension PVC is plasticated by the high-temperature process (see Sec. 7.4.2.4.4). This material generally has a K-value [29] of about 58 to 62.
- Emulsion PVC is worked by the Luvitherm process (Sec. 7.2.4.4). It has a K-value of about 62 to 65.

Before processing, the PVC must be compounded [30], and for this purpose it is important to know which of the processes mentioned in the preceding section is to be used. Recipes for use on extrusion calendering lines are constructed around several components. A typical recipe for a transparent, high-impact PVC film has the following composition [25]:

Component	Parts by wt.
PVC powder	100
Impact modifier	12
Stabilizer – thermal	1.5 to 2
Internal lubricant	1.0
External lubricant	0.3
E or OP wax	0.8
UV stabilizer	0.3
Organic pigment	6 to 8×10^{-4} (blue or violet)

For coloration of transparent films, 0.1 to 0.6% by weight of organic pigment is used instead of the blue/violet one, and for masking color, 2 or 3% of inorganic pigment.

Soft PVC films can be produced by adding a plasticizer to the recipe, usually liquid DOP (dioctyl phthalate), instead of the impact modifier. Depending on the degree of softness required, 20 to 30 or 40 parts by weight of DOP are added, together with about 2 parts of epoxidized soybean oil.

Films of ABS/PVC blends must leave the calender entirely stress-free, because they are used in the automotive industry for thermoforming applications. The use of ABS in the blend improves the shear strength and tensile strength at high temperatures [32].

The special properties of these blends are exhibited by the following typical compound:

Component	Parts by wt.
PVC powder, K 59 to 61	100
ABS powder	50
Stabilizer	1.8
External lubricant	0.5
Internal lubricant	0.5
E or OP wax	0.8
Pigment e.g. carbon black	0.45

The remaining materials mentioned are delivered ready-compounded in pellet form to meet special properties requirements like [31]:

- high impact strength,

- good stress-corrosion resistance,
- high stiffness,
- good high- and low-temperature resistance,
- low water uptake, and
- antistatic behavior.

Notes on the most important sheet extrusion materials [32]

a) for single films and sheets:

- ABS, delivered as pellet; high impact strength and hardness over a wide temperature range, and good temperature-cycling resistance.
- PS is used only as styrene/butadiene (SB) copolymer, or as SAN copolymer, for films and sheet. SB possesses good hardness and stiffness, and good low-temperature impact strength. The good temperature-cycling resistance and heat-distortion resistance of SAN is noteworthy. While SB yields only colored opaque products, SAN films and sheets are glass-clear, transparent, or colored opaque.
- PVC can be converted into sheets and films. The recipes are similar to those mentioned earlier for calendering. Soft PVC products can be rubber-elastic; applications are many and various and depend on the kind and amount of plasticizer added. Rigid PVC films and sheets are hard and stiff; they are highly impact-resistant over a wide temperature range, and form-stable up to moderate temperatures; they are glass-clear with a yellowish tint.
- PMMA can be used to produce brilliant, crystal-clear films and sheets, with a high degree of ageing- and weather-resistance, and high strength.
- High-density polyethylene, HDPE, ($\varrho = 0.95$ cf. 0.92 g/cm^3 for LDPE) is stiff and can be processed on sheet extrusion lines. The films and sheets are transparent to opaque, and have a waxy surface.
- PP, despite its lower density ($\varrho = 0.906$), is harder and stiffer than HDPE, but its low-temperature toughness is worse. It can be used at somewhat higher temperatures. It is transparent to opaque.
- Among the numerous kinds of polyamide the 6 and 66 types can be used on sheet extrusion lines. They are characterized by great hardness, stiffness, abrasion resistance, and high-temperature form stability.

b) for filled and/or reinforced films and sheets:

- PP powder mixed with 40% pulverized chalk in the extruder. Used for films and sheets from 0.3 to 2 mm thick for food packaging [33].
- Other mineral fillers are calcium carbonate, talc, kaolin, and mica; organics are woodflour, cellulose, and cotton. Talc, chalk, and woodflour can be mixed with polyolefins in proportions up to 60% by weight, or in special cases up to 80%.
- PA and polyesters can contain talc or mica up to 40 wt.%.
- PVC compounds containing up to 1000 parts by weight of mineral fillers relative to 100 parts of PVC, can be used in the cable sector [43].

7.3.4 Special quality aspects of films and sheets

Objective judgement of product quality requires objective establishment of quality characteristics, of testing- and monitoring methods, and first-hand knowledge of the machinery and processing parameters that affect the working of a line [44 to 55].

7.3 Quality of films and sheets

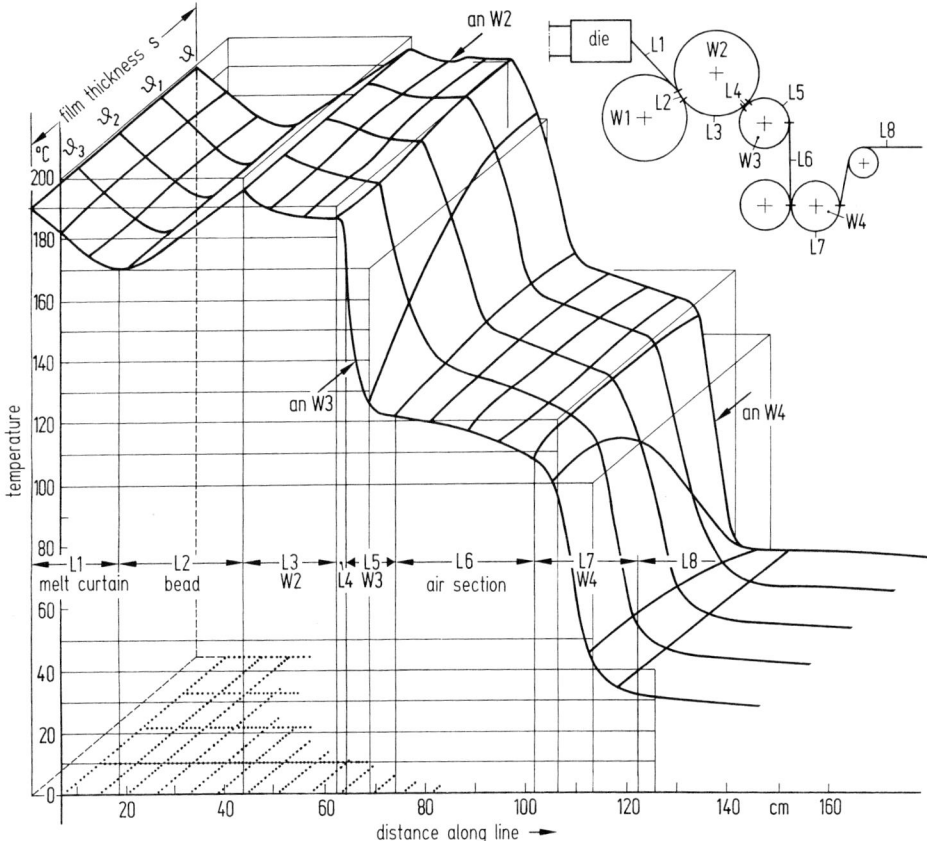

Figure 14. Temperature distribution in the thickness of a rigid PVC film during manufacture on an extrusion/calender line

One simple way of judging the process by which the films or sheets are being made is to determine their shrinkage. This is a measure of the tendency of the plastic to revert to an original state on reheating, for example during the thermoforming process [94]. For control purposes one needs to monitor the temperature and cooling programs throughout the process, the speeds of the product web in each section – on the rolls, between the rolls, in the cooling sections – and possibly the stretch ratios and relevant temperature dwell times as well [1]. The way such quantities change is shown for an extrusion/calender line (Figure 14) and a sheet extrusion line (Figure 15). The temperatures on the upper and under sides of the web at all points along the production line are determined by means of a computer program. This program takes into account all machine and process variables, and the properties of the materials being processed [4].

The difference between processing on calenders and on polishing rolls is made clear by comparing these two diagrams. For example, one can see that there are big temperature gradients across the film or sheet thickness because of the intensive cooling on the polishing rolls.

The temperature/time history determines the relaxation of the plastomer, and this, in the final analysis, determines the residual shrinkage [4, 5, 55]. Figure 16 shows how the longitudinal shrinkage normally found in films or sheets produced on extrusion lines varies with product thickness from 0.5 to 10 mm. The effect of output rate is also shown.

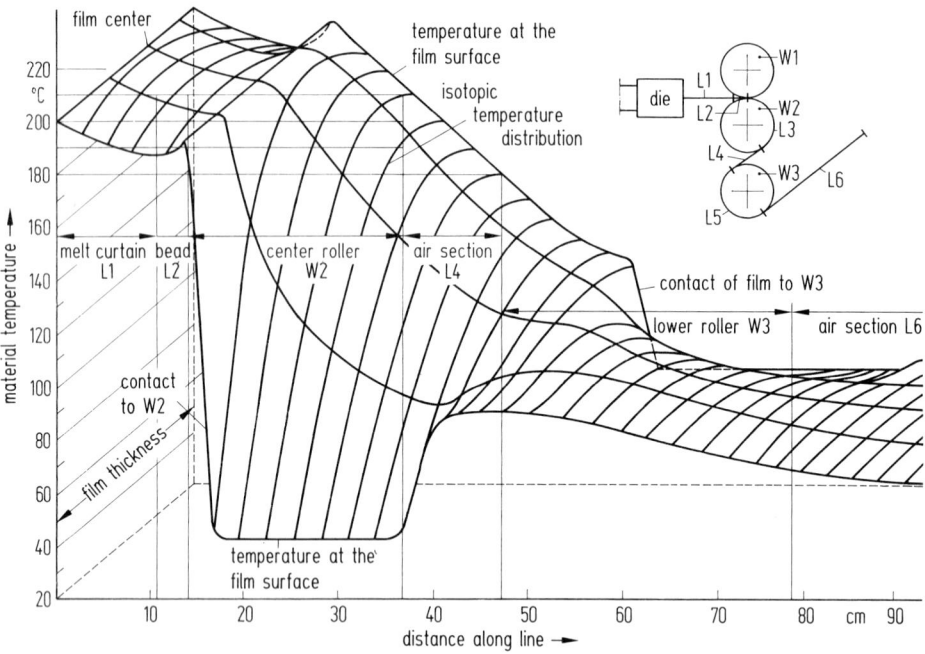

Figure 15. Temperature distribution on and within a 2 mm thick PS film during manufacture on a sheet extrusion plant ($\dot{m}$ = 560 kg/h, film width 1170 mm, roll diameter 400 mm)

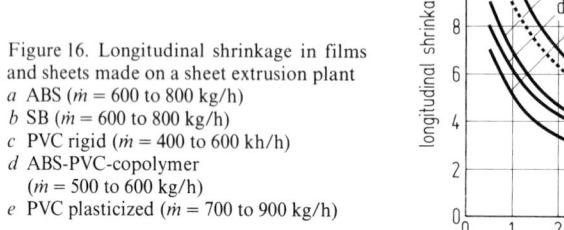

Figure 16. Longitudinal shrinkage in films and sheets made on a sheet extrusion plant
a ABS ($\dot{m}$ = 600 to 800 kg/h)
b SB ($\dot{m}$ = 600 to 800 kg/h)
c PVC rigid ($\dot{m}$ = 400 to 600 kh/h)
d ABS-PVC-copolymer
 ($\dot{m}$ = 500 to 600 kg/h)
e PVC plasticized ($\dot{m}$ = 700 to 900 kg/h)

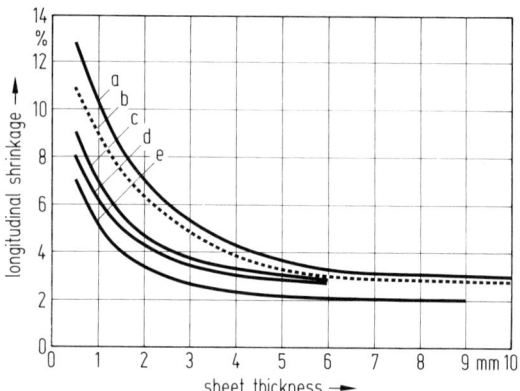

An additional quality criterion is the thickness profile [51], measured by a hand-held contact gage or, more often nowadays, by an automatic device sited on the line after the roll stack. This traverses the running web and uses ultrasound, or isotope radiation transmission to provide a signal, proportional to thickness, to a chart recorder.

As part of the effort to automate sheet extrusion manufacture, lines are now available on which thickness deviations measured are used automatically to make a thickness correction at the slot die, by adjusting the restrictor bar and/or the lips [56, 57].

7.3.5 Applications of film and sheet

The summary below mentions some of the many possible uses and special features of calendered film, and of extruded films and sheets [31, 58, 94, 99, 112].

Material	Machinery used	Thickness mm	Condition	Appearance	Application
Rigid PVC	Calender	0.03 to 0.12	unstretched	transparent or opaque; smooth or embossed	covers, separators, laminates, insulation
		0.035 to 0.07	stretched	transparent or opaque	adhesive tape
		0.025 to 0.08	low stretch ratio, impact-resistant	glass-clear	bands, covers, bags
		0.1 to 0.5	unstretched thermo-formable	glass-clear	lids, cups, blisters
		0.1 to 0.2	unstretched thermo-formable	brown-pigmented	inserts for confectionery packs
		0.25 to 1.2	unstretched thermo-formable	opaque 1-2 sides, matt-finish or one side glossy	cups, packing of fats and foodstuffs
		0.15 to 0.25	unstretched with white pinchability	colored	embossed labels
LDPE	Calender	0.05 to 0.2	high impact, good stress-crack resistance	natural color	clothing bags electr. insulation
		0.07 to 0.1	silicone-coated	colored	covering and separating film for sticking plaster decorative foils
		0.04 to 0.2	unstretched	embossed or smooth	roll film
		0.035 to 0.08	weldable	printed or unprinted high gloss	films for packaging machines, bags
		0.06 to 0.25	uni- and bi-ax. stretched	natural color	thick shrink-film for pallets, furniture, carpets, building components
		0.025 to 0.055	uniax. stretched antistatic	natural color	thin shrink-film for smaller packages
HDPE	Calender	0.09 to 0.15	good low temp.- and heat resistance	sterilizable at 120 °C	boil bags, freezer bags, bags for pre-cooked meals
		0.04 to 0.25	form stable at 100 °C	sterilizable at 120 °C	rice bags, lamination, insulation

Material	Machinery used	Thickness mm	Condition	Appearance	Application
PP	Calender	0.018 to 0.03	biaxially stretched	heat-seal coated on both sides	bags and covers
		0.04 to 0.06	biaxially stretched	embossed and surface-treated for printing	coatings, covers, separators
		0.015 to 0.06	biaxially stretched	smooth	adhesive ribbon, composites
PP	sheet extrusion plant with polishing unit	0.5 to 1.2	from unfilled granules, powder w. 40% weight talc	white	food packaging
PMMA and PC	,,	2 to 6	high-impact	embossed or smooth	sanitary wares, windows, light covers, illuminated signs
Rigid PVC	,,	1.5 to 4	smooth or embossed	transparent or colored	household wares, apparatus construction, heating and ventilation equipment, wall and exterior cladding
Soft PVC	,,	4 to 6	smooth, very rubbery	transparent or colored	strips and swing-doors, cutter backing
ABS and ABS/PVC-blend.	,,	0.5 to 3	high-impact	transparent, embossed or colored	boxes, covers in automobiles and aircraft, seat backs
PS and HIPS	,,	0.5 to 6	high-impact	transparent or colored, smooth or embossed	refrigerators, shower cabins, bathroom furniture, toys
HMPP	,,	1 to 6	stress-free	colored	tank construction
ABS	,,	3 to 10	high-impact	colored, printed, embossed	refrigerators, household wares, furniture, boats, vehicles, ski top surfaces
HDPE, LDPE, PP	,,	3 to 6	stress-free	natural or colored	covers, technical wares
HMPE-Bitumen	,,	1.5 to 2.5	plasto-elastic	black	laminated sealing webs
PC	,,	2 to 6	high-impact	glass-clear	security glass
CA and CAB	,,	0.125 to 15	high-impact	glass-clear, colored-opaque or transparent	technical products, packaging

Coextruded films and sheets [112, 113]:

- SB base film with PS glossy layer; for freezer cabinets;
- ABS base with PMMA UV-resistant layer; for boatbuilding;
- SB carrier, both sides coated with a PMMA scratch-resistant layer; for wet-cells (shower cabins etc.);
- SB carrier, tie-layers on both sides, for Frigen-resistant ABS layer on one side, and grease-resistant ABS on the other; for freezer cabinets;
- SB rework carrier layer, with white SB covering layer and colored decorative layer; for freezer cabinets;
- SB foam carrier layer with Frigen-resistant ABS layer on one side, and SB covering layer finished with a PS glossy layer on the other; for freezer cabinets.

Coated films and sheets [108]:

	Superficial weight g/m^2
– PVC coatings for:	
Leathercloth – clothing	100 to 250
Leathercloth – luggage, shoes, cushions, auto.	250 to 800
Wallpaper and bookbinding	100 to 200
Floorcoverings	800 to 1200
Conveyor belts	800 to 1600
Air structures	750 to 1250
Growing houses	300
Roof-sealing webs	1000 to 1800
Truck tarpaulins	600 to 700
Carpet backings	1000 to 2000
– PUR coatings for:	
Leathercloth – clothing	100 to 250
Leathercloth – luggage, shoes, cushions, auto.	250 to 800
Conveyor belts	800 to 1600
Embankment covers	1100
– LDPE and HDPE coatings for:	
Film-tape packaging fabrics	100 to 250
Heat-sealable materials	150 to 200
Carpet backings	1000 to 2000

7.4 Production lines for films and sheets

The construction of the sheet extrusion and calender lines illustrated in Figures 1 and 4 is described in detail below. The basic types and special versions of these lines are also dealt with.

7.4.1 Raw material compounding

The calendering and roll-stack polishing processes put different demands on the way their feedstocks are compounded. While the polishing stack of the sheet extrusion line is fed with fully plasticated, homogenized melt, the first roll gap of the calender receives strands of material that are plasticated, dispersed, but not fully molten.

On sheet lines the raw material is prepared by the extruder, meaning that it is melted, dispersed, and homogenized. If it is fed to the extruder as a powder it has usually already been through a mixing process.

The PVC and ABS/PVC recipes detailed in 7.3.3 are mixed in high-speed hot-cold mixers, like that shown in Figure 17, under precise conditions of cycle time and temperature. During the hot phase liquid components like stabilizers and plasticizers can be added. The high temperatures of around 80 °C when liquids are added, of 140 °C during the agglomeration phase, and 30 to 40 °C after cooling down, produce a dry, free-flowing, gritty pre-product for extrusion. Mixing takes about 12 minutes.

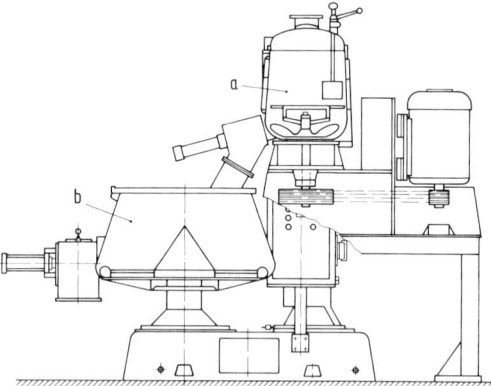

Figure 17. Heating and cooling mixer combinations:
a high-speed heating mixer (turbo-mixer),
b low-speed cooling mixer

Granular materials require only pigments, rework, or other additives to be put in. Since simple physical mixtures are sufficient at this stage, cold mixers are adequate. Nowadays pneumatically fed, dosing/mixing units are frequently used for this task [2].

It is essential to have electronic or magnetic detector-separators, to be sure of removing metallic foreign matter before it can get into the extruder and cause damage. The same kind of preparation equipment is needed for extrusion calendering lines. On the other hand, film-drawing calenders require another compounding process. The mix ingredients are preplasticated in an internal mixer, or an internal kneader, or sometimes on mill rolls; or alternatively in a Ko-Kneader, or a planet-roll extruder – under some circumstances as a cascade arrangement [2].

The standards achieved in the various methods of compounding plastomer feedstocks, and the correct tailoring of the mixture or recipe, have equally large influences on the properties to be expected in the film or sheet. The compounding operation must, of course, be carried out under conditions of careful timing and thermal control.

If fillers or reinforcements have to be mixed into the plastic, it is normally necessary to carry out the pelletizing on a separate line [2, 23].

However, some sheet extrusion lines have been built on which the filler is added to the polymer in the extruder for dispersion and homogeneous plastication. A line of this sort for producing chalk- and talc-filled polypropylene sheet is described in [43]. This kind of process avoids the need for intermediate pelletizing.

7.4.2 Construction of sheet extrusion lines

The difference between lines for sheet extrusion and those for calendering – as has already been described in relation to the roll stacks used – becomes even more evident from a consideration of the complete line. The different compounding machines are dedicated to the tasks that have been described, and are therefore not exchangeable.

In the following sections still more variants of plastication, auxiliary equipment, and shaping systems on sheet extrusion lines will appear on the scene. Cooling of films and sheets, and the final handling of the semi-finished product, round out the discussion.

7.4.2.1 Extruders and kneaders

Two basic lines can be followed, namely:

- thorough homogenization and plastication in the extruder;
- preplastication or gelling in a mixer, kneader, roll stack, strainer, Ko-Kneader, or roll extruder.

7.4.2.1.1 Extruders

There is a wide choice of extruders for producing homogeneous melts for feeding polishing stacks:

- The single-screw extruder is the simplest of all, and is used predominantly for plastication of PS, ABS, PE, PP, PC, and PA in pellet or granule form. A single-screw extruder with barrel degassing is shown in Figure 18. PS, ABS, and PC must be degassed because gases are evolved under thermal loading in the extruder. Single-screw extruders can be fitted with barrel-vacuum degassing, or with double vacuum-hopper degassing [34]. The other materials that have been mentioned do not require degassing. For each material the extruder must be fitted with a specially tailored screw having specific drive power, and a speed range to suit the shear characteristics of the plastic. Modern single-screw extruders with grooved, intensively cooled bushes in the feed zone are used for processing polyolefins (PE and PP). Grooved feed zones result in material being moved by the solids-conveying principle, in contrast to the normal molten-film-shearing principle [35, 36].

Figure 18. Single-screw extruder. (Photo: Battenfeld, Bad Oeynhausen, West Germany)

It should be noted that extruder gearboxes may be fitted with double-helical hot-ground gears for particularly quiet running. Bi-metallic barrels and screws with stellite surfaces minimize wear. The helical cooling pipes on water-cooled extruders are made from aluminum, cast integrally with the heating elements, and the distilled cooling water (no furring of pipes) circulates in a closed system [80].

- Because of their force-feeding action, double-screw extruders are very suitable for feeding powders. Thus the PVC and PVC/ABS powder compounds referred to in 7.3.3, also heavily filled PP powder and fine PMMA granules, can be plasticated well on them

[2]. A double-screw extruder fitted with a degassing section is illustrated in Figure 19. On these extruders also, choice of screw should be varied to suit the material being processed [37].

Figure 19. Twin-screw extruder.
(Photo: Reifenhäuser, Troisdorf, West Germany)

— The planet-roll- or roll extruder, whose function can be understood from Figure 20, besides being used as a preplastication unit, is frequently employed on calender lines with vacuum-hopper degassing, and also as a full plastication machine with barrel degassing on sheet extrusion lines [33]. Pellets and powders can be processed on these machines, which combine rolling the plasticated material between the central and planet spindles with delivering the melt by a single screw. Here, too, drive power and the speed of the central screw must be properly matched to the particular task [1, 38].

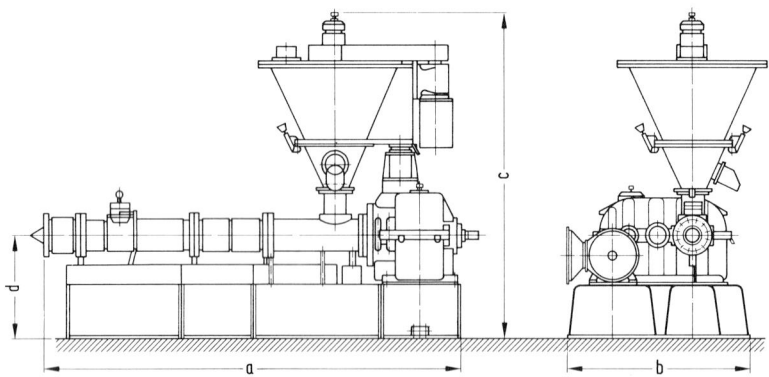

Figure 20. Planetary-roll extruder.
(Photo: Battenfeld, Bad Oeynhausen, West Germany)

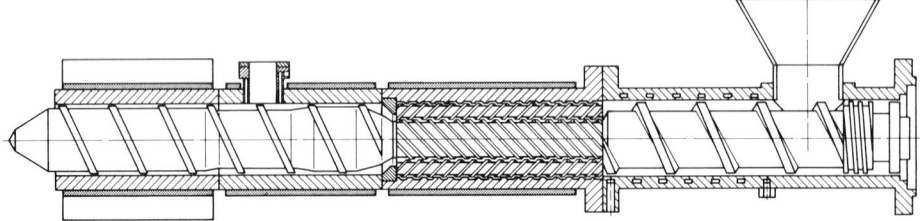

7.4.2.1.2 Mixers and kneaders

Compounding of materials for calendering can be done on the machines mentioned above; they are usually combined to make up compounding lines like the one illustrated in Figure 12. Extrusion calender lines can, of course, be fed by planet-roll- or roll extruders, using a slot die or a traversing strand-feed device. The extruder thus takes over the whole task of plastication, so that the calender is reduced from a three-roll- to a two-roll unit (cf. Figure 13) [39]. The machines in Figure 12 have the following tasks:

- The internal mixer provides discontinuous gelling of the premixed material.
- The mixing rolls turn the charges coming from the internal mixer into a continuous sheet, homogenize them, and degas them.
- The strainer, an extruder only some 6 D long, has the function of removing bubbles or undisintegrated particles; to do this it has a screen-changer in front of the strainer die.

The compounded material is transported between these units and the calender in lumps or in strand form, on conveyor belts that pass in front of the rolls.

Alternatively, calenders can be fed by continuously running screw kneaders or (planet-) roll extruders. These machines combine in one unit the steps of plastication, mixing, homogenization, dispersion, and degassing [2].

- On single-screw kneaders the screw goes through an axial oscillating motion simultaneously with rotation. The well-known Ko-Kneader operates in this way [28].
- The twin-screw kneader (TSK) has two corotating intermeshing screws with a self-cleaning profile. To assist kneading and mixing this machine is, additionally, fitted with several kneading disks set behind one another on each screw shaft [40].
- A special compounding technique for making filled sheet is provided by the corotating twin-screw extruder [43]. Up to 80% by weight of fillers can be incorporated with this device, including glass fibers and rovings, chopped-strand glass, and glass ballotini, with 0.3 to 3% by weight of pigments [97, 98].
- Often these kneaders are used as high-speed devices in combination with a slow-running single-screw extruder. The dispersion and homogenization is carried out by the kneader, while the extruder screw produces the melt pressure to overcome the flow resistance in the slot die (cascade extruder). These machines are particularly suitable for mixing fillers into PVC [43].

7.4.2.1.3 Combined extruders

There are many combinations of the machines that have been described; for example there are the so-called tandem arrangements:
- Twin-screw extruders as plasticators and single-screw extruders as output units, with degassing in the transfer adaptor and the same drive for the two extruders.
- Two single-screw extruders of different diameters and different screw lengths in series, with separate drives; degassing also in the connecting adaptor [60].
- Planet-roll extruder as plasticator with a single screw as the output device. Also vacuum-degassed in the connecting channel; both machines driven separately.

Or vertical-horizontal combinations like:
- Perpendicular single-screw extruder feeds a horizontal single-screw unit.
- Stuffing unit forcibly feeds flow-resistant plastic mixes into the planet-roll extruder.
- Dosing units feed measured quantities to the twin-screw extruder.
- Combined dosing/mixing/(stuffer) units for all kinds of extruder [66].

The high-performance compounder that can both compound and extrude raw material mixtures has been developed to follow the maxim: "From commodity plastics into engineering products" [92, 95, 111].

7.4.2.2 Screen changers

The plasticated mass normally passes through a screen-changing device between the extruder and the slot die, where it is shaped. The function of changers is to collect undispersed particles, foreign bodies and dirt in the screens they carry. In addition they can control the back pressure on the screw according to the mesh size of the screen.

The pressure drop across the screen and the support plate should be very small. In order that screens may also be used with thermally sensitive materials like PVC, dead spots in the melt flow channels must be totally avoided. Furthermore the screen support plate must have sufficient strength to carry the melt pressure and offer little flow resistance [67]. Figure 21 shows a hydraulically driven screen changer.

Figure 21. Extruder with screen changer and flat sheet die.
(Photo: Ramisch-Kleinewefers, Krefeld, West Germany)

- Screen changers with hydraulic sealing: the screen carrier is made tight against the melt pressure of some 300 to 400 bar by high surface pressure on the seal ring. The changeover time can be 0.5 to 1.5 s, depending on the size of the unit [67].
- The screen-cleaning block is recommended for all materials, except for those that are thermally sensitive; this device operates without disturbing production, by switch-venting the melt. A contaminated screen can be cleaned by controlled backflow of melt in a few seconds [72].
- The Auto Screen unit uses a band of mesh that is drawn continuously through the melt [73], at a rate controlled by the melt pressure. The band is 18 m long and is therefore usable for many thousands of hours.
- Cassette screen changers, similar to the one shown in Figure 21, can be used with PE, PS, PP, PA, PVC, and NR. For thermally sensitive materials, prefilling and vacuum arrangements are available [74, 80].
- Two or more rapid-change cassette units can be operated with a hydraulic unit and an accumulator. The corresponding number of distributors and switchover valves must be installed for this. Very fast cassette changeover is accomplished by the use of a hydropneumatic type of accumulator [75].

- Screen changers with a rotating, self-cleaning cage work fully automatically. The melt flows radially through the screen cylinder from the outside over a large surface area. The filter is cleaned automatically when the pressure in the extruder exceeds a preset limit. The cleaning procedure lasts about one to five seconds [76].

Instead of screens, adaptors with throttle rings and throttle valves are frequently used in the processing of PMMA and PC [78]. There are also extruders with fixed perforated plates supporting filter mesh. At the end of a production run a hinged flange provides access to the plate for exchanging the filter [79]. The melt pressure and temperature are measured before and after the filter plate to monitor the state of the screen.

7.4.2.3 Slot dies

Different materials flow differently in the molten state because of their structural viscosity: meaning that their viscosity is dependent on melt pressure and temperature, and on the shear stress in the flow channels of the die. Since, additionally, many different end-product thicknesses and widths are wanted, there can be no universal slot die. The cylinder of melt delivered from the adaptor is shaped in the die into flat web. There are a number of different construction methods:

- standard dies in many forms, with or without restrictor bars,
- single-channel coextrusion dies on a multilayer feed block (so-called black box) for certain materials, and multichannel dies with separate flow channels for all polymers.

Die bodies and die lips are made from tool-steel alloys. The surfaces of the flow channels are hard-chromed at least 25 µm thick and polished, and on some models the outer surface is chrome-plated [68]. An opened-up slot die and a sectional view are shown in Figure 22. The die body is heat-treated to give a surface hardness of from 48 to 54 Rockwell C. The surface roughness in all flow channels is about 0.1 µm. In one special process,

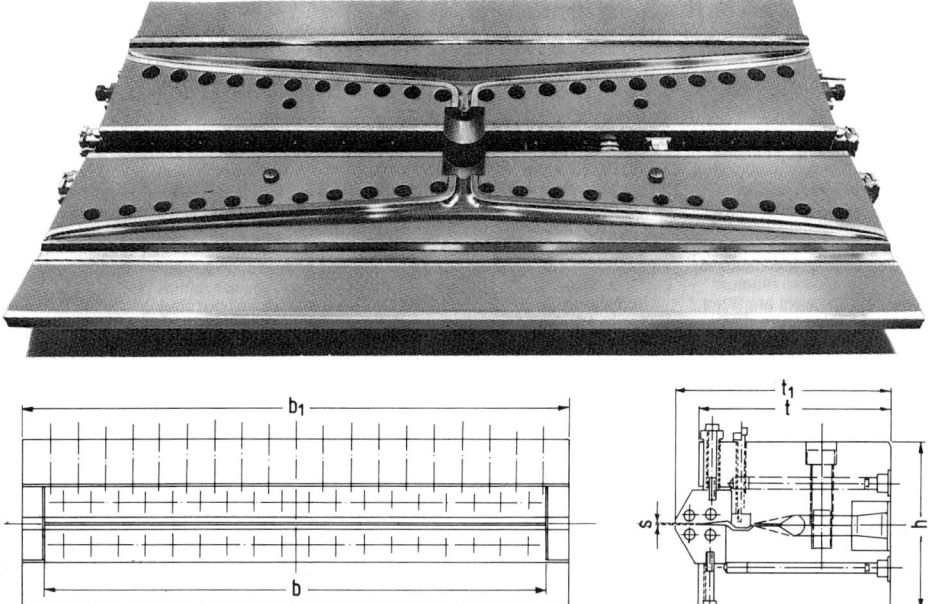

Figure 22. Opened-up flat sheet die with sectional drawing. (Photo: Cincinnati Milacron Austria AG, Wien, Austria)

fluorocarbons are embedded in the micropores in the chrome layer, thus reducing the friction coefficient and the wettability of the channel surface. These properties are of significant advantage when processing sticky materials like PVC [102].

Most film slot dies use coathanger-type manifolds designed with the help of rheological theory. To reach the die outlet from the manifold feed channel the melt flows through a flat, slit-shaped channel over a raised area called the pre-land (or island). The function of the island is to balance the different large pressure drops along the individual stream lines, so that the pressure drop along all paths is the same. The requirement is that the velocity of the melt emerging from the die should be the same at all points across the width. Thus, the flow resistance in the manifold and across the island must be the same for each component of the melt stream. To make it possible to balance out differences in these factors across the die width, a restrictor bar is placed in the manifold just before the island (see Figure 22).

Using this basic model one obtains the following equation for calculating the pressure drop in the slot die:

$$\Delta p = (8\,\dot{V}_0/\pi L) \int_{l}^{L} \frac{[x\,\bar{\eta}_R(x)]}{R^4(x)} \cdot dx + \left(\frac{12\,\dot{V}_0\,\bar{\eta}_s}{LH^3}\right) y(l) \tag{11}$$

$\dot{V}_0$ volume of melt delivered to one half of the die
L half the die width
l any stream line at point l
x coordinate across the die width
R radius of the manifold
$\bar{\eta}_R$ mean viscosity in the manifold
$\bar{\eta}_S$ mean viscosity in the slit region of the island
y thickness at x

The fundamentals of slot die design calculations can be found in [69] and [70].

When designing a melt distribution system for a slot die, the material having the greatest slope to the viscosity function should always be used as the basis of calculation. Dies for PMMA and PC, for example, are frequently needed, and since the flow curve of PMMA has the greater slope, the die would be designed rheologically for PMMA.

With proper manifold design the restrictor bar needs only minimal adjustment in the edge regions. With PS and ABS, for example, the bar must be adjusted in the middle. This will only amount to a few tenths of a millimeter, following the procedure mentioned, to correct the melt distribution to give a uniform melt flow from the die exit [103].

The standard dies mentioned include those with elastically adjustable lips, known commercially as Flex-Lip or Ultraflex dies:

- Dies with a flexible upper lip and no restrictor bar are especially popular for thermally sensitive materials like PVC. The lower lip can be moved as a whole over a range of 1 mm, and gaps of 0.5 to 3 mm can then be set flexibly by the upper lip. Their external shape allows them to be brought close to the roll nip. Since the adjustment screws work against the melt pressure and the elasticity of the steel, they are free of play [82].
- For larger thicknesses of extrudate, from 3 to 12 mm, flex-dies are given longer lips, and a restrictor bar chamfered in the flow direction.

If pressures in excess of 250 bar are likely to be developed during plastication, as occurs when processing thermoplastic PUR (up to 700 bar) or for very wide sheets or films, dies with heavier dimensions can be provided. This also applies to sheet dies.

- Sheet dies are designed for a wide range of thicknesses, 0.25 to 20 mm, but are not very suitable for heat-sensitive polymers, because of the head-on attitude of the restrictor bar

7.4 Production lines for films and sheets

to the melt stream (see Figure 22). The adjustable upper lips provide a maximum opening and adjustment range of 14 mm (or 20 mm in large versions) for lip lengths of 80 to 160 mm (or 240 mm), depending on gap size.

On dies that cover the thickness range 0.25 to 20 mm, the die lips have to be changed in the following steps [11]:

Die gap 0.5 to 1.5 mm:	one set of lips for 0.5 to 1.5 mm
Die gap 1.5 to 3 mm:	one set of lips for 1.5 to 3 mm
Die gap 3 to 10 mm:	one upper lip for 3 to 10 mm
and I:	one lower lip for 3 to 6 mm
II:	one lower lip for 6 to 10 mm
Die gap 8 to 20 mm:	one upper lip for 8 to 20 mm
and I:	one lower lip for 8 to 13 mm
II:	one lower lip for 13 to 20 mm

According to [114], manufacturers offer slot dies with bodies made from hardened steel, and die lips and restrictor bars from stainless steels. The flow-contact surfaces and flow channels are highly polished but not chromed.

Die bodies are heated by cartridge heaters, and the lips by oil that is thermostatted in its own closed heating/cooling system. This control of lip temperature ensures that the surface finish on the sheets is of high quality.

Film or sheet width can be altered by means of movable deckling bars on the die face, which cover up to 30% of the gap [79].

A die with a 1500 mm slot width may weigh around 1200 kg.

- The Auto-Flex die is automatically adjustable [71]; the lip adjustment screws have the facility for length changes of up to 0.25 mm by heating, and can provide particularly sensitive regulation of the lip-gap.
- Each screw sits in a block which permits it to be both electrically heated and air cooled. Coarse adjustment of lips can be carried out conventionally by means of differential push-pull screws [78, 79].

7.4.2.4 Polishing stacks and calenders

The rheological processes occurring in viscoelastic melts in the roll gaps of polishing rolls and calenders were dealt with in 7.2.1 and 7.2.2. Polishing stacks and calenders are distinct as individual machines, as was explained in 7.2.3; and, because the processes they carry out are different, they belong to lines that are quite differently designed (compare Figures 1 and 13). The polishing stack on a sheet extrusion line generally has three rolls, all with roller bearings in the two side walls. The polishing stack is of welded construction with cross-beams, moving gear, and unwind stations for laminating films.

7.4.2.4.1 Deformations of the polishing stack

Particular importance has to be attached to the design of the walls, because their deformation under load can be the cause of changes in the roll gap, in roll separating loads, and therefore in the thickness of the product web. Reproducible correction of this widening is possible only by means of calibrated force/deformation measurement equipment. For this it is essential that the deformations should be registered directly at the roll journals, so as to exclude elastic connecting links [6].

The deformations occurring in single-wall cast pillars for polishing stacks like that shown in Figure 23, have been demonstrated with the help of the finite element method (FEM) in [59]. Centrally driven and individually driven rolls were investigated.

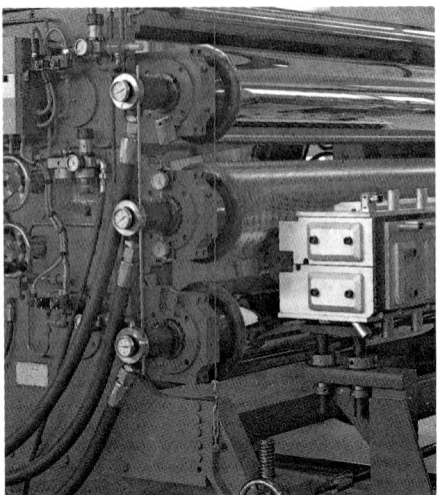

Figure 23. Polishing stack with solid pillars.
(Photo: Omipa, Varese, Italy)

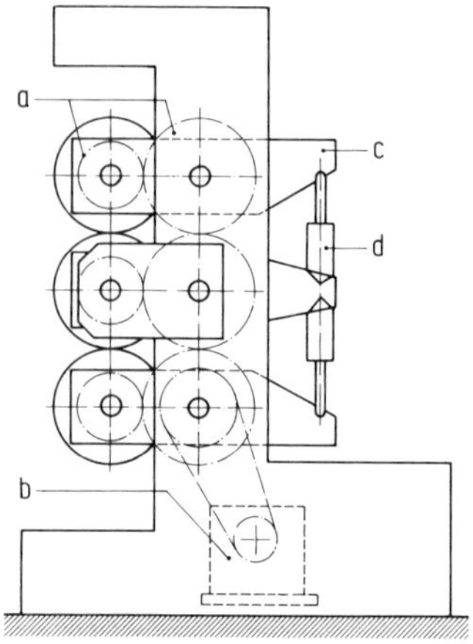

Figure 24. Polishing stack with central drive (see also Figure 1)
a gears for torque transmission,
b drive with reduction gear,
c lever,
d hydraulic cylinder

Figure 25. Polishing unit with individual drives.
(Photo: Reifenhäuser, Troisdorf, West Germany)

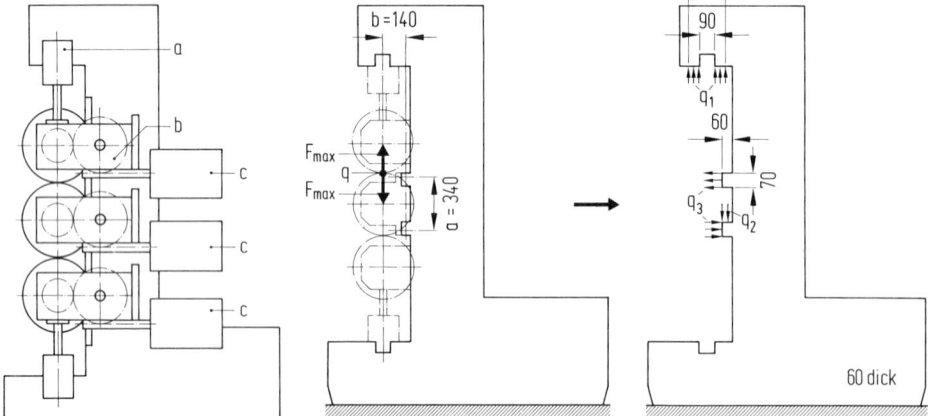

Figure 26. Polishing unit with individual drives, with load in the pillars from material fed into the upper nip
a lead-screw drive, b reduction gear, c electric motor

7.4 Production lines for films and sheets

On roll stacks with central drive, the rolls are adjusted only with a lever system, as explained in Figure 24. Polishing stacks with individual drives to the rolls can be adjusted both by lever systems, as in Figure 25, and by means of leadscrew boxes ("screwdowns") (Figure 26).

FEM studies show that the deformations in the pillar walls by lever systems are unfavorable for recording of measurement, because the levers themselves deform outside the walls.

In order to avoid the disadvantage of lever bending, the roll journals in a new design are pressed against the end stops on both sides, by means of a closed scissors system. The scissors are operated by double-action, tandem pneumatic cylinders [22]. The separation

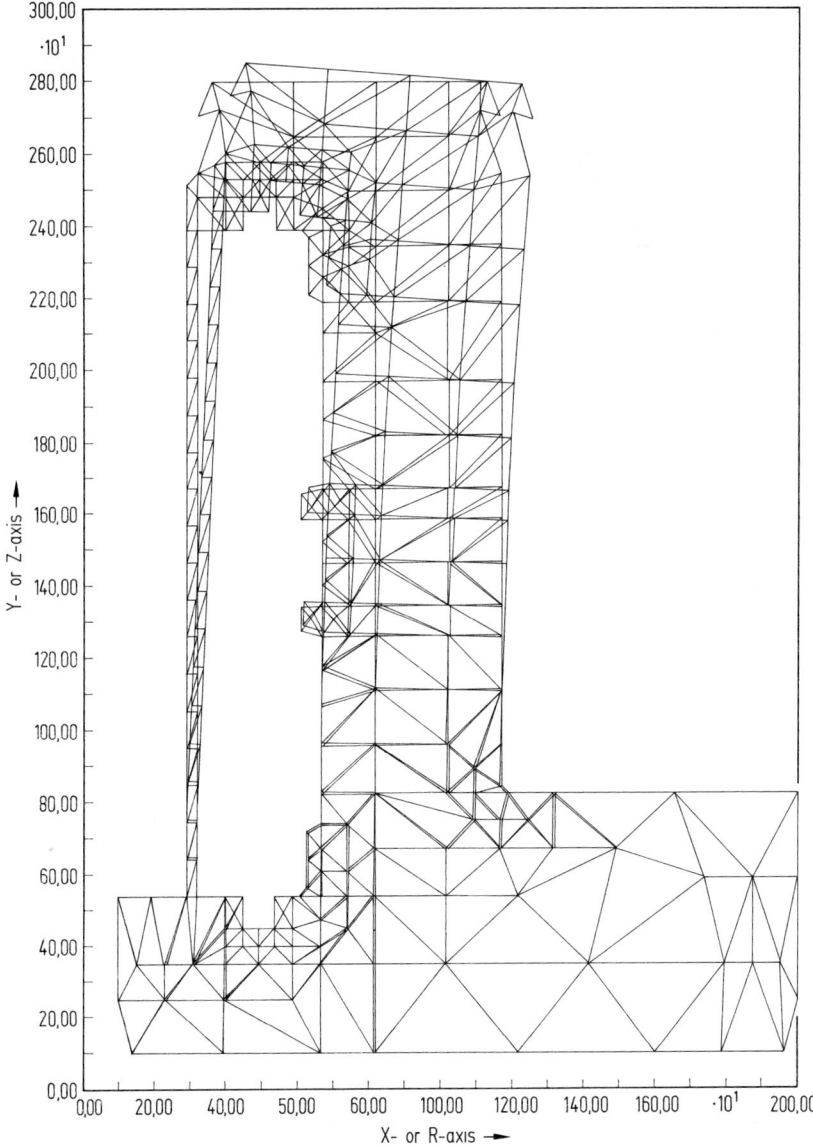

Figure 27. Element network for the pillar of Figure 26 showing deformations due to the load in the roll gap

between the rolls is adjusted by a lead-screw-driven wedge. Deformations are smallest when, as shown in Figure 27, a tensioning strap is mounted on the pillar in front of the roll bearings, and the rolls are adjusted by means of screw-downs. These straps of course make it difficult to dismantle the rolls.

For a separating load between the rolls of 800 N/cm, and with the 2000 mm-long rolls both loaded, the widening of the upper gap is 0.042 mm and of the lower one 0.04 mm. The associated deformation of the frame, which has 60 mm-thick pillars, can be clearly seen on the diagram (Figure 27), scale 1:10.

If only the upper gap is working against melt pressure, it experiences a widening of 0.062 mm, while the mid-point of the lower roll is still shifted some 0.007 mm upward. On the other hand, if the melt web is run through the lower roll-pair the gap here is widened by 0.058 mm, while the midpoint of the upper roll is shifted about 0.014 mm downwards.

Figure 28. Polishing stack with pillar guidance of the rolls.
(Photo: Battenfeld, Bad Oeynhausen, West Germany)

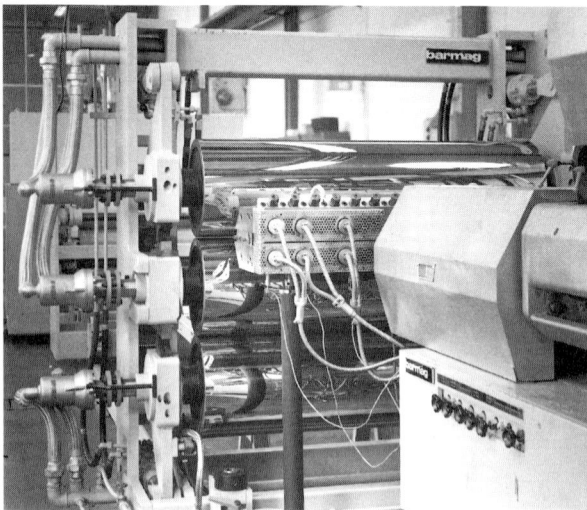

Figure 29. Polishing unit with double-wall pillars.
(Photo: Barmag, Remscheid, West Germany)

In addition to the method of roll-stack construction using floating roller bearings, shown in Figure 23, there is the technique illustrated in Figure 28 employing column guidance, and the one in Figure 29 that has two-side frame guidance where the roll journals are connected direct to the piston rod of the pneumatic cylinder. There is also a method of construction in

which the levers carrying the upper and lower rolls are mounted in a double-walled fashion, pivoted on both sides of the pillar walls. This eliminates lateral bending moments [81].
Less frequently one finds double-walled roll-mill frames of cast or welded design [60, 77].
Pneumatic adjustment of rolls, which has been the usual method until recently, is increasingly being replaced by lead-screw systems, which make it possible to track the displacement by digital or other methods of measurement, and even to reduce the roll separations mentioned [61].

7.4.2.4.2 Polishing-stack drives

The question of which kind of roll drive is preferable will be answered by referring to practical results. Thin webs, and films up to about 2 mm thick, can be run on centrally driven mills, because the change in wrap-round from roll to roll does not cause a very great change in the extension gradient within the film. Torque transfer from the central drive to the rolls can be carried out direct by a toothed chain, or by toothed-wheel gearing from the middle roll, which is driven by toothed belt or chain [80, 81]. With thick sheets the flexural change on transfer from roll to roll is much greater, as Figure 3 makes clear. Reduction of this stretching can only be achieved by adjusting the roll speed. The different roll speeds required for this can most simply be provided by the use of individual thyristor-controlled DC shunt-wound drives on each roll. It is also possible with central drive to fit each roll with a differential gearbox. A Schmidt type coupling [88] is installed between the output shaft and the roll journals.
The polishing stack is synchronized with the haul-off, from the drive. Speed matching of $\pm 0.5\%$ between the two drives is achieved. Additionally, the haul-off speed can be varied some 10% with respect to the polishing rolls.
Since the haul-off machine produces the tension in the web, it can be arranged that the roll stack runs tension-free. The roll-drive then wastes no power. To ensure that the roll torque does not become unstable, and the set speed is held precisely irrespective of winder tension, there is a strong preference for using four-quadrant drives for polishing mills. These provide synchronized control to a precision of $\pm 0.5\%$ [62, 80, 87].
Attention also has to be paid to the stability of the height adjustment of the traveling gear on the polishing stack. On modern units leadscrew gearboxes on each wheel are centrally driven to make uniform height adjustment possible [63].

7.4.2.4.3 Special roll arrangements

According to the latest findings, the processing conditions and quality characteristics of films and sheets can be significantly improved by alternative arrangements of the three rolls of the polishing stack (see Figure 11).
With the rolls vertically over one another, one can run in from above or beneath. If the material web is fed into the lower roll gap, good surface quality is obtained by double polishing in the two gaps. When film or sheet is embossed on the middle roll the lower roll should be fed, because then the embossed side will be uppermost during cooling, and will not be scratched and can be inspected easily.
If it is required to produce thick sheet, the melt is fed into the upper gap, so that the plastic mass clings to the middle roll without trapping air bubbles, which would hinder heat transfer and lead to surface imperfections and local stresses. Good contact of thicker webs is also favored by their own weight acting downwards on the roll.
When lamination is carried out in the upper gap, adhesion can be improved by preheating the film well. If the axes of the rolls lie in the 45° plane, as shown in Figure 11, the wrap angle of the web around the roll is reduced from 360° to 320°. The reduced dwell time,

which may be advantageous with thermally sensitive materials, does not affect the surface quality. A further advantage is that the slot die can be brought even closer to the roll gap, and the length of melt curtain, and thus the amount of air entrained, is reduced. Since feed can optionally be from below or above at this point, the method can be used equally for film and sheet.

The third variant of roll arrangements (Figure 11) also brings advantages with large sheet widths. For strength reasons already mentioned, a large roll diameter has to be used. Since with this arrangement the wrap angle is only 240°, the time of contact of the melt with the rolls can also be reduced without the surface quality suffering (compare 7.4.2.6). For processing PC, even a fully horizontal arrangement of the polishing stack with individual roll drives has proved to have advantages [96].

Such variations are not possible with calenders, because of the large scale on which they are built.

7.4.2.4.4 Thermal processes on polishing stacks

Extrudate webs are cooled or thermally conditioned on the rolls of the polishing stack. Those of partially crystalline materials are strongly cooled so as to prevent crystallization and the possibility of undesirable haziness. Amorphous polymers are annealed at relatively low roll temperatures to carry out stress relaxation. For the same purpose, partially crystalline polymers are held for a fixed time after cooling at a temperature somewhat below that of crystallization. The boundary conditions for cooling calculations can therefore vary greatly. And in addition, the production speed depends on the cooling capacity of the rolls, on the dwell time required in a suitable temperature range to obtain a specified quality, and on the uniformity of cooling and the temperature gradient in the product web.

The temperature change $\Delta\theta$ in time t can be calculated by solving the Fourier differential equation:

$$\frac{\partial \theta}{\partial t} = a(\theta) \left[\frac{\partial^2 \theta}{\partial x^2} + \frac{\partial^2 \theta}{\partial y^2} \right] + d(\theta) \left[\left(\frac{\partial \theta}{\partial x} \right)^2 + \left(\frac{\partial \theta}{\partial y} \right)^2 \right], \tag{12}$$

involving the temperature-dependent materials properties:

$$a = \lambda/\varrho\, C_p = \lambda \cdot V_p/C_p;$$

a is the temperature conductivity in cm^2/s, λ the thermal conductivity in W/m K, V_p the specific volume in cm^3/g and C_p the specific heat at constant pressure in Ws kg^{-1} K^{-1}. In the second term of Eq. 12, d stands for an expression of the form

$$d = (V_p/C_p)\, d\lambda/d\theta \quad [4, 5, 84, 85].$$

In the processing of partially crystalline polymers, the cooling rate and the stretching of the melt during cooling on polishing roll stacks affect the development of the structure, the thickness and the elastic modulus of the film or sheet. If the cooling rate is reduced, the modulus increases and the spherulite diameter becomes greater. It has been established that the more the melt is stretched between the die and the lower roll, the higher is the value of density measured on the film, in spite of unchanged cooling-roll temperatures [86].

Roll change can be carried out rapidly on most lines. Facilities such as roll journals with rotating, self-tightening, snap-fit water connections are usually fitted [80].

The knead-stock illustrated in Figure 5 is generated during the so-called high temperature calendering process. In the lower temperature Luvitherm process, the PVC melt moves into the nip without back-flow or swirling [64]. As a consequence of the relatively low processing temperature of only about 180 °C on the calender, the film is not produced from a homogeneous melt, but by sintering. By subsequent Luvithermization, a flash heat treatment at up to 220 °C, good-quality films can be produced. This is the essential difference between the Luvitherm process and roll-kneading or high-temperature processes [65].

7.4.2.5 Embossing, laminating, and coating

Three-roll polishing stacks are also suitable for embossing of film and sheet. The middle roll is exchanged for an embossing roll, for example a roll with a matt surface for one-side matt film or sheet.

Sheet extrusion lines can also be used for making composite films and sheets. One possibility is to let a molten film run perpendicularly into the upper gap of the roll stack, from a slot die on an auxiliary extruder (Figure 30). If a film of melt is laid down on a carrier or on an existing melt web, one speaks of extrusion laminating. If a solid film from the roll, like that shown in Figure 30, is combined in the roll gap with the main extruded melt web, one speaks of extrusion coating. Two-sided- as well as single-sided coating is possible. This is carried out by feeding an additional film to the underside of the lowest polishing roll from an unwind station beneath the roller table [78].

Complementarily, doubling is the joining of two solid webs, often textiles, possibly with the addition of an adhesion promoter. Such composites are produced on a doubling calender.

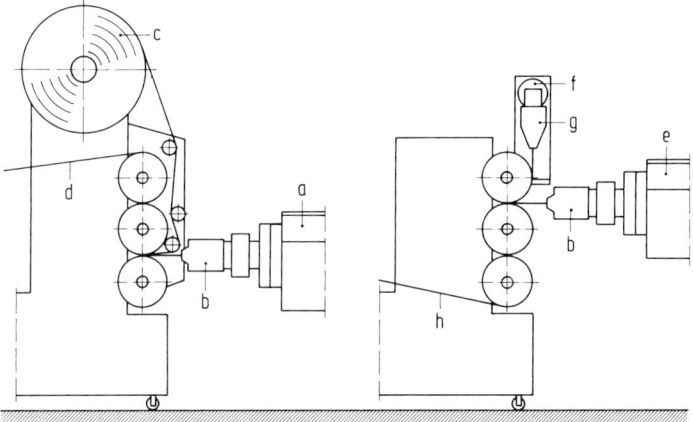

Figure 30. Polishing stacks with arrangements for melt laminating and melt coating
a extruder, *b* slot die, *c* carrier web, *d* coated sheet, *e* main extruder, *f* auxiliary extruder, *g* casting die, *h* laminated film or sheet

7.4.2.6 Cooling sections

Cooling of thicker films and sheets is usually by means of the support air on the haul-off roller track. Roller tracks may be 4.7 and 10 m long [87], and are normally fitted with chromed steel rollers with full-width axles.

Thin films need the intensive cooling of water-fed rolls, because of the high production speeds. If the film has to be wound up at room temperature it is recommended that two to five large-volume cooling rolls be installed between the polishing stack and the wind-up. These rolls are made of surface-hardened steel and are highly polished to a surface roughness of 0.1 to 0.2 µm; for processing PVC they are chromed to give corrosion protection [96]. Water flow at 1.5 to 2.5 m/s is achievable with a pump capacity of 100 to 250 l/min [89].

The Fourier differential equation for thermal conductivity given in the earlier section [12] has been used with the help of computers to determine temperature profiles in films and sheets. Results are presented in Figures 14, 15, and 31. Figures 14 and 15 refer to amorphous polymers, while Figure 31 is based on crystalline PP.

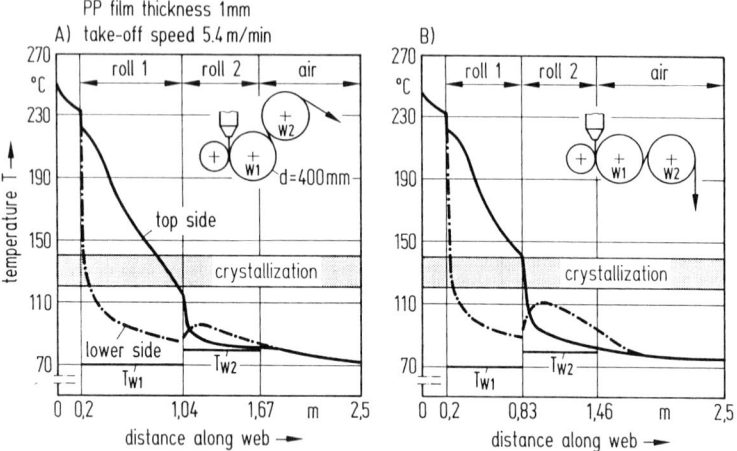

Figure 31. Cooling of a film web with different roll arrangements

For example, in Figure 31 one sees that in the crystallization range the cooling rates are very different on the lower and upper sides of the film. Before the upper side contacts the roll, the underside has already been cooled below the crystallization temperature. Thus the upper and lower sides of the film have different mechanical properties. If these films are reheated and thermoformed, they will distort. In order to remedy this disadvantage, either the temperature of the rolls must be reduced or the contact time with the first roll shortened in order to equalize the cooling of the upper and lower sides. At a preset production speed, however, the change of temperature influences the target cooling rate. The alternative is to change the dwell time on the first roll, which can be done by altering the wrap angle, as Figure 31 illustrates [93]. This kind of requirement is motivation for the design features explained in Section 7.4.2.4.3 and Figure 11. By solving Eq. 12 one obtains the following equation for the contact time, t_K, of the film on the roll [85]:

$$t_K = \frac{s^2}{\pi^2 a} \ln\left(\frac{8}{\pi^2} \cdot \frac{\theta_M - \theta_W}{\theta_E - \theta_W}\right). \tag{13}$$

It can be seen that the cooling time required increases with the square of the web thickness. It is also clear that, in order to shorten the cooling time, the temperature (θ_E) of the web as it leaves the last roll should be as high as possible, while the roll surface temperature should be as low as possible. The melt temperature (θ_M) as it enters the roll gap should also be as low as possible. But these requirements cannot be met, for the reasons already explained, and so compromises have to be made. Practically speaking, it is good policy in designing a sheet extrusion line, to fix the cooling factor, defined as [93]:

$$\theta_K = (\theta_E - \theta_W)/(\theta_M - \theta_W) \tag{14}$$

for each polymer in relation to the quality requirements. This provides a key value for use in the design and operation of sheet extrusion lines.

7.4.2.7 Take-off units with cutting equipment

The take-off as a rule has two rubberized, centrally driven rolls pressed together by pneumatically operated levers. The roll drive is electrically coupled to the roll-stack drive.
The tension on the web can be adjusted by hand with a potentiometer, or by preprogramming. The tension produced in this way, or by additional pendulum or dancer rolls, must be sufficient to ensure that the web can be cut cleanly.
Lengthwise cutting devices are smoothly adjustable for width of cut; there are various systems:

- Industrial razor blades fixed to the roller table; these are used for ABS, PP, and PE films and sheets; up to 12 mm thick they are still soft enough after leaving the roll mill to accept a blade cut.
- Scissors cut with rotating circular knives, for well-cooled ABS and HIPS films; light versions for films 0.5 to 4 mm thick, heavier ones up to 8 mm; knives are often driven by the haul-off.
- Circular saws, with swarf suck-off, mounted on the haul-off; for thick sheets – 6 to 20 mm – of tough polymers like PMMA, PS, and PC.
- Circular saws, with swarf suck-off, located ahead of the haul-off; for thick sheets – 10 to 30 mm – of the HDPE and PP.
- Hot-wedge separating after the haul-off; for brittle materials like PMMA and PS, sheets 2 to 6 mm thick; the sheets are melted along the cut-line and later broken apart by the double haul-off unit [112]. Such a haul-off has a pulling force of 10 kN.

Edge trim from film less than 0.4 mm thick is evacuated and conveyed by an injector line powered by air from a radial fan. Above 0.4 mm film edge strips are wound at speeds of up to 50 m/min.

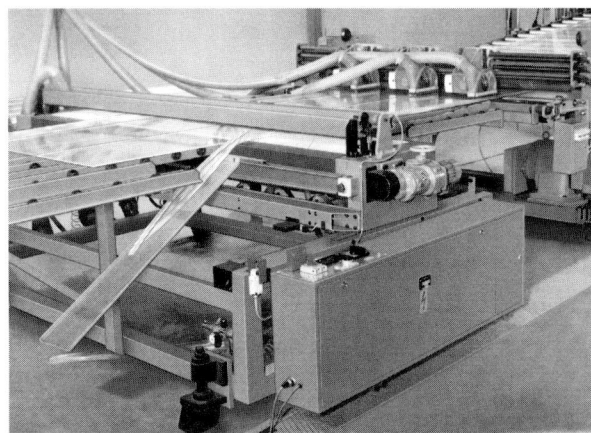

Figure 32. Traversing cutter unit uses circular saws for brittle sheet materials. Swarf sucked away.
(Photo: Reifenhäuser, Troisdorf, West Germany)

Sheet cross-cutting is carried out, under photocell control of the format length, by a guillotine, up to a maximum thickness of 1.5 mm. Impact-resistant plastics like PP, HDPE, HIPS, rigid PVC, and ABS can be cut in this way. With brittle materials like PS and PMMA, and especially tough ones like PC, or for greater thicknesses, traveling circular saws with swarf removal are used [109]. An example of such a unit is shown in Figure 32 [78, 87]. However, since saws produce noise, swarf, and dust they are largely being replaced by devices that use heated industrial cutting blades. These are set to leave a thin skin of material, and the prepared sheet is fed into a haul-off with two roll-pairs having an arched roller track between them. The enforced bending breaks the sheet at the cut-line [80].

A protective film can be introduced from a separate off-wind before the haul-off to protect the surface of the sheet when it is stacked [109].

7.4.2.8 Sheet off-loaders

Nowadays there is no handling. Instead, program-cut sheets are automatically removed sideways from the production line by suction-cup lifters, and stacked on pallets. Equipment of this kind is illustrated in Figure 33.

Figure 33. Suction-cup lifter/stacker removes format-cut sheet from the line.
(Photo: Reifenhäuser, Troisdorf, West Germany)

7.4.2.9 Film winders

Films up to 2 mm thick can be wound. Film winders available range from single-station, stationary winders to fully automatic multiple-turret winders. Winders usually have an additional winding station as the blank for continuous roll change.

The requirement laid on all types is the same: to produce flat, straight-edge rolls. To achieve this objective, a number of sheet extrusion lines have the extruder mounted horizontally on an oscillating turntable. Any thickness variations across the film are spread out by this arrangement [78]. In another situation a pneumatic oscillation generator with hydraulic damping is fixed to the winder to obtain flat-wound film [83].

The film can be automatically transferred to the winding core up to a web speed of 610 m/min. The maximum speed for manual roll change is 50 m/min.

Devices for keeping the web tension constant work very precisely if the winding ratio (roll/core diameter) does not exceed 6:1 [80]. Modern winding technology offers lines embodying all the units needed for roll transport, cutting and winding. The basis of this is the turret winder with automatic roll change.

The following methods for keeping the web tension constant are available [107, 110]:
- The tension-measurement roll with a non-driven, finely adjustable contact roll, controls the DC shunt-wound motor of the winding shaft.

- In contact winding the winding shaft is driven by means of a speed-regulated contact roll. The growing roll of film is moved horizontally on free-running sleeve-bearing guides. A pneumatic system determines the contact pressure.
- The hyperbolic winder controls the torque in relation to the speed of the winding shaft, to give a constant tension on the film web.

7.4.3 Factors affecting production and quality

7.4.3.1 Scale of production

The production range of sheet extrusion lines extends from thicknesses of 0.2 to 25 mm, and from widths of 600 to 3600 mm. A compilation of the production output levels to be expected from sheet extrusion lines for different sheet dimensions and polymers used, is given in Figure 34. Using the haul-off speeds specified and equations 12 and 13, it is possible to judge the technical quality of the program available for producing films and sheets.

Figure 35 gives a similar compilation for films over the thickness range from 0.2 to 2 mm, which can be produced on sheet extrusion lines down to this lower limit.

Specific energy input (kWh/kg) is one merit characteristic of screw geometry. Another significant one is specific output per revolution; with LDPE, for example, typical values are:

        80 mm screw − 0.04 kg/rev.
       150 mm screw − 0.074 kg/rev.

and for PP:

        80 mm screw − 0.044 kg/rev.
       150 mm screw − 0.076 kg/rev.

7.4.3.2 Energy usage

The quantities displayed in Figure 36 are also interesting for drawing up energy balances. The control of roll-surface speed covers a range of 1:100 in two steps of 1:10 by means of polarity switching: 0.5 to 5 m/min and 5 to 50 m/min for example.

The two-roll haul-off has a power range from 1.1 to 2.7 kW [78]. One can see from Figure 36 that in the lower roll-width range, up to about 1800 mm, the overall consumption for the central drive is approximately the same as that for the system using individually driven rolls. As the roll width increases, however, the power consumption of individual drives increases much more sharply than with central drive.

It can also be seen that lines with central drive are available up to a roll width of some 2700 mm, and those with individual drive up to about 3600 mm.

The ratios between production-line roll widths and diameters, compiled in Figure 37, tend to become larger the wider the roll.

The total drive power of a medium-size sheet extrusion line with 1800 mm-wide rolls, without the extruder and screen changer, can reach the following level [78, 81]:

Slot die	1750 mm	41 kW	
Polishing stack	1800 mm	3 kW	
Two-roll haul-off	1800 mm	1.6 kW	
Circular saw	1800 mm	3.7 kW	
Subtotal:		49.3 kW ≈ 50 kW	
Heating for one roll 12 kW, thus for three polishing rolls		36 kW	
Heat transfer fluid pump		4 kW	40 kW
Total:			90 kW

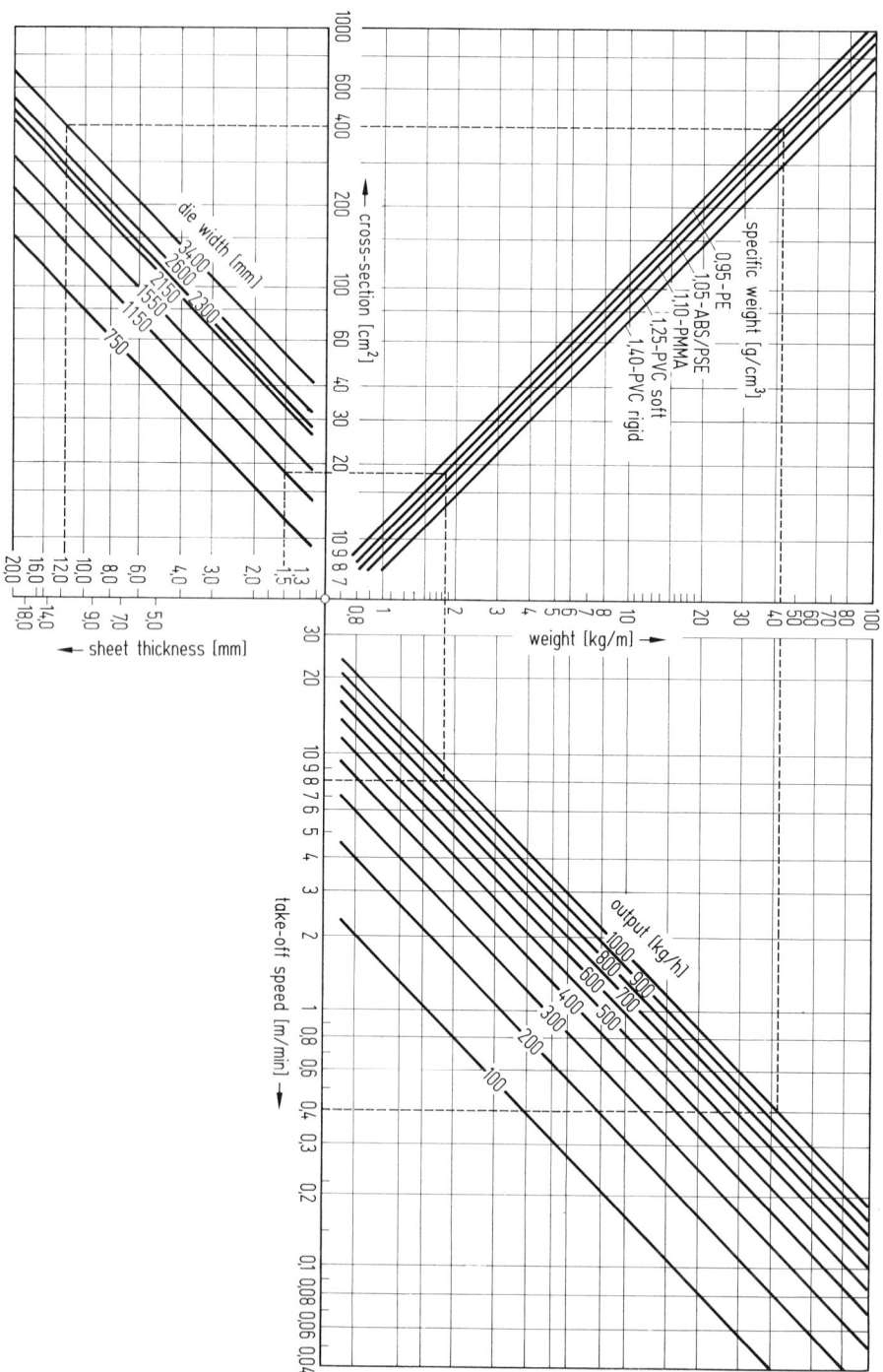

Figure 34. Production performance of extrusion lines for various materials and sheet sizes [78]

7.4 Production lines for films and sheets

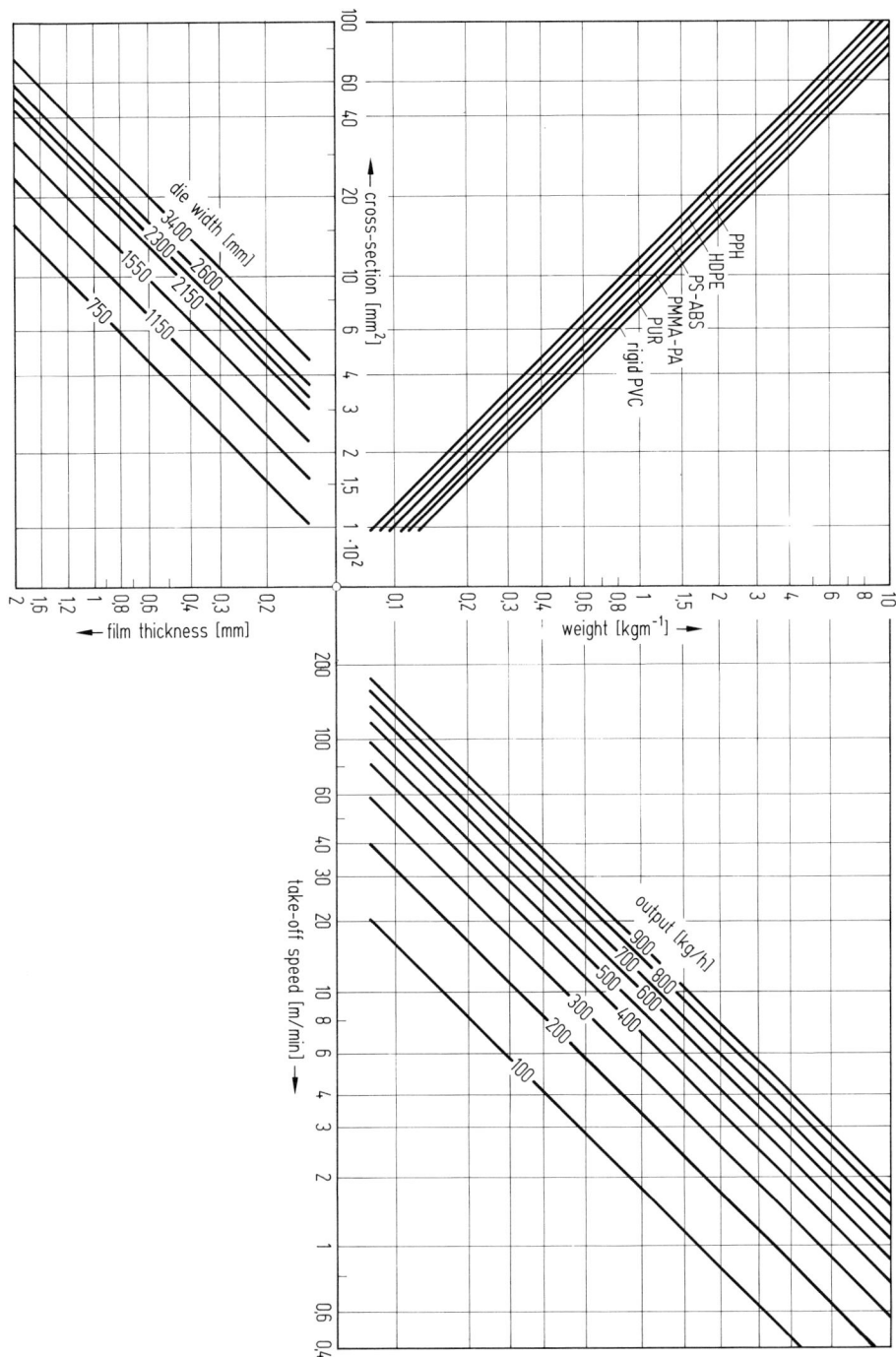

Figure 35. Production performance of extrusion lines for various materials and film dimensions

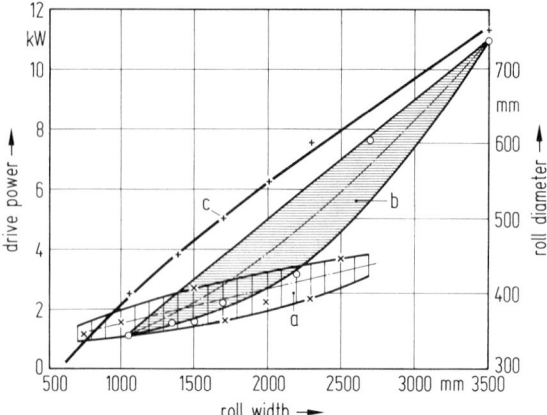

Figure 36. Roll dimensions and drive powers
$a \cong$ power consumption with central drive, $b \cong$ power consumption with individual drive, $c \cong$ roll dimensions

The loading on three-roll polishing stacks of different sizes is deducible from Figure 37, where the linearity between the torque required for rolling and the line loading on the rolls is obvious. The relevant roll dimensions are indicated for each of the points plotted.

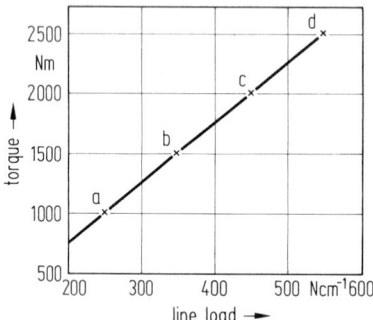

Figure 37. Loading of the rolls on a polishing stack
$a \cong$ diameter/length ratio 300/800,
$b \cong$ diameter/length ratio 400/800 to 1800,
$c \cong$ diameter/length ratio 600/160; 490/2400,
$d \cong$ diameter/length ratio 600/3000

7.4.3.3 Production limitations

Limitations in the processing of thermoplastic melts on sheet extrusion- and calendering lines are to be seen:

1. In the shear tolerance of thermally sensitive materials (like PVC) in the extruder.
2. In the shaping of sticky polymers in the slot die, above all in the balance of the melt distribution in wide dies.
3. In the possibility of applying the line loadings necessary for rolling-out in the nip.
4. In the shortest permissible dwell time of the product web on the roll surfaces. Possible adjustment for higher circumferential speeds by increasing the roll diameter appropriately, is then limited by the flatter angle of entry into the roll gap. The result is that the roll-separating force is increased to such a degree that the rolls are greatly deflected and, under some conditions, pushed apart.
5. In the thermal treatment period on the rolls that results from the low thermal conductivity of the plastic, and the subsequent cooling period required. On sheet extrusion

lines the product web is cooled by the supporting air; thus, appropriate design effort has also to be put into the cooling section. On calender lines film webs run over a series of large-volume, water-cooled rolls.
6. In the roll-length-dependent bending and the uniformity of the temperature distribution on the rolls. Because of this, the roll length is limited and dependent on general requirements.
7. In the drying process, especially with hygroscopic polymers. Drying often requires several hours, and depends on the polymer type, granule structure and granule size [41, 42]. Increasing capacity by using larger units, or several dryers (dry-air systems) will largely be limited by product-related costs.
8. In the requirement for greater clarity in calendered films. Best clarity is obtained by using the Luvitherm process. To achieve this, the temperature of the feedstock and the roll temperature must be low, and the roll speed high. When the melt temperature is relatively high, to get good clarity the dwell time on the rolls must be increased by reducing the roll speed [64].

7.4.3.4 Influences on quality

The quality of the extrudate can be affected from the beginning by the raw material compounding systems, as already described in 7.4.1. But the method of plastication is important too. Use of the co-rotating twin-screw extruder is taken as a particularly important example.

The quality of the sheet is essentially determined by uniform dosing of fillers, effective degassing of the melt so as to wet the fillers properly with the melt, and good dispersion and homogeneous mixing of the fillers.

Packing density, particle size distribution, pourability, and tendency of the fillers and the polymer to fluidize, are also influential factors. The fillers can be worked into the plastic in the solids zone or the melt zone of the kneader. If large shear forces are needed for dispersion, the fillers are added to the solids zone. With fine-grain fillers no further size reduction is needed; they can thus be dispersed homogeneously in the melt zone. If large quantities of fillers are required, they can be added to the solids- and the melt regions [43].

The quality of the film or sheet is strongly affected by:

- the distance between the slot die and the roll gap,
- the uniformity of the contact line parallel to the axis of the middle roll, between the product web and the roll,
- the uniformity of cooling both sides of the product web, which determines whether or not symmetrical morphology develops from outside to inside.

7.4.4 Special designs of line

Some special units have been developed from the sheet extrusion lines whose basic construction was described earlier. Their essential features are outlined in the following section.

7.4.4.1 Coextrusion lines

As a rule these units differ from the normal sheet extrusion lines only in the arrangements ahead of the polishing stack: instead of one extruder there are two to four. Each one is purpose-built in size and function for its particular job [99, 105]. It is even possible for one extruder to contribute two layers to the multilayer composite.

There are three variants:

- layer formation in an adaptor, the so-called black-box process: the extruders deliver their melts to the adaptor input together to create the layers.
- layer formation within the slot die: the extruder feeds the die through a separate adaptor. The melt streams are kept apart within the die until they are united just before the lips.
- separate melt streams within a multichannel die, which are united immediately after the die exit: in this case each extruder is connected to the die by a separate adaptor.

Demand for coextruded film has arisen to meet requirements that cannot be met by single-layer films. For example, long-life food packs must be completely neutral in food contact (PE or PP), provide an oxygen barrier (PVC), and have good thermoforming characteristics (SB or PS) [115].
In this case a five-layer film has to be produced using five extruders; one extruder each for PE, PVC, PS, and two small extruders for the tie-layers between the different polymer films [104].

Figure 38. Calender for lamination with melt.
(Photo: Ramisch-Kleinewefers, Krefeld, West Germany)

The shelf life of packed foodstuffs can be greatly increased by using a PVDC barrier layer in a five-layer composite with PE and SB; or in six-layer film, where the additional one is of SB rework. The Dow feed-block process, which is used to make these films, spreads out the individual streams into several thin layers; these are then shaped as multilayer melts in normal single-layer dies. It is essential for perfect operation that the viscosities of the polymers should be similar at processing temperature. Then it is possible to produce multilayer sheets and films in thicknesses from 0.15 to 15 mm with up to 15 layers [105, 106].
So as to obtain the desired individual layer thicknesses, the various extruders are fed by dosing devices. This allows the throughput of each extruder to be controlled in relation to the layer thickness and haul-off speed [127]. With i extruders the total throughput of the line is:

$$\dot{m}_{tot} = \dot{m}_1 + \dot{m}_2 + \ldots + \dot{m}_i \tag{15}$$

7.4.4.2 Coating lines

The method of construction of these lines departs significantly from that of sheet lines. Coating lines take carrier layers of paper, aluminum, films or textiles and coat them from roll to roll on laminators with one or more molten polymer films on top of one another [90]. But it is also possible for carrier layers to be coated with E- or S-PVC melts on three-roll calenders (see Figure 38). Roll widths used are between 1350 and 3500 mm.
Carrier layers like non-wovens or textiles are preheated by infrared radiators [15].
Such solvent-free coating is carried out with thermoplastics, elastomers, and synthetic rubbers using dry blends, pellets, agglomerate, or granules. The middle roll has a smaller diameter than the two outer rolls, which reduces the roll line-loading. In addition, the lower roll supports the middle roll and diminishes its distortion. Thus even high-viscosity thermoplastics and elastomers can be processed [108].
Coat thicknesses run from 0.05 to 1 mm. Carrierless films in thicknesses from 0.05 m to 0.8 mm can also be produced.

7.4.4.3 Ribbed hollow-profile lines

Lines used for manufacturing ribbed hollow profiles from PMMA, high-impact PMMA, and PC can be regarded as special versions of sheet lines. Single-screw extruders are used for plastication in association with special slot dies.
These slot dies have a range of exchangeable divider inserts that can be mounted in the die mouth, or in the land and pre-land area. They vary in size and number depending on the dimensions of the sheet to be produced. Figure 39 shows several examples of cross-sections of hollow-profile sheets, which can be up to 2000 mm wide and 22 mm thick.

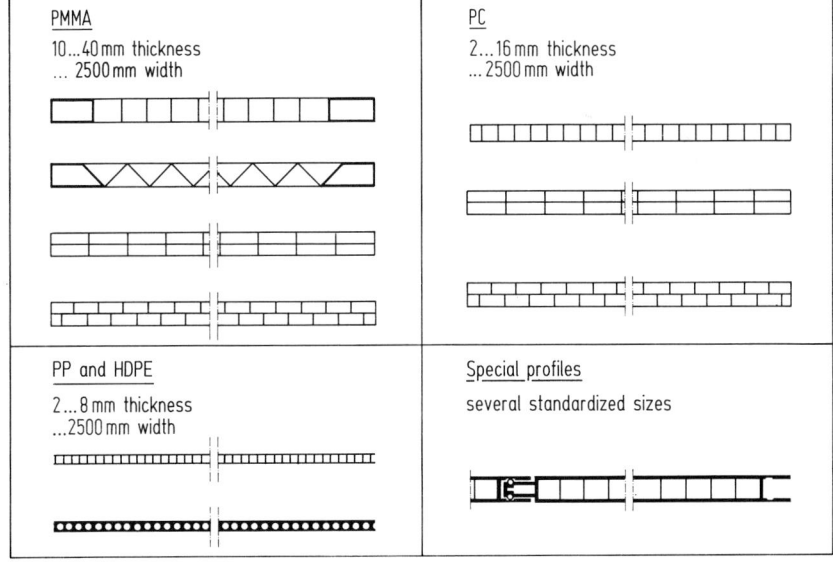

Figure 39. Cross-sections of double-wall ribbed profiles

The hollow profile is polished on both sides and stabilized after it leaves the die, in a contact calibration device. Cooling and vacuum zones alternate in the calibrator.
In order to overcome the large resistance forces that occur during calibration without damaging the sheet, the haul-off has six steel-on-rubber roll-pairs. The lower, rubberized

rolls are driven, while the upper ones are held in contact vertically by pneumatic pressure. A PE film is fed in on both sides to protect the sheet surface from damage.

Internal stresses in the sheet are then relaxed, in-line, under infrared radiation. Next come transverse saw-cutting with digital setting of sheet length, and the hydraulic intermediate storage table.

Applications of hollow profiles:

- for growing houses, shed roofs, and insulation – profiles of HIPMMA 10 to 40 mm thick and PC 2 to 16 mm thick, 2000 to 2500 mm wide, are used. These sheets are distinguished by good light transmission and a low heat-transmission coefficient,
- for solar radiation collectors – PMMA sheets 5 mm thick and 2500 mm wide have internal water channels 4 mm in diameter,
- for cartonage – hollow-profile PP and HDPE sheet 2 to 8 mm thick and 2500 mm wide is used [100].

7.4.5 Auxiliary equipment for sheet extrusion lines

Stretching of sheets

For biaxial stretching the films or sheets are heated to a particular optimum temperature, somewhat below the crystalline melting point, and are stretched at a constant rate in the lengthwise and transverse directions. Biaxial stretching lines work either separately using roll-film feedstock, or integrated with the sheet line. Heat left in the extrudate on in-line stretching units is useful for the process [116]. During stretching it is essential to hold the optimum temperature precisely; this should be some 2 or 3 °C below the crystalline melting region. If the temperature is too low the film or sheet will tear away from the clips, and if it is exceeded, holes will appear in the web [117].

The elastic modulus of PP can be raised to some 1500 N/mm^2, double the value of normal material. This was achieved [117] with an area stretch ratio of 1:38 (length increase over 1:6 in each direction) and a gripper speed in the lengthwise and transverse directions of 10.8 m/min.

Removal of electrical charge from film and sheet

Electrostatic charge is generated on the product surface during manufacture by rubbing against the rolls, and during wind-up and stacking. This charge can make it difficult to print on the surface, and it attracts dust.

This disadvantage can be avoided if the surface is electrically neutralized. Many kinds of ionizers are available on the market for this purpose. They all work on the same principle: a high-voltage generator feeding a semi-conductor ion generator. The highly charged positive or negative ions produced are transferred to the film or sheet surface by means of conductive metal brushes, or by an airstream.

The positive ions neutralize the surface and the negative ones the dust particles, which can then be removed by blown air [118, 119, 120]. The ozone created by ionization of the air is evacuated. The generator current can be 3 to 16 A at normal voltage, depending on the size of the unit; its output can be between 1.2 and 20 kVA at 8 to 17 kHz.

Safety devices

Tug lines that run all round the roll stack on the operator side, and which can be reached from all positions, make it possible to activate emergency cut-out switches. These cause the roll-mill drive to be stopped at once and the gap opened to 100 mm [80, 96].

7.4.6 Monitoring systems and operator comfort

Automation of sheet lines is difficult, because one is considering a continuous thermal process in which the flow behavior of the various polymers differs. The aim of automatic operation must be to reach target product dimensions as quickly and cheaply as possible after start-up or change. It is worth while to have operator-free running – meaning fully automatic self-monitoring and initiation of actions when faults occur.

Microprocessors are used for the task of open- and closed-loop control, even of sections of the extrusion line, for continuous monitoring of the most important measurement and control points, and for recording the production process.

In [121] a microprocessor control system is presented that calls up the production target values stored on a cassette tape, and passes them to the relevant functional elements. The microprocessor monitors the line, controls it in relation to production data, and reports disturbances.

At the end of a production run the microprocessor shuts down all moving sections of the line, brings down the extruder barrel temperatures under control, and then switches the extruder off.

Programmed data are printed out and/or displayed on a monitor screen (VDU). The control of program running should be by means of a program language that is easy to understand and learn.

Now that data storage costs have come down, there is a move to decentralized microprocessors. The microprocessor is taking over the function of the control cabinet. For example, it is moving into the place of the analog controller, and operates digitally. The DD controller (DDC) which operates in this way has switchable control structures, meaning that it carries out different dedicated control functions on start-up and during production running [123].

The temperature control on the rolls, the extruder, and the slot die is important for material saving, less waste and good quality. Therefore, following the ideas discussed in [96], the digital potentiometer System 301^{TM} is being used instead of the usual PID controller. This gives digitized signals to the microprocessor, which are also displayed on the VDU and/or printed out. The complete sheet line can be optimized with the FM-3 computer, and be kept in a state of steady production.

Rolling processes, too, can be automatically reproduced. The speed of the rolls, the gap during plastication, mixing and take-off, and the time involved in individual steps are preset digitally. Subsequently the complete process is controlled automatically [122].

An example of microprocessor-controlled sheet extrusion is described in [124]. The speeds of the twin-screw extruder and of the dosing unit are measured digitally, and the relative speeds of the three-roll stack with individual roll drive and of the take-off drive are digitally controlled.

The sheet extrusion line is fully computer-controlled by three interfaced System Automate 35 microcomputers. All working data can be displayed on a VDU and can be altered during production by means of a program keyboard. Instructions for various raw material recipes can be stored on magnetic tape.

Manual intervention is possible during start-up, and any altered settings are automatically recorded and programmed by the 18 K-Byte RAM computer. It is possible to use variations in the product thickness as quantities for controlling process parameters. This is done by the Profitmaster System 5001 [125]. It combines the functions of measurement, monitoring and correction, and optimization of the film or sheet thickness against throughput.

The corrections are carried out at the slot die by automatic adjustment of the die bolts [125, 129]. The process-control section of the system provides print-outs for each finished lot, and for a full shift.

The combination of memory-programmed control and an S5-210A computer is applicable to sheet extrusion lines, as well as to injection molding- and blow molding machines [126].

This system has Bit-processors and microprocessors working together to provide open- and closed-loop control, and to compute and process such data as temperatures, displacements, speeds, and pressures. The line is controlled by a monitor screen on which target- and actual values, production data, graphical functions, and the position of faults are displayed.

The thicknesses of layers in coextruded films of certain materials, including LDPE/PS, can be measured individually. The normal radioisotope and infrared methods are combined for this purpose. The isotope-based device measures the superficial mass to give the overall thickness of the composite film, and the infrared equipment is used to obtain that of the transparent covering layer, which could be PS [127].

Further developments of infrared methods have been carried out, in view of the importance of coextrusion. A minimum sampling width of 2 mm and measurement time of 20 m/s now make precise film-profile measurements and analysis possible. By the use of filters for different wavelengths, even the thicknesses of individual layers of a composite film can be measured [128].

Microcomputers are also being used extensively in extrusion coating. In addition to controlling process variables, these systems are employed to record the coating thickness dis-

```
EGAN MACHINERY COMPANY COMPUTER CONTROL                           CMR 1000    7/7/81    16:47:14
                              ***** END OF ROLL REPORT *****
     NO  EXT A      SET  ACT  STATUS       NO   EXT B      SET   ACT   STATUS
      1  BARRL 1    240  240  *FAULT       17   BARRL 1    240   240    6C
      2  BARRL 2    300  300   55H         18   BARRL 2    310  *999  *TC BK
      3  BARRL 3    300  300   11H         19   BARRL 3    310   310    6H
      4  BARRL 4    300  300    1C         20   BARRL 4    310   310    7H
      5  BARRL 5    300  300    9H         21   BARRL 5    310   310    6H
      6  FLANGE     310  310   19H         22   FLANGE     320   320   20H
      7  ADAPTER    310 *308 * 85H         23   ADAPTER    320   319   40H
      8  PIPE 1     310  310   18H         24   PIPE       320   320    5H
      9  PIPE 2     310  310   21H         25   COMCHAM    315   315   14H
     10  PIPE 3     310  310   26H         26   DIE 1      320   320   62H
     11  PIPE 4     310  310   11H         27   DIE 2      315   315   31H
     12  PIPE 5     310  310   14H         28   DIE 3      315   315   12H
     13  PIPE 6     310  310   12H         29   DIE 4      315   315   15H
     14  PIPE 7     310  310   10H         30   DIE 5      315   315   17H
     15  PIPE 8     310  310   15H         31   DIE 6      315   315   28H
     16  PIPE 9     310  310   17H         32   DIE 7      320   320   24H
           EXT A    EXT B    LINE
TGT SPD    214.0    153.0     260      43 HEATER KW
SET SPD    200.0    143.0     243         3 KW-HR            196.0 SCAN      0.0 CTR
ACT SPD    200.0    143.0     243      7/7/81  16:41:43            SCANNING
% OF TGT      93       93      93                             AUTOMATIC CONTROL
AMPS         324      267                                       TARGET    ACTUAL
KW           140      135              PROFILE CONTROL OFF     20.00 AVG   20.02
KWHR          14       13                                      21.00 MAX   20.36
HI PRESS     600      600              12 PROFILING GAIN       19.00 MIN   19.70
LO PRESS     200      200              60 SEC STABILIZATION   521.00 CAL
PRESSURE     468      423               5 SCAN AVERAGE
MELT TEMP    310      320                                       BOLT     GAUGE
KG/REV      .1460    .0694                                        22      20.05
TOT KGS      430      118

     PRESENT   PREVIOUS   TOTAL    ROLL      ROLL    AVG      MAX      MIN     NO OF
     METRES    METRES     METRES   START     NO      GAUGE    GAUGE    GAUGE   SCANS
      8798      9055      45747   16:11:33    5      20.03    20.54    19.36    34
```

Figure 40. Production conditions of a coating line displayed on the monitor of the control computer [129]

tribution (by isotope or infrared methods), the melt pressure and temperature, the surface quality of the rolls, the material quality at the off-wind station and of the coated web at the wind-up; and the number of checking sequences.
Start-up and fault detection are automatic; and with expected further development the process computer will even be in a position to provide self-diagnosis. Thus it will be possible to make the necessary corrections very quickly.
An example of the display of manufacturing conditions that appears on the monitor of the computer system is given in Figure 40 [129].

References for Chapter 7

For the convenience of the reader the English titles of all publications other than English are shown in parentheses.

[1] *Beuer, H.:* Neues Konzept zur Herstellung von Folien aus plastomeren Kunststoffen (New Concept for Producing Films and Sheets from Thermoplastic Materials). Plastverarbeiter 29 (1978), pp. 113/122.
[2] *Schaab, H., Stoeckhert, K.:* Kunststoff-Maschinen-Führer (Plastics Machinery Guide). Carl Hanser Verlag, München, Wien, 1979.
[3] Extrusionsanlagen zum Herstellen von Tafeln (Extrusion Lines for Producing Sheets). Company brochure Reifenhäuser GmbH & Co., Troisdorf, West Germany.
[4] *Breuer, H.:* Schrumpf als Folge von Orientierungen in Folien und Tafeln (Shrinkage Resulting from Orientation in Sheets and Films). Dissertation RWTH Aachen, 1979.
[5] *Michaeli, W.:* Zur Analyse des Flachfolien- und Tafelextrusionsprozesses (Analysis of the Flat Film and Sheet Extrusion Process). Dissertation RWTH Aachen, 1975.
[6] *Kwa, L. J.:* Konstruktion und Berechnung eines 3-Walzen-Glättwerkes zur Herstellung von Folien und Tafeln aus thermoplastischen Kunststoffen (Design Calculations for a Three-roll Polishing Stack for Manufacturing Films and Sheets from Thermoplastics). Final year study Fachhochschule Niederrhein, 1981.
[7] *Ardichvili, G.:* Über die Dimensionierung von Kalanderwalzen (Dimensioning of Calendering Rolls). Kunststoffe 54 (1964) 8, pp. 520/521.
[8] *Unkrüer, W.:* Druckverlauf und Fließvorgänge im Walzenspalt (Pressure History and Flow Processes in Roll Gaps). Kunststoffberater (1971) 9–12, and (1972) 1.
[9] *Laczynski, B.:* Vorkommende Größen beim Walzen von Kunststoffen (Roll Sizes Used for Roll-milling Plastics). Kunststofftechnik 10 (1971) 8, pp. 273/280.
[10] *Dierkes, A.:* Berechnung von Eigenspannungsprofilen und des Fertigungsverzuges extrudierter Tafeln aus thermoplastischen Kunststoffen (Calculation of Internal Stress Profiles and Distortion in Extruded Plastic Sheets). Dissertation RWTH Aachen, 1980.
[11] *Kopsch, H.:* Kalandertechnik (Calender Technology). Carl Hanser Verlag, München, Wien, 1978.
[12] *Elden, R. A., Swan, A. D.:* Calendering of Plastics. Iliffe Books, London, 1971.
[13] *Rautenbach, R.:* Das Fließverhalten von Kunststoffen im Walzenspalt, untersucht am Beispiel von Polyäthylen (The Flow Behavior of Plastics in Roll Gaps: a Study of Polyethylene). Dissertation RWTH Aachen, 1961.
[14] *Takserman-Krozer, R., Schenkel, G., Ehrmann, L.:* Non-Newtonian Fluid Flow Between Rotating Cylinders. Rheol. Acta 16 (1977), pp. 240/247.
[15] *Potente, H.:* Schmelzkaschiertechnik (Melt Lamination Techniques). Discourse at the Ramisch-Kleinewefers Symposium at K 83 in Düsseldorf, West Germany.
[16] *Vlachopulos, J., Kiparissides, C.:* A Study of Viscous Dissipation in the Calendering of Power-law Fluids. Poly. Eng. Sci. (1978) 3, pp. 210/214.
[17] *Dubbel:* Taschenbuch für den Maschinenbau (Mechanical Engineering Handbook). Springer-Verlag, Berlin, Heidelberg, New York, 1983.
[18] *Bauhaus, F.:* Auslegung von Kalanderwalzen (Design of Calender Rolls). Final year study Fachhochschule Niederrhein, Krefeld, 1983.
[19] *Schuller, R.:* Neuere Entwicklungen im Bau von Folienkalandern, Teil 2 (New Developments in the Construction of Film Calenders, Part 2). Kunststoffe 61 (1971) 2, pp. 89/98.

[20] *Reitmeier, W.:* Neuere Entwicklungen im Bau von Folienkalandern, Teil 3 (New Developments in the Construction of Film Calenders, Part 3). Kunststoffe 61 (1871) 3, pp. 161/166.
[21] *Hofbauer, L.:* Entwicklung einer Kalanderausformtheorie und beispielhafte technische und wirtschaftliche Erprobung an PVC-Folien (Development of a Calendering Theory with Model Technical and Economic Studies on PVC Films). Dissertation TU Berlin, 1980.
[22] Verbesserte Tafelqualitäten durch neue Glättwerk-Konzeption (Improved Sheet Quality with a New Polishing Stack Design). Reifenhäuser Nachrichten 1983, Nr. 8, pp. 10/11.
[23] *Saechtling, H.:* Kunststoff-Taschenbuch (Plastics Handbook). Carl Hanser Verlag, München, Wien, 1974.
[24] Schmelz-Kaschier-Anlagen (Melt Laminating Lines). Company brochure, Ramisch-Kleinewefers, Krefeld, West Germany.
[25] *Breuer, H.:* Anlagen zur Herstellung von Folien und Platten aus verschiedenen thermoplastischen Kunststoffen (Lines for Manufacturing Films and Sheets from Various Thermoplastic Materials). Plastverarbeiter 25 (1974) 1, pp. 1/8.
[26] Komplette Anlagen für Flachfolien und Platten (Complete Lines for Flat Films and Sheets). Company brochure, Extrusionstechnik Battenfeld, Bad Oeynhausen, West Germany.
[27] Kalanderanlagen zur Verarbeitung von Kunststoffen (Calendering Lines for Processing Plastics). Company brochure, Berstorff, Hannover, West Germany.
[28] Ko-Kneter (Ko-Kneaders). Company brochure, Buss AG, Prattelen, West Germany, 1983.
[29] Kunststoff-Physik im Gespräch (Discussion on Polymer Physics). BASF, Ludwigshafen, West Germany.
[30] *Lutterbeck, K.:* Aufbereitungstechnik für Kunststoffe (Plastics Compounding Technology). Kunststoffberater (1984) 3, pp. 23/26.
[31] 4 P Folie (4 P Sheet). Company brochure, Forchheim GmbH, Forchheim, West Germany, 1983.
[32] *Carlowitz, B.:* Kunststoff-Tabellen (Plastics Sheeting). Schiffmann-Tabellen-Verlag, Bensheim, 1973.
[33] Extruderkombination für gefülltes Polypropylen-Pulver (Extruder Combinations for Filled Polypropylene Powders). K.-Plastik-Zeitung, 10. 2. 1984.
[34] *Breuer, H.:* Die Bedeutung der Vakuumtechnik für die Extrusion unter besonderer Beachtung der PVC-Bearbeitung (The Importance of Vacuum Techniques in Extrusion, with Particular Reference to PVC Processing). Plastverarbeiter 28 (1977) 5, pp. 233/240.
[35] *Hegele, R.:* Untersuchungen zur Verarbeitung pulverförmiger Polyolefine auf Einschneckenextrudern (Studies of the Processing of Polyolefin Powders on Single-screw Extruders). Dissertation RWTH Aachen, 1972.
[36] *Dalhoff, W.:* Systematische Extruder-Konstruktion (Systematic Design of Extruders). Krausskopf-Verlag, Mainz, 1974.
[37] *Reitemeyer, P.:* Rohre aus thermoplastischen Kunststoffen (Pipes from Thermoplastic Materials). Reifenhäuser, Troisdorf, West Germany, August 1979.
[38] *Lahn, W., Kurth, D.:* Planetary Roll Extruders in Plastics Fabrication. Maschinenmarkt 80 (1974) Nr. 63, Aug., 6.
[39] *Kopsch, H.:* Kalandrieren und Beschichten (Calendering and Coating). Kunststoffe 71 (1981) 10, pp. 743/746.
[40] Zweizellen-Schneckenkneter ZSK (Twin-screw Kneaders ZSK). Company brochure, Werner und Pfleiderer, Stuttgart, West Germany.
[41] *Kornmayer, H.:* Trockenluft-Trockner für die Kunststoffverarbeitung (Dry-air Dryers for Plastics Processing). Kunststoffberater (1984) 3, pp. 29/31.
[42] *Fischer, W.:* Stand der Technik bei Trocknern für die Kunststoff-Verarbeitung (The State of Drying Technology for Plastics Processing). Final year study Fachhochschule Niederrhein, Krefeld, 1983.
[43] *Hess, K.-M.:* Fortschritte beim Herstellen gefüllter Kunststoffe (Progress in the Manufacture of Filled Plastics). Kunststoffe 73 (1983) 6, pp. 282/286.
[44] VDI Guidelines – 2020: Erstellen von Werkstoffblättern und Hinweise für ihre Anwendung (Preparation of Material Information Leaflets and Advice on their Use). Beuth Verlag, Berlin, 3/1972.
[45] VDI Guidelines – 2021: Temperatur-Zeit-Verhalten von Kunststoffen (Temperature/Time Behavior of Plastics). Beuth Verlag, Berlin, 1/1970.

[46] DIN 53363: Prüfung von Kunststoffolien, Weiterreißversuch an trapezförmigen Proben mit Einschnitt (Testing of Plastics Film – Tear Propagation Test on Notched Trapezoidal Specimens). Deutscher Normenausschuß.
[47] DIN 53373: Prüfung von Kunststoffolien, Durchstoßversuch mit elektronischer Meßwerterfassung (Testing of Plastics Films: Penetration Test with Electronic Recording of Results). Deutscher Normenausschuß.
[48] DIN 53374: Prüfung von Kunststoffen, Hin- und Herbiegeversuch (Testing of Plastics Films: Bend Reversal Test). Deutscher Normenausschuß.
[49] DIN 53490: Prüfung von Kunststoffen; Bestimmung der Trübung von durchsichtigen Kunststoffschichten (Testing of Plastics: Determination of Haze in Transparent Plastics Layers). Deutscher Normenausschuß.
[50] DIN 53371: Bestimmung der Kältebruchtemperatur von Folien aus PVC weich (Determination of the Low-temperature Brittle Point of Soft PVC Films). Deutscher Normenausschuß.
[51] DIN 53370: Bestimmung der Dicke durch mechanische Abtastung (Thickness Measurement by Contact Methods). Deutscher Normenausschuß.
[52] DIN 53369: Isolierfolien der Elektrotechnik; Bestimmung der Schrumpfkraft (Electrical Films: Determination of Shrinkage Force). Deutscher Normenausschuß.
[53] DIN 53365: Entwurf von Kunststoff-Folien-Verbunden und Beschichtungen; Bestimmung der flächenbezogenen Masse (Flächengewicht) der Einzellagen (Make-up of Composite Plastics Films and Coatings: Determination of Superficial Weight of Individual Layers). Deutscher Normenausschuß.
[54] DIN 53455/VII: Reißfestigkeit, Reißdehnung (Tensile Strength, Extension at Break). Deutscher Normenausschuß.
[55] DIN 53377: Bestimmung der Maßänderung (Schrumpf) (Determination of Dimensional Changes (Shrinkage)). Deutscher Normenausschuß.
[56] Control Systems for Extrusion-coating Plant. Company brochure, Bone Markham Ltd., Wembley HAO, Great Britain, 1983.
[57] Extrusion Dies. Company brochure, Extrusion Dies Inc., Chippewa Falls, Wisconsin, USA, 1983.
[58] Maschinenangebot und Know-how für die kunststoffverarbeitende Industrie (Supply of Machines and Know-how for the Plastics Processing Industry). Company brochure, Battenfeld Extrusionstechnik, Bad Oeynhausen, 7/1983.
[59] *Ernst, B., Vogt, R.:* Auslegung der Ständer eines 3-Walzen-Glättwerkes mit Hilfe der Methode der Finiten Elemente (Design of the Pillars of a 3-roll Polishing Stack by the Method of Finite Elements (FEM)). Final year study Fachhochschule Niederrhein, Krefeld, 1984.
[60] Plastics Machinery. Company brochure, Barmag, Remscheid, West Germany, 1983.
[61] *Taleb-Bahmed, B.:* Grenzen des Einsatzes von Schraub- und Schneckengetrieben zur Verstellung der Walzen als Kalandern (Limits to the Use of Screw- and Mandrel-gears for Adjustment of Calender Rolls). Final year study Fachhochschule Niederrhein, Krefeld, 1981.
[62] *Weiß, H.-W.:* Entwicklung und Berechnung des Antriebes eines 3-Walzen-Glättwerkes zur Herstellung von Folien und Tafeln aus thermoplastischen Kunststoffen. (Development and Design Calculations for the Drive of a 3-roll Polishing Stack for the Manufacture of Plastics Films and Sheets). Final year study Fachhochschule Niederrhein, Krefeld, 1982.
[63] *Tirschler, U.:* Konstruktion eines 3-Walzen-Glättwerkes (Design of a 3-roll Polishing Stack). Final year study Fachhochschule Niederrhein, Krefeld, 1982.
[64] *Meinel, G., Leidl, J.:* Vorgänge beim Kalandrieren von PVC-Folien nach dem Luvitherm-Verfahren (Operations Involved in Calendering PVC by the Luvitherm Process). Kunststoffe 73 (1983) 8, pp. 398/401.
[65] *Kopsch, H.:* Neuere Entwicklungen im Bau und Betrieb von Folienkalandern (Recent Developments in the Construction and Operation of Film Calenders). Kunststoffe 61 (1971) 1, pp. 18/26.
[66] Förder-, Misch- und Trockeneinrichtungen für die Kunststoffindustrie (Conveying, Mixing, and Drying Units for the Plastics Industry). Company brochure, Motan, Isny, West Germany, 1983.
[67] *Uhland, E.:* Siebwechselsysteme mit hydraulischer Abdichtung (Screen-change Systems with Hydraulic Seals). Company brochure, Berstorff, Hannover, 1983.

[68] Sheet Dies. Company brochure, Cincinnati Milacron Austria AG, Vienna, 1983.
[69] *Görmar, H.:* Beitrag zur verarbeitungsgerechten Dimensionierung von Breitschlitzwerkzeugen für thermisch instabile Thermoplaste, insbesondere PVC-hart (On the Dimensioning of Slot Dies for Processing Thermally Unstable Thermoplastics, Especially Rigid PVC). Dissertation RWTH Aachen, 1968.
[70] *Wortberg, J.:* Werkzeugauslegung für Ein- und Mehrschichtextrusion (Design of Dies for Single and Multilayer Extrusion). Dissertation RWTH Aachen, 1978.
[71] US-PS 3.940.221. Welex Inc., Blue Bell, Pennsylvania, USA.
[72] Kleen-Screen. Extruders Screens Cleaned During Continuous Operation by Back-flushing. Company brochure, Welding Engineers Ltd., Genf, Suisse.
[73] Auto-Screen: A New System of Screen Changing. Company brochure, Process Developments Ltd., London, Great Britain, 1983.
[74] Siebwechsler nach dem Kassettenprinzip (Cassette Screen Changers). Company brochure, Bolton-Emerson SA, Lausanne, Suisse, 1983.
[75] Rapid Screen Changer. Company brochure, Omipa, Morazzone, Italy, 1983.
[76] Melt Filtration Users Guide. Company brochure, Cresta, Waterlooville, Great Britain.
[77] Sheet and Flat Film Extrusion. Company brochure, Rulli-Davis Standard, Sao Paulo, Brasilia, 1983.
[78] Extrusionsanlagen zum Herstellen von Tafeln (Extrusion Lines for Manufacture of Sheet). Company brochure, Reifenhäuser GmbH & Co., Troisdorf, West Germany, 1983.
[79] Extruder. Company brochure, Erwepa, Erkrath, West Germany, 1983.
[80] Extruders for Plastics and Sheet Take-Off Systems. Company brochure, Welex Inc., Blue Bell, Pennsylvania, USA, 1983.
[81] Sterlex EX-M-PLAR Series Sheet Systems. Company brochure, Sterling Extruder Corporation So., Plainfield, New Jersey, USA, 1983.
[82] Leesona Report: Flex Sheet Die. Company brochure, Johnson Plastics Machinery, Chippewa Falls, Wisconsin, USA, 1983.
[83] Indispensable for the Conversion Process. Company brochure, Daniels Engineering Ltd., Gloucestershire, Great Britain, 1983.
[84] *Predöhl, W.:* Technologie extrudierter Kunststoffolien (Technology of Extruded Plastics Films). VDI-Verlag, Düsseldorf, 1979.
[85] *Breuer, H.:* Gemeinsame Verarbeitung von Gummi- und Kunststoffabfällen (Combined Processing of Rubber and Plastics Wastes), in: Seng, H. J.: Entsorgung von Kunststoffabfällen durch Verwertung (Commercialization of Plastics Wastes). Expert Verlag GmbH, Grafenau, 1982.
[86] *Menges, G., Winkel, E.:* Einfluß der Abkühlgeschwindigkeit auf die Morphologie, die Dichte und den Elastizitätsmodul von extrudierten Folien und Tafeln aus Polyproplyen (Effect of the Cooling Rate on the Morphology, Density, and Elastic Modulus of Extruded PP Films and Sheets). Kunststoffe 72 (1982) 2, pp. 91/95.
[87] Komplette Anlagen zur Herstellung von Platten (Complete Lines for Sheet Manufacture). Company brochure, Battenfeld Extrusionstechnik, Bad Oeynhausen, West Germany, 1983.
[88] Anlagen zur Herstellung von Folien und Platten (Lines for Film and Sheet Manufacture). Company brochure, Weser Lenze GmbH & Co., Hameln, West Germany, 1983.
[89] *Kähler, M.:* Herstellung einschichtiger Flachfolien auf Extrusionsanlagen (Production of Single-layer Flat Films on Extrusion Lines). Reprint, Stahlkontor Weser Lenze KG, Hameln, West Germany, 1983.
[90] Hochleistungsmaschinen und Anlagen für die Herstellung und Veredelung von Folien, Papier und Karton (High-performance Machines and Lines for the Manufacture and Conversion of Films, Paper and Carton Board). Company brochure, Erwepa, Erkrath, West Germany, 1983.
[91] Nonwoven Technology. Company brochure, Ramisch-Kleinewefers, Krefeld, West Germany, 1983.
[92] Compounder gleichzeitig Extruder? (Compounder and Extruder Simultaneously?). K-Plastic- u. Kautschuk-Zeitung, 31. 10. 1983, p. 20.
[93] *Menges, G., Winkel, E., Gross, H., Masberg, U., Wortberg, J.:* Extrudieren, Thermoformen und Streckblasen von Polypropylen (Extrusion, Thermoforming and Stretch-blowing of Polypropylene). Kunststoffe-German Plastics (1983) 11–12, pp. 31/39.

[94] *Verse, N., Jarrasch, W., Bayer, H. G.:* Thermoformverhalten modifizierter ABS-Folien mit unterschiedlicher Einfärbung (Thermoforming Behavior of Modified ABS Films with Various Pigmentation Systems). Kunststoffe 73 (1983) 8, pp. 407/411.
[95] Vom Compound zur Platte (From Compound to Sheet). Prodoc (1983) Sept./Oct., pp. 496/498.
[96] Sheet Extrusion Systems. Company brochure, David-Standard Division of Crompton & Knowles Corp., Pawcatuck, CT, USA, 1983.
[97] Kunststofftechnik, kontinuierliche Aufbereitung technischer Kunststoffe (Plastics Technology, Continuous Compounding of Technical Plastics. Technical Information Leaflet, Werner u. Pfleiderer, Stuttgart, West Germany, 1983.
[98] *Cormont, J., Meijer, H., Herrmann, H., Herres, N.:* Wirtschaftliches Herstellen gefüllter Polypropylen-Platten (Efficient Production of Filled PP Sheet). Kunststoffe 73 (1983) 10, pp. 599/602.
[99] Tafel- und Flachfolienanlagen (Sheet and Flat Film Lines). Company brochure, Kuhne GmbH, St. Augustin, West Germany, 1983.
[100] Extrusion Plants for Hollow Plastic Profiles. Company brochure, Omipa, Varese, Italy, 1983.
[101] Flexible Walzen verringern Produktionskosten (Flexible Rolls Reduce Production Costs). Kunststoffe 73 (1983) 10, p. 608.
[102] *Trompler, S.:* Werkzeugoberflächen mit verringerter Haftneigung (Tool Surfaces with Reduced Sticking Tendency). Kunststoffe 73 (1983) 10, p. 596.
[103] *Wortberg, J., Tempeler, K.:* Breitschlitzwerkzeug mit großer Verarbeitungsbreite (Very Wide Slot Dies). Kunststoffe 73 (1983) 8, pp. 404/406.
[104] Coextrusionsanlage für Fünf-Lagen-Mehrschichtfolien (Coextrusion Line for Five-layer Composite Films). Kunststoffe 73 (1983) 8, p. 406.
[105] Das Dow „Feedblock"-Verfahren; Grundlage der Erwepa-Dow-Coextrusionsanlagen (The Dow 'Feedblock' Process – Basis of the Erwepa-Dow Coextrusion Lines). Coating (1977) 10.
[106] *Ast, W.:* Coextrusion von Folien, Tafeln und Beschichtungen (Coextrusion of Films, Sheets and Coatings). Coating (1977) 12, pp. 314 ff.
[107] Wickelmaschinen für die Kunststoff-Industrie (Winders for the Plastics Industry). Stahlkontor Weser Lenze GmbH & Co., Hameln, West Germany, 1983.
[108] Beschichtungsanlagen nach dem Bekalex-Prinzip (Coating Lines Using the Bekalex Principle). Company brochure, Berstorff, Hannover, West Germany, 1983.
[109] Extrusion Plants for Laminated Plastics. Company brochure, Omipa, Varese, Italy.
[110] Zweifach-Automatik-Wickelwendemachine (Automatic Twin-turret Winders). Company brochure, Berstorff, Hannover, West Germany, 1983.
[111] *Rottenburger, H.:* Compoundieren und Extrudieren in einem Arbeitsgang spart Kosten (Compounding and Extrusion in a Single Process Saves Costs). Werner u. Pfleiderer, Stuttgart, West Germany, VKS-T2.
[112] *Predöhl, W., Herres, N.:* Extrudieren von Flachfolien, Tiefziehfolien und Tafeln (Extrusion of Flat Films, Thermoforming Films, and Sheets). Kunststoffe 71 (1981) 10, pp. 659/665.
[113] Sheet and Film Lines. Company brochure, SMTP Kaufman, Le Havre, France, 1983.
[114] Sheet Dies and Film Dies. Company brochure, Cincinnati Milacron Austria AG, Vienna, 1983.
[115] Information Service Nr. 24. Barmag Barmer Maschinenfabrik AG, Remscheid, West Germany.
[116] *Hennig, J.:* Schrumpfkinetic von biaxial gerecktem Polymethylmethacrylat (Shrinkage Kinetics of Biaxially Stretched PMMA). Kunststoffe 67 (1977) 7.
[117] *Menges, G., Winkel, E., Gross, H., Madberg, U., Wortberg, J.:* Extrudieren, Thermoformen und Streckblasen von Polypropylen (Extrusion, Thermoforming, and Stretch-blowing of Polypropylene). Kunststoffe-Plastics (1983) 3, pp. 7/15.
[118] Kompaktanlage beseitigt Staub und elektrostatische Auflagungen (Compact Line Gets Rid of Dust and Electrostatic Charge). Kunststoffe-Plastics (1984) 1, p. 36.
[119] Koronastationen für Folien, Formteile und Platten (Corona Stations for Films, Moldings and Sheets). Company brochure, Van Leeuwen, Dachau, West Germany, 1983.
[120] Dispositifs de traitement corona à semi-conducteurs (Corona Equipment with Semi-conductors). Company brochure, Sherman Treater Ltd., Thame, Great Britain, 1983.

[121] Automatisierung von Extrusionsanlagen (Automation of Extrusion Lines). Reifenhäuser-Nachrichten 1981, Nr. 5.
[122] Vollautomatisches vorprogrammiertes Walzen (Fully Automated, Preprogrammed Rolling). Kunststoffe 73 (1983) 7, p. 352.
[123] *Menges, G.:* Trends bei der Steuerung und Regelung von Kunststoffanlagen (Trends in Open- and Closed-loop Control of Plastics Processing Lines). O + P ölhydraulik und pneumatik 27 (1983) 9, pp. 593/595.
[124] Processor-controlled Extrusion Plant. K-Plastic- u. Kautschuk-Zeitung, 26. 8. 1983, p. 3.
[125] Für die Flachfolienherstellung (For Flat Film Manufacture). K-Plastics- u. Kautschuk-Zeitung, 26. 8. 1983.
[126] Mikrocomputer mit Bildschirm (Microcomputer with Monitor Screen). K-Plastic- u. Kautschuk-Zeitung, 26. 8. 1983.
[127] *Hensen, F., Hessenbruch, R., Bongaerts, H.:* Entwicklungsstand bei Coextrusion von Mehrschichtblasfolien und Mehrschichtbreitschlitzfolien (The State of Development in Coextrusion of Multilayer Blown Film and Multilayer Flat Film). Kunststoffe 71 (1981) 9, pp. 530/538.
[128] Infrarot-Dickenmeßanlage für die Folienherstellung (Infrared Thickness Gages for Film Production). K-Plastic- u. Kautschuk-Zeitung, 26. 8. 1983.
[129] Egan CMR Microcomputer Systems. Company brochure, Egan Machinery, Sommerville, New Jersey, USA.

8 Manufacture of oriented films

F. Hensen

8.1 Introduction

The manufacture of oriented film begins with the extrusion casting of a primary film from a slot die or an annular-gap ('tubular') die. This film is then oriented by stretching.

The following chapter describes the development of the process and explains the process and the equipment with reference to the morphology of the polymers used. The different techniques and line concepts, and the units from which the lines are built, are described and compared.

8.2 Development of oriented films and their commercial importance

Even before *Staudinger* described the oriented molecular form of polymeric materials in the mid-thirties, the increase in strength by orientation was utilized in the production of man-made yarns for the textile industry.

Biaxially oriented films were already being produced in Germany during the thirties, from the amorphous thermoplastics polystyrene and polyvinyl chloride. It was realized that it was possible to improve physical properties like strength, elongation, shrinkage, elastic modulus, tear resistance, splitting tendency, rigidity, and transparency by orienting the film, and to utilize these improved properties in many applications.

The orienting process was further developed in the fifties, for making films from crystalline polyamides and polyesters, and increased in importance when low-pressure (high-density) polyethylene was developed in 1955, and isotactic polypropylene in 1957. The industrial utilization of the production process for oriented films on a large scale began after 1960 and experienced high growth rates when, throughout the sixties, cellophane packaging was replaced by PP, and photobase cellulose acetate by polyester. To stretch, or 'draw', the films in the longitudinal direction (machine direction, MD), roll stretching systems were used; and for transverse (TD) drawing, stenter (or tenter) frames, used in the textile industry, were employed.

In order to achieve an operationally reliable orienting process nowadays, one uses the principle of optimization of individual process steps [17], in which each of the steps, if possible, is carried out in a dedicated line element. This increases the number of independently controllable factors, and the overall process itself becomes more controllable. This principle demands an extremely high standard of design, and a large number of units connected in series to form the complete line.

In the interest of efficiency, such lines are designed for high output rates. With appropriate film dimensions, outputs of 1000 to 5000 kg/h are achieved. For this reason oriented film production lines rank among the largest of the plastic processing operations both in terms of physical dimensions and in level of investment.

8.3 Theory of the orienting process

The molecular structure and the 'texture' of amorphous or partially crystalline thermoplastics determine the way in which the physical properties can be influenced and altered, in particular by means of thermomechanical treatment at temperatures below the melting point. In a stretching process the macromolecules are given an orientation determined by the direction of draw; this orientation, thanks to better utilization of the valence forces, increases the strength and elastic modulus, and reduces the elongation at break; it also improves blocking properties. The practitioner knows from experience, and the theory of the drawing process described later makes it clear, that the product quality achievable and the efficiency of the process depend absolutely on the properties of the polymeric raw material used, and its treatment during melting, homogenization, and molding.

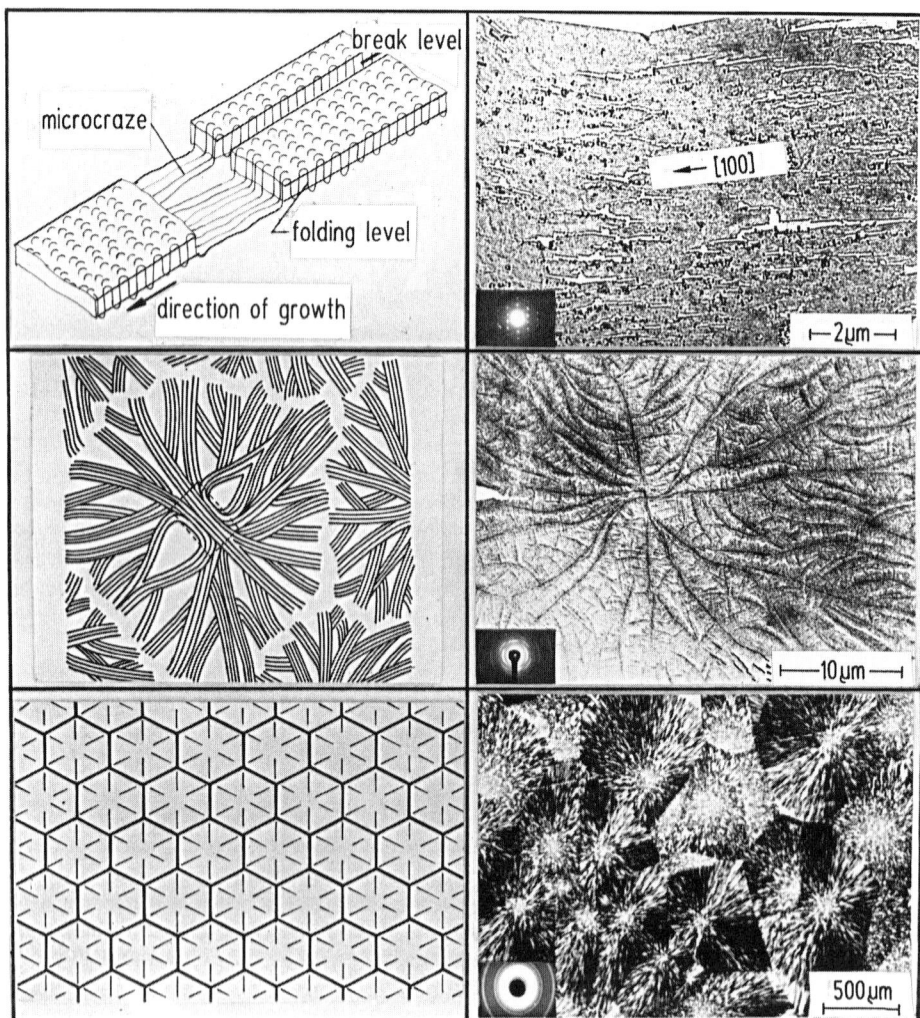

Figure 1. Microscopic structure of partially crystalline polypropylene
l.h.: schematic diagrams
r.h.: photo-micrographs
top: folded lamellae as basic elements of a crystalline structure [8]
center: growth of folded lamellae into a spherulitic superstructure [3, 10]
bottom: spherulites growing together from the melt

Oriented films are, for the most part, made from the partially crystalline thermoplastics polypropylene, PET, low-pressure polyethylene and polyamide; in addition small quantities of oriented films are made from amorphous thermoplastics such as polystyrene and polyvinyl chloride.

For a better understanding of the orienting process and the line concept developed to carry it out, one needs to know the structure of the partially crystalline materials which are subjected to it. A number of models of the structure of partially crystalline thermoplastics have been proposed and these are backed up to a great extent by microscopic examinations [1 to 6]. Figure 1 shows this structure. The basic element of morphology is the folded lamella (Figure 1, top), which is created by regular, repeated folding of the molecular chain. The X-ray diffraction pattern shows that these plate-like folded lamellae have a crystalline structure. During cooling they arrange themselves into crystalline superstructures (Figure 1, center), with spherulitic growth being most frequently observed. After formation of a crystal nucleus, which may be due to spontaneous folding of individual molecular segments, or may occur at the surface of an impurity particle or on an intentionally introduced nucleant, one first obtains a spherulite nucleus, in which individual folded lamellae are combined into packages of the same orientation. Only after a spherical growth front has formed with the molecular chain axes oriented almost perpendicular to the spherulite radius, do lamellae come together. The extreme legs branch off and enclose parts which cannot be crystallized. Upon complete solidification of the melt the growth fronts collide, with different parts of one molecule possibly being built into two spherulites, thus tying them together. At the same time there is an accumulation between the spherulite boundaries of material that cannot crystallize, which is pushed along ahead of the growth front (Figure 1, bottom) [7 to 10].

The orienting ability of the partially crystalline polymer is determined by the spherulite structure. When tension is applied, there is first an orientation of the amorphous parts between the folded lamellae and the spherulite boundaries. Then the randomly arranged platelets are converted into a highly oriented structure in which the chains are preferentially aligned in the drawing direction.

An amorphous phase and defects in the crystalline regions (connecting molecules, chain ends, entanglements) seem to be important requirements if a polymer is to be orientable.

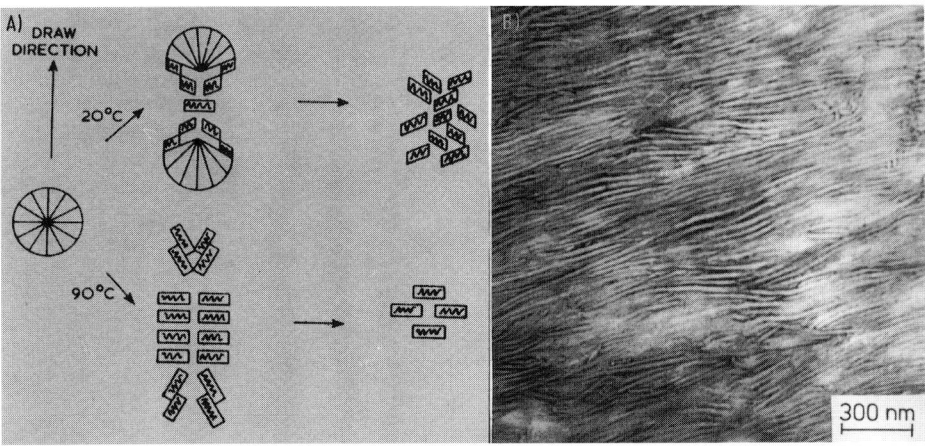

Figure 2. Deformation behavior of partially crystalline polymers during drawing
l.h.: deformation behavior of a spherulite after *Ilev* and *Keller*
r.h.: ultrathin cut of oriented polyethylene – contrast by the *Kanig* method (vertical drawing) [5]

By using a new sample preparation process (Figure 2) *Kanig* [11] made the orientation of the folded lamellae directly visible by transmission-electron-microscopy. Using HDPE, he demonstrated that lamellae are tilted by shearing processes. With further stretching the lamellae are destroyed and small crystalline blocks are created. These are connected to each other by tie molecules, and this results in a network of small blocks connected in all directions by tie molecule sections.

During drawing, energy conversion takes place. After drawing, entropy effects tend to return the aligned structure to its original configuration, causing the film to shrink. To avoid this, the film is held at constant area and heated up nearly to melting temperature. During this process the lamellae grow larger and the crystallinity becomes more perfect [12]. In the amorphous regions tensions are reduced and energetically more favorable configurations are adopted by the chain so that the free enthalpy is reduced and the system is stabilized. The form stability achieved by maintaining the web under tension while increasing the temperature, causes the molecules in the amorphous regions to rearrange and curl up again. This rearrangement is possible because of sliding and diffusion mechanisms promoted by higher temperatures; and the reversion tendency, and consequently film shrinkage, are reduced after heat treatment [13].

A description of the drawing process will show clearly what is important for producing orientation:

- A high degree of crystallization should be prevented as far as possible by the composition of the thermoplastic and by control of the process.
- The molecular weight distribution should be as uniform as possible.
- In the case of PP, fast crystallization should be avoided by raising the atactic content, and/or adding a different type of polymer, or by copolymerization.

As far as the line is concerned, these requirements can be met by means of homogeneous thermal treatment, even metering, intensive mixing, proper filtration and fast cooling.

As a result of cooperation between raw material producers, machine manufacturers and processors, natural products have been replaced by oriented polymer plastic films in some fields of application. The advantages of oriented films with regard to price, chemical resistance, strength and raw material availability accelerated substitution and led to new applications. Following the extraordinarily high growth rates in the use of oriented films, large new industrial sectors for manufacturing and converting them have come into existence.

Table 1 shows the processes used, fields of product application, preferred raw materials, and the normal film thicknesses; it also lists the most important property requirements.

To achieve the most important of the properties mentioned, oriented films are increasingly made with several layers. The layers consist either of the same basic material into which special additives have been mixed for the different layers, or of different materials. The additives or the different materials are intended to improve the target properties, for example weldability, adhesion, barrier effect against gases and liquids, splitting tendency, and so on.

The thickness range is extremely large. Dielectric films must be as thin as is technically possible. Film thicknesses of 2 µm are already being achieved, and development work for even smaller thicknesses is being done. X-ray films are produced in thicknesses of up to 300 µm. The large variety of raw materials being used, the wide thickness range (2 to 800 µm) of oriented films and film tapes – and the variety of very important properties required – show clearly that, in order to achieve optimum properties with the different processes and fields of application, it is necessary to have special lines and line elements which are optimized for each application, or to produce special properties. Such lines and elements have now become available.

8.3 Theory of the orienting process

Table 1. Drawing processes, fields of application, preferred raw materials, film thicknesses and most important properties of oriented films and film tapes

Drawing process	Fields of application	Raw materials (preferred)	Film thickness (drawn) µm	Premium properties
Biaxial (simultaneous or two-stage)	packaging	PP (PET; PA 6)	10 to 40	transparency, weldability, printability, machinability
	drawing-office film	HDPE	20 to 100	homogeneity, flatness
	photographic film	PET	100 to 300	transparency, purity, profile precision
	sound recording film	PET	4 to 8	strength, homogeneity
	electric insulating film	PET, PP, PS	3 to 200	homogeneity, purity, profile precision
	condenser film	PET, PP, PS, PC	2 to 10	homogeneity, purity, profile precision
Uniaxial (shrinkage-reduced)	adhesive tapes	PP, HDPE	50 to 200	tensile strength, profile precision, adhesion
	electric insulating film	PC	2 to 50	homogeneity, purity, profile precision
	warp-beam film	HDPE	25 to 30	tensile strength/tenacity
Uniaxial	packaging tapes	PP (PET, PA)	200 to 800	tensile strength, shape stability, profile precision, elongation
	sack fabric tapes	HDPE (PP)	25 to 35	tenacity, adhesion
	basic carpet fabric tapes	PP (PET)	35 to 50	tenacity, low shrinkage
	baler twines	PP/PET	50 to 80	tenacity, elongation
	split fibers	PP	25 to 35	splitting tendency, tenacity

Table 2 gives a comparison of the film properties that are achieved by simultaneous biaxial drawing of PP, PA, PET, and PS [16]. For comparison purposes a common thickness of 25 µm was used. In the case of PP only, an additional comparison is given: with the properties achieved by two-stage biaxial drawing, and for 40 µm as well as 25 µm film. The draw ratio depends on the polymer, and was, in each case, that normally used in production.

With PP, a comparison can be made between the three principal stretching methods – uniaxial, two-stage biaxial, and simultaneous biaxial – and the improvements in properties achieved with them. With all of the methods the physical properties were improved – tensile strength, extension at break, tear-propagation resistance, puncture resistance, and E-modulus. It is noticeable in this context that nominally undrawn film shows differences in properties in the longitudinal (MD) and transverse directions (TD); these are due to orientation of the primary film during production. In the two-stage biaxial drawing process, different values are obtained in the two directions, with the order of drawing having an effect. In the case of uniaxial drawing these differences are extremely large, because the transverse orientation is only very slight. Simultaneous biaxial drawing produces nearly equal properties in both directions.

The barrier properties of PP film are considerably improved by orientation, and the low-temperature impact resistance is also increased. The four right-hand columns of Table 2 show, for comparison, the values obtained with PP, PA, PET, and PS simultaneous biaxial drawing. It should be remembered that the overall draw ratio (area draw ratio) varies from material to material.

More recent results [29] have shown that by nucleating PP considerable strength increases are achieved under the same drawing conditions. Electron-irradiated PP was used as a

Table 2. Most important properties of oriented and unoriented films (after [21] to [25]). Examples of commercial films

Film properties	Standards DIN	Unit	PP undrawn	PP uniaxially drawn	PP biaxially drawn 2-stage		PP biaxially drawn simultaneous	PA biaxially drawn simultaneous	PET biaxially drawn simultaneous	PS biaxially drawn simultaneous
Thickness	–	μm	25	25	25	40	25	25	25	25
Area stretch ratio	–	–	–	1:5.5	1:10	1:10	1:10	1:8	1:6.5	1:10
Tensile strength – LD	53455	N/mm^2	50	250	140	130	200	300	200	70
– TD	53455	N/mm^2	40	40	270	250	200	300	220	70
Break elongation – LD	53455	%	430	10	140	143	80	70	130	10
– TD	53455	%	540	700	40	43	80	70	110	10
Tear-propagation resistance – LD	53363	N	7.6		0.25	0.4				
– TD	53363	N	12		0.45	0.7				
Puncture resistance	53373	N	23		200	360				
E-modulus – LD		N/mm^2	500	2500	2500	2500	3000		4300	
– TD		N/mm^2	900		4000	4000	3000		4300	
O_2 permeability 23 °C, 75% RH	53380	$cm^3/(m^2 \cdot 24\,h \cdot bar)$	2500		1000	750	800			
Water-vapor permeability 23 °C, 85% RH	53122	$g/(m^2 \cdot 24\,h)$	2.5		1.5	0.9	0.8			
Low-temperature resistance	–	°C	0		–50	–50	–50			

nucleating agent at the 1% level in the melt, the radiation dose being 8 Mrad of 3 MeV electrons. The radiation causes simultaneous chain scission and crosslinking, and the crosslinked PP serves as a nucleant. Some 20 to 50% increase in strength is achievable with PP, depending on the drawing method used.

8.3.1 Processes for the manufacture of oriented films

High growth rates and technological competition have led to the simultaneous development of various drawing processes. The differences lie essentially in the method of transmission of stretching forces and in their time sequence.
To assist understanding of the process, Figure 3 uses as an example the two-stage drawing process, and is a flow diagram showing the steps required for the complete process, and the factors operating at each of the stages identified on the material flow line. The dried pellets are supplied metered and mixed to the extruder. In the extruder the pellets are conveyed, melted, compressed and homogenized, and discharged as melt in metered fashion.
Metering can be done by the extruder itself, or by use of a gear pump, or with an additional extruder. The melt next passes through a large-area filter to remove contaminant and gel particles before discharge through a film die onto the casting roll. It is cooled, and solidifies as it is carried around the chill roll in good contact with the surface. It passes through a thickness gage and is stretched, first longitudinally between two rollsets and then transversely in an oven. This procedure is followed by further thickness measurement and trimming of the edges. The edge trim is chopped up at once and fed back to the extruder inlet pneumatically. The oriented film is then cut into usable widths and wound up. A look at the most important process variables will help to give a clearer picture of the production process. These factors show that the process is very complex and that there are many possibilities for influencing the product properties.
This becomes particularly clear when one considers the melt or film temperatures required at different stages of the process. The curve shown in Figure 4 applies to the biaxial drawing of PP, but its shape is similar for all thermoplastics, with the temperatures differing in relation to their crystalline melting points.
The pellets are preheated before they are fed to the extruder, in order to improve its efficiency. The preheating temperature should be very slightly below the crystalline melting point. After melting, homogenization in the extruder, metering, filtration and casting, the temperature must be lowered to allow the stretching force to be transmitted to the film. In the orienting zones (Figure 4d, f and g) there is further heating by roll contact or hot air up to almost the crystalline melting point so that orientation at high stretching speeds can be carried out reliably. With increasing orientation, operating temperatures can be higher. In the setting zone it is therefore possible to carry out film relaxation at a still higher temperature. Ahead of the take-up unit the film is cooled to room temperature to ensure that a low-tension, even package is formed.
For uniform film quality and production efficiency, the temperature at different stages is important, and TD temperature uniformity is particularly so. The need for a uniform temperature puts special demands on the control system and on the design and finish of the temperature-conditioning rolls and hot-air ovens.

8.3.2 Lines for the production of oriented films

Although all film drawing lines are based on the same principle, the various process steps can be carried out in different ways.

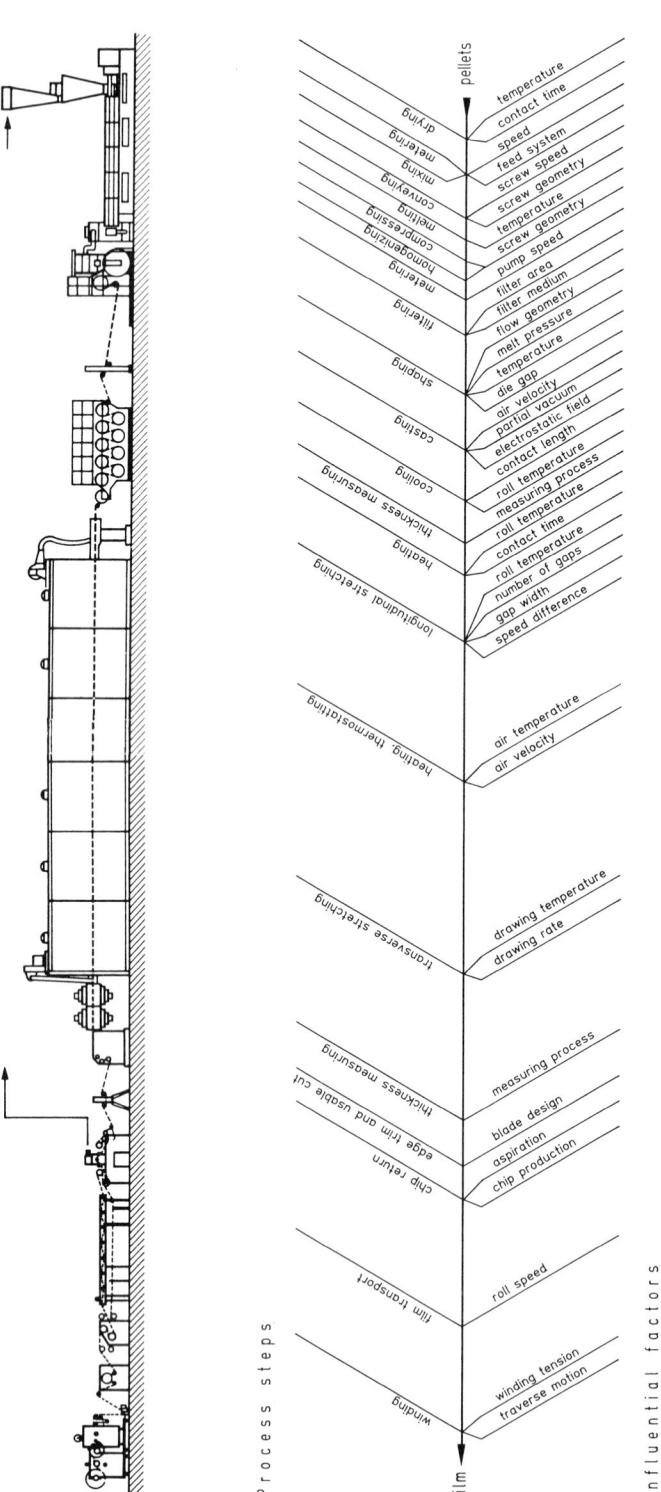

Figure 3. Principle of the production of biaxially oriented film (two stages)

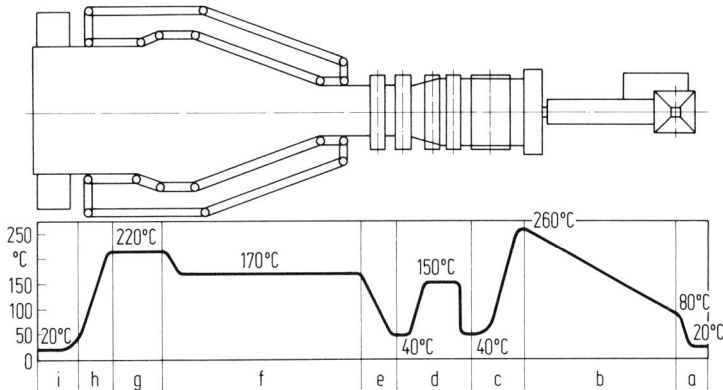

Figure 4. Temperatures used in the production of biaxially oriented films from PP
a preheating, *b* extruding, *c* cooling, *d* longitudinal stretching, *e* heating, *f* transverse stretching, *g* heat setting, *h* cooling, *i* wind-up

The components of film drawing lines are shown in tabular form in Figure 5, diagrammatically associated in the correct sequence with the various process steps. The most important technical data are given in the table. Commercially available film lines are built from these elements, some of which consist of several individual components. Such units have been developed to meet the special requirements of the drawing process, and may be chosen to meet individual needs with regard to design detail and performance.

Extrusion unit

In the extrusion unit the thermoplastic must be melted, homogenized, possibly mixed with additives, and filtered, and evenly extruded as single-layer film or, if several extruders are used, as multilayer film. Optimum results with regard to film quality and process efficiency cannot be obtained with a single extrusion unit alone in the case of larger lines; the individual process steps must be carried out by separate, individually optimized units arranged in series, such as melting extruder, mixer, filter, metering pump or homogenizing extruder, preferably with separate drives and temperature control devices [17]. The extruders used are almost exclusively single-screw extruders. The screw diameter depends on output and may be 60 to 300 mm, with screw lengths of 27 to 33 D, for outputs of up to 2000 kg/h with PP, or 4000 kg/h with PET (Table 3) [16].

Figure 6 shows the installation of a cascade extruder with 300 mm screw diameter and a total length of 30 plus 17 D, for the production of biaxially oriented PET.

From a theoretical consideration of the orienting process it became clear that the internal structure of the solidifying thermoplastic should be as evenly, finely crystalline as possible. This requires that the material be melted without excessive local stress in the shearing gap under conditions of uniform temperature and dwell time, and extruded as a homogeneous melt with regard to component and temperature distribution. Based on the experience that has been accumulated with more than 7000 extruders for drawn thermoplastic products − which have been improved during more than 20 years of continuing development − we now have extruders at our disposal which produce an optimum primary film. Features of these extruders are combinations of materials, special geometry and temperature conditioning in the feed zone, screw geometry and temperature treatment devices on the barrel. The metering precision for maintaining the film dimensional tolerances is improved by a metering gear pump attached to the extruder. When there is a requirement for high output and special quality, the pump is replaced by a melt-fed extruder. Thus the homo-

Figure 5. Schematic diagram of the elements of film stretching lines

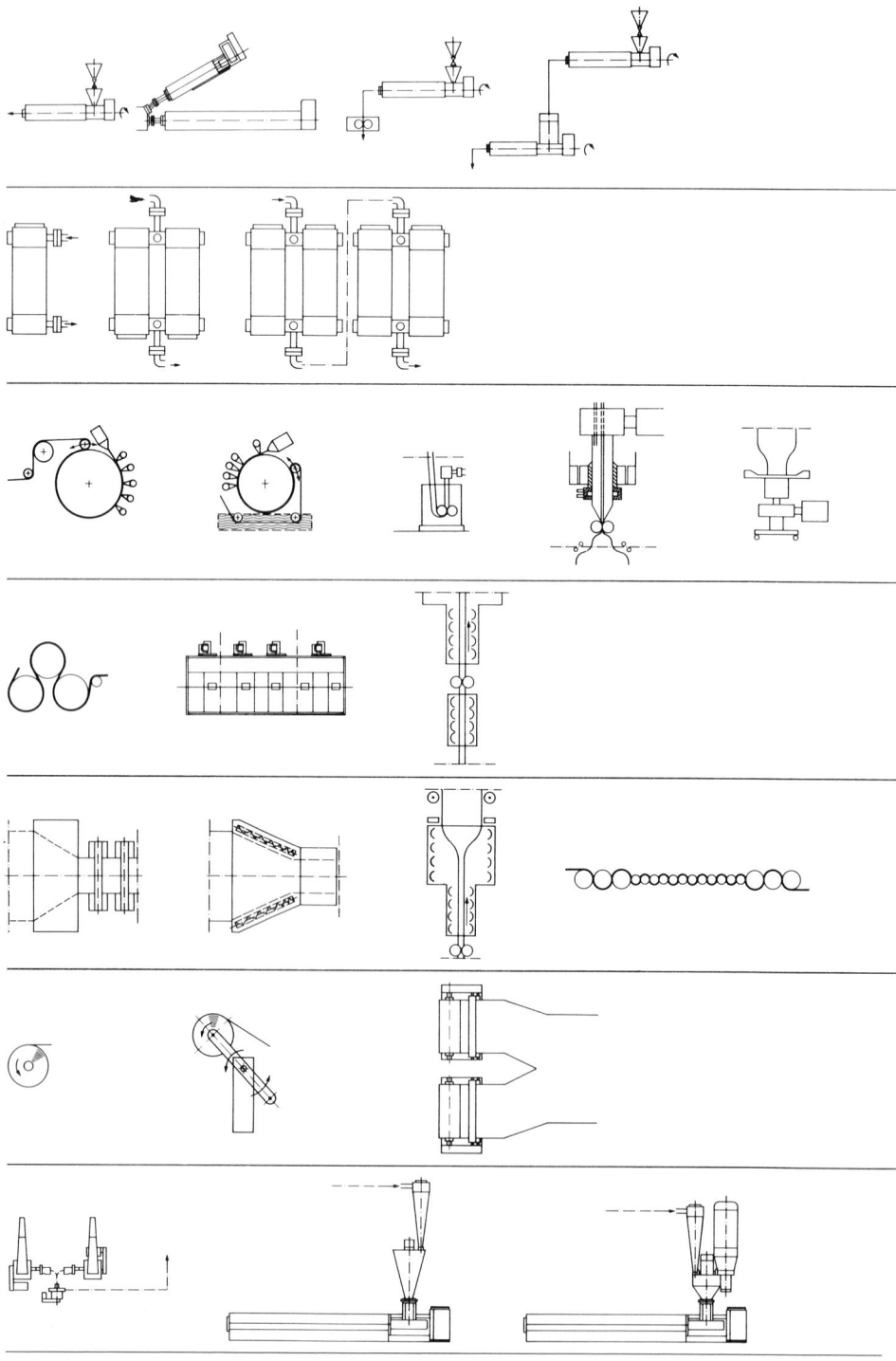

8.3 Theory of the orienting process

Process step	Designation	Technical data
Extruding Coextruding Metering	single-screw extruder parallel extruder and laminar adaptor metering gear pump cascade extruder	screw dia. 45 to 300 mm output $\leq$ 4500 kg/h $V_{pu} < 2500$ cm^3/rev $\Delta P \leq 0.3$ bar $\Delta Ts \leq 0.2\,°C$
Filtering	long-life filter twin-chamber filter cascade filter	filter area 0.1 to 46.5 cm^2 mesh width 10 to 20 µm $\Delta p = 20$ to 100 bar service life 1 to 3 weeks
Casting Cooling	cast film die, chill roll (cooled), airmeter blowing head, cooling ring, water bath	die width 450 to 1150 mm roll dia. 500 to 2500 mm take-off speed to 100 m/min film thickness to 3 mm
Temperature conditioning	thermostatted rolls (oil-heated) hot-air furnaces heat radiators	$T = 180$ to $220\,°C$ $\Delta T \leq 1\,°C$
Stretching	biaxial lines (two-step or simultaneous) uniaxial lines (multiple-gap)	stretch ratio: longitudinal 1:3 to 1:6 transverse 1:3 to 1:10 $V_2 > V_1 = 100$ to 400 m/min
Winding	central winder automatic turret winder multipurpose winder	package dia. to 600 mm package width 4000 mm
Waste return	chopping mill cyclone waste-metering units	additive to 50%

genizing and metering steps are carried out in a separate unit [16, 17]. Just as with the metering pump, this arrangement increases the conveying uniformity. The melt-fed extruder can also be used for additional homogenization, and makes it possible to recycle up to 30% waste, which arises as trimmed undrawn edge strips.

Table 3. Extruder output ranges for PP and PET as a function of screw diameter

Screw diameter	Output ranges with	
	PP	PET
mm	kg/h	kg/h
90	220 to 250	200 to 300
120	300 to 350	350 to 450
150	450 to 500	750 to 1000
200	700 to 850	1500 to 1800*)
225	1000 to 1200	2000 to 2300*)
250	1300 to 1500	2400 to 2800*)
300	1700 to 2000	3400 to 4000*)

*) cascade extruder

Figure 6. Extrusion unit with melting extruder ($D = 300$ mm, $L = 30$ D) and liquid-fed extruder ($D = 300$ mm, $L = 17$ D) (Photo: Barmag, Remscheid/West Germany)

For many fields of application it is necessary to process several thermoplastics simultaneously to make a single film. The materials may be fed into the extruder as mixed pellets, or mixed in a melt form, or coextruded to produce a multilayer film. Mixing in melt additives, and coextrusion, require additional extruders arranged in parallel, of a size to suit the number of additional components.
The mean thickness of the coextruded single layer is regulated by microprocessor speed control as a function of the continuously weighed granule throughput. Continuous measurement of pellet throughput at the extruder hopper ensures that, even with smaller layer thicknesses, no faulty areas will occur, and the thickness ratio will be maintained. In the case of three extruders the total throughput, $\dot{m}_{tot}$, is equal to the sum total of the throughputs of the individual extruders $\dot{m}_i$.

$$\dot{m}_{tot} = \dot{m}_1 + \dot{m}_2 + \dot{m}_3$$

8.3 Theory of the orienting process

with the throughputs of the individual extruders being dependent on the layer thicknesses.

$$\dot{m}_i = \frac{\dot{m}_{tot}}{\delta_{tot}} \cdot \delta_i.$$

From this follows

$$\dot{m}_{tot} = \dot{m}_{tot} \left(\frac{\delta_1}{\delta_{tot}} + \frac{\delta_2}{\delta_{tot}} + \frac{\delta_3}{\delta_{tot}} \right).$$

Uniformity of the individual layer thicknesses transverse to the extrusion direction is achieved by special melt homogeneity, symmetrical melt distribution and high dimensional precision of the melt channels.

When additives have to be mixed into a melt they must be divided into minute droplets and evenly distributed into the matrix of the base material. A dynamic mixer that uses the shearing groove principle was developed for this purpose; this divides or distributes the additives in droplet sizes of fractions of a micrometer. Figure 7 shows the construction of the mixer and the way it works [18]. For the layer thickness, d, of a drawn-out flow line, the following applies:

$d = W/(N\,n)$ with W flow velocity,
N number of grooves on the circumference,
n speed.

The number A of distributions taking place results as

$$A = N^{2K}$$

with K = number of longitudinal grooves. If data for practical designs are inserted into the formulae, one obtains layer thicknesses in the µm range in every section of the melt flow. With every rotor revolution the number of sections of partial flows lies between 40 and 100 with the exponent 22, which explains the special mixing effect. The mixer is used either as an extruder extension or as a separate unit with its own drive.

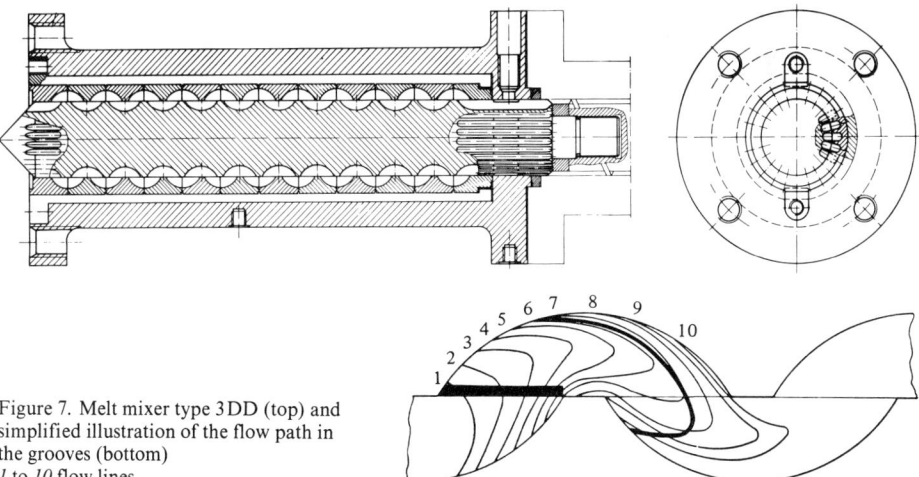

Figure 7. Melt mixer type 3DD (top) and simplified illustration of the flow path in the grooves (bottom)
1 to *10* flow lines

The morphological changes going on within the cast material during drawing, which have been described above, mean that large foreign particles or aggregations of amorphous material will lead to the formation of cracks and consequently to film breakage. To avoid this they must be removed from the melt, which can be done by passing the melt through a filter screen of 10 to 20 mesh (80 Nm pore width) at speeds of only a few µm/s and pres-

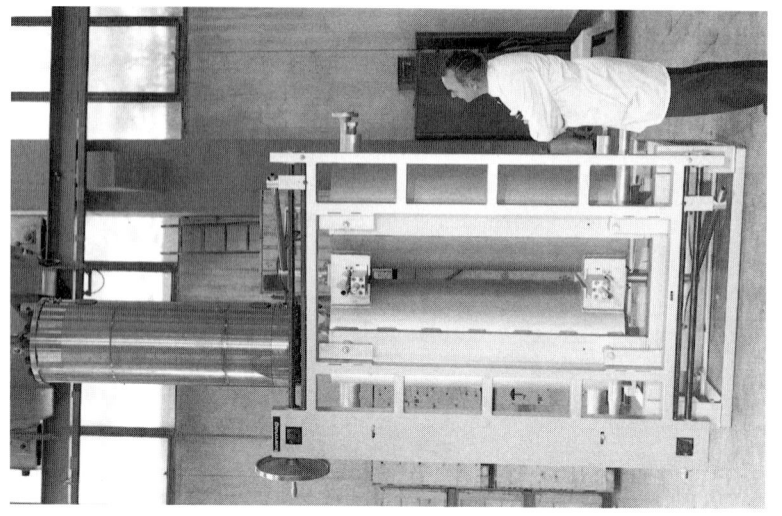
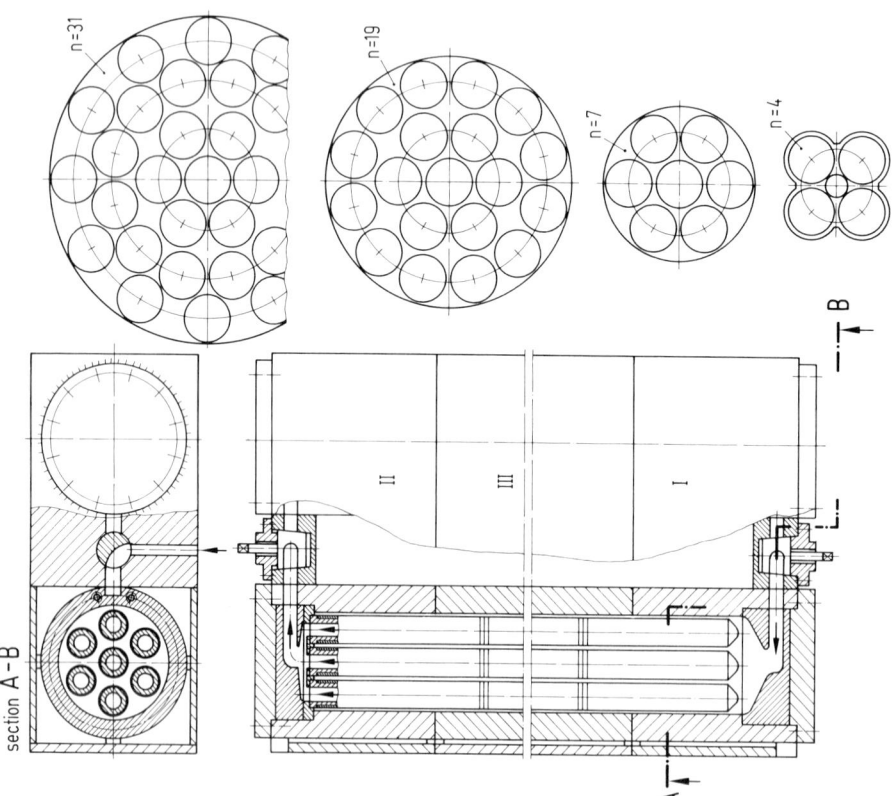

Figure 8. Non-stop large-scale melt filter of twin chamber design with switch-over valves
n number of filter candles per chamber
(Photo: Barmag, Remscheid/West Germany)

sure drops within a range of 20 to 100 bar [19]. In order to achieve this, large-scale twin-chamber filters with filter surface areas between 0.1 and 46.5 m² have been developed. These provide the maximum surface area-to-chamber volume ratio by using a battery of filter candles or disks (Figure 8). These filters make it possible to recycle undrawn edge waste, and to produce thin films at high take-off speeds; and they improve the film quality. Thanks to the twin-chamber design, contaminated filter inserts can be changed without interrupting production. The melt pipes are designed for equal dwell times and rheologically favorable flow paths.

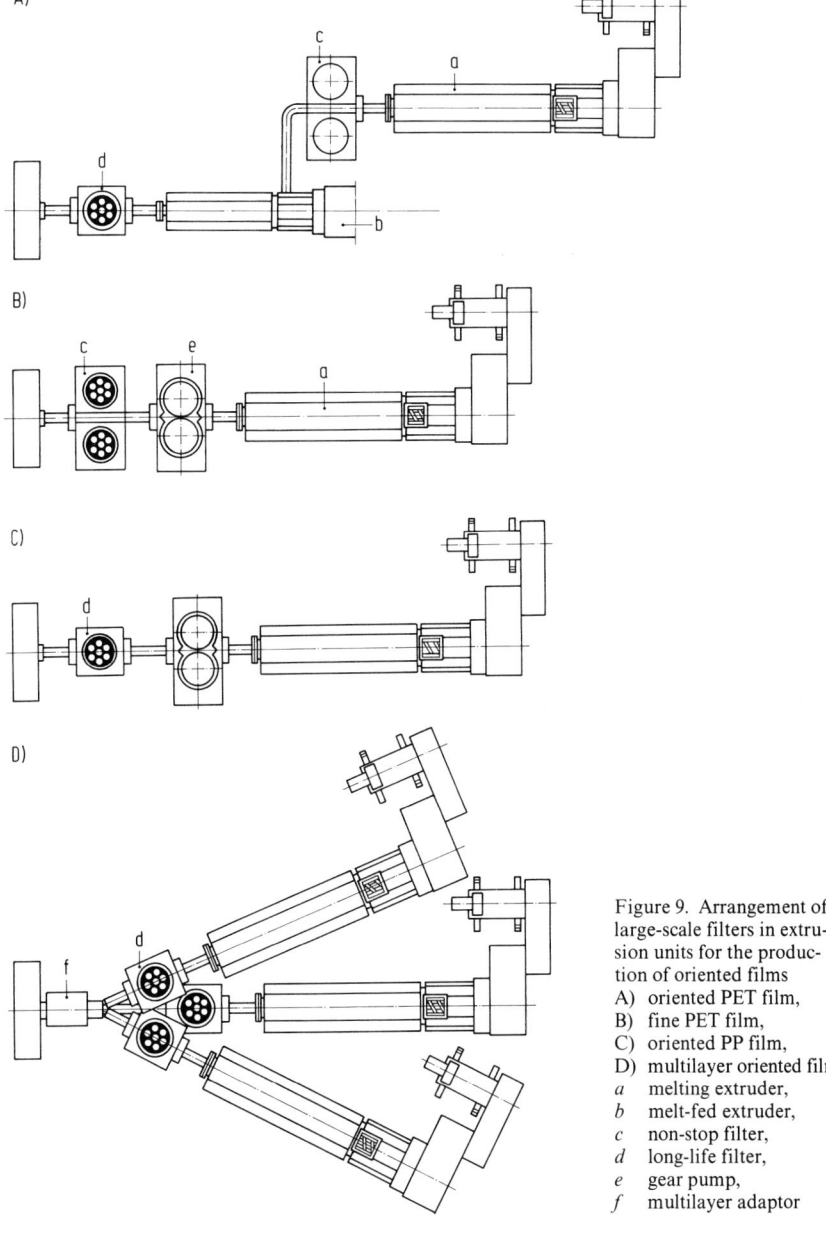

Figure 9. Arrangement of large-scale filters in extrusion units for the production of oriented films
A) oriented PET film,
B) fine PET film,
C) oriented PP film,
D) multilayer oriented film
a   melting extruder,
b   melt-fed extruder,
c   non-stop filter,
d   long-life filter,
e   gear pump,
f   multilayer adaptor

Figure 5b shows in consecutive diagrams the three most commonly used arrangements of large-scale filters: as a long-life filter; with two chambers and switch-over valve; and as a cascade arrangement of non-stop filters. As shown in Figure 9, the filters are inserted either between two extruders or between the melt metering unit and the extrusion die, depending on the film type. Insertion between two extruders permits filter changes during production without affecting film quality. This is of special advantage if large amounts of edge trim have to be recycled. The long-life or non-stop filter in cascade arrangement behind the extruder has a safety function and increases the melt purity. If melt metering is carried out by a volumetric displacement melt pump, the filter is inserted between the melt pump and the die, so that the filter resistance, increasing with dirt accumulation in the filter, has no influence on dwell time, or on shear stress on the melt, in the extruder. The filter inserts are cleaned after they have been taken out. Depending on the type of thermoplastic, cleaning is done in a furnace, solution bath or salt bath, with appropriate aftertreatment in neutralizing baths and ultrasonic baths. While the housings and filter inserts can be cleaned any number of times, the re-usability of the filter media is limited [19].

8.3.3 Extrusion dies and cooling unit for the production of oriented film precursor

The films may be formed by the flat film- or the blown film principle.
In the case of flat film, the melt is extruded as a film onto a cooled casting roll and solidified. For this purpose, a divisible cast film die heated with heating cartridges and having a flexible, deformable die lip is used; the cross-section of such a die is shown in Figure 10. The thickness range is 15 to 500 µm. Larger thicknesses are possible with larger die gaps. Flow distribution is effected by a rheologically well designed "coat-hanger" channel with a subsequent "dam zone".

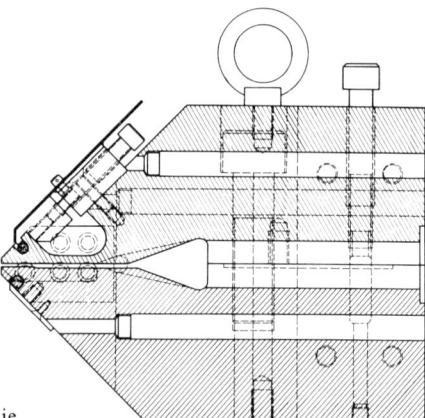

Figure 10. Cross-section of the slot die

Coextruded films can be produced either in multichannel dies or by the adaptor (or feedblock) method. In the case of the multichannel die the individual melt components are separately formed into films in the die and come together only when they reach the die gap.
Polymers possess enhanced laminar flow characteristics on account of structural viscosity. Because of this, and the fact that they do not become turbulent in the Reynolds sense even at very high shear rates, different polymers can flow in layers next to each other and change shape in the same way without mixing when the melt channel changes its shape. This behavior permits coextrusion by means of a multilayer adaptor and a single-layer

die. The adaptor system is generally preferred for production of multilayer films. With correct rheological design of the adaptor inserts, and small viscosity differences between the coextruded film components, it is possible to achieve layer thickness ratios of 1:20 and thickness deviations within the layers of ± 10%. The processes are described in more detail in Chapter 7.

Figure 11. Chill roll unit. (Photo: Brückner, Siegsdorf, West Germany)
a slot die, *b* air knife, *c* casting roll, *d* take-off roll

Before the film stretching process starts a crystallizable material must be cooled to below its melting point. The cooling must be fast and even in order to obtain an amorphous or a homogeneous crystalline structure. The chill-roll principle has proved to be the best method for this purpose. For most efficient removal of heat from the melt, the casting roll may have a diameter of from 500 to 2500 mm, depending on the film thickness and the take-off speed (Figure 11). In a double-jacket chill roll there are two helical cooling channels separated by ridges, through which are passed countercurrents of cooling fluid so that the surface temperature remains constant. In order to achieve good film/roll contact, the film is pressed down by an air 'knife', from a unit which is arranged parallel to the line of contact. In order to define the cooling contact length accurately the film is taken off by an adjustable peel-off roll after it has passed round the casting roll. The surface of the casting roll is hardened, chrome-plated and polished. It is kept at a temperature between 15 and 150 °C depending on the thermoplastic and the type of film, with a maximum deviation of ± 1 °C. Thick film is additionally cooled by means of a water bath, in the case of PET to minimize crystal growth, and with PP to promote nucleation and inhibit spherulite growth. A primary film produced in this way will behave reliably on the drawing line at high take-off speeds [20].

If cast film is to be made into film tapes, rope yarns or split fibers, it is subjected to uniaxial drawing. Since there are no strict requirements regarding surface quality, it is pos-

sible to utilize the advantage of intensive double-sided water cooling, by casting the film from a slot die into water (Figure 5 C, centre).
In the case of tubular films, which also can be uniaxially or biaxially drawn, the primary film tube is formed in an annular die gap extruding vertically upwards or downwards, depending on the subsequent process. In the case of downward film discharge the film tube may be enclosed by a weir chamber which provides for spraying and cooling of the film with water. In addition there is tubular film extrusion using air cooling, for producing supply film for the orienting process.

8.3.3.1 Temperature control of the film

Before drawing, the film must be heated to its optimum orienting temperature. This is carried out (Figure 5 D) by contact with oil-heated rolls, or by air transfer in the heating oven, or by means of heat radiators.
The requirements for temperature uniformity in longitudinal and transverse directions are very exacting.

8.3.4 Stretching units

In order to achieve the required film properties, the primary film is stretched by a tension applied in the required direction. In this procedure the film changes shape. When the film length is increased by a given factor, the draw ratio, the thickness decreases by the same ratio. If drawing is carried out as usual, longitudinally then transversely, using drawing ratios R_L and R_T respectively, a volume element of the film changes as illustrated in Figure 12.

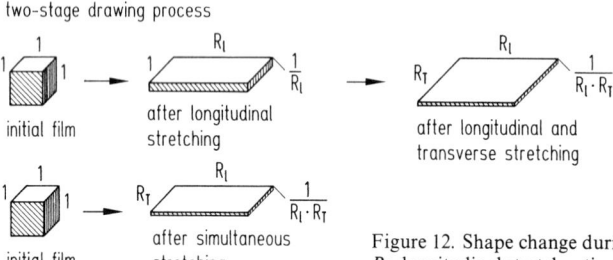

Figure 12. Shape change during stretching
R_L longitudinal stretch ratio, R_T transverse stretch ratio

If the primary film is slot-die cast, either it is first stretched longitudinally and then transversely, or it is stretched simultaneously in both directions. Stretching is carried out as the film travels between cooling roll and take-up unit. The stretching tension is applied by draw rolls or by a combination of draw rolls and sliding caliper clips on a tenter frame. During the drawing process, shown in Figure 4, the film is heated to temperatures below the crystalline melting point of the relevant raw material, by roll contact or hot air.
The design principle of a biaxial drawing unit for flat film is explained with the help of the perspective illustration in Figure 13. Two endless belts or chains arranged symmetrically in relation to each other carry film-clamping devices ('clips') which grasp the edges of the film and transport it. In the section where the clips diverge, the film is heated by hot air, and transversely stretched to a multiple of its initial width. If the film is stretched longitudinally before entering the transverse drawing unit, the spacing of the clips in the transport direction remains constant during transverse stretching.

8.3 Theory of the orienting process 275

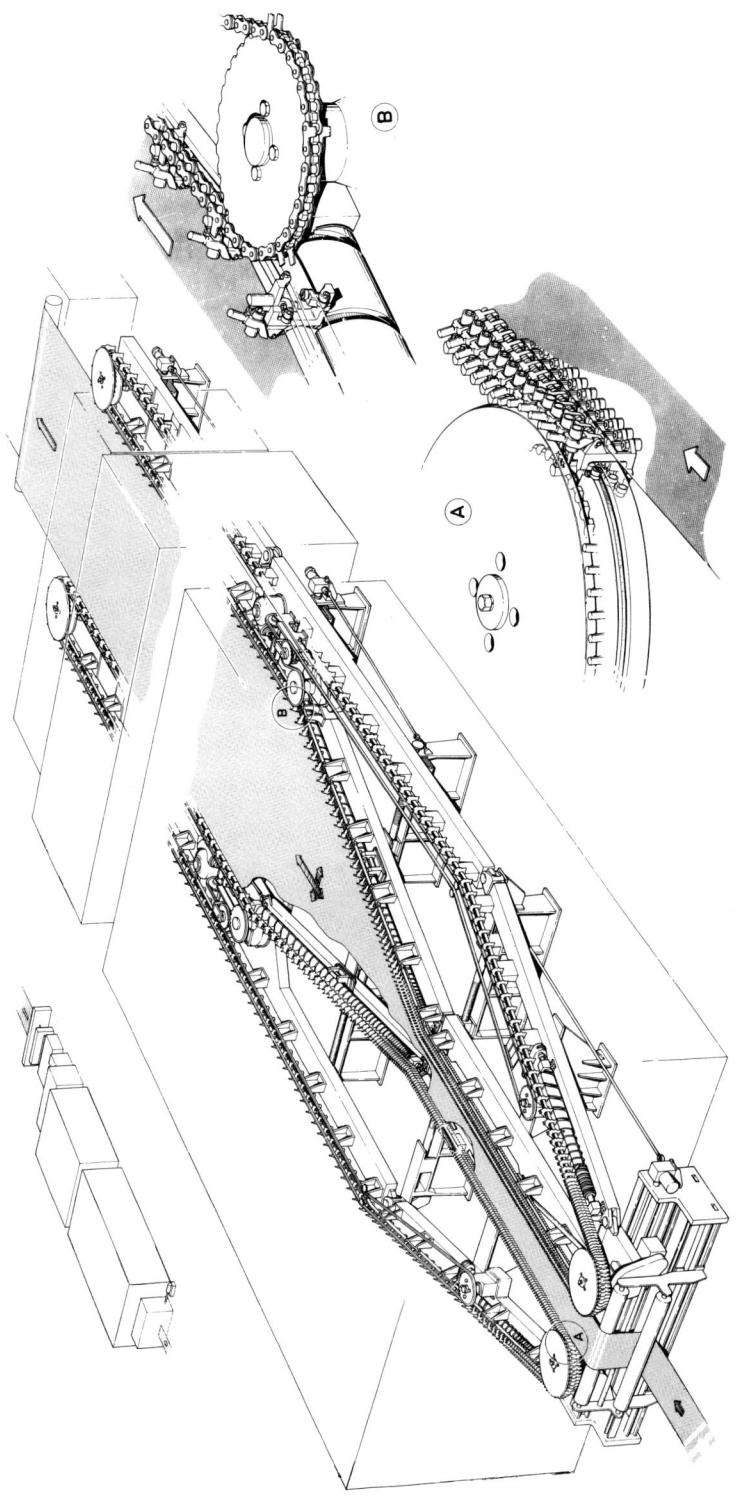

Figure 13. Unit for simultaneous biaxial stretching. (Photo: Kampf, Bielstein, West Germany)

If transverse and longitudinal drawing are to be carried out simultaneously, the clips are driven by lead screws with a progressive pitch during transverse drawing, so that the clips stretch the film transversely and longitudinally as it advances.

Figure 14 shows a unit for transverse drawing. After the required width has been reached, the spacing of the clip guides remains constant, or is reduced by a predetermined amount to relax the film (Figure 4g) so that the film dimensions can be stabilized in the hot-air oven while tension is maintained.

Figure 14. Unit for transverse stretching of films (viewed from the film inlet side).
(Photo: Dornier, Lindau, West Germany)

In Figure 15 the two common drawing processes for cast films are compared. The upper half shows the two-stage process in side elevation and in plan. Longitudinal stretching takes place between heated rolls which are driven at different circumferential speeds that increase progressively from roll to roll.

For longitudinal stretching between rolls, different roll arrangements are used for different thermoplastics and different film dimensions.

In the case of polypropylene film, the film is first heated on a group of slowly driven, heated rolls with chrome-plated surfaces; it then goes to a group of four to six stretching rolls, which can be driven at different speeds and changed in their positions relative to each other. Stretching occurs over a short distance between pairs of rolls. The rolls can be heated by steam, oil or water.

In two-stage drawing of polyester film, longitudinal stretching is preferably carried out in several consecutive stretching gaps. Between slow- and fast-driven roll sets there are one or more idler rolls which are driven by the film. The result is that the film is stretched gradually in several gaps with the stretching ratio being divided among the gaps. Because roll spacings are small, lateral film shrinkage ('neck-in') is low.

The multigap, longitudinal drawing system is also used for polypropylene films if large throughput rates are required. For high-quality films, the longitudinal stretching equipment is supplied in insulated cabinets which can be heated.

Particular features of this two-stage, biaxial drawing process:

– different LD and TD properties, adjustable,
– simple mechanics of the line,
– high maximum stretching speed and
– demarcation of separate process steps.

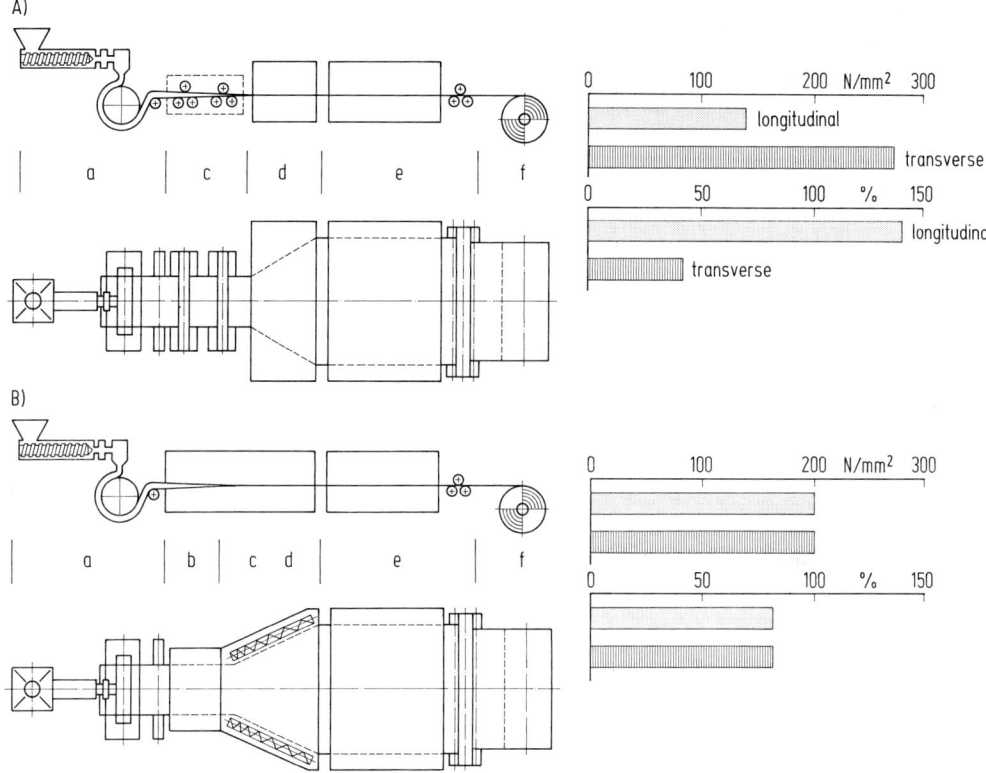

Figure 15. Comparison of the two-stage drawing process (A) and the simultaneous drawing process (B)
a production of the primary film with compression of the melt, filtration, casting and cooling, *b* heating the film to stretching temperature, *c* longitudinal stretching, *d* transverse stretching, *e* dimensional stabilization by heating the film, *f* processing, winding, distribution

Production example:

Using the two-stage process, 25 µm PP film was made using an area-stretching ratio of 1:10. Tensile strength of 130 N/mm² LD and 270 N/mm² TD with elongations of 140% (LD) and 40% (TD) are obtained.
The lower half of Figure 15 shows the principle of the simultaneous drawing process.
Particular features of the simultaneous orienting process:

– produces isotropic film structure,
– prevents intermediate crystallization,
– non-contact stretching of the drawn area of the film, and
– relatively low energy usage.

Production example:

Using the simultaneous process to make PP film of 25 µm final thickness under the same conditions, the LD and TD tensile strengths were both 200 N/mm² and the correspondimg elongations 80%.
Both processes are employed on a large scale in industry. Economic considerations, related to the features of the two processes mentioned, and to the product requirements, determine which process is used.

In the production of 'tensilized' polyester film, which has preferential longitudinal orientation for video tapes and magnetic tapes, additional longitudinal stretching is required after initial LD and TD orientation. This longitudinal stretching is also carried out between rolls, in one or several narrow gaps.

For longitudinal stretching of coextruded polypropylene films, a special machine is required, since the polymers in the different film layers have different melting points. Stretching is carried out between heated and cooled rolls, to prevent the film layers from sticking to the rolls. Heating of the film layer with the higher melting point is augmented by hot-air blowers or infrared radiators.

The clips for holding the film during transverse stretching are designed for each specific thickness of film and each material to be handled. As the film enters the unit the clips are opened by means of levers fixed to the clips, which slide across a stationary cam profile as the chain circulates. After the film has been fed in, the clamps are closed by spring action (Figure 16). Reaction to the force created by the film during stretching increases the clamping effect. To enhance the frictional force generated by clip closure, the clamping surface is profiled.

Figure 16. Clips for film stretching.
(Photo: Brückner, Siegsdorf, West Germany)

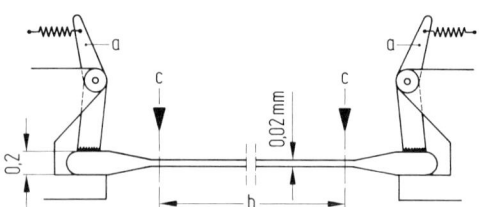

Figure 17. Principle of transverse stretching of flat film
a clip, b usable width, c blade

The principle of transverse drawing of flat film is shown in Figure 17.

For thinner films up to 75 µm max., final film widths of 4000 mm and stretching speeds of up to 120 m/min, sliding clips which are simple to service are given preference.

Polyester films for photo-base with thicknesses of up to 350 µm, final film width of up to 5000 mm and stretching speeds of up to 200 m/min require clips running in multiple-roller bearings. For high-speed lines for PP and polyester packaging films operating at stretching speeds of up to 350 m/min, and with final widths of 8000 mm, special clips are available which are guided on up to 10 rollers on special tracks, known as monorail tracks.

For economic and efficiency reasons, lines for manufacturing biaxially oriented flat film are generally designed for high production rates. Figure 18 shows the layout and principal dimensions of a normal production line for biaxially oriented PP film with an annual output of 15 000 tons [27]. Other data on this line:

Output: 2000 kg/h of 25 µm film, not edge trimmed.
Thickness range: 0.012 to 0.050 mm
Film width: 5100 to 6000 mm (without edge trimming)
Stretching ratio: 1:1 to 1:12 longitudinally
1:3 to 1:12 transversely
Winding speed: 8.3 to 250 m/min.

8.3 Theory of the orienting process

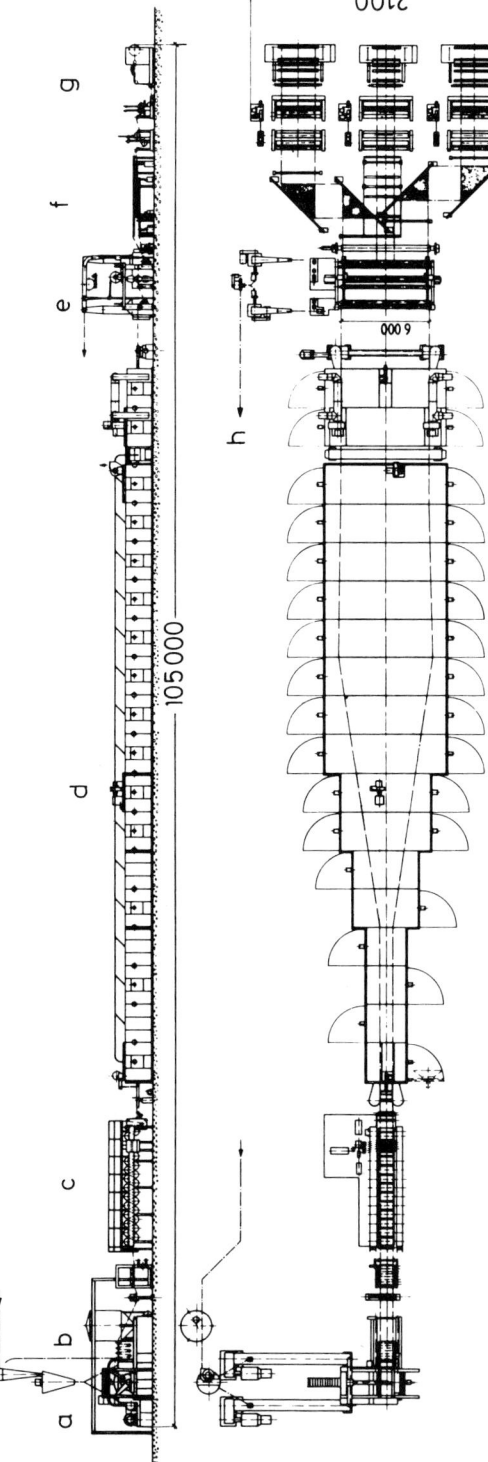

Figure 18. Line for the production of biaxially oriented PP film with a throughput of 3,500 kg/h
a extrusion unit, *b* chill roll equipment, *c* longitudinal stretching machine, *d* transverse stretching machine, *e* take-off unit with thickness measuring instrument, *f* guiding system, *g* wind-up unit, *h* automatic edge-trim recycling

Lines for manufacturing biaxially oriented PET, with an output of 3500 to 4000 kg/h, have been built.

Biaxially oriented film can also be produced by the blown-film method (Figure 19). In this case, a heated tubular cast film is continuously inflated by gas pressure while being stretched longitudinally, so that the oriented product has properties similar to those obtained by simultaneous biaxial drawing of flat film. The tube drawing processes became known as the ICI, the Bexphane and the Pirot processes (Bemberg Co.). They are preferred for use with PVC, PP and PET, particularly for films of 6 to 40 μm in thickness, and differ mainly in the method of cooling the film tube. In order to obtain a primary film that is essentially amorphous, crystalline thermoplastics require the use of a water-cooling system, or of contact cooling by means of an internal mandrel. Longitudinal stretching is carried out between the two pairs of trip rolls 1 and 2 (position d and h, Figure 19) while transverse stretching takes place simultaneously by means of air inflation. The drawing process is stable if the stretching tension in the film balances the excess pressure in the film bubble. The film is heat-set by passing it through a hot-air oven.

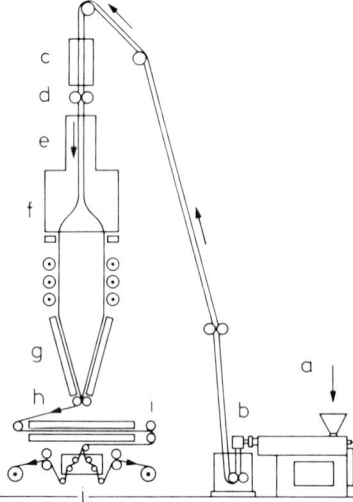

Figure 19. Blown-film stretching process
a extrusion, *b* water bath, *c* preheating, *d* nip-roll pair 1, *e* preheating system, *f* stretching system, *g* collapsing frame, *h* nip-roll pair 2, *i* heat setting, *j* wind-up unit

For many applications it is sufficient, or preferred, to orient the film only in one direction. Uniaxial stretching can be carried out in one or more gaps between rolls. The stretching force is generated exclusively by means of friction on the surfaces of the drive rolls.

As the film width must be maintained during stretching, the spacing of the stretching rolls is made small enough for the friction forces to prevent shrinkage of the film across its full width. Only the film edges shrink, and the film thickness over most of the web is reduced by stretch ratio R, so there is a usable width of more than 90% of the initial film width possessing the same properties [14].

The heat is transferred to the film by contact. In order to reduce shrinkage stresses, orienting is effected in stages, with the roll temperature being set to suit the raised thermal form stability induced by increased orientation.

Figure 20 shows the schematic layout and a general view of a uniaxial drawing line using several stretching stages, and primary film produced on a blown film unit.

Use of friction to prevent lateral shrinkage during MD stretching gives rise to a transverse tensile stress in the film, which causes transverse orientation. Although this is low in relation to the main orientation, it is sufficient to prevent splitting of the film, which can result from high anisotropy.

8.3 Theory of the orienting process

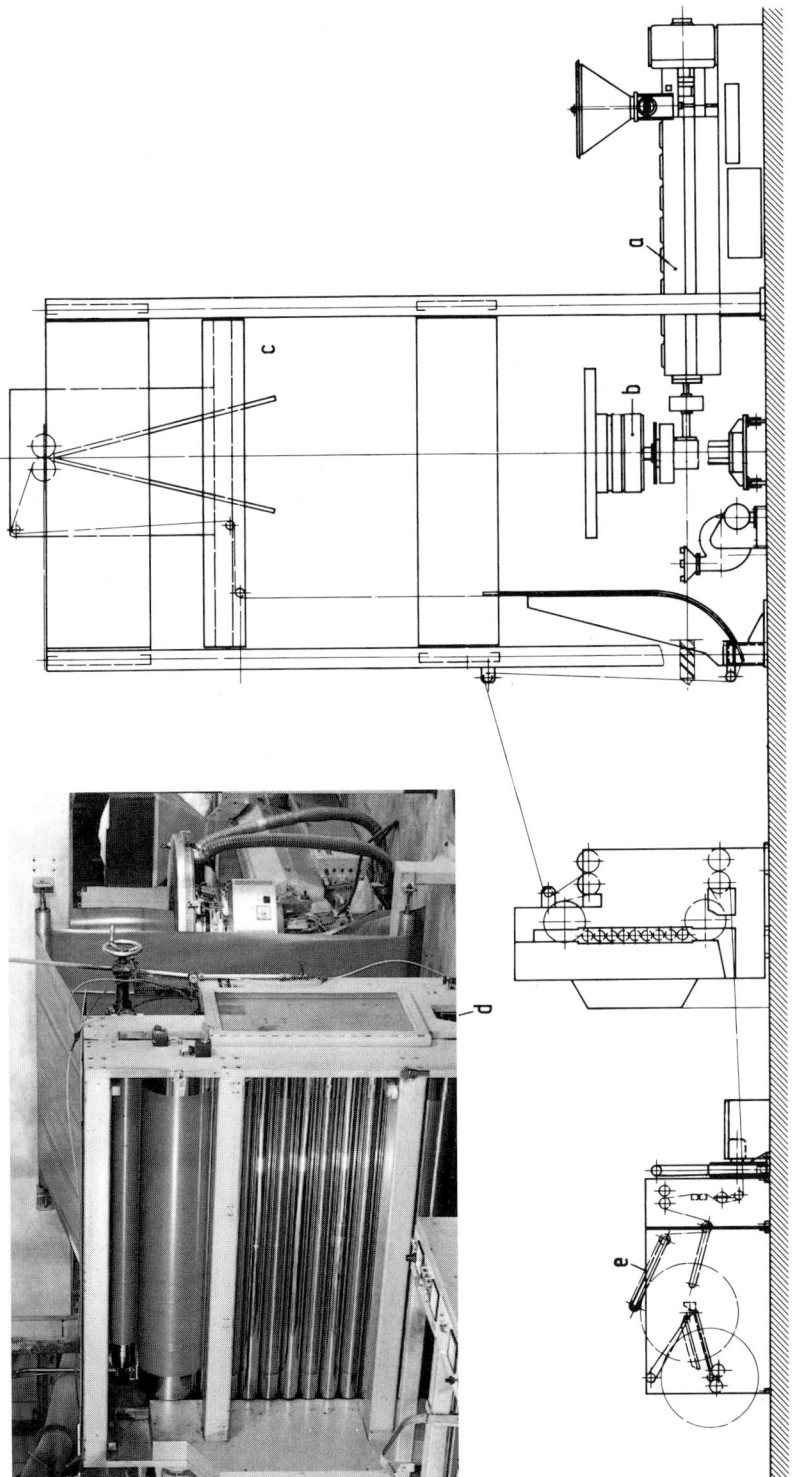

Figure 20. Line for making biaxially oriented film by the multigap process for neck-in-reduced drawing
a extruder, *b* blown-film tube head, *c* collapsing frame, *d* neck-in-reduced drawing unit, *e* wind-up
(Photo: Barmag, Remscheid, West Germany)

The blown film can be introduced into the stretch gap either unfolded or folded double. When double-layer films of polyolefins are stretched, the two layers weld together because of the rise in temperature and the normal force which arises in wrapping round the roll. Uniaxial lines with several stretching stages are employed for making warp-beam films, among other products, which eliminate the winding and beaming of film tapes.

The films are wound the way they are on flat-film lines, with centrally driven winders (Figure 21). These are turret winders with automatic transfer devices for roll change without loss. As the film width is usually larger, owing to transverse drawing, than that required for later use, it is slit into usable widths and wound separately.

Figure 21. Wind-up units for biaxially oriented film, with usable width of 2 m
(Photo: Brückner, Siegsdorf, West Germany)

References for Chapter 8

For the convenience of the reader the English titles of all publications in languages other than English are shown in parentheses.

[1] *Uhlmann, D. R., Kohlbeck, A. G.:* Scientific American 12 (1975), pp. 96/106.
[2] *Hornbogen, F., Friedrich, K.:* Gefüge von Kunststoffen (Morphology of Plastics). Sonderband Praktische Metallographie Nr. 9 (1978), pp. 143/170.
[3] *Schäfer, K., Friedrich, K.:* TEM-Untersuchungen der Morphologie und des Deformationsverhaltens von Polypropylen-Filmen mit unterschiedlichen ataktischen Anteilen (TEM Studies of the Morphology and the Deformation Processes of Polypropylene Films with Different Atactic Contents). Beitr. elektronenmikroskop. Direktabb. Oberfl. 12/1 (1979), pp. 125/132.
[4] *Kanig, G.:* Kolloid. Z. u. Z. Polymere 251 (1973), pp. 782/783.
[5] *Friedrich, K.:* Praktische Metallographie 16 (1979), p. 321.
[6] *Patel, G. N., Patel, R. D.:* J. Polym. Sci. Part A-2, 8 (1970), pp. 47/59.
[7] *Keith, H. D., Paddon, F. J. Jr.:* J. Appl. Physics 35 (1964), p. 1270.
[8] *Friedrich, K.:* Dissertation, Ruhr-Universität Bochum, 1978.
[9] *Schäfer, K.:* Diplomarbeit, Institut für Werkstoffe, Ruhr-Universität Bochum, 1978.
[10] *Ward, I. M.:* Mechanical Properties of Solid Polymers. Wiley Interscience, London, New York, 1971.
[11] *Kanig, G.:* J. Crystal Growth 48 (1980), pp. 303/320.

[12] *Petermann, J., Gleiter, H.:* Colloid & Polymer 254 (1976), pp. 247/248.
[13] *Schäfer, K.:* Molekulare Vorgänge während eines Verstreckprozesses in teilkristallinen Kunststoffen (Molecular Processes in Partially Crystalline Polymers During a Stretching Process). Technical Information, Barmag, 1981.
[14] *Hensen, F., Braun, S.:* Textilindustrie (1979), pp. 844/848.
[15] *Hensen, F., Braun, S.:* Kunststoffe 68 (1978), pp. 221/229.
[16] *Hensen, F., Bongaerts, H.:* Plastverarbeiter 31 (1980), pp. 541/549 u. 615/621.
[17] *Hensen, F., Gathmann, E.:* Kunststoffe 64 (1974), pp. 343/349.
[18] *Pahl, M. H.:* Dispersives Mischen mit dynamischen Mischern (Dispersive Mixing with Dynamic Mixers). VDI-Verlag, Düsseldorf, 1978, pp. 177/196.
[19] *Hensen, F., Siemetzki, H.:* Kunststoffe 70 (1980), pp. 753/758.
[20] *Hensen, F.:* Kühlen von Tafeln, Tiefziehfolien und Reckfolien (Cooling of Sheets, Thermoformable Films, and Stretch Films). In: Kühlen von Extrudaten (Cooling of Extrudates). VDI-Verlag, Düsseldorf, 1978, pp. 81/96.
[21] *Lange, M.:* Biaxiales zweistufiges Verstrecken von Flachfolien (Biaxial Two-stage Stretching of Flat Films). In: Extrudierte Feinfolien und Verbundfolien (Extruded Thin Films and Composite Films). VDI-Verlag, Düsseldorf, 1976, pp. 89/105.
[22] *Hutzenlaub, A.:* Recken von Kunststoffolien mit besonderer Betrachtung des monoaxialen Reckens von Flachfolie (Stretching of Polymer Films with Particular Consideration of the Uniaxial Drawing of Flat Films). In: Extrudierte Feinfolien und Verbundfolien (Extruded Thin Films and Composite Films). VDI-Verlag, Düsseldorf, 1976, pp. 71/87.
[23] *Wellenhofer, P.:* Monoaxiales Recken mit besonderer Berücksichtigung des biaxialen Flachfolienstreckens (Uniaxial Stretching, with Particular Consideration of the Uniaxial Drawing of Flat Films). In: Extrudierte Feinfolien und Verbundfolien (Extruded Thin Films and Composite Films). VDI-Verlag, Düsseldorf, 1976, pp. 65/91.
[24] *von Harnier, A.:* Kühlen von Flachfolien (Cooling of Flat Films). In: Kühlen von Flachfolien (Cooling of Extrudates). VDI-Verlag, Düsseldorf, 1978, pp. 97/132.
[25] *Kuhlmann:* Herstellung von Folien durch Verstrecken (Production of Films by Stretching). Company brochure, Kampf, Bielstein, West Germany, 1981.
[26] *Kuhlmann:* Simultaneous Biaxial Film Orientation. Company brochure, Kampf, Bielstein, West Germany, 1981.
[27] General Layout for a POPP Plant. Company brochure, Brückner, Siegsdorf, West Germany.
[28] *Hensen, F.:* Anlagen zur Herstellung von Folienbändchen und Monofilamenten (Lines for Manufacturing Film Tapes and Monofilaments). In: Foliengewebe, Vliesstoffe aus PP (Film Textiles – PP Nonwovens). VDI-Verlag, Düsseldorf, 1979, pp. 107/149.
[29] *Menges, G., Kirch, D., Nordmeier, J., Winkel, E., Wortberg, J.:* Qualitätsverbesserung von teilkristallinen Kunststoffen durch gezielte Nukleierung am Beispiel von Polypropylen (Quality Improvements in Partially Crystalline Polymers by Selective Nucleation, Using Polypropylene as an Example). Kunststoffe 73 (1983) 5, pp. 258/260.
[30] *Hensen, F., Hessenbruch, R., Bongaerts, H.:* Entwicklungsstand bei der Coextrusion von Mehrschichtblasfolien und Mehrschichtbreitschlitzfolien (State of Development in Coextrusion of Multilayer Blown Films and Multilayer Slot-die Films). Kunststoffe 71 (1981) 9, pp. 530/538.

9 Extrusion of film tapes

F. Hensen

9.1 Introduction

Film tapes are uniaxially oriented thermoplastic semi-finished products with a high width-to-thickness ratio, which can be converted into twines, ropes, and knitted or woven fabrics.

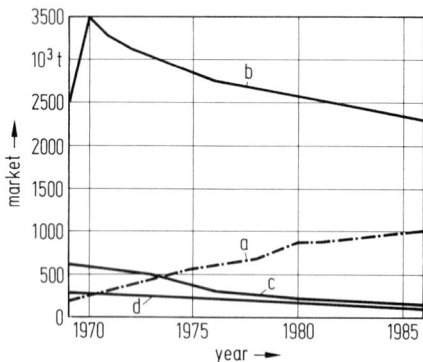

Figure 1. World production of hard fibers and polyolefin film tapes (estimated)
a PO film tapes, *b* jute, *c* sisal, *d* hemp and others

Ever since industrial extrusion of film tapes began in 1956, their production has been steadily increasing. As shown in Figure 1, production of the natural fibers jute, sisal and hemp has decreased during the same period; and they have almost completely been replaced by film tapes in some fields of application. Such fast growth can be explained as follows:

– the techniques and production lines in use are more cost-effective than those used with natural products. And are, furthermore, ecologically harmless and energy-saving.
– the raw materials employed are polypropylene, polyethylene, polyester or polyamide. In comparison with the natural products used for the same applications (jute, sisal and hemp) they have better properties, more stable prices, and are freely available.
– newer applications include geo-textiles, concrete reinforcements and asbestos substitutes.

9.2 Development and market significance of film tape extrusion

The advantages in processing technology already mentioned, and the product properties and efficiency, have led to the growth of a separate new sector of industry, with large growth rates, within a very short time. The patents and literature references listed in the appendix to this chapter give a chronological summary of the development of film tape extrusion.
In 1936, long before the development of the thermoplastics used today, a patent application by the then I. G. Farben Company described fiber-like products that could be produced by means of mechanical treatment of paper, film strips or tubelets. Subsequent patent applications by others described surface treatment of slit film tapes by rolling, twisting, rubbing, scratching, cutting, profiling or incorporating splitting additives, in order to make them similar in appearance and processability to natural products used previously.

The extrusion of polymer film tapes gained large-scale industrial significance in the mid-sixties, when the techniques of using the favorable geometry of continuous tapes for the economic production of surface-covering products by the weaving process, such as sacks, technical fabrics, carpet backing materials or awnings, became widespread.

At the same time the economic usefulness of wider tapes for the production of twines, ropes and hawsers, or as a substitute for or complement to steel tyre cords, was discovered.

9.3 Methods of manufacture of film tapes

A process for making oriented tapes for certain fields of application is shown in Figure 2. The first step is to produce flat or blown film, either smooth or profiled, from polymer pellets. Cooling methods include the use of a water bath, chill rolls or air jets. The film is then drawn, either uncut or slit lengthwise into strips, and with or without mechanical aftertreatment wound into spools or film packages. Depending on the intended application, the tapes may be twisted and cabled or woven, knitted, tufted, or felted without being twisted. The field of application determines the type of raw material, the tape dimensions, the stretch ratio, and any subsequent mechanical treatment that may be required.

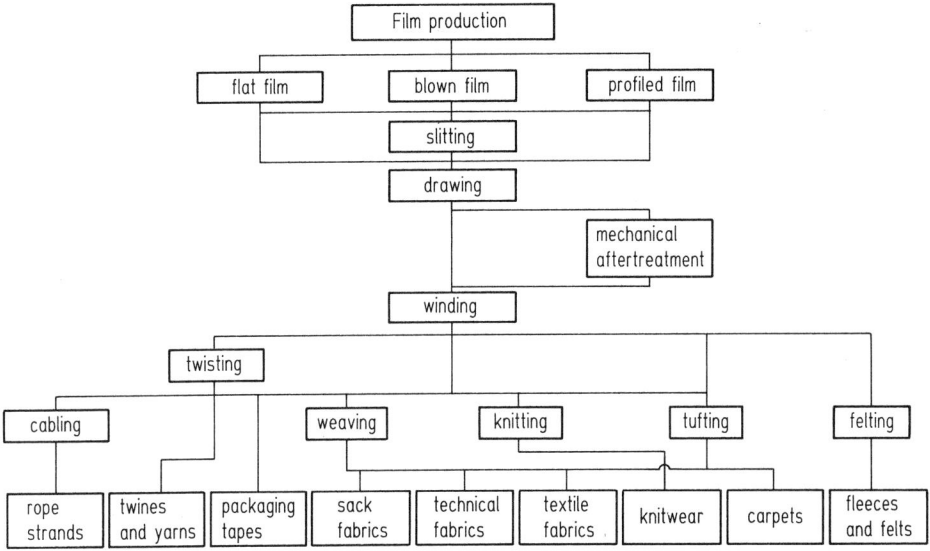

Figure 2. Production procedures with film tapes for different fields of application

The properties of the tapes, in particular tenacity, elongation and hot shrinkage, depend on the type and intensity of cooling after extrusion and on the degree of orientation of the molecules achieved during drawing. The morphology of the polymer types used and the film thicknesses required for the end product, determine the cooling process needed, with air, water or roll contact being used according to circumstances. In order to achieve molecular orientation, the film or the group of tapes is progressively elongated to many times the original length. If this orientation takes place uniaxially and without contact over a sufficiently large distance between godet rolls running at different speeds, width and thickness will decrease by the same ratio (Figure 3, top).

$$b_2 = \frac{b_1}{\sqrt{\text{stretch ratio}}}, \quad d_2 = \frac{d_1}{\sqrt{\text{stretch ratio}}}$$

b_1 initial width
b_2 final width
d_1 initial thickness
d_2 final thickness

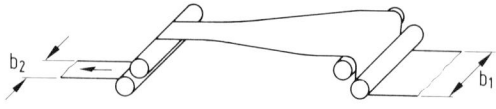
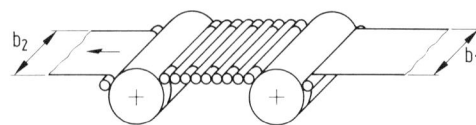

Figure 3. Principle of uniaxial stretching (top) use of a short gap for reduced "neck-in" (bottom)

Under these circumstances the anisotropy of the tenacity in longitudinal and transverse directions is so great that the tapes may split under transverse stress. The effect can be utilized to achieve a fibrous character in the tapes. If this is to be avoided, some orientation of the molecules transverse to the stretching direction is required, and can be achieved by preventing "neck-in" during drawing (Figure 3, bottom). Thanks to continuous roll contact and drawing over an extremely short distance between the rolls, the reduction in width is small and limited to the marginal zones, whilst the initial thickness of the strips is reduced by the stretch ratio.

$$b_2 = \left(b_1 - \frac{x \cdot b_1}{100}\right)$$

x % lateral shrinkage

Drawing with reduced neck-in is utilized for the production of warp-beam films, which are supplied instead of a warp-tape beam to the loom or knitting machine. In this case the oriented film is slit into tapes when entering the machine.

9.4 Characteristics of film tapes

In the course of the development of film tapes, the cross-sections best suited for certain fields of application were worked out. Figure 4 shows the most commonly used types of cross-section.
In the case of weaving tapes the ratio of width to thickness ensures that the tapes lie flat during the weaving process and permits high weaving speeds with low material consumption.
Knitting tapes are made thinner than weaving tapes to allow for the special requirements of the knitting process.
The thickness and width of twine and rope strands are made such that the flexibility and yarn weight per unit length match those of the common hard-fiber yarns made of natural products used previously. This means that existing twisters and cablers can be used. The fiber character is favorably influenced by longitudinal profiling and possibly by splitting the film.
The dimensions of strapping tapes depend on the application requirements with regard to tenacity, and product protection and handling during packaging.
Rot resistance, chemical resistance, lightness, low water absorption and ease of change of tenacity, elongation and shrinkage values have accelerated the introduction of oriented tapes into the market, at the expense of natural products in their respective fields of ap-

9.4 Characteristics of film tapes

plication. Table 1 lists the common fields of application and their most important requirements, the normal draw ratios, usual dimensions and most commonly used polymers. The data given have intentionally been limited to the most important fields of application.

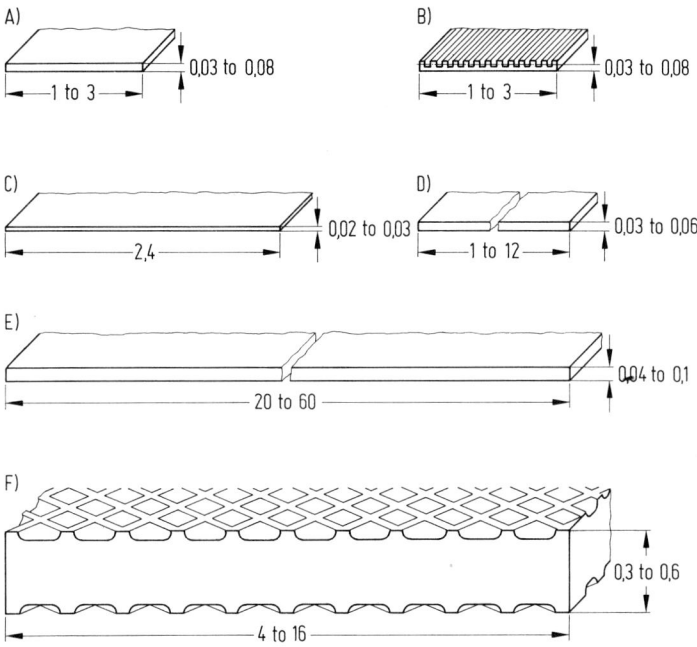

Figure 4. Cross-section dimensions of film tapes
A) weaving tapes, B) weaving tapes (profiled), C) knitting tapes, D) tufting tapes (outdoor carpets), E) twine and hawser strands, F) strapping tapes (embossed)

Table 1. Fields of application and important requirements

Fields of application	Requirement	Stretch ratio	Dimensions (width b in mm, thickness d in mm and titer T_t in dtex)	Thermoplastic
Basic carpet fabric	low shrinkage high tenacity temperature resistance specific splitting tendency matt surface	1 : 7 1 : 5	b = 0.8 to 2.5 d = 0.037 to 0.05 T_t = 270 to 1125	PP
Tarpaulins	high tenacity	1 : 7	b = 2.4 d = 0.04 T_t = 850	PP PE
Sacks	high tenacity high friction specific elongation weather resistance	1 : 7	b = 3.0 d = 0.03 T_t = 800	PP PE
Ropes	high tensile strength specific elongation good splitting tendency	1 : 9 to 1 : 11 (15)	b = 20 to 60 d = 0.04 to 0.1 T_t = 15,000 to 50,000	PP

(Continuation of Table 1)

Fields of application	Requirement	Stretch ratio	Dimensions (width b in mm, thickness d in mm and titer T_t in dtex)	Thermoplastic
Twines	high tensile strength high knot strength	1:9 to 1:11	b = 30 to 60 d = 0.03 to 0.06 T_t = 14,000 to 30,000	PP PP/PE blend
Separating fabric	high tenacity	1:7	b = 2.1 d = 0.04 T_t = 750	PP
Filter fabric	low shrinkage abrasion resistance	1:7 1:5	b = 1.0 to 2.0 d = 0.04 T_t = 350 to 700	PP PET
Reinforcing fabric	low shrinkage specific elongation temperature resistance	1:7 1:5	b = 2.0 d = 0.03 T_t = 550	PP PET
Wallpaper and home textiles	UV-resistance low static charge even dyeing textile feel	1:7	b = 1.2 to 3.0 d = 0.035 T_t = 350 to 900	PE
Outdoor carpets	low shrinkage wear resistance weather resistance good recovery even dyeing defined splitting	1:7 1:6	b = 1.0 to 12 d = 0.03 to 0.06 T_t = 300 to 3500 b = 1.0 to 3.0 d = 0.02 T_t = 300 to 1000	PP PET
Decorative tapes	effect surface low specific weight	1:6	b = 4 to 12 d = 0.03 to 0.05 T_t = 800 to 3000	PP with blowing agent
Knitted tapes, sacks and other packaging	high knot strength low splitting tendency suppleness UV-resistance	1:6.5	b = 2.4 d = 0.025 T_t = 550	PP PE
Strapping tapes	high tenacity low splitting tendency	1:9 1:7	b = 5 to 16 d = 0.3 to 0.6	PP PET
Fleeces	fiber properties	1:7	b = 20 to 300 d = 0.025 T_t = 10 to 70	PP and polymer blends
Geo-textiles	high tenacity low splitting tendency specific elongation UV-resistance	1:6.8	b = 2.0 to 2.5 d = 0.025 to 0.085 T_t = 450 to 1920	PP PE
Containers	high tenacity specific elongation weather resistance	1:11 1:8	b = 2.7 to 3.5 d = 0.070 to 0.090 T_t = 1700 to 2840	PP PP/PE blend

The (uniaxial) stretch ratio used in production determines the tenacity and other properties such as elongation, shrinkage or splitting tendency. Different stretch ratios are set for different fields of application.

$$R = \frac{v_2}{v_1}$$

R stretch ratio
v_1 tape speed before stretching
v_2 tape speed after stretching

The maximum possible stretch ratios — and the maximum tenacities — can be considerably increased by means of melt alloys, addition of nucleating agents or by achieving particular crystalline structural conditions during cooling or temperature treatment of the film tapes. Figure 5 shows the dependence of tenacity on stretch ratio. The shaded tape width is explained by the variation in process data. With increasing stretch ratio, tenacity is increasing owing to better orientation, and at the same time the break elongation is reduced in proportion. With regard to reliability in later processing operations and the requirements of the end product, optimum stretch ratios are chosen for special fields of application. Ratios of up to 10 are routinely possible with suitable polymers on normal extrusion lines. For higher stretch ratios, it is necessary to achieve a suitable crystalline structure during extrusion and subsequent thermal treatment. This is obtained either by intensive cooling or by nucleation of the melt or melt alloy, depending on the type of polymer and the tape density.

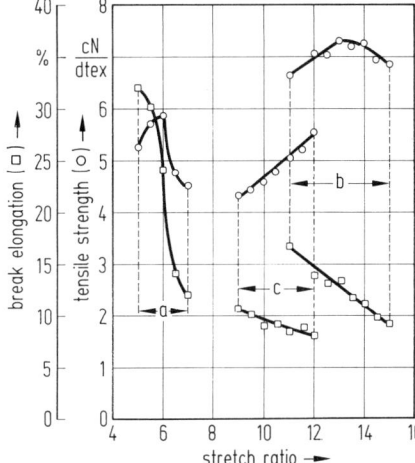

Figure 5. Dependence of tenacity on stretch ratio during PP film tape production
a weaving tapes, b high modulus tapes (polymer blend with LDPE), c twines and hawser strands

The splitting tendency of PP increases with increasing uniaxial stretch ratio. This splitting tendency is desirable with ropes and twines, only conditionally permissible with carpet backing fabrics and carpet-pile yarn, and not at all permissible with packaging fabrics or knits. If tenacity and elongation require a stretch ratio which leads to splitting, this splitting tendency can be reduced by melt additives of related polymers such as PE or PB to the PP melt. If the splitting tendency of PP tapes is to be increased with low stretch ratios, this can be achieved by mixing in small quantities of an incompatible thermoplastic such as PS.

Some fields of application require low shrinkage values, because tapes will be subjected to heat treatment in a later processing operation. The shrinkage tendency is determined by the quality of orientation achieved during the stretching process. Low values of shrinkage can be achieved by careful process control during stretching and by subsequent heat-setting under tension. Figure 6 shows shrinkage values that are obtained with common extrusion processes and with special stretching and heat-setting processes.

In addition to defined tenacity, break elongation, splitting tendency and shrinkage, certain fields of application require the tapes to have a high E-modulus. This is considerably influenced by morphological structure. For this reason, high-modulus tapes are produced under specific cooling conditions on a special extrusion unit utilizing a chill-roll system.

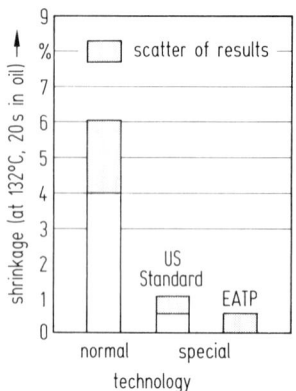
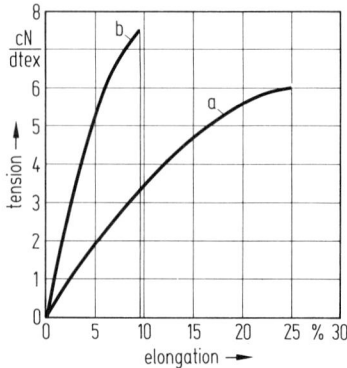

Figure 6. Shrinkage values of film tapes with normal and special stretching technology

Figure 7. Comparison between normal (a) and high E-modulus (b) film tapes, illustrated by tension/elongation diagrams

Figure 7 shows the different load/elongation behavior of normal PP film tapes and of the high E-modulus product attainable with appropriate line design.
Many line concepts have been developed using the modular construction principle, and the units can be combined optimally to produce film tapes with defined properties.

9.5 Film tape production methods

For a better understanding of the design of tape lines, the production process is illustrated in Figure 8. All line designs described later, regardless of the choice of the individual units shown in Figure 8, are based on the following process:

- Tape grade thermoplastic pellets are fed to the extruder from the storage container either by hand or pneumatically.
- In the extruder the pellets are compressed by the effect of the screw geometry and speed, melted by heat input and shearing energy, homogenised in a special screw section, or mixed with additives, and extruded as film or a bundle of tapes at constant mass throughput against the resistance of melt filter and casting die.
- The film or bundle of tapes is solidified by cooling in contact with a chill roll or in the cooling agent. If the tape width has not been determined by the casting die, the film is cut into a group of tapes with the blade spacing determining tape width and width deviation; other process data remain unchanged.
- The group of tapes is stretched between godet roll-sets which run at increasing surface speeds. The power transmission is by driven godet rolls. In order to improve orientation the tapes are heated in an airstream.
- Setting then takes place with the bundle of tapes being heated under defined tension.
- The annealed tapes are wound individually or in pairs or groups into cross-wound cheeses. The package structure is determined by yarn tension and traverse ratio. The traverse ratio is the number of revolutions of the chuck per traverse cycle of the thread-guide.
- The edge trim strips are sucked away after the first godet unit, chopped and fed into the extruder hopper.
- On start-up, the tapes are temporarily evacuated after the last godet unit to simplify the lacing of individual tapes onto the winding heads. The vacuum guns are also used to remove tapes that, in rare cases, break during stretching.

9.5 Film tape production methods

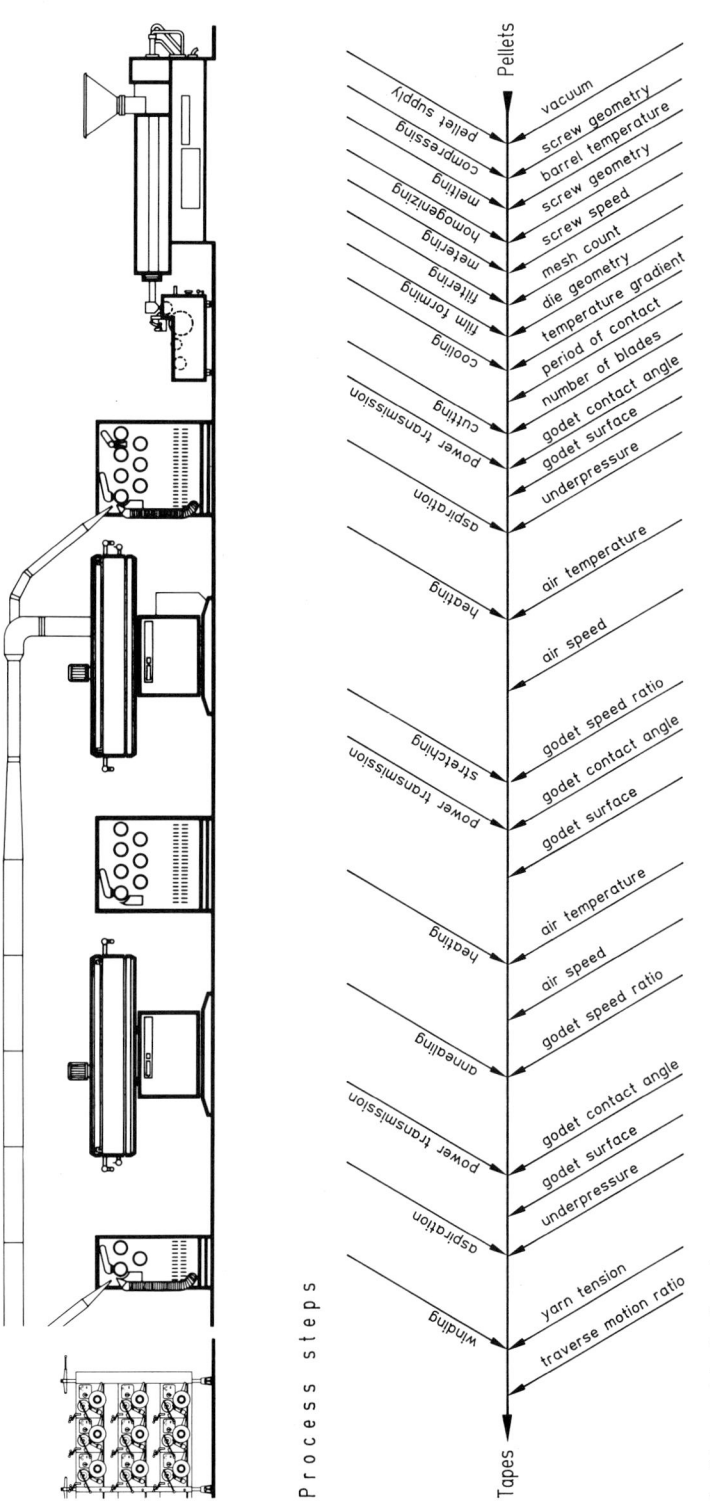

Figure 8. Process for manufacturing film tapes. Flow diagram showing process steps and influential factors

For a better understanding of this relatively complex process, Figure 8 shows only the most important line components, process steps and process parameters. It is clear to the expert that for improved process operation the size of the individual units, as well as the number of process steps and associated influencing factors, can be expanded. This is explained below.

9.6 Description of tape lines

A variety of designs based on the fundamental concept have been produced to meet the differing demands of various applications. They differ in the number, order and design of the individual units which make up the line.

The film tape lines are designed on the principle of technically optimizing separate process steps. Following this concept, each individual step is, if possible, assigned to an individual unit adapted in size and design to the application.

Figure 9 shows the line components in common use today for carrying out the main process steps, and lists some technical relevant data.

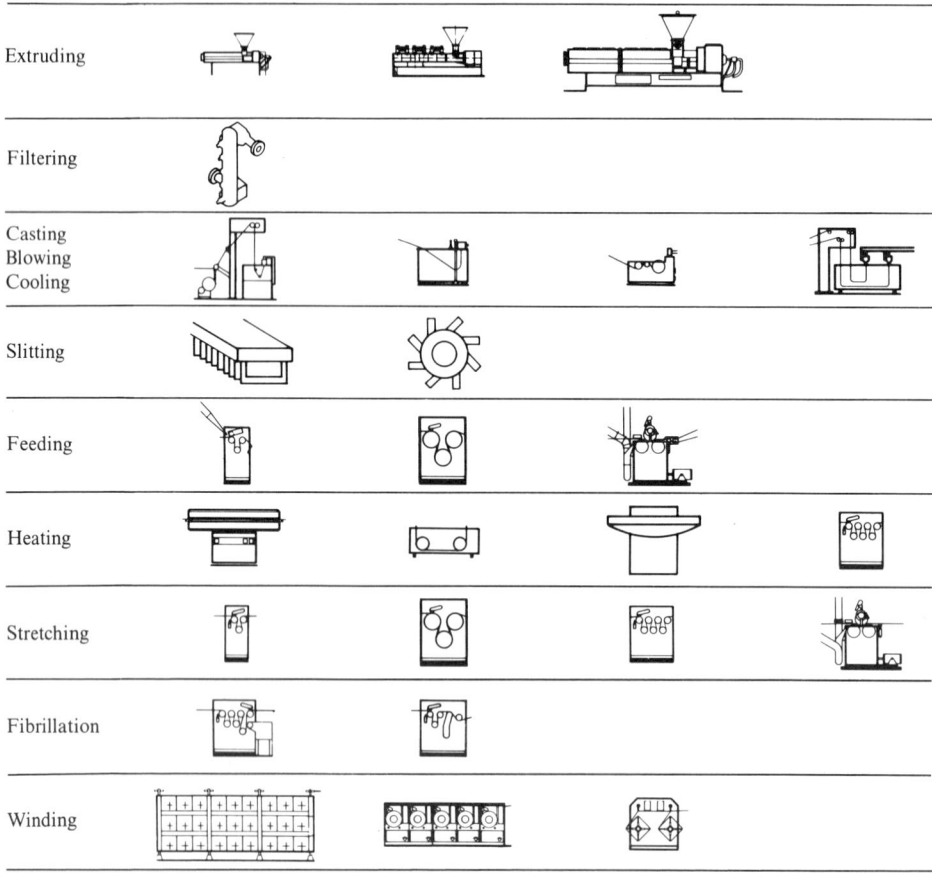

Figure 9. Elements of the film tape line and associated technical data

9.6.1 Extruding

For the preparation of the melt, single-screw extruders of horizontal design are used. They have the task of creating a gas-free melt from the pellets, homogenizing it and extruding it steadily. Throughput rate depends on the choice of screw diameters and lengths given in Figure 9. Extruder designs differ in the type of barrel temperature control, barrel insulation, type of drive and design of the feed zone. When processing polyolefins, the conveying effect of the feed zone can be enhanced if it is externally cooled, and longitudinally grooved. The extruder screws are in most cases of single-start design with 1 D constant pitch, and have special shearing and mixing elements in the discharge zone. Thanks to these, an even distribution of melt additives such as stabilizers and colorants, and a good distribution of the melt flow behind the extruder are achieved. Even temperature distribution is of special importance for uniformity of tape properties, since it strongly influences the melt structure during film formation and consequently the quality of orientation in the stretching process.

Figure 10 compares the temperature distributions achieved by three screws, using polyester as the test material. All other conditions being constant, the uniformity of the melt

		screw diameter: 45, 60, 75, 90, 120, 150 mm screw lengths: 24, 27, 30 D output: up to 300 kg/h for weaving tapes up to 700 kg/h for hawser strands
		standard: LLf 1, filter area 1000/5000 cm²
		flat film dies: 460 to 1660 mm blown-film die: 150 to 600 mm cooling: air, water, surface contact
		single blade cutter multiblade cutter blade change without interrupting production
		working widths: 400 up to 1200 mm at present godet can be heated and cooled
		heating via convection, contact, self-heating
		stretching force up to 10 000 N
		interim or subsequent addition possible
		stroke: 2 × 115, 100, 150, 200, 250, 300

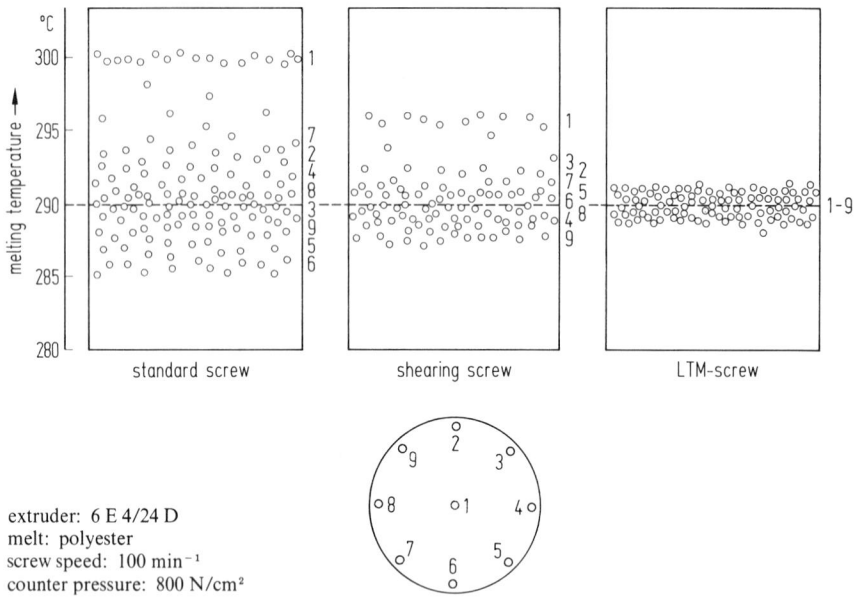

extruder: 6 E 4/24 D
melt: polyester
screw speed: 100 min⁻¹
counter pressure: 800 N/cm²

Figure 10. Melt temperature distribution in the adaptor between extruder and film die

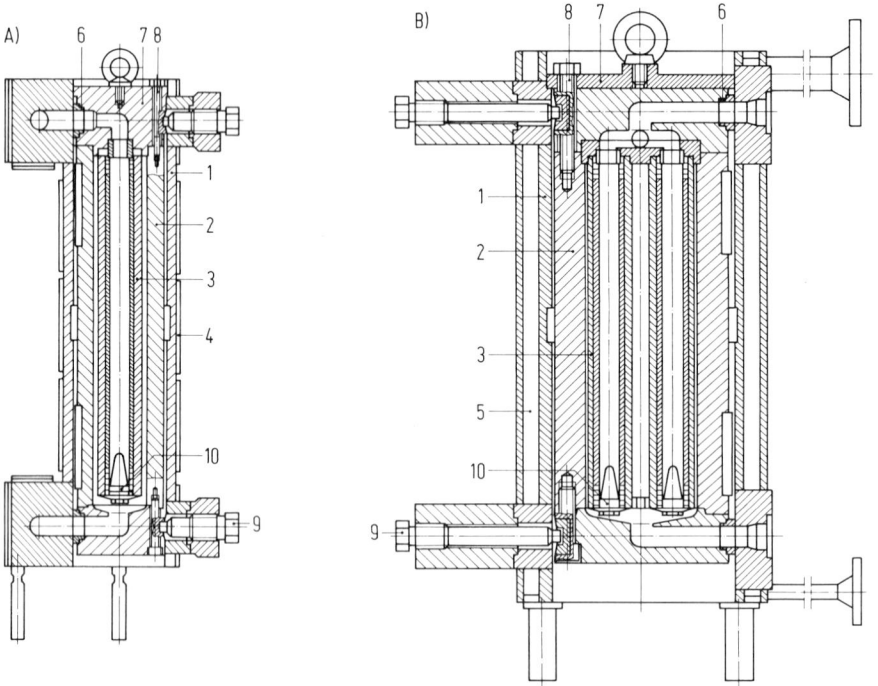

Figure 11. Long-life filter with one or several filter candles (legend)
1 filter housing, *2* filter insert, *3* filter rod, *4* heating band, *5* heating jacket, *6* seal, *7* deflector, *8* assembly screw, *9* pressure screw, *10* displacement tip

temperature is improved by using a cylindrical shearing component and an LTM mixing component with cylindrical mixing pins on the root of the screw. The temperature spread was observed by means of nine thermocouples distributed along the screw cross-section as indicated in the drawing. The best temperature distribution is achieved with the mixing screw. This screw prevents the temperature being higher in the center of the melt channel than in the marginal zones, which is common with other arrangements.

9.6.2 Filtering

The melt must not only be homogeneous; it must also be pure. It must not have any large inclusions, as these lead to breaks during the stretching process or to a reduction in tenacity. For this reason a melt filter is always inserted between the extruder and the extrusion die. In order to keep pressure losses small even with fine-mesh filters, large-surface candle-type filters are preferred; their effective areas are 40 to 400 times larger than disk filters inserted in front of a breaker plate at the barrel end.

Figure 11 shows the cross-section of the long-life filter. The filter candle consists of a tubular support element which carries the filter medium. It is located in a cylindrical housing, with the melt flowing from the outside to the inside and the filtered melt being discharged through the inside of the support element. The filter housing is arranged in a heated jacket which is firmly installed between the extruder and the extrusion die, and can easily be exchanged.

In order to avoid stoppage of production during filter change, the melt filter can also be designed as a 'non-stop' device. In that case a three-way valve is arranged between two filter chambers at the inlet and outlet, to permit a diversion of the melt flow from one chamber to the other.

9.6.3 Film formation

The film (Figure 9) is formed either by casting from the flat film die into a water bath or onto a chill roll, or by using the film blowing process to obtain tubular film. Cooling is effected directly with water or by roll contact, or, in the case of blown film, by air. With the same polymer, the cooling speed and the pre-orientation of the melt by draw-down from the die gap determine the crystalline structure, and consequently the properties achievable during stretching. Because the different end-products require different starting film thicknesses (Figure 4) and in view of the different crystallization behavior of the various types of polymer, different film-forming methods are chosen, according to purpose (see Table 2).

Table 2. Film forming methods for different products

Film forming method	Product Thickness range (initial thickness) µm	Type of tape
Water-bath process	60 to 250	PP and HDPE weaving tapes low-shrinkage PP weaving tapes PP twines and PP cable strands
Chill-roll process	> 200	PP high-modulus tapes HDPE weaving tapes
Blown-film process	50 to 70	HDPE knitting tapes

Since the width is reduced during stretching it is possible to cast two flat films simultaneously for better utilization of the line width. These enter the first godet unit in a double layer just as the tubular film does. The uniformity of film thickness transverse to the extrusion direction is determined by the shape of the extrusion die and the temperature homogeneity of the melt, and can be evened out by local adjustments to the die gap.

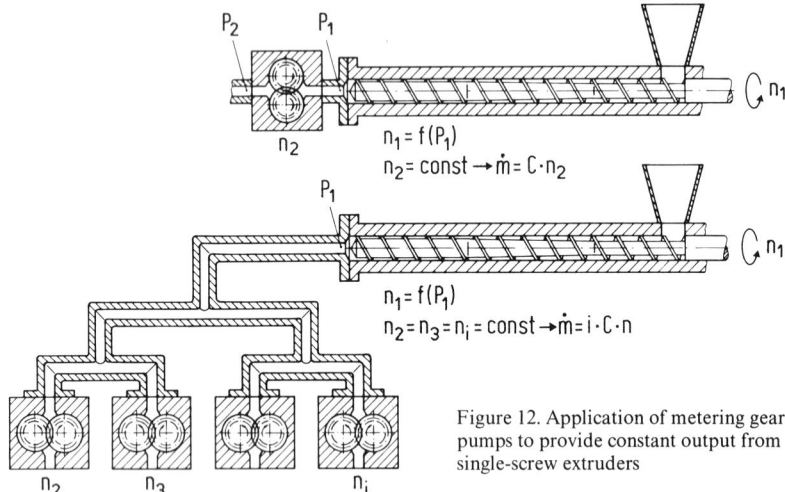

Figure 12. Application of metering gear pumps to provide constant output from single-screw extruders

The thickness uniformity in the extrusion direction depends on the metering precision of the extruder. Although good metering uniformity can be achieved with proper screw geometry and effective temperature controllers on the extruder barrel, it can be improved by using a volume-displacement gear pump. Figure 12 shows a schematic arrangement of the pumps when one (top) or more (bottom) extrusion dies are used behind the extruder. With a correctly designed and manufactured pump the leakage losses between pump housing and gear wheels are negligible, so that the quantity conveyed in time Q is proportional to the pump speed n_2; in contrast, the metering constancy of the single-screw extruder is dependent on pressure and viscosity. If a melt filter is used at the same time, the pump may be inserted either ahead of or following the filter. Insertion of the pump ahead of the filter has the advantage that as the melt pressure increases with dirt accumulation in the filter, this does not influence the extruder screw speed, since the metering rate of the pump is independent of pressure. Insertion between filter and film-casting die offers the advantage of the gear pump being protected against hard melt inclusions which can generally pass the extruder but lead to stoppage and damage in the gear pump. Figure 13 shows the insertion of a pump (b) between long-life filter (c) and film-casting die (a).

9.6.4 Slitting

In the most frequently used production process the quenched films are cut into a group of individual tapes prior to stretching. Exceptions are the production of simultaneously extruded individual tapes, as in the extrusion of monofilaments, and the orientation of the film without slitting into tapes. When the film is cut, a large number of blades arranged next to each other slit the pre-tensioned film. The blades are separated by small accurately placed spacers to form a slitter bar. In order to increase their service life, the blades can be moved up and down in the cutting direction. If there is excessive wear on the blades,

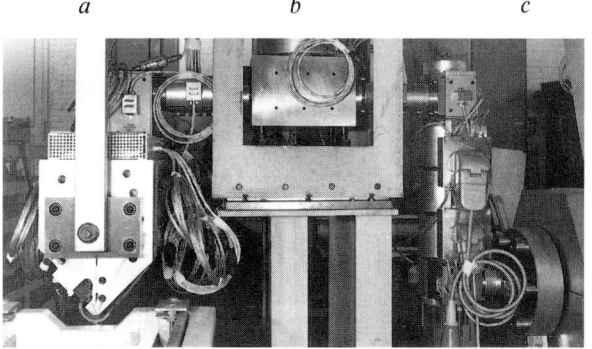

Figure 13. Arrangement of metering gear pump *b* between long-life filter *c* and film casting die *a*. (Photo: Barmag, Remscheid, West Germany)

which may be caused by abrasive melt additives, multiple slitter bars are available which can be switched over during cutting, without interrupting production in order to disengage the worn row of blades and engage a new one. Figure 14 shows the arrangement of the cutting device as the film enters the first godet unit.

Figure 14. Application of cutting device prior to film entry into the first godet unit. (Photo: Barmag, Remscheid, West Germany)

9.6.5 Feeding

Just as in the production of oriented films, the film tapes must be stretched to many times their original length in order to achieve the required properties, and thus their shape is changed permanently. In the nature of the process, the stretching force must act on the bundle of tapes while they are running. It is generally transmitted by driven rolls and arises from the speed differences between the first and second godet units. The first unit serves to feed and hold the tapes against the stretching tension applied by a second godet unit whose speed exceeds that of the first by the stretching factor. For convenience of operation, driven rolls of cantilever design are chosen. They are arranged in series to form a gripping or drawing unit and have slightly increasing or decreasing circumferential speeds to compensate for thermal expansion or shrinkage, respectively.

The force transmitted is given by the following equation:

$$F_1 = F_2 \, e^{\mu \alpha}$$

with F_1 stretching force, μ friction coefficient,
 F_2 holding force, α godet contact angle.

In order to increase the friction coefficient, the roll surface is either chrome-plated and polished, or provided with a hard metal plating, for simultaneous reduction of damage risk.

To supply the holding force, three singly supported feed rolls are generally sufficient. In special cases, fully supported rolls may be used, and a pneumatically operated, rubberized contact roll pressed onto them to increase the holding and/or stretching force transferred in the nip.

9.6.6 Heating

The stretching force and the tendency of the oriented tapes to shrink are influenced by the stretching temperature. To control temperature the tapes are heated externally to augment the self-heating that occurs during stretching. The temperature is brought as close as possible to the crystallite melting point, without causing the tapes to stick to the godet rolls. Figure 9 shows the different methods of heating, diagrammatically. Hot-air ovens, hot-water baths, hot plates, and internally heated godet rolls have all proved equally successful. Hot-air ovens permit relatively long contact periods even at high line speeds and they are easy to operate. Hot-water baths are preferred when high stretch ratios are required. Hot plates and heated rolls are easy to operate and permit heat treatment of the tapes in stages, to obtain specific properties. The temperature of the heat-transfer medium (air, water or oil) is thermostatically controlled.

9.6.7 Stretching

The stretching unit is similar in its arrangement to the feed unit (Figure 15). The speed range and the drive power for the stretching rolls are set to give the required stretch ratio and stretching force. Available stretching forces may be as high as 10,000 N with roll surface speeds up to 300 m/min.

Figure 15. Tapes passing through drawing system.
(Photo: Barmag, Remscheid, West Germany)

If the film is to remain wide, using the principle shown in Figure 3 (bottom) the rolls are arranged above each other, or next to each other, with a small gap between them, and can be individually heated in steps. Because the gap is small the film is effectively in constant contact with the roll surfaces and is prevented by friction from necking-in.

9.6.8 Fibrillating

In some fields of application the fiber-like character of the natural products previously used, jute, sisal or hemp, is required. The production process for film tapes and the anisotropy of tenacity achieved with uniaxial stretching of PP, offer several methods for achieving this fiber-like character. The processes shown in Figure 16 were developed in parallel with each other. In the Barfilex process (top) profiled PP tapes (a) are highly oriented (b) during take-off from the profiled die, being drawn with such a high stretching ratio (1:10 to 1:11) that they split along thin lines in the profile. A similar effect is achieved in the roll-embossing process (Figure 16, center). With this method, smooth tapes (a) are embossed by means of profiled rolls with a castellated profile (b) and split into individual filaments during stretching. The character of natural fibers is approached particularly well if the stretched tapes are fibrillated by means of needle, pin, or knife rollers which slice into the oriented, tensioned tapes as shown in Figure 16, bottom. The result is a net-like split tape (Figure 17). The length of the connecting fibrils is determined by the relative speed between needle or knife tips and the tape.

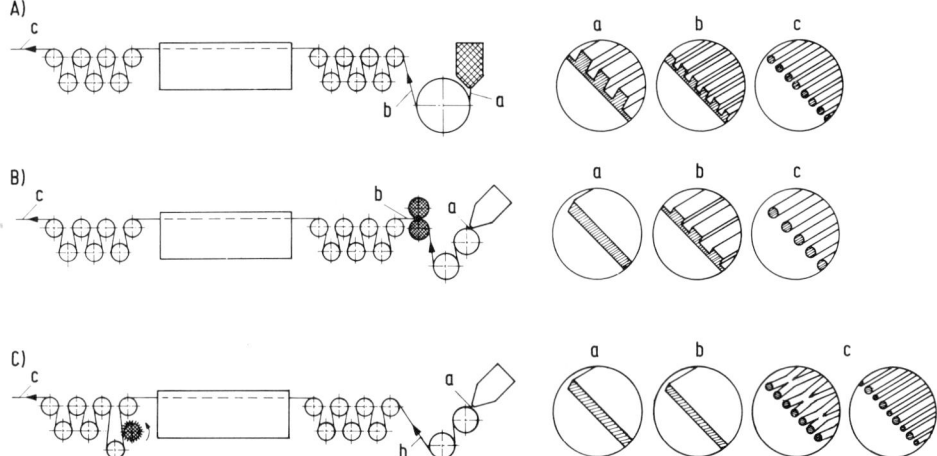

Figure 16. Principles of split fiber production
A) barfilex, B) roll embossing, C) fibrillating with blades and needles

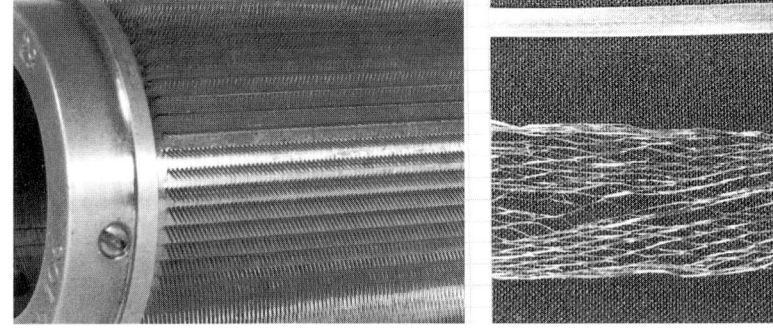

Figure 17. Needle roll (left) tapes before needling (right, top) and after needling (right, bottom). (Photo: Barmag, Remscheid, West Germany)

9.6.9 Wind-up

For economic reasons and to allow for ease of further processing, the film tapes are wound onto flangeless wind-up tubes. During wind-up, the tapes must be wound across the entire package width so that crossing layers create a firm package, with as few cavities as possible, that can be unwound easily. Because the tapes are supplied to the winding head at constant speed, the spindle speed must decrease with increasing package diameter. In practice, the problem is solved by the package being driven either at the surface by friction, or by means of a variable-speed motor.

As shown in Figure 18, in friction winding the friction roller and the yarn-traverse mechanism are driven separately. The winding angle remains constant. This leads to the formation of what is called "ribboning", giving a poor-looking package. The appearance can be improved by means of cyclic alteration of the speed of traverse by the control gear, the "antiribboning device".

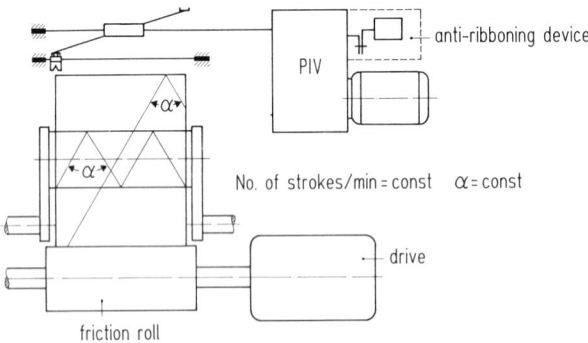

Figure 18. Principle of friction winding

In practice, the cross-winding method explained in Figure 19 has been more widely adopted. With this method each individual wind-up position is driven by its own control motor, the speed of which determines both the spindle speed and the yarn-traverse motion.

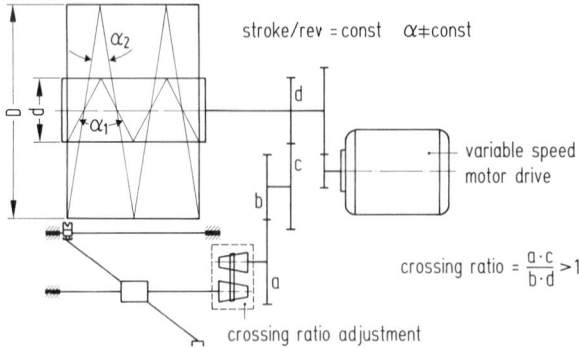

Figure 19. Principle of precision cross-winding

The choice of the number of teeth of reduction gears, a, b, c, and d determines the crossover ratio. The latter is chosen as a function of tape dimensions so that a solid, inherently stable package is built, which will allow troublefree take-off. In special cases, where required, the cross-over ratio can be altered during production by means of an additional variable-speed gear. The tape tension is kept constant during wind-up. It is monitored by means of a compensator arm and serves as measurement of the wind-up drive speed.

In addition to components already described for carrying out the main process steps, operation of the tape line requires some auxiliary equipment, including:

- edge-trim recycling unit with aspirator,
- chopping mill with air feed,
- aspirator for start-up waste and broken tapes,
- air (suction) guns for tape thread-up.

Figure 20 shows a conventional tape line in production; this uses all the units mentioned, and the auxiliary equipment required.

Figure 20. Film-tape line in production. (Photo: Barmag, Remscheid, West Germany)

9.7 Design variations of tape lines

It would be neither technically possible nor economically justifiable to produce all the types of tapes mentioned in Table 1 and Figure 4 on one type of line. For most tape types and/or tape properties the line components shown in Figure 9 are combined to create the design variations shown in Figure 21. The required output is determined by the wind-up speed, usable line width, and the available drive and heat-treatment power. Certain properties such as low shrinkage rate, high elastic modulus, high tenacity, fibrillation or softness can be achieved economically only by correct arrangement of the various line components, or by using auxiliary equipment.

Tenacity and elongation values (Figure 21) are determined on normal tenacity-measuring instruments. As there is no suitable method available for measuring residual shrinkage, film tape producers have agreed to use the EATP (European Association of Textile Polyolefines Producers) method shown in Figure 22. The heat treatment is done in silicone oil.

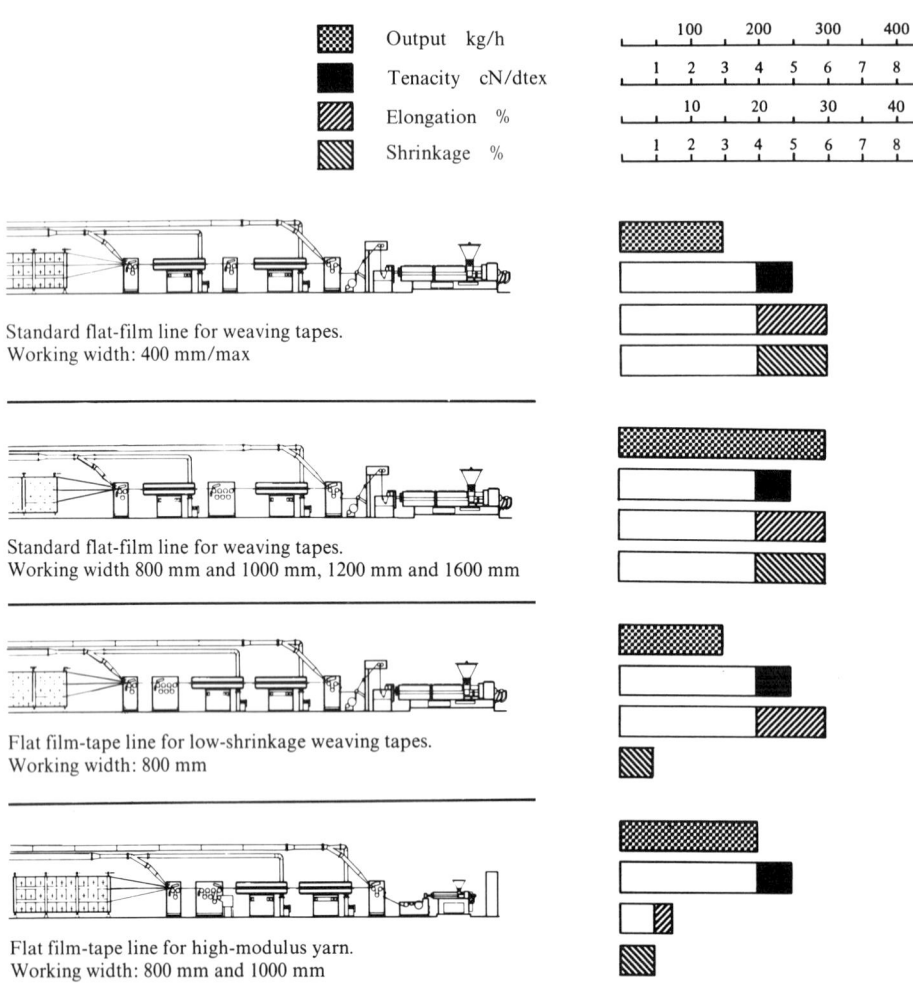

Figure 21. Design variations of film-tape lines with respective outputs and tape properties achievable. (Photo: Barmag, Remscheid, West Germany)

Figure 22. Shrinkage measuring method for film tapes (after EATP) ▶
A) marking, B) measuring, C) shrinking, D) measuring

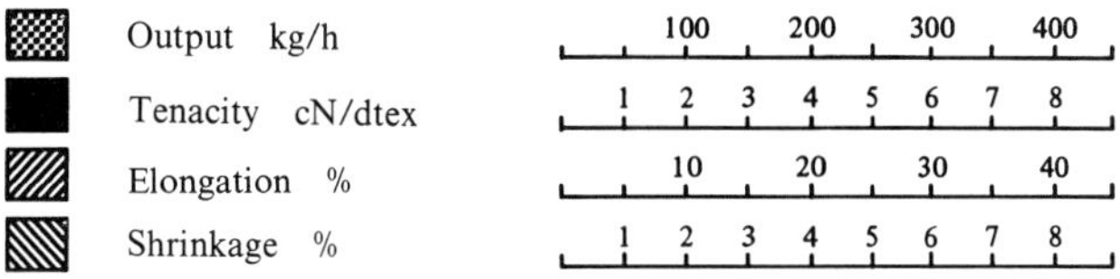

9.7 Design variations of tape lines

■ Output kg/h
■ Tenacity cN/dtex
▨ Elongation %
▧ Shrinkage %

```
100   200   300   400
 1  2  3  4  5  6  7  8
   10    20    30    40
 1  2  3  4  5  6  7  8
```

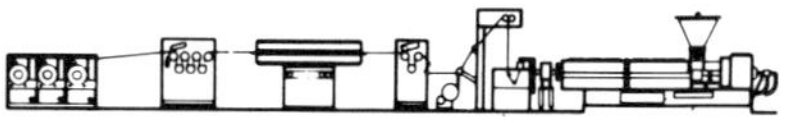

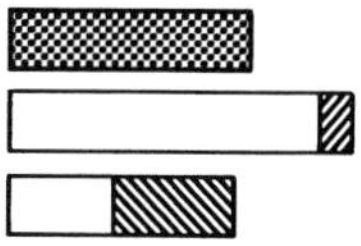

Standard flat-film line for hawser strands.
Working width: 800 mm, 1000 mm and 1600 mm

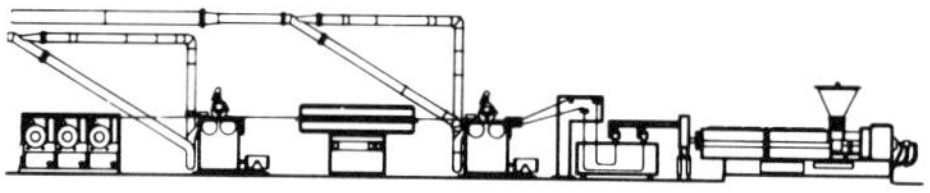

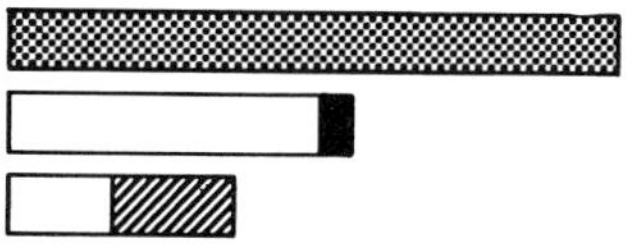

Flat-film hawser strand line (trio).
Working width: 1150 mm

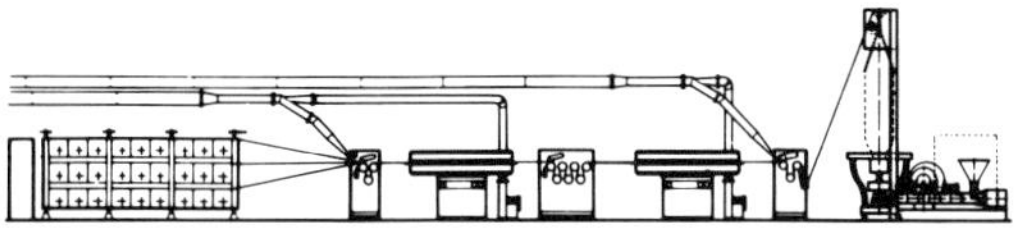

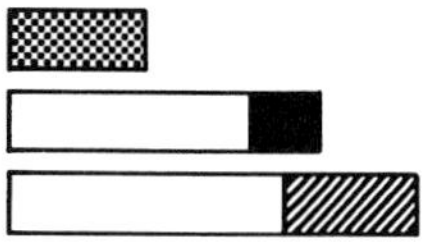

Standard blown-film line.
Working width: 800 mm max

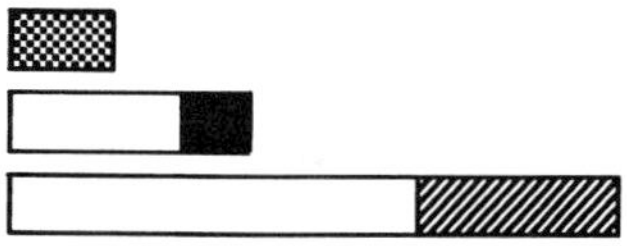

Longitudinal film-stretching line.
Working width: 1500 mm max

Figure 21 (continued)

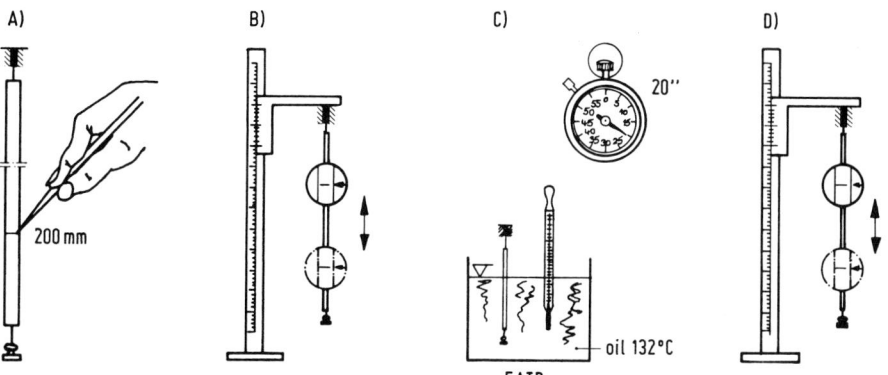

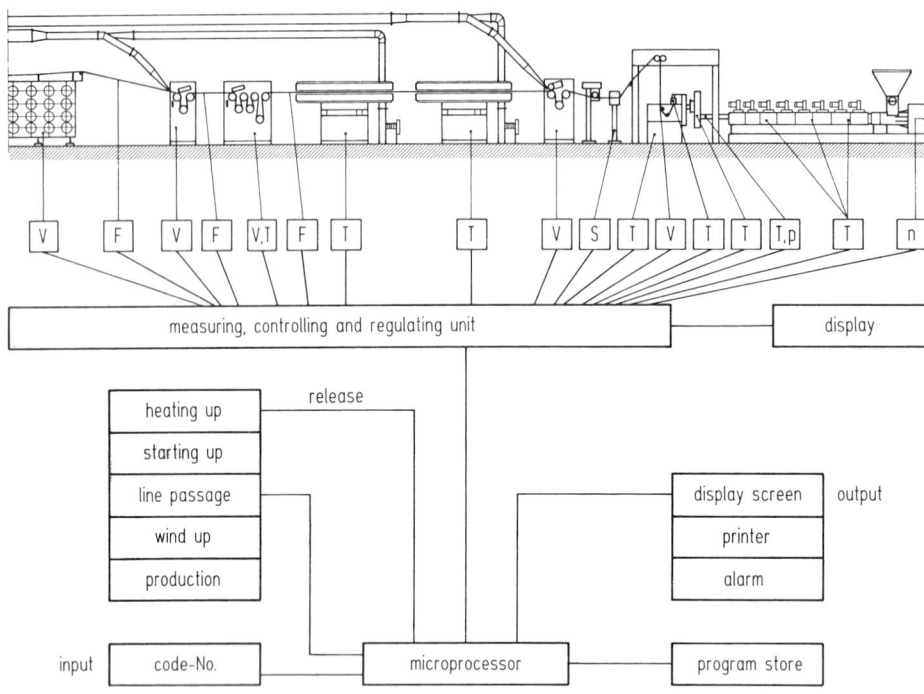

Figure 23. Scheme for automatic process control of film-tape lines

Figure 24. Data-recording installation for film-tape lines with computer, monitor and printer
(Photo: Barmag, Remscheid, West Germany)

9.8 Automation of tape lines

Maintaining the required product properties requires constant control of all relevant process parameters. It is demanding too much to expect operators to constantly monitor all temperatures, speeds, pressures and so on — but even a small fault on an individual unit noticed too late may result in faulty product and substantial losses. Therefore an automatic process monitoring system, offered as standard equipment for film-tape lines, should be used. Figure 23 is the schematic diagram, and Figure 24 the photograph of a practical design. The control system continuously monitors the process parameters, stores them for a pre-set time and generates an alarm signal if a permissible tolerance is exceeded. In addition to process monitoring, the control system also stores all machine settings and carries out the entire setting operation, as well as controlling the start-up and shut-down procedures. An additional task is to determine operating data for statistical analysis and for controlling and/or improving the execution of orders.

References for Chapter 9

For the convenience of the reader the English titles of all publications in languages other than English are shown in parentheses.

[1] *Steinen, P.:* Entwicklungsstand bei Bändchengeweben (The State of Development of Woven Tape Fabrics). Verpackungsrundschau (1967) 4, pp. 374/378.
[2] *Hensen, F., Klawonn, G.-W.:* Über die Herstellung und Verarbeitung von monoaxial gereckten Folienbändchen (On the Manufacture and Processing of Uniaxially Drawn Tapes). Textil-Praxis (1967) 5, pp. 329/331 and 410/413.
[3] *Hensen, F.:* Herstellung von Monofilen und monoaxial gereckten Folienbändern (Manufacture of Monofilaments and Uniaxially Drawn Film Tapes). VDI Bildungswerk, 16th Contribution, Seminary 1967/68.
[4] *Klust, G.:* Zwirne aus Polypropylen-Folienbändern (Polypropylene Film-tape Twines). Textil-Praxis (1968) 6, pp. 361/363.
[5] *Hensen, F.:* Winding of Monoaxially Stretched Film Tapes. Conference Plastics Institute, Manchester, 05/04/1968.
[6] *Penker, H.:* Fäden aus Folie (Filament from Film). Textilindustrie 70 (1968) 11, pp. 774/781 and (1968) 12, pp. 867/874.
[7] *Moorwessel, D., Pilz, G.:* Ein neues Polypropylen für den Grobtextilsektor (A New Polypropylene for the Coarse Fiber Industry). Kunststoffe 59 (1969), pp. 205 and 539.
[8] *Fischer, P.:* Synthesefasern aus Folie (Synthetic Fibers from Films). Textilindustrie 71 (1969), Nov.
[9] *Hensen, F.:* Herstellung von Fäden aus Folie (Manufacture of Filaments from Film). Industrieanzeiger (1969) 10.
[10] *Hensen, F., Hessenbruch, R.:* Extrusion von Mehrschichtblasfolien (Extrusion of Multilayer Blown Film). Plastverarbeiter (1971) 5.
[11] *Nott, R. E. et al.:* Report, Shirley Institute and Plastics Institute, Textiles from Film. Textile Month (1971), Sept./Oct.
[12] *Lennox-Kerr, P.:* Forschungsarbeiten auf dem Gebiet der Folienbändchen in Dänemark (Research in Denmark on Film-tape Technology). Meliand Textilberichte (1971) 3, pp. 267/269.
[13] *Harms, J., Krässig, H., Saßhofer, F.:* Fäden und Fasern aus Folien (Filaments and Fibers from Films). Lenzinger Berichte (1971) Dec.
[14] *Balk, H.:* Herstellung von Folienfasern und Garnen, Anwendung in Textilerzeugnissen, Entwicklungstendenzen (Manufacture of Film Fibers and Film Yarns, Application in Textile Production, Development Trends). Chemiefasern/Textilindustrie 74 (1972) 22, p. 236.
[15] *Hensen, F.:* Neue Anlagen zer Herstellung von Fasern aus Polyolefinen (New Lines for the Manufacture of Fibers from Polyolefins). Chemiefasern/Textilindustrie (1973) 7.

[16] *Feuerböther, D.:* Herstellung von Bindefäden aus Polyolefinen (Production of Binder Twine from Polyolefins). Textiltechnik 23 (1973) 2, pp. 93/96.
[17] *Kautz, G.:* Herstellen monoaxial gereckter Webbändchen und Monofolie aus Polyolefinen (Manufacture of Uniaxially Drawn Tapes and Monofils from Polyolefins). Kunststoffe 63 (1973) 10, pp. 682/686.
[18] *Berger, W., Schmack, G.:* Untersuchungen zum Fibrillieren von Folien aus Polymermischungen (Studies on the Fibrillation of Mixed-polymer Films). Textiltechnik 24 (1974) 1, pp. 36/41.
[19] *Hensen, F., Gathmann, E.:* Fortschritte in der Extrusionstechnik durch Optimieren getrennter Verfahrensschritte (Progress in Extrusion Technology by Optimizing Individual Process Steps Separately). Kunststoffe 64 (1974) 7, pp. 343/349.
[20] *Schreiner, L. L.:* Slit Film/Monofilament Extrusion. Plastics Technology 23 (1977) 2, pp. 53/59.
[21] *Hensen, F., Braun, S.:* Herstellen von Folienbändchen (Manufacture of Film Tapes). Kunststoffe 68 (1978) 4, pp. 221/229.
[22] *Hensen, F.:* Kühlen von Extrudaten (Cooling of Extrudates). VDI-Verlag, Düsseldorf, 1978, pp. 81/96.
[23] Neue Technologien und Anlagen zur Herstellung von Spinnfasern und Filamentgarnen aus vorzugsweise PP (New Technology and New Lines for Manufacturing Spun Fibers and Filament Yarns, Chiefly from PP). Chemiefasern/Textilindustrie 80 (1978), pp. 36/40.
[24] *Hensen, F.:* Verbesserte Extrusionstechnik durch Großflächenschmelzefiltration (Improved Extrusion by the Use of Large-area Melt Filtration). Kunststoffe 70 (1980) 11, pp. 753/758.
[25] *Krässig, H.:* Maschinensysteme zur Herstellung und Verarbeitung von Polymerfolien (Machinery for Producing and Converting Polymer Films). Lenzinger Berichte (1980) May, Nr. 49.
[26] *Hensen, F.:* Extrudieren von gereckten Folien und Folienbändchen (Extrusion of Drawn Films and Film Tapes). Kunststoffe 71 (1981), pp. 643/652.
[27] *Slack, D.:* Orientation of PP Monofilaments and Tapes. Fiber Producer (1983) June, pp. 58/60.
[28] DRP 667234 (01. 07. 1936) IG Farbenindustrie: Herstellung gereckter Folienbändchen (Manufacture of Stretched Film Tapes).
[29] DPES 1127535 (09. 04. 1953) Hermann Becker: Falten von Folienbändchen beim Verstrecken (Folding of Film Tapes During Stretching).
[30] DAS 1111339 (19. 11. 1954) Dynamit Nobel: Verstrecken von Folienbändchen (Stretching of Film Tapes).
[31] DEPS 1108420 (01. 12. 1955) Dynamit Nobel: Verstrecken von Folienbändchen (Stretching of Film Tapes).
[32] DEPS 1175385 (11. 06. 1958) Du Pont: Fibrillierte Folien (Fibrillated Films).
[33] GBPS 1035657 (1962/63) Courtaulds: Lengthwise Ribbed Tapes.
[34] USAS 2853741 (27. 05. 1954) Dow Chemical: Fibrillation with Brushes.
[35] CHPS 415940 (06. 09. 1963) Rasmussen: Fasern durch Nitscheln (Fibers from Friction Cones).
[36] GBPS 1134243 (23. 03. 1966) Courtaulds: Fibrillation by Twisting.
[37] DEPS 1660231 (21. 04. 1967) Barmag: Barfilex.
[38] CHPS 482033 (05. 09. 1967) Barmag: Barfilex.
[39] DEOS 1917822 (08. 04. 1969) Shell: Roll Embossing.

10 Monofilament extrusion

S. Braun

10.1 Introduction

The first monofilament lines were developed immediately after the commercial introduction of the types of thermoplastic polymers which achieve high tensile strength by stretching. Compared with present-day equipment, these lines were of simple design and of low output.
The use of monofilaments has increased steadily as a substitute for natural products and for a wide range of industrial applications. There has been a consequent acceleration in the development of process technology and machine design. At the present time, monofilament lines with up to 200 individual ends and production speeds of up to 160 m/min are in operation.
Because of the wide range of applications for monofilaments, these lines have a significant market impact. This will become even more important as the use of blends and co-polymers, and specially engineered polymers, increases and opens up new applications.

10.2 Theory

Monofilaments are wire-like polymer strands whose tensile properties are achieved by stretching. They usually have a circular cross-section, and – as shown in Figure 1 – parallel orientation of their molecular chains has been produced by drawing.

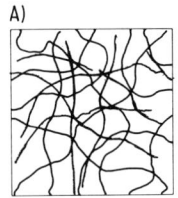
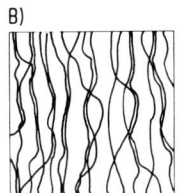

Figure 1. Diagram of molecular chains of amorphous undrawn (A) and drawn (B) polymer

This orientation of the molecular chains results in considerable enhancement of certain properties of monofilaments. Monofilaments are produced in a melt spinning process.
The polymer is melted in the extruder, compressed, homogenized and conveyed under pressure to the monofilament spinning unit, either direct or with the assistance of a metering gear pump.
Polyester or nylon are virtually Newtonian fluids at process temperature; thus, using a single-flight screw of pitch $t = D$ and a flight width of $0.1 \cdot D$, the extruder output can be calculated approximately from the equation:

$$\dot{m} = \frac{0.9 \, \pi}{12} \sin^2 \varphi \, \varrho \left(\frac{6 \, \pi}{\tan \varphi} D^2 \, h \, n - \frac{\Delta p \, D \, h^3}{\eta \cdot L} \right) \tag{1}$$

$\dot{m}$ mass throughput,
ϱ melt density,
D screw diameter,
h pitch in the metering zone,
n screw speed,
Δp melt pressure drop in metering zone,
η melt viscosity,
L length of the metering zone,
φ helix angle.

In the spinning die, the melt is forced through filter packs and die-plate holes. Depending on the shape of the holes, the monofilaments emerge with circular, oval, delta-shaped or cruciform cross-sections. To freeze these cross-sections and achieve the solid state required for further processing, the monofilaments are pulled through a quenching fluid and cooled to a temperature below the crystallite melting point.

The amount of heat, Q, to be removed during this cooling process is determined from the following equation:

$$Q = m(h_2 - h_1). \tag{2}$$

m is the mass throughput of the melt through the quenching fluid, h_2 the specific enthalpy of the melt, and h_1 the specific enthalpy of the cooled monofilaments. The enthalpy values can be found in graphs which show the specific enthalpy as a function of the material temperature.

After emerging from the quenching bath, the monofilaments are reheated and drawn to several times their original length. Proper choice of temperature is crucial to allow the *Van der Vaals* forces to achieve their full effect during drawing.

Application of heat treatment before and during drawing also ensures good runnability and, for certain polymers, is absolutely essential to achieve successful drawing. The drawing force for monofilament bundles is determined from the following equation:

$$F = \frac{d^2 \cdot \pi}{4} \cdot i \cdot \sigma_s. \tag{3}$$

In this equation, d is the diameter of the undrawn monofilaments, i the number of monofilaments, and σ_s the specific draw tension. The specific draw tension must be determined during production or in trials. It depends largely on draw ratio, draw temperature, and draw speed used.

To reduce the residual retractive forces still present in the monofilaments after drawing, the monofilaments are heat-set by reheating and allowing them to shrink longitudinally. In special cases it may be necessary to pass the monofilaments through a quenching bath after heat-setting to ensure that they are cold before wind-up.

Monofilament production requires a large number of process steps as the monofilament properties are determined by many factors, including: the type of polymer, how it melts in the extruder, the actual melt temperature, the design of the spinning die and die-plate holes, the draw-down of the monofilaments after they emerge from the holes, the length of the heated draw zone, the temperature of the quenching bath and the manner in which the monofilaments pass through it; and the draw-temperature, -ratio and -speed, the heat-setting temperature, setting ratio and time, and, finally, the cooling of the monofilaments.

10.3 End product

The following table shows the most important end uses for monofilaments, specific properties required for various uses and the types of polymers used.

End uses	Required properties	Type of polymer used
ropes	tensile strength, wear resistance, UV resistance, elongation	polypropylene (PP) high density polyethylene (HDPE)
nettings	tensile strength, knot strength, wear resistance, UV resistance	polyamide 6 (PA 6) polyamide 6.6 (PA 6.6) copolyamides, PP, HDPE
awning cloth	UV resistance, wear resistance	PP, HDPE
fishing lines	tensile strength, knot strength, flexibility, weight	PA 6 copolyamides
sewing thread	tensile strength, flexibility, lubricity, heat resistance, narrow diameter tolerance	PA 6.6 polyethyleneterephthalate (PET)
filters, conveyor belts, drying belts	narrow diameter tolerance, dimensional stability, wear resistance, hydrolysis resistance, low heat-shrinkage, high elastic modulus, chemical stability, high-temperature resistance, soiling resistance	PET polybutyleneterephthalate (PBT) ethylene/tetrafluorethylene (ETFE) copol.
zip fasteners	narrow diameter tolerance, wear resistance, dimensional accuracy, formability, dyeability	PET PA 6, PA 6.6
industrial fabrics	tensile strength, wear resistance, low heat-shrinkage	PA 6, PET, HDPE
bristles	dimensional stability, low heat-shrinkage, ease of splitting	PA 6, PA 6.6, PA 6.10, PET, copolyamides, PP, HDPE, polystyrol (PS), polyvinyl-chloride (PV)
carpets, floor mats, non-wovens	wear resistance, low heat-shrinkage, dimensional stability, dyeability, soiling resistance	PA 6, PP
industrial filaments as single- and two-component monofilaments (core/sheathing)	tensile strength, wear resistance, narrow diameter tolerance, hydrolysis resistance, low heat-shrinkage	PA 6, PET
optical fibers as two-component monofilaments (core/sheathing)	optical transmission elastic deformation	polymethylmethacrylate (PMMA) PS
tennis racket strings	tensile strength, wear resistance, elastic deformation, E modulus	PA 6 copolyamides

10.4 Process

The production process essentially consists of the following steps: extrusion, forming, drawing, heat-setting, and winding.
In the upper part of Figure 2 an arrangement to provide these process steps is shown. In this concept, drawing takes place in two steps — the first in the water bath and the second in the hot-air oven.
The lower part of Figure 2 relates process steps to the material flow and relevant parameters.

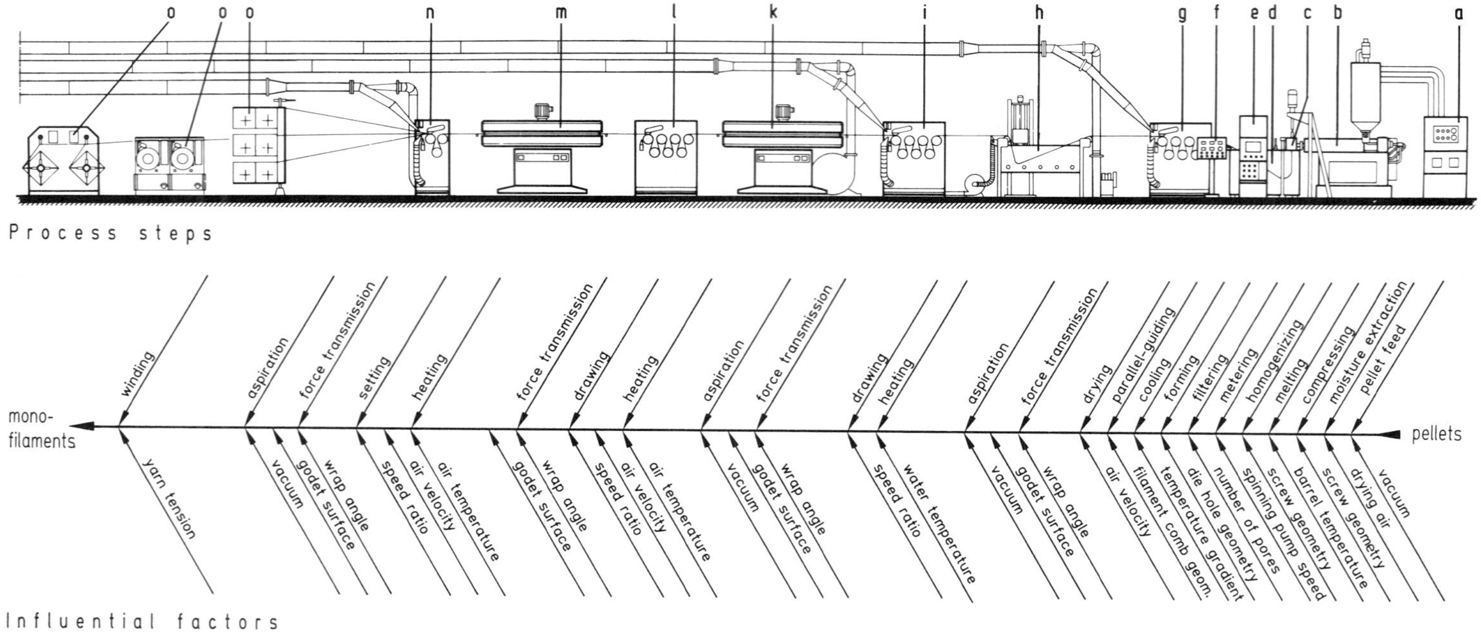

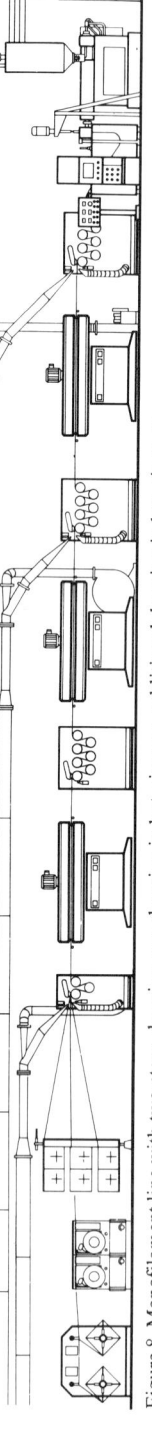

Figure 2. Monofilament line with two-stage drawing, pre-drawing in water bath, secondary drawing in hot-air oven. Flow chart with process steps and relevant influential factors
a drying cabinet, *b* extruder, *c* monofilament die, *d* quenching bath, *e* process control system, *f* control panel, *g* draw unit I, *h* water bath, *i* draw unit II, *k* hot-air oven II, *l* draw unit III, *m* hot-air oven III, *n* draw unit IV, *o* wind-up unit

Figure 8. Monofilament line with two-stage drawing, pre-drawing in hot-air oven, additional drawing in hot-air oven

10.5 Lines

10.5.1 Line concepts

10.5.1.1 Monofilament line with two-stage drawing: pre-drawing in a water bath – secondary drawing in a hot-air oven

The arrangement shown in Figure 2 is used chiefly for the production of monofilaments from polyethyleneterephthalate. Graphs 3 to 7 show the tensile strength, tear elongation, knot strength, boil shrinkage, and heat shrinkage of 0.2 mm-diameter PET filaments as a function of pre-drawing, total draw ratio, and temperature in the hot-air oven II (secondary drawing).

Figure 3. Tensile strength of PET monofilaments 0.2 mm diameter as a function of pre-drawing in the water bath, total draw ratio, and temperature of hot-air oven II (secondary drawing)
a 200 °C, $\lambda = 5$; b 200 °C, $\lambda = 6$; c 225 °C, $\lambda = 6$; d 250 °C, $\lambda = 6$; e 250 °C, $\lambda = 7$

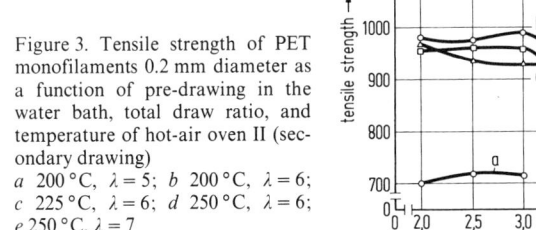
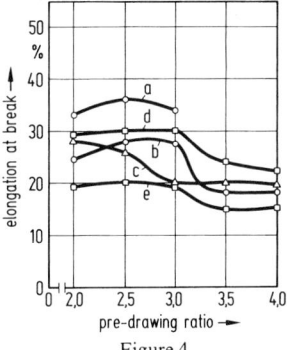

Figure 3

Figure 4

Figure 4. Elongation at break of PET monofilaments 0.2 mm diameter, as a function of pre-drawing in the water bath, total draw ratio, and temperature of hot-air oven II (secondary drawing)
a 200 °C, $\lambda = 5$; b 200 °C, $\lambda = 6$; c 225 °C, $\lambda = 6$; d 250 °C, $\lambda = 6$; e 250 °C, $\lambda = 7$

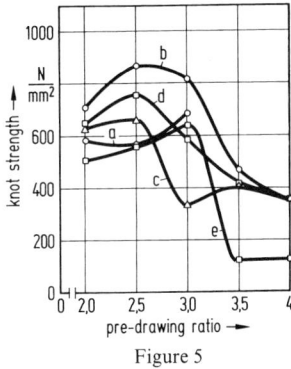
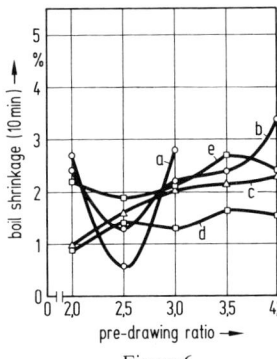
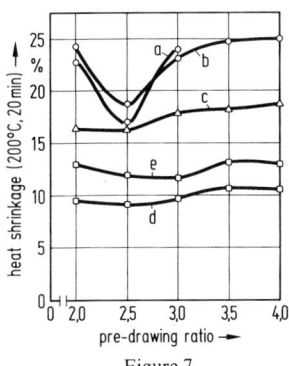

Figure 5

Figure 6

Figure 7

Figure 5. Knot strength of PET monofilaments 0.2 mm diameter, as a function of pre-drawing in the water bath, total draw ratio, and temperature of hot-air oven II (secondary drawing)
a 200 °C, $\lambda = 5$; b 200 °C, $\lambda = 6$; c 225 °C, $\lambda = 6$; d 250 °C, $\lambda = 6$; e 250 °C, $\lambda = 7$

Figure 6. Boil shrinkage of PET monofilaments 0.2 mm diameter, as a function of pre-drawing in the water bath, total draw ratio, and temperature of hot-air oven II (secondary drawing)
a 200 °C, $\lambda = 5$; b 200 °C, $\lambda = 6$; c 225 °C, $\lambda = 6$; d 250 °C, $\lambda = 6$; e 250 °C, $\lambda = 7$

Figure 7. Heat shrinkage of PET monofilaments 0.2 mm diameter, as a function of pre-drawing in the water bath, total draw ratio, and temperature of hot-air oven II (secondary drawing)
a 200 °C, $\lambda = 5$; b 200 °C, $\lambda = 6$; c 225 °C, $\lambda = 6$; d 250 °C, $\lambda = 6$; e 250 °C, $\lambda = 7$

The most important process parameters, all of which are kept constant, are:

- extruder temperature,
- die temperature,
- melt temperature,
- pump inlet pressure,
- quench bath temperature,
- gap between die plate and water,
- temperature of water bath,
- speed of draw unit III,
- temperature of hot-air oven III,
- speed of draw unit IV.

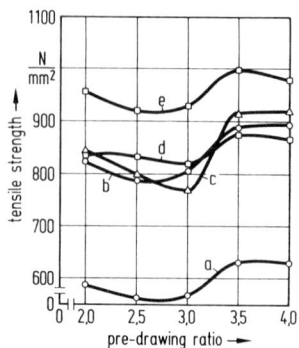

Figure 9. Tensile strength of PET monofilaments 0.2 mm diameter, as a function of pre-drawing in hot air, total draw ratio, and temperature of hot-air oven II (secondary drawing)
a 200 °C, $\lambda = 5$; b 200 °C, $\lambda = 6$; c 225 °C, $\lambda = 6$; d 250 °C, $\lambda = 6$; e 250 °C, $\lambda = 7$

10.5.1.2 Monofilament line with two-stage drawing – pre-drawing in a hot-air oven – secondary drawing in a hot-air oven

The set-up shown in Figure 8 is primarily used for the production of monofilaments from polyethyleneterephthalate and polyamide (Figure 8).
The monofilament line in this illustration differs from the one shown in Figure 2 in that the drawing process does not take place in the water bath but in a hot-air oven. Graphs 9 to 13 show the monofilament properties achievable with this line. A comparison with graphs 3 to 7 shows the different effects of water bath and hot-air oven in achieving the monofilament properties.

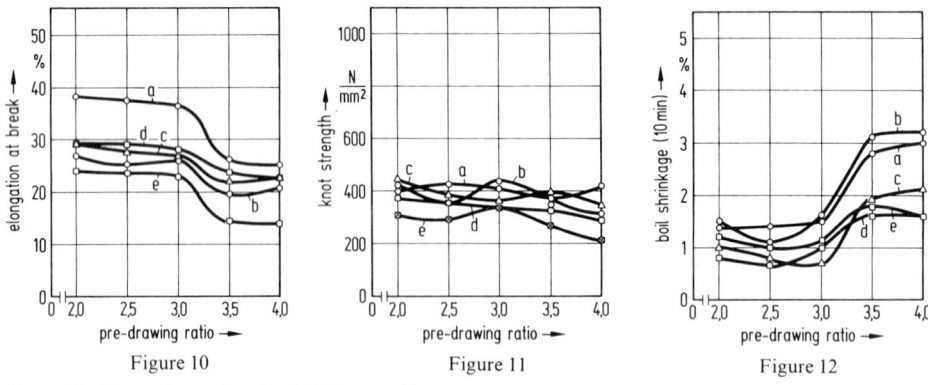

Figure 10.   Figure 11.   Figure 12

Figure 10. Elongation at break of PET monofilaments 0.2 mm diameter, as a function of pre-drawing in hot air, total draw ratio, and temperature of hot-air oven II (secondary drawing)
a 200 °C, $\lambda = 5$; b 200 °C, $\lambda = 6$; c 225 °C, $\lambda = 6$; d 250 °C, $\lambda = 6$; e 250 °C, $\lambda = 7$

Figure 11. Knot strength of PET monofilaments 0.2 mm diameter, as a function of pre-drawing in hot air, total draw ratio, and temperature of hot-air oven II (secondary drawing)
a 200 °C, $\lambda = 5$; b 200 °C, $\lambda = 6$; c 225 °C, $\lambda = 6$; d 250 °C, $\lambda = 6$; e 250 °C, $\lambda = 7$

Figure 12. Boil shrinkage of PET monofilaments 0.2 mm diameter, as a function of pre-drawing in hot air, total draw ratio, and temperature of hot-air oven II (secondary drawing)
a 200 °C, $\lambda = 5$; b 200 °C, $\lambda = 6$; c 225 °C, $\lambda = 6$; d 250 °C, $\lambda = 6$; e 250 °C, $\lambda = 7$

10.5 Lines

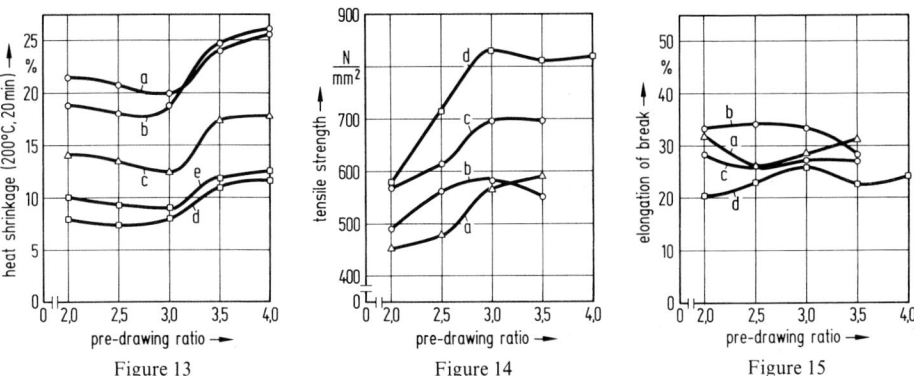

Figure 13   Figure 14   Figure 15

Figure 13. Heat shrinkage of PET monofilaments 0.2 mm diameter, as a function of pre-drawing in hot air, total draw ratio, and temperature of hot-air oven II (secondary drawing)
a 200 °C, $\lambda = 5$; b 200 °C, $\lambda = 6$; c 225 °C, $\lambda = 6$; d 250 °C, $\lambda = 6$; e 250 °C, $\lambda = 7$

Figure 14. Tensile strength of PA 6 monofilaments 0.2 mm diameter, as a function of pre-drawing, total draw ratio, and temperatures of hot-air oven I (pre-drawing) and hot-air oven II (secondary drawing)
a 110/180 °C, $\lambda = 4$; b 110/200 °C, $\lambda = 4$; c 110/200 °C, $\lambda = 4, 5$; d 140/220 °C, $\lambda = 5$

Figure 15. Elongation at break of PA 6 monofilaments 0.2 mm diameter, as a function of pre-drawing, total draw ratio, and temperatures of hot-air oven I (pre-drawing) and hot-air oven II (secondary drawing)
a 110/180 °C, $\lambda = 4$; b 110/200 °C, $\lambda = 4$; c 110/200 °C, $\lambda = 4, 5$; d 140/220 °C, $\lambda = 5$

Graphs 14 to 18 show tensile strength, elongation at break, knot strength, boil shrinkage, and heat shrinkage of PA6 monofilaments 0.2 mm diameter as a function of pre-drawing, total draw ratio, and temperatures of hot-air oven I (pre-drawing) and hot-air oven II (secondary drawing).

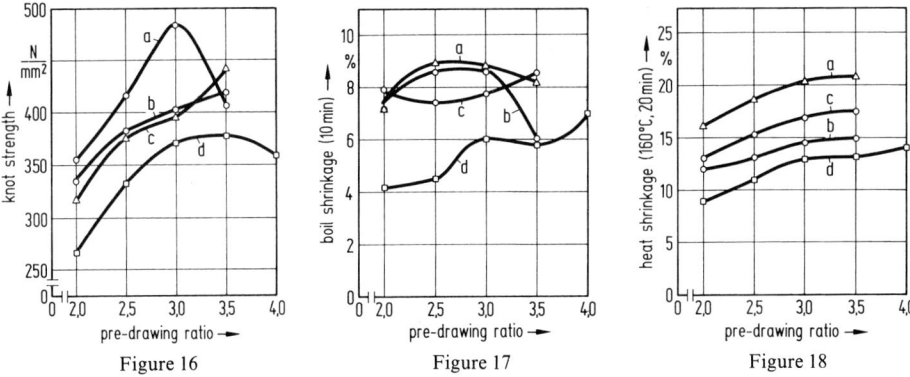

Figure 16   Figure 17   Figure 18

Figure 16. Knot strength of PA 6 monofilaments 0.2 mm diameter, as a function of pre-drawing, total draw ratio, and temperatures of hot-air oven I (pre-drawing) and hot-air oven II (secondary drawing)
a 110/180 °C, $\lambda = 4$; b 110/200 °C, $\lambda = 4$; c 110/200 °C, $\lambda = 4, 5$; d 140/220 °C, $\lambda = 5$

Figure 17. Boil shrinkage of PA 6 monofilaments 0.2 mm diameter, as a function of pre-drawing, total draw ratio, and temperatures of hot-air oven I (pre-drawing) and hot-air oven II (secondary drawing)
a 110/180 °C, $\lambda = 4$; b 110/200 °C, $\lambda = 4$; c 110/200 °C, $\lambda = 4, 5$; d 140/220 °C, $\lambda = 5$

Figure 18. Heat shrinkage of PA 6 monofilaments 0.2 mm diameter, as a function of pre-drawing, total draw ratio, and temperatures of hot-air oven I (pre-drawing) and hot-air oven II (secondary drawing)
a 110/180 °C, $\lambda = 4$; b 110/200 °C, $\lambda = 4$; c 110/200 °C, $\lambda = 4, 5$; d 140/220 °C, $\lambda = 5$

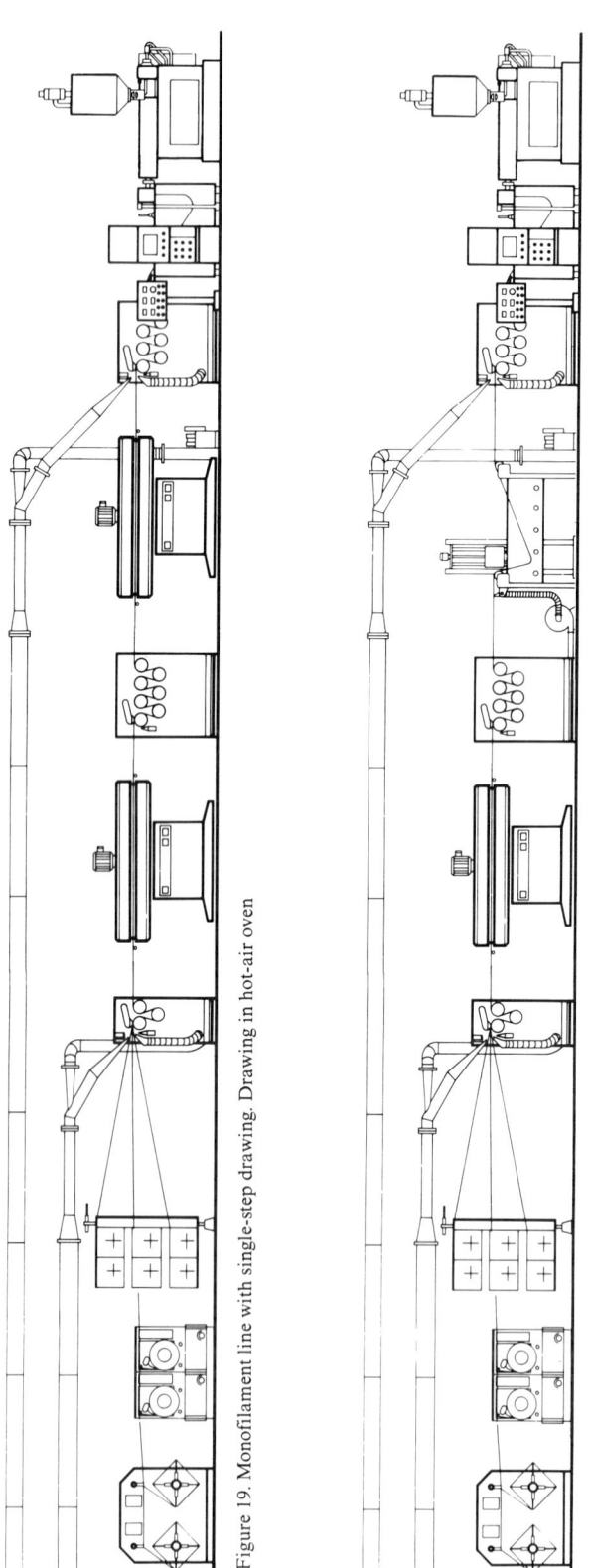

Figure 19. Monofilament line with single-step drawing. Drawing in hot-air oven

Figure 24. Monofilament line with single-step drawing. Drawing in water bath

10.5.1.3 Monofilament line with single-stage drawing – drawing in a hot-air oven

The line shown in Figure 19 is mainly used for the production of monofilaments from polypropylene.
Graphs 20 to 23 show tensile strength, elongation at break, knot strength, and boil shrinkage of PP monofilaments 0.2 mm diameter, as a function of draw ratio and temperature of hot-air oven I. Process parameters and speed of draw unit III were kept constant.

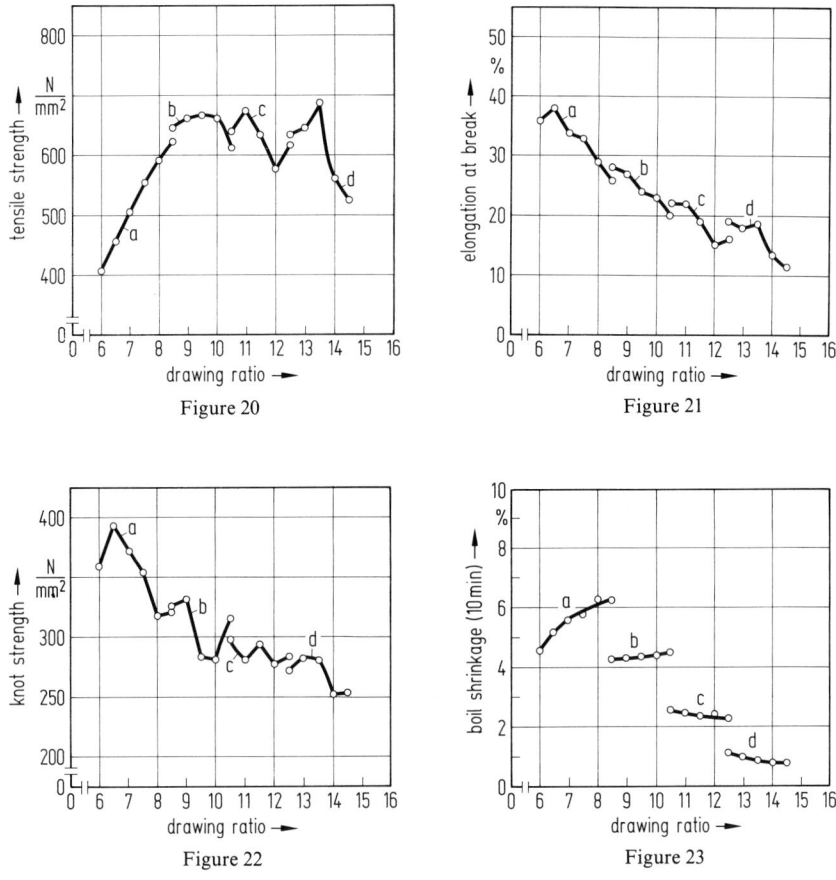

Figure 20. Tensile strength of PP monofilaments 0.2 mm diameter, as a function of draw ratio and temperature of hot-air oven I
a 120 °C, b 140 °C, c 160 °C, d 180 °C

Figure 21. Elongation at break of PP monofilaments 0.2 mm diameter, as a function of draw ratio and temperature of hot-air oven I
a 120 °C, b 140 °C, c 160 °C, d 180 °C

Figure 22. Knot strength of PP monofilaments 0.2 mm diameter, as a function of draw ratio and temperature of hot-air oven I
a 120 °C, b 140 °C, c 160 °C, d 180 °C

Figure 23. Boil shrinkage of PP monofilaments 0.2 mm diameter, as a function of draw ratio and temperature of hot-air oven I
a 120 °C, b 140 °C, c 160 °C, d 180 °C

10.5.1.4 Monofilament line with single-stage drawing – drawing in water bath

This arrangement is primarily used for the production of monofilaments from high-density polyethylene. The line layout is shown in Figure 24 (see page 314).
Graphs 25 to 28 list the values for tensile strength, elongation at break, knot strength, and boil shrinkage of HDPE monofilaments 0.2 mm diameter as a function of draw ratio and gap between die plate and water. Curves a show the results with a die-plate spacing of 50 mm, curves b those with a die-plate spacing of 70 mm.

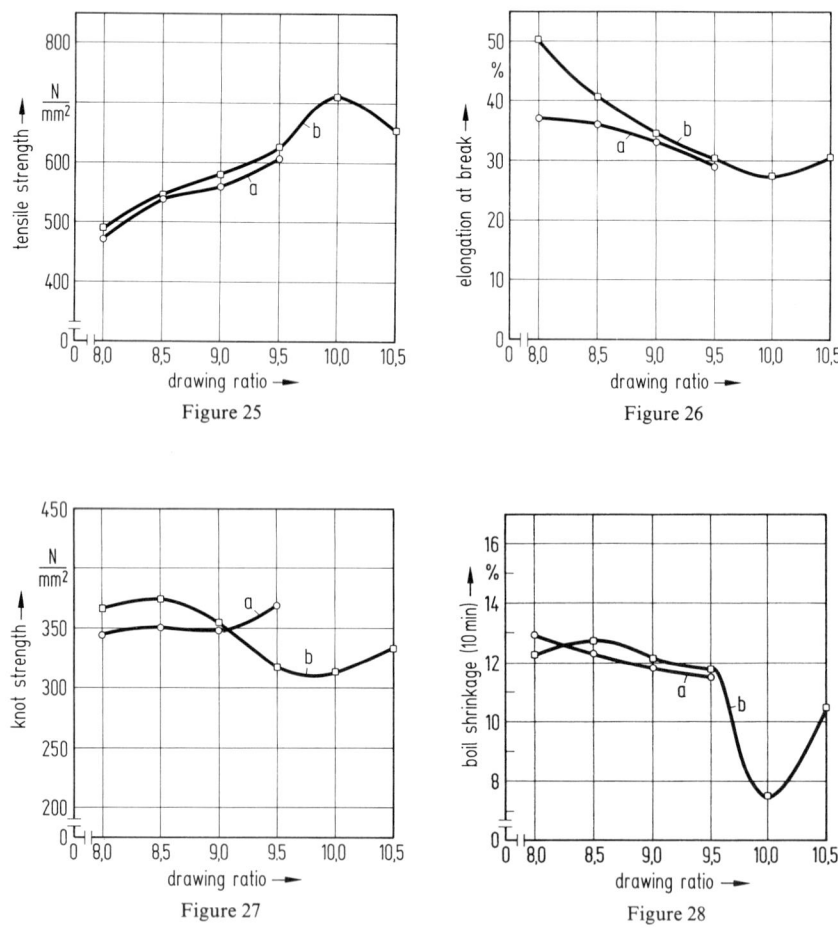

Figure 25. Tensile strength of HDPE monofilaments 0.2 mm diameter, as a function of draw ratio and gap between die plate and water ($a \cong 50$ mm, $b \cong 70$ mm)

Figure 26. Elongation at break of HDPE monofilaments 0.2 mm diameter, as a function of draw ratio and gap between die plate and water ($a \cong 50$ mm, $b \cong 70$ mm)

Figure 27. Knot strength of HDPE monofilaments 0.2 mm diameter, as a function of draw ratio and gap between die plate and water ($a \cong 50$ mm, $b \cong 70$ mm)

Figure 28. Boil shrinkage of HDPE monofilaments 0.2 mm diameter, as a function of draw ratio and gap between die plate and water ($a \cong 50$ mm, $b \cong 70$ mm)

10.5.2 Line components

10.5.2.1 Polymer feed

The method of feeding polymer to the extruder depends on the output capacity of the line, the number of lines, and the properties of the polymer to be processed.

Lines with low output, whch are used primarily for the production of monofilaments with a diameter range from 0.1 to 0.2 mm, can be fed manually. In lines with higher output, the polymer is fed in pneumatically, either by means of a pellet conveyor placed on top of the polymer container beside the extruder, or, if several lines processing the same polymer are arranged side by side, by means of a central polymer-feed system. Hygroscopic materials must often be dried or re-dried before or during feeding.

Vacuum tumble dryers are used for those materials which are dried before feeding. With this type of dryer, the material, in the form of pellets, is heated in a rotating container under vacuum to extract the moisture. This method has the advantage that no oxygen is present during drying; thus surface oxidation is prevented.

If the material is dried or re-dried during feeding, heated dry air is blown through the container above the extruder. The dry air circulates through the container and a drying cabinet. In the drying cabinet the air passes through several drying cartridges, is heated, and then passed through the container again as dry air. This drying cycle operates continuously, as the cabinet is equipped with a large number of cartridges. Several of them are always connected to the air circuit while the rest of them are being dried again by heating.

If hygroscopic materials such as polyethyleneterephthalate or polyamide are extruded with too high a moisture content, chemical degradation and simultaneous loss of viscosity occurs, and results in considerable deterioration of the monofilament properties. When processing polyamide, difficulties occur during spinning as a result of bubbles forming in the melt. The maximum moisture content should therefore not exceed 0.005% for polyethyleneterephthalate and 0.05% for polyamide.

10.5.2.2 Extruder

The functions that the extruder in a monofilament line must perform are basically the same as those of an extruder in a spinning line. For this reason, monofilament lines are usually equipped with so-called spinning extruders. The extruder barrels are 24 to 30 D long. They are electrically heated and, as a rule, have no cooling zones.

In order to ensure the greatest possible dimensional stability and precision-finished surfaces in production, the extruder barrels and screws are made of highly wear-resistant materials. In most cases the barrels are bimetallic, precision-ground and honed. The screws are nitrided and have armor-plated flights. They are precision-ground and in some cases chrome-plated.

The extruder takes in the feed material without oxidation, conveys it and melts it evenly, compresses it free of gas inclusions, homogenizes it, and finally, feeds the melt under pressure and without pulsations to the spinning pump or the spin pack.

How well these functions are fulfilled depends largely on the screw geometry. It is determined by the material to be processed, its viscosity, additives that may be present, required throughput and the properties of the monofilaments to be produced.

Use of different screw designs when processing different feed materials is absolutely essential with high output and stringent monofilament quality requirements.

In addition to functional design of the individual zones, the screws are almost always equipped with shearing and mixing components in order to ensure perfect homogenization of the melted material and uniform melt temperature.

In extruders used in monofilament lines, the screws are equipped with a gas seal on the gear-box side so that hygroscopic thermoplastics can be protected against moisture absorption by introducing inert gas, if necessary. As a rule, the extruders used in monofilament lines have screw diameters of 45, 60, or 90 mm.

10.5.2.3 Monofilament die

In the monofilament die, the material melted by the extruder is filtered, refiltered, if required, and formed into monofilaments. In the design and manufacture of monofilament dies very exacting demands have to be met to achieve monofilaments which are consistent in shape and dimensions, with the narrowest possible tolerances in relation to each other as well as in a longitudinal direction, and the least possible deviations in roundness.

The melt must be fed to the spin pack centrally from the top. Melt distribution with filtration and subsequent transport to the die-plate holes must be designed in such a way that there are no stagnation points in the flow and the melt emerging from all die-plate holes has the same residence time in the spin pack. Residence time in the die and the spin pack should be as short as possible. The heating method, and design of die and spin pack, must ensure that all melt channels reach the melt temperature.

The monofilament die consists of a die block that has a spinning pump on top of it or an adaptor in its place, and with the spin pack on its under side.

Easy and quick exchangeability of the spin pack is essential for troublefree operation. A large contact surface between die block and spin pack ensures optimum temperature equalization between the two components.

The die block can be heated by two different systems. The most frequently used and least expensive is an electrical system using either heater bands attached to the exterior surfaces, or heating cartridges inserted into the die block. If heating cartridges are used, it is advisable to insulate the outer surfaces of the die block.

The area affected by the heater bands or heating cartridges is provided with temperature sensors which are connected to temperature controllers by electric cables. The temperature controllers monitor and control the set temperatures.

A rather expensive system used in isolated cases is the Dowtherm heating system. To heat the die, the die block is provided with cavities through which the temperature-controlled Dowtherm flows either as a liquid or as vapor. This type of heating requires insulation of the exterior die-block surfaces.

The Dowtherm heating system is expensive because it requires an additional electrically heated and temperature-controlled boiler. Boiler, Dowtherm pipes, and die block must conform to safety regulations. Operation of the line also requires adherence to safety regulations.

The spin pack is usually heated electrically, independently of the heating system used for the die block, and its temperature is controlled by a separate controller. In addition, there are monofilament die designs in which the spin pack is heated and kept at temperature merely by contact heat.

In the spin pack the melt is filtered and forced through the die-plate holes. The filters are located above a support plate and are held in place by it. They consist of several layers of medium-fine, fine, and superfine screen fabric with pore sizes ranging from 0.150 to 0.250 mm (medium-fine), 0.080 to 0.112 mm (fine), and 0.030 to 0.060 mm (superfine). Optionally, sand filters may be used, with a layer of sand of a certain grain size between the upper and lower layers of screen fabric. It is also possible to use filters made of sintered metal.

The distribution of the holes on the die-plate surface, which comes in contact with the material, is of special importance for achieving uniformity in volume throughput per die hole.

10.5 Lines

For circular die plates, it is recommended that the holes be arranged on a single pitch circle. However, if there are more than 80 holes, it is necessary to arrange them in two or three circles. Because of the melt-pressure load it is not possible to extend the die-plate diameter indefinitely. Furthermore, with large die-plate diameters there is the danger that the temperature will not be uniform on the surface; this would lead to uneven distribution of the melt across the die plate thus causing greater fluctuations in the monofilament diameters in relation to each other.

Shape, surface quality, and dimensional precision of the die holes are of the greatest importance in achieving monofilament uniformity. Figure 29 shows the cross-section of a die plate for the production of monofilaments.

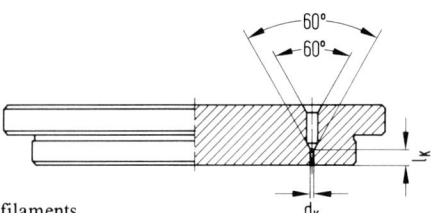

Figure 29. Die plate for the production of monofilaments

Diameter d_K of the capillary holes is machined to a tolerance of ± 0.001 mm and length L_K to ± 0.01 mm. The surfaces of the die hole and on the material outlet side of the die plate have a precision finish. The service life of the die plates depends on the quality of these surfaces and on the sharp-edged, burr-free finish of the die-hole edges on the material outlet side.

The capillary hole diameter d_K is determined from the following equation, based on monofilament diameter d_M and taking into consideration the draw ratio and hot-drawing of the monofilament:

$$d_K = d_M \cdot \sqrt{\lambda_1} \cdot \sqrt{\lambda_2}. \tag{4}$$

The draw value λ_1 is defined as the speed ratio of the last to the first draw unit of the monofilament line. The hot-draw ratio λ_2 represents the speed ratio of draw unit I to that of the melt flow in the capillary die-plate holes, or the ratio of the capillary hole cross-section to that of a monofilament quenched in the water bath.

In order to minimize diameter fluctuations among the monofilaments, the hot-draw ratio used should be as small as possible. On the other hand, this ratio, which is dependent upon the viscosity of the material, should be large enough to ensure that the draw force applied to the monofilaments between draw unit I and the die plate is large enough to prevent the monofilaments from passing through the water bath erratically, and causing extreme diameter and longitudinal fluctuations from filament to filament.

The capillary hole length l_K is determined by equation (5)

$$l_K = k \cdot d_K. \tag{5}$$

Factor k is an empirical value and, as a rule, lies between 1.5 and 5. In order to increase the melt pressure in the spin pack and optimize distribution over the die plate, a factor k of 3 to 5 is used when determining the capillary hole length for die-plate holes with larger capillary hole diameters.

If necessary, monofilament dies can be equipped with two or more spin packs. In that case, either the spin packs are fed by twin or multiple spinning pumps, or each spin pack is fed by a separate spinning pump with separately controlled drive.

In addition to the most frequently used spin packs with a circular cross-section, there are rectangular dies for use in special situations. The advantage of rectangular dies is that the monofilaments lie side by side and can therefore be quickly and easily guided through the water bath at the start of the spinning process, and placed in the comb guide at the outlet from the water bath. Design and manufacture of rectangular spinning dies must, however, meet special requirements with regard to even melt distribution and heating.

In order to ensure perfect functioning of the die and the extruder, and to monitor this function continuously, melt pressure sensors are installed in the adaptor between extruder and die, and between spinning pump and spin pack. Using a signal from the melt pressure sensor between the extruder and the die, the extruder screw speed is regulated by a closed-loop controller and a downstream pressure speed-control device, in such a manner that the spinning pump, regardless of its speed or output, is always supplied with a constant melt pressure, which is, however, adjustable at the controller. This melt pressure or spinning pump inlet pressure is usually set within a range of 6 to 12 MPa.

The controller, which also works as a melt pressure indicator, is fitted with an adjustable pressure trip-switch. If, for some reason, the melt pressure rises to the set maximum, the extruder drive is switched off. Thus, the entire extrusion unit is protected against damage by excessive melt pressure.

The melt pressure between spinning pump and spin pack is transmitted to an indicator by the pressure sensor located in that section. This pressure is a combination of the pressure required to overcome the flow resistance of the die plate and that needed to overcome the flow resistance of the filter. As dirt accumulates in the filter and increases its flow resistance, the melt pressure ahead of and within the spin pack increases continually until a value is reached which requires the spin pack to be replaced.

10.5.2.4 Spinning pumps

Spinning pumps are required to convey the melted material evenly, independently of changing counterpressures. In order to meet these requirements as far as is possible, gear pumps are used. Gear pumps work on the forced-conveying principle. Their conveyed

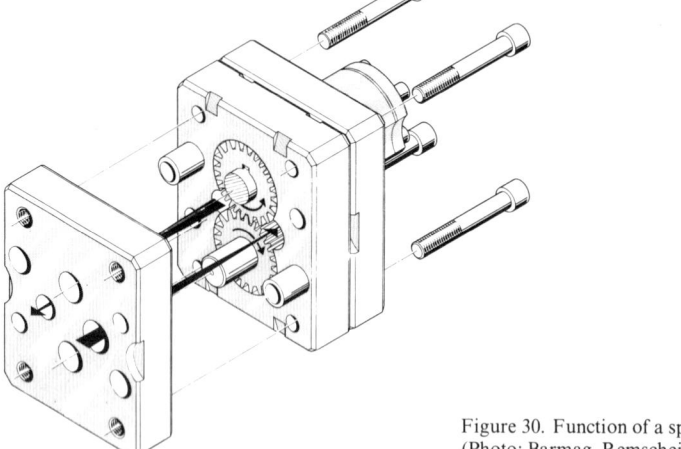

Figure 30. Function of a spinning pump.
(Photo: Barmag, Remscheid, West Germany)

volume is less than the swept volume only by the very small amount of leakage that is needed to build up and maintain a lubricating film between rotating and stationary parts. Figure 30 shows how a spinning pump works.

Spinning pumps are precision units. They are made of special types of hardened steel. Their manufacture is subject to the most exacting standards of manufacturing precision and surface quality. Specifying the clearance between rotating and stationary components takes into account the materials to be conveyed. If, for example, spinning pumps dimensioned for polyamide or polyethyleneterephthalate are used for polypropylene, they jam after a short operating period because the clearances are not sufficiently large for polypropylene. Spinning pumps dimensioned for polypropylene, on the other hand, have too much leakage flow when conveying polyamide or polyethyleneterephthalate.

The pump drives, like the spinning pumps themselves, require utmost precision. Synchronization has to be precise within $\pm 0.1\%$. In order to achieve this precision, in most cases electronically controlled DC motors, which drive the spinning pumps by means of reducing gears, are usually used. Between the reducing gear and the spinning pump there is a positive overload coupling which protects the pump against total destruction if jamming or blocking occurs.

Spinning pumps are operated in the braking mode if the melt pressure in the spin pack is more than 3 to 4 MPa lower than the pump inlet pressure, when using die plates with large capillary-hole diameters.

The spinning-pump inlet pressure should be set in the range 6 to 12 MPa to ensure reliable filling of the pump. Operation at higher pump inlet pressures increases the extruder counter pressure and therefore the melt return flow in the extruder which, in turn, results in better homogenization of the melt.

Figure 31. Monofilament die with quenching bath.
(Photo: Barmag, Remscheid, West Germany)

If the monofilaments produced do not have to be of high quality, particularly in regard to diameter fluctuations in the longitudinal direction, or if monofilaments are to be produced from materials which are not suited for conveying by means of spinning pumps, the spinning pumps are replaced by bypass elements. In that case the extruder must be operated without pressure/speed control and the screw rpm must be manually adjusted to the value that corresponds to the required monofilament diameter. As the extruder throughput is dependent upon pressure, the monofilament diameter must be checked from time to time. If necessary, the screw speed must be increased to compensate for the drop in output due to increasing dirt accumulation in the filter.

Figure 31 shows a monofilament die with the associated quenching bath.

10.5.2.5 Quenching bath

The monofilaments leaving the die plate in a molten state are cooled in the quenching bath to a temperature below the crystallite melting point.
The cooling medium is generally water. As the properties and dimensional precision of the monofilaments depend to a large extent on the cooling speed, the uniformity of the cooling process, and the temperature of the cooling medium, it must flow through the quenching bath evenly, without turbulence, and must always enter at a constant temperature.
With quenching baths using water as cooling medium, these requirements are often met by using a twin-chamber system. With this system, a mixing chamber in which the water is heated if necessary, is connected to the quenching bath; or, if cooling is required, cold water flows in, regulated by a controller, and is mixed with the water that is already in the chamber. A pump removes the temperature-adjusted water from the mixing chamber and transports it to the quenching bath. The bath is equipped with an overflow device through which the water is discharged either by way of the mixing chamber or direct to the outside. The twin-chamber system can also be replaced by a separate adjustable, thermostatted device which is connected to the quenching bath by hoses. With the help of the thermostatted device it is possible to control the quenching bath to temperatures within a range from 5 to 80 °C. The monofilaments, which during the spinning process are immersed vertically in the quenching bath, are deflected by elements installed in the quenching bath. They are pulled from the bath in an upward direction by draw unit I. The deflecting element is vertically adjustable for the purpose of cooling large monofilament cross-sections, and in order to prevent faults in roundness and diameter fluctuations due to deflection forces. Large numbers of monofilaments are best conveyed through double- or triple-deck deflection cages and appropriately divided, in order to prevent the monofilaments from touching each other or sticking together.
At the quench bath outlet, there is a comb guide into which the monofilaments are sorted. The way the monofilaments are guided through the comb creates the conditions for parallel passage through the downstream line and causes the monofilaments to pass through the water separately, consequently carrying along less water. A light installed at the front of the quenching bath below the water surface makes it possible to observe the monofilaments, and simplifies sorting them into the filament comb.
Quenching baths in monofilament lines are usually designed to be height-adjustable so that the space between water surface and die plate required for optimum monofilament quality can be quickly adjusted. The quenching baths can be moved sideways to facilitate installation of the spin packs and to simplify the start of the spinning process at the dies, and the cleaning of the extruders.

10.5.2.6 Monofilament drying

Excellent surface quality and uniform monofilament cross-sections can only be achieved if the monofilaments enter the drawing process in hot-air ovens absolutely dry. For this reason, in monofilament lines a drying section is incorporated following the quenching bath and, if necessary, following the drawing bath. The drying section consists of several components. Depending on the material used and the process parameters employed, however, not all components need necessarily be utilized. Most of the water carried along by the monofilaments is wiped off by sponge-wipers or deflecting rods immediately following the quenching bath. The remaining water is removed by one or more suction nozzles. In some cases, where the monofilaments are not yet completely dry, it may be necessary to remove the residual moisture by evaporation with the help of hot-air blowers and suitable blower nozzles.

10.5.2.7 Draw units

The physical properties of the monofilaments, such as tensile strength, elongation, elastic modulus, shrinkage, shape retention, wear resistance and flexibility, are largely determined by the degree of orientation imparted to the filament-like macromolecules by the drawing process. For the processing of materials that are drawn in a single-step operation – like high-density polyethylene and polypropylene – the draw zone of a monofilament line consists of draw unit I, a drawing bath or hot-air oven, and draw unit II. For materials that are drawn in two steps, such as polyamide and polyethyleneterephthalate, the draw zone consists of the units already mentioned and an additional hot-air oven and draw unit III. Drawing of the monofilaments takes place within the draw zone composed of these two or three draw units, with the speed ratio of the last and first draw units being the draw ratio λ.

For cost-efficient, continuous production of monofilaments in sheets, draw units with seven godets running in cantilevered bearings have proved to be the most suitable of all draw units. In this draw unit, seven godets are arranged side by side on two horizontal lines running one above the other with a space between them. With this godet arrangement, the monofilaments running side by side wrap themselves around each godet at an optimum wrapping angle.

The draw forces acting on the individual monofilaments are transmitted to the godets by means of friction according to equation (6). With this type of force transmission there is no damage to the monofilament surfaces due to clamping or pressing.

$$F_1 = F_2 \cdot e^{\mu \alpha}. \tag{6}$$

For application of tensile force, we have:

F_1 draw force at the entry side,
F_2 draw force at the outlet side of a godet.
       If the draw unit operates in the braking mode, the flow of force is reversed.
μ friction coefficient between monofilament and godet,
α angle of wrap around the godet.

The wrapping angle is determined by the way the godets are arranged in relation to each other. The angle is limited by the necessity of maintaining a certain space between the godets to ensure ease of operation and proper function of the draw unit. The godet design determines friction coefficient μ. The surfaces are such that the friction coefficient is as large as possible without the monofilaments being damaged while passing over the godets. Draw force F_2 acting upon the monofilaments when they leave the individual godets is a function of the speed ratio of each succeeding godet to the preceding godet. For the last godet in the draw unit, F_2 depends upon the speed ratio to the first godet of the following draw unit.

Depending on the use of the draw units, i.e. whether they have to supply drawing or restraining forces, the speed ratios of the godets in relation to each other either increase or decrease, in increments that ensure slippage-free transmission of the monofilament draw forces to the godets of the individual draw units.

As a rule, the godet diameter is between 180 and 250 mm and, depending on the number of monofilaments required to pass over the godets side by side, the godet length is between 450 and 1050 mm. The godets are provided with hard surfaces, to ensure wear resistance and insensitivity to damage. Figure 32 shows a draw unit with seven godets and a working width of 800 mm. The draw units are driven either by three-phase AC motors with steplessly adjustable mechanical gears or by electronically controlled DC motors with the help of reducing gears.

Figure 32. Draw unit with seven godets.
(Photo: Barmag, Remscheid, West Germany)

The production programs planned for the monofilament lines determine the drive capacity to be installed in the draw units; this depends on the draw and setting forces to be generated, and the speeds and the positions of the individual draw units within the monofilament line.

In draw units equipped with steplessly variable mechanical gears, the torques generated, and consequently the draw forces, are in inverse proportion to the speeds used. This inverse proportion of the torque to the speed of the drive shaft of such adjustable mechanical gears is why they are eminently suitable for use in draw units: because for drawing large monofilament cross-sections at low speeds, large draw forces are required, whereas drawing of small monofilament cross-sections at high speeds requires low draw forces.

In draw units equipped with DC motors, this favorable draw force/speed ratio does not apply despite electric and electronic control features. For reasons of control engineering, the motors must therefore be overspecified in relation to the actual power consumption of the draw units. The motors, however, have the advantage of being easier to control, especially with synchronous control systems. The control system is virtually free from wear and tear. Draw units equipped with DC motors work at lower noise levels than draw units with steplessly adjustable mechanical gears.

The individual draw-unit godets are usually driven by toothed gears or a flat-belt drive. Toothed gears allow slippage-free operation but have higher noise levels than flat-belt drives, particularly at higher speeds. Flat-belt drives have the advantage that increases or reductions in the speeds of individual draw unit godets relative to each other, can be made quite easily by sliding different pulleys of appropriate diameters onto the godet shafts. The speed increase or reduction of the godets in relation to each other is necessary in order to reduce the draw tensions from godet to godet in steps, taking into consideration the elastic elongation of the monofilament.

In draw units equipped with toothed gears, the required speed differences between the individual godets are achieved by means of different godet diameters. If, for reasons of process engineering, a change in the speed gradations of the individual godets in relation to each other is required, this can easily be achieved by exchanging the pulleys, if draw units with flat-belt drives are used. In draw units with toothed gears, however, the godets must be exchanged.

If the monofilaments are subjected to heat setting (to stabilize the orientation and reduce residual recovery or shrinkage tensions) the setting unit is subject only to the comparatively low setting tensions. It operates at a speed that is up to 10% lower than that of the last draw unit of the draw zone, and, because of the low forces to be transmitted, it is equipped with only three godets.

10.5.2.8 Drawing bath

The water drawing bath was found to be successful for drawing monofilaments made from materials which, for reasons of process stability or to achieve special characteristics, must be drawn or pre-drawn at temperatures below 100 °C. This drawing bath is 1.5 to 2 m long and is made of stainless steel. Its heating system consists of heating elements which are evenly distributed directly above the bottom of the bath. The desired bath temperature is set by means of a temperature controller which controls the functions of the heating elements.

The loss of water in the drawing bath due to evaporation and water being carried along by the monofilaments, is compensated by a continuous supply of fresh water controlled by a level-sensing device. In order to keep the loss of water as low as possible, hinged cover plates are installed directly above the drawing bath so that the water condensate can flow back into the bath. Furthermore, the monofilaments leave the drawing bath without touching each other, in vertical or at least nearly vertical fashion, so that they carry only a very small amount of water with them. Guide plates attached to the outlet side collect water spray, or water that has been wiped off the monofilaments, and direct it back into the drawing bath.

Non-driven rolls with a grooved profile are installed to suit the working height of the draw units, one on the inlet and one on the outlet side of the drawing bath, to deflect the monofilaments into the temperature-controlled water. Proper allocation on these deflecting or grooved-profile rolls ensures that all monofilaments entering and leaving the drawing bath maintain the same distance in relation to each other. This is important in order to obtain high-quality monofilaments, ensure process stability, and keep to a minimum the amount of water being carried along by the monofilaments.

Figure 33. Drawing bath in operation. (Photo: Barmag, Remscheid, West Germany)

One or two cylindrical guide rolls with non-profiled surfaces can be moved electromechanically into the drawing bath next to the profiled deflecting rolls. If only one guide roll which can be lowered or lifted is installed, it is mounted next to the deflecting roll on the outlet side of the drawing bath. During start-up of a monofilament line equipped with a drawing bath, the monofilaments are fed in over the deflecting rolls and under the guide rolls, these being in their uppermost position. With this method, the draw unit following the drawing bath operates at virtually the same speed as the draw unit ahead of the drawing bath. As the speed ratio of the two draw units in relation to each other increases, the guide rolls move into the drawing bath and the monofilaments pass through the temperature-controlled water, where they are drawn.

Figure 33 shows a drawing bath with a working width of 800 mm in operation.

10.5.2.9 Hot-air oven

The unit most frequently used for heating the monofilaments, for drawing as well as for heat-setting, is the hot-air oven.

As a rule, hot-air ovens for monofilament lines are designed for a working width between 450 and 1000 mm. They are used in lengths from 2.5 to 5.0 m. In order to achieve rapid, even heating from all sides during the heating phase of the monofilaments, most hot-air ovens operate with a dual-cycle air circulation system. With this system, one half of the total air circulation takes place in the upper portion, and the other half in the lower portion of the hot-air oven.

On the inlet side, the two airstreams contact the monofilaments from top and bottom at speeds of up to 50 m/s to give optimum heat transfer to the monofilaments. By approaching the monofilaments at an angle of 30 to 40° — with the main flow component pointing towards the upper and lower suction openings at the outlet side of the hot-air oven — the air in the drawing oven itself moves in the same direction as the monofilaments.

The air temperature in the hot-air oven is monitored and controlled by an open-loop device. The upper temperature limit is usually 300 °C.

In order to achieve identical properties for all monofilaments produced in one process, it is necessary to keep the temperature and air velocity fluctuations as low as possible over the entire width of the hot-air ovens. With proper hot-air ovens, the temperature fluctuations over the entire width are below $\pm 1 \,°C$, and those of the air velocity less than ± 0.5 m/s.

To string up the monofilaments during start-up or for restringing broken filaments, the hot-air ovens are opened 8 to 10 mm on the operating side. For cleaning, the top part of the oven can be swung open by about 300 mm.

10.5.2.10 Winding

Depending on the end product or on the subsequent process, monofilaments are wound either individually or in strands. Strand winding is used if the monofilaments are subsequently processed into cord, ropes, or hawsers or into bristles. All other applications require individual winding.

Monofilaments for the manufacture of bristles are wound on reels or reeling machines, from which they are taken off in bundles. Monofilaments used for cord, rope, or hawser manufacture are either parallel-wound on flanged bobbins or cross-wound on wind-up tubes into cylindrical or biconical packages. Package weights usually are within the range 10 to 40 kg.

For strand winding, each wind-up spindle with its associated traverse-motion mechanism is driven separately. The required wind-up tensions can be adjusted by means of electrical control circuits.

For winding of individual filaments, either the wind-up spindles are driven collectively with the help of magnetic couplings, or each spindle has its own drive. The monofilaments are guided to and deposited on the bobbins of the individual winding positions by means of traverse motion mechanisms which work in conjunction with the associated

spindles or by means of a collective traverse motion mechanism equipped with its own drive.
The winding tension is adjusted mechanically for spindles with collective drive, and electrically for spindles with individual drives. If spindles with individual drives and compensating-arm control are used, the pre-set tension is controlled electromechanically or electronically.
If monofilaments are wound individually — especially monofilaments that have to meet the most stringent requirements regarding precision of shape — the parallel-winding system onto flanged bobbins is selected. As a rule, flanged bobbins conforming to DIN 46399 are used for this purpose.
Both strand winding and individual winding can be done on flangeless tubes on which the individual monofilaments are cross-wound into cylindrical or biconical packages. This type of package build, however, is only suitable for monofilaments with a diameter range below 0.3 mm. However, shape precision need not be of primary concern in this case, as the monofilaments deform easily at crossing points.

10.5.2.11 Aspirator unit

As an aid in start-up and in order to remove monofilaments broken during the drawing process or left behind in the last draw unit after doffing, aspirator cones are attached to the outlet of the individual draw units. They are slightly wider than the winding or working width of the draw units and, depending on how they are mounted, they convey the monofilaments in aspirator ducts which are arranged either above the monofilament line or behind it directly above the floor. In most cases, the vacuum required to operate the aspirator is produced by an injector nozzle fed by a fan. The monofilaments are blown into a collecting bin through a pipe connected to the injector nozzle.
A different aspirator design, which operates without an injector nozzle, consists of an aspirator cone followed by a suction or transport pipe and vacuum vessel. An exhaust fan is connected to the suction side of the vacuum vessel, which has a volume of 5 to 15 m^3, depending on the output of the extruder to which it is assigned.
Compared with aspirator units working with injector nozzle, the aspirator unit equipped with a vacuum vessel has the disadvantage of having to be taken out of operation for short periods of time while the monofilaments are being removed from the vacuum vessel. On the other hand, it uses less energy.

10.5.2.12 Process control systems

Process control systems equipped with microprocessors are being used increasingly on plastics processing lines.
Because of their efficiency in areas of digital control, continuous process monitoring, operating- and diagnostic aids, these systems raise productivity; and on monofilament lines, too, they improve the reproducibility of processing conditions and keep them constant to ensure production of high-quality monofilaments to narrow tolerances.
The advantages of digital control are found primarily in the ability to reproduce set values and control parameter settings exactly, and work without drift over long periods of time, even if temperatures fluctuate. Furthermore, once the optimal settings have been established, they can be placed in data banks, like magnetic tape cassettes or floppy disks, or Electrically Erasable Programmable Read Only Memory banks (EEPROMs), and can be recalled at any time. A change to different products can thus be made quickly and without mistakes.
Digital control simplifies semi-automatic start-up and setting of specific draw ratios.
Continuous process monitoring makes possible constant checking of all process parameters

to ensure that they remain within their limit values. An alarm signal is triggered whenever they move outside the tolerance band, and a warning signal is switched on and the deviating measured value conspicuously displayed on a color monitor, often winking in red. In addition, an alarm record is printed showing date, time, and serial number, as well as location and type of trouble.

By accumulation of measured values of critical line parameters and tracking of trends, it is possible to diagnose trouble at an early stage and take corrective measures to forestall deterioration in monofilament quality, or even damage to the line.

As all process data are stored for a certain length of time, in case of trouble it is possible to obtain a print-out on the history of that type of trouble. Numerical outputs and graphs of measured values are displayed on a color monitor, and can be printed out with date and time, either on request or automatically at specified intervals. It is also possible to obtain print-outs of the actual and the set-point values of the line.

Instruments that measure filament diameter by means of laser beams can be incorporated in the process system. In this case, the measuring head of the instrument is mounted ahead of the last draw unit. It moves across the monofilament sheet, measuring either one individual monofilament after another separately or, alternatively, combined monofilament groups. The results are displayed on the monitor. If required, a print-out of the record can be produced.

By using the calculated average diameter of all monofilaments, the speed of the melt-metering pump is controlled in such a manner that the pre-selected diameter is maintained. The extruder screw speed is controlled by a melt-pressure/speed control system to keep the pre-set pressure ahead of the melt-metering pump constant.

Precise logging and analysis of the doffing times at the individual wind-up positions under production conditions, i.e. keeping constant all process parameters required for a specific monofilament quality, makes it possible at all times to obtain comprehensive information on the output and quality, efficiency, set-up, start-up, and downtimes of the line. This procedure is important for planning, and expediting orders.

It is now possible to design process control systems in such a way that, aside from stringing the monofilaments through the line and on to the wind-up bobbins ('string-up'), the lines are fully automatic.

After input of a certain code number assigned to the monofilament diameter, feed material, and monofilament quality, these lines function automatically from the beginning of the heating process, through start-up at low production speeds (placing the monofilaments on the wind-up bobbins) all the way up to operation at high production speeds.

By maintaining uniformly high product quality, reducing downtimes, eliminating waste, by material savings as a result of narrower tolerances and making the lines easier to operate, process control systems today pay for themselves within approximately two years. This payback time will undoubtedly soon become even shorter.

10.6 Layout of the line

The cost effectiveness of a monofilament line is determined by the quality of the monofilaments it produces and by its output capacity and efficiency. For new investments it is therefore important to choose the optimum line layout, taking into consideration the future production program.

Key points are the choice of proper extruder size and screw geometry, whether single or two-step drawing is to be used and what type of heating system is to be used for drawing and heat-setting of the monofilaments, how long the heat-setting zone should be, what working width the line should have and how large the number of monofilaments should be.

10.6 Layout of the line

One of the main design criteria for a monofilament line is the output capacity $\dot{m}$. Its dependence upon monofilament diameter d, number of monofilaments i, production speed v and density ϱ of the material being processed, is expressed in equation (7):

$$\dot{m} = \frac{d^2 \cdot \pi}{4} \cdot i \cdot v \cdot \varrho. \tag{7}$$

This relationship is represented as a nomogram for some familiar situations in Figure 34.

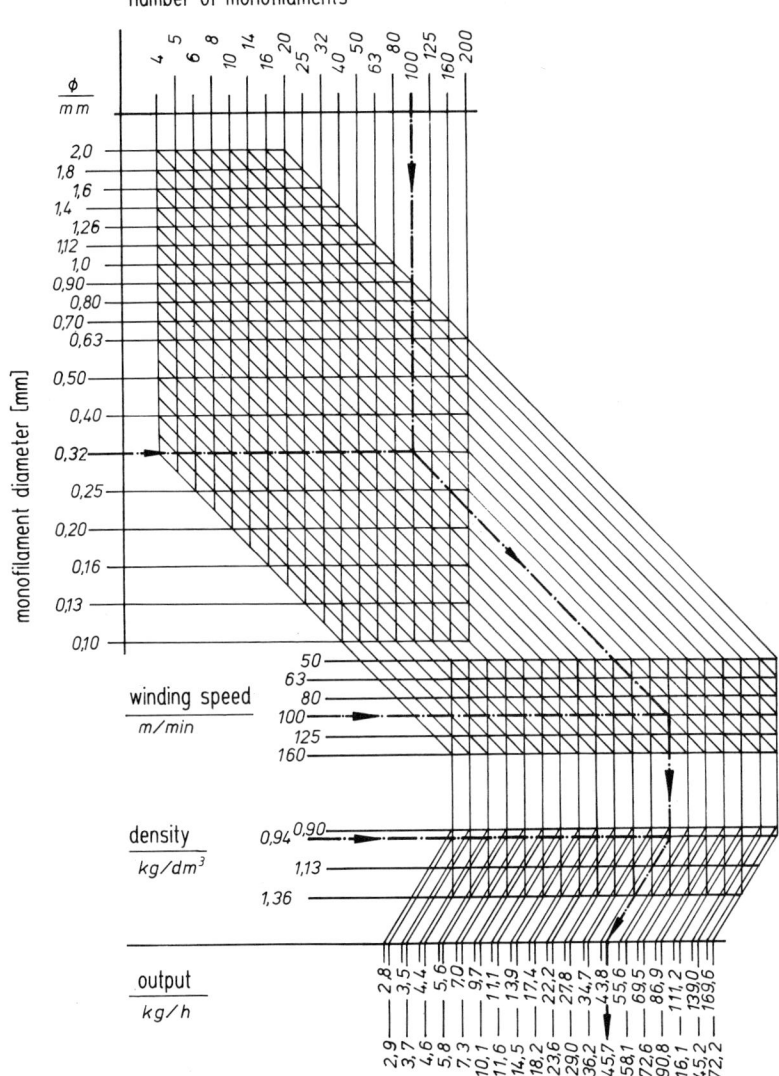

Figure 34. Nomogram to determine the output capacity as a function of monofilament diameter, number of monofilaments, wind-up speed, and material density

The extruder size is established on the basis of the maximum output capacity required for a given range of monofilament diameters and numbers. When establishing the production range, care should be taken that the ratio between minimum and maximum output does not exceed 1:6, since otherwise the residence time of the material in the extruder component will be too long when producing small diameter monofilaments, and thermal damage may result.

As a rule the working width of monofilament lines is between 400 and 1000 mm. It is determined by the maximum number of monofilaments to be produced.

For the production of high-quality monofilaments, it is necessary to guide them parallel through the entire line, always maintaining a certain space between them. If this space is too small, i.e. if the number of monofilaments in relation to the working width of the line is too large, the string-up and sorting process of the monofilaments during line start-up, or in case of troubleshooting during production, becomes too complicated and time-consuming. There is too much waste and too much production time is lost, diminishing the efficiency of the line.

Densities of 16 to 20 monofilaments per 100 mm have proved successful under production conditions and should not be exceeded in production of high-quality monofilaments.

Greater densities are used in the production of monofilaments for bristles, ropes, and hawsers, where several monofilaments are usually combined into one strand and the individual strands are guided through the line parallel to each other and with a space between them. With this method it must be kept in mind that certain concessions are required regarding maximum achievable draw ratio and uniformity of monofilament properties, since the monofilaments — when guided in strands — are not so evenly heated in the draw- and heat-setting zones as they are when each monofilament is conveyed separately. For this reason, no more than five to eight monofilaments should be combined into one strand, otherwise a reliable drawing process can no longer be ensured.

The additional design criteria mentioned — screw geometry, whether single or two-step drawing is used, which heating system should be selected for drawing and heat-setting, and how long the drawing and heat-setting zones should be — require decisions by the line manufacturers based on their experience and know-how on the various materials to be processed, and on the required monofilament properties and production speeds.

Reference for Chapter 10

For the convenience of the reader the English title of this German publication is shown in parenthesis.

[1] *Jakobi, H. R., Cegla, J.:* Auslegung von Extrudern für quasi-Newtonsche Kunststoffschmelzen (Design of Extruders for Quasi-Newtonian Plastic Melts). Kunststofftechnik 12 (1973) 8, pp. 213/217.

11 Extrusion coating and laminating

E. Schöllkopf, K.-F. Roesch

11.1 Paper, film and foil converting

The improvement of the characteristics of various materials in web form occupies a large part of the paper, film and foil converting industry's efforts.
By combining two or more materials, by coating or lamination, their individual characteristics are enhanced.
Such enhanced properties may concern the material surface only, i.e. coating or laminating with another component improves the appearance of the basis substrate, its printability and its resistance against mechanical and chemical damage. Coating a substrate with another material component or laminating two substrates improves the structural properties of the resultant material. Thus it can have better dimensional stability, improved stiffness, impact strength and durability, temperature resistance, machinability and sealability. However, the packaging industry uses the coating and laminating processes mainly to improve the barrier properties of a basic substrate. The most important barrier properties include resistance against light, grease and vapor as well as resistance against the migration of gases and flavoring oils.
Representative material combinations are listed and explained in Section 11.6.

11.2 Coating and laminating

Coating and laminating are two distinctly different converting processes. In the coating process a layer of identical or dissimilar material is put onto a prefabricated substrate. This includes the lacquering of foil, and wax-, hot melt- and extrusion coating or coextrusion. A coated substrate is already a multilayer material.
In the laminating process two or more prefabricated webs are combined by using an adhesive. Laminating media include solvent-based and solventless adhesives, waxes and hot melts as well as extruded or coextruded thermoplastic materials. Coating and laminating by extrusion or coextrusion should not be considered a substitute for other coating and laminating processes in paper, film and foil converting. In certain cases, however, it is more and more recognized as an economical alternative to the dry and wet laminating processes. In comparing similar or identical end products, extrusion coating or laminating should, as such, be more economical than the conventional coating and laminating processes. Prefabricated film of minimal thickness does not have to be adhered to the substrate, nor do the solvents of a wet film have to be evaporated, for the solids to form a film.
Other advantages in using an extruder for paper, film and foil converting are:

- very often, higher production speeds;
- the elimination of an adhesive and the related energy consumption and machinery cost;
- no need for a prefabricated film, i.e. the elimination of a complete processing step;
- the possibility of applying extremely thin layers of film by direct extrusion, giving material savings and eliminating problems connected with the handling of thin films.

Despite the many positive characteristics of the extrusion process, one should not forget the relatively high investment needed and the disadvantages inherent in the process itself.

For instance the extruder screw must not be stopped from rotating while the extruder is at process temperature, which means that material is extruded even during product changeover time. While changing the resin the entire plant must be shut down for a relatively long period of time, especially when the screw has to be changed. For these reasons extrusion coating plants can very often only be used economically for medium and large size production runs. An extrusion coating plant should run for several shifts before the changeover to the next product is scheduled.

The prime application of the extrusion coating process is the manufacture of packaging materials, including both simple and complex structures.

11.3 Material combinations

The properties and the performance of laminated material can be influenced and controlled by the combination of two or more single webs (paper, board, plastics films, metal foil), since the properties of the structure are determined by the specific properties of the individual substrates used. The specific properties of extrudable plastics materials for paper-, film- and foil converting are listed on data sheets of raw material suppliers and converters. The DIN reference numbers of the most important material testing methods are listed in Table 1.

Table 1. Test methods

Properties	Test methods
Density	DIN 53420
Melt index	DIN 53735
Tensile strength	DIN 53455
Impact resistance	DIN 53448
Water absorption	DIN 53428
Moisture vapour transmission rate	DIN 53122
Gas permeability	DIN 53380

11.4 Adhesion

One of the quality-determining parameters in the production of multilayer materials is the adhesion at the boundary layers. Adhesion can be described as the total of all forces that resist the delamination of two layers at their common boundary. Thus adhesion depends critically on the quality of the surface, and its polarity, affinity and wettability. These conditions combine to determine the surface tension, which is measurable. Exactly why adhesion between two extruded plastics films, or between extruded plastics films and aluminum foil, paper or film occurs, has been the subject of many theories, which can be found in the literature.

11.4.1 Adhesion values

No direct method is available for measuring the adhesion between different materials. For testing the adhesion level it is necessary to delaminate a structure: i.e. to measure the forces needed for separating two layers or for peeling one layer from the other. According to the DIN 53357 test method, material samples 15 mm wide are separated at a speed of

100 m/min. The force required is recorded and can be used to compare the adhesion levels of various material combinations. As soon as the adhesion forces between two layers exceed the cohesion forces in either layer, the weaker layer will be damaged by mechanical stress before delamination takes place.

11.4.2 Improving the adhesion

Material structures with extruded or coextruded layers are used mainly for protective purposes (packaging, covering, wrapping). External influences put severe demands on the adhesion. The adhesion levels of a material structure must be chosen to suit its ultimate use. Insufficient adhesion between extruded layers may be caused by an inadequate manufacturing process, unsuitable storage and finishing, wrong choice of materials. It may also be caused by the product packed in it.
Most extrusion, coextrusion coating, and laminating products need adhesion promoters. Adhesion promoters are called for when two materials without basic affinity have to be stuck to each other. They must also be used when higher adhesion levels, higher web speeds, lower extrusion temperatures, must be achieved, and whenever the final structure should have better chemical and thermal resistance.
The most important adhesion promoters are:

− flame pretreatment
− corona discharge pretreatment
− ozone shower
− primers.

11.4.2.1 Flame pretreatment

This pretreating method is often used in high-speed extrusion coating of paper and board to achieve good adhesion levels. The air supply to the burner must be precisely adjustable to obtain an oxidizing flame. The flame oxidizes the web surface, and bridging COOH groups develop which improve the adhesion between the substrate and the extruded film.
The additional heat build-up in the substrate and the elimination of protruding fibers and of surface humidity also improve the adhesion.

11.4.2.2 Corona discharge

High-frequency corona discharge is a physical surface treatment process to improve the adhesion of extruded coatings on plastics films, paper and metal foils.
An alternating voltage of approx. 12 to 20 kV is generated in a high-frequency generator (20 to 40 KHz). The energy generated is transferred to the substrate by a system of electrodes. The air in the gap between the electrodes is ionized and nascent ozone and oxygen are created. This causes oxidation of the substrate surface with the development of adhesion-promoting groups. Applying a corona discharge to aluminum foil will burn off slip lubricants and other surface impurities.
The voltage that can be applied is limited by the puncture voltage of a given substrate. Higher voltages will damage the web. High-speed extrusion coating machines are therefore equipped with multiple electrodes, from which the necessary energy can be transferred to the web at lower voltages. The amount of energy transferred by the electrodes per unit of web surface determines the degree of change in the characteristics of the web surface, measurable as surface tension [2].

11.4.2.3 Ozone showers

For practical reasons, flame and corona pretreatments are only applied to the surface of the substrate to be coated. It is, however, known that oxidation of the extruded plasticated film prior to contact with the substrate to be coated will improve the adhesion. The oxidation of the extrudate can be achieved by blowing ozone against the contact surface (for laminating on both sides) of the plasticated film [3]. Ozone is one of the strongest oxidizing agents known.

Oxidation is the result of the high rate of decomposition of ozone, enhanced by the heat of the extruded film. Although an ozone shower will noticeably improve the adhesion, its effectiveness will always be influenced by the other extrusion process parameters (e.g. melt temperature). Ozone showers are very often used in addition to the above-mentioned pretreatment methods.

11.4.2.4 Primers

Primers promote adhesion between the extrudate and the substrate to be coated, but themselves need a suitable substrate surface to stick to. Carrier webs of PA or non-polar polyolefins (PP, PE) require pretreatment with a flame or by corona discharge.

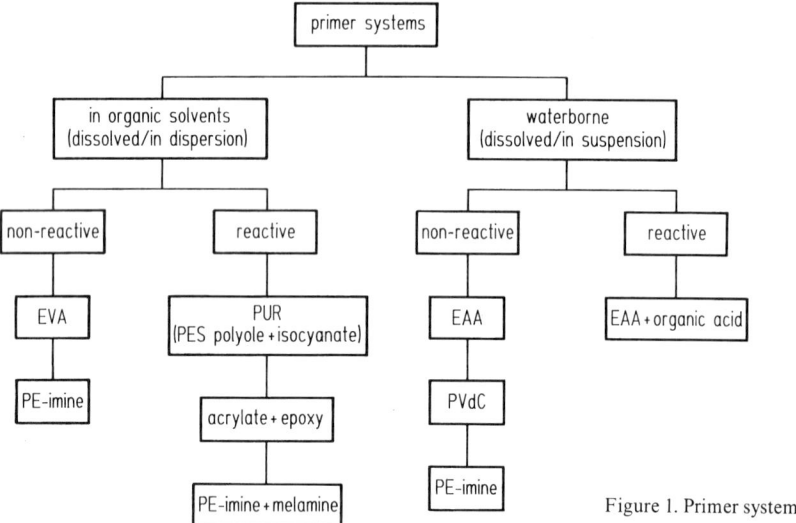

Figure 1. Primer systems

The primer is applied in a roll-coating station, which is an integral part of most modern extrusion coating and laminating equipment. The selection of a suitable primer depends on the final use of the laminated material. Figure 1 shows the most important raw materials used for the preparation of primers [4]. The organic solvents of solvent-based primers and the water of water-based primers must be evaporated in a drying tunnel which follows the primer station. Solvent-based primers can only be applied on explosion-proof machinery. The choice between reactive and nonreactive primers depends very much on the properties which the final laminate must have. Nonreactive systems are easier to work with in production. The final adhesion level is reached immediately after extrusion coating, and the material can immediately be used for further production and finishing operations.

Reactive primer systems need up to several hours of curing time, depending on the type of primer and the ambient temperature. Further processing steps must be delayed accordingly. Premature handling, or processing of such materials prior to the completion of the primer-curing process may drastically reduce or completely destroy the adhesion. These processing disadvantages are more than offset by the advantages which reactive primers offer, such as improved chemical resistance and better thermal and mechanical resistance of the final structure [4]. Primer coating weights vary a lot. Low-weight primers, which work properly only at molecular layer thickness, are applied at dry-coating weights of 0.01 g/m² or less. Other primers are applied at dry-coating weights of 0.5 g/m² and more [1].

11.4.3 Adhesion-related coating- and laminating parameters

Figure 2 shows the most important parameters that influence the adhesion level in extrusion coating and laminating. As PE was the first polymer to have its extrusion behavior thoroughly examined, it has been taken for reference to explain the various effects shown in the table.

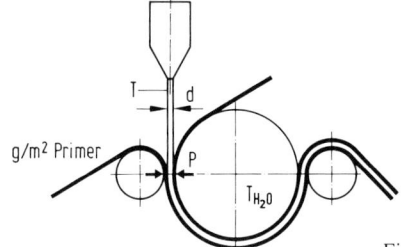

Figure 2. Adhesion-related coating and laminating parameters

11.4.3.1 The temperature of the extruded PE film

The adhesion of the extruded PE film to the carrier web improves with increasing extrusion temperatures T (°C), especially when the film is exposed to an oxidizing environment. Figure 3 shows the level of adhesion of extruded PE on corona-pretreated and untreated

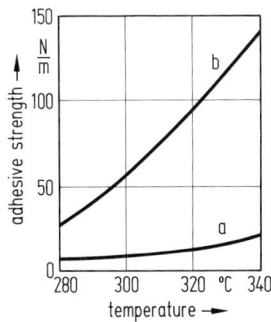

Figure 3. Adhesion values of extruded LDPE on aluminum foil as a function of the extrusion temperature. Web speed 60 m/min, coating weight 30 g/m².
Curve a: untreated aluminum, curve b: corona pretreated aluminum

aluminum foil, as a function of the extrusion temperature. The diagram shows that a noticeable increase in adhesion at higher extrusion temperatures can only be achieved when the LDPE film is being extruded onto pretreated foil [5]. Higher extrusion temperatures, however, may cause polymer decomposition and the development of oxidation products, which in turn may have a negative influence on stability, odor, reel blocking and heat sealability.

11.4.3.2 Surface conditions of the carrier web and the laminating web (Dyne/cm, g/m² primer)

Adhesion-promoting treatment of the surfaces of substrates makes it possible to lower the temperature of the extruded film, thus eliminating the disadvantages listed above (11.4.3.1) and producing an equal or even better adhesion value. The heat developed in the substrate by corona discharge or flame pretreatment will improve the adhesion. Pretreating and primer stations, however, necessitate higher investment and result in higher operating costs. Use of primers may cause odor problems due to residual solvents in the final laminate, and environmental problems associated with evaporated solvents, unless sufficient drying capacity and solvent recovery or incinerating systems have been installed.

11.4.3.3 The chill-roll temperature

The usual chill-roll surface temperature T_{H_2O} (°C) is between 15 and 40 °C. Most machines operate at 18 to 25 °C. Higher chill-roll temperatures usually result in higher adhesion levels. At higher chill-roll temperatures there is an increased tendency for the extrusion-coated film to stick to the chill-roll surface, causing delamination between coated film and carrier web, or even a web break when the web is peeled off the chill roll.

11.4.3.4 Nip pressure between chill roll and impression cylinder

Usually the nip pressures range from 150 to 250 N/cm (Cellophane/PE 100 to 200 N/cm, paper/PE 150 to 300 N/cm, aluminum/PE 150 to 200 N/cm and in special cases up to 500 N/cm).
The evaluation of various test runs indicates that increasing the nip pressure from 0 N/cm to a certain point will increase the adhesion. Exceeding this point will not further improve the adhesion significantly [6].

11.4.3.5 The thickness of the extruded film

With other process parameters remaining constant, the adhesion level will fall with decreasing film thickness. The reason for this is the amount of heat energy in the extruded film and heat-energy-related reactions. A thin film will cool down much faster after leaving the extrusion die and there will, therefore, be less oxidation at the surface of the film.

11.4.4 Coextrusion as adhesion promoter

Coextrusion, the extrusion of a multilayer film, offers the possibility of combining an almost unlimited variety of plastics materials. It follows that the coextrusion technique not only offers the possibility of upgrading the barrier properties of a laminated structure but also of improving the adhesion. Many extrusion-coated barrier materials do not sufficiently adhere to other plastics materials and do not have sufficient heat sealability. By coextrusion of suitable adhesion promoters it will be possible to embed such barrier layers in heat-sealable laminates. The coextrusion of adhesion promoters, as shown in Figure 4, will in many cases eliminate the need for primers, corona discharge pretreatment and an ozone shower. A typical example is the coating of a coextruded PE/PE extrudate onto a carrier web. The PE layer which adheres to the carrier web can be extruded as adhesion promoter at a relatively high melt temperature; this will result in a higher degree of oxidation and, therefore, in better adhesion. The coextruded PE top layer can have a much lower melt temperature, resulting in better heat sealability and an acceptable odor level.

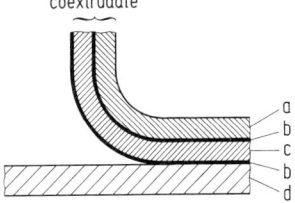

Figure 4. Coextrusion as adhesion promoter
a LDPE for food contact and for sealability,
b adhesion promoter,
c Saran* as gas and vapor barrier,
d substrate
* Trademark Dow Chemical Company

11.5 Machine design considerations

Besides the above-mentioned parameters, economic and design criteria play an important part in the project phase of a (co-)extrusion coating and laminating machine.

11.5.1 Web widths

Usual web widths range from 500 to 2200 mm. Most packaging materials are between 1000 and 1400 mm wide. For special extrusion coating applications, e.g. the coating of cloth, web widths of up to 3000 to 4000 mm are used.

11.5.2 Web speeds

Production speeds are determined by:

– the conditions which govern the cooling process of the extruded film (chill roll diameter, distance between the die and the nip),
– the corona discharge pretreating equipment (number of electrodes) or primer equipment (drying capacity),
– the machine design (winding speed, controllable speed range, etc.),
– the capacity of the extruder (see section 5.4).

Web speeds usually range from 100 to 500 m/min; higher web speeds are possible for special applications. Cloth is usually extrusion coated at web speeds of 30 to 100 m/min.

11.5.3 Coating weights of extruded films

In German-speaking areas the quantity of extrudate (the extruded film) or the coating is expressed as weight per unit area, for instance 0.5 g/m² primer or 15 g/m² PE. The carrier webs are defined by their web thicknesses, e.g. 9 µm aluminum foil or 25 µm polyester film, with the exception of paper, which is specified by weight per unit area, for instance 60 g/m² Kraftliner.

Usual extrusion coating weights range from 10 to 100 g/m², and in some special applications up to 200 g/m². Very low extrusion coating weights of 5 g/m² can be obtained at high melt temperatures with raw materials of very low viscosity. On specially designed low-speed machinery it is possible to reach extrusion coating weights of up to 500 g/m².

The weights of the individual layers of coextruded coating films are controlled by other parameters, and will not be discussed here.

11.5.4 Extruder size

Extruders for extrusion coating machinery are equipped with various screw diameters depending on the desired web widths and web speeds. Usual screw diameters range from 90 to 150 mm. The screw diameters of coextrusion machinery are usually smaller, and range from 45 to 60 mm.
Table 2 shows typical extruder outputs at various screw diameters.

Table 2. Extruder outputs for various screw diameters

Web width	mm	840	1040	1280	1430	1670
Screw diameter	mm	90	105	120		150
Output
LDPE
HDPE
PP
EVA
Ionomer	kg/h
kg/h
kg/h
kg/h
kg/h
kg/h
280
150
200
200
200 |
380
200
260
260
260 |
500
260
370
370
370 | |
750
380
560
560
560 |

11.6 Machine layout criteria

Layout criteria of extrusion coating machinery are discussed in the following paragraphs.

11.6.1 The coating principle (see Figures 5 and 6)

The die (d) as described in Chapter 7 is arranged above the laminating rolls (a) and (b). The thermoplastic film (e) (PE, PP or other thermoplastic materials) leaves the die and is coated onto the carrier web (f) in the nip between chill roll (a) and impression roll (b). A three-layer laminated structure is combined when a second substrate (g) is fed into the nip.
The contact point of the extruded film can be varied, as shown in the picture, to enable adjustments which can influence the quality of the lamination to be made. Adjustment can be achieved by moving either the extruder with the die, or the chill roll/impression roll assembly.
The lapse of time between the film leaving the die and contacting the chill roll or the carrier web is also variable. The distance between die and nip can be adjusted according to product specifications.
Chill roll diameters vary widely. For low production- and cooling capacities chill roll diameters range from 350 to 500 mm, or up to 1200 mm for high-capacity machinery.
The impression roll (b) is exposed to severe working conditions. In order to prevent the temperature of the rubber roll cover from rising, the impression roll is water cooled from the inside, and from the outside by a contact cooling roll. The temperature is kept between 15 and 20 °C.
The contact cooling roll (c) also transfers the nip pressure, generated by pneumatic cylinders, via the impression roll against the chill roll.
Toward the edges the extruded film is always thicker, which puts some additional load on the impression roll. When entering the nip, the extruded film is always wider than the sub-

strate, and the protruding edges of the hot film make direct contact with the rubber covering of the impression roll. Various types of rubber coverings are used. Silicone rubber is dominant. In order to improve the service life of these rubber roll covers, running ®Teflonized belts are used or ®Teflonized adhesive tape is put around the edges of the impression roll.

Dr. Ing. *G. Schenkel* [7] describes an ideal solution for this problem; however, it is seldom used because the web width changes frequently. The contraction of the extruded film between the die and the coating nip presents a further problem [8].

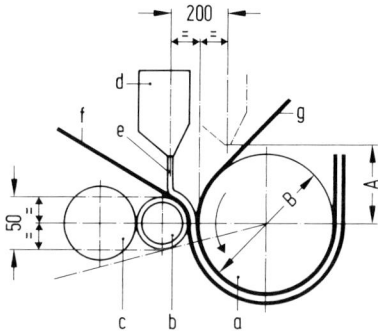

Figure 5. The extrusion coating principle
a Chill roll, *b* impression roll, *c* contact cooling roll, *d* die, *e* extruded film, *f* carrier web, *g* laminating web, A: 130 to 430 mm, B: 350 to 1200 mm

Figure 6. Extrusion coating and laminating unit. (Photo: Barmag, Remscheid/West Germany)

11.6.2 Extrusion coating and laminating machinery

In its early days the extrusion coating technique was mainly used on paper.
Further process refinement and the combination with film- and foil-converting machinery enormously widened the range of products that can be manufactured on modern extrusion coating and laminating machines.
The most basic extrusion coating and laminating machine, shown in Figures 7 and 8, comprises the following components:

a unwind for the carrier web,
b extruder with die,
c chill roll,
d impression roll,
e rider roll,
f edge trimmer,
g pull roll assembly,
h thickness gage,
i unwind for the laminating web,
k rewind,
l web tension control.

The types of unwinds and rewinds, their web tension ranges, and the controls have to be specified in relation to the products to be manufactured on such extrusion coating and laminating machines.

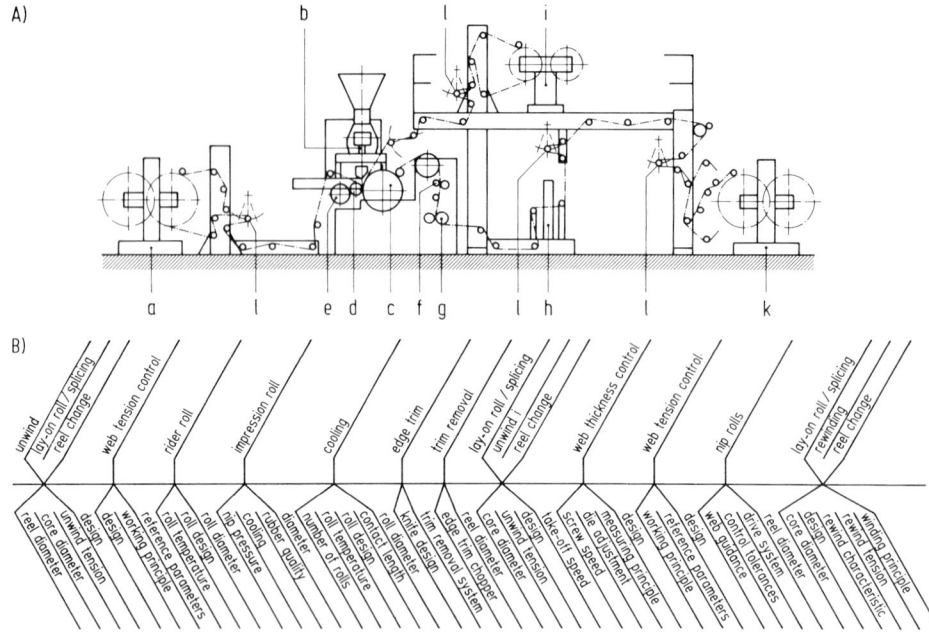

Figure 7. Extrusion coating and laminating machine
a unwind for the carrier web, *b* extruder with die, *c* chill roll, *d* impression roll, *e* rider roll, *f* edge trimmer, *g* pull-roll assembly, *h* thickness gage, *i* unwind for the laminating web, *k* rewind, *l* web tension control

Product example:

Unwind (*a*): – paper, from 30 g/m² to cardboard up to 300 g/m² or
– aluminum foil, 20 to 60 µm or
– prefabricated laminates, e.g. glue or adhesive laminates with a total weight of up to 300 g/m².

Unwind (*i*): – aluminum foil, 8 to 20 µm or
– oriented, printed polyethylene (LLDPE), 20 to 40 µm, etc.

The carrier web from unwind (*a*) can either be coated with an extrudate or be combined with the second substrate from unwind (*i*) using the extrudate as a laminating medium.

Laminate structure: LLDPE, 20 µm
                     printed and metallized
                     LDPE extruded, 12 µm
                     paperboard, 300 g/m²

End use: multipack for beverage cans.

Using preprinted LLDPE for this laminate makes it necessary to provide an extremely sensitive web-tension measuring and control system, in order to avoid stretching of the web and mechanical distortion of the print. For specific detail see Section 11.7.

Since the extruder must run continuously, it is advantageous to install extrusion coating and laminating machines with flying-splice unwinds and rewinds, which permit automatic reel changes at full production speed.

An extrusion coater must be shut down for product change. For this purpose the extruder, with die attached, has to be laterally withdrawn from the machine. The extruder screw continues to run at a reduced stand-by speed with the extruded material falling into a catch pan. Figure 8 shows the side view of the withdrawn extruder.

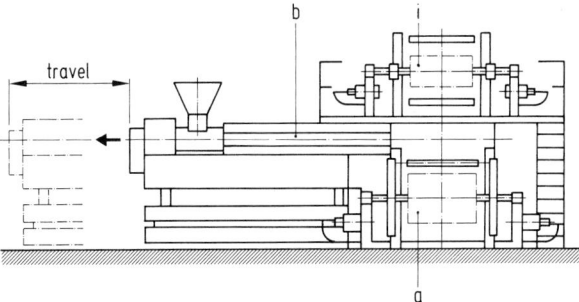

Figure 8. Side view of withdrawn extruder (*explanations see Figure 7*)

11.6.3 Coating machine with primer coating station

Figure 9 shows a coating machine with integrated primer coating (*a*) and pretreating stations (*b*) [9]. Very often the web width of the final product does not correspond to the coating width, and to save one extra production step (slitting and rewinding), the web can be slit inline (*c*) prior to being rewound. The primer coating station can be bypassed on the underside, i.e. the substrate can be threaded from the tension-controlled unwind to the corona discharge pretreating station and from there direct to the coating unit.

The primer is usually applied by the gravure coating method; multiroll systems are seldom used. The coating machine described is used to extrusion-coat paper, light paperboard and aluminum foil with PE or other thermoplastic materials.

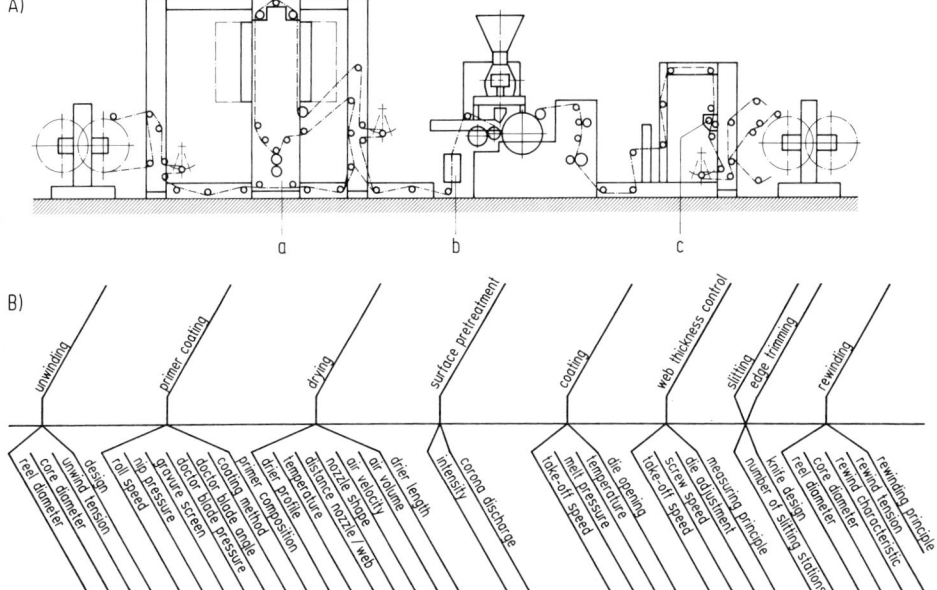

Figure 9. Extrusion coating machine with integrated primer coating station

End uses for such materials are:

- PE-coated folding carton for the packaging of deep-frozen fish and vegetables;
- decorated shopping bags with a PE coating on the outside for attractive promotional purposes;
- PE-coated paper (inner surface) for the packaging of hygroscopic products;
- aluminum foil with double-sided PE coating (2 passes) for the cable industry.

11.6.4 Coating and laminating machine with primer station for the laminating substrate

Figure 10 shows an extrusion laminating machine with corona discharge pretreatment on the carrier web, and solvent-based primer coating on the second (laminating) substrates, on which the following webs can be coated or laminated:

- aluminum foil   9 to 40 µm
- cellophane   30 to 80 g/m²
- polyester film   12 to 23 µm
- paper   30 to 150 g/m²
- oriented polypropylene 12 to 30 µm

Multiple-pass production is possible, e.g. laminating in the first pass and coating or again laminating in a second pass. This enables the manufacture of a wide variety of laminated structures.

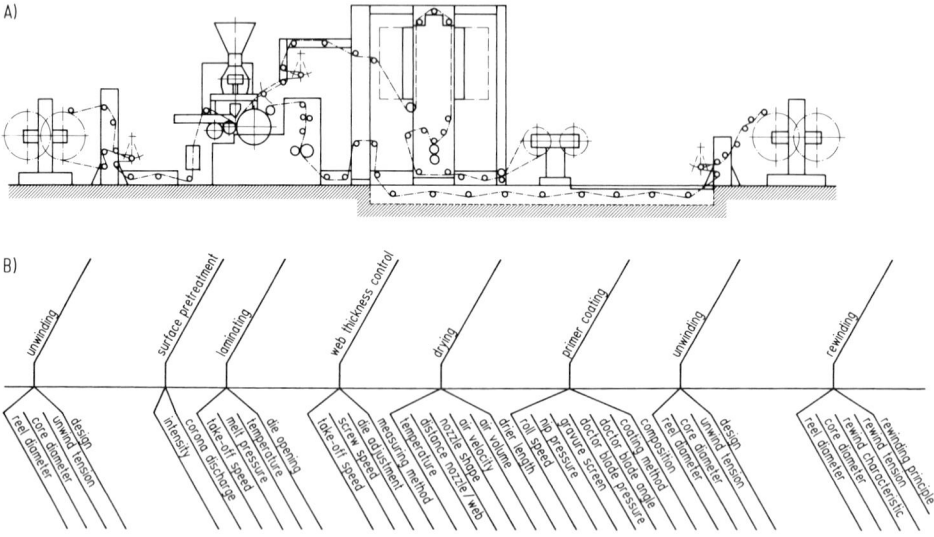

Figure 10. Extrusion coating and laminating machine with integrated primer coating station for the laminating web

11.6.5 Extrusion coating and laminating machine with primer coating and overlacquering station

The machine as shown in Figure 11 comprises the following main components:

a main unwind,
b primer coating station, which can also be converted into a one-color rotogravure or flexo-printing station,
c coating and laminating extruder,
d unwind station for the laminating web,
e web turner-bar unit,
f overlacquering station, which could be converted into a coating station,
g rewind.

The following substrates can be converted on this machine:

− aluminum foil,
− paper and light paperboard,
− cellophane,
− polypropylene,
− polyamide,
− polyester,
− laminates of above-mentioned basic substrates within the given total thickness range.

So as to be able to apply an overlacquer or coating to either side of the web, a turner-bar unit has been incorporated, which will turn the web 180° whenever necessary. The coating weights applicable are as follows:

− primer coating   0.5 to 1.5 g/m^2 dry weight,
− overlacquering   2.0 to 6.0 g/m^2 dry weight.

11.6.6 Tandem extrusion coating and laminating machine

Figure 12 shows a tandem extrusion coating and laminating line with primer coating possibility for the carrier web.
The machine comprises the following components:

a unwind,
b primer coating station,
c coating and laminating extruder I,
d unwind for the laminating substrate,
e thickness gage I,
f turner-bar unit,
g edge-guide control,
h coating extruder II,
i thickness gage II,
k rewind.

The machine comprises two extruders. The first extruder can be used for extrusion coating or laminating, whereas the second extruder can, in this application, only be used for extrusion coating. The turner-bar unit has been installed in front of extruder II in order to allow for extrusion coating onto either side of the web. Such installations are very useful for the vast variety of possible paper, film and foil combinations.

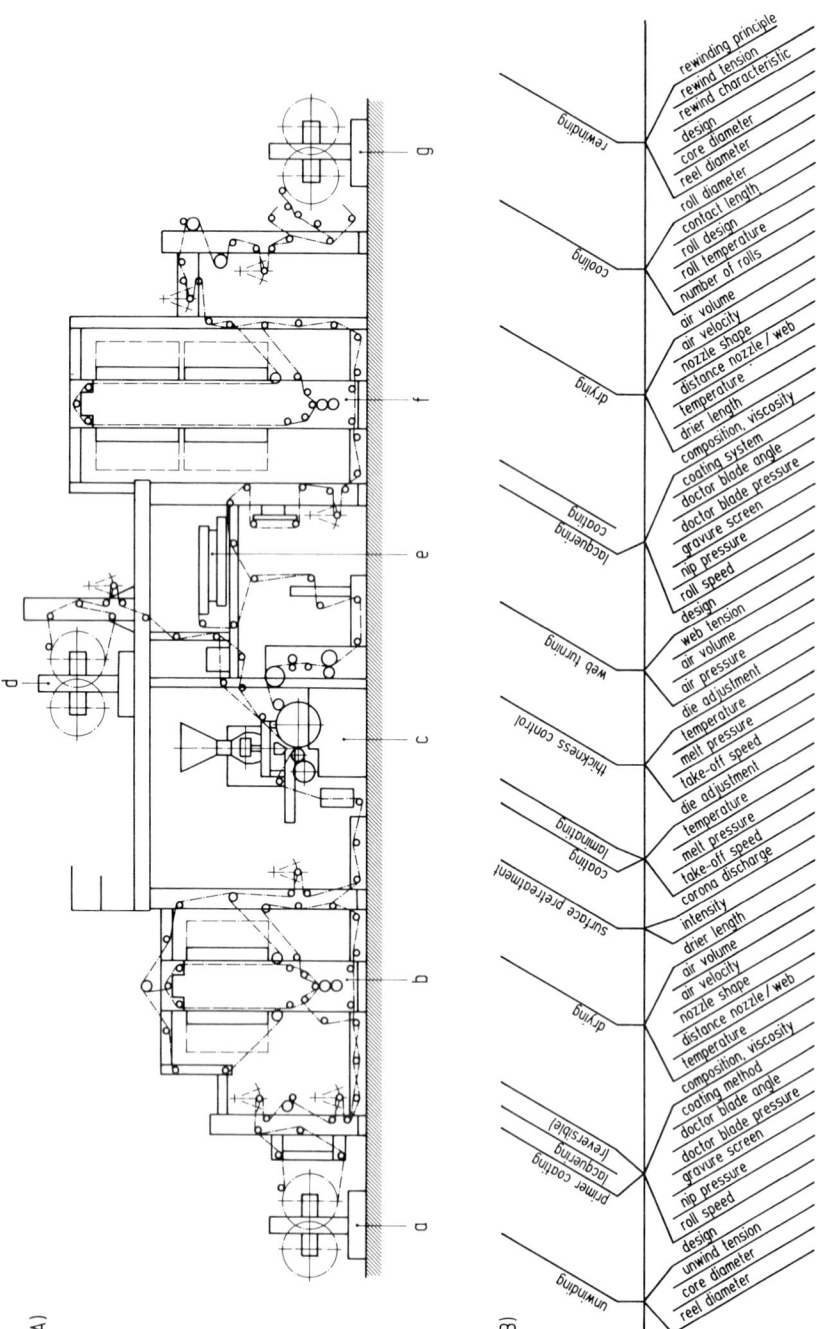

Figure 11. Extrusion coating and laminating machine with primer coating and lacquering stations. *a* main unwind, *b* primer coating station, which can also be converted into a one color rotogravure or flexo-printing station, *c* coating and laminating extruder, *d* unwind station for the laminating web, *e* web turner-bar unit, *f* overlacquering station, which could be converted into a coating station, *g* rewind

11.6 Machine layout criteria

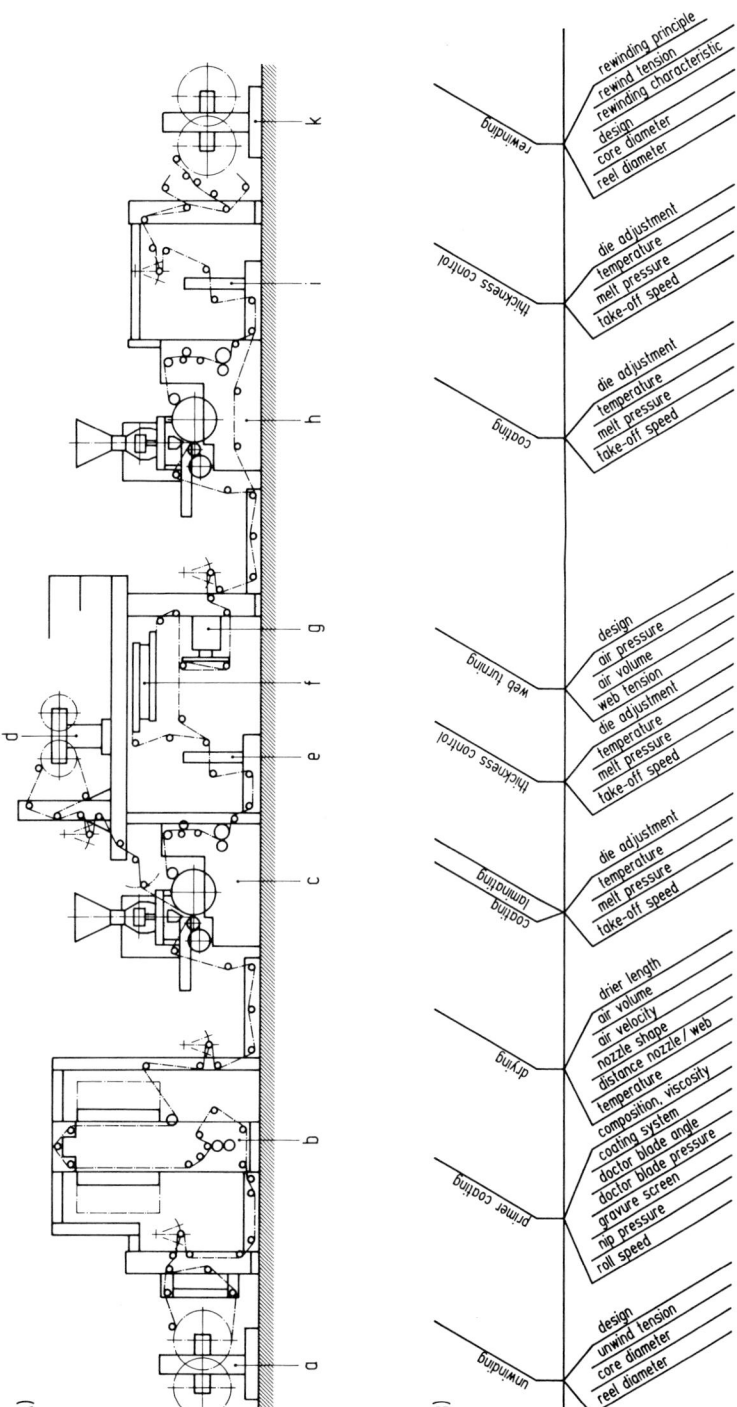

Figure 12. Tandem extrusion coating and laminating machine
a unwind, *b* primer coating station, *c* coating and laminating extruder No. 1, *d* unwind for the laminating substrate, *e* thickness gage No. 1, *f* turner-bar unit, *g* edge-guide control, *h* coating extruder No. 2, *i* thickness gage No. 2, *k* rewind

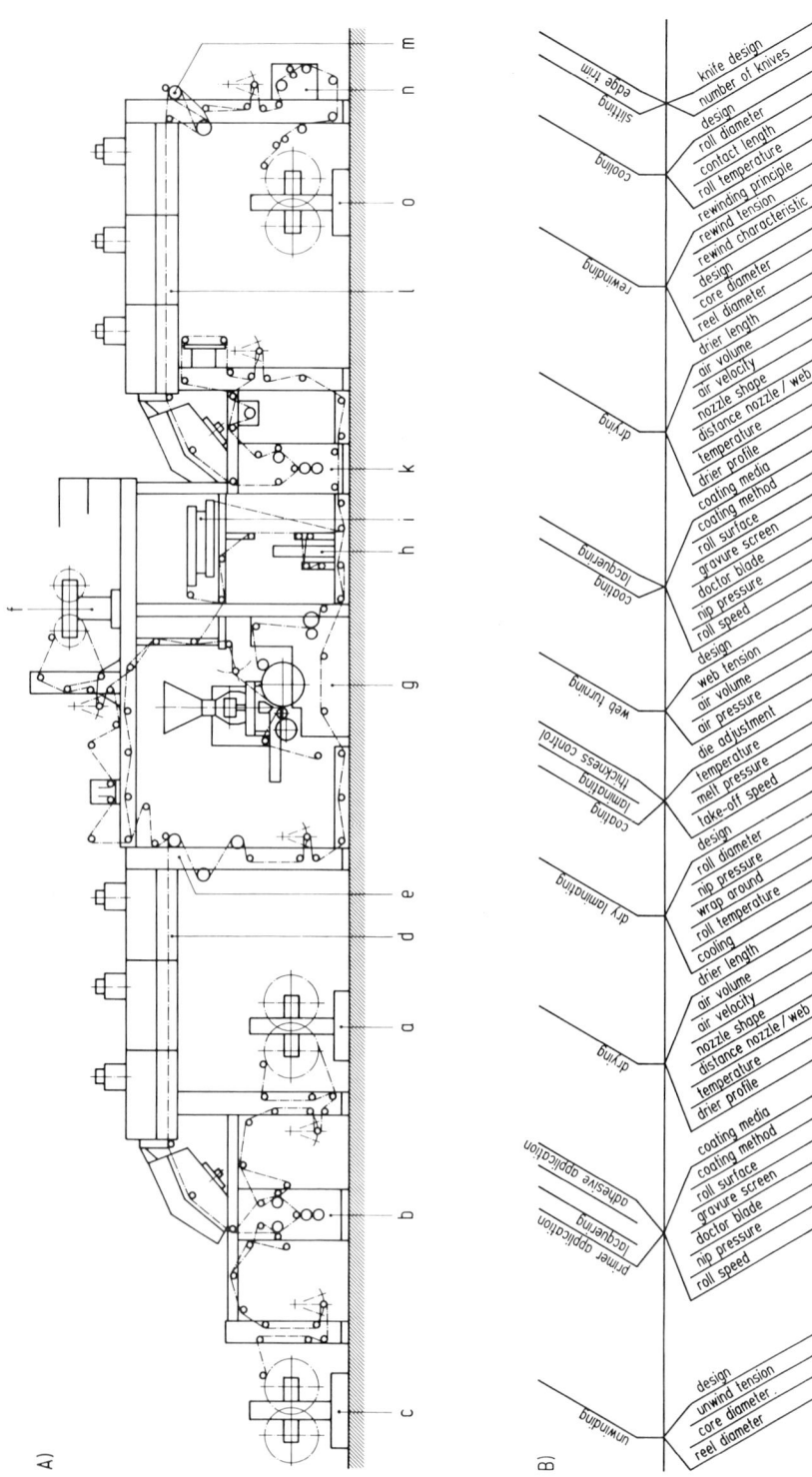

Figure 13. Multipurpose extrusion coating and laminating machine
a main unwind, *b* multipurpose coating station, *c* unwind for wet laminating substrate, *d* drying tunnel, *e* dry-laminating station, *f* unwind for extrusion or dry lamination, *g* coating or laminating extruder, *h* thickness gage, *i* turner-bar unit, *k* multipurpose coating station, *l* drying tunnel, *m* cooling cylinders, *n* edge-trimming and slitting station, *o* rewind

11.6.7 Multipurpose extrusion coating and laminating machine

Figure 13 shows a multipurpose extrusion coater and laminator, which comprises the following components:

a main unwind,
b multipurpose coating station,
c unwind for wet-laminating substrate,
d drying tunnel,
e dry-laminating station,
f unwind for extrusion or dry lamination,
g coating or laminating extruder,
h thickness gage,
i turner-bar unit,
k multipurpose coating station,
l drying tunnel,
m cooling cylinders,
n edge-trimming and slitting station,
o rewind.

With the addition of certain equipment on the extruder and an air knife at the chill rolls, it is possible to use this machine to produce cast film.
This cast film is threaded to the dry-laminating station (*e*) where it can be laminated with a substrate which has been coated with an adhesive at the coating station (*b*).

Example 1: polyester, 12 µm
 print (on back side)
 adhesive
 LDPE cast film, pigmented white

This laminate is used for the packaging of milk powder.

Example 2: overlacquer, 1 g/m^2
 rotogravure or flexo printing
 base lacquer, 1.5 g/m^2
 aluminum foil, 8 µm
 glue, 2.5 g/m^2
 paper, 40 g/m^2
 polyethylene, 20 g/m^2

This laminate is used for the packaging of biscuits, and it is manufactured as follows:

1. The lacquered and printed aluminum base foil is unwound at *a* and is coated with glue in coating station *b*.
2. The paper is unwound at *c*, wet-laminated with the aluminum foil after the coating station *b* and the laminate dried in the drying tunnel *d*.
3. The paper side of the laminate is then extrusion coated with PE.
4. After cooling, the laminate is trimmed, and passes the thickness gage *g* and the turner-bar unit *i*. It then goes to the coating station *k* where the aluminum side is overlacquered; the lacquer is dried in the drying tunnel *l* and cooled on the cooling cylinders *m*.
5. The final laminate is trimmed and slit into individual web widths at station *n*, and finally rewound on the rewind *o*.
6. Since flying-splice reel-changing equipment is installed, the machine need not be stopped for reel changes.

11.6.8 Tandem lacquering and laminating machine with extrusion coating- and laminating facility

Figures 14 and 15 show a combination of a lacquering and laminating machine with an extrusion coating machine. The two machines can be operated either separately for optimized scheduling and utilization of the individual machines, or in-line together as one machine.

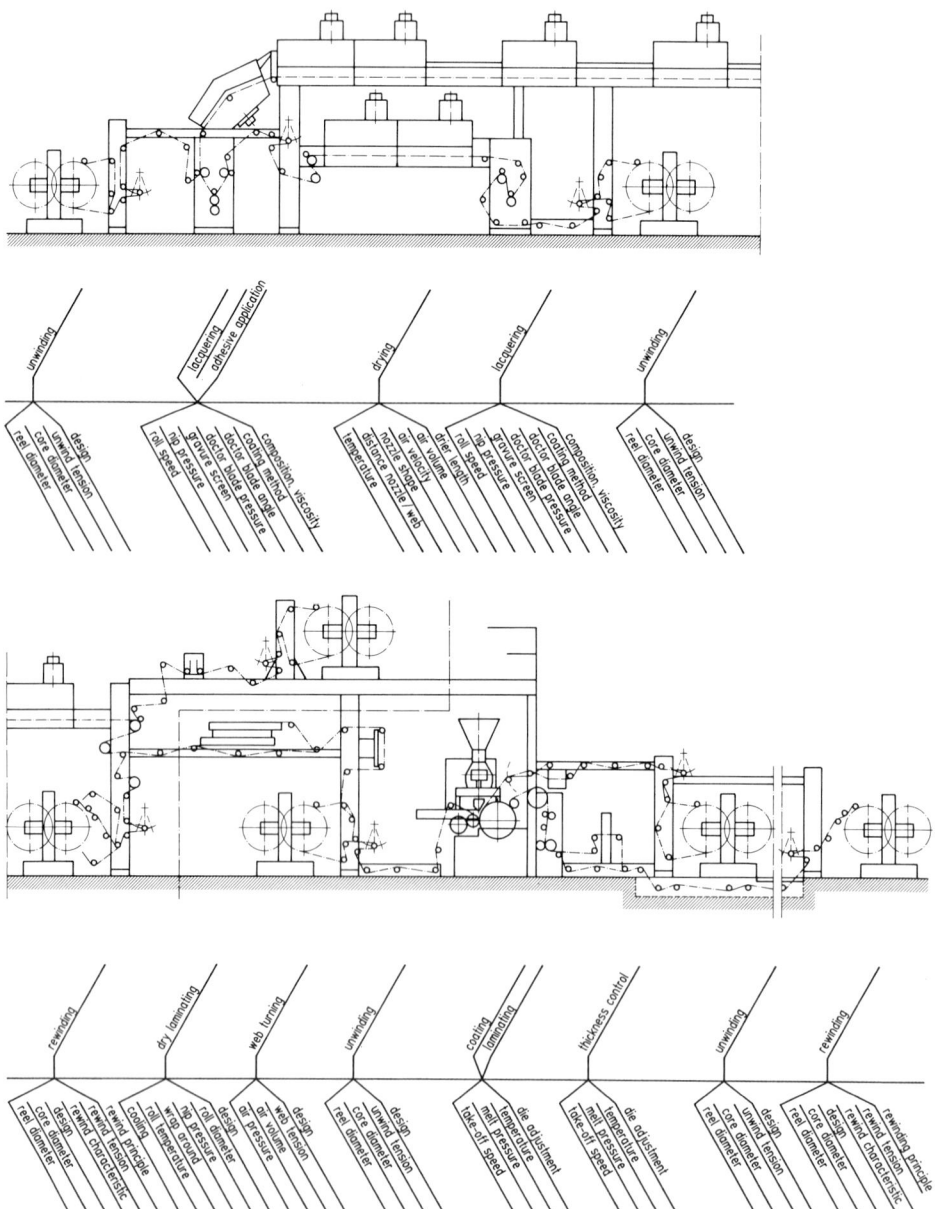

Figure 14. Tandem lacquering and laminating machine with extrusion coating and laminating facility

11.6 Machine layout criteria

Figure 15. Extrusion coating and laminating machine. (Photo: Polytype, Fribourg, Switzerland)

The lacquering and laminating machine on the left is used for lacquering, adhesive laminating, or dispersion coating of paper, plastic films and aluminum foil. Depending on the product, the machine uses two or three unwinds and one rewind. The extrusion coating and laminating machine on the right can be operated separately, or in-line with the other machine. This part of the complete line is equipped with two unwinds and one rewind so that it can be run separately. A turner-bar unit between the two machines enables either side of the substrate from machine 1 to be extrusion coated in machine 2.

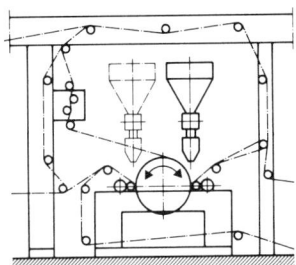

Figure 16. Extrusion laminating unit with impression and support rolls on either side for reversible chill-roll drive applications

The turner-bar unit can be eliminated if the extruder can be moved to extrude on either side of the chill roll assembly, necessitating a set of impression and support rolls on either side, and a reversible drive for the chill roll assembly (see Figure 16).

11.6.8.1 Product examples – lacquering/laminating machine

Example 1: lacquer, 2 g/m^2
aluminum, 10 μm
glue, 2.5 g/m^2
paper, 90 g/m^2

Example 2: lacquer, 2 g/m²
aluminum, 12 µm
adhesive, 3 g/m²
printing
polyester, 12 µm

11.6.8.2 Product examples – extrusion-laminating machine

Example 1: LDPE, 15 g/m²
paper, 150 g/m²

Example 2: paper/board, 60 to 200 g/m²
PE, 20 g/m²
aluminum, 9 to 15 µm

11.6.8.3 Product examples – tandem operation of both machines

Example 1: lacquer, 1 g/m²
aluminum, 9 µm
glue, 2.5 g/m²
paper, 70 g/m²
PE, 25 g/m²

Example 2: PE, 25 g/m²
primer, 1 g/m²
aluminum, 9 µm
glue, 2.5 g/m²
coated paper, 70 g/m²

Example 3: polyester, 12 µm
rotogravure print
adhesive, 3 g/m²
aluminum, 9 µm
PE, 50 g/m²

The different uses of these machines illustrate the versatility of extrusion coating and laminating machinery layouts. Modular machine design makes it possible to adapt converting machines to ever varying product needs.

11.7 Machine components

This section covers the most important components of an extrusion coating and laminating machine.

11.7.1 Unwinds

Almost all extrusion coating and laminating machines are, for technical and economic reasons, equipped with flying-splice unwinds. This type of unwind permits flying reel changes for all substrates at full production speed.

Figure 17 shows a flying-splice unwind which comprises the following elements:

a turret,
b reel carrier 1,
c reel carrier 2,
d central shaft and pivoting point for the reel carriers,
e levers with splicing lay-on roll and splicing knife,
f tachometer roll,
g reel loading device.

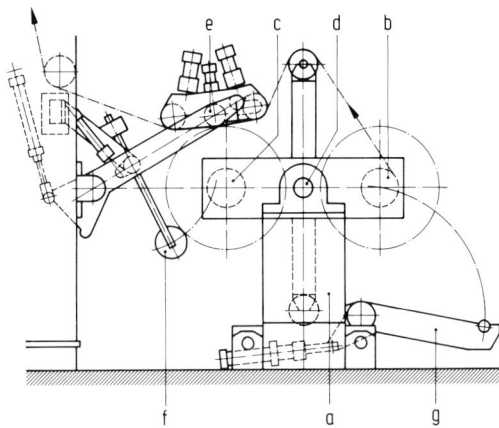

Figure 17. Flying-splice unwind

A typical flying-splice reel change can be broken down into 8 phases as shown in Figure 18.

A) Machine start-up with the unwinding reel in carrier 1.
B) A new reel is loaded into carrier 2 while the machine is running.
C) The lay-on roll is withdrawn.
D) 90° rotation of the turret.
E) The lever with the lay-on roll and the splicing knife is swung into stand-by position and held at a distance of approx. 10 mm from the new reel.
F) At the same time the tachometer roll is put onto the reel and the circumferential speed of the new reel is matched with the speed of the old reel. Splicing.
G) The tachometer roll is lifted off the reel.
H) The core of the old reel is taken out of the reel carrier.

The unwinding substrate must be fed into the machine at a certain web tension.
The amount of web tension depends on the substrate itself and is built up by unwind brakes such as

– drum brakes,
– disk brakes,
– magnetic powder brakes,
– regenerative electric motors.

In order to maintain the web tension constant as the reel diameter diminishes, it is necessary to alter the braking power accordingly. This can be achieved by a control system:

– sensing the roll diameter,
– sensing the web tension by a static-force transducer,
– sensing the web tension with a dancer roll.

Most machines are equipped with regenerative drives, since a wide variety of thin, thick, extensible and nonextensible substrates must be handled on the same unwind. Regenerative drive systems are based on 4-quadrant controls. These permit the driving or braking of the motors, which becomes a necessity as soon as extremely thin substrates like aluminum foil, 6 to 7 μm, or thin extensible plastic films must be handled.

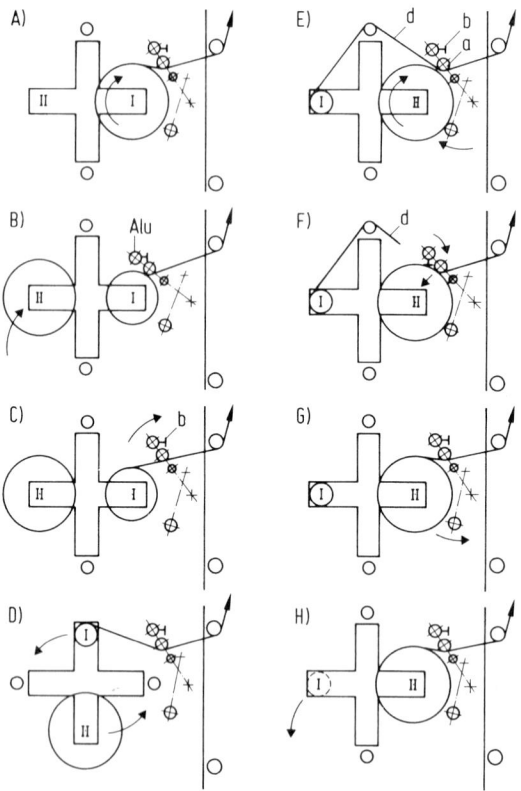

Figure 18. Working sequence of a flying-splice unwind
A) Machine start-up with the unwinding reel in bearing I. *B)* A new reel is loaded into bearing 2 while the machine is running. *C)* The lay-on roll is withdrawn. *D)* 90° rotation of the turret. *E)* The lever with the lay-on roll and the splicing knife is swung into stand-by position and held at a distance of approx. 10 mm from the new reel. At the same time the tachometer roll is put onto the reel and the circumferential speed of the new reel is matched with the speed of the old reel. *F)* Splicing. *G)* The tachometer roll is lifted off the reel. *H)* The core of the old reel is removed

Typical unwind web tensions

Aluminum foil   6 to 8 N/mm² cross-section
Polyester film   5 to 7 N/mm² cross-section
Polypropylene film 4 to 6 N/mm² cross-section
Polyethylene film 2 to 3.5 N/mm² cross-section
Paper   3.5 to 7 N × G
                   (G = weight in g/m²)

Winding cores are usually made of board, steel or aluminum.

11.7.2 Rewind

Figure 19 shows a rewind with a flying-splice attachment similar to the one used on the unwinds; this permits automatic reel changes at full production speed. Rewinds comprise the following elements:

a turret,
b reel carrier 1,
c reel carrier 2,
d central shaft and pivoting point for the reel carrier,
e levers with splicing lay-on roll and splicing knife,
f lay-on roll,
g secondary lay-on roll (during turret rotation),
h reel unloading device.

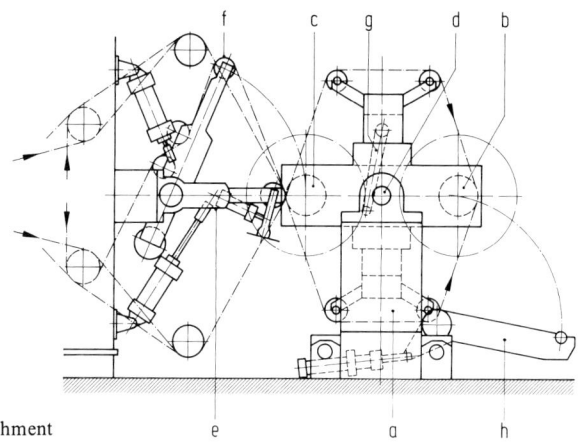

Figure 19. Rewind with flying-splice attachment

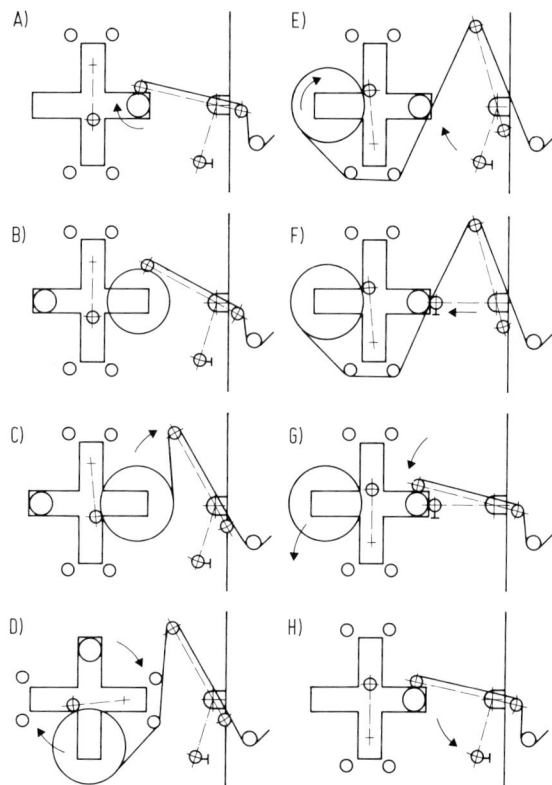

Figure 20. Working sequence of a flying-splice rewind.
A) Machine start-up with lay-on roll on. *B)* The lay-on roll follows the increasing reel diameter with constant pressure. *C)* The secondary lay-on roll is swung in and the lay-on roll is lifted off. *D)* 90° rotation of the turret. *E)* The levers with the splicing lay-on roll and the splicing knife are swung into stand-by position as the new core starts rotating, and its circumferential speed is matched with that of the reel. *F)* The web is spliced to the new core and cut off the rewound roll. *G)* The lay-on roll is put on the reel and the secondary lay-on roll is withdrawn. *H)* Withdrawal of the splicing unit

The various phases of a flying-splice reel change on a rewind are shown in Figure 20.

A) Machine start-up using lay-on roll.
B) The lay-on roll follows the increasing reel diameter with constant pressure.
C) The secondary lay-on roll is swung in and the lay-on roll is lifted off.
D) 90° rotation of the turret.
E) The levers with the splicing lay-on roll and the splicing knife are swung into stand-by position as the new core starts rotating, and its circumferential speed is matched with that of the reel.
F) The web is spliced to the new core and cut off the rewound roll.
G) The lay-on roll is put on the reel and the secondary lay-on roll is withdrawn.
H) Withdrawal of the splicing unit.

Rewinds are driven by DC motors. Their tension controls incorporate an automatic taper which decreases the web tension as the diameter of the reel increases. This is especially important for the rewinding of high-slip substrates. Most rewind controls have a fixed taper, e.g. 20%. In addition, and depending on the substrate to be handled, the rewind can be preset to reach a minimum of 7% of the original web tension.

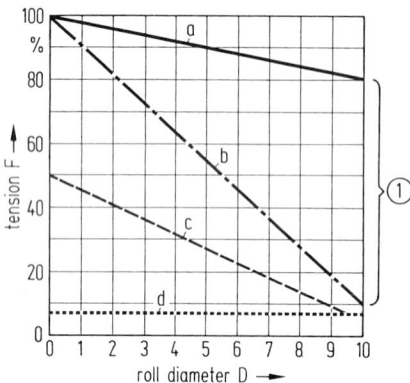

Figure 21. Winding tension control and taper
a upper limit line by setting 100% starting tension and 0% decrease of web tension, b lower limit line by setting 100% decrease of web tension, c slope for 50% original tension and 100% decrease of web tension, minimum tension is not reached, d absolute minimum tension, e any values may be set. For example: reel $d = 100$ mm (min. $\varnothing$)
splice $D = 1000$ mm (max. $\varnothing$)

The graph in Figure 21 shows the maximum range of tension-taper controls. Depending on the webs, the rewind tensions are between 30 and 50 percent higher than the unwind tensions. The above-mentioned rewinds are so-called center rewinds, in which the reel is driven through the shaft. On other rewinds the reels are driven by friction on the circumference (Pope-system). Still others have combined circumference/center drive systems. The contact line between the reel and the driving roll on circumferentially driven rewinds must be as continuous as possible; this calls for concentrically running rolls and substrates with very even thickness profiles across the web.

11.7.3 Web-tension controls

The web tension can be controlled by dynamic or static control equipment. This is used to control the web tension between the various machine components, and on winding equipment. Flying-splice unwinds and rewinds are usually equipped with dancer rolls, because of their web accumulating capacity.

11.7.3.1 Dynamic control system

Figure 22 shows a dancer roll. The angular position of the pivoting shaft is sensed by a potentiometer and the signal of the potentiometer is used for tension control. The desired web tension is applied to the dancer by pneumatic cylinders. The web tension produced by the drive motor works against the air pressure on the dancer and the dancer moves into a balance position. Higher air pressure on the cylinders must be countered by higher web tension from the drive motor, to keep the dancer roll in its balance position.

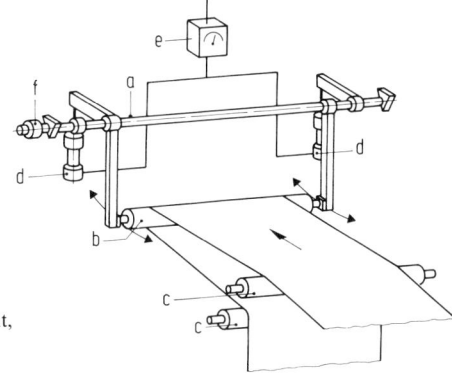

Figure 22. Dynamic tension-control system with dancer roll
a pivoting shaft, *b* dancer roll, *c* guide rolls, *d* pneumatic cylinder, *e* set-point adjuster unit, *f* potentiometer

11.7.3.2 Static control system

In Figure 23 a measuring roll is shown with an integrated static-force transducer, measuring the web tension on both sides. The web tension is measured via strain gages in these high-precision instruments, necessitating a sturdy overload protection.

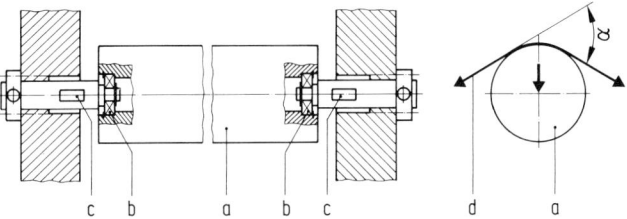

Figure 23. Static tension-control system with force transducer
a measuring roll, *b* carrier, *c* shaft with strain gage, *d* web angle α: wrap-around angle

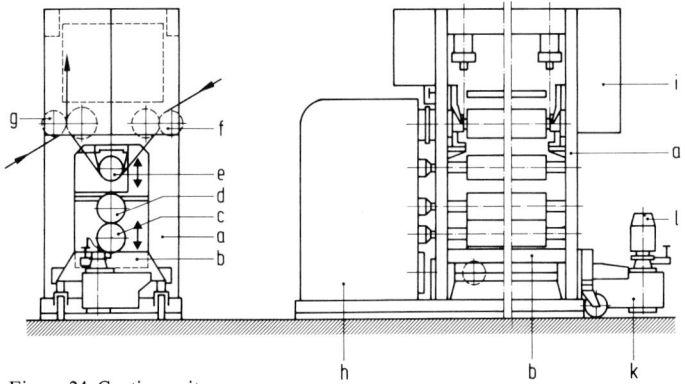

Figure 24. Coating unit
a machine frame, *b* removable trolley, *c* dip roll, *d* coating roll, *e* impression roll, *f* in-feed draw-roll nip, *g* wet-laminating station, *h* drive, *i* control panel, *k* tank for coating media, *l* circulating pump with viscosity control

11.7.4 Coating unit for primers, lacquers, glues and dispersions

Figure 24 shows a coating unit for gravure- or smooth-roll coating applications. This type of coating unit is used for solvent or water-based coating media. The most important coating methods are shown in Figure 25.

A) Smooth-roll coating, direct and reverse,
B) Fountainless gravure coating with reverse-angle doctor blade, direct and reverse,
C) Same as B) but with smoothing bar,
D) Same as B) but as kiss coating,
E) Gravure-roll coating with trailing doctor blade, pre-wetting device and dip tray.

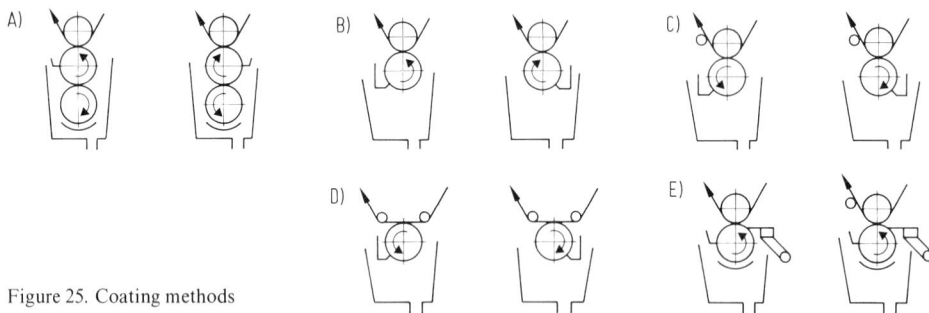

Figure 25. Coating methods

Modern coating units are equipped with removable trolleys for quick production changeover. Doctor blades [10], impression rolls, fountains [11], [12], [13] are designed for quick change too. Since coating weights must be kept within close tolerances, special media infeed- and recirculating systems with viscosity controls are provided, as shown in Figure 26. Coating units are usually equipped with DC multimotor drives which are adjustable within wide rpm ranges.

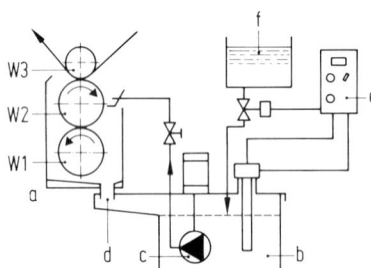

Figure 26. Lacquer circulating system with viscosity controls
a coating unit, b tank, c circulating pump, d return, e viscosity control, f solvent tank

11.7.5 Drying tunnels

In contrast to the technique with extrusion coating (plasticated melts), it is necessary to install drying tunnels for the evaporation of solvents or water from primers, lacquers or dispersions. Convection driers are the most widely used for solvents, and impingement driers for water. The drying air is blown at high velocity at a right angle against the web surface to be dried, resulting in high energy- and mass transfers [14]. A number of individual drier sections are usually assembled to form a drying tunnel, as shown in Figures 27 and 28. One of the drier sections in Figure 29 is shown with the drier lid opened for maintenance or inspection. The aerodynamics of a well designed drying tunnel must ensure a careful and

11.7 Machine components

even drying process. The drying intensity must be precisely adjustable to suit the substrate to be dried. Figure 30 shows the evaporation profile of a 5-section drying tunnel.

Energy consumption and environmental aspects must be considered carefully. Driers should be constructed to work in combination with solvent recovery or exhaust air combustion plants, the latter including energy recovery equipment.

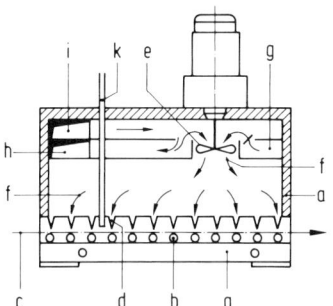

Figure 27. Convection drier

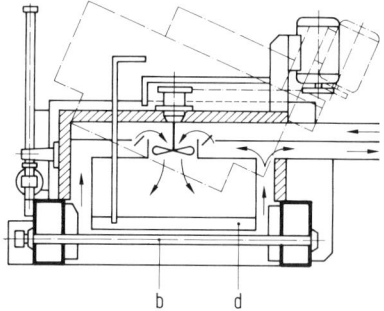

Figure 28. Convection drier, cross-section

a longitudinal section, *b* guide rolls, *c* film web, *d* nozzles, *e* ventilator, *f* drying air, *g* recirculated air, *h* exhaust air, *i* fresh air, *k* sensor for vapor alarm

Figure 29. Convection drier with lid open

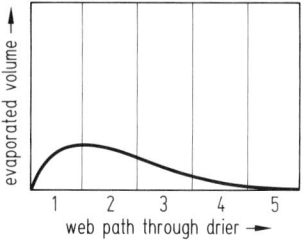

Figure 30. Drier evaporation profile

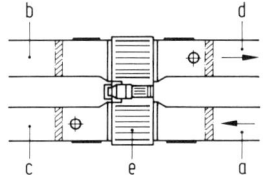

Figure 31. Rotary-wheel heat exchanger

a exhaust air (warm), *b* fresh air (cool), *c* exhaust air (cooled down), *d* fresh air (warmed up), *e* rotary-wheel heat exchanger

The exhaust air can, for this purpose, flow through rotary-wheel (Figure 31) or other types of heat exchangers in order to preheat the fresh air that is used for the drying process and/or the heating system of the building.

Working with solvent-based coating media necessitates compliance with safety regulations, such as the regulations to prevent explosions in tunnel driers [15]. Since safety regulations vary from country to country, contact with the local authorities is necessary.

Declining energy supply and environmental considerations led to the development of new drying or curing techniques, such as the curing of web coatings by exposure to accelerated electrons.

There are two types of electron accelerators in industrial use

Type 1: the scanner type,
Type 2: the linear-cathode type.

Scanner-type devices are usually used for systems with ratings of 300 kV and more, whereas the linear-cathode type is most frequently used for voltages between 150 and 300 kV. The voltage necessary to cure a coating material depends on the density and the thickness of the layer. Figure 32 shows a machine layout with a scanning beam unit [16]. The total length of the curing chamber is 1200 mm, but the window for electrons is only 25 mm wide. The electron-beam curing unit as shown in Figure 33 is used for the curing of layer thicknesses up to 500 µm.

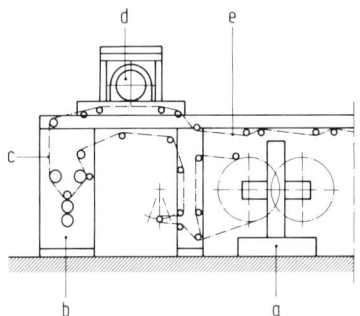

Figure 32. Machine with electron-beam drier
a unwind, *b* coating unit, *c* coated substrate, *d* electron-beam generator, *e* substrate with cured coating

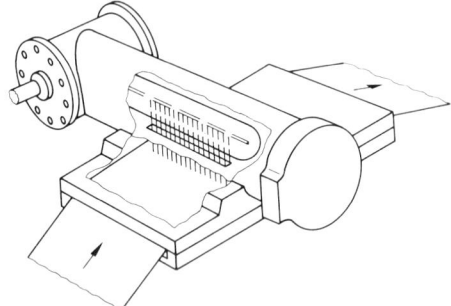

Figure 33. Electron-beam unit

Electron-beam accelerators must have integrated shielding. The radiation curing chamber must also be shielded and be equipped with safety interlocks to comply with safety regulations.

Coating media for electron-beam radiation curing contain only solids and, therefore, have no volatile components to be evaporated. The availability of EB curable coating media is still limited, although the chemical industry is seriously working on the development of a wider range of these types of materials.

The most important advantages of this curing process are:

– very low energy consumption,
– absence of solvents, which eliminates the need for solvent recovery or combustion systems, and also eliminates environmental pollution and an inherent explosion risk,
– polymerization at room temperature: the substrate temperature is only raised by a few degrees, and temperature-sensitive substrates can be handled without thermal damage,
– appreciably reduced machine area.

11.7.6 Drying and dry laminating

Extrusion coating machines are very often equipped with dry-laminating stations as shown in Figure 15. A typical dry-laminating station is shown in Figure 34. The substrate is coated with an adhesive, and combined with another substrate in a heated laminating nip. The laminating substrate can be in-line extruded cast film, or blown film. Laminating very often requires high nip pressures, and it is therefore often necessary to install deflection-com-

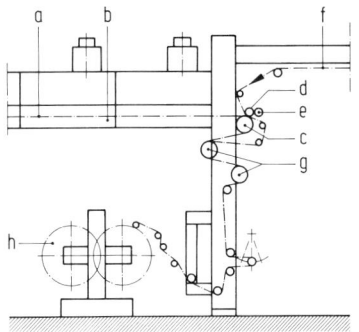

Figure 34. Dry-laminating station
a substrate with adhesive coating, *b* drier, *c* laminating cylinder, *d* impression cylinder, *e* cooling roll, *f* laminating web, *g* cooling cylinders, *h* rewind

pensating roll systems. Laminating-roll systems without deflection compensation will exert less laminating pressure on the laminate towards the middle part of the roll, as the rolls forming the nip deflect in opposite directions, as shown in Figure 35. Even nip pressure at any pressure level across the entire nip is ensured by deflection-compensating devices [17] as shown in Figures 36 and 37.

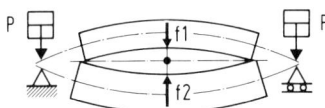

Figure 35. Roll deflection

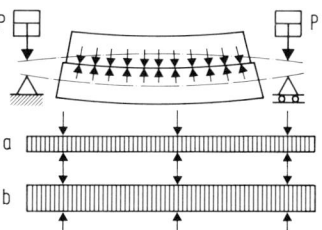

Figure 36. Deflection-compensating roll
a contact zone at low pressure, *b* contact zone at high pressure

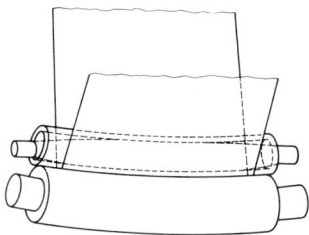

Figure 37. Laminating nip with deflection-compensating roll

11.7.7 Web turner-bars

Turner-bars are needed whenever the web has to be turned for subsequent in-line operations. Turner-bars have been known to the printing industry for decades. The equipment has now been improved and is no longer restricted to the handling of papers. Practically all types of films and foil can now be handled (see Figure 38).

11.7.8 Edge-guiding equipment

Despite the use of good web-tension control systems it is possible that a substrate will show a tendency to move sideways, especially on machines with very long web paths. Extrusion coating and laminating machines are, therefore, often equipped with automatic edge guides.
Edge guides can be installed at unwinds and rewinds, but they may also become necessary between the various machine components or converting steps, to sense the lateral position of the web and guide it back to the center line of the machine whenever needed.

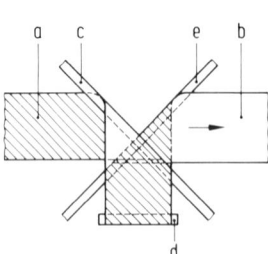
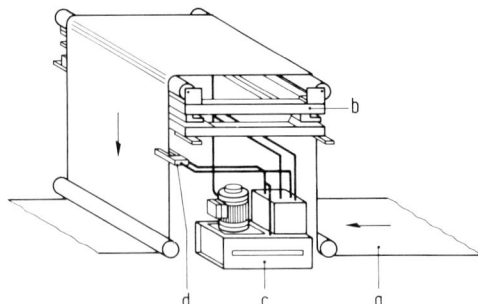

Figure 38. Turner-bar unit
a incoming web, *b* turned web, *c* turner bar *1*, *d* returning roll, *e* turner bar *2*

Figure 39. Edge guide
a incoming web, *b* deflector frame, *c* hydraulics, *d* sensor

The position of the edge is sensed by means of photoelectric or pneumatic sensors without contacting the web. The lateral web position can be sensed at both edges, or at one edge only. Double-sided edge guides shift the web to the center line of the machine. One-sided edge sensing keeps one edge of the web in the desired lateral position.

The lateral web-shifting mechanisms are controlled by the signals of the sensors. Depending on their position in the machine, camber-roll or deflector frame systems are used as web-shifting mechanisms. The shifting is actuated by hydraulics or by electric motors. Figure 39 shows a deflector frame assembly.

11.7.9 Drive and control systems

Modern high-performance production equipment calls for the installation of reliable drives and sophisticated control equipment. Precise DC motor drives and microprocessor controls ensure constant and reproducible product qualities.

The individual machine components are therefore equipped with regenerative drives with thyristor units (SCR, silicon-controlled rectifier) and DC shunt motors.

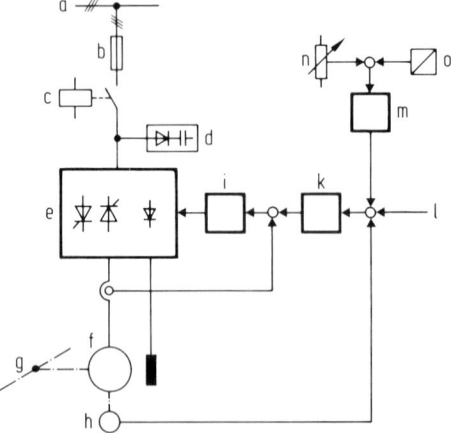

Figure 40. Schematic diagram of a tension-controlled drive system
a main power supply, *b* fuse, *c* main contactor, *d* surge suppressor, *e* SCR unit for regenerative drive, *f* DC-shunt motor, *g* mechanical drive connection, *h* tachogenerator, *i* current controller, *k* rpm controller, *l* rpm reference, *m* tension controller, *n* tension reference, *o* actual tension value

11.7 Machine components

Figure 40 is the basic schematic diagram of a web-tension controlled drive [19]. Extrusion coating and laminating machines have several similar components, equipped with programmable logic controllers, called PLCs, or microprocessors. PLCs have no hardwired relaymatic or logic control elements, but they comprise a processor with input and output interfaces, a central processing unit, a software memory and a memory for the applicator program. PLCs offer the following advantages:

- modular design,
- software instead of hard-wiring,
- changes in the program and machine extensions can easily be realized,
- simple diagnosis and testing facilities,
- wear-resistant equipment.

11.7.10 Process reference parameters

Complex machines for highly sophisticated production programs are often connected to process control systems, see Figure 41.

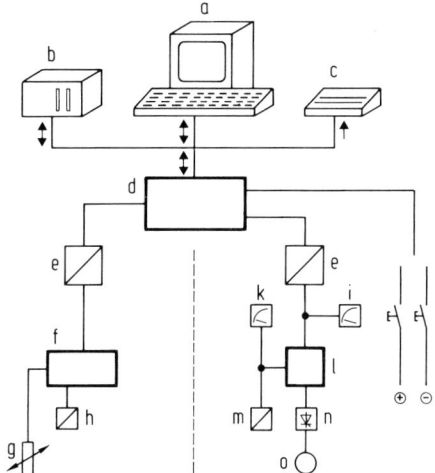

Figure 41. Process control system, e.g. temperature and tension control
a video terminal, b floppy disk station, c printer, d central unit with memory, e digital/analog converter, f temperature controller, g temperature sensor, h regulating unit, i tension reference, k actual tension value, l tension controller, m tension measuring unit, n thyristor unit, o DC drive motor

The process parameters are controlled by separate analog devices such as:

- rpm/tension controllers for drives,
- temperature controllers for dryers,
- air pressure controllers for impression cylinders.

The process control system provides only the process reference parameters.
This offers the possibility of operating the machine in manual mode, just like any machine without a process control system. The individual control loops operate autonomously and guarantee precise control, especially of drives with heavy dynamic loads. The process reference parameters must be compatible with the PLCs installed.

The following process reference parameters can be used:

- machine speed,
- web tension at various machine sections,
- nip pressures at impression rolls, which are important for the converting process,
- thicknesses of applied coatings as indicated by thickness gages [20],
- rpm of rolls at primer and lacquering stations,
- dryer and roll temperatures, and so on.

Process-data recording and the use of digital signals for the indication of faults at individual controllers, of web breaks, of faulty operation and inadequate wind-up, are additional facilities available with process control systems.

References for Chapter 11

For the convenience of the reader the English titles of all publications in languages other than English are shown in parentheses.

[1] *Kettenbach, F.:* Extrusionskaschierung und deren Bedeutung auf dem Verpackungssektor (Extrusion Laminating and its Importance in the Packaging Industry). Adhesives and Film Conversion Seminar, Fachhochschule Munich, 1979.
[2] *Prinz, E.:* Neue Erkenntnisse über die Koronabehandlung (Recent Findings about Corona Discharge Treatment). Adhesives and Film Conversion Seminar, Fachhochschule Munich, 1979.
[3] DEOS 3.045.413 (1980), Windmöller & Hölscher.
[4] *Spikmann, H. H., Wiedeyer, H. R.:* Primern von Polymerfolien (Priming of Polymer Films). Adhesives and Film Conversion Seminar, Fachhochschule Munich, 1981.
[5] *Vasko, M., Majling, J.:* Adhäsion von Polyethylen zur Alufolie beim Extrusionsauftrag (Adhesion of Aluminum to Polyethylene during Extrusion Coating). Papier + Kunststoff-Verarbeiter (1981) 10, pp. 12/18.
[6] Kaiser Aluminium & Chemical Sales, Inc. In: Kaiser Aluminium Foil, Kaiser Center, Oakland, Ca., USA, 1958.
[7] *Schenkel, G.:* Kunststoff-Extrudertechnik (Polymer Extrusion Technology). Carl Hanser Verlag, München, 1963.
[8] GB-OS 8.108.486 (1981), Du Pont, Canada Inc.
[9] DE-OS 2.752.052 (1977), K. Kalwar, Steinhagen, West Germany.
[10] CH-PS 498.720 (1970), Maschinenfabrik Winkler, Fallert & Co. AG, Bern, Suisse.
[11] SE-PS 7.714.432 (1977), Polytype AG, Fribourg, Suisse.
[12] SE-PS 7.808.278 (1978), Polytype AG, Fribourg, Suisse.
[13] CH-AS 624.526 G (1981), Polytype AG, Fribourg, Suisse.
[14] *Dosdogru, G.:* Prallstrahltrockner im Vergleich untereinander und mit anderen Trocknern (Comparison of Various Baffle-dryers and Other Types of Dryer). Deutsche Papierwirtschaft (1982) 3, pp. 200 f.: (1982) 4, pp. 174 f.; (1983) 1, pp. 78 f.
[15] Explosionsschutz an Durchlauftrocknern (Explosion Prevention on Continuous Dryers in the Paper Industry). Hauptverband der gewerblichen Berufsgenossenschaft. Oct. 1982. Carl Heymanns Verlag KG, Köln.
[16] US-PS 530.942 (1974), Energy Sciences Inc., Bedford, Mass., USA.
[17] CH-PS 456.649 (1968), Maschinenfabrik Winkler, Fallert & Co. AG Bern, Suisse.
[18] Erhardt u. Leimer KG, 8900 Augsburg, West Germany.
[19] *Flükiger, E.:* Mikroprozessorgesteuerte Beschichtungsanlagen (Microprocessor-controlled Coating Lines). Adhesives and Film Conversion Seminar, Fachhochschule Munich, 1983.
[20] *Menges, G., Bolder, G., Bergweiler, E., Breil, J., Esser, K., Ramm, H. F., Wortberg, J.:* Rechnergesteuerte Extrusions-Automatisierungskonzepte des IKV (Concepts for Computer-controlled Automation of Extrusion from the IKV). Kunststoffe (1982) 11, pp. 29/43.

12 Extrusion blow molding

H.-G. Fritz

12.1 Introduction

12.1.1 Techniques of blow molding

At every stage in the development or modification of the blow molding process in the last 50 years, the criteria and standards for a technically and economically optimum process have been repeatedly discussed; and each time extrusion blow molding has maintained an undisputed lead position. Almost 90% of worldwide annual production of plastics hollowware is manufactured by this process. All other blow molding concepts must be measured against the technical capabilities and productivity of extrusion blow molding [1].

The processes practiced nowadays can be classified according to the method of preform manufacture and/or the essential steps followed in the blow molding process itself. Thus, the various methods are: extrusion blow molding, injection blow molding, stretch blow molding and dip blow molding. Each method:

- requires the development of machinery tailored to the needs of the particular blow molding process,
- employs process engineering that is specific to the process sequence, to preform manufacture and conditioning, and to the final shaping technique,
- makes it possible to obtain a product that is tailored precisely for particular fields of application and defined by the raw material, the proportion of waste, the welding seams, the number of layers, and the product geometry.

Extrusion blow molding is economically the most important, the most widely adopted, and hence the best known process in the field of blow molding. It is the only process discussed extensively in this chapter. One reason for this dominant position is the almost unlimited variation that is possible in the geometry of the blown part: unsymmetrical articles, hollow articles with several openings and/or handles, and even containers with inserts can be manufactured. Extrusion blow molding and rotational molding are the only processes that offer this facility. The other blow molding processes mentioned can be employed only if the end product has simple geometry and, as a rule, only one opening. Even the slanted or eccentric openings of a container may create problems if a process other than extrusion blow molding is employed. There is no doubt that the efforts to reduce costs and the amount of material used, and to lower the total expense of production, while still achieving the quality demanded of the product — especially of packaging evolved for "quality products" — have been very fruitful for the continuing development of these adaptable, modern blow molding processes. The upward trend during the past 10 years in the stretch blow molding sector bears this out most strongly.

12.1.2 The historical development of extrusion blow molding

A new chapter in blow molding technology, based on long-established tradition in glass blowing, began in the last century. The starting point was the development of thermoplastic, moldable synthetic materials, or their forerunners — materials which could very easily be formed into hollow bodies at lower temperatures than those needed for glass [2].

As early as 1851, a patent application by *S. T. Armstrong* [3] described an improved method of manufacture of hollow bodies from gutta percha, in which a tubular parison was made and subsequently shaped to form a hollow end product by application of internal pressure. Towards the end of the 19th century, the first patents describing plants and processes to manufacture blow-molded hollow products from small celluloid tubes were published [4, 5]. At that time, articles destined primarily for the technical sector − the forerunners of today's technical blow-molded parts − and toys were manufactured from gutta percha, natural rubber and celluloid. It is interesting to note that various possibilities for using blow molding technology with thermoplastics were already envisaged in those patent specifications.

Between 1930 and 1945 blow molding technology received a new stimulus with the development and marketing of PE and PVC. Much effort went into producing parisons correctly designed for both the material used and the blow molding process, and into automation of the manufacturing process [2]. The following patent specifications exemplify this trend in development:

- US patent No. 2,222,461 (1938):
  Manufacture of moldings from celluloid or other moldable materials using a tubular parison. Applicant: *W. J. de Witt* and others (New York).
- US patent No. 2,288,454 (1938):
  Manufacture of a bottle neck by injection molding followed by the extrusion of a tubular parison subsequently molded using the initial heat content. Applicant: *J. B. Hobson* (Plax Corp.)
- US patent No. 2,260,750 (1938):
  Manufacture of a parison and blow molding it using the initial heat content. Applicant: *W. H. Kopitke* (Plax Corp.)
- US patent No. 2,298,716 (1942):
  Process and equipment for injection blow molding of hollow bodies. Applicant: *T. Moreland* and others (Owens, Illinois).

The techniques which would be classified today as injection and dip blow molding at first clearly led the field [1]. The industrial use of such processes began in the late 1930s, when well-known glass makers like Owens Illinois in the USA started to manufacture and market thermoplastic hollow articles for packaging. Since this American know-how was not available in Europe − and especially not in Germany − because of the Second World War, a completely isolated, self-reliant development began in Germany after the war, and proceeded along paths completely different from those followed in the USA 20 years before.

This modified technology, later called extrusion blow molding, closely resembled the technology of continuous extrusion. Tubular parisons made from various thermoplastic materials by continuous extrusion were shaped, with the help of compressed air, into hollow articles, in a single-stage process, using the initial heat content of the parison. In comparison with injection blow molding, the extrusion blow molding process, even during its initial development phase, already offered several advantages and greater freedom with respect to the geometry and size of the blow-molded product, together with a theoretical possibility of extension to making technical blow-molded parts. The adoption of this approach, not least, was the reason that US industry lagged behind Europe in plant and process development for extrusion blow molding until the end of the seventies. Extrusion blow molding machines with a number of advanced technical features (e.g. continuous tube extrusion, vertically arranged blow molds, calibration by the blow mandrel), which are still found on the most modern blow molding lines, were first built between 1950 and 1952, with the firm Kautex-Werke, Reinhold Hagen GmbH blazing the trail. For the first time impact-resistant and chemical-resistant packaging and technical hollow articles could be manufactured on a large scale with these machines. The products were eagerly accepted by the market, and the plastics raw material industry, encouraged by this response, began to develop polymer

12.1 Introduction

types suitable for blow molding, the polyolefins and rigid PVC being the plastics of the greatest interest at that time. Similarly, since the plastics processors were supported by an enthusiastic market reaction, they, too, made efforts to find new fields of application for blow-molded products. From this beginning in the early 1950s until the middle of the 1960s, most of the fundamentals on which the market and the technology still thrive today were established [2]. The aims of the intensive research and development work of the polymer producers and machinery manufacturers were to design extruder systems, tube extrusion dies and blow molds for particular materials, to improve calibration techniques, and to construct highly productive machinery for large-scale output. Moreover, the manufacture of PVC containers on extrusion blow molding lines of conventional design – despite many special developments in the two-stage process – was steadily improved and optimized, but without involving the development of high-speed Ferris-wheel type machines like that then taking place in France to meet the special market conditions there. Reviewing the past, it should be emphasized that PVC in particular, with its severe demands on process engineering, has given strong impetus to the development of extrusion blow molding, and provided many incentives to carry out thorough technical analyses of the rheological/thermodynamic phenomena in the plasticating unit and the extrusion die [2].

Because of the steadily increasing use of extrusion blow-molded hollow articles, and also harder competition in the market, awareness of quality and productivity became more and more intense in the late sixties. The result was optimization and a certain degree of specialization, both of the machinery and of the process engineering, all of which needed continuing new developments, summarized below:

- use of wall-thickness controls for producing hollow articles with a wall-thickness distribution optimized for the type of end use;
- modification to calibrating techniques for precision forming of container openings;
- mechanization of the process sequence, and automatic scrap removal from inside and outside the blow mold;
- use of multiple extrusion dies to increase the output rate of machines;
- development of modified machinery concepts for producing large containers such as drums and heating-oil tanks;
- first experiments to produce multilayer hollow articles by coextrusion of parisons.

Many different kinds of machine were thrust upon the market in the early years, designed for particular processes and products, very often to overcome patent restrictions. Only at the beginning of the 1970s was there some rationalization of machine design, at least in the case of continuous tube extrusion systems, initiated by market forces or the expiry of patents. However, many technically distinct machine designs continued to exist, particularly those with special melt-accumulator and die designs for blow molding large containers. Of the many different concepts tested, the annular-piston accumulator head has now been adopted as standard equipment for blow molding machines of all sizes.

High-molecular-weight polyethylene (HMW-PE) in the form of powder or pellets was introduced into the market at the end of the sixties, and although additional costly development effort was required, this material opened up new areas of application for large extrusion blow-molded containers and technical parts.

The processing technologies and plant developed and introduced then specially to process HMW-PE have been established as the state of the art today. The key points of these innovations can be summarized as follows:

- introduction of single-screw extruders equipped with a grooved section for forced feeding of powders; and shearing and mixing elements on the screw for melt homogenization;
- optimization of the flow geometry in extrusion dies;
- introduction of sensitive, rapid-response wall-thickness control systems, with integrated tube-length control;

– development and production of precision-made large blow-molded articles such as heating-oil tanks, diesel-fuel tanks, and canisters and drums for transporting dangerous chemicals.

Extensive research and development work carried out in the seventies, equally by the raw material producers, the machine manufacturers and the processing industry, lay behind these developments, which are merely the salient features of extrusion blow molding development. Since 1980, an important focus of development has remained the open- and closed-loop control of extrusion blow molding plants. The enormous increase in the performance of microprocessors has made it possible to replace entire control systems, including the individual control components, with a system of hierarchically interconnected microprocessors. This has been achieved without overall change in costs. Thus a fully controlled, self-monitoring extrusion blow molding plant should be possible soon.

12.1.3 The economic importance of extrusion blow molding

Blow molding, more than 90% of which is extrusion blow molding, has become one of the most important plastics processes, next to injection molding and extrusion. This state of affairs applies as much to the manufacture of blow molding machinery as to the production of the hollow products themselves.

Figure 1 shows that in West Germany alone blow molding equipment to a value of 243 million DM was sold in 1983 [6]. More than 70% of this plant was exported, despite the very large gap between the high costs of German industry and those of manufacturers in weak-currency countries. This could be interpreted as an indication that German machinery manufacturers still have a substantial technological advantage in the field of blow molding, one that has brought them unchallenged worldwide success in the market for almost three decades.

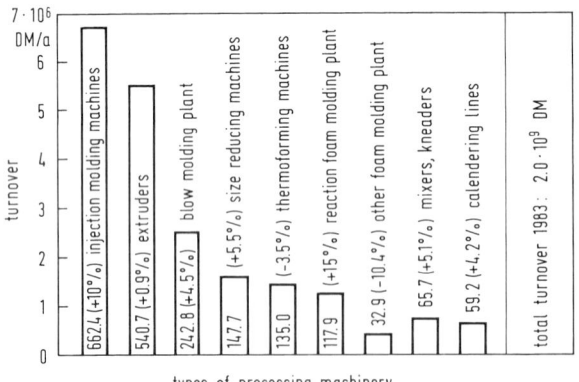

Figure 1. 1983 sales of plastics-processing machines (million DM); values in parentheses: change compared with 1982

In this context, it is also interesting to note how the individual market shares (in millions of DM sales) held by injection molding machines, extrusion lines and blow molding machines, changed between 1970 and 1982 (Figure 2). Whereas the percentage share held by extrusion equipment has remained almost constant over the last 15 years, the injection molding machine sector had to bear a drastic setback when economic recession following the two oil crises in 1973 and 1979 hit profits badly. In contrast to the molding technologies that do not use a preforming stage, the market share, in money terms, held by blow mold-

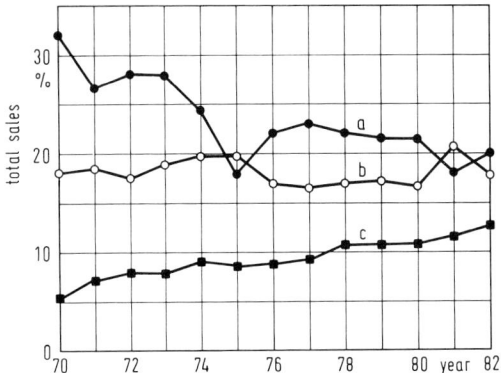

Figure 2. Percentage shares, by process, in sales value of machines
a injection molding machines, *b* extrusion plant, *c* blow molding machines and thermoforming equipment

ing plant has steadily, if not remarkably, increased since 1970. This phenomenon can be attributed to continuous expansion of manufacturers' product ranges and to further development and modifications of the process. The stretch blow molding process, which is not discussed further here, is a widely acknowledged example of this trend (Figure 3). Increased precision of blow-molded parts manufacture has led, in some cases, to the adoption of blow molding in place of injection molding.

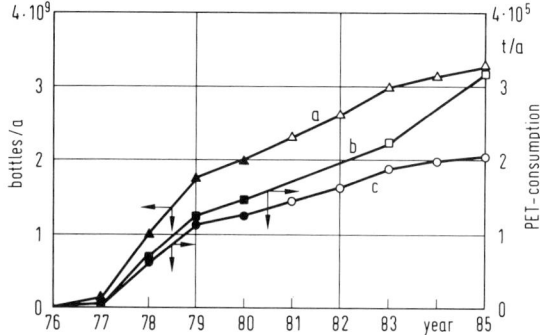

Figure 3. Growth to 1980 and market projections for stretch blow-molded PET bottles
a PET bottles in USA, *b* material consumption for PET bottles, USA, *c* world usage of PET for bottles

12.2 Polymers for extrusion blow molding

The importance, properties and fields of application of the thermoplastics most commonly used in extrusion blow molding are briefly described in this section. They are classified according to the chain reaction by which they are synthesized — i.e. polymers, polycondensates or polyadducts.

In 1980, blow-molded commodity plastics such as PE, PVC, PP and PS/PB accounted for some 12 to 13% of the total consumption of plastics. The clear lead position was held by PE, with almost 70% of the blow molding sector, followed at a long distance by PP (9% share), PVC (8%) and PS/PB (2%). Except for PET, which has attained importance since the development of stretch-blow molding, all remaining thermoplastics lag far behind, having only minute shares of the market.

Polyethylene

Being odorless and tasteless, physiologically non-hazardous and having good water-barrier properties, PE is an ideal material for the blow molding of food packaging.
Squeeze bottles and liners for multicomponent packages are frequently made of LDPE. On the contrary, HDPE is a preferred material for all types of canisters and drums, since it is distinguished by higher rigidity and melt viscosity, and high chemical and stress-crack resistance. With ethylene copolymers it is possible to manufacture storage tanks of 600 to 10,000 liters capacity for fuel oil and other materials. The development of the plastics gasoline tank for automobiles has opened up a new field, which was expected to consume about 45,000 metric tons in 1986 [6]. Compounds filled with active carbon black can be used to blow mold containers with a surface resistance below 10 MΩ for safe transport of fuels.

Polypropylene

A wide range of propylene homopolymers, block copolymers and random copolymers are used for blow molding today. PP bottles and PP cans are quite suitable for the packaging of hot-filled juices, syrups and sauces, and also for oils, cosmetics and pharmaceuticals [7]. On a larger scale, polypropylene is specified for making blow-molded engineering products used in several industrial sectors: for example, the automobile industry (air-ducting, containers for brake fluid and windshield-wiper fluid, overflow tanks for cooling systems, elbows for air filters etc.), and the household-appliances sector (parts for washing machines, dishwashers and coffee-makers, components for clothes driers, and tanks for electric water heaters); also for use in laboratories, hospitals, and plumbing generally.

Polyvinylchloride

Compounds consisting mainly of rigid PVC with a K-value of between 57 and 60, and other components such as an impact modifier (8 to 12% of an ABS polymer or a methylmethacrylate/butadiene copolymer) and a stabilizer (either tin stearate or calcium stearate), are used for making glass-clear bottles (0.1 to 2 l) and canisters (2 to 5 l). Polyvinylchloride was at one time discredited for this use because of its VC-monomer content, but in recent years – thanks to intensive research and development work done on polymer mass production, compounding and processing – PVC has become a physiologically unobjectionable material (Figure 4). Even so, considerable long-term efforts will be necessary to overcome the deep-rooted aversion in a number of countries to the use of this polymer for food packaging [8].

Graft copolymers and other styrene copolymers

Today, pure PS is only used for a few products like small hollow articles for cosmetics packaging. Graft copolymers of polybutadiene and ABS are tough and impact-resistant, and suitable for blow molding furniture components and even small pieces of furniture.

Polycarbonate

Special features of this material are high transparency, impact toughness, acceptance of sterilization, and scratch resistance. Babies' bottles, lighting covers, drinking-water tanks (volume ranging from 15 to 25 l), and even double-walled computer-cabinet doors are blow-molded in high-viscosity polycarbonate.

Other blow-moldable thermoplastics for special products

Polyacetal (or polyoxymethylene, POM) and its copolymers are highly regarded, because of their impermeability to gases and their high mechanical strength, and can be used to

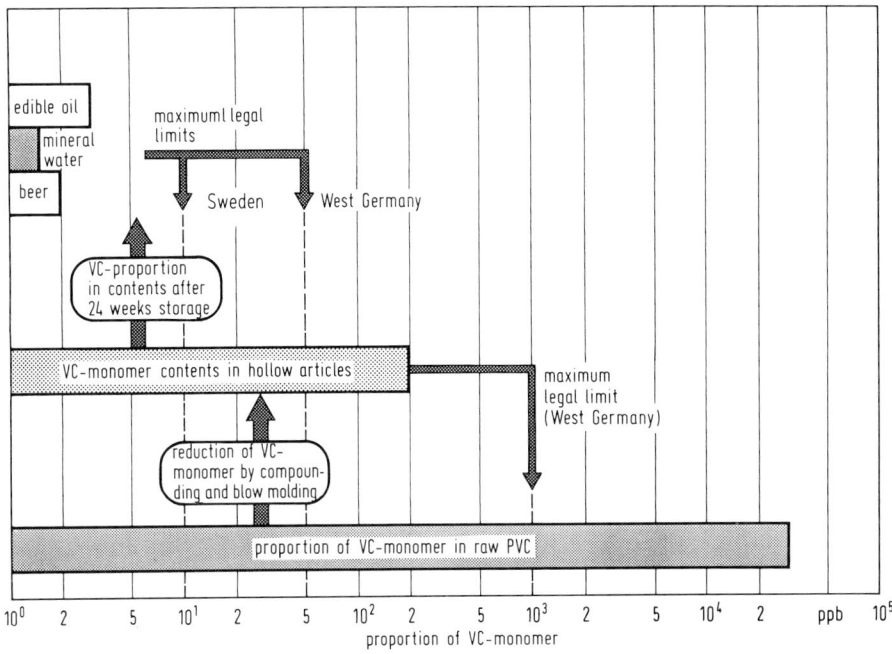

Figure 4. Change in VC-monomer content as a function of processing and storage conditions

blow mold pressure-resistant aerosol packs and larger pressure containers. The large containers must be installed with impact protection to compensate for the brittleness of the material.

Modified polyphenyleneoxides (PPO) offer good high-temperature form stability, favorable low-temperature impact strength and ductile, non-shattering behavior. These characteristics make them suitable for use in applications like blow-molded automobile components (rear spoiler, glove-compartment cover etc.) [9].

Highly viscous polyamide types (PA) possess sufficient melt strength to support blow molding into tanks for fuels and hydraulic fluid, on an accumulator-head extruder, if certain precautions are taken against the tendency of the melt to oxidize.

Thermoplastic (linear) polyurethanes (PUR) and their modifications are ideal materials for bellows.

12.3 Classification of extrusion blow molding processes

As the name suggests, classic extrusion blow molding comprises the following steps: tube extrusion, the blow molding operation and form stabilization of the hollow product by cooling. Blow molding with *continuous* tube extrusion is distinct from that based on *intermittent* tube output. In both cases, the parisons may be either single-layered or multilayered (coextrusion blow molding). The tube sections are extruded with either single-channel or multichannel dies and molded without intermediate treatment into the finished product with blown air only, while the tubular parisons are still in the thermoplastic state. The extruded-parison type process is characterized by the existence of a welding seam on the product, constancy of dimensions and weight within a narrow tolerance range, and a certain amount of scrap, which may be reprocessed.

12.3.1 Continuous tube extrusion blow molding

Sizes of typical hollow products made by the process can range from 10 ml to 30 l in volume, or from a few grams up to 1.5 kg in weight, depending on the material used. Since the development of high-molecular-weight HDPE polymers, even big drums of up to 120 l capacity can be blow molded [10]. The limiting factor in this case is not the melt viscosity but cooling of the long parison during extrusion; cooling affects the quality of the pinch weld adversely. Extrusion speeds in the continuous process are low compared with the intermittent process; for rheological reasons this requires a modified die design (cf. Figure 41) if a comparable surface quality on the blow-molded product is to be achieved. This process is now being used effectively with HDPE for producing canisters of up to 30 l capacity, with extrusion times of about 70 s.

Continuous tube extrusion requires either bypassing clamp units or — unusually — transport of the parison into the blow mold by means of grippers (Figure 5 D). If the blow molds beneath the extrusion die are moved in a horizontal direction only (Figure 5 A, B) then the extrusion speed has to be drastically reduced to avoid undue overrun and deformation of the parison [11, 12]. On the other hand, if the molds are moved vertically the outcoming tube is not disturbed by the lifting and lowering operations. Although the latter method makes for the shortest possible idle times, there are dimensional limits to the mold mass that can be moved.

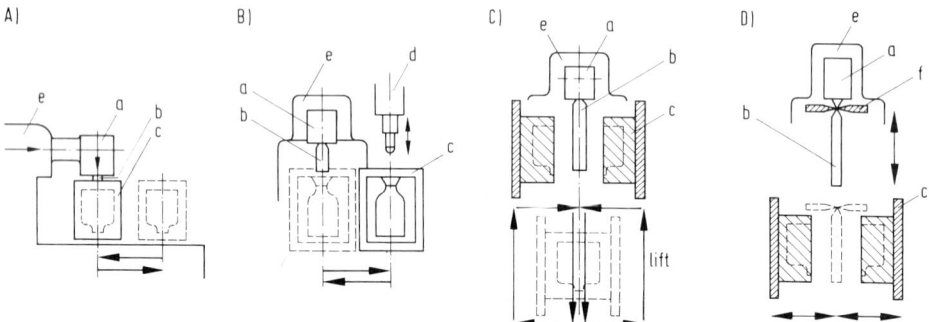

Figure 5. Movement of the mold or the parison during continuous tube extrusion
a extrusion mold, *b* tube, *c* mold, *d* blowing mandrel, *e* extruder, *f* transport grippers

On modern commercially available blow molding plant, the clamp unit either moves on inclined rails or swings through 45° from the extrusion and tube transfer position into the blow molding and calibration position. The motion in this case may occur either laterally or in the direction of the extruder axis (Figure 6). Moving the mold along a circular path has the advantage that during the first phase of its lowering motion the closed mold moves away from the extrusion die rapidly.

Alternatively, a continuously extruded tube may be cut off by a succession of 'book' molds on a vertical wheel or chain system, with the wheel or chain speed and number of cavities matched to the extrusion speed.

12.3.2 Extrusion blow molding with intermittent tube extrusion

Intermittent tube extrusion is used for the manufacture of hollow bodies of large volume, like a tank, or great length, like a surfboard, requiring parison weights between 1.5 and 250 kg. The melt is taken from one or more extruders operating continuously, collected in

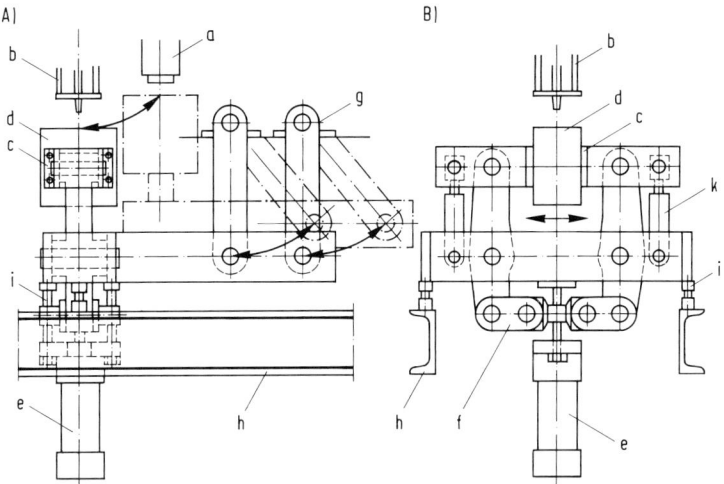

Figure 6. Oscillating motion of the clamp system along the extruder axis
A) side view, B) front view
a extrusion die, *b* blowing mandrel, *c* tool platen, *d* mold, *e* hydraulic cylinder, *f* force transfer element, *g* mold swing, *h* tie bar, *i* clamp support, *k* parallel guide linkage

an accumulator system, and intermittently extruded rapidly through an annular-gap die to obtain the parison. In the intermittent-type process, an electronic system of wall-thickness control is always employed nowadays; the result is that the distribution of axial wall thickness is matched to the shape of the blow-molded product, and savings are achieved in the cooling time and the amount of material used. The extruded tube section is pinched off as the mold halves are closed automatically by a clamp unit, which may be of the stationary or the indexing type, sometimes without tie-bars. Now enclosed in the mold cavity, the parison in its viscoelastic state is blown by means of a blowing mandrel or a blowing pin. Depending on the product shape and the number of openings and their contours, the use of a parison-stretching device, pre-weld grippers and/or a thread-forming device may be required.

12.4 Extrusion blow molding lines

In this section the structure and function of the important components of typical extrusion blow molding lines are explained and the fundamentals of their design described. In each of the sub-sections important technical data on the commercially available components are compiled in tables and diagrams. Finally, proven designs of plant constructed from these components are briefly outlined.

12.4.1 Description of individual units

12.4.1.1 Extruder systems

Extruders used for extrusion blow molding the materials mentioned in Section 2 are almost exclusively single-screw types. Contra-rotating twin-screw extruders are used only for a few special materials and/or for very high levels of output. The intermittent plunger-screw type of unit once used on German blow molding lines is nowadays found only on lines built by French and American manufacturers.

A number of limiting factors must be considered when choosing and designing single-screw extruders, and when making comparisons with other plasticating units, as follows [13]: The melt has to move through the die passage against resistance that is by no means constant with time. But during both the intermittent storage stage and the extrusion stage with accumulator systems, and when the extrusion is continuous, the material conveying- and homogenization functions of the extruder have to remain undisturbed, in spite of the drastic changes in flow impedance caused by adjustments to the die gap by the thickness control system. In addition, the melt temperature must be as low as possible, so as to keep the cooling time short and avoid unacceptable lengthwise stretching ("sag") of the parison. Also, the extruder must be readily capable of dispersing various kinds of pigments and additives, and varying amounts of regrind, in the virgin material.

The range of materials that has to be processed is very wide — from easy-flow LDPE, through highly viscoelastic HMW-PE, to PVC, which is very sensitive to shear and thermal stressing.

And if one recognizes, further, that for many of these requirements simply lengthening the extruder does not provide a general solution, because the L/D ratio of extruders on blowing lines affects the price of the overall line more strongly than with conventional extrusion lines, it can be appreciated why people continue to rely on the various single-screw designs available for the different kinds of blow molding operations.

There are two distinct types of plasticating unit:

– plasticating units with smooth barrels (conventional single-screw extruders);
– plasticating units with a grooved bush in the feed zone (extruders with feed zones optimized for conveying, or grooved extruders).

Before going into the details of their different characteristics, it is worth while noting the *common features* of the two types of unit.

On blow molding lines extruders are usually mounted horizontally, but the choice between the vertical and the horizontal arrangement still depends on the gearbox construction and the design of the line. Vertical extruders ($D \leq 50$ mm) are used primarily on bottle blowing machines and occupy a minimum of floor space. They are generally equipped with hydraulic drive connected either direct to the screw, or through a reduction gearbox.

On the older types of blow molding machines that still have vertically mounted extruders, drive is by means of shuntwound AC motors (speed adjustment range 4:1 to 6:1), or by DC motors (adjustment range 10:1), usually through three-step helical-gear reduction boxes. If necessary these can be designed as speed-change boxes, with two speed ranges.

There are many kinds of drive system available for use with horizontal extruders — the arrangement preferred on the newly developed blow molding machines: AC motors with adjustable belt- and reduction gear drive, for smaller units with screw diameters up to 60 mm; shunt-wound AC motors with reduction gearboxes, sometimes variable ratio, for medium size extruders ($60 \leq D \leq 90$ mm).

Both variants use piggyback mounting of gearbox and motor (Figure 7A, B); and for special purposes both versions can be fitted with the more expensive DC motors. For single-screw extruders with diameters of between 90 and 160 mm, a parallel arrangement of the drive motor and a two-stage box is often preferred (Figure 8). Here coupling is by means of vee-belt drive, which has the facility for providing a wide range of screw speeds by rapid exchange of the drive pulleys.

Although the options of AC and DC motors are offered for 90 and 100 mm extruders, units with screw diameters between 120 and 160 mm are fitted exclusively with thyristor-controlled DC motors, for space-utilization and cost reasons.

To reduce noise emission to below legal limits the machines are fitted with optimized fresh air blowers and high-precision helical gears. In this respect vee-belt drives, which generally operate at lower speeds, also contribute to the noise reduction.

12.4 Extrusion blow molding lines

Figure 7. Extruder drive by A) three-phase AC motor and belt-shift-type reduction gear, B) commutator motor and reduction gear

Figure 8. Extruder drive by commutator motor and reduction gear mounted parallel to the extruder axis

Figure 9. Motorized height adjustment of the extruder platform

The extruder of a blow molding machine is generally mounted on a platform that is part of the line. The height can be adjusted by motorized or manually operated leadscrews (Figure 9). Also for smaller, platform-mounted extruders, at least, self-supporting designs are available in which the platform height can be adjusted in relation to the blow mold by means of vertical column guides and set screws (see Figure 7). The extruder itself may be mounted on the platform in such a manner that it can be swung round and/or shifted parallel to its own longitudinal axis for easy access or variation of the processing technique.

Modern extruder screws and barrels are made from high-alloy nitriding steels such as 34CrAlNi7, X35CrMo17 or 31CrMoV9, surface-hardened to more than 900 HV-30 in an NH_3 gas stream. With screws for PVC, usually only the flight lands are hardened, while the basic screw and the faces of the flights are hard-chrome-plated to a depth of 15 to 20 µm, and highly polished. Another method of surface treatment is ion implantation [14].

Abrasive wear on the barrel and screw of a grooved extruder can be severe in the conveying zone, where the material is not molten, especially with compounds containing inorganic pigments [15, 16]. To reduce the extent of such wear, barrel components are centrifugally cast from Fe/Cr/Ni/B or Fe/Ni/B alloys, or machined from tool steel; and screws are made from X155 CrVMo 121 high-speed steel (1.2379) and fully hardened, or are surfaced only on the screw flights with a layer of a Co/Cr/W or Cr/B steel alloy [17].

In this connection it is worth mentioning that an elegant and cost-effective way of minimizing wear can be to modify the color-concentrate carrier material [18].

12.4.1.1.1 Conventional single-screw extruders

Basic relationships

The conventional plasticating extruder generally consists of a three-zone pressure generator (the barrel) connected direct to a pressure utilizer (the die) as shown in Figure 10. Thus it is a multifunctional machine that feeds the solid material and operates as a plasticating unit, a homogenizing system and a pressure-generating unit. The various tube dies to which it is coupled have flow impedances related to their internal geometry and the rheological properties of the material being processed. In order that these various functions can be carried out, the screw in particular must have certain design features and be operated in a specified manner. The design detail and operating conditions of the screw depend, as described below, on the raw material and the finished-product quality required. It is charac-

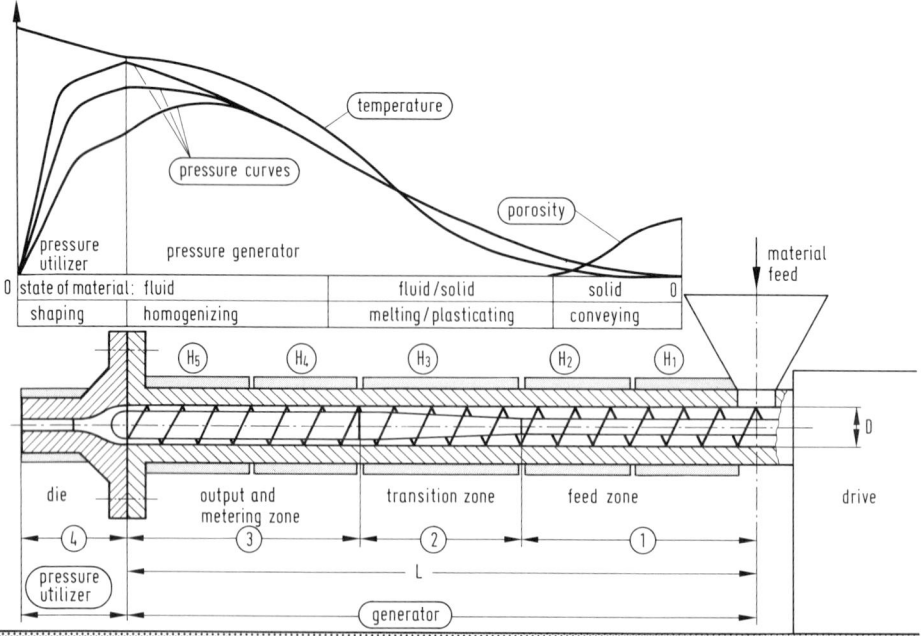

Figure 10. Assembly sections and functional zones of a conventional single-screw extruder

teristic of this type of extruder that throughput $\dot{m}$ is determined by the conveying zones. Changes in the die resistance affect the throughput, and thus react directly on the melt quality or homogeneity. These interactions can be followed easily with the help of schematic performance diagrams like that shown in Figure 11 for a conventional single-screw extruder [19].

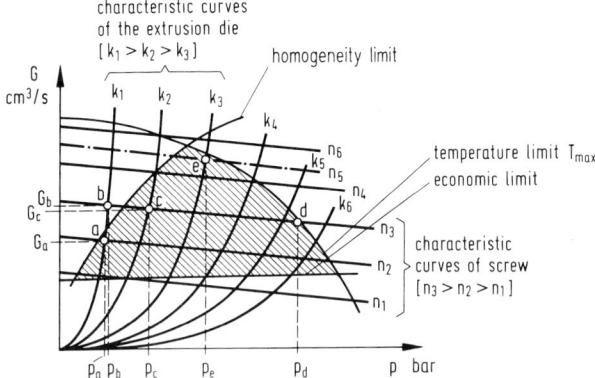

Figure 11. Performance diagram of a conventional single-screw extruder

In recent years the conventional extruder has lost ground to the grooved feed-zone type, because of the way polymers have developed. The grooved extruder is now established as standard equipment on advanced blow molding machines. Extruders with smooth barrels and screws of 30 to 100 mm diameter, or preferably 50 to 90 mm diameter are primarily used for processing PVC alone, but, fitted with the appropriate design of screw, they can be used to process both PVC and LDPE or, alternatively, HDPE of low or medium molecular weight. They are also still used economically for plasticating polycarbonate, which is very sensitive to shear and thermal stresses, and exceedingly hard-grained in the solid state.

Extruders for processing PVC

The screw/barrel systems of conventional single-screw extruders have an L/D ratio of between 20 and 24, with the trend of development distinctly toward higher L/D values. Depending on its size, the barrel is heated in three or four adjacent zones by ceramic resistance elements. If barrel cooling is needed it is carried out by individual air blowers for diameters between 35 and 60 mm; but for larger-diameter extruders copper tubing is often rolled into the grooved outer surface of the barrel to carry the heat exchange medium, preferably an organic liquid. Each of the heating/cooling zones is controlled by a three-term (PID) controller; or all are collectively controlled by a microprocessor, with the advantage that optimum values of the control parameters for each zone can be set independently.

The screws of these units are bored out so that they can be effectively cooled by compressed air, particularly in the tip region. They have two or three sections, three being preferred for blow molding machines. The design of these three-section screws is diameter-dependent, but they generally consist of a feed zone of length $L_1 = 0.18$ L, a second zone of length $L_2 = 0.64$ L – the compression zone – and a homogenization zone of length $L_3 = 0.18$ L (Figure 12). The plasticating/compression zone has to be the longest, because of the broad softening range of PVC. The ratio of screw-channel depth in the feed section to that in the homogenizing section (h_1/h_3) must be chosen to suit the K-value of the raw PVC to be processed. The ratio varies from 1.8 to 2.2 as the molecular weight of

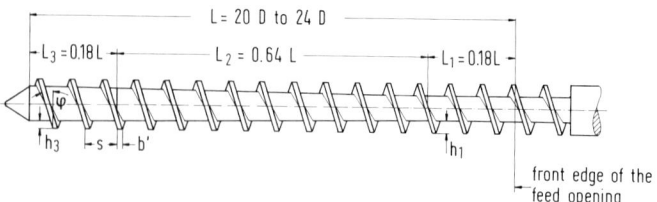

Figure 12. Geometric layout of a typical triple-section screw for plasticating PVC
screw pitch: $s = 1.0\,D = 17°40'$, channel depth: $h_1/h_3 = 1.8$ to 2.2, flight width: $b' = 0.10\,D$

the PVC increases. Figure 13 shows values of channel depths h_1 and h_3 now in use with various values of screw diameter, for a constant value of the ratio $h_1/h_3 = 2.0$. If screws are designed with ratios other than 2.0, depth h_1 should be kept the same, and only h_3 be changed. The pitch of the screw, s, is generally made equal to the diameter D (pitch angle $\varphi = 17.76°$) and the land width $b' = 0.1\,D$. The minimum radial clearance δ between the screw and the inner surface of the barrel lies between 0.1 and 0.15 mm, depending on the screw diameter.

To intensify the plasticating action and to improve the melt homogeneity in terms of temperature distribution and degree of mixing, some special homogenizing sections are also built into the screw. These, typically, consist of two to four disks carrying radial pins with the clearance between adjacent pins being 1.0 to 1.3 mm. The section is part of the metering zone of the screw. Certain types of extruder in addition allow for axial displacement of the screw, so that the conical ring-gap between the screw tip and the die inlet can be varied, and the plasticating and homogenizing action improved.

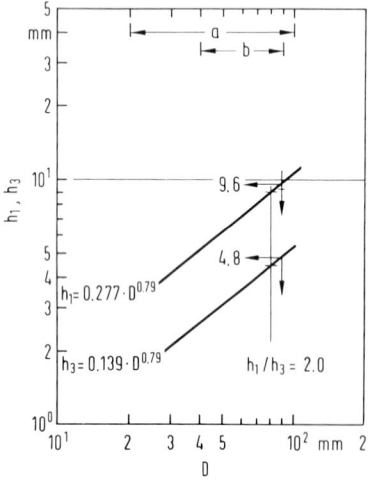
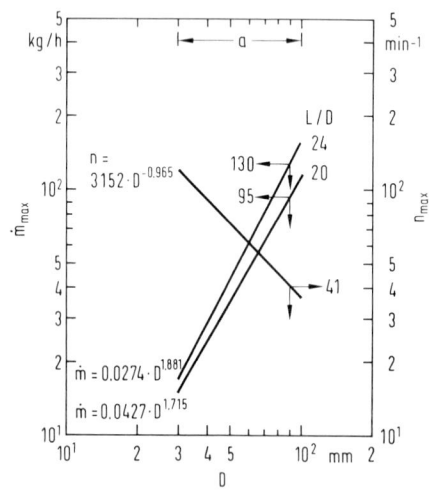

Figure 13. Screw channel depths $h = f(D)$ for PVC molding extruders of $L/D = 20$
a PVC extruder, b popular sizes

Figure 14. Maximum throughput rates and screw speeds for processing PVC with conventional extruders
a PVC extruder

To avoid mechanical or thermal overstressing of sensitive compounds, the maximum peripheral speed (v_u) of the screw must be limited to 0.2 m/s. But nowadays a 90 mm extruder satisfying this limiting condition can produce 95 kg/h at maximum screw speed for $L/D = 20$, and 130 kg/h for $L/D = 24$ (Figure 14). Specific melt throughputs $\dot{m}/n$ that can be achieved with extruders of different sizes are shown in Figure 15. The installed drive power N_A required on PVC extruders, which depends on the screw diameter (D) and the

12.4 Extrusion blow molding lines

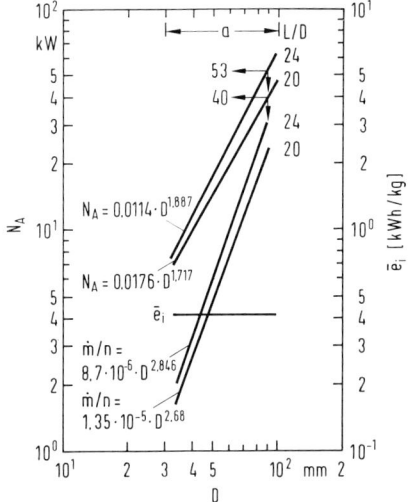

Figure 15. Installed drive capacities and specific throughput rates of PVC extruders
a PVC extruder

L/D ratio, is also shown in the same figure. These data allow an average value of the installed specific drive power to be determined: $\bar{e}_i = N_A/\dot{m} = 0.415$ kWh/kg = const. The following equations can be helpful in designing and developing a series of extruders for processing PVC powder mixes that may also contain certain proportions of regrind:

maximum screw speed:
$$n_{max} = 3152\, D^{-0.965}\ \text{min}^{-1}, \tag{1}$$

maximum throughput rate:
for $L/D = 24$: $\dot{m}_{max} = 0.0274\, D^{1.881}$ kg/h, (2a)
for $L/D = 20$: $\dot{m}_{max} = 0.0427\, D^{1.715}$ kg/h, (2b)

specific throughput rate:
for $L/D = 24$: $\dot{m}/n = 8.7 \cdot 10^{-6} \cdot D^{2.846}$ kg min/h, (3a)
for $L/D = 20$: $\dot{m}/n = 1.35 \cdot 10^{-5} \cdot D^{2.680}$ kg min/h, (3b)

installed drive power:
for $L/D = 24$: $N_A = 0.0114\, D^{1.887}$ kW, (4a)
for $L/D = 20$: $N_A = 0.0176\, D^{1.717}$ kW. (4b)

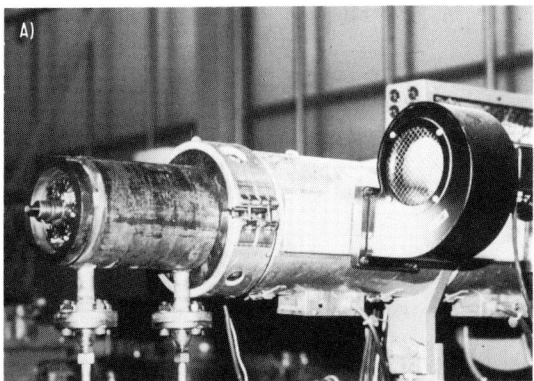

Figure 16. Planetary-screw homogenizing section of a high-output, grooved PVC extruder, with $L/D = 20$
A) planetary-screw homogenizing section (general view), B) with screw tip removed

This information should in no way obscure the fact that PVC powder/regrind mixtures and PVC pellets can also be compounded optimally with grooved extruders. It has been shown that a grooved 90 mm extruder with an additional planetary screw homogenizing section, having its own oil-based temperature-control system (Figure 16), can be used to plasticate PVC pellet into a blowable melt at a rate of almost 230 kg/h [20]. It is clear that further development in this sector will be energetically pushed ahead, and that, in particular, it should be possible to design screws for specific materials and to use only the homogenizing elements that are absolutely necessary.

Extruders for processing polyolefins

The barrel configuration used for PVC can be combined with particular screw geometries and screw-speed ranges to readily process medium-molecular-weight (MMW) grades of LDPE and HDPE, polystyrenes and also – with certain limitations – polypropylenes. These polymers can be worked at screw surface-speeds (v_u) of between 0.3 and 0.35 m/s, so the ranges of screw speeds for these materials are different from those for PVC (see Figure 17). Hence, if PVC has to be processed on the same extruder, it is advantageous to install the gearbox mentioned earlier, so that two different screw speeds can be set, and thus make possible optimum plastication of two different kinds of polymer with one extruder.

The screws used for this purpose generally have three sections ("zones"), whose lengths, irrespective of screw diameter, are as follows:

feed section: $L_1 = 0.44\ L$,
compression (transition) section: $L_2 = 0.15\ L$,
homogenizing and metering section: $L_3 = 0.35\ L$,
mixing elements: $L_M = 0.06\ L$.

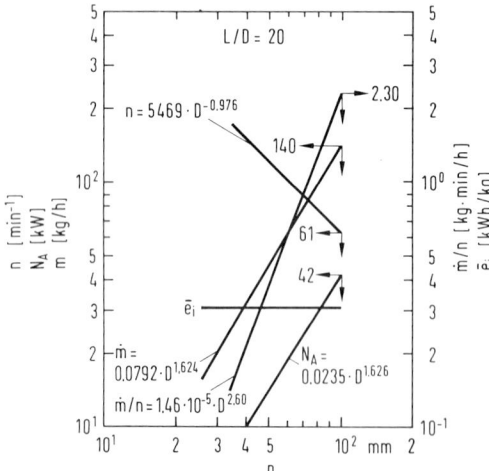

Figure 17. Design data for conventional PE extruders ($L/D = 20$)

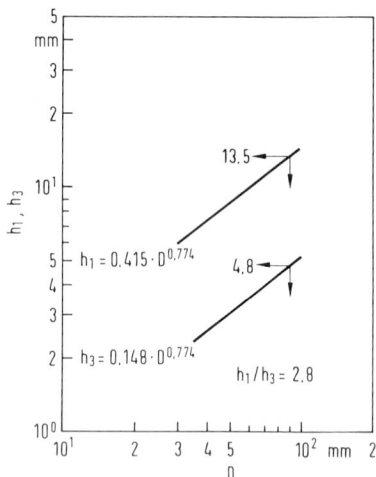

Figure 18. Flight depths of PE screws on conventional extruders

As with PVC screws, the effective L/D ratio of these screws is 20 in most cases today, though there is a clear trend in this sector also towards L/D values of 24 or 25. The ratio of screw-flight depths h_1/h_3 is approximately 2.8. Figure 18 shows typical values of h_1 and h_3 for various screw diameters. The values of screw pitch, land width and the radial gap are the same as those given for the PVC screw design. To improve the homogeneity of mixing and temperature distribution, polyolefin screws are generally equipped with a screw section 1.0 to 1.5 D long, having homogenizing elements of the rhombic or slotted-disk type, which drop melt pressure a little; these are generally fitted at the screw tip (Figures 19, 20).

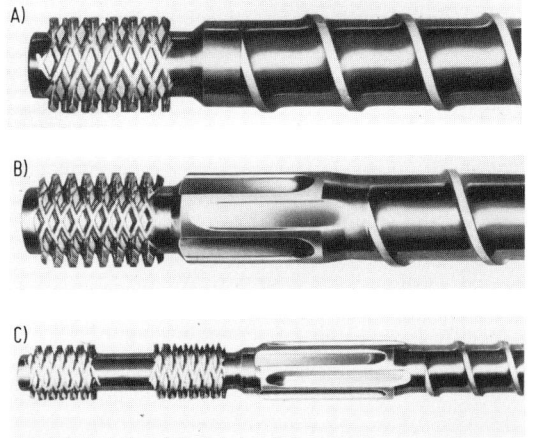

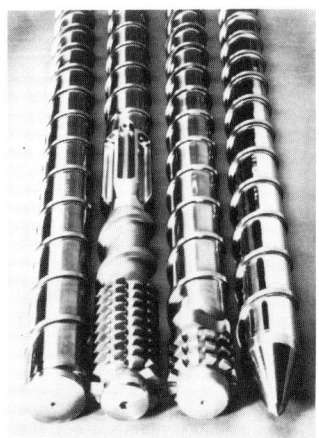

Figure 19. Screw designs for conventional and grooved extruders
A) conventional extruder, PE screw with rhombic mixing section,
B) grooved extruder, screw for HM-HDPE,
C) grooved extruder, PE screw for special mixing tasks

Figure 20. Assortment of screws sold for use on blow molding extruders

Using the maximum screw speeds shown in Figure 17, the raw material can be converted into a homogeneous, blowable melt on a 90 mm extruder at a rate of about 125 kg/h for $L/D = 20$, or about 140 kg/h for $L/D = 24$. The specific throughputs and drive power required may be obtained by reference to Figure 17, from which, also, the following mathematical relationships can be extracted for designing a series of extruders for processing polyolefins:

- maximum screw speed: $\quad n_{max} = 5469\, D^{-0.976}\, \text{min}^{-1}$, (5)
- maximum throughput rate: $\dot{m}_{max} = 0.0792\, D^{1.624}\, \text{kg/h}$, (6)
- specific throughput rate: $\dot{m}/n = 1.46 \cdot 10^{-5} \cdot D^{2.6}\, \text{kg min/h}$, (7)
- installed drive power: $\quad N_A = 0.0235\, D^{1.626}\, \text{kW}$. (8)

The average value of the installed specific drive power ($\bar{e}_i$) given by Equations (6) and (8) is $\bar{e}_i = N_A/\dot{m}_{max} = 0.3$ kWh/kg.

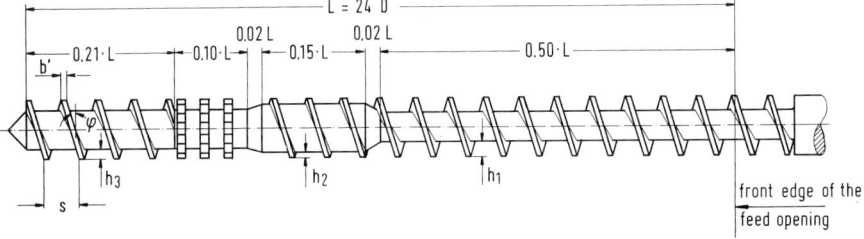

Figure 21. Polyolefin screw with a deep-flight metering section and a slotted-disk homogenizing section for use on conventional extruders
screw diameter: $D = 60$ mm, channel depths: $h_1 = 10.0$ mm, $h_2 = 3.5$ mm, $h_3 = 5.9$ mm, screw pitch: $s = 1.0\, D$
$\varphi = 17°40'$, land width $b' = 0.1\, D = 6.0$ mm

In addition to three-zone screws for processing polyolefins, there is also a modified design of screw occasionally used for the same purpose, mostly in combination with longer barrels: the 60 mm plasticating screw shown in Figure 21 is an example. The comparatively shallow plasticating zone, and the intermediate homogenizing zone equipped with slotted-disk homogenizing elements, is followed by a deep-cut discharge zone which carries out the task of pressure generation and metering the melt in an ideal fashion.

12.4.1.1.2 Extruders with a forced-feed section (grooved extruders)

The grooved extruder was originally developed to plasticate HDPE of very high molecular weight, usually obtained in powder or granular form from the polymerization reactor [21 to 23]. This type of extruder has now become the standard plasticating unit on modern blow molding machines. It can be used to plasticate and homogenize the usual varieties of LDPE and HDPE, and polypropylene powders, mixed with their own regrind, as well as thermoplastics like ABS, PUR and POM (acetals), which, however, are seldom used for blow molding. By virtue of its higher specific throughput rate (Figure 22) and the consequent low temperature of the outcoming melt, and because its conveying capacity is largely independent of die resistance and it can be easily adapted to diverse homogenizing requirements, the grooved extruder is distinctly superior to conventional plasticating units with smooth barrels. The obviously higher machine construction costs of the grooved extruder are of secondary importance compared with the advantages it offers in processing.

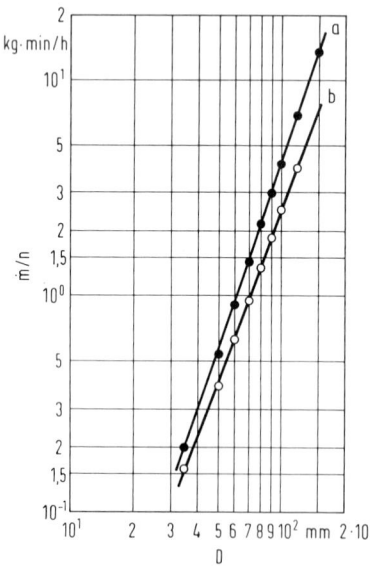

Figure 22. Comparison of specific throughput rates $\dot{m}/n$ of conventional and grooved extruders (material: HDPE, Lupolen 5021 D) a grooved extruder $\dot{m}/n = 6.6 \cdot 10^{-6} \cdot D^{2.9}$, b conventional extruder $\dot{m}/n = 3.1 \cdot 10^{-6} \cdot D^{2.7}$

Special construction features

A grooved extruder differs from a conventional single-screw extruder in the design detail of the feed section (cf. Figure 23). The casing of the speed-reduction gear is flanged onto the barrel feed section (*a*). The feed bush (*b*) has a cylindrical bore and can be directly and intensively cooled. The bush is tightly fitted or shrunk into the barrel of the feed section, and has conically tapered axial grooves (*c*) on its inner surface. Near the feed opening (*d*) the feed bush is enlarged ("feed pocket" (*e*)) for more effective filling of the screw channels. The combined action of a back-flow check flight (*f*) on the shaft and a sealing

12.4 Extrusion blow molding lines

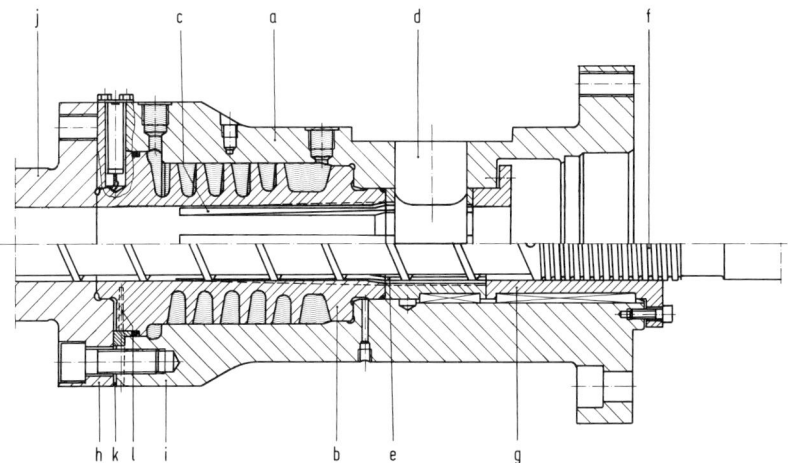

Figure 23. Grooved feed section for improved conveying of solid material

bush (g) prevents the leakage of powdery material at the drive end of the extruder. With suitably designed connecting flanges (h, i), a very good heat barrier (k) is obtained between the water-cooled grooved bush and the main barrel (j) (the barrel is hardened by gas-nitriding). Temperature control of the barrel is achieved by three to five combined heating/cooling zones, depending on the extruder size, using resistance-type ceramic heating elements and individual cold-air blowers. The effective length of the cylinder varies between 20 D (standard) and 25 D, with a distinct trend towards the use of longer conveying sections. The temperature of the barrel inner wall at the groove run-out is generally measured with an Fe-constantan thermocouple (1). The indicated temperature provides information about feed stability and the instantaneous operational condition of the feed section. Standardized design data, as compiled in Table 1, now exist for the grooved bushes in the extruders of blow molding machines.

Table 1. Design data for grooved feed bushes on blow molding extruders

Number of axial grooves:	$z_N \approx D/8$ (D in mm)
Groove width:	b_N = 12 to 14 mm
Groove length:	l_N = 3 D
Groove depth at front edge of feed opening:	t_N = 4 mm
Diameter of feed pocket:	D_T = D + 4 mm
Angle of intersection between feed pocket and cylinder bore:	$\alpha_T \approx 7°$
Groove shape:	Rectangular cross-section

Recent investigations have shown that the much-discussed spiral (helical) grooves [24] never give higher throughput rates than axial grooves, but they do produce a lower specific rate of energy exchange in the feed section, and thus reduce the loss element in the total energy balance. This energy saving is possible, apparently, because the rate of shear between the solid material compacted in the grooves and that in the screw channel is somewhat reduced (see Figure 24). As a result of an exponential rise in pressure $p_M(L)$ in the feed zone, the bulk density of the material $\varrho(p_M)$, and hence the feed angle ω of the solids packet, change continuously along the screw length L (Figure 25); consequently, to minimize the rate of energy loss the grooves should start with a helix angle of 90° (i.e. as axial grooves) that reduces nonlinearly to 30° following the $\omega(L)$ curve. Since grooved bushes with a changing helix angle are very complicated to manufacture, the compromise of machining the grooves with a constant helix angle of about 55 to 60° is suggested.

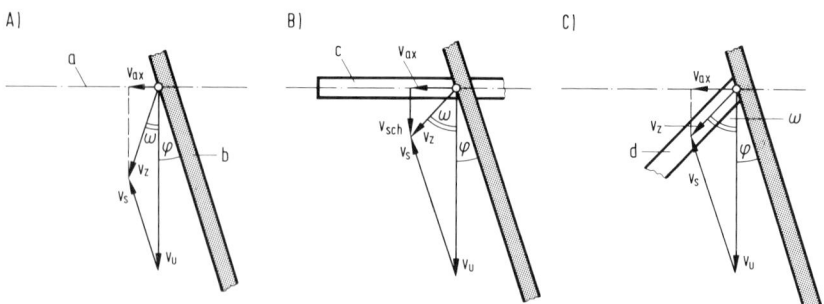

Figure 24. Velocity components in single-screw solids conveying systems
A) smooth barrel, B) axial grooves, C) helical grooves
ω = conveying angle, φ = screw pitch angle, v_{sch} = shearing velocity,

$$\dot{m} = A \cdot \varrho \cdot v_{ax} \cdot \Omega \cdot \frac{D}{Z} \cdot \frac{\tan \varphi \cdot \tan \omega_0}{\tan \varphi + \tan \omega_0}, \quad \omega_0 = f(\mu_a, \mu_{eff}, h, D, \varrho)$$

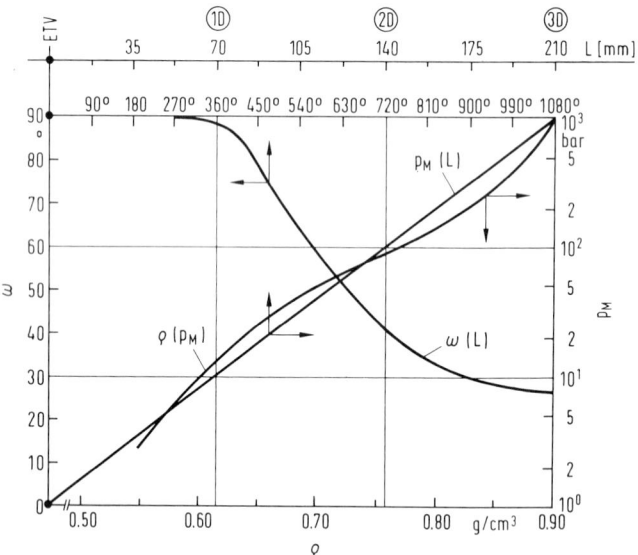

Figure 25. Changes in bulk density, melt pressure and conveying angle in the grooved feed section. 'ETV' Front edge of the feed opening

Conveying mechanism and operational behavior

The throughput rate of the extruder is determined not only by the geometry of the screw and the grooved bush, but also by material parameters such as bulk density ϱ_S, the coefficients of internal friction (μ_i) and external friction (μ_a), and the shear strength and deformation resistance of the raw material. When the transport phenomenon in the grooved feed section is being investigated, an average coefficient of friction ($\bar{\mu}_z$) [25] or a locally variable, equivalent coefficient of friction $\mu_e(l)$ [26] with $\mu_a \leq \mu_e(l) \leq \mu_i$ should be used for interaction between the block of solid material and the machine surfaces in the feed zone, because local (internal) friction arises (plastic on plastic) between the compacted material moving in the grooves and that in the screw channel.

The fact that $\bar{\mu}_z$ or μ_e is bigger than μ_a leads to the effective conveying angle ω at the end of the grooved zone being larger than is inherently possible with the same screw working

12.4 Extrusion blow molding lines

in a smooth barrel, under the same operational conditions. (The angle ω is that between the velocity vector of material movement and the reference plane perpendicular to the screw axis – see Figure 24.) The result is an accumulation of material, and an intensive densification and compacting process that raises the bulk density nearly to the full material density over a three to four D length of screw, between the forward edge of the feed opening and the run-out from the grooves. This conveying mechanism results in a pressure maximum in the neighbourhood of the heat barrier and has the additional consequences that all the following sections of the screw, and the mixing and shearing sections, are force-fed and work as pressure droppers rather than pressure generators [27]. And by comparison with the conventional extruder, the axial pressure distribution is of a basically different type (Figure 26). For characterizing the degree of force-feeding the dimensionless quantity:

$$a = G_p/G_s = 1 - G_{eff}/G_s \qquad (9)$$

is used in which, in general, $-0.4 \leq a \leq -0.2$. The pressures that are found at the outlet end of the grooved zone p_{EZB} are a function of the die flow impedance w, the screw geometry as a whole, the screw speed n, the throttling quotient a, and the viscosity function of the material being processed.

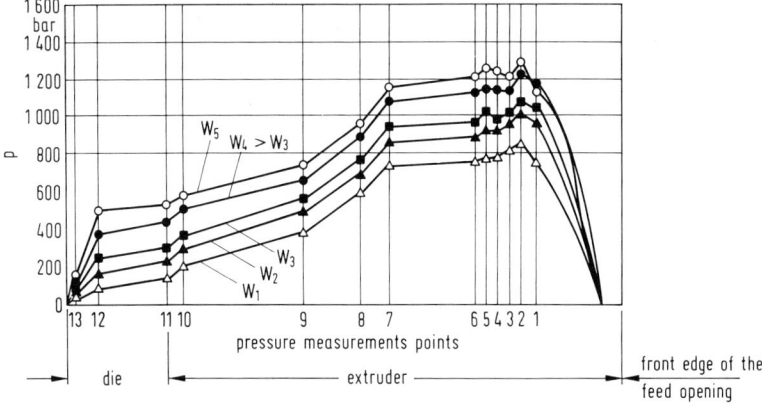

Figure 26. Pressure distribution in a grooved extruder with $D = 120$ mm, material: HMW-HDPE, polymer type II (granule/regrind mix), $n = 36$ min^{-1}

The design and optimization of grooved feed zones, and the description of operational behavior, require the target melt throughput rate $\dot{m}$, and drive torque Md to be expressed in terms of the relevant geometry, and of material and operational parameters. These relationships can usually be obtained from a physical/mathematical model, or by means of similarity theory in combination with model studies [28] (Figure 27). Computing models developed over the last few years started, initially, from the force balance of a small element of length in the screw channel [25, 29, 30]. With this approach, so-called pressure-anisotropy coefficients had to be introduced, which were not, in fact, pure material parameters. To get around this difficulty in another model approach, the force balance of a volume element was calculated [31, 32]. The important results of these model studies, which were in substantial agreement, can be summarized as follows:

- The melt pressure increases exponentially in the conveying direction. Because the density of the material is pressure-dependent, the conveying angle ω continuously decreases from the maximum value ω_0 at the hopper opening. The value of ω_0 is fixed by the screw geometry and the frictional behavior, and determines the melt throughput rate $\dot{m}$.

- To ensure that conveying behavior is independent of the back pressure, this pressure p_{EZB}, which is fixed by the flow-resistance coefficients of the die and the homogenization zones, must be reduced to the value p_0 at the hopper opening, which is less than the flow pressure p_F of the material in the hopper. The feed zone length required to achieve this effect, $L_{EZ,\,eff}$ is $3\,D$ to $5\,D$, depending on screw and bush geometries. Up to now feed zones have been designed almost entirely with a standard length $L_{EZ} = 3\,D$ as can be seen in Table 1.
- The drive torque needed in the region of the feed zone is independent of screw speed, and increases linearly with increasing back pressure p_{EZB}. This pressure-dependence also holds for the amount of cooling needed to remove frictional heat from the grooved bush; however with p_{EZB} constant, the cooling capacity must be raised approximately linearly with screw speed n, so as to avoid the unacceptable formation of a film of melt at the surface.
- If, for design reasons, a high back-pressure is required, the use of shallow flights in the feed zone is recommended. The same result can be obtained by reducing the pitch, s, of the screw, because the pressure-raising component of the frictional force on the barrel is increased by this means. Of course, the conveying angle, ω_0, and the specific throughput $\dot m/n$ are reduced under otherwise constant conditions.

The results summarized here, which are only the key points resulting from the model studies quoted, are also confirmed by similarity theory methods, based on extensive model studies.

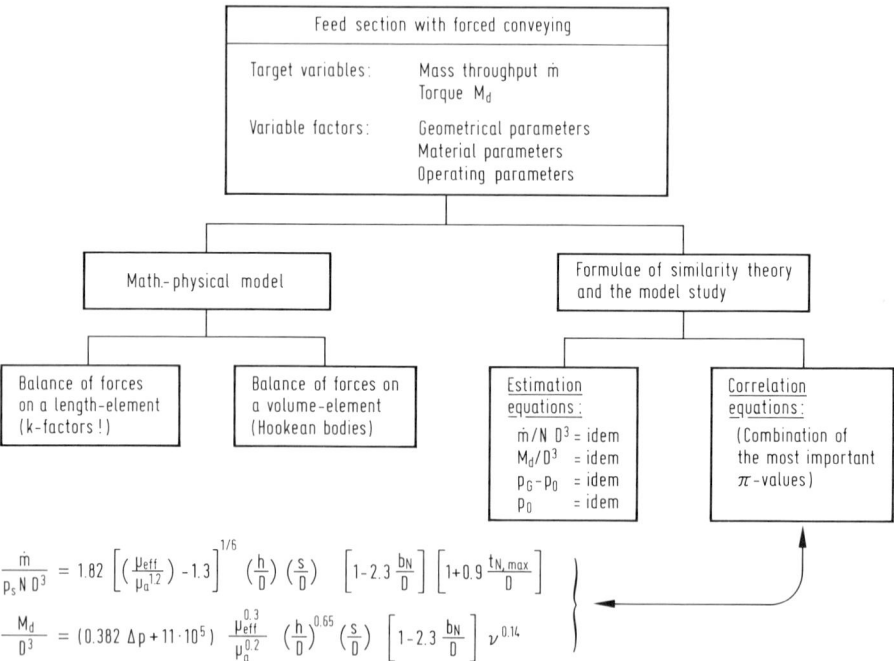

Figure 27. Procedures for calculating the behavior in grooved feed zones

From the relationships demonstrated it is clear on energy grounds, as much as for reducing wear effects, that one should try to reduce the level of pressure p_{EZB} in the region of the heat barrier. One way that this can be done is to increase the channel depth at the beginning of the feed section of the screw, which may be $5\,D$ or $6\,D$ long. This raises the inherent conveying capability of the following zones, but results in a worsening of the heat transfer and plastication functions. A more efficient solution is to choose a larger pitch angle φ for the following sections [33], with the optimal value φ_{opt} dependent on the specific throughput rate $\dot{m}/n$.

Because of the nature of the conveying mechanism described, the melt throughput rate of grooved extruders is independent over wide ranges of the die impedance and the geometrical configuration of the homogenization zones built into the screw. The performance curves of these screws run almost horizontally – in contrast to those of conventional screws (Figure 28). Only if the energy generated in the shear planes of the grooved bush becomes so great that softening, melting, and/or shear failure of the packets of solid material in the grooves occur, do the lines show a spontaneous kink. This break point moves in the direction of smaller values of die impedance with increasing screw speed.

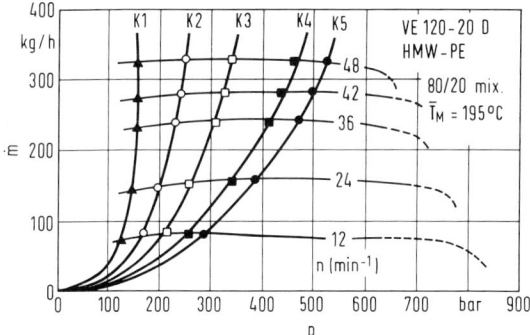

Figure 28. Melt throughput rate as a function of back pressure in grooved extruders $D = 120$ mm, $L/D = 20$ (pellet/regrind mix)

With system behavior of this kind it is hardly to be expected that the thermal and general homogeneity of the melt will be improved by raising the die impedance (throttling). For this reason, screws designed for grooved extruders embody special functional zones, such as tearing, shearing, and mixing sections, to which particular compounding functions are assigned, depending on their position. In the following section the shaping and positioning of these elements will be discussed.

Screw designs, throughputs, drive power

To accommodate the various material properties and homogenization tasks, grooved extruders nowadays are usually fitted with variable stacking-element- or screw-together modular screws (Figure 29). In many cases, extruder screws with an L/D ratio of 20 are constructed with a screw section 16 or 17 D long, a 2 to 2.5 D shearing section, and a diamond-faceted mixing section of the same length. At the end is either a smooth screw tip or one provided with slotted disks. For less difficult homogenization tasks, the familiar shearing and mixing sections can be exchanged for elements possessing shearing and mixing rings and a simple threaded part. For a screw with an $L/D = 20$ and a conveying screw 16 to 17 D long it is recommended that the pitch $s = 0.8\ D$ ($\varphi = 14°17'$), and that the expression

$$h = 0.293\ D^{0.674} \tag{10}$$

be used to obtain the channel depth (see Figure 30). These screw sections can be readily employed up to a diameter of $D \leq 80$ mm, with a flight width $b' = 0.1\ D$. To reduce the surface pressure on the flight faces, and therefore the adhesive wear on them, an extruder with D greater than 80 mm should have a double-start screw with $b' = (0.05\ D + 2)$ mm, which extends over a length $L_E = 6$ to $7\ D$ and then becomes a single-start screw. From this point on, even a slightly larger channel depth or screw pitch can prove advantageous for reducing the pressure p_{EZB}. The task of the adjacent blind-grooved shearing section with a shear gap of between 0.4 and 0.7 mm is to hold up unmelted particles, and thus exercise a form of filtration. It has turned out to be an advantage, at least with polyolefins, to machine the inflow and outflow grooves relatively deep, because this reduces the pressure drop and at the same time increases the mean dwell time of material in the extruder. Because of the high degree of melt-stream subdivision and the high shear deformation it causes, the diamond mixing section produces an excellent mixing action, but certainly also exhibits high specific flow resistance.

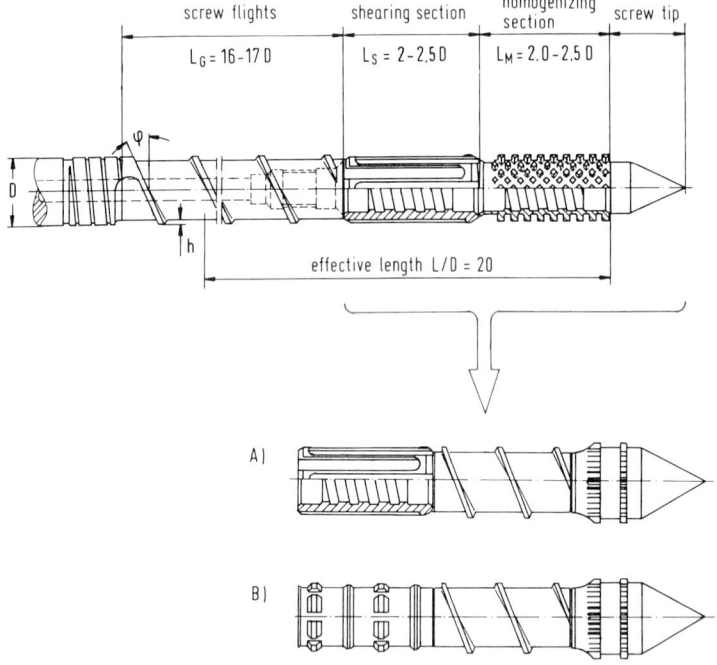

Figure 29. Screw built from standard modules for grooved blow molding extruders
A) combination of shearing section and homogenizing tip, B) combination of shearing/homogenization section and homogenization tip

As well as the mixing and shearing sections that have been mentioned, many modifications are found in the patent literature and in industrial operations [34], which differ with regard to performance and difficulty of manufacture.

The maximum achievable throughput rate $\dot{m}$ and specific rate $\dot{m}/n$ in processing mixtures of virgin and regrind high-MW polyethylene vary with diameter according to the equations:

$$\dot{m} = 0.0984\ D^{1.703} \tag{11}$$

and

$$\dot{m}/n = 1.64 \cdot 10^{-5}\ D^{2.703}. \tag{12}$$

It can be recognized from these data that it is very often possible to specify the next-smaller diameter grooved extruder, in place of a conventional extruder, to obtain a given melt output rate. The steps in screw speed and the installed drive power, N_A, for each extruder size can be extracted from Figure 30.

12.4 Extrusion blow molding lines

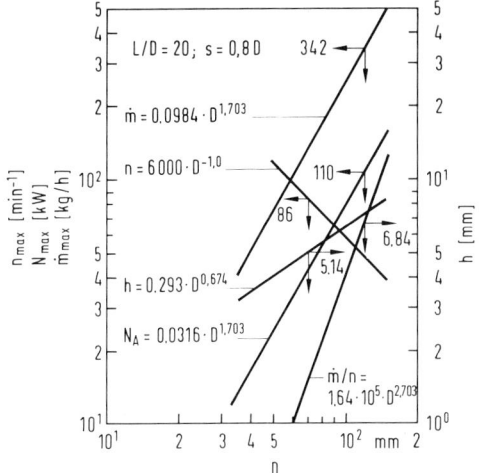

Figure 30. Design data for a series of grooved extruders

The trend to greater L/D ratios with grooved extruders, also, is obvious. By increasing the pitch ($s = 0.8\,D$ to $s = 0.9\,D$), and simultaneously raising the screw and/or the channel depth, it is possible to obtain considerably higher throughputs with longer extruders ($L/D = 24$ or 25), as shown by the following practical example: in an automobile gas-tank manufacturing operation it was possible with a 120 mm extruder ($L/D = 24$, $s = 0.9\,D$, $h = 8.5$ mm) to achieve perfect plastication of HMPE at an output rate $\dot{m}$ of at least 480 kg/h with a screw speed of 64 min^{-1} ($\dot{m}/n = 7.5$ kg min/h).

Any increase in specific output implies an increase of length in the plasticating zone. If in future it is required to increase extruder performance still further, keeping the quality constant or even improving it, and with a fixed L/D ratio of 25, particular effort will have to be devoted to forcing the melting and plasticating processes, by building in special sections. In detail this can involve (Figure 31):

- Using elements that tear the solid material apart: the melt flowing between them and the solid material, and the increased area for heat transfer intensify the plastication. It is essential that melt containing such islands of solid material should be passed through a subsequent shearing section that acts somewhat like a filter (Figure 31 A);
- Having screw sections with polygonal cross-section to provide additional kneading for intensifying the plastication (Figure 31 B);
- Employing kneading elements of good conveying efficiency, combined with radial pins [35] to form a comb system in which the material is plasticated by continuous shearing and dividing action (Figure 31 C);
- Installing a screw section for separation of solids from the melt — known as a barrier screw. The original concept of this was embodied in the Maillefer screw, but it has since been modified frequently in many details [36, 37]. The essential design feature of the Maillefer screw is the interleaving of two screws of different pitch (Figure 31 D). A considerable increase in plasticating rate can be achieved with such barrier sections [38, 39], which have a constant width of channel for solids, but a depth that decreases in the direction of material movement (Figure 31 E). The main advantages of such a modified design, which may have increased importance in the future, can be summarized as follows:

- structural and functional zones always coincide, whatever the operational conditions; no clusters of solid material;

- almost constant specific surface area for heat transfer along the length of the solids channel; simultaneous increase of the Fourier index in the conveying direction;
- the spectrum of dwell times, and hence the lengthwise mixing action, can easily be affected by varying the depth of the melt channel;
- the average temperature of the plasticated material and the thermal stress on it can be kept low.

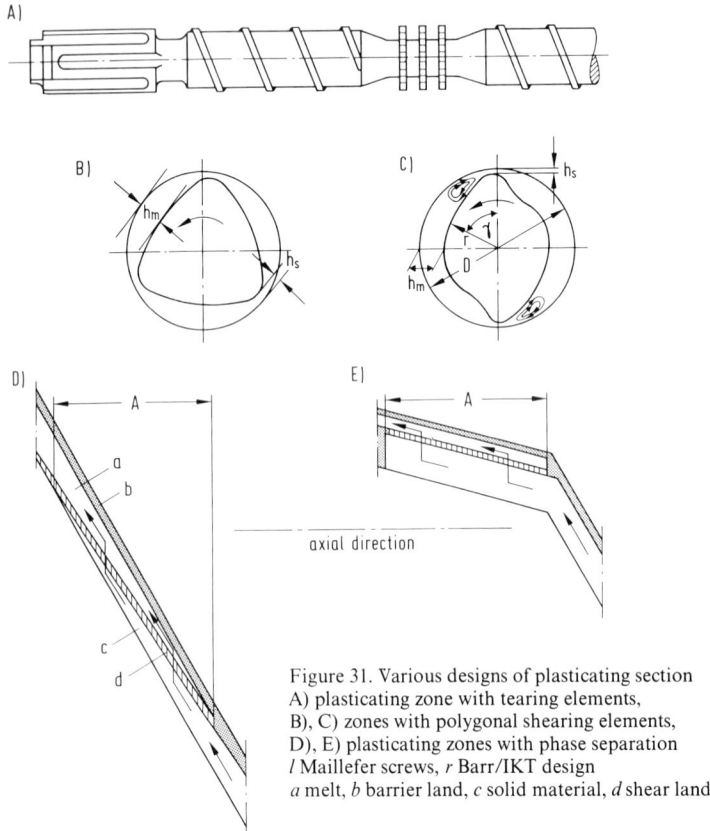

Figure 31. Various designs of plasticating section
A) plasticating zone with tearing elements,
B), C) zones with polygonal shearing elements,
D), E) plasticating zones with phase separation
l Maillefer screws, *r* Barr/IKT design
a melt, *b* barrier land, *c* solid material, *d* shear land

The specific advantages of barrier screws can only be fully exploited, however, when the optimal geometry in the plasticating zone is established by means of an analytical model [40]. It then has to be decided, with a length fixed by line design considerations, and for a specified throughput, whether a given minimum pressure drop over the length of the intensive melting zone can be achieved (Figure 32). Resulting from this are minimum values for drive torque, feed-zone cooling, and wear − and here the wheel comes full circle to the previous design methods. Figure 33 shows such a multichannel intensive melting section, which could be installed between the feed zone and the succeeding mixing section.

12.4.1.2 Dies

Dies for continuous and intermittent extrusion of tubes are dealt with in the same systematic way as in section 12.3.

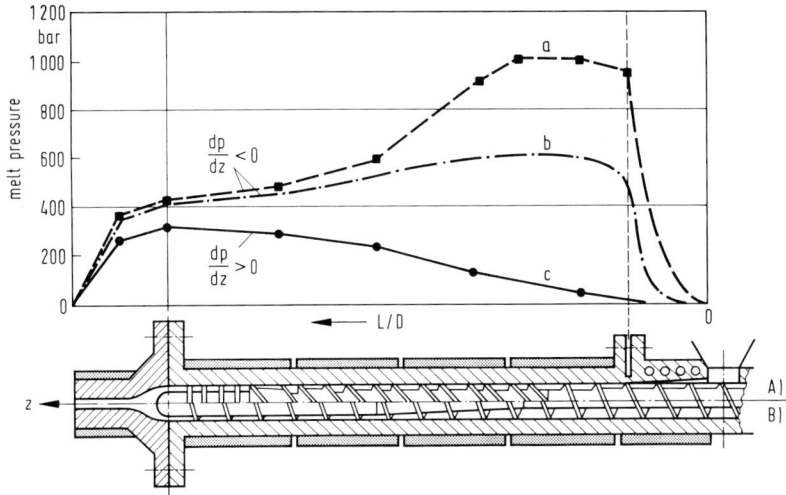

Figure 32. Pressure profiles along the barrel in grooved and conventional extruders
n = const., w = const., material: HDPE
A) grooved extruder, B) conventional extruder
a grooved feed zone, plasticating zone with tearing, shearing and mixing, *b* grooved feed zone, optimized plasticating zone with melt separation, *c* smooth feed zone

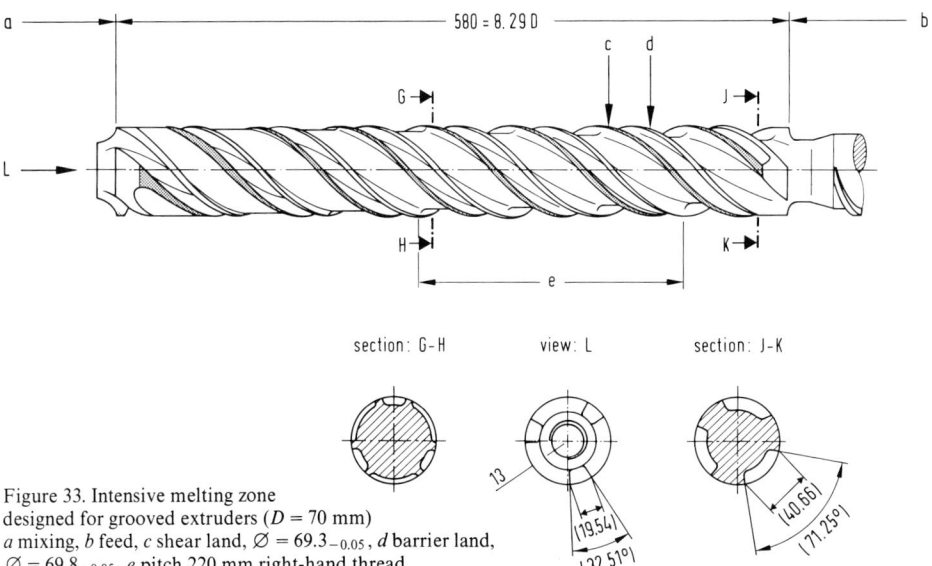

Figure 33. Intensive melting zone
designed for grooved extruders (D = 70 mm)
a mixing, *b* feed, *c* shear land, $\varnothing$ = 69.3$_{-0.05}$, *d* barrier land,
$\varnothing$ = 69.8$_{-0.05}$, *e* pitch 220 mm right-hand thread

12.4.1.2.1 Tube dies for continuous production of parisons

Two basic designs of die are used for the continuous production of tube nowadays, which differ in the way the mandrel is supported: there are *side-fed dies* and *spider-mandrel parison dies*. Both types are available as simple versions, and as those that produce multiple, overlapping flows. It has not yet proved possible to use spiral-mandrel distributors and breaker-plate supported mandrels, well known in blown film and pipe extrusion, for blow molding.

The construction principles of pipe dies, the advantages and disadvantages, and the most important criteria for their structural design are compiled in the sections following.

Side-fed dies

The key task of a side-fed die (Figure 34) is to form the radially fed sheet of melt from the extruder into a tube that leaves the die gap with the same mean velocity, and the same thickness, at all points around its circumference.

The solution to this problem is achieved by giving the supply manifold − consisting of a melt distribution channel and a flow restrictor (or island) zone − the well-known coat-hanger shape, appropriately dimensioned to take into account the processing conditions and the rheological characteristics of the melt. The analysis of the melt distribution problem can be handled two-dimensionally if the ratio of the thickness of the distribution channel to the mandrel diameter is sufficiently small. Model studies of such manifold systems, which assume thin-wall pipe flow in the melt distribution region and flat-slot flow in the restrictor area, can be found in [41 to 44]. A modified design strategy, one in which similar flow elements (flat-slot flow) are combined and which, in theory at least, leads to a solution that is independent of material and flow rate [45, 46], is explained at the end of this section. Side-fed dies designed using this principle have proved satisfactory for processing low- and medium-molecular-weight HDPE (MFI 190/2.16 $\geq$ 0.15 to 1.25 g/10 min.).

A device frequently used in the past for melt distribution consisted of the combination of an annular groove of large cross-section and correspondingly small flow resistance, followed by a parallel flow section with a large flow resistance (the overflowing weir principle [47]. Such designs do not satisfy today's quality criteria.

The side-fed die, with its simple construction and uncomplicated components, is inexpensive, and is still widely used. Because of the slender construction, the internal dimensions of multitube dies can be small, and the forces needed for adjusting the die gap low. This is because the torpedo can be moved, by means of a tension rod fixed axially to it, unlike spider-type dies, where the whole die must be moved. Furthermore, the side-feed concept makes it easy to introduce support air into the tube and to provide mandrel temperature control on larger dies.

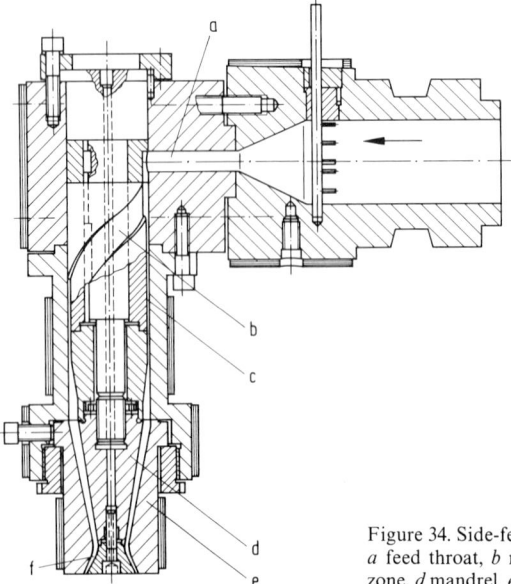

Figure 34. Side-fed die − schematic
a feed throat, b melt distribution channel, c throttling zone, d mandrel, e die body, f exit gap

There are also a number of significant disadvantages to side-fed dies. For one, the tube obtained with a rheologically correct design of melt distributor can show a thickness distribution that is uniform overall initially; but because of the orientation in the flow lines (knit-lines) generated where the divided melt streams come together, it stretches more here than elsewhere and this leads to a thin spot in the blown article, which becomes more apparent with increasing stretch ratio. This difficulty can largely be eliminated by using two melt distributors coaxially and concentrically arranged at 180° to each other: this design will be explained more fully in connection with annular-piston accumulator die heads.

As to the future of side-fed dies, it remains to be seen whether it will be possible to use the calculation methods mentioned, and gradually to optimize these simple melt distribution systems to meet today's quality requirements. In this connection it must always be remembered that only the viscosity functions of polymers are considered in these models. To be able to take into account the effect on melt distribution of the elastic behavior of higher-molecular-weight extrudates with increased melt strength, these models will have to be improved still further.

Analytical design of a melt distribution system for a side-fed die

In the concept presented here, similar flow patterns occur in the distributor channel and the restrictor zone, namely flat-slot flow (Figure 35 B), in contrast to conventional manifolds (Figure 35 A).

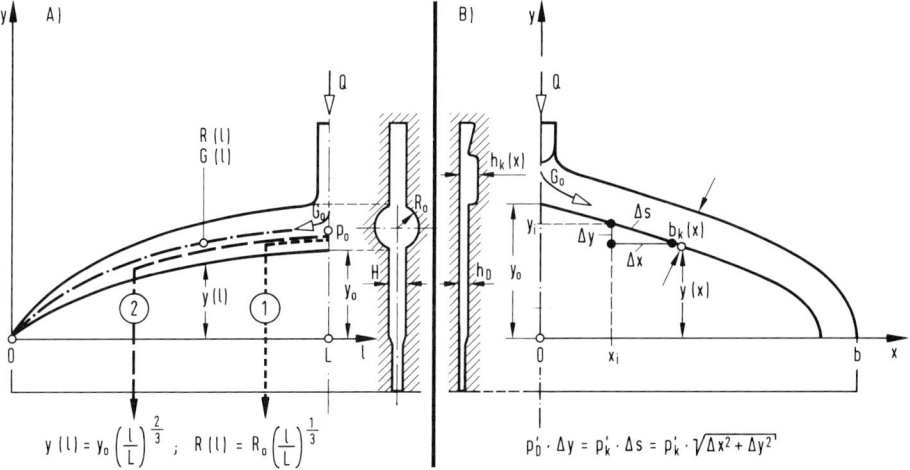

Figure 35. Schematic diagrams of two systems of melt-flow distribution

In order to obtain an extrudate with uniform wall-thickness with this die system, it is a basic requirement that the geometrical tuning of the distributor channel and the restrictor zone should ensure that the *mean exit-velocity around the whole of the die gap is the same*. This requirement is fulfilled if the total pressure drop in the die, taken as the sum of the drop in the distributor and that in the restrictor gap, is independent of flow path (compare Figure 35 A). The requirement for constant average velocity at the die exit determines, automatically, that the volume displacement $G(x)$ at the width coordinate x decreases linearly in each half of the manifold, from the maximum value $G_0 = Q/2\varrho$ to zero (Figure 35 B). Q (in kg/h) is the total rate of melt flow arriving radially from the extruder. The assumptions and boundary conditions underlying the design concept explained below are formulated thus:

a) the flow patterns in the restrictor zone and in the distribution channel should be the same, namely laminar slot flow;
b) the balance of local pressure drops at any point (x_i, y_i) of the manifold contour (Figure 35 B) can be expressed in the form:

$$p'_D \cdot \Delta y = p'_K \cdot \Delta s = p'_K \cdot \sqrt{\Delta x^2 + \Delta y^2} = p'_K \cdot \Delta x \sqrt{1 + \left(\frac{\Delta y}{\Delta x}\right)^2} \qquad (13\,\text{a})$$

which contains the orthogonality condition that the lines $y = $ const. are isobars.

It can be seen at once from Eq. (13a) that the expression used in various studies:

$$p'_D \cdot \Delta y = p'_K \cdot \Delta x \qquad (13\,\text{b})$$

only applies if $(\Delta y/\Delta x)^2 \to 0$.
Because of the requirements stated below this is not generally the case.

c) the structural viscosity of the melt can be described by the Ostwald/de Waele power law:

$$\dot{\gamma} = \phi \tau^m, \qquad (14\,\text{a})$$

or rewritten, using

$$\tau = \tau_0 \cdot \left(\frac{\dot{\gamma}}{\dot{\gamma}_0}\right)^{1/m} \quad \text{and} \qquad (14\,\text{b})$$

$$\eta = \eta_0 \left(\frac{\dot{\gamma}}{\dot{\gamma}_0}\right)^{1/m - 1}, \quad \text{as} \qquad (14\,\text{c})$$

$$\phi = \frac{1}{\eta_0 \cdot \dot{\gamma}_0^{m-1}}, \qquad (14\,\text{d})$$

where τ_0, $\dot{\gamma}_0$, and η_0 are reference values.
d) critical shear rates (material-dependent) shall not be reached:

$$\dot{\gamma}_W < \dot{\gamma}_{1,\,\text{crit}}. \qquad (15)$$

Additional conditions specified are:
e) the optimal melt distribution function shall always be independent of the throughput rate and of the rheological functions of the material.
f) the average dwell time of the melt in the manifold shall be identical for all flow paths.
g) in addition, it is assumed that the flow in the distribution channel and the restrictor gap is the stationary, one-dimensional, isothermal streamline flow of an incompressible polymer melt sticking to the channel walls.

Determination of the manifold contour

The melt delivered by the extruder at rate Q, after a 90° change of direction, enters the distributor system. On symmetry grounds it is sufficient to investigate the half-width $b = B/2$ of the system, which is associated with $G_0 = Q/2\varrho$. Because of the basic requirement that the mean flow velocity $\bar{v}_D$ at any point x_i in the restrictor zone shall have the same value, the volume flow rate in the manifold must decrease linearly with the x-coordinate:

$$G(x) = \frac{Q}{2\varrho} \cdot \left(1 - \frac{x}{b}\right). \qquad (16)$$

12.4 Extrusion blow molding lines

For the average flow velocity in the restrictor gap and manifold we arrive at:

$$\bar{v}_D = \frac{G_0}{b \cdot h_D} = \frac{Q}{2\varrho \cdot b \cdot h_D}, \tag{17}$$

$$\bar{v}_K = \frac{G(x)}{A(x)} = \frac{Q\left(1 - \frac{x}{b}\right)}{2\varrho \cdot h_K(x) \cdot b_K(x)}. \tag{18}$$

It may be noted here that the channel width $b_K(x)$ is measured perpendicular to the tangent of the channel contour. From condition f) it follows that:

$$\frac{\Delta y}{\bar{v}_D} = \frac{\Delta s}{\bar{v}_K} = \frac{A(x)}{G(x)} \cdot \sqrt{\Delta x^2 + \Delta y^2}, \tag{19}$$

from which, using the relationships (13a), (16), (17), and (18) one obtains the ratio of the pressure gradients:

$$\frac{p'_D}{p'_K} = \frac{\Delta s}{\Delta y} = \frac{\bar{v}_{K'}}{\bar{v}_D} = \frac{h_D(b-x)}{h_K(x) \cdot b_K(x)}. \tag{20}$$

Since slot flow should occur both in the square-section manifold and in the restrictor zone, for the general case:

$$G = \frac{b \cdot \dot{\gamma}_0 \cdot h^2}{2(m+2)} \cdot \left[-\frac{h \cdot p'}{2\eta_0 \cdot \dot{\gamma}_0} \right]^m \tag{21}$$

or $\quad G = b \cdot h \cdot \bar{v} \tag{22}$

and we obtain for the pressure gradients in these flow channels:

$$p'_K = -\frac{2\eta_0 \cdot \dot{\gamma}_0}{h_K(x)} \cdot \left[\frac{2(m+2) \cdot \bar{v}_K}{h_K(x) \cdot \dot{\gamma}_0} \right]^{1/m}, \tag{23}$$

$$p'_D = -\frac{2\eta_0 \cdot \dot{\gamma}_0}{h_D} \cdot \left[\frac{2(m+2) \cdot \bar{v}_D}{h_D \cdot \dot{\gamma}_0} \right]^{1/m}. \tag{24}$$

From (20), (23), and (24):

$$\left[\frac{\bar{v}_D}{\bar{v}_K} \cdot \frac{h_K(x)}{h_D} \right]^{1+1/m} = 1, \tag{25}$$

which is satisfied when:

$$\frac{\bar{v}_K}{\bar{v}_D} = \frac{h_K(x)}{h_D}. \tag{26}$$

Combining the relationships (20) and (26) produces the equations:

$$\boxed{h_K(x) = h_D \cdot \sqrt{\frac{b-x}{b_K(x)}} \quad \text{and} \quad b_K(x) = (b-x) \cdot \left(\frac{h_D}{h_K(x)} \right)^2} \tag{27}$$

which in generally valid form connect the geometrical quantities $h_K(x)$, $b_K(x)$, h_D, and b. They show that at the point $x = b - b_K$ the depth of the manifold h_K is equal to h_D, the restrictor-zone gap. The increase in the manifold contour is obtained from (13a), (20), and (26) as:

$$\frac{p'_D}{p'_K} = \frac{\Delta s}{\Delta y} = \sqrt{1 + \left(\frac{\Delta x}{\Delta y}\right)^2} = \frac{\bar{v}_K}{\bar{v}_D} = \frac{h_K(x)}{h_D}$$

or in differential equation form:

$$\frac{dy}{dx} = -\frac{1}{\sqrt{\left(\frac{h_K(x)}{h_D}\right)^2 - 1}} = -\sqrt{\frac{b_K(x)}{b - x - b_K(x)}} \qquad (28)$$

from which the manifold contour $y = f(x)$ results.
The initial slope of the function $y = f(x)$ and its slope at the point $x = b - b_K$ are given by the following equations:

$$\left(\frac{dy}{dx}\right)_{x=0} = \frac{1}{\sqrt{\frac{b}{b_K(0)} - 1}}, \qquad (29)$$

$$\left(\frac{dy}{dx}\right)_{x=b-b_K} = -\infty, \qquad (30)$$

which means that the manifold finishes with a vertical tangent (see Figure 35 B).

Geometrical design of the manifold

There are two possibilities for dimensioning the rectangular distribution channel (Figure 36):

— use a constant channel width:

$$b_K(x) = b_K = \text{const.}$$

— keep the profile quotient — the ratio of channel width to channel depth — constant:

$$h_K(x)/b_K(x) = \beta = \text{const.}$$

Case I: constant channel width

Using the dimensionless quantities

$$\eta = y/b, \quad \xi = x/b, \quad \alpha = b_K/b, \qquad (31 \text{ a to c})$$

and by integration of differential Equation (28), one obtains the manifold contour; this is shown in Figure 37 with the relative channel width α as parameter. Depth as a function of position is given by:

$$h_K(\xi) = h_D \cdot \sqrt{\frac{1-\xi}{\alpha}}. \qquad (32)$$

Finally the relationships between the largest value of restrictor-gap length and the channel width at the point $\xi = 0$ may be noted:

$$\eta_0 = \frac{y_0}{b} = 2\sqrt{\alpha} \cdot \sqrt{1-\alpha} \qquad (33)$$

$$h_K(\xi = 0) = h_D \cdot \sqrt{1/\alpha}. \qquad (34)$$

12.4 Extrusion blow molding lines

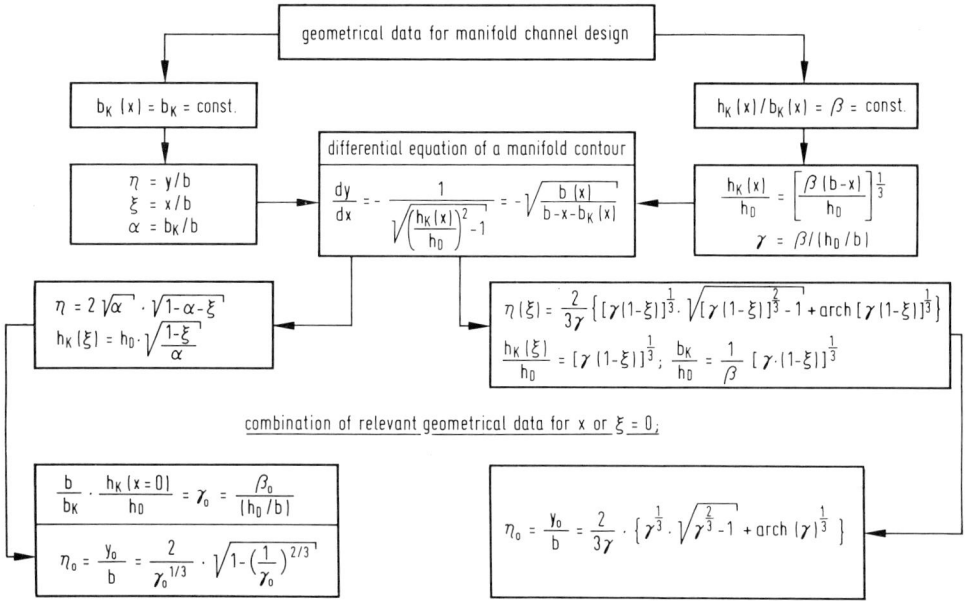

Figure 36. Formulae used in the geometric design of melt distribution channels (manifolds)

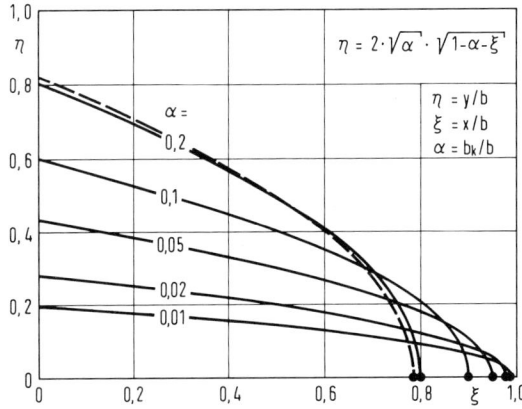

Figure 37. Manifold contours for various channel widths α (case I: b_k = const.)

It can be seen from Eq. (33) that the maximum length of the restrictor zone η_0 increases very rapidly with increasing channel width.

For x and $\xi = 0$, a connection between the relevant geometrical dimensions, from which the area of application for this design approach can also be deduced, is given by the correlation function:

$$\eta_0 = \frac{y_0}{b} = \frac{2}{\gamma_0^{1/3}} \cdot \sqrt{1 - \left(\frac{1}{\gamma_0}\right)^{2/3}} \tag{35}$$

which is shown in Figure 38.

In specific design situations, b and Q and the rheological functions of the melt are given. By taking into account structural and strength considerations, and on the basis of a given restrictor zone width h_D, an appropriate geometrical design for the manifold can be arrived at with the help of Figure 38.

The maximum permissible pressure drop Δp_{zul} and/or a critical shear rate $\dot{\gamma}_{1,\text{krit}}$ can be taken as the criterion for determining the width h_D of the restrictor zone (Figure 39).

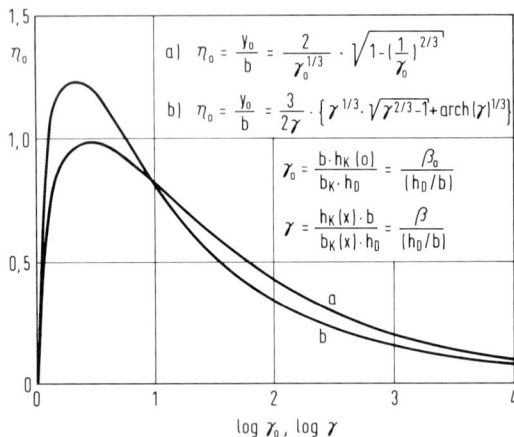

Figure 38. Maximum length η_0 of the flow-restricting gap as a function of the parameter γ_0 or γ

Case II: profile quotient constant

With $\beta = \text{const.}$; using the relationships

$$\frac{h_K(x)}{h_D} = \left[\frac{\beta(b-x)}{h_D}\right]^{1/3} \tag{36}$$

and

$$\gamma = \frac{\beta}{(h_D/b)}, \tag{37}$$

and by integration of Eq. (28), which has general validity, we arrive at the definitive equation for the manifold contour in the form:

$$\eta(\xi) = \frac{3}{2\gamma} \cdot \{[\gamma(1-\xi)]^{1/3} \cdot \sqrt{[\gamma(1-\xi)]^{2/3} - 1} + \text{arch}\,[\gamma(1-\xi)]^{1/3}\}. \tag{38}$$

The function $\eta(\xi)$ with $\dot{\gamma}$ as parameter is plotted in Figure 40. The relationships

$$\frac{h_K(\xi)}{h_D} = [\gamma(1-\xi)]^{1/3} \tag{39a}$$

and

$$\frac{b_K(\xi)}{h_D} = \frac{1}{\beta}[\gamma(-\xi)]^{1/3} \tag{39b}$$

provide the relevant depths and widths of manifolds of rectangular cross-section. The maximum restrictor-zone length for $\xi = 0$ turns out as:

$$\eta_0 = \frac{y_0}{b} = \frac{3}{2\gamma}\{\gamma^{1/3} \cdot \sqrt{\gamma^{2/3} - 1} + \text{arch}\,(\gamma)^{1/3}\}. \tag{40}$$

12.4 Extrusion blow molding lines

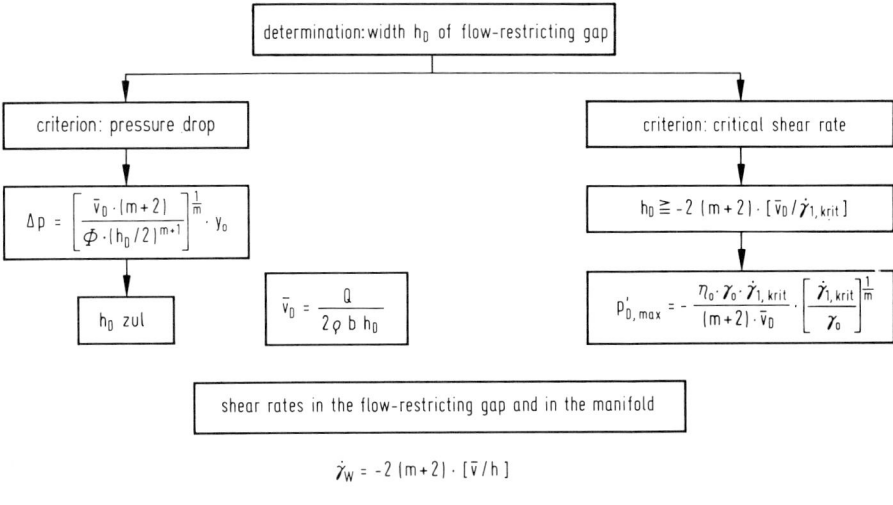

Figure 39. Formulae for determining the width h_D of the flow-restricting gap

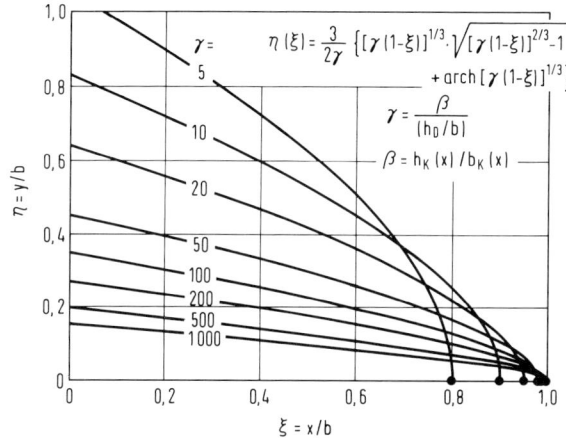

Fig. 40. Manifold contours for various values of γ (case II: β = const.)

It can be seen from the plot of the function $\eta_0(\gamma)$ in Figure 38 that short lengths can be achieved only for large values of γ, which result from either small ratios of h_D to b, or from the largest possible values of the profile quotient β ($0.2 \leq \beta \leq 1.0$). Rectangular cross-sections with side dimensions in such ratios cannot, however, be treated as flat slots: furthermore the hold-up action of the sides of the channel itself requires that the modified correction factor, f_p, which is related to the shear sensitivity ("structural viscosity") of the melt, should be taken into account [45]. A wide selection of side-fed dies (summarized in Figure 41) have been dimensioned using the design approach just explained, automatically machined on a numerically controlled multiaxis end-mill, and experimentally checked. It was shown that a manifold contour with b_k = const. (Case I) gave the best results for tube-wall thickness uniformity. The same method was equally successful in the profiling of manifolds for 250 mm-wide profile dies (Figure 41 B), of multilayer blown film dies, and wire-coating dies.

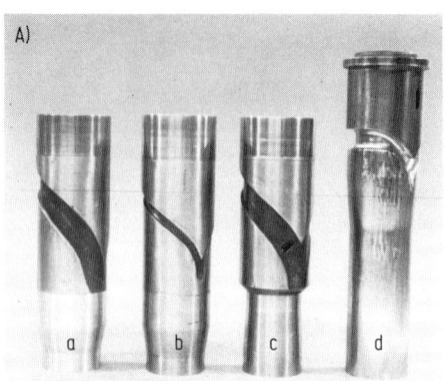

Figure 41 A. Side-fed dies designed according to various strategies
a case I, b case II, c case I modified, d empirically determined

Figure 41 B. Profile-extrusion die combined with a manifold system (case I)

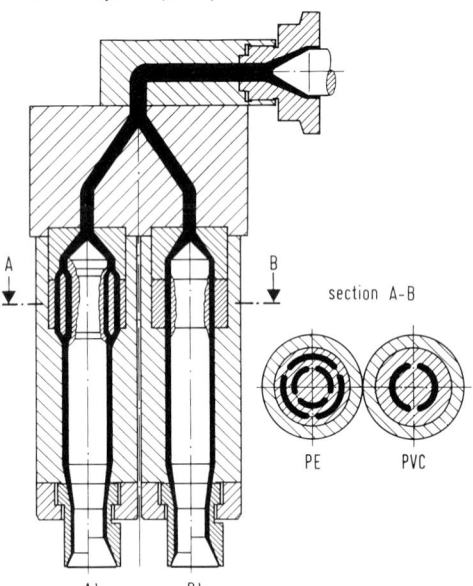

Figure 42. Schematic arrangement of spider-type extrusion dies with simple mandrel support B and staggered-leg support A

Spider-mandrel dies

The spider-mandrel dies shown diagrammatically in Figure 42 come into use when side-fed dies are pushed to their limits because the self-cleaning time is too long, or the dwell time spectrum is too wide. The center feeding of the spider assures uniform axial flow velocity at all points around the circumference of the tubular parison, and therefore a good uniformity of wall thickness.

The knit lines associated with the radial supports are visible on the tube wall after blowing, to an extent that depends on the melt relaxation time. The effect can be diminished somewhat by using offset spiders (Figure 42 A).

Single-spider heads

It is advisable to use single-spider heads (Figure 43) with supports at 180° when processing polymers like PVC, PC, PET and PAN, which react sensitively to high temperatures or long dwell times ("thermal overload"), and which also have short relaxation times.

Proper design generally allows the two supports ("spiders") to be placed in the parting plane of the mold, irrespective of the design of the rest of the machine or the mold arrangement. The consequence is that knit lines originating from the spiders are hardly visible on the finished part.

The melt feed to these heads goes through a streamlined 90° elbow to the spider, and the tip of the spider expands the melt curtain to form a tube. Dimensioning of the melt channel as a whole can be carried out by the analytical methods indicated below, which can ensure that dwell times are not too great, and that operational shear rates and pressure gradients cannot reach unacceptably high levels.

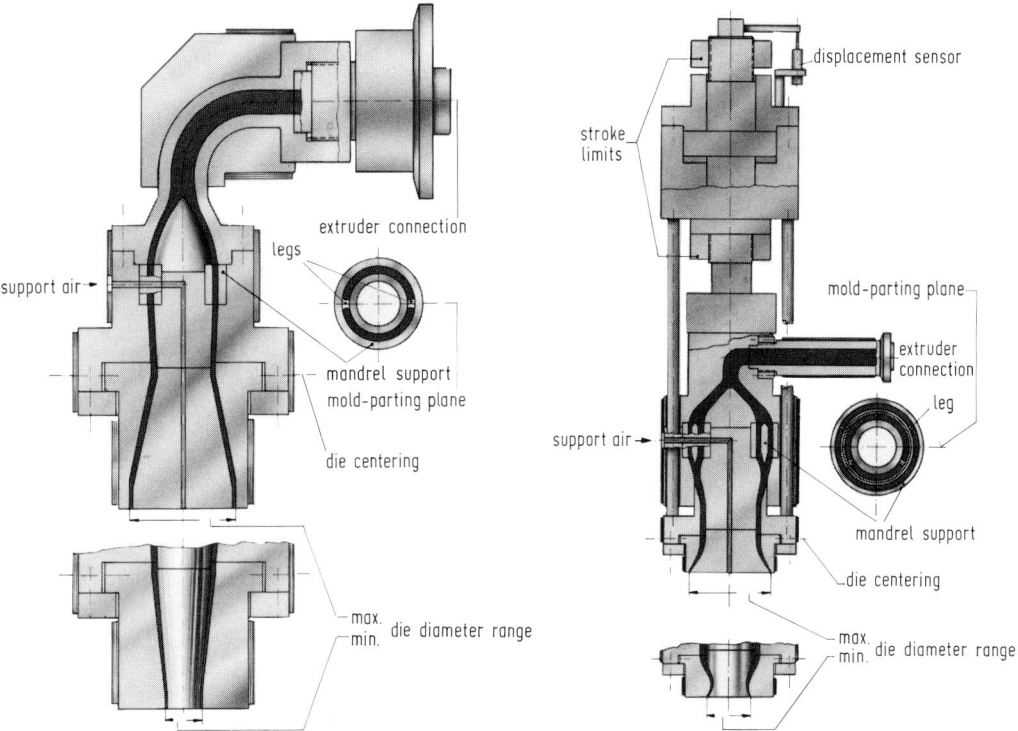

Figure 43. Spider-type parison die for PVC

Figure 44. Staggered-leg spider-type parison die for HMW-HDPE and PP

Offset-spider heads

The problem of formation of thin spots in the blown part can essentially be eliminated by use of offset-spider heads to give overlapping melt streams (Figure 44). These are certainly expensive to make, needing spark erosion for the holes; and they complicate the supply of tube-support air and require large forces for die adjustment, which is carried out hydraulically by displacing the die ring with externally mounted push-pull bolts. In spite of this expensive method of construction, offset-spider heads are in widespread use today for processing PP and high-molecular-weight HDPE; and for all other thermally insensitive thermoplastics and elastomers they also offer advantages of principle in comparison with single-spider heads [48].

The offset spider (Figure 45) is the heart of such extrusion dies, and has a big effect on product quality. Its dimensions are fixed by the melt throughput required and the rheo-

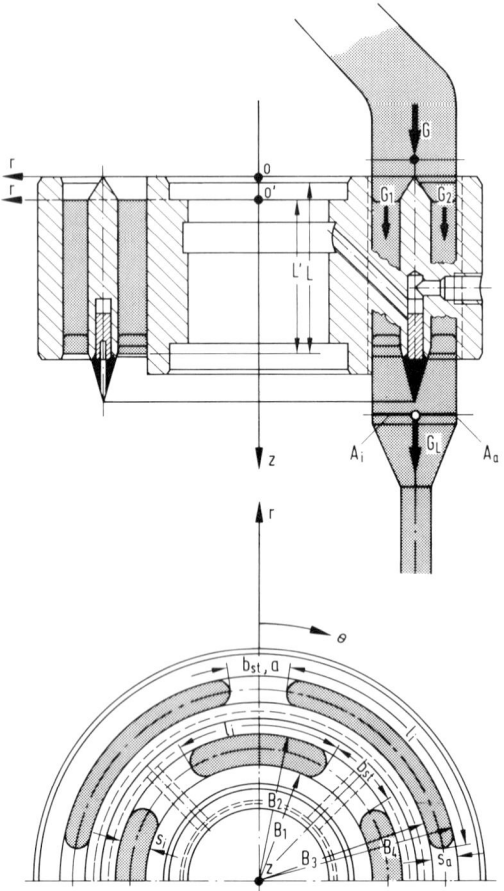

Figure 45. Staggered-leg mandrel support for a tube die

logical properties of the material. The sizes of the openings must be such that the maximum shear rate locally does not exceed 30 s^{-1}. In order to keep the elastic (or reversible deformation) effects to a minimum when processing HM-HDPE, these shear rates should be reduced further to $10 \leq \dot{\gamma}_{max} \leq 20$ s^{-1} [49, 50]. The number of radial supports required is determined by the size of the head and the associated question of mechanical strength and stiffness. There are die designs on the market with spiders having 2×2, 2×3, 2×4, 2×6, and 2×8 radial legs, depending on the size of the die. It is interesting that the number of legs also has direct reactions on the cross-sectional geometry of the tube: because of the deformations that the melt undergoes in passing through the spider, a 2×4 spider gives a four-sided profile, and a 2×6 spider one that is like a ring with six flats on it (Figure 46).

It was realized quite early on that if the inner and outer openings of offset spiders had the same gap width ($s_i = s_a$), an excess of melt would be supplied to the peripheral region. This is caused especially by the greater influence of the flanks of the inner radial legs. If now the mean flow velocities across the annular surfaces A_i and A_a after the spider are to have the same value $\bar{v}$, s_i must be greater than s_a. The following analytical method shows how the determination of these two quantities is carried out.

In the rheological treatment of the problem, the adjacent parallel openings can be considered approximately as plane slits. The effect of the side-faces of the radial legs is taken

12.4 Extrusion blow molding lines

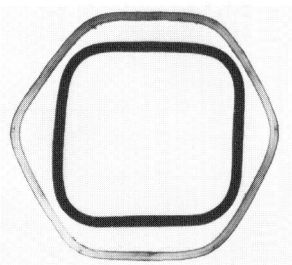

Figure 46. Changes in tube cross-section caused by deformation of the melt in the mandrel support system

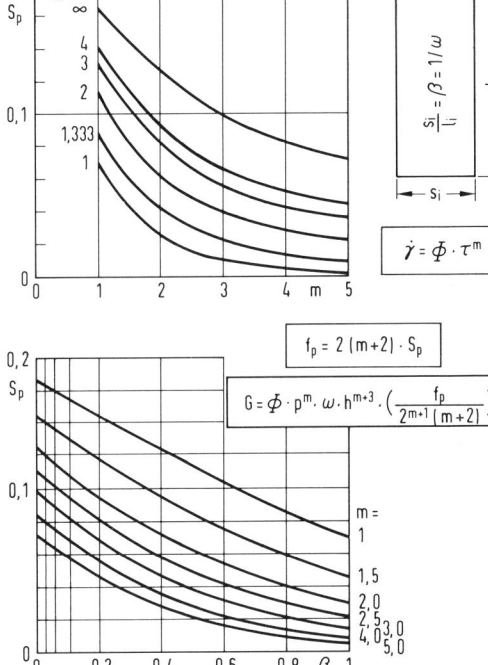

Figure 47. Effect of walls on volume throughput, G, for openings of rectangular cross-section $s_i \cdot l_i$: plots of S_p and f_p for Ostwald power-law melts, and formulae for G

into account by a correction factor, f_p, in the equation describing flow through a hole. These factors are dependent on the exponent m of the flow power law, and on the s/l ratio, as shown in Figure 47.

It can be shown that the ratio of surface areas A_a/A_i in the flow channel just after the spider is described by the equation

$$\frac{A_a}{A_i} = \frac{4B_4^2 - (B_2+B_3)^2}{(B_2+B_3)^2 - 4B_1^2} . \tag{41}$$

And it follows from Figure 45 that the widths of the holes are:

$$s_i = \tfrac{1}{2}(B_2 - B_1) \tag{42}$$

$$s_a = \tfrac{1}{2}(B_4 - B_3). \tag{43}$$

It is now required that $\bar{v}_i = \bar{v}_a = \bar{v} = $ const., whence:

$$\frac{A_a}{A_i} = \frac{G_a}{G_i} \tag{44}$$

where G stands for the volume flow rate.
Using the die equation $G = \phi\, p^m / w$, which incorporates the *Ostwald/de Waele* power law, it follows that

$$\frac{G_a}{G_i} = \frac{w_i}{w_a}, \tag{45}$$

in which, quite generally, for a non-planar rectangular gap we have

$$w = \frac{s^{m-1}(m+2) \cdot L^m}{l \cdot s^{m+2} \cdot f_p}. \tag{46}$$

From (45) and (46) it follows that

$$\frac{G_2}{G_1} = \frac{A_2}{A_1} = \frac{l_a \cdot s_a^{m+2} \cdot f_{p,a}}{l_i \cdot s_i^{m+2} \cdot f_{p,i}}. \tag{47}$$

For the hole widths it is clear that

$$l_a = \frac{\pi}{2k}(B_3 + B_4) - b_{St,a}, \tag{48}$$

$$l_i = \frac{\pi}{2k}(B_1 + B_2) - b_{St,i} \tag{49}$$

(k = number of radial legs).

Finally, from Eqns. (41) and (47) we obtain:

$$s_i' = s_a \left(\frac{A_2 \cdot l_i \cdot f_{p,i}}{A_1 \cdot l_a \cdot f_{p,a}} \right)^{-\frac{1}{m+2}}. \tag{50}$$

Eqn. (47) can also be solved for s_a', and the geometry of the outer channel determined. The new value of s_i' provides fresh spider dimensions with which another iteration step can be carried through: this process becomes rapidly convergent.

Each of the diameters B_1, B_2, B_3, and B_4 can be changed at will by use of Eqn. (50). Physical alteration of B_1 or B_2 on an existing spider is possible by machining, but changes to B_3 or B_4 can only be made during initial manufacture.

With all tube-extrusion dies, and particularly with the offset-spider dies predominantly used for HM-HDPE, the distributions of velocities, shear rates and dwell times are of interest, as are the pressure gradient and the pressure drop along the melt channel. The values of maximum shear rate occurring at the walls are of special importance therefore, because HDPE shows pronounced flow anomalies, as can be seen from the relevant flow curves (Figure 48). Above a critical shear rate $\dot{\gamma}_{krit}$, which depends on molecular weight and melt temperature, first of all slip-stick effects and then slip processes occur, which affect the tube quality adversely [51] or can even make troublefree blowing impossible. From this point of view, at the very least, it seems useful to dimension melt channels and dies with the help of simple equations.

Actually, multidimensional flow patterns occur in a number of sections along the channel, which can be described properly only by finite-element methods [52]. Nevertheless the equations and inferences set out below can be useful aids in the design and choice of tube dies.

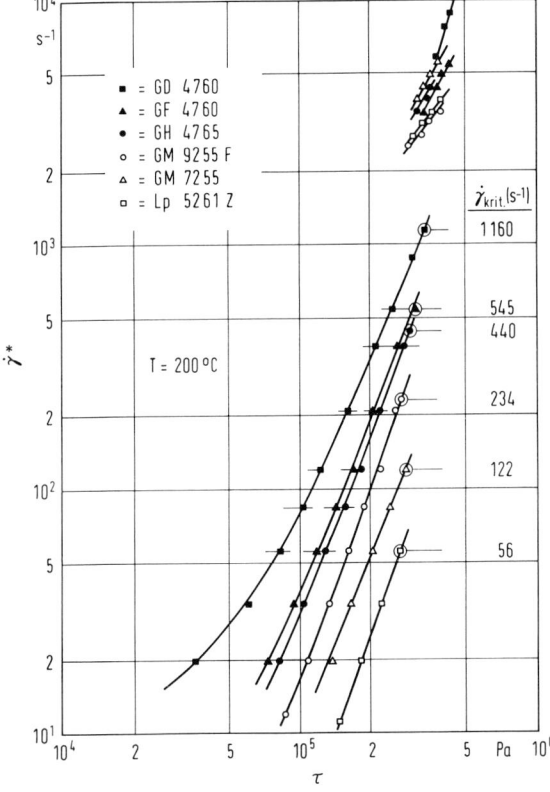

Figure 48. Flow curves for HDPE grades of various molecular weights

The melt flow between the melt inlet and the die exit in the dies described earlier, goes through channel segments of different lengths; these may be cylindrical or conical bores, or have the form of cylindrical, divergent or convergent annular channels with a constant gap, or one that changes along the direction of flow. For a flow geometry consisting of a succession of n channel sections, the *die equation* formulated for an individual element

$$G = k \cdot \phi \cdot p^m \tag{51}$$

must be modified. In the case of an incompressible polymer melt, because the continuity equation applies

$$G = G_1 = G_2 = G_n = \text{const.} \tag{52}$$

The values of pressure loss p_i associated with the geometry of individual channel sections make up the total pressure drop p_{ges} between the melt inlet and the die exit

$$p_{\text{ges}} = \sum_{i=1}^{n} p_i. \tag{53}$$

Correspondingly, for individual regions Eq. (51) gives

$$p_i = \left(\frac{G}{k_i \cdot \phi}\right)^{1/m}. \tag{54}$$

If Eq. (54) is inserted into Eq. (53) one obtains

$$G = \phi \cdot p_{\text{ges}}^m \left[\sum_{i=1}^{n} \left(\frac{1}{k_i}\right)^{1/m}\right]^m. \tag{55}$$

The flow impedance of the extrusion tool as a whole, w_{ges}, for the situation with all sections in series is

$$w_{\text{ges}} = \left[\sum_{i=1}^{n} (1/k_i)^{1/m}\right]^m = \left[\sum_{i=1}^{n} w_i^{1/m}\right]^m. \tag{56}$$

If flow-channel impedances work in parallel in a particular section of the tool — a typical example is the spider mandrel — the melt conductance value of this zone as already noted is given by

$$k_{\text{ges}}^* = \sum_{i=1}^{n^*} k_i, \tag{56a}$$

$$\frac{1}{w_{\text{ges}}^*} = \sum_{i=1}^{n^*} \frac{1}{w_i}. \tag{56b}$$

For *channels whose cross-section is not constant along the direction of flow* — typical examples being conical bores and conical annular gaps — it is in fact no longer valid to use the differential form of the basic equations, discussed earlier, for flow in infinitely long channels of constant cross-section. This is because the flow is *not* one-dimensional. However, if the change in cross-section is kept within certain limits, one can use the differential form of the infinite-length equation at each point of such a channel section with good approximation. In this procedure the geometrical quantities that vary with z — like diameter or gap width — are expressed as a function of the length coordinate. The flow impedance of such a channel of varying cross-section turns out to be a generalization of the series-coupling of elements of impedance, thus

$$w_i = \left[\int_{z=0}^{L} (w'(z))^{1/m} \cdot dz\right]^m. \tag{57}$$

For the pressure profile we obtain

$$p(z) = \int_{z}^{z=L} \frac{\partial p}{\partial z}(z) \cdot dz \tag{58}$$

and the *dwell time per unit of length*, which corresponds to the reciprocal of the local average flow velocity $\bar{v}(z)$, is given by

$$\frac{dt_v}{dz} = \frac{1}{\bar{v}(z)} = \frac{A(z)}{G}. \tag{59}$$

These particular calculations have been carried out for the channel geometries in most frequent practical use. The resulting equations of state for the most important mold-dimensioning parameters are compiled in Tables 2 to 4. They provide the basis for approximate calculation of pressure drop, shear stress, shear rate, and of mean dwell times along the length of the melt channel. An example will now be given of a model design calculation for the centrally fed, offset-spider-mandrel tube die shown schematically in Figure 49, modified slightly in the die feed region. The following values have to be determined for the particular channel geometry:

- the pressure drop along the length of the flow channel $p(z)$, and the total pressure drop p_{ges};
- the maximum shear stresses τ_W at the channel walls, and maximum shear rates $\dot{\gamma}_W$ along the channel;
- the mean dwell time per unit of length (Eq. 59) along the channel, and the mean overall dwell time.

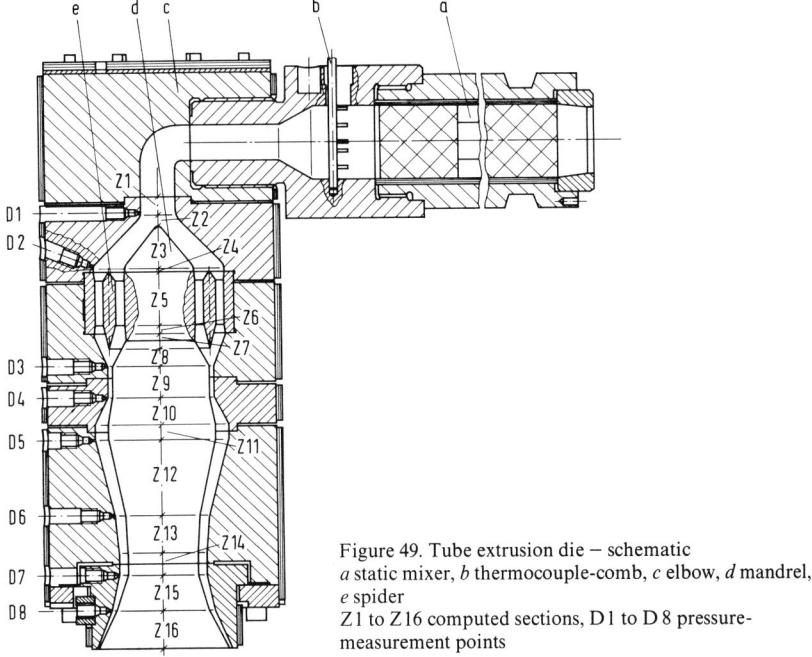

Figure 49. Tube extrusion die – schematic
a static mixer, b thermocouple-comb, c elbow, d mandrel, e spider
Z1 to Z16 computed sections, D1 to D8 pressure-measurement points

From a consideration of these quantities, conclusions can, if necessary, be drawn about stagnation zones, about shear rates that could lead to flow anomalies, and about regions of unnecessary pressure drop.

Basic operational and material parameters used are:

Melt throughput rate	$\dot{m}$	= 100 kg/h
Melt temperature	T_M	= 210 °C
Material		Hostalen GF 7740 F
with		
flow exponent	m	= 1.79
fluidity	ϕ	= 35.8 bar$^{-m}\cdot$s^{-1}
	$\varrho_{210°}$	= 0.77 g/cm^3.

The calculation of these quantities is carried out by a data processor, using a computer program based on the equations of state compiled in Tables 2 to 4.

Table 2. Simple design calculations for flow channels of circular cross-section

Flow channel geometry / Parameter	Cylindrical bore	Conical bore	Circular cross-section, optional profile $D = D(z)$
Eqns. of the contour	$D(z) = D_E = $ const.	$D(z) = D_E + (D_A - D_E) \cdot \frac{z}{L}$ $\alpha = \arctan \frac{D_A - D_E}{2L}$; $\beta = \frac{D_E}{D_A}$	$D(z) = D_A \cdot f(\zeta)$ $\zeta = \frac{z}{L}$; $\beta = \frac{D_E}{D_A}$
Cross-sectional area	$A(z) = \frac{\pi}{4} \cdot D_E^2 = $ const.	$A(z) = \frac{\pi}{4} \cdot D^2(z)$	$A(z) = \frac{\pi}{4} \cdot D^2(z)$
Overall flow impedance	$W_{ZB} = \frac{2^{2m-3} \cdot (m+3) \cdot L^m}{\pi \cdot D_E^{m+3}}$	$W_{KB} = W_{ZB} \cdot \left(\frac{m}{3}\right)^m \cdot \beta^m \cdot \left(\frac{1-\beta^{3/m}}{1-\beta}\right)^m$	$W_{ges} = W_{ZB} \cdot \beta^{m+3} \cdot \left\{ \int_0^1 \frac{d\zeta}{[f(\zeta)]^{\frac{m+3}{m}}} \right\}^m$
Pressure profile	$p(z) = \left(\frac{G \cdot W_{ZB}}{\Phi}\right)^{1/m} \cdot \left(1 - \frac{z}{L}\right)$	$p(z) = \left(\frac{G \cdot W_{ZB}}{\Phi}\right)^{1/m} \cdot \beta^{\frac{m+3}{m}} \cdot \frac{m}{3} \cdot \frac{1}{1-\beta} \cdot \left[\frac{1}{[\beta+(1-\beta)\cdot\frac{z}{L}]^{3/m}} - 1\right]$	$p(z) = \left(\frac{G \cdot W_{ZB}}{\Phi}\right)^{1/m} \cdot \beta^{\frac{m+3}{m}} \cdot \int_\zeta^1 \frac{d\zeta}{[f(\zeta)]^{\frac{m+3}{m}}}$
Shear stress at the channel wall	$\tau_w(z) = \left(\frac{G \cdot W_{ZB}}{\Phi}\right)^{1/m} \cdot \frac{D_E}{4 \cdot L} = $ const.	$\tau_w(z) = \left(\frac{G \cdot W_{ZB}}{\Phi}\right)^{1/m} \cdot \frac{D_E}{4 \cdot L} \cdot \left(\frac{\beta}{\beta + (1-\beta)\cdot\frac{z}{L}}\right)^{3/m}$	$\tau_w(z) = \left(\frac{G \cdot W_{ZB}}{\Phi}\right)^{1/m} \cdot \frac{D_E}{4 \cdot L} \cdot \left(\frac{\beta}{f(\zeta)}\right)^{3/m}$
Shear rate at the channel wall	$\dot{\gamma}_w(z) = G \cdot W_{ZB} \cdot \left(\frac{D_E}{4 \cdot L}\right)^m = $ const.	$\dot{\gamma}_w(z) = G \cdot W_{ZB} \cdot \left(\frac{D_E}{4 \cdot L}\right)^m \cdot \left(\frac{\beta}{\beta + (1-\beta)\cdot\frac{z}{L}}\right)^3$	$\dot{\gamma}_w(z) = G \cdot W_{ZB} \cdot \left(\frac{D_E}{4 \cdot L}\right)^m \cdot \left(\frac{\beta}{f(\zeta)}\right)^3$

12.4 Extrusion blow molding lines

Table 3. Simple design calculations for flow channels of annular cross-section, with gap width $s = \text{const.}$

Parameter	Flat slot and cylindrical annular gap (Gap width constant along z)	Conical annular gap (Gap width constant along z)
Flow-channel geometry	Gap width — Flat slot: $b \gg s$; ($s \hat{=} h$) Cyl. annular gap: $b = \pi \cdot \bar{D}_E$, $\bar{D}_E \gg s$, $\bar{D}_E = \text{const.}$	
Eqns. of the contour		$\bar{D}(z) = \bar{D}_E + (\bar{D}_A - \bar{D}_E) \cdot \dfrac{z}{L}$ $\alpha = \arctan \dfrac{\bar{D}_A - \bar{D}_E}{2 \cdot L}$; $\beta = \dfrac{\bar{D}_E}{\bar{D}_A}$
Cross-sectional area	$A(z) = \pi \cdot \bar{D}_E \cdot s = \text{const.}$	$A(z) = \pi \cdot \bar{D}(z) \cdot s$
Overall flow impedance	$w_{zR} = \dfrac{2^{m+1} \cdot (m+2) \cdot L^m}{b \cdot s^{m+2}}$	$w_{kR} = w_{zR} \cdot \left(\dfrac{m}{m-1}\right)^m \dfrac{\beta}{(\cos\alpha)^m} \dfrac{\beta^{1/m}}{\cos\alpha} \cdot \left(\dfrac{1-\beta^{\frac{m-1}{m}}}{1-\beta}\right)^m$; $m \ne 1$
Pressure profile	$p(z) = \left(\dfrac{G \cdot w_{zR}}{\Phi}\right)^{1/m} \cdot \left(1 - \dfrac{z}{L}\right)$	$p(z) = \left(\dfrac{G \cdot w_{zR}}{\Phi}\right)^{1/m} \dfrac{m}{m-1} \left(\dfrac{\beta}{\beta+(1-\beta)\frac{z}{L}}\right)^{1/m} \dfrac{1-\left[\beta+(1-\beta)\frac{z}{L}\right]^{\frac{m-1}{m}}}{1-\beta}$; $m \ne 1$
Shear stress at the channel wall	$\tau_w(z) = \left(\dfrac{G \cdot w_{zR}}{\Phi}\right)^{1/m} \cdot \dfrac{s}{2L} = \text{const.}$	$\tau_w(z) = \left(\dfrac{G \cdot w_{zR}}{\Phi}\right)^{1/m} \cdot \dfrac{s}{2L} \cdot \left(\dfrac{\beta}{\beta+(1-\beta)\frac{z}{L}}\right)^{1/m}$
Shear rate at the channel wall	$\dot{\gamma}_w(z) = G \cdot w_{zR} \left(\dfrac{s}{2L}\right)^m = \text{const.}$	$\dot{\gamma}_w(z) = G \cdot w_{zR} \cdot \left(\dfrac{s}{2L}\right)^m \cdot \dfrac{\beta}{\beta+(1-\beta)\frac{z}{L}}$

Table 4. Simple design calculations for flow channels of annular cross-section, with gap width $s = s(z)$

Flow-channel geometry Parameter	Cylindrical annulus: gap varying linearly with z	Annulus: optional variation of gap, s and radius R along z
Eqns of the contour	$\bar{D}(z) = \bar{D}_E =$ const.; $s(z) = s_E + (s_A - s_E) \cdot \frac{z}{L}$ $\gamma = \frac{s_E}{s_A}$	$\bar{D} = \bar{D}(z)$; $s = s(z)$ $\bar{R}(z) = \bar{D}(z)/2$
Cross-sectional area	$A(z) = \pi \cdot \bar{D}_E \cdot s(z)$	$A(z) = \pi \cdot \bar{D}(z) \cdot s(z)$
Overall flow impedance	$w_{zRS} = w_{zR} \cdot \left(\frac{m}{2}\right)^m \cdot \gamma^m \left(\frac{1 - \gamma^{2/m}}{1 - \gamma}\right)^m$	$w_{ges} = w_{zR} \cdot \bar{D}_E \cdot s_E^{m+2} \cdot \left[\int_0^L \frac{\sqrt{1 + \left(\frac{d\bar{R}(z)}{dz}\right)^2}}{\bar{D}^{1/m}(z) \cdot s^{\frac{m+2}{m}}(z)} dz\right]^m \cdot \frac{1}{L}$; $w_{zR} = w_{zR}(s_E, \bar{D}_E)$
Pressure profile	$p(z) = \left(\frac{G \cdot w_{zR}}{\Phi}\right)^{1/m} \cdot \frac{m}{2} \cdot \gamma^{\frac{m+2}{m}} \cdot \frac{1}{1-\gamma} \cdot \left[\frac{1}{[\gamma + (1-\gamma) \cdot \frac{z}{L}]^{2/m}} - 1\right]$	$p(z) = \left(\frac{G \cdot w_{zR}}{\Phi}\right)^{1/m} \cdot (\bar{D}_E \cdot s_E^{m+2})^{1/m} \int_z^L \frac{\sqrt{1 + \left(\frac{d\bar{R}(z)}{dz}\right)^2}}{[\bar{D}(z)]^{1/m} \cdot [s(z)]^{(m+2)/m}} dz$
Shear stress at the channel wall	$\tau_w(z) = \left(\frac{G \cdot w_{zR}}{\Phi}\right)^{1/m} \cdot \frac{s_E}{2L} \left(\frac{\gamma}{\gamma + (1-\gamma) \cdot \frac{z}{L}}\right)^{2/m}$	$\tau_w(z) = \left(\frac{G \cdot w_{zR}}{\Phi}\right)^{1/m} \cdot \frac{s_E}{2L} \left[\frac{\bar{D}_E}{\bar{D}(z)} \cdot \left(\frac{s_E}{s(z)}\right)^2\right]^{1/m}$
Shear rate at the channel wall	$\dot{\gamma}_w(z) = G \cdot w_{zR} \cdot \left(\frac{s_E}{2L}\right)^m \left(\frac{\gamma}{\gamma + (1-\gamma) \cdot \frac{z}{L}}\right)^2$	$\dot{\gamma}_w(z) = G \cdot w_{zR} \cdot \left(\frac{s_E}{2L}\right)^m \cdot \frac{\bar{D}_E}{\bar{D}(z)} \cdot \left(\frac{s_E}{s(z)}\right)^2$

Input data for cylindrical and conical bores are the diameters D_E, and D_A, at the inlet and outlet, and the respective channel lengths L (compare sections 1 and 2 in Table 5). The input data for the various annular gap geometries are the inner and external diameters of the channels at their inlets and outlets, which are shown in Table 5 as $D_{E,i}$ and $D_{E,a}$; and $D_{A,i}$ and $D_{A,a}$, respectively, together with the channel lengths. The quantities $\bar{D}_E$, $\bar{D}_A$, $\bar{D}(z)$, s_E, s_A, and $s(z)$ are computed before the main calculation, using the contour equations for each channel in a sub-program.

The overall pressure drop calculated in this case was 141.5 bar and the average melt dwell time $\bar{t}_v$ came to 30.1 s.

In Figure 50 the functions $p(z)$, $\tau_W(z)$, $\dot{\gamma}_W(z)$, and $dt_v/dz = 1/\bar{v}$ are shown, as well as the channel geometry, from which the following details can be recognized:

- large pressure gradients exist in the land zone and in the die-exit gap. The constriction following the spider, by causing higher extensional and shear forces, should improve the welding together of the divided streams produced by the spider legs;
- the highest shear rates ($\dot{\gamma} \approx 495$ s^{-1}) occur at the end of the die, but do not exceed the critical value $\dot{\gamma}_{krit}$ which is 1000 s^{-1} for GF 7740 F;
- immediately before and after the spider the mean specific dwell times become rather large, because of the large cross-section of the channel.

One can arrive at a relatively simple and rapid statement about the mechanical/thermal energy interchange taking place during flow through such extrusion die systems, by combining the physical equation of state $i = i(p, v)$ with the thermal equation of state, after *Spencer* and *Gilmore*:

$$i_2 - i_1 = c_p(T_2 - T_1) + b^*(p_2 - p_1). \tag{60}$$

For the case of adiabatic throttled decompression $i_2 - i_1 = 0$, so that the mean temperature rise, $\Delta T = T_2 - T_1$, is obtained as the sum of the shear heating and the cooling caused by adiabatic cooling:

$$\Delta \bar{T} = b^* \cdot p_{ges}/c_p. \tag{61}$$

The temperature rise so calculated is to be understood as a mean value over the cross-section of flow; the local temperature in some places is higher, in others lower. The highest temperatures occur at the channel walls in the adiabatic process, and along the center line of each channel section in the isothermal case.

The structural viscosity of the melt was written into the above calculation by means of the *Ostwald/de Waele* power law. This change from Newtonian to shear-rate-dependent viscosity happens in some types of polymer, for example PMMA and PC, only at relatively high rates of shear – see $\eta(\dot{\gamma})$ for PMMA in Figure 51. In this case the flow curve or viscosity function can be more precisely described by means of the *Carreau* expression, a modified form of the power law:

$$\tau = \frac{a \cdot \dot{\gamma}}{(1 + b \cdot \dot{\gamma})^c}, \tag{62}$$

$$\eta = \frac{a}{(1 + b \cdot \dot{\gamma})^c}, \tag{63}$$

$$\eta_0 = \lim_{\dot{\gamma} \to 0} \eta = a. \tag{64}$$

By using the *Carreau* formulae, simple stationary streamline flow in typically contoured channels can be ideally described for PMMA or PC, as shown in [53].

Table 5. Geometrical input-data for a model design calculation on a tube-extrusion die

Input data for designing a tube-extrusion die

Total number of flow-channel sections: 17

Section 1: bore (cross-sectional shape)
DE = 32.00 MM
DA = 32.00 MM
L = 46.00 MM

Section 2: bore
DE = 32.00 MM
DA = 44.00 MM
L = 6.00 MM

Section 3: annular gap
DEI = 0.00 MM   DEA = 44.00 MM
DAI = 62.00 MM DAA = 118.00 MM
L = 37.00 MM

Section 4: annular gap
DEI = 62.00 MM DEA = 118.00 MM
DAI = 62.00 MM DAA = 116.00 MM
L = 14.00 MM

Section 5: mandrel zone
with staggered leg spider:
D1 = 62.00 MM   D2 = 78.00 MM
D3 = 101.57 MM D4 = 116.00 MM
L = 39.00 mm
No. of radial legs: 6
Inner width of radial leg = 14.00 MM
outer width of radial leg = 21.00 MM

Section 6: annular gap
DEI = 62.00 MM DEA = 116.00 MM
DAI = 62.00 MM DAA = 118.00 MM
L = 10.00 MM

Section 7: annular gap
DEI = 62.00 MM DEA = 118.00 MM
DAI = 62.00 MM DAA = 115.00 MM
L = 5.00 MM

Section 8: annular gap
DEI = 62.00 MM DEA = 115.00 MM
DAI = 88.00 MM DAA = 95.00 MM
L = 22.00 MM

Section 9: annular gap
DEI = 88.00 MM DEA = 95.00 MM
DAI = 88.00 MM DAA = 95.00 MM
L = 28.00 MM

Section 10: annular gap
DEI = 88.00 MM DEA = 95.00 MM
DAI = 96.00 MM DAA = 120.00 MM
L = 23.50 MM

Section 11: annular gap
DEI = 96.00 MM DEA = 120.00 MM
DAI = 96.00 MM DAA = 120.00 MM
L = 15.00 MM

Section 12: annular gap
DEI = 96.00 MM DEA = 120.00 MM
DAI = 67.00 MM DAA = 84.00 MM
L = 67.00 MM

Section 13: annular gap
DEI = 67.00 MM DEA = 84.00 MM
DAI = 62.00 MM DAA = 78.00 MM
L = 34.00 MM

Section 14: annular gap
DEI = 62.00 MM DEA = 78.00 MM
DAI = 62.00 MM DAA = 78.00 MM
L = 20.00 MM

Section 15: annular gap
DEI = 62.00 MM DEA = 78.00 MM
DAI = 30.00 MM DAA = 40.00 MM
L = 32.00 MM

Section 16: annular gap
DEI = 30.00 MM DEA = 40.00 MM
DAI = 30.00 MM DAA = 40.00 MM
L = 15.00 MM

Section 17: annular gap
DEI = 30.00 MM DEA = 40.00 MM
DAI = 46.00 MM DAA = 50.00 MM
L = 18.00 MM

Computed results:

$P_{ges} = 141.6$ [bar]
$\bar{t}_v = 30.1$ [s]
$p(z), \tau_W(z), \dot{\gamma}_W(z), \dfrac{dt_v}{dz}$
(see Figure 50)

12.4 Extrusion blow molding lines

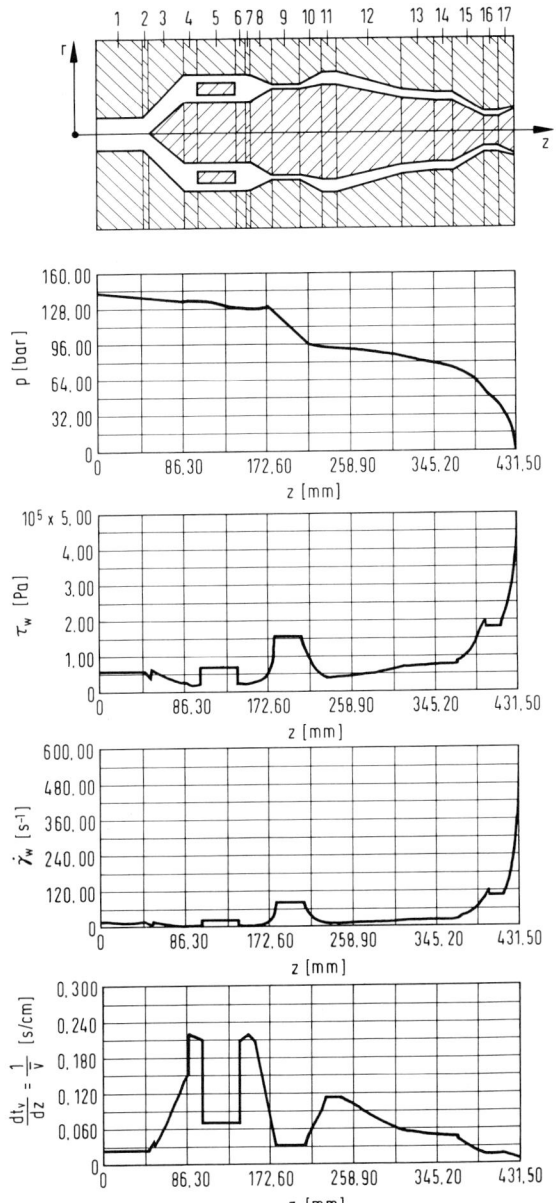

Figure 50. Operating parameters for a centrally fed tube die

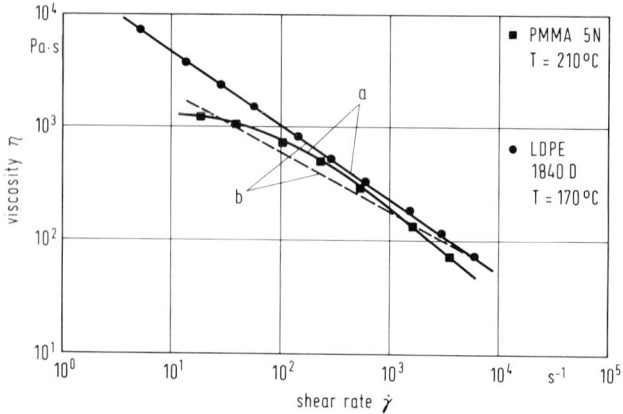

Figure 51. Flow curves of PMMA and LDPE melts matched to the power law and the Carreau formula
a Carreau fit, *b* power-law fit

12.4.1.2.2 Dies for continuous multilayer tube extrusion

Uses, polymer combinations

The parisons for making multilayer hollow bodies by extrusion blow molding are provided by coextrusion dies. Multilayer hollow bodies are now in a position, because of their properties profile, to displace traditional glass, aluminum, and tinplate packaging. The properties improvements and product modifications that can be achieved by coextrusion blowing are the following:

- multilayer bodies can show outstanding barrier properties against gases (O_2, CO_2) and water vapor, as well as good aroma retention. The barrier properties and the related increase in shelf-life that can be achieved, depend on the kind of barrier material, the thickness of the layer, and the environmental conditions − temperature and pressure differences, air humidity. Barrier materials in use today include EVOH, PVDC, PA, PAN, and even PET.
- the sequence of layers and the polymers used can be such that the material in contact with the package contents is physiologically harmless, and behaves neutrally with respect to the contents.
- with proper choice of material, and if the state of orientation in the blown part is very low, it is possible to hot-fill, e.g. with juices, syrups.
- HDPE hollow bodies can be made with a highly polished or special scratch-resistant surface that is printable and dust-repellent. By using a surface layer of polyethylene with a high carbon black content − a relatively expensive material − cost-effective gasoline canisters with a low surface resistance can be blow molded.

A decisive criterion in choosing materials to meet such product specifications is good adhesion between the individual layers (peel strength 5 to 10 N/cm), because delamination in the weld or pinch areas would make the blown article useless. In combinations of related materials, for example HDPE with LDPE for high surface gloss, one can always achieve adequate adhesive bonding. But combinations of dissimilar materials require the use of adhesion promoters; EVA copolymers, ionomers, and EVA terpolymers (e.g. Admer, Plexar, CXA) are used for this purpose. The *two-layer process*, using a main extruder and an auxiliary unit for each, can be used to make composite films from similar and dissimilar material pairs [54]. Preferred material combinations for the latter type of product are LDPE/PA and PP/PA, with the polyolefin, to which adhesion promoter is added before extrusion, forming the inner, carrier layer. The external PA layer provides the essential barrier properties, as well as the abrasion resistance, printability, and surface gloss that are also needed. Because a large amount of adhesion promoter is needed, and the adhesion achieved between materials is frequently unsatisfactory, use of the two-layer technique for dissimilar composites has almost disappeared from the scene. Some years ago its place was taken by *three-layer coextrusion*, in which a layer of adhesion promoter (tie layer), supplied by a second auxiliary extruder, unites the carrier layer and the barrier layer. Such three-layer structures can be advantageously produced for packaging foodstuffs, using the layer sequence HDPE/TL/EVOH with thicknesses 400 to 600/50/100 µm. The relatively thick layer of EVOH serves as a barrier against CO_2 loss.

Restrictions on the proportion of regrind that could be used because the carrier layer comes in contact with the package contents, led quickly to the *four-layer technique*, in which a layer consisting entirely of regrind (RG) is coextruded. Since this compounded neck and bottom waste is up to 90% carrier material, it sticks readily to the carrier layer without the need for adhesion promoter. It is positioned as the penultimate inner or outer layer, so that a layer sequence of the type HDPE/RG/TL/EVOH is obtained.

Since the permeation coefficient of the barrier layer can increase under some circumstances, because of the uptake of water, five- and six-layer structures have been developed in which the barrier layer is embedded between two polyolefin layers coated with adhesion promoter. A hot-fill ketchup bottle that was developed using this concept, with the wall structure: PP/TL/EVOH/TL/PP, 370/20/20/20/370 µm, proved a sweeping success in the market [55, 56]. The regrind generated can be used again, as with the four-layer structure, as the penultimate outer or inner layer (layer sequence: PP/TL/EVOH/TL/RG/PP, 140/20/20/20/100/100 µm).

Relatively expensive extrusion dies are required for making these multilayer tube sections. Their basic construction is explained below.

Structural design of coextrusion dies

An important distinguishing characteristic of the coextrusion dies available today is the melt feeding arrangement. Material feed for the relatively thick carrier material, the adhesion promoter, and barrier layer can be carried out by twin or single manifolds of the coat-hanger type, by spider or strainer-plate mandrels and by spiral mandrels (Figure 52). The arriving melt streams are led individually into the various analytically designed distributor systems and adjacent land areas, and then formed into closed tubes with uniform wall thicknesses, and successively overlaid on one another. Rigid, high-precision components ensure that thickness uniformity Δs is $\leq 10\%$, so that it is not necessary to incorporate rings for adjusting the thickness distribution of the individual layers. This makes even three-layer heads a great deal easier to set up. The channel walls must be heated separately in the die region and in the section after the individual tubes become a single multilayer tube; and the dimensions of the channels must be such that large pressure gradients and appreciable average dwell times are realized. Both points have decisive effects on weld quality and peel strength.

Single and twin-head dies are available commercially for both three- and five-layer technology. The axial thickness profile can be pre-programmed in both systems, which work on the assumption that the thicknesses of all layers change in the same proportion for a given adjustment to the mandrel or the die gap, depending on design.

In operating with such multilayer dies one necessarily has to use extruders with grooved feed zones whose specific advantages – back-pressure-independent conveying at essentially constant rate, and easier control of melt temperature – come fully into play.

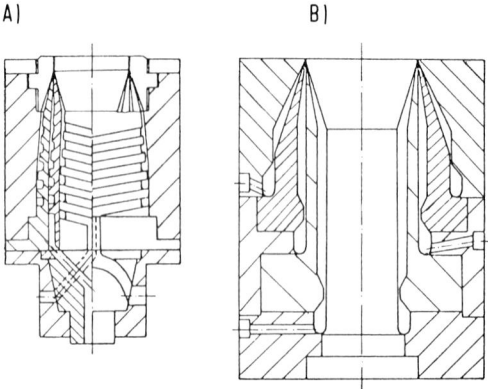
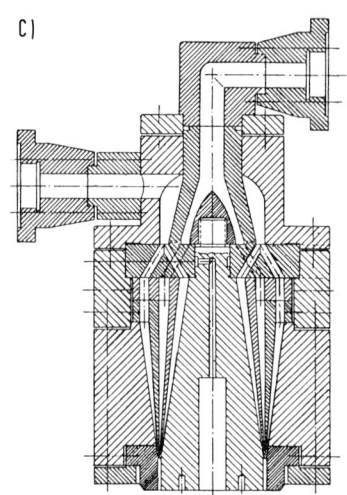

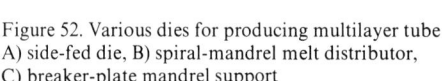

Figure 52. Various dies for producing multilayer tube
A) side-fed die, B) spiral-mandrel melt distributor,
C) breaker-plate mandrel support

When processing polymers with hard pellets, like PC and PA, intense cooling is used in the grooved zone. The number of extruders required and their size is determined by the method of production and the material combinations prescribed. Thus a five-layer parison can be produced with three, four, or even five extruders.

From these descriptions it is abundantly clear that the cost of the machinery for plasticating and shaping the polymer is extremely high. It therefore remains to be considered whether one should use such expensive units only with high-performance blowing-mold systems.

12.4.1.2.3 Dies for intermittent extrusion of tube

Requirements

Dies combined with melt accumulators are used for the following reasons:

- to prevent the tube lengthening too much under its own weight (parison droop), which occurs with lower viscosity melts (LDPE, PA, ABS) especially;
- to exclude cooling at the tube end, which would affect the quality of the seam weld;
- to avoid oxidative damage to the polymer;
- to minimize dead times in blow molding machines with fixed mold-clamping systems, on which extrusion can take place only when the mold is open.

The demands such accumulator systems have to meet can be formulated in relation to the storage and extrusion cycles, as follows [57]:

- The flow resistance (impedance) of the connecting channels between accumulator and extruder must be low, but the need to find a compromise in relation to self-cleaning action has to be borne in mind. Highly symmetrical, concentric filling of the accumulator space must be aimed at.
- To hold dwell times to acceptable limits, no melt must be left in the storage system after each extrusion stroke. The melt that goes in first must come out first — the FIFO or "first in–first out" principle. There must be no build-up of material on the walls of the accumulator or on the piston head. The sliding guides between the piston and the storage cylinder should allow controlled flow leakage, such that troublesome deposits of oxidized or crosslinked particles (black streaks) on the surface of the tube are avoided.

12.4 Extrusion blow molding lines   415

- The hydraulic drive for the piston should, in general, provide for adjustable, very short extrusion times. And the channels connected to the accumulator must permit melt to flow concentrically, without subdivision.
- With regard to the structural requirements of the machine, first of all the calculation of the accumulator volume should be independent of the dimensions of the extruder, especially of diameter. Accumulator volumes, head sizes and die diameters that can be accommodated must be matched to each other so that a wide range of use with adequate overlap is available.

Of the many accumulator systems that were available at the beginning of the 1970s, only a few could meet this broad range of requirements. These were, exclusively, the so-called tubular-ram accumulator heads, or tubular-ram accumulators, which are now standard components on modern blow molding lines. Before these types are described in detail, some notes on a few of the older, now superseded accumulator systems, will be given for the sake of completeness.

Older accumulator systems

A comparison of the accumulator systems itemized here with the performance specification formulated earlier for such components, provides a sound means of assessment. The following discussion is arranged according to the position of the accumulator between the plasticating unit and the die. Reciprocating screw and push-out ram accumulators, and accumulators mounted separately between the extruder and the parison head are dealt with.

Reciprocating-screw accumulators

We are concerned here with the ram screw, well known from injection molding. In blow molding, the ram screw has the disadvantages of varying effective screw length and differences in melt homogeneity linked to this. The grooving of the feed zone gives rise to pronounced difficulties, since the forces required to displace the screw and for parison extrusion become excessive. Of all the accumulator construction systems the ram screw is the one tied most closely to the extruder dimensions: in general the diameters of the storage and plasticating cylinders are identical, so that the prospective storage volume is always small. The FIFO principle is certainly guaranteed, but high melt-flow velocities and large deformations occur in the stream-dividing sections, and affect parison quality adversely.

Push-out ram accumulators

In contrast to reciprocating screws, push-out ram accumulator dimensions can be largely independent of extruder dimensions. A grooved extruder of constant effective screw length pumps the melt into a storage cylinder, whose inner diameter is slightly larger than the outer diameter of the plasticating cylinder. The plasticating cylinder and screw moving together axially function as the extrusion piston; the stroke, and thus the storage volume, are variable over a wide range. The earlier comments about the FIFO principle and high flow velocities in the spider region apply here too. Disadvantages seen for the system are the large weights of metal that must be moved, and the scarcely controllable sealing problem on the sliding surfaces of the accumulator.

Separate accumulators

Such accumulators can be designed independently of the extruder and the parison head, and be built into the blow molding machine in quite different ways. An arrangement with the accumulator parallel to the extruder axis, or vertical, is often preferred. The extrusion piston can be designed either as a recuperator piston or as a plunger. However, in both cases the piston-head profile must be shaped to ensure that the channel between the extruder and the accumulator remains open during parison extrusion. The FIFO principle is not always realized — in contrast, however, to undesirably large deformation rates in the adjacent spider region, that are.

Annular-piston accumulator heads

Figure 53 is a schematic diagram of the melt-channel system of an annular-piston accumulator. The hydraulic die-gap adjusters and the band heaters have been left off in the interests of clarity. Common to all the construction methods in current use is the plunger version of the annular piston. All designs have melt feed and spider sections that bring about the overlapping of melt streams. These produce tubes in which the seam welds do not run right across the thickness; this ensures that the thickness distribution in the blown part will be satisfactory. The partial streams are united again immediately after the spider section into two tubular streams; after passing through separate annular channels these feed into a common annular channel formed between the mandrel and the inner wall of the annular piston. All the machine parts in which there is divided flow lie between the extruder and the accumulator space [13]. The flow velocities here are low, and thus the deformation rates are small, and the resulting orientation in the extruded material is minimal. The FIFO principle applies to all such units. When the die exit is closed the incoming melt pushes the annular piston upwards.

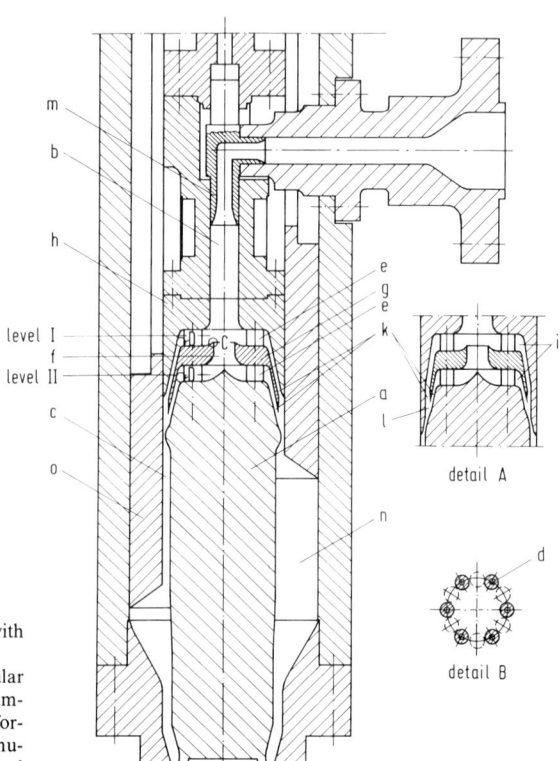

Figure 53. Tubular-ram accumulator head with bolt-type mandrel support
a mandrel, *b* central feed channel, *c* annular gap, *d, e* axial bolts, *f* axial channel, *g* umbrella flow divider, *h* inlet body, *i, k* tube formation zones, *l* land zone, *m* elbow, *n* accumulator chamber, *o* annular piston, *I* mandrel support level 1, *II* mandrel support level 2

12.4 Extrusion blow molding lines

There are significant differences between annular piston heads in the way the melt is fed in. Such differences in this region of the feed channel, which have a large effect on the parison quality, are considered in the following section.

Battenfeld design

The accumulator design shown in Figure 54 A, which was arrived at more than 15 years ago particularly in connection with the manufacture of gasoline tanks, uses a simple annular channel for melt feed. With the die shut, the incoming melt pushes the ring piston vertically upwards. An improved version provides two annular channels separated by an adjustable sleeve (Figure 54 B), whose feed points are set opposite to each other to achieve the required 180° of melt stream overlap. The annular piston slides on the mandrel shaft. The whole mandrel is moved to adjust the die gap.

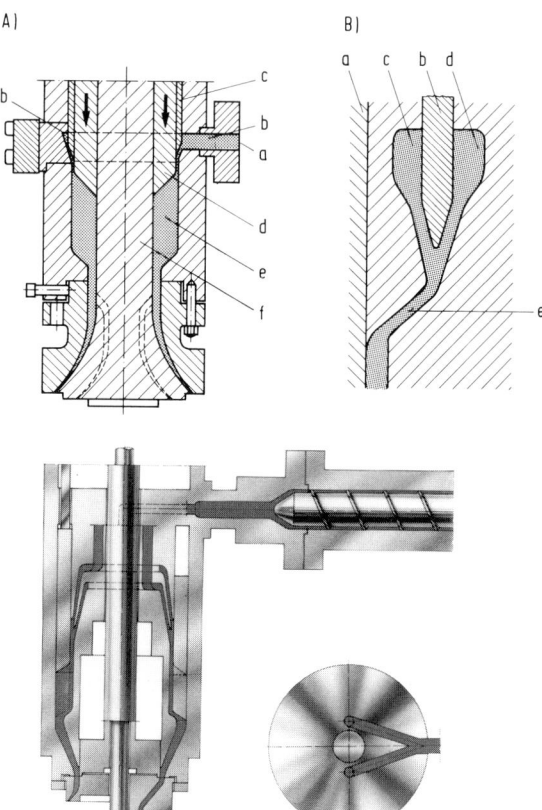

Figure 54. Tubular-ram accumulator head with ring feed
A) single ring
a extruder connection, *b* annular channel, *c* adjustable sleeve, *d* annular piston, *e* accumulator chamber, *f* axially adjustable mandrel support
B) overlapping rings
a tubular-ram outer wall, *b* adjustable sleeve, *c* inner annular channel, *d* outer annular channel, *e* adjustable throttling gap

Figure 55. Tubular-ram accumulator head with two-stage heart-groove feed system (Battenfeld-Fischer design)

Battenfield-Fischer design

The accumulator head design shown in Figure 55 is characterized by the so-called two-stage heart-curve feed system. Here we have a development of the central flow, bolt supported mandrel [58] — see Figure 53 — in which the mandrel is supported by a number of bolts parallel to the die axis in planes I and II. The axial bolts, which produce a considerable number of knit lines, have been replaced in an improved version by a flat melt distributor system in each plane, each fed through a separate channel. The individual tubes formed in these planes and in the adjacent screen-like annular channel zones, are overlaid so that the knit lines in the parison resulting from the heart curves are 180° displaced from each other. A generously dimensioned central bore in the accumulator head allows control shafts to pass through it for positioning inserts, and/or manipulation of the tube.

Bekum design

The melt from the extruder is divided into two streams by impinging on a torpedo. These then flow through 90° around the torpedo in two horizontal channels (Figure 56), and are then turned in its axial direction to reach two coat-hanger manifolds that each extend round 180° of the torpedo circumference. The tubes are therefore formed from two half-shells with two knit lines at 180°, across the full wall thickness. A directly adjacent ring channel now divides the melt into two annular streams, one within the other [59, 60]. Diagonal fins on the wall of the inner channel (Figure 57) deflect the two original weld lines around the axis of flow; thus a possible line of weakness running through the wall of the parison formed by overlapping the two annular flows is avoided. The outer surface of the annular piston slides in the die housing, and forms an annular channel of variable length with the torpedo. Die-gap changes are made by shifting the torpedo axially by means of a tension rod.

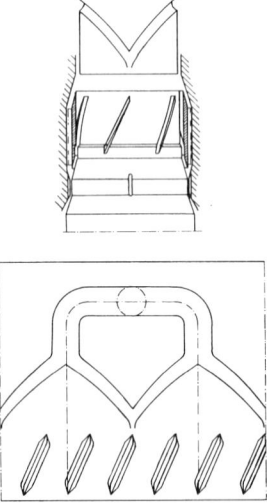

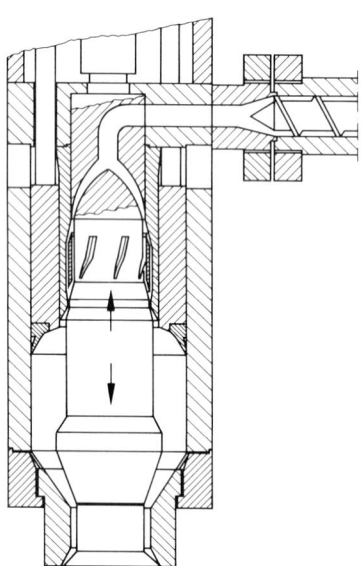

Figure 56. Melt distributor systems and a flow-guiding ring make it possible to overlap partial streams and so avoid knit lines through the full thickness of the tube wall (Bekum design)

Figure 57. Tubular-ram accumulator head with side-feed, and flow-guiding ring

Krupp-Kautex design

Two concentric coat-hanger manifolds, housed one inside the other, separately form tubular elements (Figure 58). The feed points of the two manifolds are 180° offset from each other, like the weld lines of the two tubular structures created [61, 62]. These two are brought together, and the melt stream now flows through a land (or throttling) zone to reach the accumulator by way of the channel formed by the annular piston and the mandrel. Because of the way the melt is directed there are separate overlapping streams, and therefore no full-thickness weld lines. It should be noticed, incidentally, that with the Krupp-Kautex and Battenfeld-Fischer methods the parison does not have to be made from concentric tubular elements. Identical circumferential thickness differences between the two tubular preproducts generally result in a substantially constant wall thickness, because of the 180° offset arrangement of the manifold system. There is also the advantage that the thickness of the weld line is often only a fraction of the full thickness of the tube wall. In this type of accumulator head, also, the die gap is varied by shifting the mandrel.

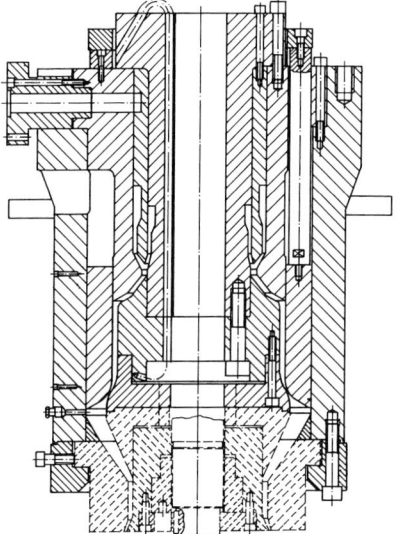

Figure 58. Tubular-ram accumulator head with overlapping heart-shaped grooves. (Krupp-Kautex design)

The third and fourth mandrel-support and melt-feed systems described are also used on tube dies for continuous extrusion; these, too, produce parisons of very high quality. The foregoing description of the complexities of melt channel design makes one realize that the optimization of such systems is possible only by analytical methods. Just as with feed systems, flow channels in the die region require product-related, computer-aided design to arrive at short extrusion times, with extrusion rates up to 1.5 kg/sec. This is an essential precondition for obtaining good surface quality, and for the manufacture of very long parisons; for example, like those now required for producing surfboards [63].

Dimensions of annular-piston accumulator heads

Annular-piston accumulators are now built in a range of standardized volume steps, which allow the stored volume to be varied from 1 dm^3 to 400 dm^3. Accumulator heads with 1/2.5/4/6.3/10/25/40/63/100/250 and 400 l storage capacity are on the delivery programs of manufacturers of blow molding lines. There is a correspondingly broad range of die diameters – from 40 to 1300 mm. An increasing number of the large extruders used in the manufacture of fuel oil tanks and storage containers with volumes up to 10,000 l, are working with annular-piston accumulator units.

These arrangements mean that reliable plasticating units with screw diameters from 120 to 160 mm can be brought into use without the need for costly development work.

12.4.1.2.4 Wall-thickness control

Uniform wall thickness lengthwise and circumferentially is generally sought in blow molded parts. Because of the differing stretch ratios, this can be achieved only if the parison wall thickness is influenced in a carefully controlled way during extrusion. Efficient thickness control systems make it possible to produce hollow articles with higher compressive strengths from a given weight of material, or to produce a part to a given performance specification with less material, in a shorter cooling time, which in addition can have better stress-crack resistance. Because these advantages are available, hardly a blow molding machine, apart from PVC bottle-blowing units, is delivered today without electronic wall-thickness control.

At the beginning of the seventies, systems were developed with which the axial thickness profile of the parison could be determined ideally in relation to the geometry of the blow-molded part. Then as now, the target thickness profile could be simply and reproducibly preset on a control device. Depending on whether tube extrusion is continuous or intermittent, the extrusion time, or the "push-out" time, of the annular piston is divided into a maximum of 64 intervals, for each of which a percentage change in gap can be set by means of a potentiometer. Such preset target values are compared in a control loop with the actual values for the die and mandrel positions, and adjusted until they agree. Hydraulic servo-drives with built-in position indicators are used for this purpose. As these details show, we are concerned here with closed-loop control of the die or mandrel position or open-loop control of the axial thickness profile of the tube.

In attempting to hold the performance specification of the part with ever decreasing weight, process engineers over many years have tried to improve the perimeter wall-thickness distribution (a hollow body of rectangular cross-section is a typical example) by means of die/mandrel profiling in the exit-gap region. Optimal die-gap geometry is still usually arrived at empirically, and it is certainly possible to make changes in the wall thickness locally. But these apply only for the particular material rheology and melt throughput, with the given part geometry. One way out of this dilemma is provided by adjustable die/mandrel profiling (Figure 59), using a static flexible die ring (SFDR die) [65, 66]. With the help of this device the machine setter can carry out the necessary profile adjustments in a very short time without dismantling the die and without metal cutting processes. An

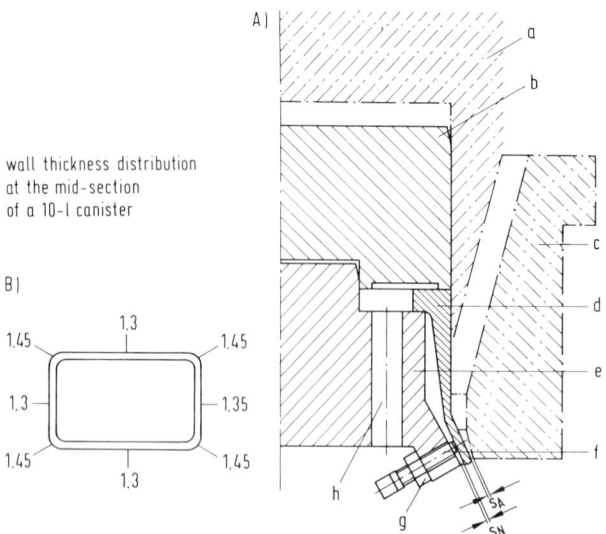

Figure 59. Schematic layout of a semi-flexible die ring for a center-fed die (A) and the wall thickness distribution attainable with rectangular containers (B)
a mandrel housing, *b* mandrel, upper, *c* die body, *d* SFDR, *e* mandrel, inner, *f* set screw, *g* lock-nut, *h* melt-leakage path

additional big advantage is that these adjustments are all reversible. If this were not so it might be restricting to have to work with the single profile chosen, for a range of extrusion operations. The partial wall-thickness control system (PWTC system) that has been available since 1979 solves this problem by making the actual die mouthpiece as an elastic ring (Figures 60, 61) that can be deformed to an elliptical contour by a pair of programmed positioning cylinders working in opposition. The result is that over certain lengths the tube cross-section can have two pairs of strips of greater and smaller thickness, with like pairs opposite each other. By combining the conventional axial method with the partial radial technique, an extremely flexible and effective system for precision control of parison thickness is obtained that can be used for both continuous and discontinuous extrusion of tubes. This unified PWTC system, shown in Figure 60, is built into the accumulator head machine. It consists of a special die with a deformable die ring (DFDR), two wall thickness

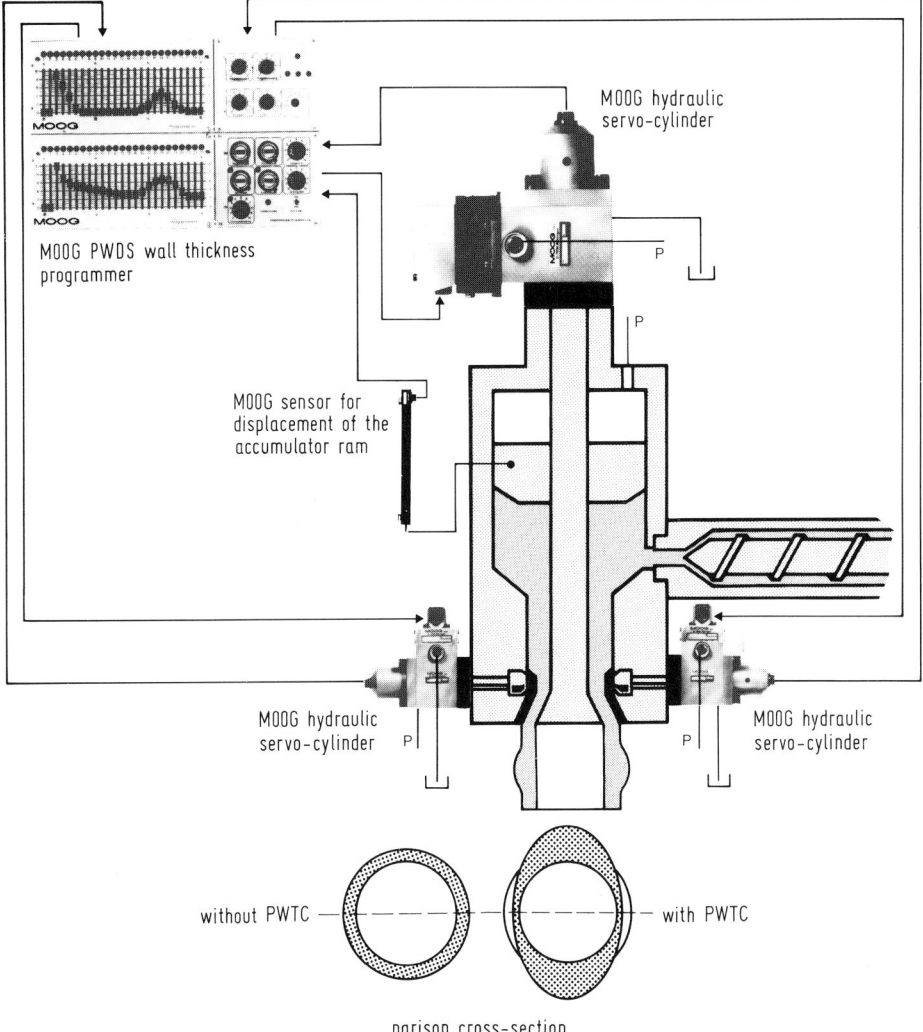

Figure 60. Schematic diagram of a programmable wall-thickness control system for an accumulator-head blow molding machine

programming devices for separate setting of the axial and perimeter wall-thickness profiles, two hydraulic servocylinders with their own displacement indicators for deforming the ring, and another cylinder for changing axial thickness. A displacement sensor on the accumulator provides a recording of the movement of the annular piston.

If the unified PWTC system is used with an SFDR die (see Figure 61) the tube wall thickness can be changed in almost any way desired, and the thickness of the blown part made remarkably uniform, as is evident from Figure 62.

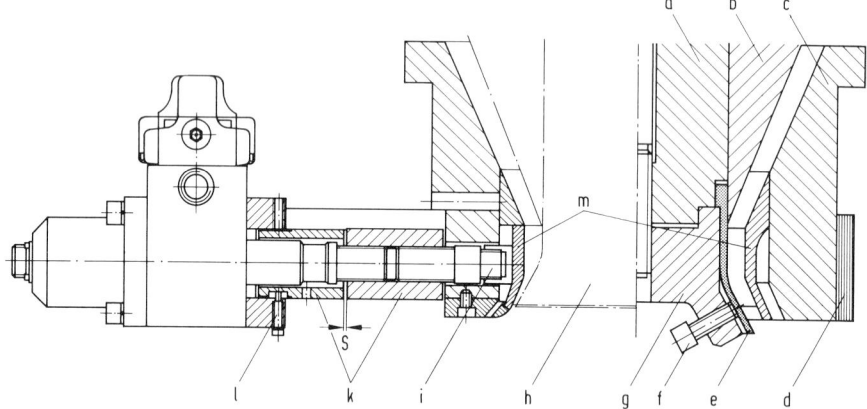

Figure 61. Partial control of parison wall thickness by deformation of an elastic die ring
a mandrel, upper, b mandrel housing, c die body, d heater band, e SFDR, f set screw, g mandrel, inner, h mandrel, i adaptor, k stops, l set screw, m nozzle ring, s deformation stroke

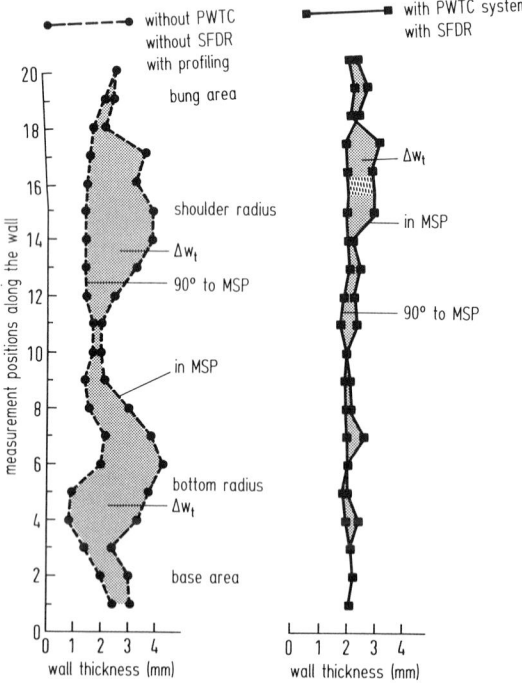

Figure 62. Wall-thickness distribution in a 25-l canister with and without radial wall thickness programming (constant weight of material = 1040 g). MSP = mold split plane

12.4.1.2.5 Mold clamping units

Depending on the size of the part to be blow molded, and on the method of extrusion of the parison tube, there are basic differences in the methods of construction and drive of mold clamping units.

Clamps for use on continuous-tube extrusion units

Newer clamp units now move between the extrusion and blowing positions along circular arc guides (see Figures 6 and 11 et seq.). This motion can take place along or across the extruder axis. Diagonal downward movements are very seldom used, and where they are, predominantly on older blow molding machines. Drive is hydraulic, with electronically controlled proportional valves, a technique that combines high speeds with an optimal acceleration characteristic.
Electromechanical crank drives allied to transmission brake motors are in the process of vanishing from the scene. High speeds and short dry-cycling times require the very lowest moving masses, which is a fundamental reason for making press platens and other clamp system components from special aluminum alloys. At their lower ends clamp systems are generally supported mechanically, to make it possible to accommodate the forces exerted by hydraulic calibrating devices without deformation.

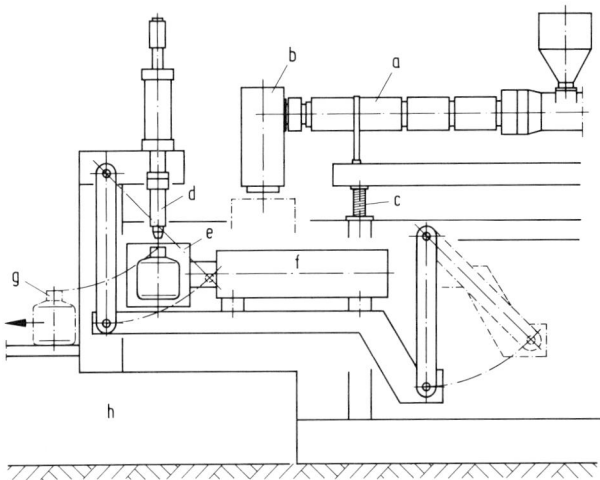

Figure 63. Swing-arm system for actuating the clamp unit on a blow molding machine
a extruder, *b* parison die, *c* height adjustment, *d* blowing mandrel, *e* mold, *f* clamp unit, *g* finished product, *h* free space

Clamp units which have swiveling arrangements are designed without tie bars, or have them placed well down to ensure good accessibility to the mold, and unhindered transfer of finished parts to downstream stations. Even without tie bars, parallel platen motion is guaranteed and the presses can be installed either vertically (Figure 6) or horizontally (Figure 63); the latter version certainly requires long support arms and heavy lever closing mechanisms, but it does allow plenty of space under the mold for special combinations of mandrels and stretching devices. To reduce the dead time the mold opening stroke can be limited to the absolute minimum required for removing the blown part. A particular advantage of many of these newly developed clamp systems is that molds of various heights can be accommodated. Large, hydraulically applied closing forces make certain that the seam welds are of very high quality and that the waste material is easy to remove.

Clamping systems for lines producing tube intermittently

Modern clamp units, whether fixed or linearly movable between the tube extrusion position and the blowing position, are all capable of providing very large closing forces and high speeds, thanks to their powerful hydraulic systems. Depending on whether the welding edges of the blow mold are formed as pressure bars or as pinch-off areas [67, 68], specific closing forces of between 2000 and 4000 N/cm are needed to ensure easy removal of flash. Maximum mold closing speeds are around 200 mm/s, reducing to 5 to 10 mm/s during the pressurizing phase. Modern large blow molding units often include press-speed control systems that can ensure a defined cycle of movement free of external influences during the blowing process, by which constant weld quality is again guaranteed.

Most of the clamp systems currently available do not have tie bars, and the closing forces are applied either by a frame or by a horizontal C-clamp. The synchronization of the movements of the mold platens is ensured by means of racks or roller chains, depending on the machine dimensions.

12.4.1.2.6 Tube cutting devices

When extrusion of the tube is complete, the two halves of the mold close around the parison and form the weld seams. Uniquely on accumulator-head systems, the tube is cut by closing the die/mandrel gap. On continuous extrusion systems, depending on the diameter of the tube and the material, hot wires or hot band cutters (PE, PC, PP), pneumatically operated spring steel daggers (PVC) or impact knives (HDPE, PVC), are used for the cutting process. If, because of the geometry of the blow mold, it is necessary to expand the parison, it is advantageous to make use of so-called cold-tube shears (Figure 64), which weld the tube and at the same time make a cold cut.

Figure 64. Bottom view of parison-die outlets, cold-cutting mechanisms, and the welding device used on twin-mold machines

12.4.1.2.7 Calibration and blowing station, stretching and pre-weld devices

Insertion of the cooled blowing pin into the projecting tube section that has been widened by support air, and calibration after closing the mold, is the most widely practiced method of calibration and blowing. Depending on the machine dimensions, the tube extrusion process, and the construction of the clamp unit, calibration can be effected from above or below. Large calibration forces can be produced by the hydraulically operated calibration and blowing mandrels, which even with thick-walled parts produce flash-free calibrated surfaces and easily removable flash tails. The insertion speeds are steplessly adjustable, and are also controlled to some extent. The calibration and blowing devices, which can all be operated with flushing air or refrigerated air, are vertically or horizontally adjustable for flexibility, and in individual cases are also swivelable to make it possible to calibrate diagonally oriented openings as well.

With technical parts – particularly automobile gas-tanks, consoles, and spoilers, and double-wall hard-shell suitcases – there is generally no opening in the parting line of the mold. Such parts are shaped by closing the tool halves around the welded tube, already expanded by support air, before a hydraulically operated hollow steel needle punctures the bubble and blows it to form the end product. This is the needle-blowing process.

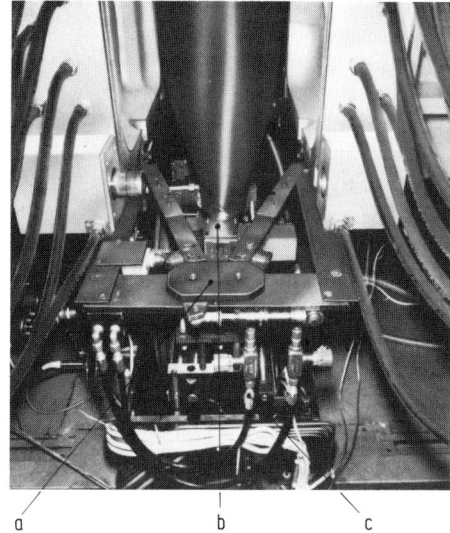

Figure 65. Device for combined stretching and welding of the parison
a welding tongs, b stretch mandrel, c stretching device

To obtain the best possible wall-thickness uniformity with somewhat flat parts like solar absorbers, automobile gas-tanks, concealed wiring junction boxes, etc., it is necessary to more-or-less flatten the extruded tube by using a stretching device (Figure 65). Blowing is then carried out direct by the stretching mandrel, or by means of an additional blowing mandrel. A further way of normalizing wall thickness is to weld the ends of tube before closing the mold round it and to inflate it while the tool is still open. This can be accomplished with the help of the pre-weld and tube shears mentioned earlier, or for large parts by means of pre-weld forceps.

12.4.2 Complete blow molding lines

Blow molding units with one or two clamp systems, and accumulator-head blow molding machines will now be discussed, following the process classification used in 12.3.

12.4.2.1 Automated blow molding machines

Blowing units equipped with swivelable or linearly movable clamp systems are characterized by the fact that the parts are oriented when they leave the blowing station and are transported, standing up, on a conveyor belt to downstream devices for leak testing, pressurizing, filling, and packing (Figure 66). After blowing, a transfer system that may or may not be linked to the mold halves seizes the blown parts and transports them cyclically through

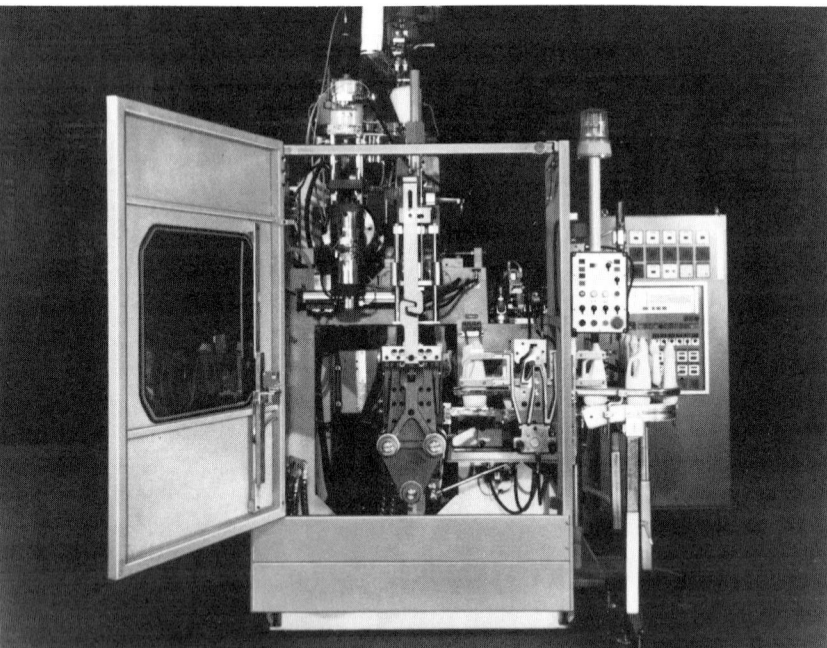

Figure 66. Single-station blow molding machine – Bekum system

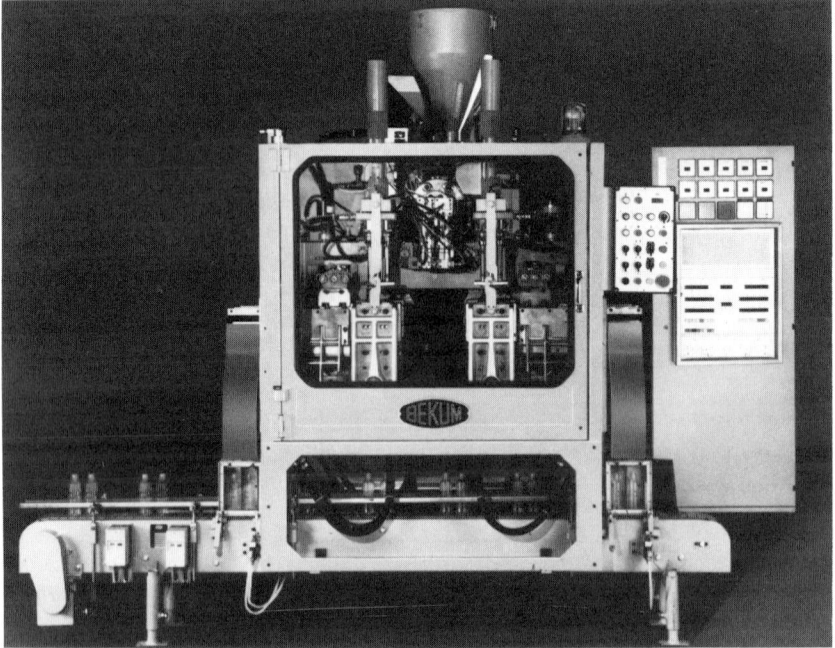

Figure 67. Two-station blow molding machine – Bekum system

the post-cooling and flash removal station. As a direct result of integrating additional post-cooling devices – simple, coolable auxiliary tools on which the blown parts can still be pressurized – it has been possible quite recently to raise the production performance of such blowing machines very significantly. If removal of waste is not accomplished through an interlocking rotary tube mounted on the blowing spigot, it is done by means of propellors, pneumatically operated strikers, or integrated flash removers. When blown parts have handle pinch-off areas, these are preferably removed by shear-punch devices, while wide-necked jars are finished by rotary cutters or spear knives. A recent development involving placement of labels in the mold enables blow-molded bottle- and canister-like articles to be labeled during the blowing process ("in-mold-labeling") [64]; this is a crucial requirement for in-case filling of containers. These examples show how far automation and rationalization have progressed in this sector. By incorporating a second clamp system, which removes tube sections alternately with the first, the output of a machine can be almost doubled (Figure 67). Depending on the shape of the part, it is possible to have up to four cavities per clamp unit, and so obtain up to eight articles per working cycle. An increasing number of coextrusion lines using this basic concept are being offered on which the only difference, apart from the design of the extrusion dies, is whether there are three, four, or five extruders.

12.4.2.2 Accumulator-head blow molding machines

Accumulator machines used for manufacturing large-volume packages, storage tanks, and technical parts, consist of a base-frame in which the hydraulics are generally housed, the extruder, the accumulator head, and the clamp unit (Figure 68). The parts are finished on downstream punching or wideneck trimming stations controlled from the blowing machine, after passing through cooling tools usually integrated with the blowing unit.

Figure 68. Accumulator-head machine for blow molding large-volume containers and technical parts – Battenfeld-Fischer system

References for Chapter 12

For the convenience of the reader the English titles of all publications in languages other than English are shown in parentheses.

[1] *Fritz, H. G.:* Systematische Verfahrensdarstellung angewandter Blasformtechnologien (Systematic Discussion of Modern Blow Molding Technologies). In: Technologien des Blasformens, VDI-Verlag, Düsseldorf, 1977.
[2] *Holzmann, R.:* Kunststoffe 69 (1979), pp. 704–711.
[3] US Patent 8180 (1851).
[4] US Patent 237168 (1891).
[5] German Patent 112770 (1899).
[6] *Johnke, K. D., Behr, P.:* Kunststoffberater (1982) 12, pp. 13/21.
[7] *Stoeckhert, K.:* Kunststoffberater (1980) 2, pp. 24/33.
[8] *Veal, P. L.:* Recent Developments in Blowmoulding Materials. Paper AG at Conference in Advances in Blow Moulding, London, 1977.
[9] *Six, J., Wichmann, U.:* Plastverarbeiter 33 (1982), pp. 689/691.
[10] *Daubenbüchel, W.:* Plastverarbeiter 31 (1980), pp. 313/317.
[11] *Holzmann, R.:* Blasverfahren und Methoden der Hohlkörperherstellung (Blow Molding Processes and Methods of Producing Hollow Articles). VDI Bildungswerk, Paper No. BW 649.
[12] *Schneiders, A.:* Kunststoffe 59 (1969), pp. 637/642.
[13] *Schneiders, A.:* Extruder, Schlauchköpfe, Schmelzspeicher in Extrusionsblasformen (Extruders, Parison Dies, and Melt Accumulators in Extrusion Blow Molding). VDI-Verlag, Düsseldorf, 1979.
[14] *Lülsdorf, P.:* Verschleißprobleme mit Zylinder und Schnecken beim Extrudieren (Wear Problems with Barrels and Screws during Extrusion). VDI-Lehrgang 1975, paper No. BW 3019.
[15] *Fritz, H. G.:* Kunststoffe 65 (1975), pp. 176/182.
[16] *Schüle, H., Fritz, H. G.:* Kunststoffe 73 (1983), pp. 603/605.
[17] *Lülsdorf, P.:* Verschleißfragen (Wear Problems). Company brochure, Reiloy Metall GmbH, 9/1983.
[18] *Schüle, H., Fritz, H. G.:* Kunststoffe 75 (1985), pp. 399/403.
[19] *Schenkel, G.:* Kunststoff-Extrusionstechnik (Plastics Extrusion – Technology and Theory) Carl Hanser Verlag, München, Wien, 1963.
[20] *Junk, P. B.:* ANTEC 1984, pp. 897/899.
[21] *Schneiders, A.:* Plastverarbeiter 19 (1968), pp. 797/799.
[22] *Fuchs, G.:* Plastverarbeiter 19 (1968), PP. 765–771 and pp. 237–244.
[23] *Menges, G., Predöhl, W., et al.:* Plastverarbeiter 20 (1969), pp. 79/88 and pp. 188/190.
[24] *Grünschloß, E.:* Abridged Report No. 2.1 at 7th Stuttgarter Kunststoffkolloquium, 8th – 10th April 1981.
[25] *Goldacker, E.:* Dissertation, RWTH Aachen, 1971.
[26] *Grünschloß, E.:* Kunststoffe 74 (1984), pp. 405/409.
[27] *Fritz, H. G.:* Voith Forschung und Konstruktion (Research and Machine Design at Voith's). No. 23, Oct. 1975, SD 2239.
[28] *Fritz, H. G.:* Chem.-Ing.-Tech. 55 (1983), pp. 256/266.
[29] *Langecker, G.:* Dissertation, RWTH Aachen, 1977.
[30] *Grünschloß, E.:* ANTEC 1979, pp. 160/165.
[31] *Pfeiffer, H.:* Dissertation, RWTH Aachen, 1981.
[32] *Rautenbach, R., Pfeiffer, H.:* Kunststoffe 69 (1979), pp. 377/379.
[33] *Grünschloß, E.:* Abridged report No. 3.4 at 9th Stuttgarter Kunststoffkolloquium, 13th – 15th March, 1985.
[34] *Miessner, F.:* Misch- und Scherteile in Einschneckenextrudern (Homogenizing and Shearing Elements in Single-screw Extruders). Student's Report, Institut für Kunststofftechnologie (IKT), University Stuttgart, 1975.
[35] *Anders, D., Müller, W.:* Kunststoffe 72 (1982), pp. 62/65.
[36] *Maillefer, C.:* Kunstst.-Plas. (1961) 3., pp. 323/327.
[37] *Maillefer, C.:* Mod. Plast. (1963) 1, pp. 132/138.

[38] *Barr, R. A.:* Plast. Technol. (1972) 9, pp. 67/70.
[39] *Kim, H. T.:* ANTEC 1976, pp. 439/443.
[40] Final Report of the DFG research program No. Fr 562/1 at IKT, Stuttgart, 1983.
[41] *Görmar, E. H.:* Dissertation, RWTH Aachen, 1968.
[42] *Wortberg, J.:* Dissertation, RWTH Aachen, 1978.
[43] *Knappe, W., Schönewald, H.:* Kunststoffe 61 (1971), pp. 497/504.
[44] *Michaeli, W.:* Extrusionswerkzeuge für Kunststoffe (Extrusion Dies for Plastics). Carl Hanser Verlag, München, Wien, 1984.
[45] *Fritz, H. G.:* Abridged report No. 4.1 at 7th Stuttgarter Kunststoffkolloquium, 8th–10th April, 1981.
[46] *Winter, H. H., Fritz, H. G.:* ANTEC 1984, pp. 49/52.
[47] *Weekes, D. J.:* Brit. Plastics (1958) 31, pp. 156/160 and pp. 201/205.
[48] *Fritz, H. G., Meder, S.:* Kunststoffberater (1985) 4, pp. 21/25.
[49] *Boes, D.:* Rechnerische Auslegung von Schlauchköpfen und Schmelzespeichern (Design Computations for Tube Extrusion Dies and Melt Accumulators). In: Extrusionsblasformen. VDI-Verlag, Düsseldorf, 1979.
[50] *Boes, D.:* Kunststoffe, 72 (1982), pp. 7/11.
[51] *Uhland, E.:* Dissertation, Stuttgart University, 1978.
[52] *Masberg, U.:* Kunststoffe 71 (1981), pp. 15/17.
[53] *Geiger, K., Kühnle, H.:* Rheol. Acta (1984) 23, pp. 355/367.
[54] Coextrusion, Fortschritte der Blasformtechnik (Coextrusion, Progress in Blow Molding Technology). Company brochure, Bekum.
[55] *Smolik, G. R.:* Mod. Plast. Int., Dec. 1983, pp. 36/38.
[56] *Peters, G. W.:* Packaging, Jan. 1984, pp. 31/38.
[57] *Schneiders, A.:* Speichersysteme (Melt Accumulator Systems). In: Blasformen, VDI-Verlag, Düsseldorf, 1972.
[58] *Fritz, H. G., Maier, R.:* Kunststoffe 66 (1876), pp. 390/396.
[59] DE-OS 2537419 (1975).
[60] US-PS 4063865 (1977).
[61] DE-OS 2100192 (1971).
[62] *Daubenbüchel, W.:* Kunststoffe 66 (1976), pp. 15/17.
[63] Kunststoffe 73 (1963), p. 228.
[64] Verpackungsforum. Information Leaflet No. S. 106 (1984).
[65] *Voelz, V.:* Plastverarbeiter 32 (1981), pp. 326/330.
[66] *Voelz, V., Feuerherm, A.:* Kunststoffberater (1983) 1/2, pp. 17/22, and (1983) 3, pp. 29/34.
[67] *Schubbach, R.:* Plastverarbeiter 24 (1973), pp. 608/614.
[68] *Lohrbächer, V.:* Quetschnähte, Quetschzonen, Schließkräfte und Butzentrennung (Pinch Seams, Pinch-off Zones, Locking Forces and Deflashing). In: Extrusionsblasformen. VDI-Verlag, Düsseldorf, 1979.

13 Extrusion of foamed intermediate products with single-screw extruders

K.-D. Kolossow

13.1 Introduction

Given the need to use basic raw materials more wisely, the technology of extrusion of foamed intermediate products is becoming increasingly important.
In 1933 a blowing agent was injected for the first time into a polystyrene melt in a pressure vessel, and the discontinuous production of foamed polystyrene blocks became possible [1]. Later Dow Chemical applied for a patent on the continuous extrusion of foamed boards [2]. Since then foam process technology has been so greatly developed that foam processing of all the bulk thermoplastics is now possible. Because polystyrene has few problems related to mixing-in blowing agents, it retains the largest share of the thermoplastic foam market; but polyolefins, closely followed by PVC, have begun to catch up. The main reasons for the continued growth in the usage of foamed materials are a number of properties related to the reduction of density:

– low resin consumption per unit of volume,
– low thermal conductivity,
– improved mechanical and acoustical damping,
– increased bend stiffness at the same resin weight.

Conventional extrusion systems can be used for the manufacture of foamed products, and modified single-screw extruders are normally employed for this purpose. The development of single-screw technology in many cases went hand in hand with raw materials development, and this resulted in many materials formulation patents, and in carefully guarded processor know-how.

13.2 Phases of the manufacturing process

The manufacture of useful foamed intermediate products by extrusion requires careful selection of raw material, and proper coordination of the individual steps in the process. Generally, vaporising liquid ('physical') blowing agents are used for the manufacture of low-density foams; for higher foam densities, chemically decomposing ('chemical') blowing agents are used. The most important phases in the manufacture of foamed products by extrusion are:

– selection of raw materials and product recipe,
– feeding of raw material,
– extrusion,
– foaming,
– cooling and calibrating,
– take-off,
– winding, or cutting to length.

13.2 Phases of the manufacturing process

Operations carried out after extrusion include the following:

- expansion,
- laminating,
- scrap reclaiming.

These processes can be combined with the extrusion process, if desired.

13.2.1 Raw materials and recipes

The following *thermoplastic* resins are suitable for foam extrusion using chemical or physical blowing agents:

- polystyrene (PS), and polymer blends and copolymers such as high-impact polystyrene HIPS, SB, SAN, ABS, for films, boards, and profiles,
- low-density polyethylene (LDPE) and EVA copolymers for films, boards, and profiles,
- polyvinylchloride (PVC), rigid and soft, including blends for films, boards, and profiles,
- polypropylene (PP) for films, boards, and profiles.

An essential characteristic of a suitable raw material is its foamability. This is favored if the melt viscosity of the product decreases slowly with increasing temperature over the softening range. Accordingly, amorphous thermoplastic resins are easier to foam than polyolefins, because the temperature and critical viscosity range for foaming is very narrow (Figure 1) [3].

Partially crystallized thermoplastics must be crosslinked (chemically or by radiation) to achieve low foam densities.

There is a large selection of general purpose polystyrenes, but medium- and high-molecular-weight resins with a melt index (MFI 200/5) in the range 1 to 5 grams per 10 minutes are generally preferred. Low-molecular-weight PS is easy to extrude; however, its melt strength is too low for foaming, and the vapor pressure in the foam cell ruptures the cell walls and the foam collapses.

A basic recipe for foam extrusion is the following:

Component	Parts
Polystyrene	70
Re-extruded foam scrap	30
Nucleating agent	0.1 to 0.8
Physical blowing agent	2 to 12
Pigments	0.05 to 0.15
Additive (flame retardants, bonding agent, etc.)	0.1 to 6

The final recipe has to be determined with the properties of the final product in mind. Blending in of high-impact polystyrene (HIPS) in low quantities is recommended, to improve the impact strength of the final product.

Low-density polyethylene (LDPE) with a melt index (MFI 190/2.16) of 0.5 to 2 grams per 10 minutes and a density range of 0.917 to 0.922 grams per cubic centimeter can be used for foam extrusion. The basic recipe for the manufacture of a chemically foamed and crosslinked sheet is the following:

Component	Parts
Polyethylene (LDPE)	100
Chemical blowing agent (ADC)	8 to 15
Crosslinking agent (peroxide)	0.5 to 1.5

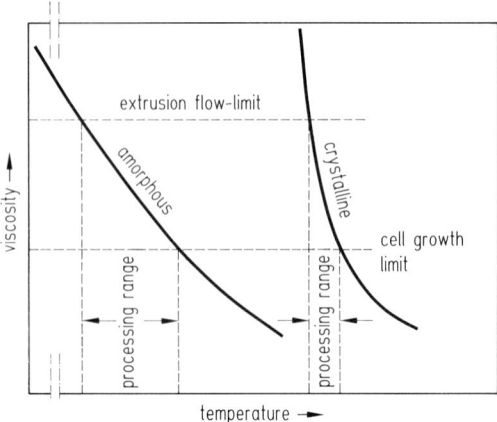

Figure 1. Schematic indication of the processing range available for foaming with thermoplastics of different structures; dependence on temperature and viscosity

The type and quantity of the crosslinking agent affect the viscosity increase during processing and determine the degree of foam expansion. The more gradual the viscosity increase, the more favorable are the conditions for production of low-density foam. The following is a basic recipe for the manufacture of PE foam sheets using the direct gassing process:

Component	Parts
Polyethylene (LDPE)	~ 70
Re-extruded foam scrap	~ 30
Nucleating agent	0.1 to 0.3
Physical blowing agent	2 to 6
Color masterbatch	0.05 to 0.4

All the basic PVC types – *emulsion (E-PVC), suspension (S-PVC)*, and *mass polymerized (M-PVC)* – as well as VC/VA copolymers and PVC/ABS blends are used for foam extrusion.
Materials with K-values of 55 to 58 process well [4]. The rheological characteristics and the limited heat stability of polyvinylchloride, require that processing aids be added, including "internal lubricants" to improve behavior in the extruder and in the die, and modifiers to increase melt extensibility.
The basic recipe for PVC foam sheet extrusion using the direct gassing process is:

Component	Parts
Rigid PVC	100
VC/VA copolymer	2
Stabilizer	0.5 to 2.5
Internal and external lubricant	0.5 to 2
Modifier	3 to 5
Filler	2 to 5
Nucleating agent	0.2 to 0.7
Physical blowing agent	4 to 8

Several PVC manufacturers already offer PVC compounds usable for foam extrusion [5].

13.2 Phases of the manufacturing process

The guidelines mentioned for PE extrusion also apply to the selection of polypropylene formulations. The viscosity range in which polypropylene foam can be extruded is very narrow, which makes the process even more difficult, because of the crystallization tendency of this polymer.

13.2.2 Nucleating agents and additives

Nucleating agents influence the foam structure and the foam density. At a given blowing agent concentration the amount of nucleating agent determines the number of cells, and therefore the cell size. Various systems, with differing modes of operation, are possible, as follows:

- fine dispersion of strongly exothermic substances in the melt, immediately before foaming; these reduce the melt viscosity and surface tension locally and promote the formation of blowing-gas bubbles [6].
- addition of gaseous or liquid substances, which increase the super-saturation of the melt by the blowing gas as the melt leaves the die, and which form minute, finely divided bubbles into which the proper blowing agent diffuses. Nitrogen and carbon dioxide are two highly effective nucleating agents. The second substance is generated, with water and sodium citrate, by the reaction of sodium bicarbonate and citric acid [7]. Because of the threefold function of this nucleating agent, citric acid and sodium bicarbonate are extensively used for the direct gassing of melts, mixed in the stoichiometric ratio of 1 to 1.33.

Figure 2 shows the effect of varying nucleating agent concentration on average foam cell diameter [8].

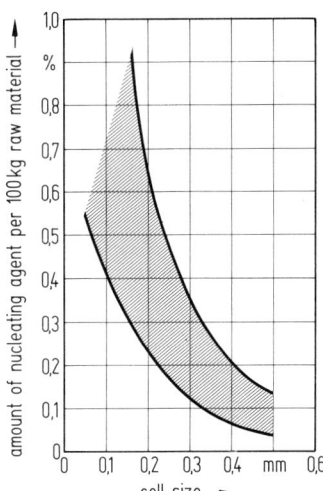

Figure 2. Average cell diameter, as a function of the amount of nucleating agent

- dispersion of very fine metal powder or metal compounds in the polymer melt. Such dispersions are known as "hot spots", because of their ability to form hot spots in the melt just before foaming, which reduce surface tension and melt viscosity locally.
- particulate solids such as talc, silica, and titanium dioxide can also be used as nucleating agents. The blowing gas is adsorbed on the surface of the particle during phase separation [8].
- other important additives are organic fillers, pigments, and flame retardants, and adhesion promoters to bond the powdered components to the surface of the pellets.

13.2.3 Blowing agents

The blowing agent selected determines the degree of foaming and the design of the foam extrusion line. Two basic types of blowing agent – 'physical' and 'chemical' – are used in foam extrusion. Physical blowing agents are liquids with boiling points below the softening points of the resins used. Chemical blowing agents are substances that decompose over a narrow temperature range to produce gases. Physical blowing agents are normally used if low foam densities are needed. An exception is the three-stage process for the production of polyolefin foam, which uses chemical blowing agents in combination with crosslinking agents.

Physical blowing agents

A physical blowing agent must have the following properties:

- relatively high solubility in the resin, without drastic change of the resin's viscosity or glass transition point.
- low diffusion tendencies.
- fast vaporization during expansion to achieve low residual gas concentration in the cell wall.

Fluoro-chlorinated hydrocarbons, aliphatic hydrocarbons and, to some extent, nitrogen and carbon dioxide, are physical blowing agents demonstrating all these properties.

There are many fluoro-chlorinated hydrocarbons, sold under various trade names: FRIGEN, KALTRON, FREON, and FLUGENE. For extrusion of foamed PS and PVC, tri-chloro-fluoro-methane (CCl_3F) called R 11, or di-chloro-di-fluoro-methane (CCl_2F_2) called R 12, or a mixture of both, is used. For foaming of polyolefin, R 114 is preferred. Their most important properties are summarized in Table 1.

Table 1. Properties of physical blowing agents

Blowing agent	Molecular weight	Density at 25 °C g/cm^3	Boiling point	Heat of vaporization kJ/kg	Thermal conductivity at 20 °C W/K m
Pentane	72.15	0.616	36.1	360	–
Trichlorofluoromethane (CCl_3F) R 11	137.38	1.476	23.8	182	0.0086
Dichlorodifluoromethane (CCl^2F^2) R 12	120.9	1.311	– 29.8	166	0.0103
Mixture R 11/R 12 (50 : 50)	129.14	1.39	– 2.5	164	0.0095
Carbon dioxide, CO_2	44.01	1.977	– 78.5	574	0.0163
Nitrogen, N_2	28.016	1.251	– 196	201	0.0258

Because of the need for environmental protection, there is a tendency to replace the fluoro-chlorinated-hydrocarbons by other physical blowing agents. Recent suggestions regarding the depletion of the world ozone layer by released fluoro-chlorinated hydrocarbons are supported by the latest test results [10].

Aliphatic hydrocarbons are further possible replacements for fluoro-chlorinated hydrocarbons, in addition to nitrogen and carbon dioxide.

Iso-pentane and *n*-pentane are also popular blowing agents, even though they are flammable and form explosive mixtures with air. Another disadvantage is the relatively high solubility of the pentanes in various resins, which reduces the degree of foaming.

Their main advantages over fluoro-chlorinated hydrocarbons are low price per unit weight and the larger gas volume per unit weight, due to their low molecular weight.

Chemical blowing agents

The main criterion in their selection is that the temperature of decomposition lie within the processing temperature range of the resin (Table 2). In addition, the rate of decomposition to gaseous products must not be too slow. It is also advantageous that the following conditions be fulfilled:

- the products of decomposition must not discolor the resin,
- the products of decomposition must not be corrosive,
- the products of decomposition themselves should act as nucleating agents.

Azo-di-carbonamide (ADC) is most commonly used. The best gas yield of 220 cm^3/g is achieved at a temperature of 210 °C. The blowing agent decomposes into solid and gaseous products, where the gaseous component is nitrogen. The temperature of decomposition, however, is too high for several of the temperature-sensitive thermoplastic resins. The decomposition temperature can be reduced by the addition of initiators (metal compounds such as zinc oxide and zinc stearate).

Table 2. Properties of chemical blowing agents

Blowing agent	Short nomen-clature	Processing temp. range °C	Gas yield cm^3/g	Concentration %	Used with
Azodicarbonamide (Azobisformamide)	ADC	165 to 215	220 (210 °C)	0.1 to 4.0	PP, PS, ABS, PE (hard), PVC
Azodiisobutyronitrile	AZDN	110 to 125	130 (110 °C)	0.5 to 6.0	PVC
p.p-oxibis (benzol-sulphonhydrazine)	OBSH	150 to 200	160 (160 °C)	0.5 to 2.0	PE (soft), EVA
Trihydrazinotriazine	THT	250 to 300	220 (270 °C)	0.1 to 1.0	PA, AC
Barium-azodicarbonate	BADC	250 to 300	200 (270 °C)	0.1 to 1.0	PVC, PA, PC, ABS
p-toluenesulfonyl semicarbazide	TSSC	180 to 210	200 (200 °C)	0.5 to 2.0	PE (hard), PP, PS, PVC

13.2.4 Feeding of raw materials

13.2.4.1 Solids

In a standard foam extrusion recipe the polymer component, apart from repelletized waste, normally amounts to more than 90% by weight. The resins are delivered in pellet or powder form and are suitable for transportation in trucks or rail tank cars. Obvious economic advantages are achievable by storing the resins in silos and using standard conveying systems to move the material to the premixing station. The dosing and premixing of the solid components require high accuracy to achieve uniform foam quality. Proven methods of carrying out this process use continuous dosing units in combination with a slow-running mixing paddle in the feed-hopper. For high accuracy, a volumetric or gravimetric dosing system to suit the condition and behavior of the solid components may be selected. The principle of volumetric dosing is based on a continuous counter, and a rotating disk provided with chambers with defined volumes. Because of the large number of such small chambers the desired accuracy can be achieved with free-flowing products. If even higher accuracy is called for, gravimetric dosing systems must be used.

13.2.4.2 Liquids

When the direct-gassing process is used, accurate dosing of the blowing agent is very important, since the quantity added determines the viscosity of the melt. The best devices for this job are multihead piston pumps with adjustable strokes. The pumping characteristic is based on a fixed-frequency reciprocating piston. The piston with cross-section A_k displaces a volume proportional to stroke length, h_s, and piston frequency, n_h.
The pump capacity can be calculated as follows [12]:

$$\dot{m} = \delta \cdot h_s \cdot A_k \cdot n_h \cdot \eta_F \tag{1}$$

$\dot{m}$ pump capacity
δ density of liquid
h_s stroke length
A_k piston cross-section
n_h piston frequency
η_F pump efficiency factor

The efficiency factor η_F is the ratio between the real volume injected into the extruder and the theoretically calculated volume. The particular parameters affecting this factor are the physical state, the temperature, and the compressibility of the liquid to be pumped, meaning that the efficiency is pressure- and viscosity-dependent.
Accordingly, to maintain an accurate and uniform pumping rate, the blowing agent is transferred from the storage tank to the dosing pump under fixed pressure; and it is also advisable to maintain a constant pressure on the discharge side of the pump, by installing a suitable back-pressure control valve.

13.2.5 Extrusion

The common feature of foam extrusion processes is that the blowing agent is always mixed into the melt in the extruder. Thus it can be seen that the most important steps in the process of foam production are premixing of resin and additives, injection of blowing agent, shaping of the foam, and post-treatment. The processing steps carried out in the extruder include the following:

- feeding of premixed resins and additives,
- surge-free conveying and compression,
- melting and homogenizing,
- achieving good, thermal homogeneity prior to injection of the physical blowing,
- homogeneous mixing of the blowing agent under pressure,
- steady conveying and cooling of the melt,
- extrusion of the melt at a higher pressure than the partial vapor pressure of the blowing agent.

Fast plastication at the beginning of the extrusion process is necessary when processing a polymer melt containing a chemical blowing agent. It is essential to control the melt temperature accurately, to avoid decomposition of the blowing agent too early. Fast plastication can be achieved by shear heating in the melt and by additional heat input, under accurate temperature control, through the extruder barrel. The purpose of this procedure is to establish a melt seal in the extruder towards the feed opening and to build up a high pressure before decomposition of the chemical blowing agent can start. Homogenizing of the melt in the extruder is a stage of the process where the melt temperature must be increased very fast in the last barrel section; but it must remain constant in the die.

When the direct gassing method is used, the procedures in the plasticating and melt zones are the same as with chemical blowers. But to obtain a homogeneous foam with this method the melt temperature and pressure have to reach their peaks before the blowing agent is injected.

For a better understanding of the processes going on in the single-screw extruder, a few equations due to Ast [3] are listed below:

Shear rate in the metering zone

$$\dot{\gamma} \approx \frac{\pi \cdot n \cdot D}{60 \cdot h} \, s^{-1}, \qquad (2)$$

Average residence time in the metering zone expressed as the ratio of screw free-volume V, and volumetric throughput, $\dot{V}$, of the extruder

$$\bar{t} = \frac{V}{\dot{V}} \approx \frac{\pi \cdot D \cdot h \cdot b (L/D) \cdot 3.6 \cdot \varrho}{\cos \varphi \cdot \dot{m} \cdot 10^3} \, s, \qquad (3)$$

Temperature increase ΔT_s by heat dissipation in the metering zone

$$\Delta T_s \approx \frac{\eta \cdot \dot{\gamma}^2 \cdot \bar{t}}{\varrho \cdot c} \approx \frac{\eta \cdot \pi^3 \cdot D^3 \cdot n^2 \cdot b \cdot (L/D)}{h \cdot \cos \varphi \cdot \dot{m} \cdot c \cdot 42.7 \cdot 10^3} \, °C, \qquad (4)$$

Temperature change, ΔT_L, of the melt by heating or cooling through the extruder barrel

$$\Delta T_L = \frac{\dot{Q}}{\dot{m} \cdot c} \approx \frac{\alpha \cdot \pi \cdot D \cdot L \cdot (\bar{T} - T_z)}{\dot{m} \cdot c \cdot 10^3} \, °C. \qquad (5)$$

D screw diameter in mm
L screw length in mm
h screw channel depth in mm
b screw channel width in mm
φ screw flight pitch
n screw speed RPM
$\dot{m}$ throughput in kg/h
ϱ melt density in g/mm^3
c specific heat in cal/g $\cdot$ °C
η melt viscosity in kp $\cdot$ s/mm^2
α heat transfer coefficient between melt and extruder barrel in cal/mm$^2 \cdot$ h $\cdot$ °C
T_z extruder barrel temperature in °C
$\bar{T}$ average melt temperature in °C.

13.2.6 Foaming

The process of foaming is influenced by the viscosity and elasticity of the melt, by the gas permeability of the polymer, and by the vapor pressure, solubility and decomposition temperature of the blowing agent, all interacting with each polymer in a particular way.

Furthermore, if there is to be uniform cell growth at the die exit, the blowing agent in gaseous form must be dissolved or partially dissolved in the viscoelastic polymer melt. The origin of the gas is not important; it can be added direct to the melt through the extruder barrel, or be generated by a chemical blowing agent dry-mixed with the resin.

Because of the sudden pressure drop that occurs at the exit of the die, supersaturation of the melt by the dissolved gases results, which initiates the separation of gas from the melt. Thereafter, a large number of small bubbles are formed. These bubbles expand until an equilibrium is reached between the vapor pressure, the surface tension, the melt viscosity, and the degree of saturation of the remaining dissolved gases [6, 14, 15]. The melt temperature is lowered during this process by the expansion of the gas and through heat losses to the surroundings.

As long as the state of supersaturation persists, theoretically there will be nucleation of bubbles. At the same time migration of gas from small to larger bubbles will be taking place because of differences in pressure between them, and finally the gas will escape to the outer surface of the extrudate (see Eq. (8) below).

The gas transfer process with physical blowing agents can be described, following *Schleith* [16], using the energy equation of state for a bubble at equilibrium:

$$\sigma \cdot \Delta A_0 = p_i \cdot \Delta V \tag{6}$$

σ interfacial tension
p_i partial vapor pressure in bubble
A_0 bubble surface area

For a spherical bubble, radius r and volume $V = \dfrac{4\pi r^3}{3}$ we then obtain:

$$\sigma \cdot \{8\pi r \Delta r\} = p_i \cdot \{4\pi r^2 \Delta r\} \tag{7}$$

and therefore:

$$p_i = \frac{2\sigma}{r}. \tag{8}$$

Thus the internal pressure in small bubbles is higher than in larger bubbles. In hot conditions, this results in very fast diffusion of the gas from small cells to larger ones and finally towards the surface of the foam, where the gas escapes.

The resulting diffusion of the bubbles themselves follows Stokes's law:

$$w = \frac{g \cdot D^2}{18\eta} \cdot \Delta\varrho. \tag{9}$$

w diffusion rate
g gravitational constant
η dynamic viscosity
$\Delta\varrho$ density difference between bubble and melt
D $2r$.

The diffusion rate increases with increase of bubble size and reduction of melt viscosity. To avoid related problems in the manufacture of fine-structured foams it is advisable to create a large number of small cells by adding nucleating agent.

High shear stresses on the melt in the die during expansion can result in a coarse cell structure and open cells at the surface.

It is essential in foam extrusion to hold the pressure on the melt in the extruder and the extrusion tool above the partial pressure of the blowing agent. The partial pressures of blowing agents can be difficult to determine prior to extrusion. *Henry's* law enables the partial pressure of physical blowing agents to be calculated approximately. The law is only valid for low concentrations of blowing agents:

$$p_L = \frac{c}{S} \tag{10}$$

13.2 Phases of the manufacturing process

p_L partial pressure
c gas concentration in the melt
S solubility

The equation shows that the partial vapor pressure is proportional to the gas concentration in the melt and inversely proportional to the solubility, which is temperature-dependent. Alternatively the *Flory-Huggins* relationship can be used for approximate calculation of the partial vapor pressure, since it is not limited only to ideal mixtures [17].

$$\ln \frac{p_L}{p_0} = \ln(1 - V_x) + V_x + \chi V_x^2(\eta) \tag{11}$$

V_x Volume fraction of polymer
p_0 vapor pressure of blowing agent at operating temperature.

The problem in applying the above equation is to determine the interaction parameter χ. The experimentally determined linear relationships for the solubility of different gases and vapors in elastomers [18] is worth mentioning. These allow the solubility, S, to be calculated at the boiling point, T_s, and the critical temperature, T_c, of the blowing agent:

$$\lg S_{298} = 2.1 + 0.0123\, T_s \tag{12}$$

$$\lg S_{298} = 2.1 + 0.0074\, T_c \tag{13}$$

T temperature, °K.

The relation between the solubility and the heat of solution of gases in elastomers was found to fit the equation:

$$\Delta H_s = -2400 - 2000\, \lg S_{298} \tag{14}$$

According to *Burt* the following equation can be used to calculate the temperature-dependence of solubility:

$$S_T = S_{298} \exp \frac{\Delta H_s}{R} \left[\frac{1}{298} - \frac{1}{T} \right] \tag{15}$$

R gas constant.

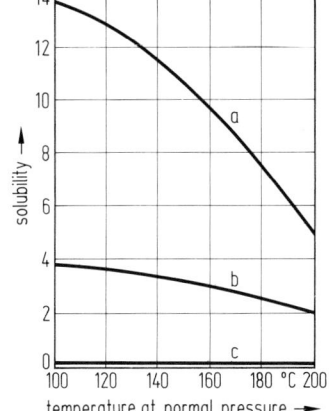

Figure 3. Solubility of a physical blowing agent in polystyrene, as a function of temperature
a trichlorofluoromethane, b dichlorodifluoromethane, c nitrogen

Assuming that the conditions of solubility of the blowing agent in the melt at the glass transition point are similar for elastomers and thermoplastics, the empirical equations (12) to (15) can also be applied to amorphous thermoplastic resins [16]. Figure 3 shows the solubility of some physical blowing agents as a function of processing temperature, using the *Schleith* equation.

13.2.6.1 Cooling, calibrating and take-off

For the manufacture of foamed intermediate products, the principle most often used is of free expansion from the die. This gives three-dimensional cell growth, and means that changes in tube circumference, wall thickness and the ratio of take-off to extrusion velocity of the extruded tube are identical.

The linear expansion ratio, or degree of foaming, F can be determined from

$$F = \sqrt[3]{\frac{\varrho_{\text{solid}}}{\varrho_{\text{foamed}}}} \,. \tag{16}$$

Following *Zingsheim* [20], the die gap, G_D, of an annular die can be calculated if foam density and foam thickness are known:

$$G_D \approx \frac{s'}{\sqrt[3]{\frac{1.4}{\varrho_{\text{foamed}}}}} \text{ mm}. \tag{17}$$

Uniform, three-dimensional expansion is very important if the foamed product is to be used for thermoforming. Foam films with low shrinkage coefficients in all three directions are best suited for thermoforming. The main influence on shrinkage is the blow-up ratio i_A defined as:

$$i_A = \frac{D_D}{D_F} \tag{18}$$

D_F diameter of foamed tube
D_D die diameter.

The effect of blow-up ratio on orientation has been determined experimentally [21]. With increased blow-up ratio, shrinkage in the extrusion direction (MD) is reduced; however, shrinkage in the transverse direction (TD) is increased. Shrinkage was least at a blow-up ratio of 1:3.2. This value is very close to what is used in production; most foam manufacturers now work with blow-up ratios of 1:3.2 to 1:4.1, depending on the composition of the blowing agent.

The effect of take-off ratio, i_1, as well as of blow-up ratio, i_A, must be determined; i_1 is the ratio of the speed of the foam sheet to that of the melt in the annular gap of the die.

$$i_1 = \frac{V_1}{V_2} \tag{19a}$$

V_1 take-off speed
V_2 melt speed in the die gap.

Then, since take-off speed

$$V_1 = \frac{\dot{m} \cdot V^*}{D_F \cdot \pi \cdot S_F} \tag{20a}$$

and extrusion speed

$$V_2 = \frac{\dot{m} \cdot V_s}{D_D \cdot \pi \cdot G_D}, \tag{20b}$$

using the blow-up ratio:

$$i_A = \frac{D_D}{D_F} \tag{18}$$

the following simplified equation is obtained for the take-off ratio:

$$i_1 = \frac{V^* \cdot G_D \cdot D_D}{D_F \cdot S_F \cdot V_s} \tag{19b}$$

where G_D die gap
          $\dot{m}$ mass throughput
          V^* specific volume at glass transition temperature
          V_s specific volume at melt temperature
          S_F sheet thickness.

Since the effect of the differences in specific volumes is negligible, the take-off ratio really only depends on the die gap, G_D, the sheet thickness, and the blow-up ratio. Since the thickness of the final film and the blow-up ratio are dictated by the shrinkage, the only remaining variable is the die gap G_D.

13.2.7 Post-expansion

One of the useful features of extruded foams is the closed-cell structure; because of it the cells still contain most of the injected blowing agent even after cooling and calibrating. If now heat is applied by steam or boiling water, the gas pressure in the cells increases, the resin softens again and the growth of the cells continues.
It is possible by post-expansion to reduce density by a maximum of 50%. Furthermore, post-expansion in combination with extrusion improves the cell size distribution which, in turn, improves the physical properties of the foam.

13.2.8 Lamination

The steps required for the manufacture of multilayered products in combination with foam extrusion can be divided in groups such as follows:

- lamination,
- extrusion coating and
- coextrusion.

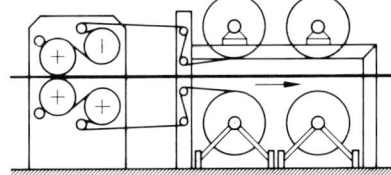

Figure 4. Principle of laminating during foam extrusion

Lamination of foamed intermediate products with paper and products like aluminum foil, fabrics, and compact extruded films with identical or similar adhesion properties, is usually achieved without the use of adhesives. Bonding can be obtained in a roll nip by applying heat and pressure simultaneously (Figure 4).

13.2.9 Scrap recovery

Economic considerations force foam manufacturers to reclaim all uncrosslinked scrap produced during start-up, edge trimming, thermoforming, and cutting. Several approaches are possible for reclaiming scrap, but continuous automatic recovery processes are best, because they are more economical and give better product quality and uniformity than discontinuous methods.

After size reduction, the foam flakes are usually re-extruded in a single-screw extruder and repelletized. Re-extrusion is preferable to densification, since melting of the voluminous waste, degassing, filtering, strand extrusion, cooling, and pelletizing are done in a very short time in a single operation. Figure 5 shows a flowsheet of a scrap-recovery line.

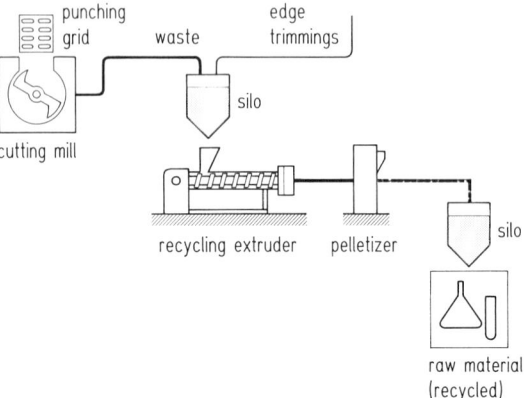

Figure 5. Schematic layout of scrap recycling (Berstorff system)

Table 3. Dimensions and densities of foamed products obtained with various blowing systems

Product	Dimension	Density	Blowing system	
	mm	kg/m^3	physical	chemical
Polystyrene sheet, laminated and nonlaminated	0.15 to 4.5	35 to 250	×	
Polystyrene board	10 to 120	32 to 80	×	
	5 to 10	250 to 600		×
Polystyrene profiles		45 to 75	×	
		250 to 500		×
LDPE sheet	1 to 10	20 to 400	×	
	0.2 to 10	35 to 600		×
LDPE board	10 to 80	30 to 60	×	
	8 to 15	32 to 400		×
LDPE pipe	ID 10 × 6 – ID 115 × 19	32 to 45	×	
LDPE profiles		250 to 600		×
PVC film rigid	1.5 to 6	90 to 280		
soft	2 to 6	350 to 500		×
PVC boards rigid	6 to 12	90 to 300	×	
soft	10 to 30	400 to 600		×
PVC rigid profiles		100 to 300	×	
		400 to 600		×
PP film	0.3 to 6	120 to 250	×	
	0.3 to 8	32 to 250		×
PP board	8 to 20	32 to 250	×	
	8 to 15	32 to 250	×	×
PP profiles		300 to 600		×

13.3 Foam products

The most important extrusion-foamed intermediate products are films, boards, profiles, tubes, and packaging materials.

Table 3 gives a summary of the dimensions and densities that can be obtained with different blowing agents.

The forms available, the properties, and main fields of application of these foamed products are listed and compared in Table 4.

Table 4. Foamed plastics products: forms, properties and applications

Polymer/ foam type	Form of product	Basic characteristics	Special features	Main areas of application
1. Polystyrene foam				
1 a. physically foamed	films	closed cell, flexible or brittle	thermoformable, laminable, printable, flockable, pressable, moistureproof and waterproof. Thermally insulating. Low water vapor- and air permeability	packaging hot meals, thermal insulation in buildings, sandwich film, linings — automobile and instrument industries
	boards	closed cell, stiff and hard	moderately good temperature form stability, high compression strength, difficult to ignite, weather-resistant, low moisture uptake, low thermal conductivity	thermal insulation and sandwich board for the building industry and for the automobile and instrument industries
	profiles	closed cell, flexible or stiff	moisture and water-resistant, good form stability, low moisture uptake, smooth surface, good insulation properties, thermoformable	building industry — interior finishing
1 b. chemically foamed	boards	closed cell, stiff and hard	solid skin, core like structural foam, easily workable, weather- and chemicals-resistant	interior finishing, furniture, roofing
	profiles	compact, hard, stiff and flexible	as for boards, 1 b above	interior finishing, furniture, skirting, frames
2. LDPE foam				
2 a. physically foamed	films	closed cell, soft, elastic	waterproof, good mechanical and acoustical damping, weldable, laminable, crosslinkable, good chemical resistance	packaging of fragile goods, acoustic insulation for buildings
	boards	closed cell, soft, elastic	as for films, 2a above	as sheet 2a furniture upholstery
	pipes, profiles	closed cell, soft, elastic	as for films, 2a above difficult to ignite	see under sheet 2a insulation for heating systems
2 b. chemically foamed	films	closed cell, tough or soft, elastic	water-repellent, good mechanical and acoustic damping, weldable, laminable, crosslinkable, good chem. resistance, thermoformable	linings automobile and instrument industries, water-proof clothing, sports articles, protective covering of sports stadiums, carrier bags
	profiles, boards	closed cell, stiff or flexible	solid skin, core like structural foam, easily workable, good chem. resistance	interior finishing, boards and profiles

Table 4. (Continued)

Polymer/ foam type	Form of product	Basic characteristics	Special features	Main areas of application
3. PVC foam				
3a. physically foamed	rigid sheets	closed cell, hard, stiff or flexible	thermoformable, laminable, printable, thermally insulating, difficult to ignite, weather-resistant, rotproof	thermal insulation and lining – building industry
	rigid boards, boards and profiles	closed cell, hard, stiff or flexible	solid skin, core uniform cell structure, good chem. resistance, rotproof, weather-resistant, difficult to ignite, easily workable	exterior and interior finishing – building industry
3b. chemically foamed	flexible film	closed cell, flexible, elastic	pressable, water-repellent, weather-resistant, rotproof	floor coverings – sanitation, agriculture
	flexible pipes	closed cell, flexible, elastic	as for soft PVC, 3b above	pipe covering
	rigid boards	closed cell, hard, stiff	solid skin, core like structural foam, chem. resistance, rot-proof, weather-resistant, ignition-resistant, easily workable	exterior and interior finishing – building industry
4. PP foam				
4a. chemically foamed	films	closed cell, flexible or stiff	good form stability at high and low temperature, thermo-formable, good mechanical and acoustic damping, weld-able, laminable, good chem. resistance	linings – automobile and instrument industries, packaging sector
	boards and profiles	closed cell, hard, stiff or flexible	as for films, 4a above – if not crosslinked	linings – kitchen equipment and sanitary ware

13.4 Processes

The most common extrusion processes for the manufacture of foamed semi-products are:
– single-stage process
– two-stage process
– multistage process.

13.4.1 Single-stage process

With this method the form and density of the product are determined in one step irrespective of the blowing agent system; it is used very extensively because of its economic advantages (see Table 3). Extrusion and foaming, together, are the most important aspects of this process. The extrusion technique may, however, vary depending on the selection of the blowing agent. Single-stage techniques using various blowing agents are schematically shown in Figure 6. The process can be carried out in four ways, viz.:

13.4 Processes 445

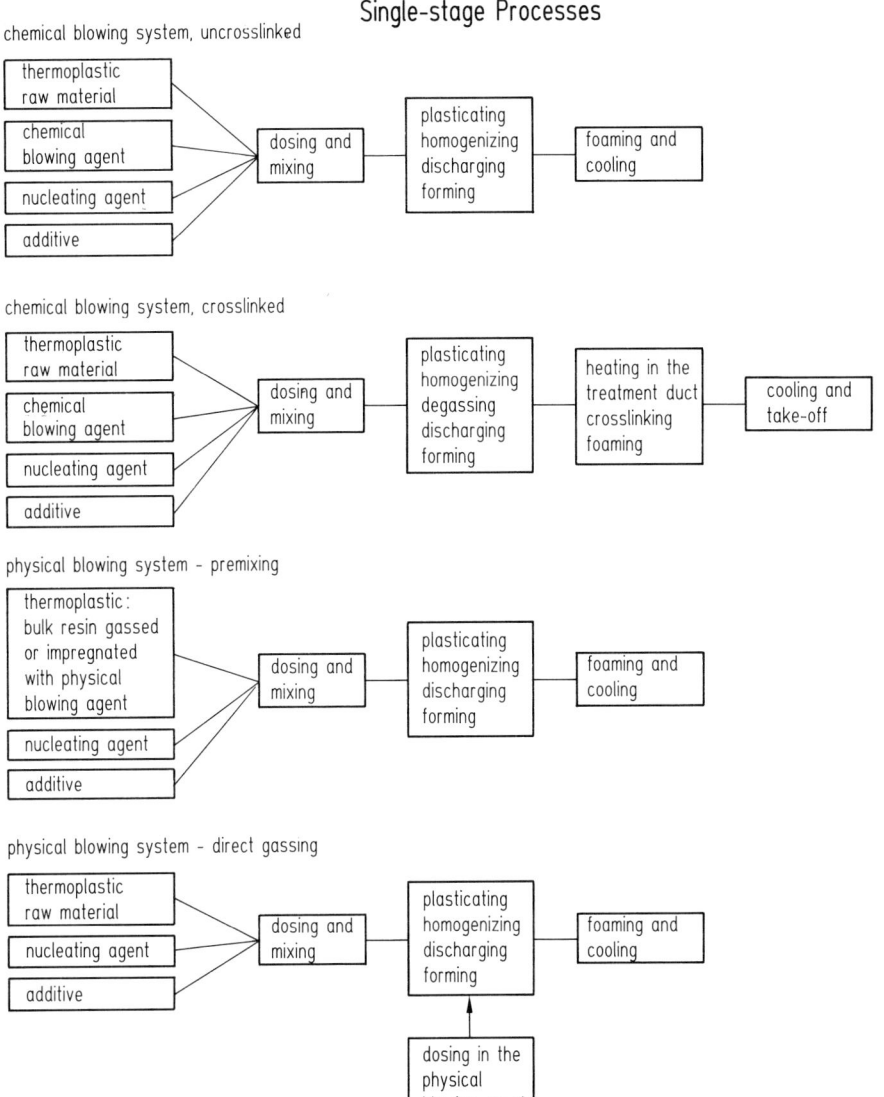

Figure 6. Use of a single-screw extruder for the single-step process.
(Berstorff system schematic)

a) Single-stage process, chemical blowing agent system, not crosslinked
  - adding of blowing agent to the raw material in the form of powder or masterbatch prior to extrusion,
  - generation of blowing gas in the extruder and during foaming.
b) Single-stage process, chemical blowing agent system, crosslinked
  - addition of crosslinking agent to the raw material and blowing agent during pre-mixing,
  - extrusion of profiles, not crosslinked, not expanded,
  - subsequent crosslinking and foaming in microwave or hot air ducts.

c) Single-stage process, physical blowing agent system
   pre-gassing:
   - feed material: a thermoplastic resin already containing physical blowing agent, incorporated in the resin during polymerization or by impregnation in an autoclave,
   - phase separation of blowing agent during foaming,
   direct gassing:
   - blowing agent injection direct into the melt in the extruder by a high-pressure injection system,
   - phase separation of the blowing agent during foaming.

Two-stage Processes

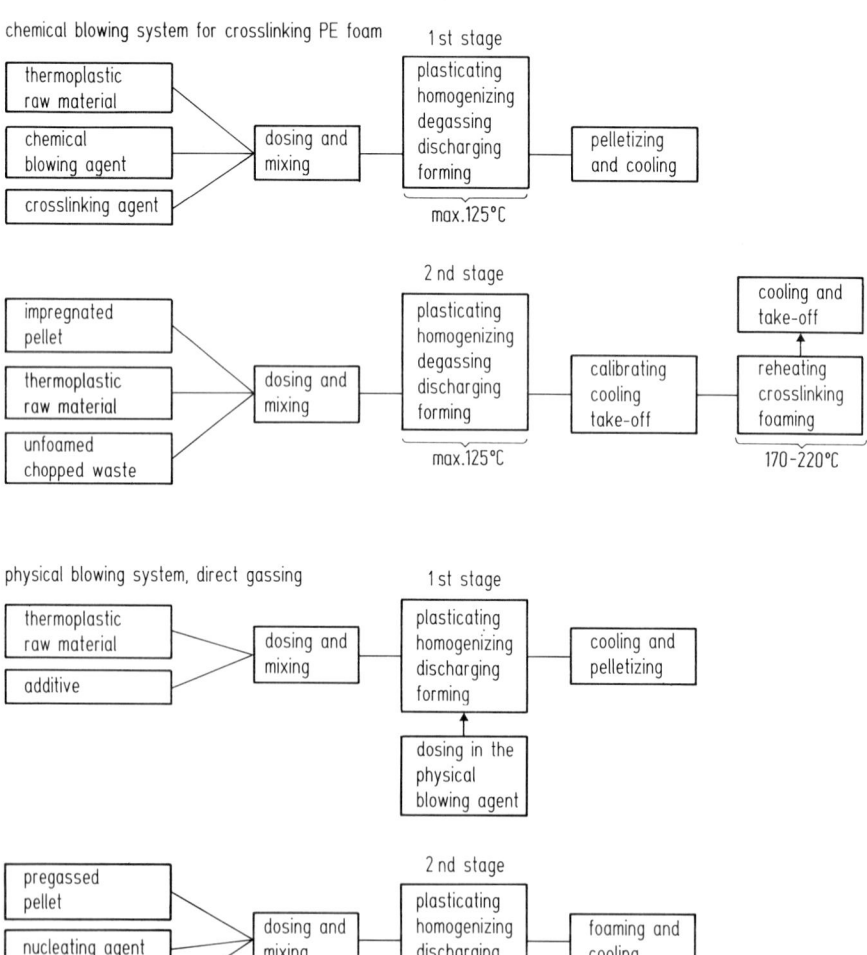

Figure 7. The predominant single-screw, two-stage process (schematic)

13.4.2 Two-stage process

The main feature of this process is the incorporation of the blowing agent into the melt in a completely separate process step prior to foaming.
The two-stage process for manufacturing crosslinking PE foams with low densities has been accepted because of the simplicity of process control. As previously mentioned in Section 13.2.1, the manufacture of low-density PE foam requires the melt viscosity to increase steadily as the material softens. It is a common practice in trying to solve this problem to add chemical crosslinking agents, whose function is to enhance the melt strength before decomposition of the chemical blowing agent. Since crosslinking in the extruder is difficult to control, crosslinking agents and blowing agents are added in large concentrations in the first stage of the process, and dispersed in the PE melt below the decomposition point of the blowing agent. The melt is then degassed and pelletized (Figure 7). The unexpanded edge trims generated in this process can be added to virgin material in the second extrusion to an extent determined by the foam density required. Extrusion and film forming are effected by the film die in combination with a polishing-roll system. Thereafter, the extruded sheet is fed direct into a heated duct in which the expansion and crosslinking take place. This is known as the *Furukawa* process and is protected by patents [22].
The two-stage process using physical blowing agents includes direct gassing and pelletizing of the melt in the first stage, as shown in Figure 7. Thereafter, the extrusion of the foamed semi-finished product is done in the second stage of the process. This process offers advantages for the manufacture of thin films.

13.4.3 Multistage process

The process shown in Figure 8 is used exclusively for the manufacture of crosslinked PE foams.
Different process steps are required with different crosslinking systems:
a) Chemical blowing agent system and peroxide crosslinking agent.

First stage
 – adding of blowing agent and crosslinking agent during premixing, before extrusion,
 – extrusion of uncrosslinked and unfoamed masterbatch.

Second stage
 – extrusion of uncrosslinked and unfoamed matrix.
Third stage
 – heating, crosslinking, foaming, and cooling of intermediate product.
b) Chemical blowing agent in combination with radiation crosslinking.
First stage
 – addition of blowing agent to the raw material during premixing, before extrusion,
 – extrusion of uncrosslinked and unfoamed matrix.
Second stage
 – radiation crosslinking.
Third stage
 – heating, foaming, and cooling of the product.

Multistage Processes

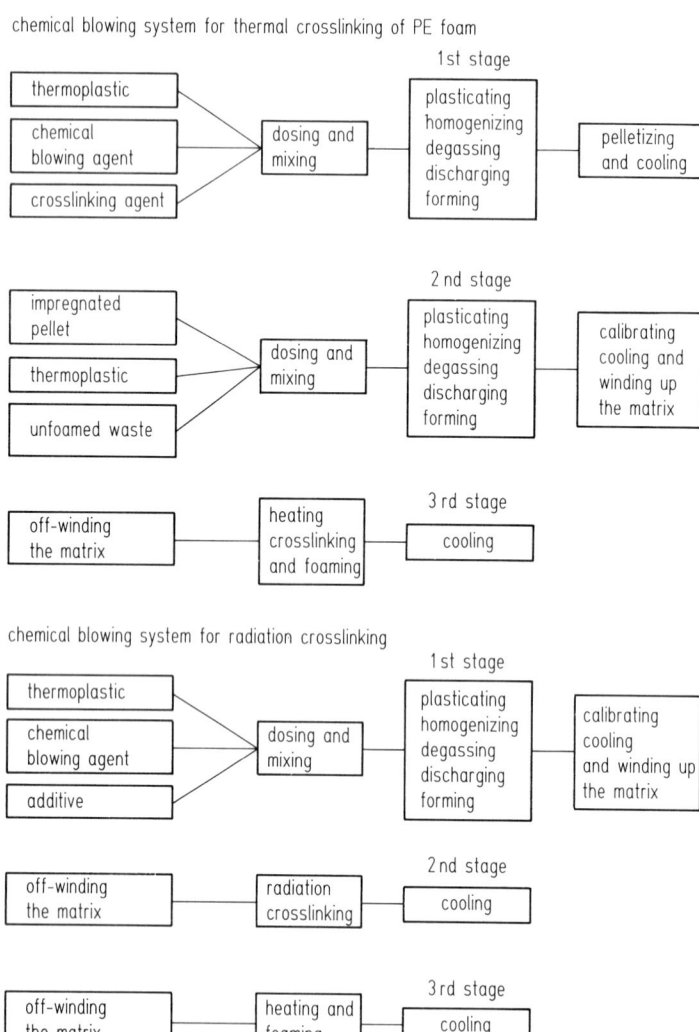

Figure 8. Steps in the single-screw multistage process

13.5 Single-screw extrusion systems

Single-screw extruders and single-screw combinations in series (tandem extrusion systems) are the usual systems employed for the manufacture of foamed intermediate products (Figure 9). For special applications the following systems are available:

- twin-screw extruder in series with single-screw extruder (twin-screw/single-screw tandem extrusion system)
- planetary-gear extruder in series with single-screw extruder (planetary-gear/single-screw tandem extrusion system).

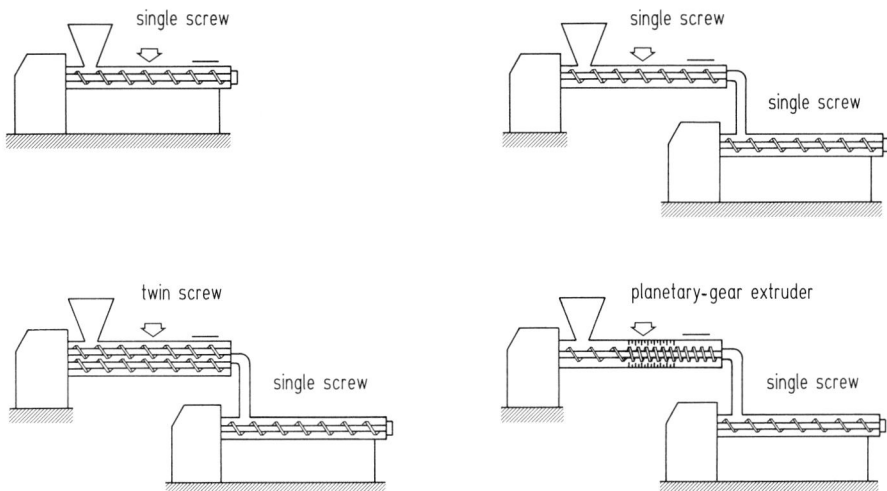

Figure 9. Schematic illustration of single-screw extruder systems for the manufacture of foamed products

13.5.1 Single-screw lines

Extruders with screw diameters of 45 to 90 mm and screw lengths of 24 to 30 D are used for the manufacture of foams with densities of 0.2 to 0.7 g/cm^3, using chemical blowing agents. The relatively small size of these extruders is dictated by their use on profile extrusion lines. The basic design of such lines is similar to that of conventional profile extrusion lines, and includes the extruder, premixing-, calibrating-, cooling, take-off-, cutting-, and dumping devices. The main differences are in the design of the die and, if necessary, of the calibrators. The established processes are results of intensive research and development, and are mostly protected by patents.

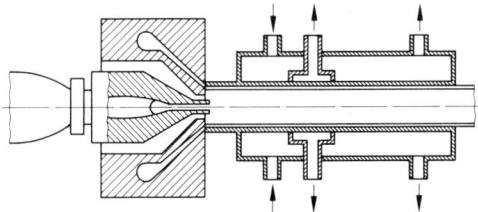

Figure 10. Coextrusion of foamed intermediate products with a foamed core from two different melt streams (after Reifenhäuser)

For instance:

- *Reifenhäuser* process [23], Figure 10: using the principle of coextrusion, the core of a solid-wall tube is filled with foam.
  This process makes it possible to combine two materials of different density or two similar resins in the die or calibrator, as well as to plasticate the resins independently.
- *Celuka* process by Ugine Kuhlman, Figure 11. The die mandrel works like a flow restrictor. In the calibrator, which is directly attached to the die, the outer skin of the melt, containing the blowing agent, is cooled intensively by contact, densified and solidified. The remaining melt expands inward and fills the profile core with foam.

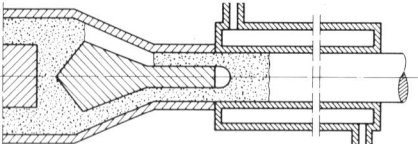

Figure 11. Extrusion of foamed solid profiles (after Ugine Kuhlmann)

— A modified *Celuka* process is shown in Figure 12. The melt, extruded in two annular die gaps, is intensively cooled in the calibrator at the inner and outer surfaces. Characteristic of the hollow profile are the skin layers with higher densities at the inner and outer surfaces [24].

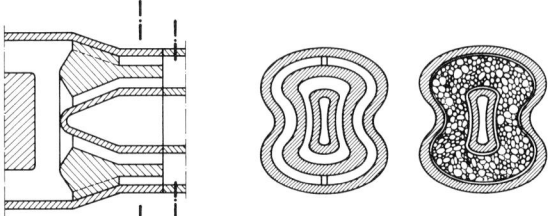

Figure 12. Extrusion of foamed hollow profiles (after Ugine Kuhlmann)

— The process shown in Figure 13 combines all the advantages of coextrusion and the use of a mandrel. It is protected by patents [25].

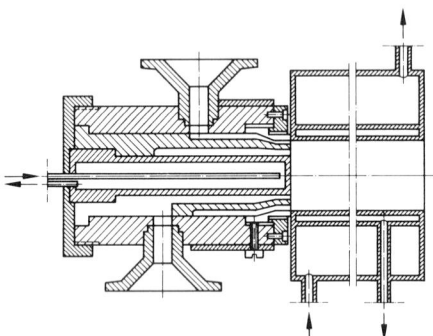

Figure 13. Coextrusion of foamed profiles from two different melt flows (after Ugine Kuhlmann)

— The BASF process (Figure 14) uses an extruder to pump the melt containing its blowing agent into a pressurized vessel. Thereafter, the melt can be released from the vessel and foamed [26].

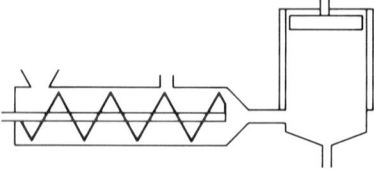

Figure 14. BASF foam extrusion process

— *Armocel* process by Société Armosig (Figure 15). This, as well as the *Celuka* process, is used for the manufacture of foamed semi-finished products, using chemical blowing agents. The design of the dies is patented [27]. A conical reduction of the flow channel before the land zone avoids early expansion in the die. The illustration shows that there is a gap between the die and the calibrator, which is not the case in the *Celuka* process. After the melt exits the die lips it can expand freely according to the die geometry until finally solidified in the calibrator.

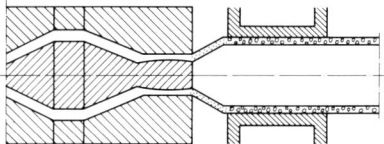

Figure 15. Extrusion of foam pipes with the Armocel process

– The Scherer & Trier process offers another possibility for producing foamed products with high-density surface layers. The melt is divided into two streams in the die in order to raise the temperature of the core stream by means of an axially located torpedo, so that it is only the core of the profile which foams.

The extrusion processes described so far are the most suitable for manufacturing foamed intermediate products, using chemical blowing agents. Chemical or physical blowing agents can optionally be used in the *Woodlite* process developed by Sekisui Plastics Coroporation Ltd., Tokio [28]. A strand die maintains the pressure in the die flow-channel as shown in Figure 16. The expansion of individual strands is limited by cooled surfaces in the calibrating channel, with the result that they flow together to form a board. Limitation of expansion along the cooled surface also creates a skin of higher density.

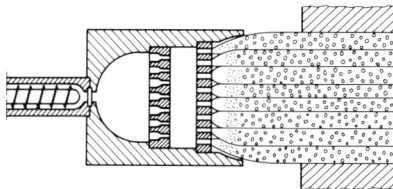

Figure 16. The Woodlite process

Figure 17 [29] shows a process developed specially for extruding PS-profiles with skins of higher density, using physical blowing agents. A lubricant is injected by a metering pump, forming a film between the melt and the surface of the tool, to prevent the low viscosity melt from sticking to the walls of the flow channel in the die. Dimensionally accurate PS-profiles can be maintained at extrusion speeds of up to 30 m/min. using this process [30].

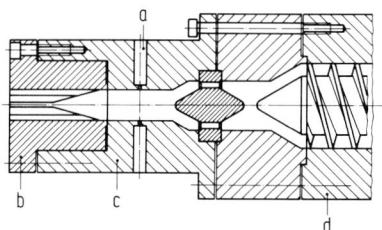

Figure 17. Extrusion process for low-density foamed profiles.
a connections for adding the slip agent, *b* die, *c* die holder, *d* extruder end or outlet cooler

Single-screw extruders are employed for the manufacture of profiles and films using the direct-gassing process with physical blowing agents. They are distinguished from conventional extruders by their greater length.

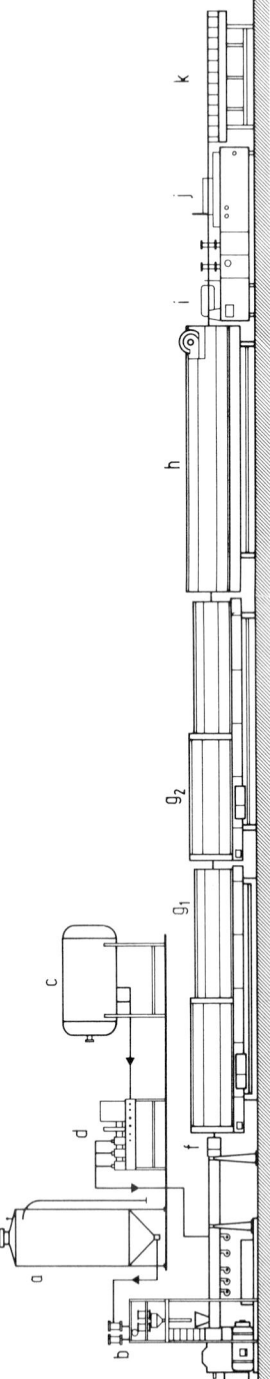

Figure 18. Single-screw direct-gassing line for the production of PE foamed pipes (Berstorff system)
a raw material silo, *b* dosing and mixing unit, *c* blowing agent tank, *d* blowing agent dosing unit, *e* single-screw extruder with direct-gassing system, *f* die, g_1 cooling bath, g_2 cooling bath, *h* drying oven, *i* take-off, *j* cutter, *k* drop-off unit

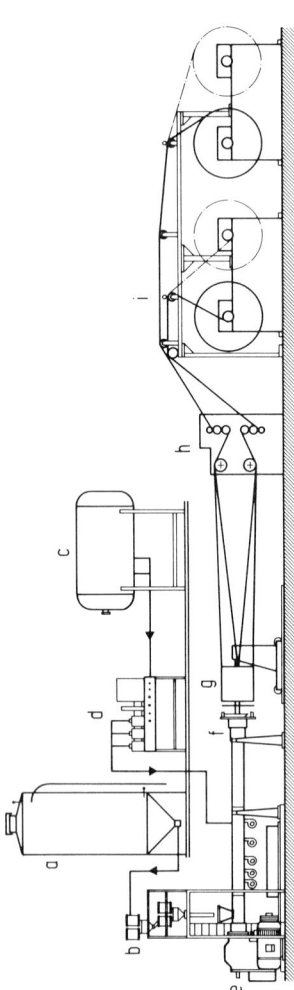

Figure 19. Arrangement of a single-screw extrusion line for the production of windable foam sheets (Berstorff system)
a raw material silo, *b* dosing and mixing device, *c* blowing agent tank, *d* blowing agent dosing unit, *e* single-screw extruder with direct-gassing system, *f* die, *g* calibrating mandrel, *h* 4-roll s-wrap, *i* center winder

13.5 Single-screw extrusion systems

Figure 18 shows a single-screw unit for the manufacture of foamed LDPE pipe insulation using the direct-gassing process. The individual raw materials are fed into a mixer by dosing devices and then fed into the extruder, which has an L/D ratio of 40:1. The direct-gassing extruder differs from the plasticating or homogenizing extruder in the design of the barrel, and in sheet die design.

The screw is designed for functions such as intake and melting of the raw materials, mixing in of blowing agent, and cooling of the melt mix to extrusion temperature. The first part of the extruder barrel is electrically heated and air-cooled. The blowing agent is injected into the extruder at 16 to 18 D from the feed opening and mixed into the melt at approximately 180 °C. The second parts of the extruder barrel and extruder screw are designed for intensive cooling by liquid heat-transfer medium; this is because of the necessity to cool the melt to approximately 130 °C before it leaves the die. The die is a modified pipe die.

The melt expands freely after leaving the die mouth, because the pressure drop is rapid.

The extrudate is led through a water-spray cooling bath and an air cooling-system by a caterpillar-type take-off, to solidify the expanded foam; it is then conveyed to a cutting device and finally to a dumping table.

The most common direct-gassing, single-screw extruders used for foam manufacturing are shown in Table 5.

Table 5. Technical data on single-screw direct-gassing extruders for producing foamed LDPE pipes

Screw diameter	Screw length	Screw speed	Installed drive power	Installed heating capacity	Installed cooling capacity	Through-put*)	Product data	
							Dimensions D Ø × S	Density
mm	L/D	min^{-1}	kw	kw	kw	kg/h	mm	kg/m^3
45	38 to 42:1	119	13	9	15	15 to 25	10 × 6 to 22 × 6	35 to 45
60	38 to 42:1	80	20	18	24	25 to 45	15 × 6 to 42 × 10	35 to 45
90	38 to 42:1	45	38	26	42	60 to 110	22 × 10 to 34 × 35	35 to 45
120	38 to 42:1	30	65	42	58	110 to 180	34 × 35 to 115 × 10	35 to 45

*) max. values achievable depend on density and thickness of product, as well as type of polymer and blowing agent.

Figure 19 shows a single-screw system for extrusion of foamed film. The extruder is equipped with a vertically oriented film-blowing die; this arrangement is different from a pipe insulation line, on which extrusion is horizontal. To obtain a smooth foamed film, the diameter of the tube at the die, determined by the die diameter, must be enlarged in a specific ratio to allow for three-dimensional cell growth during expansion (see Equation 18).

To achieve this for sheet thicknesses of 1 to 4.5 mm the tube is pulled over a temperature-controlled calibrating mandrel, stretched axially and radially, and simultaneously cooled. Air is introduced into the hollow space between die and calibrating mandrel to promote uniformity of stretching. Stationary knives are arranged immediately behind the mandrel to cut the foamed tube into two webs. The film webs are flattened by support rolls mounted at the inlet to the take-off unit, and then guided to the wind-up system. The most important technical data for single-screw sheet extruders are listed in Table 6.

The tubular-film blowing process is also used for the manufacture of thin films (0.15 to 0.8 mm). Two roller trains in a V configuration are used for flattening the blown bubble.

A two-roll take-off pulls the foamed tube away from the die and seals the blowing air inside the tube. This technique is also suitable for the manufacture of foamed sheet. The tube can be flattened and welded internally to form a sheet as it passes between the take-off and pressure rolls, by changing the take-off speed and adjusting the temperature inside the tube appropriately.

Table 6. Sizes of single-screw extruders for the production of direct-gassed foamed film

Screw diameter	Screw length	Screw speed	Installed drive power	Installed heat capacity	Installed cooling capacity	Through-put*)	Standard production data		
							Dimension sheet		Density
							width	thickness	
mm	L/D	min^{-1}	kw	kw	kw	kg/h	mm	mm	kg/m^3
45	38 to 42:1	119	13	9	15	35	2 × 500 or 1 × 1000	0.15 to 3.00	70 to 250
60	38 to 42:1	80	20	18	24	70	2 × 750 or 1 × 1500	0.15 to 3.00	70 to 250
90	38 to 42:1	45	38	26	42	135	2 × 1100 to 1 × 1800	0.5 to 3.5	70 to 250
120	38 to 42:1	30	65	42	58	215	2 × 1100 to 1 × 1800	1.5 to 3.5	70 to 250
150	38 to 42:1	22	95	63	126	310	2 × 1100 or 1 × 1800	1.5 to 3.5	70 to 250

*) max. achievable values are dependent on density and thickness of product, as well as polymer and blowing agent.

A single-screw extruder line for the manufacture of intermediate products used in making packaging cartons is shown in Figure 20. For this purpose, everything depends upon maximizing the expansion of the sheets with minimum use of raw materials prior to lamination. This can be achieved by using the *Berstorff* process. The foamed sheets coming from the extruder are immediately fed through a hot water bath or a steam-heated duct.

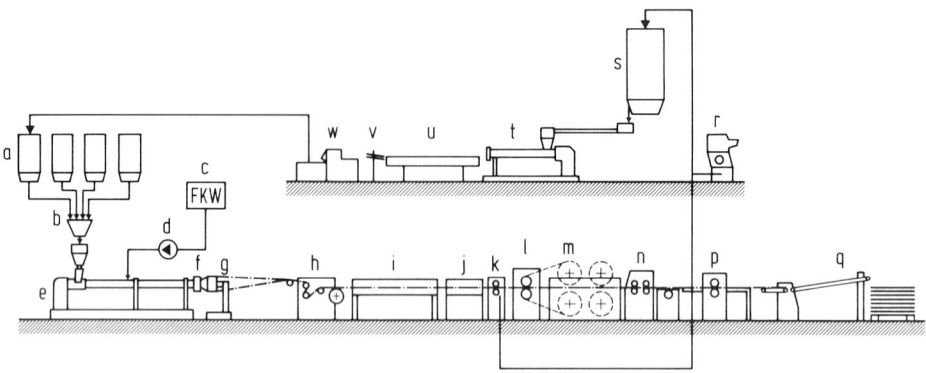

Figure 20. Single-screw extrusion plant for the production of paper laminated foam boards of low density (Berstorff system)
a raw material silo, *b* dosing and mixing unit, *c* blowing agent tank, *d* blowing agent dosing unit, *e* single-screw extruder with direct-gassing system, *f* die, *g* calibrating mandrel, *h* 2-roll s-wrap, *i* post-expansion duct, *j* dryer, *k* trimming station, *l* laminator device, *m* off-wind station, *n* take-off with grooving device, *o* slitting and cutting device, *p* cross-cutter, *q* stacker, *r* cutting mill, *s* silo for scrap reclaim, *t* reclaiming extruder with degassing system, *u* cooling bath, *v* dryer, *w* granulator

The remaining surface moisture is blown off in a dryer before the sheet is fed into the roll nip of the lamination calender. The laminator is an oil-heated two-roll calender equipped with a preheating section. This section heats the paper sheet to a temperature above the softening point of polystyrene. The surface of the foam is melted by contact with the paper and bonds to it as the two webs pass through the laminator nip together. The laminate passes through a moistening chamber after lamination to replace the moisture required to maintain the flexibility of the paper. A take-off device then transports the continuous laminate to a scoring device, longitudinal cutter, cross-cutter and finally to a stacking unit. Laminates 1600 mm wide can be produced at a line speed of 22 m/min. It is possible to manufacture unlaminated and laminated sheets in thicknesses from 2 to 12 mm.

13.5.2 Tandem single-screw lines

For production rates of more than about 320 kg/h it is economical to install two single-screw extruders in series for optimal performance. Figure 21 shows a tandem single-screw line for the manufacture of foamed intermediate products. The primary extruder has the smaller diameter, faster-turning screw, designed for raw material feeding, melting, compression, and mixing in the blowing agent. The secondary extruder with the larger diameter, slower screw, is intended mainly for cooling and mixing.

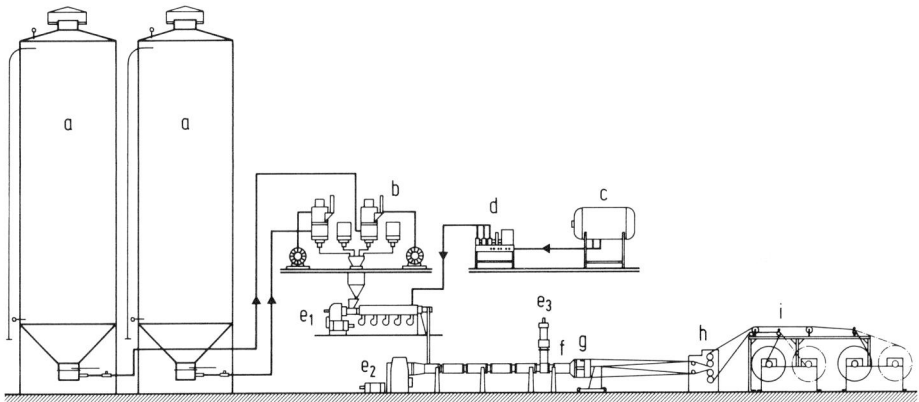

Figure 21. Single-screw tandem plant for the extrusion of foam sheets (Berstorff system)
a raw material silo, b dosing and mixing unit, c blowing agent tank, d blowing agent dosing unit, e primary single-screw extruder with direct-gassing system, e_2 secondary single-screw extruder, e_3 gear pump, f die, g calibrating mandrel, h 4-roll s-wrap, i center winder

Foam output rates of up to 750 kg/h can be achieved using this technology (see Table 7). The installation of a melt gear pump between the secondary extruder and the die can be recommended to maintain close tolerances at high output rates. The downstream equipment, such as dies, cooling devices, calibrators, take-off units and winders, is practically identical to that used on single-screw extrusion lines.

For the manufacture of heavy foam boards 30 to 100 mm thick, with densities of 35 to 40 kg/m^3, the tubular film die is replaced by a special sheet die/board calibrator combination (Figure 22). Output rates between 300 and 600 kg/h can be reached, depending on extruder sizes, with this arrangement.

Table 7. Technical data of tandem single-screw extruders for manufacturing foam products

Screw diameter		Screw length L/D		Screw speed max. min⁻¹		Installed drive power kw		Installed heat capacity kw		Installed cooling capacity kw	Through-put*) kg/h		Standard product data		
													Dimensions		Density kg/m³
													width mm	thickness mm	
Extr. I mm	Extr. II mm	Extr. I	Extr. II	Extr. I	Extr. II	Extr. I	Extr. II	Extr. I	Extr. II		**)	***)			
45	60	24 to 32:1	28 to 30:1	240 to 400	19 to 35	31	13		5	36	80		2 × 750 or 1 × 1500	0.15 to 3.00	70 to 250
60	90	24 to 32:1	28 to 30:1	180 to 300	15 to 28	90	22	14	12	54	150		2 × 1100 or 1 × 1800	0.5 to 3.5	70 to 250
90	120	24 to 32:1	28 to 30:1	120 to 190	13 to 23	90	40	18	20	80	240	265	2 × 1100 or 1 × 1800	1.5 to 3.5	70 to 250
120	150	24 to 32:1	28 to 30:1	90 to 140	11 to 20	135	60	30	38	125	400	450	2 × 1500 or 1 × 1800	1.5 to 100	38 to 250
150	200	24 to 32:1	28 to 30:1	72 to 112	10 to 17	180	115	56	100	300	620	700	2 × 1500 or 1 × 1800	1.5 to 100	38 to 250

*) max. achievable values depend on density and thickness of product, as well as polymer and type of blowing agent.
**) without melt pump.
***) with melt pump.

13.5 Single-screw extrusion systems

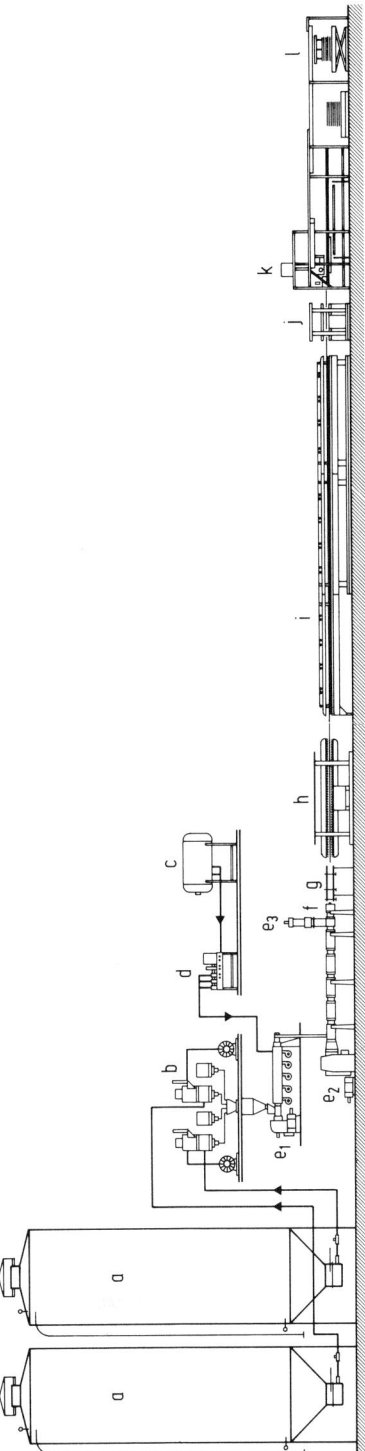

Figure 22. Single-screw tandem extrusion plant for the production of foam boards (Berstorff system)
a silo for raw material, *b* dosing and mixing unit, *c* blowing agent tank, *d* blowing agent dosing unit, e_1 primary single-screw extruder, with direct-gassing system, e_2 secondary single-screw extruder, e_3 gear pump, *f* die, *g* forming station, *h* take-off, *i* cooling channel, *j* take-off, *k* sawing station, *l* stacker

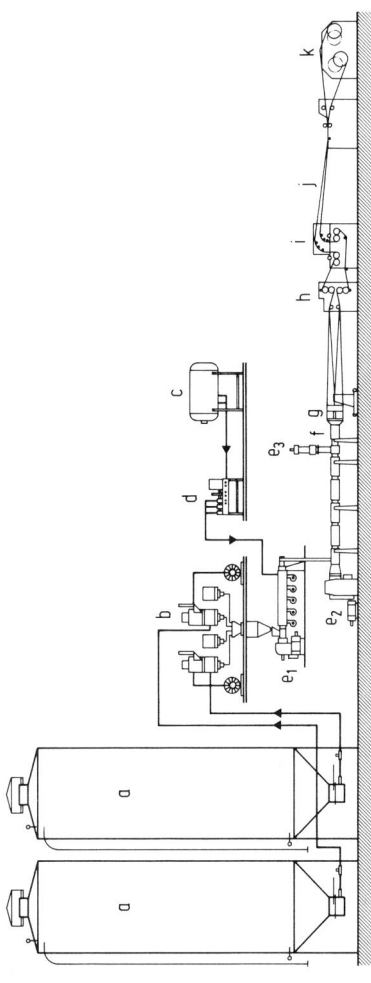

Figure 23. Tandem single-screw foam film plant combined with laminating equipment for single-sided melt coating (Berstorff system)
a raw material silo, *b* dosing and mixing unit, *c* blowing agent tank, *d* blowing agent dosing unit, e_1 primary single-screw extruder with direct-gassing system, e_2 secondary single-screw extruder, e_3 gear pump, *f* die, *g* calibrating mandrel, *h* 4-roll s-wrap, *i* laminator, *j* wind-up

Board take-off is effected by two double-roll conveyors, with a cooling device between them, and cutting and stacking units downstream. A single-screw tandem extruder system for producing coextruded foam sheet differs principally in the arrangement of the die. In contrast to a simple film extrusion line, the coextrusion die is connected direct to two extrusion systems, in such a way that the melt stream containing the blowing agent, and the base film melt stream come together in the tool in the shortest distance possible. This type of system is used mainly in the USA, for the manufacture of films up to 200 µm thick. The unfoamed component is some 10 to 15 µm thick. Such coextruded film is widely used for wrapping glass containers [32].

Single-side coated foam sheets can also be manufactured by extrusion coating, using a slot die and a chill roll to apply the compact film after solidification of the foam web (Figure 23). A pellet-fed single-screw extruder/slot-die combination, separate from the foam extruder, is used for the purpose.

In the manufacture of two-side coated foamed products, for technical reasons one can work with only one foam web. Accordingly, the extruded foam tube must be slit only once, at the lowest point after the calibration mandrel, and flattened. After take-off, the foam web passes through two extrusion coating units in series in which the films are applied to both sides. This technology is suitable for the manufacture of laminates 3 to 7 mm thick. The thickness of the applied films can vary between 50 and 250 µm. The finished width of such double-coated sheets may vary between 1400 and 1600 mm, with output rates of 250 to 400 kg/hr. This product is distinguished by excellent physical properties, including high stiffness. A major application for such products can be found in formed parts for the automotive industry [33].

Foam-in-place filler material, foam PS with a very low packing density (4 kg/dm^3) for the packaging industry, is another growing application. Unexpanded pellets containing blowing agent are the basic product. The impregnated pellets are expanded only at the point of use, in special automatic EPS foaming ovens. Shipping solid pellets reduces costs.

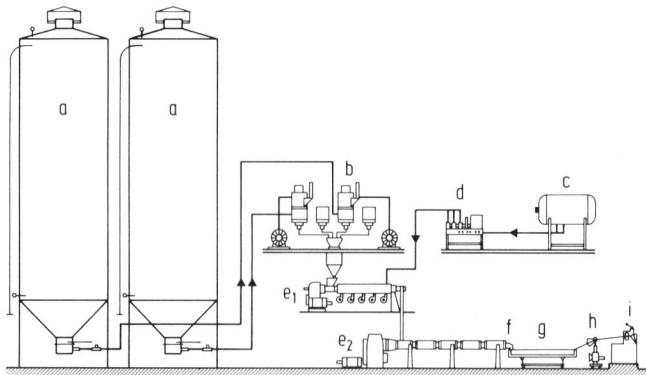

Figure 24. Tandem single-screw extrusion plant for production of pregassed pellets (Berstorff system)
a silo for raw material, b raw material dosing and mixing unit, c blowing agent tank, d blowing agent dosing unit, e_1 primary single-screw extruder with direct-gassing system, e_2 secondary single-screw extruder, f tool, g cooling bath, h dryer

A single-screw tandem extrusion system for the manufacture of foam-in-place impregnated pellets at high output rates is shown in Figure 24. This system differs from those previously discussed, in the design of the die, the pelletizing and the cooling systems. The required pellet form, and concentration of blowing agent determine the choice of pelletizing system. Die-face cutters or strand pelletizers are commmonly used. Single-screw extruders with

13.5 Single-screw extrusion systems

60 to 90 mm diameter screws are employed to meet the large requirements of packers. The foaming of the product can be carried out directly after pelletizing. Typical output rates vary between 40 to 110 kg/h.

13.5.3 Single-screw scrap recovery lines

The repelletizing of foam scrap for re-use is preferably carried out alongside foam extrusion operations, and therefore is mentioned here. Figure 25 shows a single-screw extrusion line for repelletizing foam scrap such as that produced during start-up, edge trimming, and thermoforming, as well as sawdust from cutting. The particle size of the foam scrap is first reduced in a grinder and the foam flakes, with a bulk density of approximately 80 kg/m^3, are conveyed to an intermediate storage silo. The flakes are then fed into a single-screw extruder for melting, degassing, filtering, and strand extrusion. The strands are led through a water bath for solidification. The surface moisture of the strands is removed by an air knife before they are cut in a pelletizer. Conveyor systems transport the repelletized material straight back to the new material supply line of the foam extrusion system. The form and bulk density of the repelletized material and virgin raw material are similar. Technical data on the most commonly used repelletizing systems are summarized in Table 8.

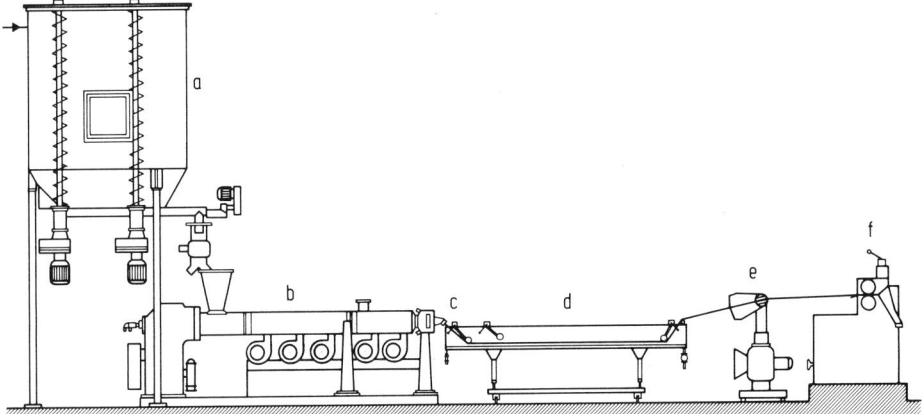

Figure 25. Single-screw extrusion line for recycling of foam scrap (Berstorff system)
a silo for scrap reclaim, *b* reclaim extruder, *c* tool, *d* cooling bath, *e* dryer, *f* pelletizer

Table 8. Sizes of single-screw extruder for scrap foam recycling plants

Screw diameter	Screw length	Screw speed	Installed drive power	Installed heating capacity	Through-put*)	Product data	
						Pellet dimensions $Q \times S$	Packing density
mm	L/D with degassing	min^{-1}	kw	kw	kg/h	mm	kg/h
60	32:1	240	32	10	85	ca. 2.5 × 2.5	0.58 to 0.62
90	32:1	200	35	24	200	ca. 2.5 × 2.5	0.58 to 0.62
120	32:1	150	90	38	320	ca. 2.5 × 2.5	0.58 to 0.62

*) indicative values for PS-foam scrap of packing density 0.1 kg/dm^3.

13.6 Plant components

13.6.1 Storage, conveying, and mixing equipment for raw materials

The main components of a foam extrusion recipe are thermoplastic resins, usually available in pellet form. Deliveries by resin manufacturers are made in bags, tankers, or bulk railcars. Storage of the raw material at the processor's plant is normally in silos or on pallets. Pneumatic or vacuum conveyors are used to transport the raw materials within the plant. The kind of system installed depends on the conveying distance and conveying capacity required.

Vacuum conveyors are recommended for transporting chopped foam scrap because of the long distances between individual generation sites and the waste treatment unit.

Figure 26. Dosing and mixing unit for single-screw foam extrusion.
(Photo: Berstorff, Hannover, West Germany)

The materials are transported from storage silos into a dosing and mixing unit mounted directly above the extruder feed hopper (Figure 26). This premixing system consists of a number of volumetric feeding devices and mixing chambers. A modular building block principle allows for the arrangement of several dosing and mixing units on top of each other. Accurate dosing is achieved by counting out volume packets of the individual components. The agitator in the mixing chamber is synchronized to disperse the materials uniformly. Level switches monitor the quantities in the dosing hoppers, mixing chambers, and extruder feed hopper, and the control system automatically demands multiple repetition of dosing and mixing cycles for continuous operation of the extruder. The dosing and mixing systems used for pellets and powders also have facilities for mixing-in liquid additives. Foam extrusion lines with provision for continuous feeding and premixing of eight components are already in operation.

These conveying, dosing, and mixing devices can be remotely controlled and operated from the main control room, providing ease of monitoring and supervision.

13.6.2 Extruder components

A single-screw extruder designed for foam extrusion consists of the following main components: drive motor, reduction gear, feed section, barrel, extruder screw and temperature control unit.

Shunt-wound DC motors are most frequently used. Power transmission from drive motor to reduction gearbox is by vee-belts.

The *reduction gearbox* on single-screw extruders is designed for high torque at low screw speed. Helical teeth, case-hardened and ground gears, and low speeds result in long service life and low noise levels. Gearbox and axial thrust bearing are force-lubricated by an external lubrication system with a separately driven oil pump.

Friction between the feed material and the screw surface, and the feed material and the inner surface of the barrel, determines the efficiency of conveying of the raw material in the extruder feed section. The friction along the barrel surface must be high to sustain effective conveying. A grooved, liquid-cooled feed section several diameters long ensures high, constant output rates even at low screw speeds.

Extruder barrels for single-screw extruders, as well as those for single-screw tandem extrusion systems, are designed to provide a desired sequence of processing steps. Typical of these are: melting, homogenizing of raw material mix, injection and mixing of blowing agent, and cooling of the melt to the temperature required for extrusion. Figure 27 shows barrel and screw designs schematically. The barrel is made of solid metal, electrically heated and air cooled up to the end of the mixing zone. The injection port for direct gassing of the melt on single-screw extrusion units is located at 16 D to 20 D from the feed end. The position of the blowing-agent addition port on tandem screws depends on the extruder barrel length, and lies in the range of 20 to 27 D.

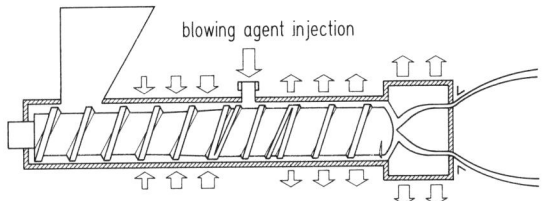

Figure 27. Barrel and screw arrangement of a single-screw extruder for direct gassing (Berstorff system schematic)

In the area of intensive cooling of the melt — needed to delay blowing — the extruder barrel is of jacketed design using a liquid heat-transfer medium; the cooling efficiency is high and can reduce the melt temperature from 210 °C to approximately 135 °C in the case of polystyrene.

The extruder barrel may be equipped with aluminum clamp-on shells with cast-in electrical heaters, and cooling channels; the detail design depends on the amount of heat to be exchanged and the heat-transfer medium used. For extended service life the barrel should be made from nitrided steel, and given a special deep hardening process.

The extruder *screw* is divided into three zones (feeding, mixing and cooling), as there are many process steps in foam extrusion.

The effective screw length should be a minimum of 22 D to accomplish raw material intake, pressure build-up, melting, homogenizing, and mixing of blowing agent efficiently. Primary extruders for tandem systems are, typically, 25 to 32 D long.

The flight depths and zone lengths on three-zone screws are designed so that melting and homogenization of the melt are complete before addition of the blowing agent. The ratio of flight depth between metering and feed sections can vary between 1:1 and 1:3.

Screw configuration in the cooling section of the extruder must be designed for optimal heat transfer. By appropriate design of conveying and mixing elements the melt can be made to circulate, forcing hot melt layers from the screw core region to contact the cooled barrel surface, bringing about intensive, and uniform cooling.

Secondary extruders, used for melt cooling in tandem systems, have screw lengths of 28 to 32 D, to achieve high output rates.

Most of these screws are manufactured from nitrided steel and are surface-hardened.

Temperature control units (TCU) are designed to heat up the extruder to operating temperatures within a set time and to keep them constant during processing with minimum variation. Extruders are usually equipped with electrical ceramic heaters to provide additional heat during initial melting of the material. Typical heat densities vary from 5 to 8 watt/cm^2, depending on the extruder diameter. Each heating section is connected to an air blower to reduce temperature peaks created by shear heating in the melt. The heating/cooling system uses a temperature sensor, located in the extruder barrel, connected to a controller.

A highly efficient temperature-control unit is required to reduce the temperature of the melt, with its blowing agent, to that for extrusion.

Water as a cooling agent offers the best conditions for heat removal, although open circuit systems may cause calcareous deposits in the cooling channels of the extruder, thus reducing the efficiency of cooling. Closed, pressurized temperature-control units are recommended for this reason. Figure 28 shows a flow sheet of such a temperature-control unit.

Oil-operated temperature-control units are recommended if it is necessary to control the polymer melt in the range above 180 °C.

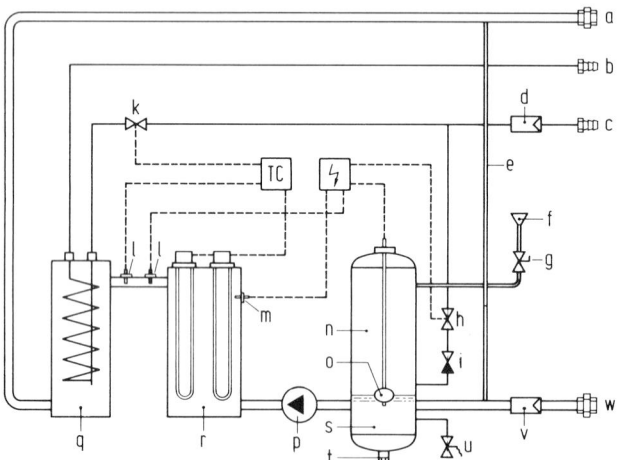

Figure 28. Pressurized water heating/cooling – schematic (single system)
a outflow, *b* cooling water outlet, *c* cooling water inlet, *d* dirt/dust collector, *e* bypass (guarantees min. flow volume if flow-line is blocked), *f* top-up valve (mandrel), *g* shut-off valve, *h* magnetic level valve, *i* return valve, *k* magnetic valve (cooling), *l* temperature sensor, *m* flow control, *n* expansion vessel, *o* float switch, *p* pump, *q* flow cooler, *r* heater, *s* degassing system, *t* drain, *u* safety valve, *v* dirt/dust collector, *w* return flow

All temperature-control units are connected with the extruder barrel section by pipes. High-performance pumps circulate the heat-transfer agent smoothly within the closed system. Each system is provided with a heat exchanger as well as with a heater, for heating and cooling the heat-transfer agent, and with a temperature controller. The temperature is measured by an accurate temperature sensor located in the liquid circuit. If the temperature to be controlled is high, cooling is effected indirectly by water through a shell-and-tube type heat exchanger. Resistance heating elements are introduced in the circuit for intensive heating.

13.6.3 Screen changer

Virgin material mixed with certain quantities of densified or agglomerate waste is used for foam extrusion. Foreign particles from such mixtures may contaminate the melt and cause uneven flow in the die and die gap. To avoid this, the melt must be filtered through a screen changer. The screen changer is typically installed between the extruder and the die if a single extruder unit is used (Figure 26), or between the primary and secondary extruders with a tandem extrusion system. Several systems for changing the screens during production have been developed.
Screen changer designs using slide plates or cassettes in which the screen packs are supported by a breaker plate, are satisfactory for this application. There is always one screen pack in the polymer stream. A second screen pack is easily accessible for exchange and cleaning. Screen changeover is initiated automatically after a preset melt pressure is reached, and on some commercially available systems can be done in less than one second. A drop in melt pressure will be noticed after the change, which means that the slide plate system should be used for tandem extrusion systems only.
Continuous screen changers guarantee a constant melt pressure, and therefore a continuous melt stream. The screen, in the form of an endless belt, is moved continuously across the melt stream, driven by the melt pressure, and has a special temperature-control system at the exit of the screen. Mechanical means of screen feeding is, therefore, not required [34].

13.6.4 Dies and calibrating devices

The manufacture of the large variety of foam products on the market can be carried out with well-known forms of die, adapted to the foam extrusion process. Extrusion dies with torpedoes are used for the extrusion of foamed films and pipes. The torpedoes are fixed in position by one or more spiders, or by a circular breaker plate. The die housing, the die lip section, and the die pin are individually heated by a liquid heat-transfer agent, under precision temperature control.
The profile-shaping part of the die must be designed to avoid any expansion before the melt leaves the die. It is customary to center and adjust the die lips radially with set screws to achieve an even thickness around the entire circumference. Axial displacement of the torpedo is used to set the die gap. Both settings must be adjustable during operation.
If unexpanded surfaces are required on the tubular film or the pipe, it must be supported externally and internally as it leaves the die, on temperature-control contact-rings or air cooling rings. In a system used in the USA, the tube is extruded vertically into a water bath and intensively cooled inside and outside simultaneously [35]. The tube is then pulled over a temperature-controlled aluminum calibrating mandrel, stretched to an accurate internal diameter, and further cooled. Uniform stretching is aided by maintaining a controlled air pressure in the tube between the die and the calibrating mandrel. To make it possible to optimize the distance between the die and the mandrel, the mandrel is fixed to a movable carriage. Slitting equipment on the downstream side of the mandrel cuts the extruded tube, once at the low point of the mandrel to obtain a single sheet (Figure 29), or twice, at the 3 o'clock and 9 o'clock positions, to give two sheets, depending on the subsequent use of the foamed sheet. Industrially, two sheets up to 1,500 mm width each and single sheets up to 1,800 mm wide can be produced using this technique. Die diameters of 75 to 250 mm are required, depending on output capacity and sheet widths specified.

Figure 29. Single-sheet foam extrusion. (Photo: Berstorff, Hannover, West Germany)

The dies used for the manufacture of foamed pipes are small versions of blown film dies. The use of dual extrusion heads is recommended for efficient production of small-diameter pipe (ID 12 mm × 6 mm wall thickness) Figure 30. In a twin-head unit, the melt stream is divided into two after the screw tip and fed into two tube dies. The dual-head design incorporates means of adjustment of polymer flow to each die independently.

Figure 30. Single-screw extruder with double-head die for the production of PE-foam (Photo: Berstorff, Hannover, West Germany)

For the manufacture of foamed boards 30 to 100 mm thick and 600 mm in width, specially designed dies are used in which the melt stream has to spread out laterally.
The main design difficulty is achieving an even flow profile over the entire width of the die, but two die designs which solve this problem, and are industrially proven, are the fishtail die and the coat-hanger die. With both designs a sizing-plate system is directly attached to the die. This consists of two liquid-heated plates, adjustable for width and thickness, to control the board gage. The sizing-plate surfaces in contact with the foam must be coated with a slip agent.

13.6 Plant components

Extrusion crossheads like those used for wire- and cable coating are suitable for the continuous coating of various materials with extruded foam. A good bond between foam and base material can only be achieved if the coating takes place within the die.

Crosshead dies are used to make pellets incorporating a blowing agent. These are similar in design to a film blowing die or a fishtail die, and fitted with a multi-hole die plate, which is horizontally mounted, close and parallel to the water surface in the cooling bath. The strands are extruded vertically downwards into the water, led through the bath on rolls, cooled, dried, and fed into a strand pelletizer [36].

For pelletizing with a center-fed die, the holes are arranged in a circle around the die plate and discharge vertically downwards. The strands are cut at the die face by a rotating knife assembly driven concentrically with the die axis, or eccentrically, and thrown centrifugally into a water circulation system for cooling. The knife speed can be adjusted steplessly.

13.6.5 Storage, conveying, and dosing systems for the blowing agent

The most important component in the dosing system for the liquid blowing agent is the high-pressure pump. Membrane and piston pumps are the most suitable for accurate dosing of the blowing agent into the extruder. Triple-head pumps are widely used to achieve uniform, pulsation-free flow.

It is advantageous for control purposes to drive the injection pump by a DC motor, or by a coupling to the extruder drive-shaft. Manual adjustment of the piston stroke is recommended. Flow-measuring devices to measure the blowing agent displacement rate are installed on either the suction or the discharge side of the pump. Suitable control valves, accumulators, and coolers are installed to maintain uniform flow rates into the extruder independent of any temperature and pump variations on the suction side of the pump or in the extruder. The polymer melt is prevented from entering the blowing-agent supply line by an injection valve located at the extruder. A pressure-monitor system should be installed for safety reasons. Figure 31 [37] shows the flow sheet of a blowing-agent injection system for the direct-gassing process. Tanks of 0.8 to 20 m^3 capacity are used for storing the blowing agent. The liquid levels in the tanks are measured and indicated by level gages. The

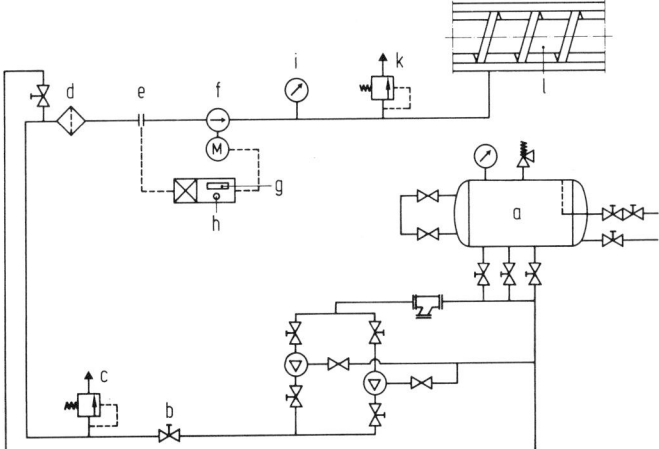

Figure 31. Schematic diagram of dosing unit for direct gassing with physical blowing agents (Berstorff system)
a blowing agent container with fill-level indicator, *b* shut-off valve, *c* safety valve, *d* filter, *e* melt-flow metering, *f* pump (regulating unit), *g* indicator, *h* setting, *i* manometer, *k* safety valve, *e* extruder

blowing agent is conveyed from the tank to the high-pressure pump under sufficient pressure to keep the blowing agent in the liquid state. The tanks must therefore be equipped with rotating displacement pumps operating at constant speed. The pumping rate of the smallest commercially available pump is approximately 10 l/min.

A manually adjustable back-flow valve is located on the pressure side of the pump to reduce the pumping rate. The setting of the back-flow valve depends on the blowing agent and the distance it has to be conveyed.

The blowing agent injection system consists of a triple-head high-pressure pump with adjustable stroke, a DC drive, flow meter, pressure valve and the controller. A rotating-float flow meter is used to control the dosing rate, by converting the position of the float into an electrical signal; this is fed to the controller and compared with the target flow-rate, to initiate an adjustment of the pump speed if necessary.

Compactly designed dosing devices are available for simple gassing tasks, consisting only of multiple-head injection pumps with manual stroke adjustment.

The *Zingsheim*-blowing agent injection system [37] is also worth noting. Since the efficiency of a dosing pump is dependent on the counter pressure, as mentioned earlier, high back pressure in the polymer melt reduces the amount of blowing agent injected. This negative effect can be eliminated with the Zingsheim system, which keeps the pump pressure and the differential pressure between injection pump and injection point at the extruder constant. A pressure-differential measuring device and a flow controller are installed downstream of the injection pump to maintain conditions constant.

13.6.6 Take-off equipment

Take-off equipment is located after the cooling section and pulls the foamed product away from the die smoothly, at a constant speed matched to the extruder output. To ensure slip-free take-off, the nip force between the take-off rolls must be carefully set. The first take-off station must provide the master speed signal for downstream equipment, which may consist of several take-off stacks. However, each unit must also be finely adjustable individually by additional controls, without affecting the master take-off speed.

Roll, belt, and caterpillar-type take-offs are used, depending on the form of the foamed product.

A four-roll S-type take-off stack is required for the manufacture of reelable foamed film (two webs – Figure 32). Each pair of rolls is mounted vertically and the film wrapped around them in an S shape. A pneumatically operated pressure roll running against the run-out roll improves the friction between the take-off rolls and the film. Two support rolls are placed on the intake side to keep the foam sheet flat.

The take-off equipment for single-web operation has only one pair of take-off rolls and the pressure roll. A pneumatically operated squeeze roll is mounted on the intake side in addition to the sheet support roll, thus facilitating threading of the foam sheet into the take-off stand.

Both designs can be equipped with knives if edge trimming or slitting into several webs is required. The edge trim can be wound up or be directly fed into a grinder.

Caterpillar and belt take-off systems are mostly used for tube and profile extrusion up to line speeds of 40 m/min.

Roll conveyor systems are particularly suitable for pulling foam boards away from sizing die plates. The special feature of this system is that the transfer of take-off forces to the foam board does not cause any unwanted deformation of the board if the diameter of the rolls is correctly chosen. Also, there is easy access for cooling air to the board surface. Normal operating speeds for this system are from 1 to 12 m/min.

Figure 32. Foam extruder with S-take-off and winding station.
(Photo: Berstorff, Hannover, West Germany)

13.6.7 Winders

The basic equipment of a foam-sheet extrusion line also includes a winding system with two winding shafts for single-web operation, or four winding shafts for two-web operation. Systems with shafts mounted in bearings on one side only are preferred, for convenience of roll handling. Rolls up to 1700 mm in diameter can be removed laterally.

Figure 33. Single-end bearing supported, double turret winder.
(Photo: Berstorff, Hannover, West Germany)

The properties of foamed film allow winding shafts designed for winding with or without cores to be used. The shafts can be expanded mechanically or pneumatically for coreless winding. Cutting the film when the roll reaches the desired diameter, and threading the empty winding shaft are usually done manually. Experienced operators can perform roll changes at line speeds of up to 30 m/min and with sheet widths of 1500 mm. The roll changing procedure can be facilitated by using a double turret winder with one-side bearings (Figure 33). The two winding shafts are mounted on an electromechanically driven rotating disk. As soon as the target roll diameter is reached the empty shaft is brought into the winding position by rotating the disk. Then the sheet is cut across and transfer to the empty winding shaft is initiated.

Each winding shaft must be driven independently by geared motors, to reduce foam scarp to a minimum during roll changes. It is essential to keep the tension and the speed of the film web constant as the roll diameter increases and the shaft speed decreases. The winder control system must be designed for the maximum shaft speed and the smallest roll diameter. The *Alquist* drive system offers a very cost-effective solution to this problem, meeting almost all requirements.

The torque characteristic of this AC drive is almost identical to the torque requirements of the winder. The torque of this motor can be adjusted by changing the magnetic field by resistors or transformers.

DC drive systems can be used in combination with standard control systems.

13.6.8 Laminators

Laminating calender, temperature-control unit, preheating system, and unwind system are the major components of the laminator.

The design of the individual system components depends mainly on the laminating specification i.e. on whether continuous single- or double-side lamination, on one or two webs is required.

The unwind system is equipped with four unwind stations from which the surfacing films are fed under constant tension into the nip of the laminating calender (Figure 34). For two-side lamination of a single web or single-side lamination of two webs, one unwind station at the top and one at the bottom are in operation. The other two stations are in the standby position, and are necessary for continuous operation of the lamination process. Manual change from one unwind station to the other when the end of the surfacing roll is reached can be carried out at line speeds of up to approximately 15 m/min.

Figure 34. Four-fold take-off unit for laminating foam sheet. (Photo: Berstorff, Hannover, West Germany)

Automatic changeover devices must be used for higher line speeds. Each unwind station is equipped with a braking arrangement linked to a pneumatic web-tension control system, which provides constant web tension between the unwind unit and the calender.

The surfacing films are heated before being fed into the laminator. The line speed and the properties of the surfacing films determine whether radiant heating ducts, preheating rolls, or direct heating by the laminator rolls must be used. Heating ducts, mostly, are used for preheating of paper, whereas preheating rolls are sufficient for laminating with plastics films. Preheating rolls cannot be used for films less than 100 µm thick at line speeds of less than 12 m/min. Bonding between the foam and the covering material is accomplished in the nip of the laminating calender under pressure and at elevated temperature, without the use of adhesives. The laminator rolls are coated, and can be heated individually by oil up to 280 °C. The laminator roll gap can be adjusted by electric motors. The largest laminating calenders in operation have a roll width of 1800 mm and a roll diameter of 800 mm. The laminate passes through a moistening device after lamination with paper, to replace the moisture lost during heating. Replacement of the original moisture is necessary for good shaping properties.

13.6.9 Cutting and stacking devices

The design of cutting and stacking devices used is determined by the kind of foamed products to be handled.
Laminated foam sheets for the manufacture of boxes and art board are cut and scored to format in the extrusion direction continuously. Scoring is necessary for folding the laminates into boxes. Scoring and cutting tools are mounted on separate full-width shafts located above and below the sheet, perpendicular to the machine direction.
Razor blades or rotary cutters are used to cut sheets and boards up to 15 mm thickness. With blades, the sheet runs towards the stationary blades, which enter the sheet at an angle of 15 to 30 degrees. The blades are held in fixtures on shafts, and their cutting angle and the distance between them can be adjusted.
Rotary cutters, which have a longer service life, are preferred for cutting foam laminates with thick and hard surface layers.
The principle of cutting is that one cutter is keyed to a driving shaft above, and the other to one below the board, with the cutting gap defined by the cutter edges. The cutters are driven in the same direction as the moving board, at a rate related to the speed of the board. A high-speed circular cross-cutter then cuts the boards to the desired lengths with high accuracy. The lengths can be programmed in the range of 300 to 10,000 mm, depending on line speed.
A belt conveyor accepts the finished board for automatic stacking on a pallet, and, to ensure continuous operation during exchange of a full pallet for an empty one, the cut boards are temporarily stacked on the conveyor belt.
It is usual to employ circular saws with hardened blades for longitudinal and cross cutting of foam boards 30 to 100 mm thick. Saws for longitudinal cutting are manually adjustable in width and can be raised and lowered pneumatically. The traversing speed for cross cutting is determined by a calculator which uses a signal proportional to extrusion speed from a wheel touching the board. The cut boards are then lifted by suction cups and deposited on a lifting table. The dust generated during cutting is removed by suction to storage silos connected with the reclaim extruder, and the edge trim is fed into grinders by conveyor belts.
For cutting foamed pipe to length, hot cutting is suitable up to extrusion speeds of 40 m/min. Resistance heaters are used to heat the knife, which is lowered vertically to perform the cut. Take-off systems immediately accept the cut pipe to convey it to an intermediate storage table.

References for Chapter 13

For the convenience of the reader the English titles of all publications in languages other than English are shown in parentheses.

[1] AP 2023204 *Munters, C. G.* and *Tandberg, J. G.*
[2] AP 2669751 DOW; *McCurdy, J. L.* and *Delong, C. E.*
[3] *Burt, J.:* Foam Extrusion, Plastics Institute of America. Inc. Castle Point, Hoboken, NJ, June 1983.
[4] *Zingsheim, P.:* 6th International Symposium on Foamed Plastics, Düsseldorf, 1975.
[5] *Barth, H.:* 3rd International Symposium on Foamed Plastics, Düsseldorf, 1973.
[6] *Frisch, K. C., Saunders, J. H.:* Plastics Foams, New York, 1972.
[7] DE-DBP 1038275.
[8] *Kolossow, K.-D.:* Herstellung geschäumter Folien durch Direktbegasung (Production of Foamed Film by Direct Gassing). Plastics and Rubber Symposium, Sarajevo, Nov. 1975.
[9] *Reichert, U.:* 3rd International Symposium on Foamed Plastics, Düsseldorf, 1973.
[10] Company brochure, Hoechst AG, FRIGEN Information. Frankfurt, Sept. 1980.
[11] *Hurnik, H.:* Kunststoffe 63 (1972), p. 688.
[12] *Vetter, G.:* Volumetrisches Dosieren von Flüssigkeiten mit Dosierpumpen (Volumetric Feeding of Liquids with a Metering Pump). VDI-Verlag, Düsseldorf, 1981, pp. 131/167.
[13] *Ast, W.:* Der Extruder als Plastifiziereinheit (The Extruder as a Plasticating Unit). VDI-Verlag, Düsseldorf, 1977.
[14] *Benning, C. J.:* Plastics Foams. Wiley Interscience, New York, 1969.
[15] *Szekely, J. et al.:* VDI-Berichte No. 182 (1972), pp. 13/22.
[16] *Schleith, O.:* Schäume aus der thermoplastischen Schmelze (Foams from Thermoplastic Melts). VDI-Verlag, Düsseldorf, 1981, pp. 17/38.
[17] *Flory, P. I.:* Principles of Polymer Chemistry. Cornell University Press, Ithaca/NY, 1953.
[18] *van Krevelen, D. W.:* Properties of Polymers. Elsevier Publishing Company, 1972.
[19] *Burt, J. G.:* J. Cell. Plast. 14 (1978), pp. 341/345.
[20] *Zingsheim, P.:* 5th International Symposium on Foamed Plastics, Düsseldorf, 1975.
[21] *Breuer, H.:* Plastverarbeiter 24 (1973) 6, pp. 350/356.
[22] DE-AS 1694130.
[23] DE DBP 183238.
[24] FP 1498620.
[25] DE-OS 1913921.
[26] DE-OS 2000039.
[27] DE-AS 2051006.
[28] DE-OS 2038803.
[29] DE-AS 2507979.
[30] *Schröder, R.:* Schäume aus der thermoplastischen Schmelze (Foams from Thermoplastic Melts). VDI-Verlag, Düsseldorf, 1981, pp. 45/56.
[31] DE-DBP 2509252.
[32] DOS 2157940.
[33] GM 7929987.
[34] CH P 69547.
[35] US 3864444.
[36] GBP 886811.
[37] *Zingsheim, P.:* Schaumextrusion von PVC hart nach dem Direktbegasungsverfahren (Extrusion of Rigid PVC Foams by the Direct-gassing Process). 5th Intern. Symposium in Foamed Plastics, Düsseldorf, 26th/27th May 1975.

14 Extrusion of foamed intermediate products with twin-screw extruders

P. Klenk, H. P. Schneider

14.1 Introduction

Foam extrusion of PVC is a process which has been practiced since the early sixties, but in recent years there has been a considerable improvement in technology, applications and economics; its importance is now increasing. There are numerous reasons for using PVC as a foamed intermediate product, the most significant of these being:

- low thermal conductivity (good heat insulation),
- good acoustic damping properties,
- easy to fabricate by mechanical means (similar to wood, permitting use of all wood-working methods,
- highly flame-retardant (Class B1, based on DIN 4102, use in building construction),
- good chemical resistance (printing with color-containing solvents),
- excellent weathering stability (outside use),
- low material costs due to low density (a significant factor in view of the continuing increase in raw material costs).

In view of the properties just mentioned, PVC rigid foam profiles frequently compete with wood nowadays. The surface of the foam extrudates can be largely adapted to that of wood in terms of structure, color and appearance. PVC rigid foam profiles are used both inside and outside in housing construction, e.g. as skirting boards, cornices, ceiling beads, curtain rails, roller-shutter casing profiles, windowsill profiles, doorframe profiles, balcony paneling profiles as well as for wall and ceiling panelings. Foamed window profiles with aluminum reinforcement are under development.

The main uses for PVC rigid foam sheets are in exhibition stands and shop construction, as signs and panels for lettering, for air-conditioning ducting, clean and wet rooms, as top sheets for work tables and shelves, in electrical equipment and housing construction or as the core material of sandwich constructions in boat building, for tanks and refrigerated vehicle bodies. The sheets are thermoformable (even after printing), and this opens up a wide range of design and application possibilities.

Semirigid PVC foam sheets can be used as protective equipment in the sports sector, shock-absorbing bases of artificial grass playing fields, wall linings of gymnastics halls, judo and wrestling mats or upholstery in aircraft and vehicle construction. PVC rigid foam pipes have a considerable market share, particularly in France, as cable conduits, ventilation, rain- and sewer pipes. Similar uses in other countries are limited by the test standards laid down for rigid PVC.

All the semifinished products mentioned so far are manufactured on conventional extruders using chemical blowing agents. It is possible to produce PVC films on special direct-gassing extruders using physical blowing agents. Such films have a homogeneous cell structure and a smooth, unbroken surface, making them suitable for printing. Examples of their use include insulation (multilayer film sheets), packaging (deep-drawn parts) and wallpaper lining.

14.2 Structure of formulations

14.2.1 PVC types

PVC types with a K value ranging from 58 to 65 are used for making rigid foams.
Basically, suspension, emulsion and melt PVC types are suitable, with preference being given to suspension and melt PVC. PVC copolymers and graft polymers have favorable processing properties although they are scarcely ever used, for reasons of price.
Suspension PVC types with a K value of 70 to 75 have proved themselves suitable for soft foams.
The grain structure should be as porous as possible to ensure that plasticisers are properly absorbed. PVC with a K value of 75 is used for preparations with low-viscosity plasticisers and PVC with a K value of 78 with higher-viscosity plasticisers [2].

14.2.2 Stabilizers

Three stabilizing systems very frequently used for processing PVC dryblend are:

− organotin,
− barium/cadmium and
− lead stabilizers.

The stabilizer selected depends on the product, the processing machine and the range of applications (interior or exterior, food packaging, ...).
Lead stabilizers play a dominant role owing to their low cost, ease of processing and wide processing range. They are particularly used for profiles for interior applications, for pipes and sheets. Tin stabilizers are added to meet the severest demands in terms of thermal stability and transparency. There are no restrictions on the types of PVC used, or other constituents in the formulation, or the processing method. Proper selection makes it possible to manufacture highly light- and weather-resistant final products as well as physiologically safe articles [3].
Barium/cadmium systems act differently with different PVC types. They have only a limited effect with emulsion PVC [4].
Tin or barium/cadmium systems have proved suitable for profiles for outside use, as well as for pipes and sheets.
In establishing the formulation, it should also be noted that various stabilizer systems have, in addition to their actual stabilizing effect, an activating effect on the blowing agent. Other compounds with a stabilizing effect, but which can only be used by themselves, are the so-called *co-stabilizers, chelators, UV absorbers* and *antioxidants* [5]:
The principal representatives of the *co-stabilizers* are the epoxy compounds, which are used as high-molecular compounds (e.g. epoxidized polyethylene) or as low-molecular compounds, the principal − physiologically safe − representative of these being epoxidized soybean oil.
With almost all stabilizers containing metal, epoxides can act synergistically, to enhance light stability particularly.
In addition, epoxy compounds act as processing aids and enhance surface gloss.
Alkyl or acryl phosphites are used as *chelators*. They are able to deactivate the metal chlorides resulting from the separation of HCl and thus improve light and heat stability.
UV absorbers convert the energy-rich ultraviolet radiation, which is harmful to PVC, into heat and thus raise light stability. Phenols and organic sulphur compounds are primarily used as *antioxidants*. They reduce the thermo-oxidative decomposition of the PVC by trapping radicals produced by the effect of oxygen.

14.2.3 Lubricants

The purpose of lubricants is to make possible and to simplify the processing of the polyvinyl chloride on the extruder. A distinction is made between the effects of *external* and *internal lubricants:*

External lubricants reduce the friction between the melt and the metal parts (screw, barrel, mold ...) with which it comes into contact.

Internal lubricants influence the rheological behavior of the melt by altering the intermolecular friction and thus the viscosity. In addition, they have an impact on the ability of the material to soften, in other words on the dependence of plasticating time on temperature.

A factor to note in foam extrusion is that it is essential to have a melt which slides satisfactorily on the mold wall to form a smooth, unbroken surface. In addition, lubricants have an impact on surface gloss as well as on the thorough mixing and distribution of the blowing gas in the melt, and thus on its cell structure.

It is normal practice to use several lubricants in PVC formulations, since external or internal lubricants alone do not fully meet the demands made on the processing method and the properties of the semifinished products. In addition, when selecting the lubricants, consideration also has to be given to the other constituents of the formulation, particularly the stabilizer system.

14.2.4 Fillers

The fillers used in the PVC are usually insoluble, solid inorganic materials, with coated and uncoated chalk playing a major role. The chalk contained in the blowing agents acts as a pore former, and thus has an effect on the cell structure dependent on the quantity used, the grain size and grain structure. In addition, use of fillers makes it possible to achieve certain properties in the final article, such as reduced abrasion resistance or increased scratch resistance. On the other hand, however, fillers used just to lower the cost of the formulation can have a detrimental effect on the flow behavior of the mixture in the extruder or mold, and on the mechanical properties of the final product.

14.2.5 Pigments

Pigments or pigment preparations are normally used in powder form for coloring PVC. The high-speed mixers used for PVC compounding are able to incorporate the powder pigments into the mixture satisfactorily. The suitability of pigments depends principally on their resistance to temperature and migration. In addition, severe demands are made on PVC pigments in respect of light fastness particularly, and also on their weathering resistance, when used in the building sector. Consideration should also be given when selecting pigments to their interaction with additives such as blowing agents, stabilizers and lubricants. Titanium dioxide is normally used for whites, and carbon black for blacks. A semifinished product with a woodlike appearance can be achieved by adding small quantities of specific color concentrates (e.g. intensely colored, stabilized PVC with a K value at least 10 units higher than that of the PVC base material).

14.2.6 Blowing agents

A basic distinction can be made between *chemical* and *physical* blowing agents. Conventional extruders can be used if chemical blowing agents are employed, whereas physical blowing agents can only be added to the melt when pressurized on special direct-gassing extruders. Foamed PVC rigid profiles, pipes and sheets are manufactured for the most part by using chemical blowing agents. A characteristic feature of chemically blown foams, however, is a limited expansion level (< 50%). In other words, foams of high and medium density are obtained − so-called structural foams − but not light foams [6]. Low foam densities can be obtained using physical blowing agents with appropriate direct-gassing methods [7, 8] on what are significantly more sophisticated extrusion systems. The use of physical blowing agents is common in the extrusion of foamed tubular films and is likely to grow in importance in the field of profile and sheet extrusion, too. Depending on polymer, method and process control, it is possible to obtain densities < 5% of the solid material density, or in other words, an expansion of > 95%.

Chemical blowing agents are inorganic or organic compounds which decompose under the effect of heat and form gaseous breakdown products (N_2, CO_2, NH_3 etc.). The decomposition process in this case is usually exothermic and irreversible.

Of the group of azo-compounds, azodicarbonamide (ADC) is the most important chemical blowing agent and the one most frequently used (see Table 1). The gas yield of 220 ml/g is extremely high compared with other organic blowing agents. The various forms in which it is offered differ primarily in particle size, which affects the decomposition temperature (approx. 200 to 230 °C), and the rate of decomposition.

Table 1. Azo-compounds commonly used as blowing agents [9]

Chemical	Acronym	Processing temp. range °C	Gas development cm^3/g	Concentration %	Used for
Azodicarbonamide (Azobisformamide)	ADC	165 to 215	220 (210 °C)	0.1 to 4.0	PP, PS, ABS, rigid PE, PVC
Azodiisobutyronitrile	AZDN	110 to 125	130 (110 °C)	0.5 to 6.0	PVC
Barium azodicarboxylate	BADC	250 to 300	200 (270 °C)	0.1 to 1.0	PVC, PA, PC, ABS

The addition of accelerators (kickers) can have a far greater effect than the particle size in reducing the decomposition temperature, which may be too high, particularly for thermally sensitive materials such as PVC. Reduction to about 150 °C is possible. A wide range of compounds containing metal, particularly on a lead, zinc, barium and cadmium basis, have an accelerator effect. Since these metal compounds already exist in the mixture as stabilizers, it may in certain circumstances be unnecessary to provide specific kicker additives. A further chemical blowing agent, which is primarily used in the method of foaming to the inside ("Celuka" method; see Section 14.3.2), is sodium hydrogen carbonate ($NaHCO_3$). This produces slow expansion, in contrast to most organic blowing agents, making it possible to achieve a compact and smooth outer skin.

Physical blowing agents are either liquids whose boiling points are below the softening point of PVC, or substances which are already gaseous under normal conditions. Physical blowing agents include aliphatic hydrocarbons (pentane, hexane, heptane ...), chlorinated hydrocarbons (methylene chloride, trichlorethylene ...), chlorinated fluorocarbons (trichlorofluoromethane, dichlorodifluoromethane ...), carbon dioxide, nitrogen, rare gases and air. The properties of the principal physical blowing agents are presented in Table 2.

In PVC foam extrusion, only the fluorocarbons type R11 (CCl_3F) or mixtures of R11/R12 are used. In addition to their advantages of non-combustibility and non-toxicity, they also have the positive property of adequate affinity to the polymer melt [8]. To obtain as fine a cell structure as possible, so-called pore regulators, also known as nucleators, are used, particularly with physical blowing agents. The most common is a mixture of sodium hydrogen carbonate and citric acid, the reaction producing water, carbon dioxide and sodium citrate. The blowing agent then diffuses into the finely distributed tiny bubbles that are called active nucleators.

14.2 Structure of formulations

Table 2. Properties of physical blowing agent [8]

Blowing agent	Molecular weight	Density at 25 °C g/cm³	Boiling point °C	Heat of evaporation kJ/kg	Thermal conductivity at 20 °C W/K m
Pentane	72.15	0.616	36.1	360	
Trichlorofluoromethane (CCl₃F) R 11	137.38	1.476	23.8	182	0.0086
Dichlorofluoromethane (CCl₂F₂) R 12	120.9	1.311	− 29.8	166	0.0103
Mixture R 11/R 12 (50 : 50)	129.14	1.39		164	0.0095
Carbon dioxide CO₂	44.01	1.977	− 78.5	574	0.0163
Nitrogen N₂	28.016	1.251	−196	201	0.0258

14.2.7 Foaming aids

Because of the poor expandability of homopolymer PVC, it is not possible to achieve foam densities significantly less than 1.0 g/cm³, since the walls of the cells are prematurely torn open during expansion by the pressure of the blowing gas. The expandability of the melt can be considerably improved by the addition of foam modifiers such as acryl polymerizates, styrene copolymerizates etc. (e.g. [10 to 13]), to achieve foam densities far below 1.0 g/cm³. Moreover, the modifiers produce an enhancement of the sliding properties of the melt in the mold, its impact strength and surface quality.

14.2.8 Mixing technique

For processing dryblend mixtures into foamed semifinished products, twin-screw extruders are used almost exclusively. The PVC foam formulations are prepared on standard commercial heating and cooling mixer combinations. For processing on extruders, the PVC is hot-mixed (mixing process performed with the addition of heat) since it is not possible to achieve a dry, free-flowing and sufficiently homogeneous mix if the initial mixture is cold. So-called high-speed mixers are used as hot mixers, operating on the principle of converting the mechanical drive power into friction heat in the product. In the case of *PVC rigid foam formulations*, the stabilizers, lubricants, chemical blowing agents, fillers and pigments are mixed together with the PVC in the high-speed mixer. Once the mixture has reached a temperature of 85 °C, the foaming aids, gelling agents and, if necessary, kicker are added, and mixing continued until the material reaches a temperature of 110 °C. To prevent any thermal damage to the PVC, but also to enable it to be more easily transported and stored, the mixture is then rotated at low speed in a cooling mixer downstream of the high-speed mixer and cooled to a temperature of 40 °C. *PVC soft foam compounds* are processed on extruders as dryblend and as pellet. According to [14], the compound should be manufactured under the following conditions:

Dryblend: Place PVC, stabilizer and lubricant in the high-speed mixer, mix at high speed up to 60 °C before adding the plasticizer; dry-mix and gel to approx. 110 to 120 °C. Modifiers, fillers and blowing agents are not added until the cooling-mixer stage.

Granulate: Place PVC, stabilizer and lubricant in mixer; mix at high speed to 60 °C; add plasticizer, dry-mix to 80 °C. Modifier and filler are added in the cooling mixer; pelletize, coat with blowing agent. If the mixture gels at low temperatures, the blowing agent can also be worked into the pellet.

14.2.9 Formulations

The formulations detailed below are guideline formulations which require careful matching to the particular processing machine and to the final product.

14.2.9.1 Guideline formulation for PVC rigid foam profiles

	Parts
PVC, M-PVC, K value 58 to 60	100.00
Tin mercaptide (stabilizer)	1.0 to 1.5
Epoxy compound (co-stabilizer)	1.0 to 2.0
Calcium stearate (lubricant and thermal stabilizer)	0.8 to 1.2
Fatty acid ester (internal lubricant)	0.5 to 0.8
Oxidized PE wax (external lubricant)	0.4 to 0.6
Acrylic polymer (flow aid, impact modifier)	6.0 to 8.0
Chalk ($CaCO_3$, particle size > 5 µm-filler)	3.0 to 4.0
Azodicarbonamide and kicker (blowing agent)	0.5 to 0.7
Pigment as required	

14.2.9.2 Guideline formulation for PVC rigid foams

	Parts
PVC, S-PVC, K value 65 to 68	100.00
tribasic lead sulphate (thermal stabilizer)	1.8 to 2.2
dibasic lead stearate (lubricant and co-stabilizer)	0.6 to 0.8
Calcium stearate (lubricant and thermal stabilizer)	0.4 to 0.5
Stearic acid (internal and external lubricant)	0.3 to 0.4
Hydrocarbon wax (external lubricant)	0.2 to 0.3
Acrylic polymer (flow aid, modifier)	4.0 to 6.0
Chalk ($CaCO_3$, particle size > 5 µm-filler)	3.0 to 6.0
Azodicarbonamide and kicker (blowing agent)	0.4 to 0.7
Pigment as required	

14.2.9.3 Guideline formulation PVC rigid foam sheet (free-foamed)

	Parts
PVC, M-PVC, K value 58 to 60	100.00
tribasic lead sulphate (thermal stabilizer)	1.8 to 2.0
dibasic lead stearate (lubricant and co-stabilizer)	0.4 to 0.5
Calcium stearate (lubricant and thermal stabilizer)	0.8 to 1.0
Organic di-acid ester (internal lubricant)	0.6 to 0.8
Glycerine monostearate (internal lubricant)	0.6 to 0.8
Hydrocarbon wax (external lubricant)	0.6 to 0.8
Epoxy plasticizer (co-stabilizer, flow aid)	0.8 to 1.0
Acrylic polymer (flow aid, impact modifier)	4.0 to 6.0
Chalk ($CaCO_3$, particle size < 5 µm-filler)	2.0 to 4.0
Azodicarbonamide and kicker (blowing agent)	0.9 to 1.0
Pigment as required	

14.3 Theory

14.3.1 Foaming operation

The theoretical principles of foaming are described first. Only direct foaming is dealt with here, in which foaming of the melt, containing blowing agents, takes place directly after it leaves the die head. Other methods, in which the melt (containing blowing agent) is extruded at a low temperature and foamed at a higher temperature once again under the influence of heat, play a subordinate role.

The foaming process is a function of viscosity and expandability of the melt, the gas pressure of the blowing agent, and the ambient pressure. The interaction between melt and blowing agent (solubility, concentration, formation of nuclei) is the principal factor in foam extrusion.

Since the foam has poorer sliding properties at the wall than the compact material, and a semifinished product with a dense outer skin is desired, foaming must not begin until after the melt has left the die head. This requires the pressure acting on the foamable melt to be high enough at every point in the extruder and die head to keep the blowing gas constantly dissolved in the melt, this being achieved by appropriate screw design, die-head design and temperature control (Figure 1). Meeting this requirement is, inter alia, the basis of the process-and-apparatus patents listed in Section 14.4. Once the melt, containing blowing agent, has left the die head, the rapid drop in the pressure of the melt results in a supersaturation of the gas dissolved in the melt.

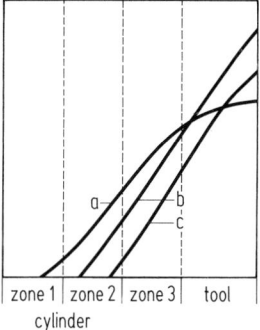

Figure 1. Ideal pattern of temperature, pressure and gas development in extrusion of foamable thermoplastics [9]
a melt temperature, *b* pressure in melt, *c* development of blowing gas

The two phases, gas and melt, separate and the gas forms a large number of bubbles almost instantaneously. The formation of the maximum possible number of cells requires the nucleants to be finely and evenly distributed throughout the melt (see also Section 14.2.6).

A distinction can be made in this respect between inactive and active gas-generating nucleants [15]. Three parallel processes take place during the formation of bubbles [16]:

– the continuous formation of new bubbles due to supersaturation.
– diffusion of the gas from smaller bubbles to larger ones, due to the larger surface tension and thus greater vapor pressure in the smaller bubbles.
– diffusion of the gas to the outer surfaces (permeation).

All three processes are basically undesirable and result in cells of different sizes being produced.

A parameter which has a significant impact on cell structure, and thus on the quality of the foamed semifinished product in terms of density and surface finish, is the temperature of the melt. This in turn, is affected by the K value of the PVC material, the structure of the formulation and the shear stress on the material during extrusion.

During foaming, the gas pressure in the bubbles acts against the melt structure. If the material temperature is too low, only an incomplete foam structure is formed, because of the higher viscosity of the melt. This results in a high extruded density. If the material temperature is too high, this causes the cells to tear open owing to the lower melt viscosity. Most of the blowing gas can escape and the cells collapse. The semifinished product displays a scarred surface and high density [17] — see also Section 14.2.7 in this connection. Foaming is also influenced by the following factors [7]:

— homogeneity of the blowing agent/melt mixture,
— degree of supersaturation of the gas dissolved in the melt,
— number, size, distribution and efficacy of the nucleants,
— heat of evaporation and vapor pressure of the blowing agent,
— solubility of the blowing agent in the melt [18],
— reduction in viscosity of the melt due to the blowing agent.

The formation of foam is terminated by calibration, which must take place immediately after the completion of foaming to prevent the foam from collapsing. The aim is to freeze the foam structure rapidly within a narrow temperature range. The density and thickness of the outer skin can be influenced by appropriate selection of the distance between calibration unit and die head, and the intensity of the cooling. The quality of the surface can be influenced by contact with the smoothing surfaces, irrespective of its hardness and the thickness of the skin [15].

14.3.2 Foaming methods

All of the methods hitherto developed for direct foaming of PVC into semifinished products can be classified into two basic methods [19]:

— free-foaming methods,
— methods for foaming to the inside.

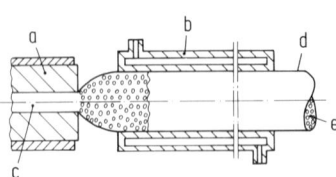

Figure 2. Free-foaming method
a die head, *b* calibrator, *c* melt, containing blowing agent, *d* outer skin, *e* foam core

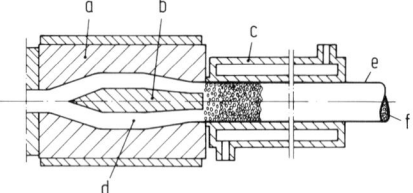

Figure 3. Method of foaming toward the core
a die head, *b* mandrel, *c* calibrator, *d* melt, containing blowing agent, *e* compact outer skin, *f* foam core

With the free-foaming method (Figure 2), the melt, containing blowing agent, expands freely right after it leaves the die head but before it obtains its final shape in the larger calibration unit some distance downstream from the die head. The entire cross-section of the semifinished product displays an approximately even density, surrounded by a somewhat denser, yet thin, outer skin which is more or less structured.
Pipes, sheets and geometrically simple profiles can be manufactured using the free-foaming method.
With the method of foaming toward the core (Figure 3), the calibration unit is positioned directly next to the die head. The calibration unit has the same outer contour as the

relevant die. As a result, the melt, containing blowing agent, undergoes rapid cooling over its entire surface as soon as it leaves the die; this prevents the formation of cells on the surface layer and any enlarging of the cross-section of the extrudate. At the same time, a mandrel in the die enables the cavity which it produces in the semifinished product to be filled by foam from the remaining melt.

In addition to pipes and sheets, it is possible to manufacture profiles with any desired geometry using this method. Such semifinished products are characterized by a compact, smooth outer skin and low density in the core zone.

The method of foaming toward the core has become known as the "Celuka" method (see Section 14.4.1.2.1).

14.4 Extrusion foaming methods

14.4.1 Methods using chemical blowing agents

14.4.1.1 Coextrusion of full profiles and sheets from two different material streams (Reifenhäuser)

This method (Figure 4) is based on the coextrusion of an outer hollow profile comprising compact material, and a foamed core. The semifinished products are thus provided with a homogeneous, smooth surface. The foam material is injected into the core of the profile as it is extruded, irrespective of the die cross-section. No uniform structure is required. Appropriate formulating and process technology make it possible to achieve a proper combination between compact material and foam. This process includes the possibility of combining two different types of plastics.

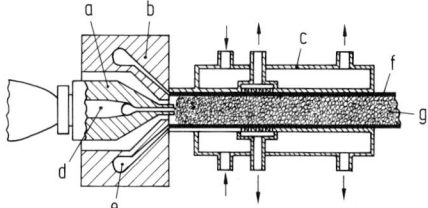

Figure 4. Method for coextrusion of solid profiles and sheets from two differential material streams [20]. *a* extrusion die, *b* extrusion head, *c* calibrator, *d* melt, containing blowing agent, *e* melt free of blowing agent, *f* hollow body, *g* foam core

14.4.1.2 Extrusion of foamed full profiles (Ugine Kuhlmann)

14.4.1.2.1 Full profiles from one material stream

This method (Figure 5) has become known as the "Celuka" method. It was one of the first patented extrusion-foam methods that made it possible to manufacture profiles with a dense, closed outer skin and a cell structure inside, from one material stream.

Principle of the Celuka method: the outer skin of the melt is compressed and cooled in the calibration unit directly adjoining the die, while a mandrel in the die head enables the remaining melt to foam freely toward the inside of the profile.

The density of the foam core and the thickness of the outer skin (0.1 mm to more than 1.0 mm) can be influenced by the intensity of cooling, the dimensions of the mandrel, and the take-off rate.

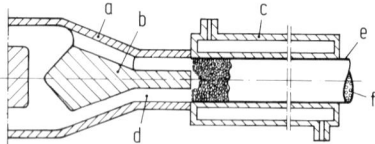

Figure 5. Method for extrusion of solid profiles from one material stream [21].
a die head, b mandrel, c calibrator, d melt, containing blowing agent, e compact outer skin, f foam core

14.4.1.2.2 Full, part-skinned profiles from one material stream

Figure 6 shows a further method for manufacturing foamed profiles. In contrast to the Celuka process, this is a method of making semifinished products having only strips of surface, which need not adjoin each other, with a smooth, skin next to the foamed core.

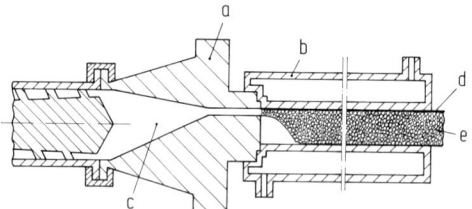

Figure 6. Method for extrusion of solid profiles from one material stream [22].
a die head, b calibrator, c melt, containing blowing agent, d compact outer skin, e foam core

This method is an example of combining the process for foaming towards the core with that of free foaming. A compact outer skin is only formed at those points where the calibration unit, which has a larger overall cross-section than the die, is flush with the die opening. The entire remaining cross-section is filled by the foaming melt, while the surface skin, which is thus formed adjoining the cell structure, is smooth and closed only in places.

14.4.1.2.3 Hollow profiles from one material stream

This method (Figure 7) represents a modification of the process described in Section 14.4.1.2.1. It is used for manufacturing hollow profiles with compact surface layers and a foamed core.

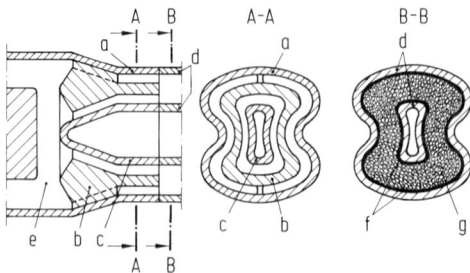

Figure 7. Method for extrusion of hollow profiles from one material stream [21].
a die, b mandrel, c inner wall, d calibrator, e melt, containing blowing agent, f compact skin, g foam core

14.4.1.2.4 Coextrusion of full and hollow profiles from two different material streams

This method (Figure 8) makes it possible, in a similar way to that described in Section 14.4.1.1, to manufacture profiles with a compact outer skin and foamed core in one operation. One of the extruders feeds the material for forming the outer skin to the outer die, which essentially has the same contour as the desired profile. The profile is provided with its final outer shape by the cooled calibration unit, which immediately adjoins the die.

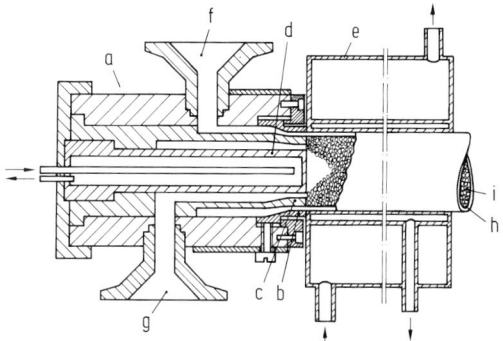

Figure 8. Method for coextrusion of solid and hollow profiles from two different material streams [23].
a die head, *b* outer die, *c* inner die, *d* mandrel, *e* calibrator, *f* melt, free of blowing agent, *g* melt, containing blowing agent, *h* compact outer skin, *i* foam core

A second extruder supplies the melt containing the blowing agent; this (in a similar way to the Celuka method) can foam freely toward the inside, with a mandrel acting as a flow brake while the melt at the same time forms a solid bond with the outer skin.
The cavity formed by the compact outer shell can be fully or only partially foam-filled in this way. This method also enables different types of plastics to be combined.

14.4.1.3 Extrusion of woodlike sheets from one material stream (Sekisui Kaseihin Kogyo)

The "Woodlite" method (Figure 9), developed with polystyrene, is used for manufacturing foamed sheets with a woodlike structure. The melt, containing blowing agent, leaves the die through a breaker plate.
Owing to the high resistance of the outlet holes, the individual strands do not foam until they reach a calibration unit or temperature-controlled rolls downstream, immediately welding together.

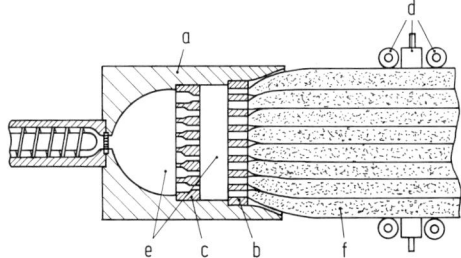

Figure 9. System for extrusion of artificial wood sheet from one melt stream [24]
a die-head tool, *b*, *c* breaker plate, *d* temperature-controlled rolls, *e* melt, with foaming agent, *f* foam strand

Each of the individual strands has a high-density surface skin and a low-density inner zone.
The medium density of the individual foam strands can be further influenced by a breaker plate fitted upstream of the outlet breaker plates. The density of the sheet surface zones can be additionally varied by appropriate cooling of the calibration unit.
For processing PVC, the inlets of the breaker plate holes must be designed in such a way as to ensure smooth melt flow.
A drawback of this method is that the flexural strength of the sheets perpendicular to the extrusion direction is very low, since the transverse strength of strands welded together in the longitudinal direction is low.
An improvement is achieved by a further method [25] patented by Sekisui.

14.4.1.4 Extrusion of full profiles from a split material stream (Scherer & Trier)

With this method, the die head is designed to split the foamable melt into at least two partial flows, enabling semifinished products to be manufactured with cross-sectional areas of different density. These areas of different density can be arranged around, next to, or above one another.

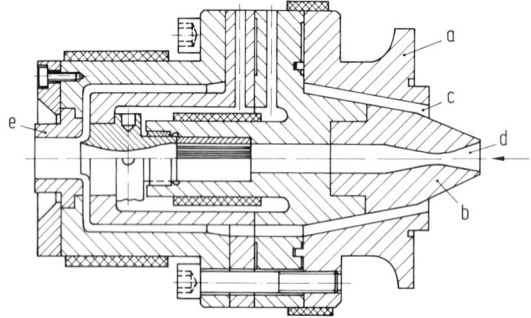

Figure 10. Extrusion die for extrusion of solid profiles from split material stream
a flange, *b* brake cone, *c* outer channel, *d* inner channel, *e* die

Figure 10 shows a possible embodiment for this patent:
- a die head for manufacturing a rod with a dense outer skin and foamed core,
- the outer material flow forming the outer shell is extruded at given temperature and provided with a smooth, pore-free surface by the adjoining calibration unit,
- the melt flowing on the inside is heated to a higher temperature before it flows out of the projecting mandrel and foams freely; this results in a foam core which is firmly welded to the outer shell.

The process can also be reversed, however, to produce profiles with a dense core and foamed outer shell.

14.4.1.5 Extrusion of pipes with foamed walls (Société Armosic)

Foamed pipes are manufactured using, essentially, the free foaming principle.
The "Armocel" method of Messrs. Armosic (Figure 11) represents a variation on this principle.
By proper design of the flow channel, an adequate pressure build-up is achieved in the die head to prevent the melt (containing decomposed blowing agent) from foaming prematurely: the annular gap has an initial tapered constriction and then continues cylindrically over a certain parallel length.

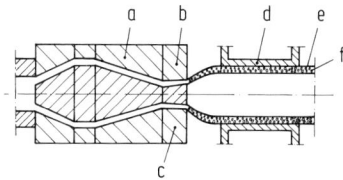

Figure 11. Method for extrusion of pipes with foamed walls [27]
a pipe die head, *b* die, *c* mandrel, *d* calibrator, *e* outer skin, *f* foam core

The extruded tube expands in diameter and in wall thickness immediately after it flows out of the pipe die head, and before passing into the calibration unit arranged a few centimeters from the die. By appropriate choice of formulation, it is possible to manufacture pipes with smooth inner and outer surfaces.

14.4.2 Processes using physical blowing agents

14.4.2.1 Extrusion of foamed PVC films

Foamed PVC films can be manufactured with directly-gassed twin-screw extruders with screw lengths of 18 to 22 D (Figure 12).
The raw material mixture is first gravimetrically metered, plasticated and gassed after approx. 10 D in the extruder. With direct gassing, the blowing agent is added continuously to the melt under pressure. Direct gassing is only possible with physical blowing agents.

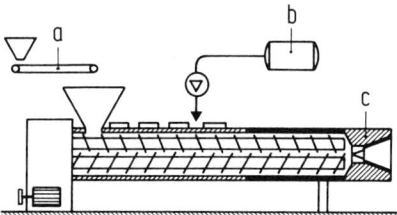

Figure 12. Twin-screw extruder with direct gassing [28]
a weighing belt, *b* blowing agent, *c* film blowing head

For PVC foam extrusion, only particular fluorinated hydrocarbon types are used [8].
After the blowing agent and melt have homogenized, the material is cooled to an extrusion temperature at which the melt viscosity is right for foaming [7].
The melt can be extruded through a horizontal blowing head.

14.5 Plant for foamed intermediate products

14.5.1 Extruder

PVC foam extrudates of medium to high density are manufactured on basically the same machines as are used for the extrusion of compact semifinished products, using powder mixtures containing blowing agent. Tightly meshing, contra-rotating systems of cylindrical and conical design have proven satisfactory in service.

Twin-screw extruders offer the following benefits in processing PVC powder mixtures:

- constant flow in the feed zone, independent of friction,
- gentle heating and plasticating of material,
- excellent thermal and material homogenizing,
- simple pressure build-up without interfering with the flow of material,
- high outputs at low speeds.

At the present time, the entire output range commonly employed in the extrusion of semi-finished products is covered by both cylindrical and conical twin-screw extruders. Since the demands made on the product in profile extrusion are greater than those in pipe extrusion, even if basically different, outputs for profiles are only approx. 50% of those attainable with the same extruder used for pipe extrusion. In view of this fact, profile extruders of somewhat simpler design in certain subareas can be used:

- enclosed screw, thermal control system in place of a more sophisticated oil system,
- air-cooled in place of oil-cooled barrel in the metering zone,
- material feed from the hopper.

If the extruder is used solely for manufacturing foamed profiles, the entire degassing unit can also be eliminated.

Pipe extruders of standard design are used for the extrusion of foamed pipes and sheets, in particular cylindrical types, although conical designs are also found. The principal features of these machines are:

- oil thermal control of barrel in the metering zone,
- oil thermal control system for screws (possibly separate control of screws for sheet extrusion),
- horizontal metering unit to handle the different bulk densities of the various materials.

Apart from an efficient temperature control system, the state of the art on today's twin-screw extruders also includes DC drives with steplessly variable speed control, and torque limiting.

PVC light foams require physical blowing agents; this necessitates the use of special gas-injection extruders. A characteristic feature of these machines is the gas-injection opening at a point in the barrel at which the PVC already exists as a melt. The pressurized liquid, or liquified blowing agent, is injected by a pump through a valve. Such machines have hitherto been primarily used for manufacturing foamed PVC films [8].

14.5.2 Screw design

Both *conical* and *cylindrical, contra-rotating twin screws* are used for processing PVC dry-blends containing blowing agent. Since the powdery or gelled material cannot hold the gas produced by the effect of heat, the mixture should be plasticated as rapidly as possible to avoid any loss of gas through the hopper opening. Only when the material is melted is it possible for the gas to dissolve. The concentration of gas dissolved in the melt depends not only on the retention time in the screws, and on the temperature control, but also particularly on the melt pressure [18].

Conical twin screws

Special screws without a degassing zone in combination with enclosed barrels have proved particularly effective for the extrusion of foamed profiles (Figure 13). A basic distinction can be made between single-conical- and twin-conical-screw systems. Whereas the depth of thread of single-conical screws is constant over the entire length, the thread depth of twin-conical screws increases smoothly from the tip of the screw to the feed zone. This makes it possible to achieve high flight-volumes with low pitch in the feed zone of twin-conical screws. The result is a higher retention time, or preheating of the material, while the large flight surface ensures good heat transmission.

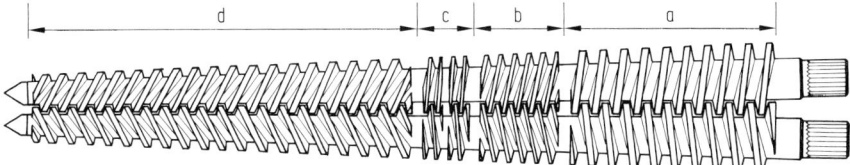

Figure 13. Design of conical twin screws for PVC foam extrusion
a inlet zone, *b* precompression zone, *c* compression zone, *d* metering zone

Elimination of the degassing zone enables the metering zone to be made very long to attain good material plastication and homogenizing. In addition, high material pressures can be kept under control. The die head can therefore be optimally designed to foaming requirements without the need to take the extruder into account.

Cylindrical twin screws

Cylindrical twin screws are employed in particular for manufacturing foamed pipes, sheets and larger profiles (e.g. window sills). Good plasticating screws have provided satisfactory service for these fields of applications. In addition to special sheet screws, standard pipe screws are used for PVC rigid foam pipe and profile extrusion (Figure 14).
Common screw diameters range from 90 to 130 mm, screw lengths from 18 to 22 D.

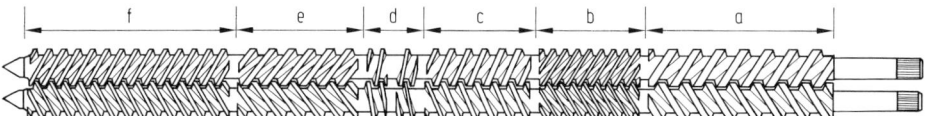

Figure 14. Design of cylindrical twin screws for PVC foam extrusion
a feed zone, *b* preheating zone, *c* precompression zone, *d* compression zone, *e* degassing zone, *f* metering zone

14.5.3 Die head

The major concern in die-head design is to develop sufficient build-up of pressure to prevent any premature foaming of the plastic melt in the head section or in the die. If the extrudate foams in the die head, this produces an uneven cell structure and rough surfaces. A certain melt pressure level can be maintained in the entire die head by an even reduction in cross-section to the die outlet; in other words, a steady increase in the geometrical die-head resistance. Since the flow processes, particularly for profile die heads, are governed by laws which have not yet been exactly determined, die-head design is primarily based on practical experience. In addition, die heads for PVC foam profiles are considerably more difficult to design and alter than those for compact PVC profiles, since consideration has to be given not only to die cross-section, flow rate, flow resistance and swelling, but also to the expansion of the extrudate. Asymmetrical profile cross-sections, in particular, require a special design of flow channel to achieve an even material flow at the die outlet. The parallel zone of profile die heads should be designed as short as possible, since premature foaming may be caused by the shear stress. In addition, a short smoothing zone promotes even foaming of the melt, since the molecules are then less oriented when they flow out of the die. Foamed rigid PVC pipes are manufactured, principally, using the free-foaming method. In such cases it is possible to use the same pipe die heads as are commonly employed for the extrusion of compact PVC pipes. The particular die diameter, die-head gaps and parallel zones are arranged as a function of the foaming level and the swelling of the material, using computer programs based on theoretical principles and practical experience.

14.5.4 Calibration unit, cooling unit

The purpose of the calibration unit is to impart its shape to the extrudate flowing out of the die head, and to fix this shape. The calibration dimensions are made larger than the final cross-section of the semifinished product by a shrinkage factor of about 0.01. Owing to the lower thermal conductivity of foamed semifinished products, both the calibration unit and the downstream cooling screw have to be designed appropriately longer. Apart from thermal conductivity, determined primarily by the foam density, the length of calibrator and cooling unit must take into account the cross-section of the semifinished product, the stability of the foam and the required take-off speed.

Vacuum water tanks have provided satisfactory service for calibrating foamed profiles, while vacuum spray tanks and spray baths are used in foam pipe extrusion. Better cooling of the extrudate is achieved in vacuum calibration, owing to the more intensive wall contact. At the same time, the vacuum prevents the foam, which has not yet solidified, from collapsing.

The surface quality and surface layer density of the foamed semifinished product are determined by appropriate selection of the distance between calibrator and die outlet, and the temperature of the calibration unit.

Large calibration distances result in an extension in the foaming time. The cell structure can form over the entire cross-section of the semifinished product. In calibration, the cells located on the surface are smoothed. This results in a coarse-cell foam structure in the surface zone, surrounded by a thin, porous outer skin with low compressive strength.

Small calibration distances, i.e. short expansion times, in combination with the sharply cooled calibrator, prevent or suppress the formation of cells on the surface of the extrudate. The semifinished products in this case display a fine-cell foam structure in the surface layer with a thicker, compact outer skin and smooth surface. If the calibrator is cooled to a lesser extent, this produces lower densities in the surface zone, likewise with a smooth surface.

A final point to note is that, because friction forces occur during calibration, the required take-off forces have to be absorbed by the solid outer layer of the semifinished product as tensile forces. Particularly in the case of profiles with a large surface and small cross-section, this can result in the tensile strength of the transition layer being exceeded and the profile strand breaking [17].

14.5.5 Take-off

The take-offs are the same as used in the extrusion of compact profiles or pipes. The tensioning force of the caterpillar take-off should not be too high, since the foamed semifinished products have relatively low compressive strength. Positive-locking rubber jaws have proven beneficial in this respect. Special pads are needed for specific profile shapes.

14.5.6 Cutting to length, punching and stacking

The fixtures used for cutting to length, punching and stacking foamed semifinished products are the same as are commonly employed in handling compact semifinished products. A particular point to note in stacking is that the products should be positioned as straight as possible, and well cooled to prevent any heat accumulation occurring in the inside. If the parts are left for a prolonged period without so doing, deformation could result.

14.5.7 Embossing, printing

Decorating the surface is particularly common for profiles made of PVC rigid foam, since these frequently compete with wood. The surface of the profile is structured by the heated embossing roller of a special machine, which applies pressure and an embossing film with the appropriate wood color and grain at the same time. Profile surfaces are normally embossed after take-off. Single- or multicolor printing machines are also used, in-line or off-line, to impart a woodlike appearance to profiles.

14.6 Selection criteria

In selecting a system, it can basically be assumed that the output of extruders used in manufacturing foamed PVC semifinished products is just as high as that obtained in manufacturing compact PVC semifinished products. Throughput, however, is determined not only by extruder capacity, but also by the shape and dimensions of the semifinished product, and the cooling efficiency of the calibrator and cooling bath.
The production data of various semifinished products are presented in Table 3.

Table 3. Performance of various systems as a function of product

Extruder	Design	Product	Density g/cm^3	Dimension mm	Weight kg/m	Take-off rate m/min	Output kg/h
KMD 50 KK	twin-conical	Wall-lining profile (PVC rigid foam)	0.6	Width: 170 Thickness: 3	0.360	2.80 to 5.55	60 to 120
KMD 90	cylindrical	Pipe (PVC rigid foam)	0.8	110 × 3.4	0.910	4.60 to 6.05	250 to 330
KMD 125	cylindrical	Sheet (PVC rigid foam)	0.7	1600 × 10	11.20	0.50 to 0.60	400 to 500

References for Chapter 14

For the convenience of the reader the English titles of all publications in languages other than English are shown in parentheses.

[1] *Morianz, E.:* Geschäumte Verbundfolie aus PVC, der Fensterrahmen der Zukunft? (Foamed Composite PVC Profiles, the Window Frames of the Future?). In: Schäume aus der thermoplastischen Schmelze (Foams from Molten Thermoplastics). VDI-Verlag, Düsseldorf, 1981.
[2] *Barth, H.:* Kunststoffe 67 (1977), pp. 674/680.
[3] *Abeler, G., Büssing, J.:* Kunststoffe 71 (1981), pp. 315/320.
[4] *Steigerwald, F.:* Plastverarbeiter 26 (1975), pp. 588/592.

[5] Solvic and Solvic-Premix für Folien (Solvic and Solvic Premix for Films). Company brochure, Deutsche Solvay-Werke GmbH.
[6] *Trausch, G.:* Physikalisch und chemisch getriebene Thermoplastschäume, Grenzen der Verfahren und Anwendungen (Physically and Chemically Blown Thermoplastics Foams — Limits of Processes and Applications). In: Schäume aus der thermoplastischen Schmelze (Foams from Molten Thermoplastics). VDI-Verlag, Düsseldorf, 1981.
[7] *Zingsheim, P.:* Schaumextrusion von PVC-hart nach dem Direktbegasungsverfahren (Extrusion of Rigid PVC Foam by the Direct Gassing Process). Paper at the 5th Int. Conference on Foamed Plastics, Düsseldorf, 1975.
[8] *Kolossow, K.-D.:* Herstellung geschäumter PVC-Folien durch Direktbegasung (Production of Foamed PVC Films by Direct Gassing). In: Schäume aus der thermoplastischen Schmelze (Foams from Molten Thermoplastics). VDI-Verlag, Düsseldorf, 1981.
[9] *Domininghaus, H.:* Hohl- und Vollprofile aus thermoplastischem Strukturschaumstoff (Hollow- and Solid Structural-foam Profiles from Thermoplastics). In: Extrudieren von Profilen und Rohren (Extrusion of Profiles and Pipes). VDI-Verlag, Düsseldorf, 1974.
[10] DE-AS 2047969 (1970), Goodrich Co., Akron, Ohio, USA.
[11] DE-OS 2016043 (1970), Japanese Geon Co., Tokio, Japan.
[12] FR-PT 1487545 (1967), ICI, Great Britain.
[13] DE-OS 2503390 (1975), Rohm and Haas Co., Philadelphia, USA.
[14] *Barth, H.:* Kunststoffe 67 (1877), pp. 674/680.
[15] *Barth, H.:* Kunststoffe 67 (1977), pp. 130/135.
[16] *Reichert, U.:* Extrusion geschäumter Thermoplaste (Extrusion of Foamed Thermoplastics). Paper at 3rd Int. Conference of Foamed Plastics, Düsseldorf, 1973.
[17] *Wanzek, R.:* Extrusion von Schaumprofilen aus Hart-PVC (Extrusion of Foamed Profiles from Rigid PVC). In: Schäume aus der thermoplastischen Schmelze (Foams from Molten Thermoplastics). VDI-Verlag, Düsseldorf, 1981.
[18] *Moritz, U.:* Kunststoffe 73 (1983), pp. 394/397.
[19] *Zingsheim, P.:* Extrusion geschäumter PVC-Hart-Profile und -Rohre (Extrusion of Rigid PVC Profiles and Pipes). Paper at 7th Int. Conference on Foamed Plastics, Fa. Krauss-Maffei, 1978.
[20] DE-PS 1183238 (1961), Reifenhäuser KG, Troisdorf, West Germany.
[21] DE-AS 1729076 (1967), Ugine Kuhlmann, Paris, France.
[22] DE-AS 2050550 (1970), Ugine Kuhlmann, Paris, France.
[23] DE-OS 1913921 (1969), Ugine Kuhlmann, Paris, France.
[24] DE-AS 2038803 (1970), Sekisui Kaseihin Kogyo K. K. Nara, Japan.
[25] DE-AS 2428999 (1974), Sekisui Kaseihin Kogyo K. K. Nara, Japan.
[26] DE-AS 2116940 (1971), Plastic-Werk Scherer & Trier oHG, Michelau, West Germany.
[27] DE-AS 2051006 (1970), Société Armosic, La Celle Saint, Clode, Yvelines, France, DE-OS 2126976 (1971, Addendum to 2501006.
[28] DE-OS 3038306 (1980), Hermann Berstorff Maschinenbau GmbH, Hannover, West Germany.

15 Crosslinking of plastics after extrusion

K. Kircher

15.1 Introduction

In plastics processing there are various processes in which polymer precursors are transformed into the final material by a chemical reaction during or after shaping [1]. In the manufacture of moldings from unsaturated polyester resins, for example, low- or medium molecular-weight polymerizable starting materials (monomers or prepolymers) are molded and cured, i.e. they are transformed into highly crosslinked macromolecules [2, 3].

And in the crosslinking of thermoplastics in a process involving extrusion, the properties of the starting material and of the end product are also different: linear, unlinked macromolecules, for the most part, are used (plus processing aids possibly) for transformation into the crosslinked end product by a chemical reaction which increases molecular size only after extrusion and shaping.

Processing operations with subsequent crosslinking are thus a combination of physical and chemical processes. The question of why the chemical part of the overall process is not carried out during the plastics manufacturing operation is easy to answer: since crosslinked high molecular-weight polymers are thermoset or thermoelastic materials, thermoplastic deformation is not possible without extensive destruction of the network. Plastic or thermoplastic shaping has therefore to be carried out on the pre-products, which are crosslinked in parallel with, or subsequent to, the shaping operation. So this sort of chemical reaction must take place in the plastics processor's works. Consequently there are more processing parameters to control and monitor: in particular temperature, dwell time, and degree of crosslinking.

In Section 13.4 the discussion of crosslinking dealt only with thermoplastics. Pure thermoplastic crosslinking is limited essentially to that of polyethylene and ethylene copolymers, especially ethylene-vinyl acetate [4], ethylene-propylene [5], and ethylene-propylene-diene [6] copolymers. The crosslinking of polypropylene [7 to 10] and several other polymers is described in the literature, but these have achieved little technical importance so far.

Current extrusion-crosslinking processes for thermoplastics were initially designed for the crosslinking of polyethylene. In the present chapter, therefore, the descriptions of crosslinking processes refer exclusively to polyethylene.

15.2 Modification of the properties of polyethylene by crosslinking

The crosslinking process alters the physical as well as the chemical structure, and the accompanying changes in some of the properties of the material result in an improved properties profile for individual applications [11 to 14]. Crosslinked polyethylene is rubber-elastic, and its properties lie between those of thermoplastics and rubbers. The degree of crystallinity in crosslinked polyethylene is lower than in the normal product, and the remaining crystallites are small and highly defective; the hardness and the stiffness are lower, and the sensitivity to stress cracking is much reduced. Above the melting point, uncrosslinked material has a relatively low viscosity melt, with hardly any mechanical strength. Crosslinked polyethylene, however, has some residual strength above the apparent melting point, which gives the material a certain form of stability; this prevents electrical conductors, for example, from moving under their own weight from their central position in an insulating sheath.

The changes in some physical properties have to be taken into account in crosslinking processes. The alteration of the physical structure affects the thermal conductivity and the water-vapor diffusion coefficient sufficiently to cause uncertainties in calculations with these quantities. Whereas the geometry of an uncrosslinked polyethylene above the melting point can be altered by a mechanical influence – for example a bend in the crosslinking tube – crosslinked material above the melting point is considerably less sensitive. Crosslinking lines have to take account of these characteristics.

15.3 Uses for crosslinked polyethylene

The outstanding electrical properties of crosslinked polyethylene, and its good stress-cracking resistance and chemical resistance, as well as adequate strength in the 120 to 150 °C range, suit it for broad application as cable insulation, particularly for high-performance cables [15 to 25]. This application area was the real impetus behind the development of the various crosslinking processes.

Another ideal area of application for crosslinked polyethylene is for pipes for underfloor heating. Also, polyethylene foam produced by extrusion is to a large extent crosslinked [26]. The technical reasons for crosslinking foamed polyethylene are that crosslinking in parallel with foaming makes the foaming process easier, and the elasticity of the foam is improved. Other application areas for crosslinking of polyethylene include the manufacture of shrink films and shrinkable tubes [27].

15.4 Crosslinking processes

The crosslinking methods that are put into practice, and the others that are still in the experimental stage, are so varied that one cannot simply speak of "the crosslinking" after extrusion. The various processes are so different that they require separate sections to be devoted to them. Also one cannot specify overall factors which influence all of the different crosslinking processes. Rather it is necessary to enumerate them separately. A summary of the various processes is given in [28]. One scheme showing the different ways of crosslinking can be set out as follows:

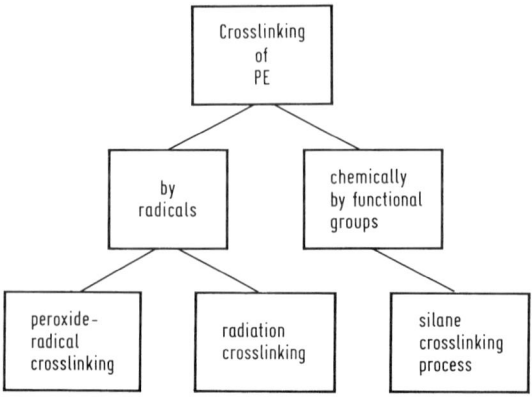

15.4 Crosslinking processes

Peroxide-radical processes are technically the most important at present. With these, also, the larger number of process variants should be borne in mind. Radiation crosslinking differs from peroxide crosslinking in some chemical aspects, and is also technically more costly. Silane crosslinking has little in common in its technical realization with radical processes, and is quite different chemically.

In almost all conversion processes crosslinking has to be considered as a separate operation, which nevertheless usually runs directly in line with the extrusion. Because of this rapid time succession, the heat energy available in the uncrosslinked thermoplastic can be used to best effect in the downstream crosslinking process.

Even though the shaping and crosslinking processes are considered as separate, there are processes in which the crosslinking is moved to some extent into the shaping tool. In all processes in which the processing temperature in the die head approaches the crosslinking temperature, part of the crosslinking takes place during shaping.

15.4.1 Chemistry of radical crosslinking

Radical crosslinking processes are based on:

− production of polyethylene radicals
− dimerization of polyethylene radicals.

Radical dimerization goes through these two reaction steps spontaneously, providing virtually no possibility for the processor to influence them. The production of chain radicals determines the course of the reaction, and these can be produced in a number of ways:

− by the direct input of energy in radiation crosslinking (electron beams, UV light)
− by radical transfer from separately produced primary radicals.

Radical production	$-\overset{H}{\underset{H}{C}}-\overset{H}{\underset{H}{C}}-\overset{H}{\underset{H}{C}}-\overset{H}{\underset{H}{C}}-\overset{H}{\underset{H}{C}}-$ $\xrightarrow{-H\cdot}$ $-\overset{H}{\underset{H}{C}}-\overset{H}{\underset{H}{C}}-\overset{H}{\underset{\ast}{C}}-\overset{H}{\underset{H}{C}}-\overset{H}{\underset{H}{C}}-$
Radical dimerization	two radical chains combining to form crosslinked structure $-CH_2-CH_2-\ \ CH_2-CH_2-$ / \ C / \ $-CH_2-CH_2-\ \ CH_2-CH_2-$

a) Electron beam crosslinking

Direct radical creation in electron beam crosslinking is possible in principle without additives; thus additive-free polyethylene can be crosslinked by electron beam irradiation [27]. The possibilities for the processor to influence the reaction are limited to energy dose levels and the temperature [29]. According to data in the literature [30 to 33], there are still many ways of affecting what goes on, by using special additives to optimize the crosslinking density produced per unit of energy input.

Radical formation occurs exclusively because of the interaction of the energy with the polymer chain, so that $-C-H$ bonds are broken, producing hydrogen and a polyethylene radical. When the radical build-up is in the presence of atmospheric oxygen, the result is that radicals created are partially saturated by the oxygen [34, 35].

b) *Radical transfer reactions*

In this reaction, primary radicals are produced in a separate step by the decomposition of an auxiliary component (usually a peroxide); then these, in turn, extract one of the H-atoms (available in large excess), thus saturating themselves and transferring the radical state to the polymer chain.

Production of primary radicals:

$$R-R \longrightarrow 2R\bullet$$

Radical generator → Radical

Radical transfer:

$$R\bullet + -CH_2-CH_2-CH_2-CH_2-CH_2- \longrightarrow R-H + -CH_2-CH_2-\overset{\bullet}{C}H-CH_2-CH_2-$$

(chain radical)

With the processes that are used nowadays for crosslinking low-density polyethylene (LDPE), the primary radicals are created by the thermal decomposition of peroxides, especially di-cumyl peroxide [36 to 41]. Common peroxides are:

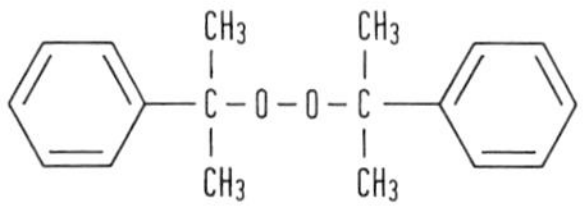

Structure	Name
Ph-C(CH$_3$)$_2$-O-O-C(CH$_3$)$_2$-Ph	di-cumyl peroxide
CH$_3$-C(CH$_3$)$_2$-O-O-C(CH$_3$)$_2$-CH$_3$	di-t-butyl peroxide
CH$_3$-C(CH$_3$)$_2$-O-O-C(CH$_3$)$_2$-Ph	butyl-cumyl peroxide

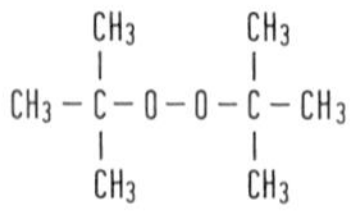

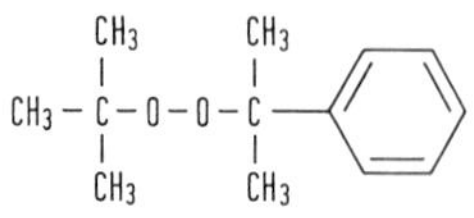

These peroxides are not sufficiently temperature-stable for crosslinking high-density polyethylene (HDPE) and would cause crosslinking in the extruder and the die at extrusion temperature (160 to 180 °C). Other peroxides are used for this purpose, for example 2,5-dimethyl-hexene-(3)-2,5-ditertiary-butyl peroxide [42].

The peroxides are either delivered ready-mixed in the polyethylene, or metered directly by a pump for mixing during extrusion. The concentration of peroxide is decided by the degree of crosslinking required [43]; it should remain constant and not exceed an upper limit. Under some circumstances the polyethylene may contain carbon black [44 to 46] or crosslink-promoting allyl compounds [47].

15.4 Crosslinking processes

For the production of crosslinked extruded LDPE, the polymer, containing the peroxide, is plasticated in an extruder and extruded as a profile at a die-head temperature of ca. 130 °C. Under these conditions the peroxide must be completely stable, otherwise early partial decomposition would lead to crosslinking in the extruder. As Figure 1 shows, di-cumyl peroxide, with a decomposition half-life of 3 to 4 hours at 130 °C, meets these requirements.

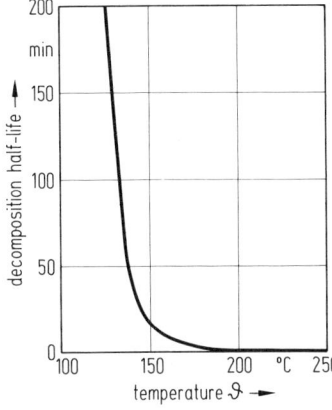

Figure 1. Decomposition half-life of di-cumyl peroxide as a function of temperature, from [49]

After leaving the die head, the extrudate is further heated, to start the decomposition of the peroxide and the crosslinking process at the same time [28, 48]. When the required temperature is reached at the points furthest from the walls, there follows a constant temperature zone to ensure the completion of the crosslinking reaction. The extrudate is then cooled.

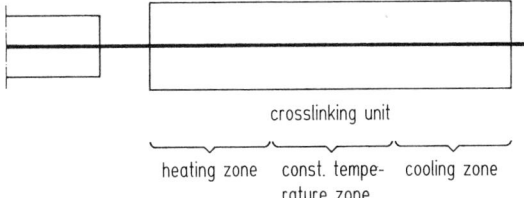

In this process the crosslinking reactor takes the form of equipment in which the extrudate is thermally conditioned. Its geometry, technical design, and performance depend on the geometry and quantity of the product to be crosslinked, and on the difference between the melt temperature at the die exit and the minimum crosslinking temperature. In order to keep the temperature difference small, and the heating time as short as possible – to make the crosslinking unit most effective – a peroxide which is completely stable at the extrusion temperature, but which decomposes rapidly with further increase of temperature, must be used.

One disadvantage of thermally initiated peroxide crosslinking is the creation of volatile decomposition products of the peroxide and the polyethylene itself [39]. These can form bubbles, which are potential failure sites and must not occur in cable insulation, for example, an important use for polyethylene. To suppress the formation of bubbles the crosslinking units operate under pressure.

If the extrudate is cooled to a temperature below 100 °C the gaseous products do not form bubbles.

```
┌─────────────────────────────────────────────────────────────┐
│ Decomposition reaction of di-cumyl peroxide                │
│                                                             │
│         CH₃   CH₃                              CH₃          │
│          |     |                                |           │
│   Ph—C—O—O—C—Ph      → 2     Ph—C—O·                       │
│          |     |                                |           │
│         CH₃   CH₃                              CH₃          │
│                                                             │
│                              O                              │
│                              ‖                              │
│                        Ph—C—CH₃ + ·CH₃                      │
│         CH₃                                   +PE           │
│          |                                        → CH₄     │
│   Ph—C—O·                                                   │
│          |                   CH₃                            │
│         CH₃                   |                             │
│                 +PE    Ph—C—OH                              │
│                               |                             │
│                              CH₃                            │
│                                                             │
│         CH₃                                   CH₂           │
│          |                                    ‖             │
│   Ph—C—OH    → H₂O +    Ph—C—CH₃                            │
│          |                                                  │
│         CH₃                                                 │
└─────────────────────────────────────────────────────────────┘
```

c) *Heat effects and kinetics of heat-activated radical crosslinking*

The heating and cooling process, apart from the heat of the crosslinking reaction, is described by the laws given in Volume 1 [50 to 56]; and the heat-up time is slightly reduced when this exothermic reaction is taken into account.
Estimation of energy states [57, 58]:

1. Positive heat effects

Decomposition of the peroxide	$H =$ 142 to 159 kJ/mol
Breakage of C–H bonds	$H = 2 \cdot (377 \text{ to } 394)$ kJ/mol
Total =	896 to 947 kJ/mol

2. Negative heat effects

Formation of alcohol –OH bonds	$H = -2 \cdot (427)$ kJ/mol
Formation of C–C bonds	$H = -344$ kJ/mol
Total = −	1198 kJ/mol

Crosslink formation involves a negative heat of reaction of 251 to 302 kJ/mol.
To what extent this simplified theoretical determination of the heat effects matches reality has to be experimentally checked. The positive heating effects certainly cannot be ignored, since temperature/time curves measured near the temperatures of peroxide decomposition, even allowing for heat transfer from external heat sources, show disproportionate increases in temperature.
The total reaction heat is not liberated suddenly, but over a certain period of time dependent on the reaction kinetics of the peroxide. In calculating the time dependence of the liberation of reaction heat, it can be assumed that the fast dimerization of chain radicals does not determine the speed of the overall reaction, and that it is only the peroxide decomposition that has to be considered. On the assumption of a first order reaction ($n = 1$) with n independent of temperature, and a final concentration of crosslinks, c_∞, one can calculate the transformation variable:

15.4 Crosslinking processes

$$x = \frac{c(t)}{c_\infty} \text{ using the relationship [50, 51]}$$

$$x = 1 - \exp\{-kt\}$$

where: $c(t)$ concentration of crosslinks at time t
         t dwell time
         k (reaction-)rate coefficient.

The rate coefficient k is determined by the equation [50, 51]

$$k = k_1 \cdot \exp\left\{\frac{A(T-T_1)}{R \cdot T \cdot T_1}\right\}$$

where: k_1 rate coefficient of peroxide decomposition at reference temperature T
         T_1 reference temperature
         T actual temperature
         A activation energy
         R universal gas constant

Assuming a specific heat for polyethylene of 2.5 J/g · K at the crosslinking temperature, the reaction heat produces a temperature rise of 22 °C.

Table 1. Phases of heat-activated peroxide crosslinking

	Object	Active Factors
Heat-up zone	Points furthest from the wall must be raised to a temperature = or > that of peroxide decomposition	– Temperature of the extrudate on entering the crosslinking zone – Section thickness – Heat transfer – Crosslinking temperature
Constant temperature zone	Completion of the crosslinking reaction	– Temperature – Additives
Cooling zone	Cooling of the extrudate to below 100 °C	– Max. temperature in crosslinked material – Wall thickness – Heat transfer – Cooling conditions
Pressure phase (normally maintained over all three zones above)	Suppression of failure-site development in extrudate by highly volatile components	– Kinds of additives – Quantity and kind of highly volatile components – Temperature

15.4.2 Silane crosslinking processes

In silane crosslinking reactions [59, 60], chemically reactive side groups on the polyethylene chain take part in a condensation reaction to form crosslinks in the following way:

$$\}\text{–OH} + \text{HO–}\{ \longrightarrow \}\text{–O–}\{ + H_2O$$

The first step involves attaching condensable side groups to the polyethylene chain. This is done by means of a radical graft reaction of trismethoxy-vinylsilane:

$$\begin{array}{c}|\\CH_2\\|\\CH_2\\|\\CH_2\\|\end{array} + CH_2=CH-\overset{\overset{O-CH_3}{|}}{\underset{\underset{O-CH_3}{|}}{Si}}-OCH_3 \xrightarrow{\text{(Peroxide)}} \begin{array}{c}|\\CH_2\\|\\CH-CH_2-CH_2-\overset{\overset{O-CH_3}{|}}{\underset{\underset{O-CH_3}{|}}{Si}}-OCH_3\\|\\CH_2\\|\end{array}$$

To produce crosslinked polyethylene the product containing the siloxane side groups, which is stable and can be stored, is first mixed with a catalyst concentrate (e.g. dibutyl tin dilaurate [61] or tetrabutyl titanium) and extruded. With the addition of water, the catalyst causes the trimethoxysilyl groups to split off, creating condensable silanol groups:

$$2\begin{array}{c}|\\CH_2\\|\\CH-CH_2-CH_2-\overset{\overset{O-CH_3}{|}}{\underset{\underset{O-CH_3}{|}}{Si}}-OCH_3\\|\\CH_2\\|\end{array} \xrightarrow[-6\ HOCH_3]{+6\ H_2O} 2\begin{array}{c}|\\CH_2\\|\\CH-CH_2-CH_2-\overset{\overset{OH}{|}}{\underset{\underset{OH}{|}}{Si}}-OH\\|\\CH_2\\|\end{array}$$

$$\xrightarrow{-H_2O} \begin{array}{c}|\\CH_2\\|\\CH-CH_2-CH_2-\overset{\overset{OH}{|}}{\underset{\underset{OH}{|}}{Si}}-O-\overset{\overset{OH}{|}}{\underset{\underset{OH}{|}}{Si}}-CH_2-CH_2-CH\\|\\CH_3\end{array}\begin{array}{c}|\\CH_2\\|\\\\|\\CH_2\\|\end{array}$$

Although the methoxy silane groups are relatively stable, the silanol groups condense forming crosslinks between chains. Crosslinking is controlled by admission of water. Polyethylene itself takes up very little water, but transmits it relatively rapidly so that even with thick walls technically acceptable crosslinking times result.

In the simplest case, extrudate crosslinking takes place on storage at room temperature, with atmospheric moisture slowly permeating the material and inducing the above reactions.

The crosslinking reaction can be accelerated by storing the material in hot water. The melting point of the polyethylene is not exceeded. The silane crosslinking process thus provides a method, as with radiation crosslinking, by which extrudate of finished shape can be crosslinked.

Table 2. Factors in silane crosslinking

Variable	Effect
– component recipe – water content on part surface – mass temperature – storage time	– determines crosslinking rate and maximum degree of crosslinking – influences water transport into the polymer – influences the diffusion rate of water – for optimum degree of crosslinking a minimum time is necessary

15.5 Checking the degree of crosslinking

Several methods for measuring the degree of crosslinking are currently in use [62, 63], among them:

a) *Equilibrium swelling:* In this method a small quantity of crosslinked polyethylene is immersed in xylol and the ratio of the weights of swollen and unswollen samples determined.

b) *Gel/sol test:* Finely disintegrated sample material is suspended in boiling xylol in a fine-mesh metal sieve for 17 hours, and subsequently dried in vacuum at 150 °C. The quantity retained, expressed as a percentage of the starting weight, gives the degree of crosslinking.
c) *Hot-set test:* The crosslinked test piece is subjected to a predetermined stress (20 N/sq.cm) and the resulting extension measured.
d) *Torsional modulus measurement* (DIN 53445): The modulus increases with increasing degree of crosslinking.

These well-known methods give various numbers that depend not only on the effective number of crosslinks, but also on the kind of crosslink that is formed. Thus they have predictive value only if practical values for optimally completed crosslinking are known.
Continuous on-line determination of the degree of crosslinking is described in [64]. The method uses a crosslink indicator added to the starting material, which allows the degree of crosslinking to be estimated by comparing the color of the crosslinked material with that of standard samples. To the extent that temperature and dwell time are thoroughly controlled, conclusions about the progress of the crosslinking reaction during production can be drawn. The determination of degree of crosslinking is indispensable for quality control and for control of start-up procedures.

15.6 Implementation of crosslinking

The different possible ways of crosslinking, viz.:

a) electron beam crosslinking (radical),
b) peroxide radical crosslinking,
c) silane crosslinking,

have the following characteristic features, respectively:

a) a device for irradiating the extrudate with high energy electrons;
b) long pressurized tubes for crosslinking, in which thermal energy is first put into the extrudate, and then extracted from it;
c) containers in which the extrudate is treated with water at high temperatures.

Peroxide radical processes are the most diversified, as will be illustrated by the example of crosslinking of cable insulation.

15.6.1 Heat-activated peroxide radical crosslinking

In the manufacture of cable insulation from crosslinked polyethylene, the conductor is covered with a layer of polyethylene in a coating head [65 to 68] (compare Chapter 3). Immediately after leaving the die, the extrudate runs into a crosslinking zone in which it is heated by high-pressure steam, by liquids or gases, by radiant heat, ultrasonics or ultra-high-frequency electromagnetic induction fields. Figure 2 shows an example of a complete line for manufacture of cables insulated with crosslinked polyethylene. A survey of the different processes is given in [28, 70].
Heating with steam is possible only up to ca. 25 bar/225 °C, because of line-design limitations. Reduction of the heating time by use of hotter heat-transfer media calls for the use of another process. Experience has shown, furthermore, that crosslinking of polyethylene with steam results in a water content that can give rise to electrical failure by "water-treeing", in which tree-like, branched, waterfilled faults within the insulator are the starting points for electrical breakdown. The various processes in current use are structured as follows:

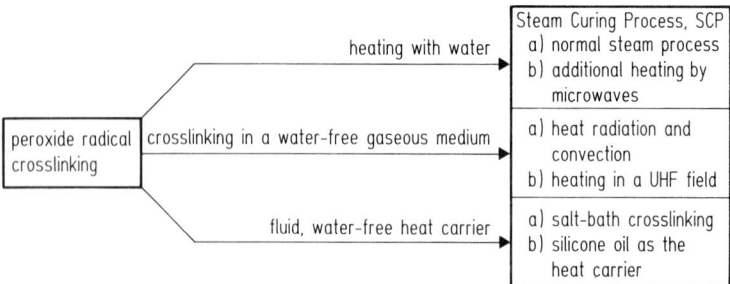

The current view is that the normal steam process, and heating in nitrogen-pressurized tubes with infrared radiation, are by far the most important processes commercially. A steady material flow in these processes is guaranteed by the fact that the conductor is pulled through the coating head, and then through all the heating and cooling zones, at constant speed and tension.

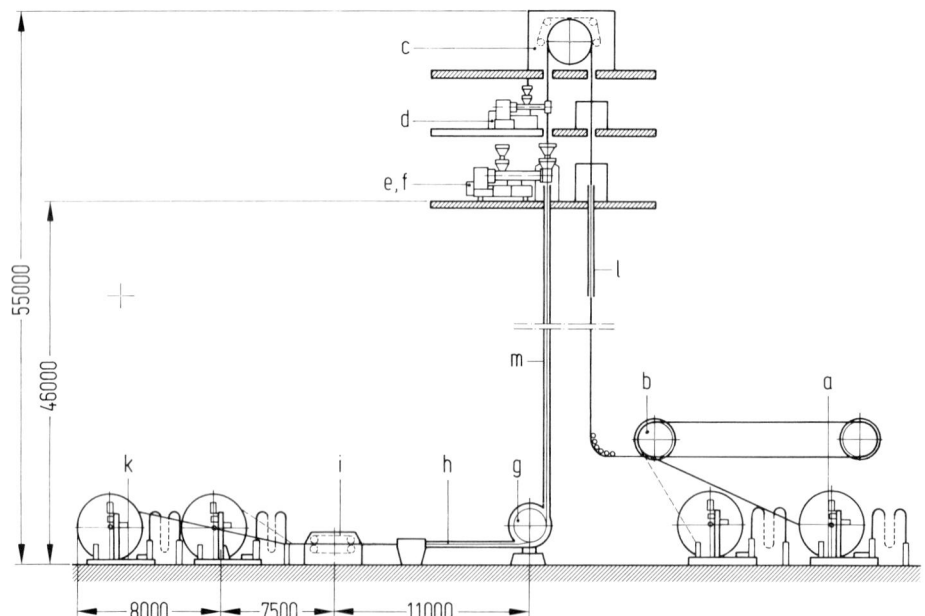

Figure 2. Vertical CV-line for manufacture of cables up to 100 mm diameter and bigger, from [49]
a driven drum-offwind, *b* horizontal dancer reservoir, *c* disk haul-off, *d* extruder for inner semi-conductive layer, *e* insulation extruder, *f* extruder for external semi-conductive layer, *g* pressurized deflecting-roll housing, *h* cooling tube, *i* haul-off unit, *h* traversing drum-windup, *l* preheating tube, *m* steam tube

15.6.1.1 Steam crosslinking processes

The different variants are distinguished by the design of the crosslinking zone (see Figure 3):

- vertical tube configurations for high-tension cables and large cross-sections,
- inclined straight tube configurations, often also called horizontal lines, for thin conductors, and
- catenary, or half-catenary configurations for cables with medium and large cross-sections.

15.6 Implementation of crosslinking

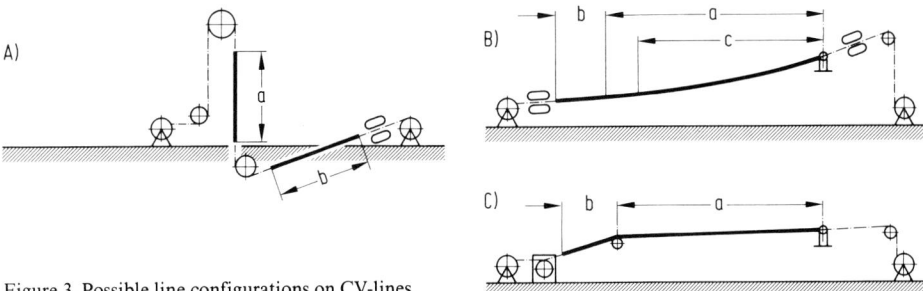

Figure 3. Possible line configurations on CV-lines. (Schematic).
A) vertical tube configuration, B) catenary tube configuration, C) inclined straight tube configuration
a vulcanization section, *b* cooling section, *c* catenary section

Steam crosslinking is often also known as the continuous vulcanization (or CV) process, largely on historical grounds related to rubber processing, and also to distinguish it from an earlier discontinuous crosslinking process. Correspondingly there are several variants, viz.:

- VCV = vertical type continuous vulcanization
- HCV = horizontal type continuous vulcanization
- CCV = catenary type continuous vulcanization.

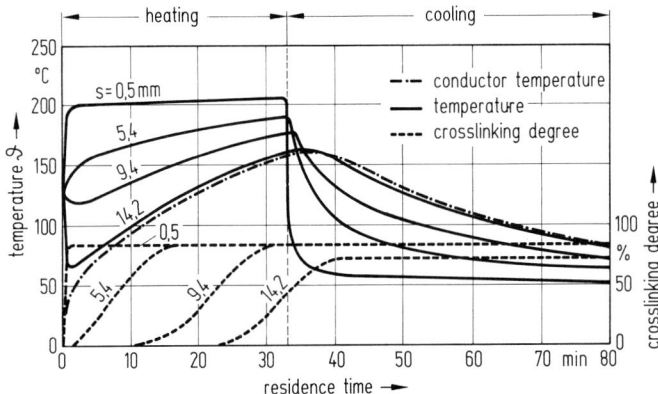

Figure 4. Dependence of temperature and degree of crosslinking on the dwell time at different layer depths, S, in a steam crosslinking tube, according to [72]. Product to be crosslinked: 60 kV cable of 500 mm² cross-section, conductor c/s 28.8 mm², total insulation thickness 14.7 mm, overall cable diameter 58.2 mm, ingoing mass-temperature 125 °C, heating-zone temperature 210 °C, mean cooling temperature 50 °C, final temperature of the conductor 85 °C

Vertical tube configuration

In this process the sheathed cable passes through a vertical crosslinking tube and a vertical cooling tube; in most such arrangements it is only at the junction with the cooling section that it passes around a deflector drum to adopt a horizontal attitude.
Vertical configuration in the crosslinking zones has various advantages; the deformation of the uncrosslinked insulator by gravity forces is negligible, even in the manufacture of extremely thick-walled cables. At the same time there is no danger, with well-centered cable guidance, of the molten uncrosslinked plastic material undergoing deformation against the tube walls, and thereby reducing the quality of the insulator surface. Control

and regulation devices are therefore not necessary for cable guidance, to avoid wall contact. Vertical orientation of the extrusion head enables several extruders to be placed behind each other at different levels on the production unit without problems. The length of the crosslinking tube is limited by the construction height available. Modern vertical lines are about 75 m high, and require appropriate tower-like buildings. According to [73] the hourly output (in kg/h) of a tube of given length increases with diameter, so that in producing thick insulation layers the crosslinking zone limitation with a vertical pipe configuration plays less of a role, and cable with an outer diameter of over 25 mm can be efficiently produced. For thin-walled cable covering, a line-length limitation means a limit on the production rate. For this reason such cables are usually crosslinked in inclined or catenary-form tubes.

Inclined straight configurations

With an inclined straight configuration no intractable building problems are encountered, and tube lengths of over 100 m can be installed without further consideration. An arrangement with the crosslinking zone above the normal production area and the placement of the extruder on a platform allows for plenty of free space on the ground floor.
Because of the accumulation of condensate water, a horizontal arrangement of the tube is not possible. According to [73], modern level-control systems for high-pressure cooling water need a 2.5 to 3% gradient to be able to work properly, so the tube end is located some 1.5 to 5 m below the level of the extruder platform, depending on the overall length. The slight inclination of the tube operates so that, depending on cable weight and tension, after leaving the die only a short stretch of cable hangs free without touching the tube wall. Later on, however, the cable touches the wall and slides along it, or is guided by rollers. Because of this contact, especially with thick insulation and stiff conductors, the surface quality is affected adversely. For this reason the inclined straight tube is used principally in the manufacture of lightweight cables.

Catenary tube configuration

With a catenary type of configuration the crosslinking zone is given the shape, more or less, of a free-hanging cable. Contactless movement of the cable through such a crosslinking zone puts severe demands on a speed regulation device. A free-hanging cable is only assured if the length of the floating section is kept constant to tight limits; this is achieved by careful control of the tension and speed of the haul-off caterpillar. The positioning of the cable in the crosslinking zone is achieved by continuous control of sag. In some cases, according to [28], at the start of the crosslinking zone the eccentricity of the cable is checked on a monitor screen. Sensors for determining the position of the cable in the tube, which touch the surface of the insulator, may only do so at points where the insulation is already crosslinked and is unaffected by contact.
The crosslinking tube is made from stainless steel tubes which are about 5 to 6 m long and 100 to 200 mm in diameter. A catenary configuration is only used in the crosslinking zone; in the cooling zone which directly follows it, the insulation surface is cooled rapidly and becomes insensitive to contact, so that a catenary form is not necessary in the cooling zone. The shape of the catenary taken up by the sagging cable depends on the weight of the cable and the applied tension. If the weight of the cable increases, the tension must be increased as well. Since the total weight of the cable under tension, which is at a relatively high temperature, is carried by the conductor alone, the strength of the conductor, particularly at joins, is a limiting factor in running the plant. Thus, coating of very thin conductors cannot be done on long catenary lines, and in general the shape of the catenary curve on a proposed new plant must be designed to accommodate the production program required.

Characteristics of steam crosslinking tubes

The steam inlet to the crosslinking tube must avoid direct jetting of the steam onto the cable. The steam pressure is normally controllable – perhaps up to 25 bar – so that the crosslinking zone can operate at different temperatures. Condensed water flows automatically to the lower part of the tube, where it mixes with the cooling water.

The heating zone runs direct into the cooling zone, and does not require mechanical separation. The cable is carried in the inclined pipe from the hot steam section straight into condensate or cooling water. The pressure of the cooling water on lines without mechanical separation from the heating and cooling zones is the same as the steam pressure, with the result that bubble formation by gaseous decomposition products is also suppressed in the cooling zone. The length of cooling zone needed depends on temperature, on the thickness of the insulation, on the cable diameter, on the heat capacity, and on the line speed; it must be long enough for the temperature of the emerging cable to be below 100 °C. Because there is no physical barrier between the heating and cooling zones, the length ratio of the two zones can be varied (for a given overall length) by control of water level, and set for each production program. A recirculating pump is necessary for controlling water temperature, to ensure that there is a constant low temperature in the cooling zone. Excess hot water is pumped off under continuous level-control and replaced by cold water.

To prevent steam from penetrating beneath the insulation, the line is run with the crosslinking tube full of water. The water level is reduced at a speed which prevents contact of the end of the conductor with the steam. During line shutdown, conversely, the water level is shifted at a speed which guarantees complete crosslinking and cooling of the cable end. Start-up and shutdown losses are minimized by this procedure.

The seals at the beginning and end of a tube length are special components. A telescopic tube against the die head serves as the steam-tube seal between heating zone and extruder. Typically, a short cylindrical pipe-fitting is mounted on the die head, and the end of the telescopic pipe, which is fitted with a seal, is pushed against it. If the effect of the steam pressure is insufficient to achieve a good seal between the telescopic pipe and the die head, mechanical assistance is necessary. The end of the cooling pipe requires a seal suitable for the dimensions of the cable; this is either made and fitted to cable dimensions, or manufactured from special flexible material which can accommodate a large range of cable diameters. Since on start-up all that is required to prevent large water losses is a seal to the metal conductor (normally the start-up cable) as it is drawn through, a small auxiliary seal is used to separate the water space from the outside. The main seal becomes operative when the coated conductor, with its larger diameter, reaches the end of the cooling zone. Once the main seal takes over, the auxiliary seal is opened. Since a single outlet seal does not totally stop the escape of water, particularly with the larger geometrical variations among the thicker cables, water escapes through the gap remaining into a collector housing, and either goes to waste or is pumped back to cool the crosslinked insulation in the cooling section of the pressure pipe.

15.6.1.2 Gas crosslinking lines (radiation methods)

One of the first designs, from Sumitomo Electric Industries, operates with radiant heat in a nitrogen-pressurized crosslinking tube to which a cooling section with pressurized water is connected. In order to avoid pressurized water getting into the insulation while it is still hot, the Nokia process operates without water even in the cooling section, and lets the cable cool thoroughly in the N_2 stream [74].

Another design, from *Hitachi* [28], dispenses with heating the polyethylene layer by radiation, and carries it out by using hot N_2 alone, the N_2 being heated in an external unit. In yet another process, developed by the same company, the N_2 is replaced by SF_6 (sulfur hexafluoride), which is also heated separately. Sulfur hexafluoride was chosen in preference to alternatives because this gas penetrates into the micropores of the insulator and is said to improve the electrical properties of the cable. The Sumitomo and Nokia processes have been well accepted as gas crosslinking lines.

Radiation crosslinking tubes with water cooling

Among possible tube arrangements available for steam crosslinking, the best known are the perpendicular, the inclined, and the catenary tube configurations. Most units on radiation crosslinking lines are identical with those on steam crosslinking lines.
On radiation crosslinking lines the crosslinking tube is made from stainless steel and is thermally insulated; it is fitted with several resistance-heating elements to allow the temperatures of individual sections to be independently adjusted. The temperature of the heating element reaches some 600 °C; the crosslinking tube itself must stand temperatures up to 500 °C and corresponding temperature variations. So as to make possible rapid cooling of the piece of cable remaining in the pipe towards the end of the run, the hot section of the crosslinking pipe is fitted with cooling facilities.
The complete crosslinking tube comprises the heating zone, the precooling zone (which is operated with cold N_2), and the cooling zone. There is no special separation between the three zones, which run one into the next. As shown in Figure 5, a stream of cold nitrogen is fed into the crosslinking tube just above the water surface, and brings about rapid cooling of the sheathing surface. During radiation crosslinking, the N_2 is heated to ca. 300 °C and leaves the radiation pipe at the front end. The gas flow through the crosslinking tube takes with it small amounts of moisture and volatile products, which are condensed during cooling of the hot N_2 from 300 °C to about 60 °C, and are removed from the tube. After passing through the precooling section, the covered cable slides into a water-cooling bath, in which it is cooled in the same fashion as with steam crosslinking, and finally goes through an exit seal to free atmosphere.

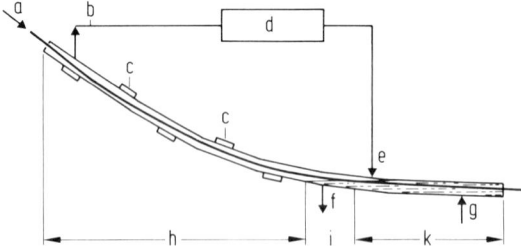

Figure 5. Schematic diagram of a radiation crosslinking tube
a cable supply, *b* hot N_2-take off, *c* heating elements, *d* N_2-cooling and separation of volatile matter, *e* supply of cold N_2, *f* outlet of cooling water, *g* inlet of cold water, *h* heating zone, *i* precooling zone, *h* cooling section

The maximum temperature of the tube in the hot zone lies between 400 and 500 °C. The heat transfer is accomplished principally by radiation, and only to a small extent by conduction from the nitrogen gas. A typical temperature profile of tube inner walls and of cable and conductor surfaces is shown in Figure 6. The heat output of each individual heating element is adjusted to ensure that an optimum temperature is reached on the surface of the cable. This must be as high as possible to reduce crosslinking time, consistent with avoiding decomposition of the polyethylene. For this reason heating control is extremely important in radiation tube crosslinking.

15.6 Implementation of crosslinking

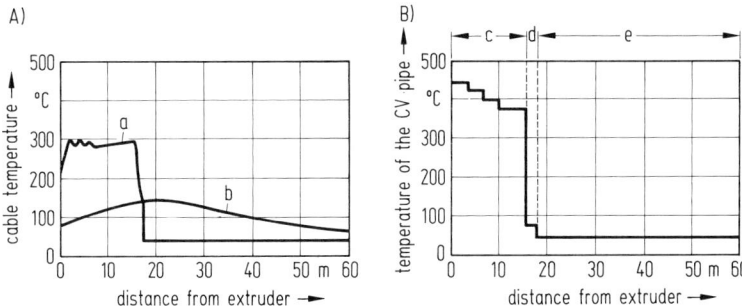

Figure 6. Temperature profile of a cable (A) and the crosslinking tube (B) during radiation crosslinking, according to [72].
a Temperature of the cable surface, b conductor surface temperature, c heating zone, d pre-cooling zone, e cooling zone

The temperature of the upper cooling region reaches a maximum of 80 °C and falls naturally to the temperature of the fresh water fed in at the supply point. A particular advantage of radiation heating is the decoupling of temperature and pressure: whilst with hot steam crosslinking the pressure is determined by the steam temperature, with radiation crosslinking the pressure can be freely chosen, and can reach 7 to 15 bar [74]. The very much higher temperatures used in radiation heating tubes shorten the time needed to reach a level of temperature at which the crosslinking reaction starts. Figure 7 explains this,

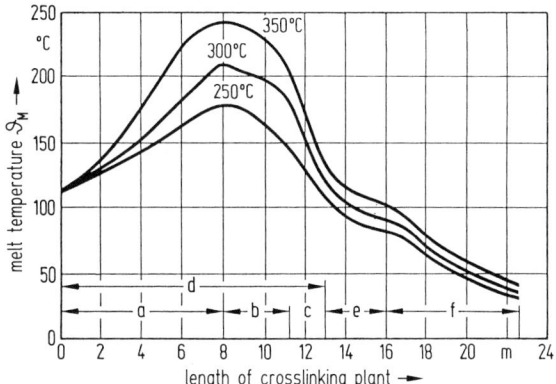

Figure 7. Examples of heating and cooling curves, measured for various wall temperatures on a laboratory radiation crosslinking line.
a heating section, b steam-cooling section, c water-cooling section, d pressurized-line unit, e bend zone, f water channel

using results obtained on a laboratory crosslinking line with three heating zones (7.2 kW, 10.8 kW, 3.6 kW) as an example. Whilst with a tube wall temperature of 250 °C a temperature of 175 °C on the inner wall of the insulation is only attained after 8 m of crosslinking zone, the same mass temperature is reached after 5 to 6 m and 4 m respectively for tube wall temperatures of 300 and 350 °C. The faster rate of heating also permits higher throughputs − requiring higher haul-off rates. If one considers a temperature of at least 175 °C at the coldest spot in the insulation to be adequate, in the laboratory line mentioned, with a wall temperature of 300 °C and 4 m/min. haul-off speed, this temperature is reached after 5 to 6 m in the crosslinking zone, or after 7 to 8 m at a speed of 10 m/min.

The gain in crosslinking efficiency that can be achieved with the higher temperatures in radiation crosslinking as opposed to hot-steam crosslinking, based on published information from plant manufacturers, is shown in Table 3.

Table 3. Ratio of line speeds for radiation crosslinking and steam crosslinking, after [72]

Thickness of insulation mm	Conductor cross-section mm^2		
	100	500	1000
20	1.39	1.36	1.60
10	1.55	1.46	1.40
5	1.66	1.48	1.31

Basis of comparison:
1. Saturated steam pressure reaches 20 bar (210 °C) in the steam tube.
2. The maximum temperature on the cable surface in the radiation crosslinking line is 300 °C.
3. The overall length of the crosslinking tube is the same for steam and radiation crosslinking.
4. The ratio of lengths of heating zones to cooling zones is chosen so that the speed of each process is the optimum.
5. The cables are produced with a semi-conductive outer sheathing.

15.6.1.3 Other gas crosslinking lines

In the Nokia process a line very similar to the Sumitomo process, with a radiation crosslinking tube, is used; it differs, however, in the method of cable cooling [74]. Whilst the Sumitomo process has only a short gas cooling section between the heating and cooling sections of the crosslinking zone, in the Nokia process the gas-cooling section is sufficiently extended not to require water cooling in the pressure tube. One motive for this arrangement is to exclude water completely from the crosslinking tube, and thus obtain higher cable quality; it does, however, require a longer cooling section.

A crosslinking process developed by Hitachi-Cable [28] dispenses with the use of heating elements on the crosslinking tube and uses externally heated nitrogen (or sulfur hexafluoride, SF_6) as the best transfer medium. In contrast to the processes mentioned above, where crosslinking is induced by radiation, in this case it takes place purely convectively. A further difference is that the nitrogen is not cooled, and volatile components cannot therefore be condensed out.

A further crosslinking method is one that uses an ultrahigh frequency inductive field (UHF field) [75 to 79]. Special peroxides which decompose in the UHF field are mixed with the polyethylene for this process. The principal difference from the methods described earlier lies in the design of the heating section, since energy input is by UHF field instead of thermally. An advantage of the process is the lower operating temperature, which is not much above the extrusion temperature. The cooling section can be made proportionately shorter and limited to a gas-cooling zone.

Whilst all the processes so far described are thickness-dependent — heating and cooling times incease with coating thickness — the UHF field penetrates the polyethylene easily and almost without loss, and acts on the peroxide simultaneously and uniformly throughout the whole layer thickness. Calculations show that to crosslink a 20 kV cable at a line speed of 10 m/min., a crosslinking zone length of only 20 m is all that is needed. So far this process has not been used commercially for crosslinking polyethylene; however, UHF crosslinking is a standard method for rubber crosslinking.

Since nitrogen pressurization also appears to be advisable in the UHF crosslinking of polyethylene, this process is also counted among the gas crosslinking methods.

15.6.1.4 Comparison of the economics of steam- and radiation crosslinking

If one compares the economics of steam crosslinking and crosslinking in a radiation tube, using current prices in Western Germany, the following results are obtained.
To produce 20 kV cable with an external surface area of 150 mm² at a speed of 10 m/min., using an extruder with an output of ca. 290 kg/h including the semiconductive layer, plants with the tabulated dimensions, investment costs, and space requirements are needed:

Table 4. Comparison of steam crosslinking and radiation crosslinking costs, according to [80]

		Steam crosslinking	Radiation crosslinking
Overall line length	m	130	110
Crosslinking zone	m	90	70
Cooling zone	m	40	40
Line	DM	2,000,000	2,000,000
Steam generator	DM	1,500,000	–
Specific crosslinking power	kW	1	1
Crosslinking energy at 290 kg/h	kW	290	290
Cooling costs:			
Cooling from: –		220 °C to 60 °C	250 °C to 60 °C
Cooling temperature interval (°C)		160	190
Cooling power requirement	kW	29	35
Total energy usage:			
Crosslinking	kW	290	290
Cooling	kW	29	35
Extrusion	kW	80	80
Total energy usage	kW	399	405
Building costs (Basis: 220 DM/m³):			
Length	m	140	120
Height	m	17	16
Width	m	10	10
Volume	m³	23,800	19,200
Cost of building	DM	5,200,000	4,200,000

The comparison between the steam crosslinking process and crosslinking in a radiation tube shows that steam crosslinking has higher plant investment costs and higher building costs, but is somewhat more economical in energy usage.

15.6.1.5 Crosslinking in liquid baths

In Sections 15.6.1.1 to 15.6.1.3 processes were described in which the principal construction elements of the overall crosslinking lines were the same or very similar; and differences in the details of crosslinking tube heating and of the cooling zone were given. All the processes use a gaseous heat carrier, heat radiation or UHF energy. Alongside these processes there are others which use fluids as heat carriers in the heating section.

a) Fujikura process (FZCV process) [28]

A high-boiling silicone oil is used as the heat carrier in this process – for cooling as well as heating. The heating and cooling sections are separated from one another by a seal. The separate heating-oil circuit is continuously reheated externally. By depressurizing and series recompression, separation of volatile components built up during the crosslinking process can be achieved in the heating-oil circuit. Because of the high viscosity of the cold silicone oil, there are few sealing problems at the end of the cooling section. In the heating zone the oil is raised to 250 °C at a pressure of some 20 bar. Pressure and temperature are separately controllable. Since the silicone oil has about the same specific weight as the cable to be crosslinked, the cable floats in the heating medium, thus counteracting deformation of the insulator.

b) Salt-bath crosslinking [28]

Salt-bath crosslinking, known from rubber technology, has also been used for cable crosslinking. This process, also known as the "Pressurized Liquid-Salt Continuous Vulcanizing System", uses as heating medium a non-aqueous eutectic mixture of:

53% KNO_3,
40% $NaNO_2$,
 7% $NaNO_3$.

The eutectic mixture melts at 145 to 150 °C and is stable up to 540 °C. The crosslinking temperatures reach a maximum of 300 °C. Since the heating medium is solid at room temperature but is water-soluble, there are no basic problems in cleaning the line.

c) Anaconda process (MDCV process) [28]

In a process developed by Anaconda Wire and Cable Co., and used by Dainichi-Nippon Cables Ltd., the covered cable is drawn through a straight, horizontal heater tube. A lubricant is used, to promote slippage and good heat transfer between the tube walls and the cable surface. The Anaconda process might thus be viewed as a fluid-bath crosslinking process; alternatively, it can be regarded as one in which the relatively short heating tube serves as an extension of the sheathing die.

After the hot uncrosslinked melt leaves the coating die, the cable is first taken through a porous tube, which is effectively an extension of the coating die. Inside this tube a lubricant is put onto the cable, which is then led into a heated tube, where the polyethylene crosslinks. Oil should not penetrate the insulation surface to an extent of more than 100 mg/cm^2 at 150 °C in 45 h [28]. The extrusion pressure is 70 to 90 bar, the heating tube length 12 m, and the heating temperature 270 to 350 °C.

Since the heating and cooling sections are separated from one another, the lubricant can be pumped off at the end of the heating section and returned via a heating unit to the beginning of the heating zone, and through its porous walls into the crosslinking tube. The difference from the Fujikura process is as follows: with the Anaconda process the crosslinking tube is externally heated electrically and the lubricant layer (only 0.2 to 1 mm thick) acts as heat carrier, whereas with the Fujikura process all the heat energy comes via the heat carrier from external heating devices. Since the internal diameter of the crosslinking tube is only slightly greater than the cable diameter, a change of cable size necessitates structural changes to the line and the provision of another crosslinking tube.

15.6.1.6 Special processes

A length reduction or total avoidance of an additional crosslinking device can be achieved if the heat energy necessary for crosslinking has already been put into the melt in the processing or shaping unit. Otherwise, thermal energy can be put into the plastic melt by heat conduction or by shear heating.

Various processes have been developed using this principle. In the Engel process the polymer is plasticated in a ram extruder with a shear gap and raised to a temperature at which the HDPE, containing ca. 0.4% peroxide, crosslinks. The process is more suitable for the production of crosslinked polyethylene tubing [81] than for that of crosslinked cable insulation. Also, the shear head from Krupp heats a polyethylene/peroxide mixture to the crosslinking temperature without a downstream heating section [82 to 84]. The shear head comprises a fixed cylinder with an internal rotor. The polyethylene melt is raised from 130 °C in this cylinder to around 200 °C, a level at which crosslinking proceeds rapidly.

In a further process [51], a temperature increase of some 50 °C is achieved in the fixed shear gap of the coating head; the wall temperature is about 200 °C and the shear gap 0.6 mm to 1.5 mm. The result is that either the crosslinking temperature is reached or at least the temperature difference is appreciably reduced.

Reduction of the heating time is also possible, in principle, by the use of inductive heating of the conductor during the crosslinking process, and this is described in [85].

15.6.2 Crosslinking by UV light

For photocrosslinking of polyethylene by UV light [60, 86], photoinitiators like benzophenone [87] or benzyldimethylketal [88] and peroxides are added to the polyethylene. The photoinitiators absorb UV light, which causes molecular breakdown and radical generation, by which crosslinking is initiated. High temperatures accelerate the crosslinking process, since the polyethylene reacts photochemically more quickly at higher temperatures, and the breakdown of the photoinitiator proceeds at a higher rate.

The process involves guiding the material to be crosslinked past a UV light source, e.g. a high-pressure discharge lamp, on a carrier or freely suspended. The effect of the UV light is to cause the polymer to heat up. The rate of crosslinking is proportional to the amount of light absorbed, and is governed by the Lambert-Beer law. The amount of crosslinking decreases the further the material is from the UV source. Since the UV light generates ozone from the air, an effective air extractor must be installed over the lamp.

UV light is suitable for the crosslinking of thin layers, especially films, but has apparently not been widely used.

15.6.3 Electron beam crosslinking

When polyethylene is irradiated with high-energy electrons (0.5 to 4 MeV), they penetrate into the plastic and interact with it. The depth of penetration of the electrons is limited, but as a rule of thumb, 2 MeV electrons penetrate about 1 cm into the polyethylene [27, 89]; a consideration here is that their energy decreases with increasing depth of penetration, resulting in a decrease in crosslinking density with depth. Because the radiation energy falls off with increasing depth of penetration, electron irradiation crosslinking is preferred for use on thin moldings.

The principal areas of use for electron-beam crosslinking are in the manufacture of shrink tubes and shrink films [27], cable insulation [90, 91], foams [92], and underfloor heating pipes.

The degree of crosslinking depends on the radiation dose and mass temperature. An important point is that although the electron crosslinking rate depends on temperature, crosslinking can be carried out even at very low temperatures. Consequently, this process can be used at any time, either directly after extrusion whilst the material is still in the plastic state, or independently of the extrusion process on cold extrudate or molded parts taken from intermediate storage. Account must certainly be taken of the fact that crosslinking of solidified partially crystalline polyethylene leads to different physical properties than with the crosslinking of a melt, since under solid conditions only the amorphous regions become crosslinked [4].

Electron-beam crosslinking has a lot of advantages over other crosslinking processes:

a) crosslinking can be carried out in the cold — i.e. in a state of optimum form stability,
b) electron-beam crosslinking is equally applicable to HDPE and LDPE,
c) even complicated extrusions like profiles can be crosslinked with an electron beam without geometrical change,
d) electron-beam crosslinking is not dependent on the use of crosslinking initiators, so extremely pure polyethylene can be crosslinked.

To raise the density of crosslinking and the effectiveness of the radiation, additives like polymerizable vinyl- and divinyl compounds can be mixed with the polymer [30, 32].
The product to be crosslinked is passed under the radiation source on a transporter. The radiation line is completely screened for protection against X-rays — for example by a concrete wall and lead sheeting. The entry and exit openings for the irradiated product must be built like canal locks, so that only greatly attenuated stray radiation can escape. Electron irradiation lines require effective air extraction equipment, to remove the ozone created by the radiation.
The realization of electron-beam crosslinking is extremely dependent on the geometry of the part to be crosslinked. The line consists of the irradiation unit, the screens, and a device for conveying the product to be crosslinked (Figure 8).

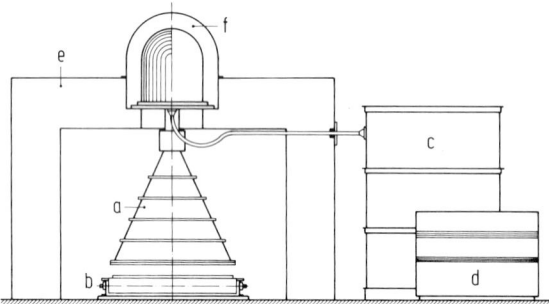

Figure 8. Arrangement of an electron accelerator in an irradiation line (schematic), according to [27].
a electron gun, *b* conveyor belt, *c* H. T. generator, *d* control desk, *e* concrete shield, *f* lead shield

According to [93] a radiation dose of 15 Mrad is required for adequate crosslinking of the polyethylene insulation on 1 kV cables; the amount of radiation required to give this level of crosslinking differs widely with cable diameter, as follows:

Conductor diameter mm	Insulation wall thickness mm	Weight of PE per km kg
1.39	0.7	4.2
17.4	1.6	83.0
25.2	2.0	156.0

The mass per unit area of polyethylene penetrated can be calculated from the following equation [93]:

$$P = \frac{\delta \cdot w}{2F} [q \cdot \cos \alpha + \sqrt{q^2 \cdot \cos^2 \alpha + 4(q+1)}] \quad (g/cm^2)$$

P   mass per unit of area penetrated
δ   density of the insulation material (g/cm³)
w   insulation wall thickness (cm)
F   correction factor for multiple irradiation
c   conductor diameter (cm)

$$q = \text{ratio:} \frac{\text{conductor diameter, } c}{\text{insulation thickness, } w}$$

$$\alpha = 180\left(\frac{n-1}{n}\right)$$

n	2 for irradiation from 2 sides	$\cos \alpha = 0$
n	3 for irradiation from 3 sides	$\cos \alpha = -0.5$
n	4 for irradiation from 4 sides	$\cos \alpha = -0.5\sqrt{2}$
n	∞ for rotating irradiation	$\cos \alpha = -1$

The particle energy needed is obtained from the following relationships:

$$P \leq 0.33 \text{ g/cm}^2 \rightarrow U(\text{MeV}) = 1.77\sqrt{P}$$
$$p \geq 0.33 \text{ g/cm}^2 \rightarrow U(\text{MeV}) = 3 \cdot P.$$

The disadvantages of electron beam crosslinking are to be found in the high handling costs and the large investment required: thus, according to [9], a total investment of some DM 7 million is required for an irradiation line with electron energies of between 0.5 and 1.5 MeV, a beam current of 50 mA max., a full beam-scan width of 2100 mm and a minimum scanned width of 1100 mm. In addition to using electron beams from accelerators for crosslinking, crosslinking lines using radio-isotope β-ray (electron) and γ-ray (X-ray) sources have been used for polyethylene [94].

15.6.4 Implementation of silane crosslinking

In the silane crosslinking process the extrudate is either given water/steam treatment immediately, or cooled and stored before treatment [95 to 100].
For thorough crosslinking of the extrudate, water must be distributed throughout the plastic in sufficient quantities to hydrolyse all the methoxy-silyl groups. Calculation of the temperature/time relationships is made difficult because the diffusion coefficient is influenced by the changing density of the polyethylene, by the Si-containing groups available, and by the changing crosslink density.

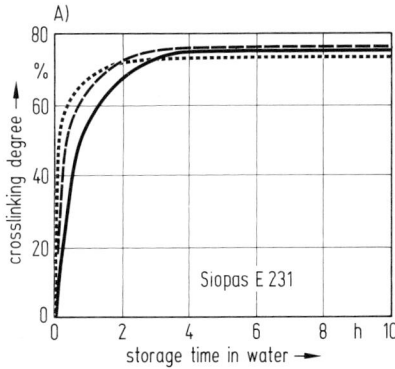
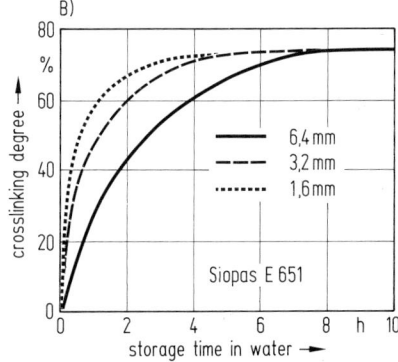

Figure 9. Dependence of the degree of crosslinking at different layer thicknesses for two different silane crosslinked products, according to [59]

Diffusion rate increases with increasing temperature, and is governed by the temperature dependence of the diffusion coefficient:

$$D = D_0 \, e^{-(E_D/RT)}.$$

Temperature variation, other than the variation in the product itself, is the one variable influencing the time needed for achieving a predetermined degree of crosslinking. The required times are experimentally determined for each crosslinkable compound; at the same time the time dependence of the degree of crosslinking at different depths for various temperatures is established (compare Figure 9).

In carrying out silane crosslinking, the extrudate is either drawn through hot-water baths – which only works for thin layers – or crosslinked discontinuously in kettle-like containers.

15.6.5 Crosslinking of polyethylene foam

Commercially available polyethylene foams are usually crosslinked [26, 101, 102]. In their manufacture a polyethylene compound containing a blowing agent and peroxide is extruded from a wide slot die and then heated (e.g. by IR radiation) in a temperature-conditioning zone. This first of all results in decomposition of the peroxide, and leads to crosslinking of the polyethylene. Decomposition of the blowing agent starts only after crosslinking has begun, with the result that the crosslinking reaction and the blowing reaction can overlap in time. Further details about this process can be found in Chapter 13.

15.7 Criteria for choosing different crosslinking processes

Which of the crosslinking processes is the most suitable for a given situation depends, first of all, largely on the geometry of the part.

Films can, in fact, be crosslinked even by the silane process, but for the most part electron beam crosslinking is the recommended procedure. Heat-activated peroxide crosslinking is not suitable for films. The peroxide process also presents problems in the manufacture of crosslinked tubes, since the only procedures that can be used satisfactorily here are those in which the molten uncrosslinked extrudate is supported internally. The most suitable for this purpose are the Engel- and silane crosslinking processes. The same considerations apply in the production of profiles, and in this case electron beam crosslinking is useful. For crosslinking cable insulation the dominant processes use heat-activated peroxides, though here too, electron beam and silane crosslinking processes can be used.

References for Chapter 15

For the convenience of the reader the English titles of all publications in languages other than English are shown in parentheses.

[1] *Kircher, K.:* Chemische Reaktionen bei der Kunststoffverarbeitung (Chemical Reactions in Plastics Processing). Carl Hanser Verlag, München, Wien, 1982.
[2] *Vieweg, R., Goerden, L.:* Polyester (Polyesters). Kunststoff-Handbuch, Bd. VIII. Carl Hanser Verlag, München, Wien, 1973.
[3] *Boenig, H. V.:* Unsaturated Polyesters: Structure and Properties. Elsevier Publishing Comp., Amsterdam, London, New York, 1964.
[4] *Köhnlein, E.:* Kunststoffe 65 (1975), pp. 583/588.

[5] *Di Giulio, E., Ballini, G.:* Kautschuk u. Gummi 15 (1962), pp. 6/13.
[6] *Ashikari, N., Kanemitsu, T., Kobayashi, S., Tajima, Y., Kawashima, T.:* Review of the Electrical Comm. Lab., Tokyo, 17 (1969), pp. 1161/1176.
[7] *Kunert, K. A.:* J. Polym. Scie., Polym. Chem. Ed. 20 (1982), pp. 2909/2914.
[8] *Chodak, I., Lazar, M.:* Angew. Makromol. Chem. 106 (1982), pp. 153/160.
[9] *Barton, J., Pavlinec, J.:* Plaste Kautsch. 15 (1968), p. 397.
[10] Gummi, Asbest, Kunstst. 36 (1983), pp. 356/357.
[11] *Jacobi, H. R., Flohr, J.:* Kunststofftechnik 9 (1970), pp. 79/85.
[12] *Tenney, K. S.:* SPE Journal 26 (1970), pp. 68/71.
[13] *Lucchesi, P.:* SPE Technical Paper 14 (1968), pp. 616/618.
[14] *Köhnlein, E.:* Kunststoffe 60 (1970), pp. 883/889.
[15] *Protassow, W. G., Baramboim, N. K.:* Plaste Kautsch. 23 (1976), pp. 267/270.
[16] *Mair, J.:* Kunststoffe 59 (1969), pp. 535/539.
[17] *Wanser, G., Wiznerowicz, F.:* Kunststoffe 67 (1977), pp. 275/279.
[18] *Kössler, L., Mair, H. J.:* Kunststoffe 62 (1972), pp. 359/362.
[19] *Domorazek, G., Falke, H.:* Elektrizitätswirtschaft 75 (1976), pp. 739/742.
[20] *Lücking, H. W., Geis, H.:* Elektrizitätswirtschaft 69 (1970), pp. 414/420.
[21] *Bergmann, J., Steckel, R.-D., Tretow, U.:* Elektrizitätswirtschaft 73 (1974), pp. 336/339.
[22] *Wanser, G., Wiznerowicz, F.:* Kunststoffe 69 (1979), pp. 105/108.
[23] *Hetzer, W., Kuhmann, H.:* ETZ-A 92 (1971), pp. 141/146.
[24] *Nowak, P., Saure, M.:* Kunststoffe 56 (1966), pp. 390/395.
[25] *Menges, G., Rheinfeld, D.:* Plastverarbeiter 23 (1972), pp. 324/331.
[26] *Kleiner, F.:* Kautsch. Gummi, Kunst. 28 (1975), pp. 149/154.
[27] *Antonetty, G.:* BBC-Nachrichten 51 (1969), pp. 274/280.
[28] *Otani, K.:* Jpn., Plast. Age 16 (1978) 2, pp. 21/31; 16 (1978) 3, pp. 33/42; 16 (1978) 4, pp. 19/29.
[29] *Robalewski, A., Wojciechowska, J.:* Int. Polym. Sci. and Technol. 6 (1979), pp. T35/T39.
[30] *Zyball, A.:* Kunststoffe 67 (1977), pp. 461/465.
[31] *Zyball, A.:* Kautsch. Gummi, Kunstst. 9 (1978), pp. 656/659.
[32] *Lee, D. W., Braun, D.:* Angew. Makromol. Chem. 68 (1978), pp. 199/211.
[33] *Chappas, W. J., Silverman, J.:* J. Polym. Sci., Polym. Lett. Ed. 17 (1979), pp. 5/14.
[34] *Kovacs, E.:* Plaste Kautsch. 8 (1977), pp. 560/563.
[35] *Schaudy, R.:* Kunststoffe 68 (1978), pp. 167/170.
[36] *Laurenson, P., Fanton, E., Roche, G., Lemaire, J.:* Europ. Polym. J. 17 (1981), pp. 989/997.
[37] *Hummel, K., Desilles, J.:* Kautsch. Gummi, Kunstst. Asbest 15 (1962), pp. 492/498.
[38] *Hulse, G. E., Kersting, R. J., Warfel, D. R.:* J. Polym. Sci., Polym. Chem. Ed. 19 (1981), pp. 655/667.
[39] *Wiedemann, R., Markert, H.:* Angew. Chem. 83 (1971), pp. 940.
[40] *Braun, D., Brendlein, W.:* Kunststofftechnik 9 (1970), pp. 275/278.
[41] *Carlson, B. C.:* SPE Journal 17 (1961), pp. 265/270.
[42] *Dorn, M.:* Gummi, Asbest, Kunstst. 11 (1982), pp. 608/611.
[43] *Kunert, K. A.:* J. Macromol. Sci.-Chem. A17 (1982), pp. 1469/1488.
[44] *Dannenberg, E. M., Jordan, M. E., Cole, H. M.:* J. Polym. Sci. 31 (1958), pp. 127/153.
[45] *Behr, E.:* Kunststoffe 53 (1963), pp. 502/509.
[46] *Ferch, H.:* Kunststoffe 52 (1962), pp. 326/331.
[47] *Kerrutt, G.:* Kautsch., Gummi, Kunstst. 4 (1971), pp. 384/387.
[48] *Otani, K.:* Japan Plastics May–June 1974, pp. 18/23.
[49] Company brochure Akzo Chemie bv, Amersfoort, NL, December 1973.
[50] *Rheinfeld, D.:* Dissertation, Inst. f. Kunststoffverarbeitung, RWTH Aachen, 1972.
[51] *Franzkoch, B.:* Dissertation, Inst. f. Kunststoffverarbeitung, RWTH Aachen, 1979.
[52] *Holzapfel, G.:* Study at the Inst. f. Kunststoffverarbeitung, RWTH Aachen, 1982.
[53] *Mirsch, D.:* Dissertation, Inst. f. Kunststoffverarbeitung, RWTH Aachen, 1982.
[54] *Rheinfeld, D.:* Plastverarbeiter 30 (1979), pp. 145/149; 30 (1979), pp. 211/216.
[55] *Stenzel, H.-D.:* Draht-Fachzeitschrift 26 (1975), pp. 600/603.
[56] *Boysen, R. L.:* IEEE, EHF Conference, Los Angeles, July 1970, pp. 926/933.
[57] *Wunsch, K., Kohlmann, G.:* Plaste Kautsch. 13 (1966), pp. 258/266.
[58] *Rheinfeld, D.:* Gummi, Asbest, Kunstst. 28 (1975), pp. 390/396, 402, 452.

[59] *Scott, H. G., Humphries, J. F.:* Modern Plastics 50 (1973), pp. 82/87.
[60] *Temin, S. C.:* JMS-Rev. Macromol. Chem. Phys. C22 (1982/83) 1, pp. 131/167.
[61] DE-OS 2.649.84 (OT 3, 1978), Kabel- u. Metallwerke Gutehoffnungshütte AG, Hannover, West Germany.
[62] *Voigt, U.:* Kautsch., Gummi, Kunstst. 29 (1976), pp. 17/24.
[63] *Hummel, K.:* Kautsch., Gummi, Kunstst. 30 (1977), pp. 807/811.
[64] *Janson, G.:* Kunststoffe 71 (1981), pp. 424/428.
[65] *Han, C. D., Rao, D.:* SPE Technical Paper 24 (1977), pp. 441/442.
[66] *Haas, K. U., Skewis, F. H.:* SPE Technical Paper 20 (1974), pp. 8/12.
[67] *Ito, K.:* Japan Plastics 8 (1974), S. 21/27; 9 (1975), pp. 14/17.
[68] *Gutfinger, C., Broyer, E., Tadmor, Z.:* Polym. Engin. & Sci. 15 (1975), pp. 381/385.
[69] *Patsch, R.:* Kunststoffe 65 (1975), pp. 89/91.
[70] *Trommer, J.:* Siemens-Zeitschrift 47 (1973), pp. 505/513.
[71] Company brochure, Troester, Hannover, West Germany.
[72] Company brochure, Maillefer SA, Ecublens-Lausanne, Switzerland: INF 6D 3.77 RBR 2000.
[73] Company brochure, Maillefer SA, Ecublens-Lausanne, Switzerland: 16D IBR 1.74 1000.
[74] *Aaltonen, M.:* Wire Journal 11 (1978), pp. 64/68.
[75] *Menges, G., Kircher, K., Franzkoch, B., Hoffacker, W.:* Kunststoffe 69 (1979), pp. 430/434.
[76] *Menges, G., Kircher, K., Franzkoch, B.:* Kunststoffe 70 (1980), pp. 45/48.
[77] DE-OS 2.611.349 (OT 22. 9. 1977), Institut f. Kunststoffverarbeitung, RWTH Aachen.
[78] DE-OS 2.803.252 (OT 2. 8. 1979), Institut f. Kunststoffverarbeitung, RWTH Aachen.
[79] DE-OS 2.904.086 (OT 7. 8. 1980), Institut f. Kunststoffverarbeitung, RWTH Aachen.
[80] Company brochure, Troester, Hannover, West Germany.
[81] *Engel, Th.:* Kunststoffe 57 (1967), p. 536.
[82] DE-PS 2.059.496, Krupp AG.
[83] *Schatz, O.:* Draht-Fachzeitschrift 12 (1974), pp. 684/686.
[84] *Anisic, L., Martin, P., Raube, M., Reckmann, H.:* Technische Mitteilung Krupp 36 (1978), pp. 19/24.
[85] *Menges, G., Hoffmanns, W., Junk, P.:* Kunststoffe 65 (1975), pp. 281/284.
[86] *Wilski, H.:* Angew. Chem. 71 (1959), pp. 612/618.
[87] *Oster, G., Oster, G. K., Moroson, H.:* J. Polym. Sci. 34 (1959), pp. 671.
[88] DE-OS 2.337.813, Ciba-Geigy.
[89] *Wilski, H.:* Polyolefine (Polyolefins). Kunststoff-Handbuch, Bd. IV, Carl Hanser Verlag, München, 1969.
[90] *Bernstein, B. S.:* SPE, East N Engl. Sect. Reg. Tech. Conf., Techn. Pap. 3 (1975), pp. 117/143.
[91] *Barlow, A., Biggs, J. W., Meeks, L. A.:* Radiat. Phys. Chem. 18 (1981), pp. 267/280.
[92] *Sagane, N., Harayama, H.:* Radiat. Phys. Chem. 18 (1981), pp. 99/108.
[93] *Bichel, H. D., Heublein, H., Weber, D.:* Research Report T81–124 of the Bundesministeriums für Forschung und Technologie, July 1981.
[94] *Terrisse, J.:* Soc. Plast. Engin., First EUROTEC, Gent, 14th/15th June 1979.
[95] *N. N.:* European Plastics News 11 (1976), pp. 31/32.
[96] *Weber, B. W.:* Bull. Schweizer Elektrotech. Ver. 69 (1978), pp. 62/66.
[97] DE-OS 2.350.876 (OT 18. 4. 1974), Dow Corning Ltd. London.
[98] DE-OS 2.255.116 (OT 17. 5. 1973), Dow Corning Ltd. London.
[99] DE-OS 2.646.080 (OT 20. 4. 1978), Kabel- u. Metallwerke Gutehoffnungshütte AG, Hannover, West Germany.
[100] DE-OS 3.210.192 (OT 14. 10. 1982), Nippon Oil Co., Ltd., Tokyo.
[101] *Hosoda, K., Shiina, N.:* Japan Plastics 2 (1968), pp. 28/34, 46.
[102] *Alfter, F. W.:* Plastverarbeiter 29 (1978), pp. 129/132.

16 Extrusion of elastomers

H. J. Gohlisch, W. May, F. Ramm, W. Rüger

16.1 Extruders

16.1.1 The task of the elastomer extruder

Extruders for processing elastomers normally provide most of the energy for heating the material by energy transfer through the screw from the drive unit. Thus, such extruders operate nearly autogenously.
These extruders are also usually equipped with external thermal control, which, because of the relatively low operating temperatures (130 °C maximum), is often carried out with pressurized water. This arrangement allows the extruder to be preheated to the required temperature before start-up.
In elastomer processing one makes a basic distinction between hot- and cold-feeding. And one also distinguishes between extrusion with and without degassing.

16.1.1.1 Hot-feeding

In hot-feeding, the rubber compound slaps, which have been stored, are reheated to 60 to 100 °C and plastified on one or more roll mills; this transforms the rubber mix into a fully flowable state by heat transfer, or input of mechanical energy. The extruder is continuously fed with strips, or discontinuously with "pigs" of this plastified mix. The extruder still has to move the material against the resistance of the forming die, during which process further plastification can occur.
As a consequence of the relatively modest demands made on the extruder, hot-feed machines have a small L/D (screw length to diameter) ratio, usually between four and six. Because of its short length, the hot-feed extruder is relatively sensitive to changes in the feed, which translate immediately into variations in the geometry of the extruded profile.
For producing blanks, discontinuous ram extruders are used, as well as hot-feed screw extruders.

16.1.1.2 Cold-feeding

In cold-feeding, the extruder is fed at room temperature with an unplasticated compound. The feed material can be strips or granules.
Since with cold-feed extrusion both heating and plastification functions must be carried out by the extruder, the length of the screw is greater than with hot-feed extruders. The L/D ratio varies with screw diameter, and lies between 10 and 18. For special tasks extruders with even greater L/D ratios have been produced.
In recent years cold-feed extruders have been adopted in almost all areas of rubber extrusion, since, by comparison with hot-feed operations, processing is significantly more economic as preheating roll mills are not required.

The functions a cold-feed extruder has to perform may be summarized as follows:

- feeding, conveying, and compacting,
- heating and plastication,
- mixing and homogenization,
- pressure build-up for extrusion.

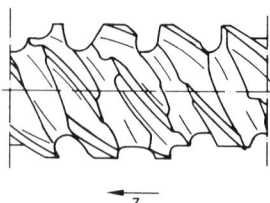

Figure 1. Schematic diagram – Troester mixing zone

In order to meet all these requirements optimally, thorough studies have been carried out in the last few years, in particular to improve the mixing and homogenization stages, and thus achieve higher output rates and lower extrusion temperatures.

As well as changing screw length, screw pitch, flight depth, and number of flights, two approaches have been adopted which have led to improvements in homogenizing action and mixing: screws with shearing zones and screws with flow-dividing mixing zones.

A complete separation of these two principles is not always possible, since even on screws with shearing zones some flow division can occur; likewise screws with flow-dividing mixing zones cause shear – even if only with a very small shear gradient (Figure 1).

16.1.1.2.1 Systems for plastication of elastomers

Screws with shearing zones

On screws with shearing zones the material is sheared to a varying extent by means of gaps and constrictions located in sections along the screw. With proper design of the barrel, some subdivision and displacement of the melt stream is ensured. The most important developments with screws having shearing zones are briefly explained in the sections following.

Extruders with EVK screws

The EVK screw works in a smooth barrel, as screws in most extruders do. The special features of this screw principle are the shearing zones and flow-dividing elements distributed along its length, which make possible intensive shearing associated with good mixing action, to an extent governed by the shear gap (Figure 2).

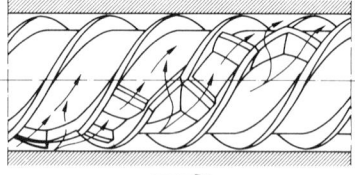

Figure 2. Schematic diagram – EVK

Though the particles in the compound must cover various path lengths, the number of locations of greatest shear loading is the same for each volume element. Because extruders with EVK screws have good mixing and homogenization action, they are used both as post-homogenization extruders after a kneader mixer and for continuous cold-feed extrusion with simultaneous forming.

Extruders with Maillefer screws (barrier screws)

Maillefer screws were originally developed for the extrusion of thermoplastics. Later, modification and optimization for the extrusion of elastomers was undertaken (Figure 3). The basic idea embodied in this principle is to separate packets of – as yet – unplastified cold material from the mass of heated and plastified particles. This separation is achieved by provision of a so-called "barrier flight" in addition to the conveying flights, with a greater pitch than the conveying flights. The barrier flight begins at the high-pressure side of a conveying flight and after a certain distance meets the low-pressure side of the conveying flight ahead of it. By this means, the channel volume in front of the barrier flight is continuously reduced and simultaneously the volume behind it increased. The flight depths before and after the barrier flight can vary, with the result that a further volume change can be caused.

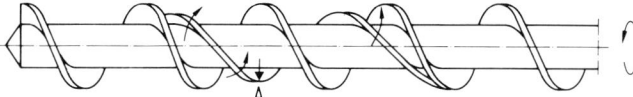

Figure 3. Maillefer screw

All the material passing into the extruder is forced by the arrangement of the barrier flight to pass through the gap between it and the barrel wall. This causes shear of the material, whose severity depends on the gap width.

Extruders with a Plastiscrew

In this case too, in addition to the conveying flights, a barrier flight is provided which has a larger pitch and a smaller outside diameter than the conveying flights. The pitch of the barrier flight can be chosen so that after several revolutions it meets a conveying flight again. By this arrangement all the particles of material – as with the Maillefer screw – are forced to transfer from the passage before the barrier flight through the gap between it and the cylinder wall into the passage on the other side of the barrier flight. The intensity of shearing of all the material particles depends upon the width and arrangement of the gap.

Extruders with Troester shearing screws

With shear screws there is a shearing section well separated from the feed zone, in which, in contrast to the shearing sections so far described, the shear gaps are axially aligned. Because of the relatively short length (3 D), four axially aligned shear gaps are provided to ensure adequate shearing of all the material, between which lie conveying flights with a very large lead (Figure 4 A and 4 B). The configuration is so arranged that the beginning of a shear gap is linked with a conveying screw land, and the end of this conveying flight is in contact with the next shear land along the screw. All particles of the material are thus forced to pass through the shear flight at least once.

Cavity mixer

Another development which (as with the Transfermix) involves the barrel as an active component influencing mixing, is the so-called "Cavity mixer" (Figure 5). With this device, an intensive exchange of volume elements within the compound takes place, and both shearing action and stream division come about. However, large-scale practical testing of the device has not yet been carried out. A disadvantage of the system is that it is not self-cleaning.

Screws with shearing zones are used for cold-feed extrusion, and are especially well suited for purposes of special applications – for example in post-homogenization. Such screws are

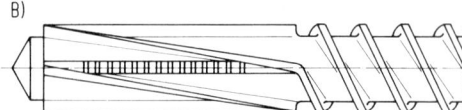

Figure 4 B). Schematic diagram – Shear section of screw

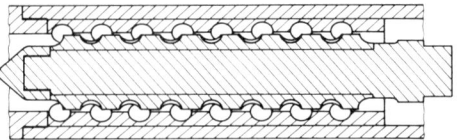

Figure 4 A). Screw with Troester shear section

Figure 5. Schematic diagram – Cavity mixer [46]

also still used in continuous extrusion, but the area of use is largely confined to smaller extruders. The reason for this is that even with narrow shear gaps, parts of the compound with different properties can pass through them relatively unchanged. This applies especially in processing elastomers. An increased number of shear gaps and a reduction of the gap width could possibly help here; however, there is then the danger of scorching by local overheating when the usual vulcanizable compounds are processed.

Screws with flow-dividing elements

On screws with flow-dividing mixing zones (Figure 6), the requirement for a mixing and dividing action with the smallest possible shear gradient has been elaborated. To achieve this, the cold core material enclosed within the screw channel is broken up to create new surfaces, so that, even with a relatively short mixing zone of this kind, there is a mechanically and thermally homogeneous compounding action. An intermediate step was the development of a screw with counter-rotating land system; this, however, has not been generally adopted, because of poor self-cleaning and the high extrudate temperature it produces. A development of this principle is found in the Troester mixing zone (Figure 1), which essentially consists of a wide conveying-screw land which is intersected many times by lands of greater pitch. This type of mixing device has previously shown that it can also perform extremely well in the feed area, and it is now used on many extruder screws of all diameters.

Figure 6. Screw with flow-dividing elements

← conveying direction

Transfermix extruder

The basic principle of the Transfermix extruder is to involve the barrel actively in the mixing process. To achieve this, the barrel is not machined smooth but, as with the screws, is provided with flights of a specific pitch (Figure 7 A). The hand of the flights in the barrel is opposite to that of the screw flights. Consequently, since the depths of the screw- and barrel flights increase and decrease along the length in a wavelike fashion, with a phase shift of 90°, the material experiences movement in the radial direction as well as being transported lengthwise (Figure 7 B). The result is that, as the screw turns, the compound is sheared in the gap between the screw- and barrel lands. The break-up of the material lying in the screw flight works in the interest of thorough mixing, and intensive subdivision of the different regions occurs as well.

16.1 Extruders

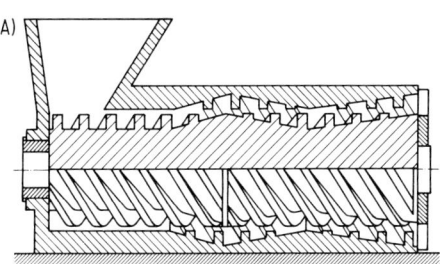

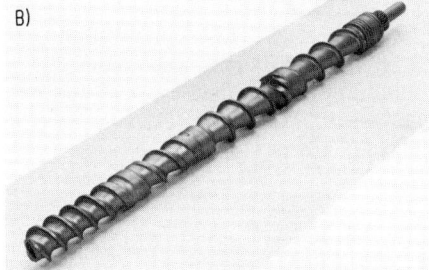

Figure 7 A). Schematic diagram of the Transfermix

Figure 7 B). Transfermix screw

A development of the normal two-flighted machine is the so-called Multicut Transfermix, which has led to an improvement in the mixing and homogenization actions and raised output, by the use of several flights in screw and barrel. The Transfermix extruder is often used as a post-homogenization extruder after the kneader, as well as for direct extrusion.

Extruders with pin barrels

Subsequent development of extruders with stream-dividing elements led to pin extruders, in which pins (penetrating the interior surface of the barrel) project into the screw channel almost to the screw core (Figure 8). The requirement for mixing with a low shear gradient is largely met by this approach. Very good mixing- and dispersion action is achieved by continual subdivision of the melt flow, and the material is not thermally overloaded by excessive mechanical stress. The material is subjected to repeated subdivision in the region around the pins and, as a result of the new surfaces created, to intensive material and heat exchange. The difference in mixing action between pin-barrel and conventional extruders can be demonstrated very clearly with specimen sections taken from screw channels after feeding the two extruder types with different-colored strips of material.

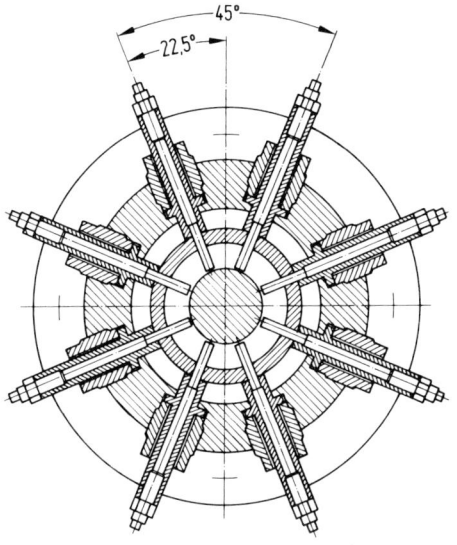

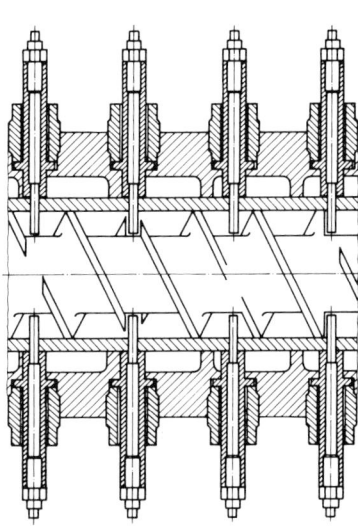

Figure 8. Schematic diagram of pin extruder

The section shown in Figure 9 was taken perpendicular to the screw flights some 3 D after the feed section. The section in Figure 10 was taken in the circumferential direction at the same radial distance, also about 3 D after the feed. With conventional extrusion, the build-up of a cold unmixed core can be clearly recognized near the middle of the channel, whilst with a pin-barrel extruder good mixing can be seen in the finely distributed bright and dark areas.

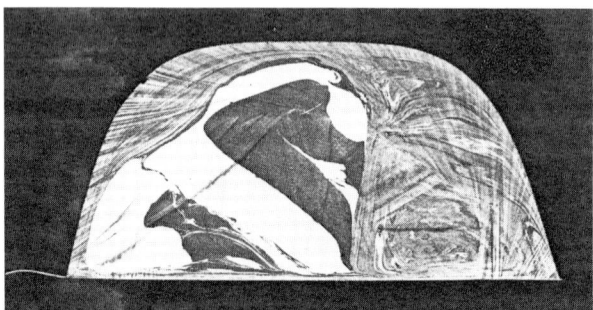

Figure 9. Channel cross-section of a simple conveying screw

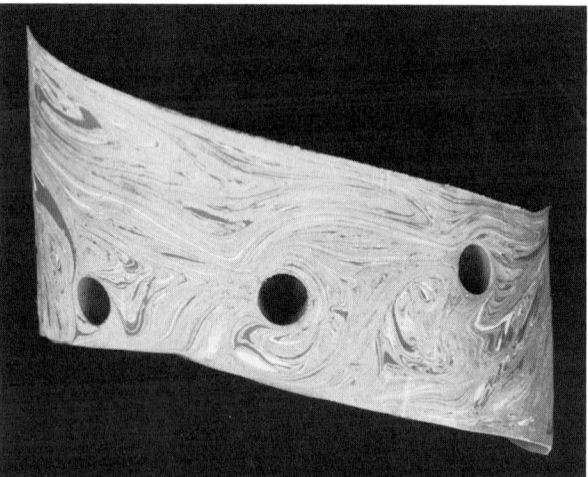

Figure 10. Flow division by the pins of the cross-flow mixing barrel

Since with this method of mixing, flow division across the conveying direction is involved, the pin-barrel extruder is also known as the cross-flow mixing or OSM extruder.
In Figure 11 the output rates of currently available pin-barrel extruders are shown as a function of extruder diameter. These rates of output are, throughout, comparable with those of hot-feed machines. Pin-barrel extruders have therefore had increasing success in recent years.
Initially, pin-barrel extruders were limited to those of large diameter ($D \geq 150$ mm), but nowadays smaller extruders with pin barrels and matching screws are also manufactured.
For especially difficult plastication and mixing tasks, the pin-barrel principle is combined with shearing and flow-dividing elements. Thus, extruders are known which have a pin-barrel in the first section and use a Maillefer screw in the second, or which feature a mixing zone in the feed section.

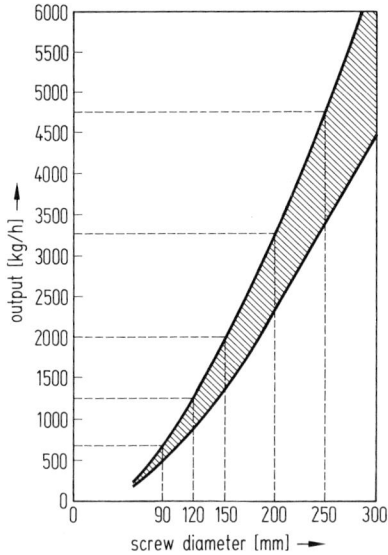

Figure 11. Discharge region of a QSM extruder

16.1.1.3 Vacuum extruders

For continuous, pressureless vulcanization of profiles and tube, an extruder is needed which removes all volatile components from the mix during the extrusion process. This has been achieved by, in effect, combining two individual machines one after the other into a single machine, in such a way that the first functions as a cold-feed extruder which passes the material through a vacuum zone to a second extruder, which then takes care of degassing and pumping the melt against the resistance of the die.

The efficiency of the vacuum zone (low pressure) is determined by the level of the vacuum, the dwell time, and the free surface of the hot mass. For best effect, therefore, extrudate is pushed from the first section of the extruder through a grooved barrier zone (Figure 12 A and 12 B), which causes the single stream to be subdivided into a large number of strands of small cross-section and correspondingly large surface area. The screw channel is relatively deep in the vacuum zone to allow depressurization of the melt.

Various ideas for achieving the highest possible degree of degassing have been examined and introduced. In deciding on the characteristics of a screw for a vacuum extruder, the series operation of a "cold-feed extruder" section and a "hot-feed extruder" section can be considered. Here, the throughput of material (against the resistance of the barrier) in the first section must be precisely matched to that of the second extruder section. If the two halves are not matched to one another, overflow in the vacuum zone or feed starvation of the output section can occur.

In addition to precise specification of the screw geometry to suit the range of tasks the extruder has to perform, precise temperature control of both screw and barrel is necessary to ensure satisfactory operation of this type of extruder. Because there are virtually two extruders connected to make a single unit, it would be desirable to be able to set temperatures independently of each other not only in the different barrel sections, but also in the various screw sections. Since this is usually not possible, a temperature is chosen which provides satisfactory operating conditions both for the "cold-feed section" (feed zone) and for the "hot-feed section" (discharge zone).

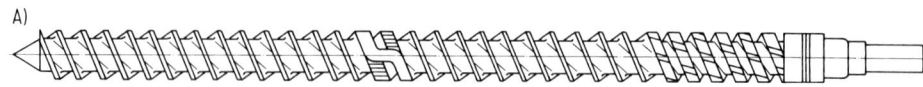

Figure 12 A). Schematic diagram of a vacuum screw

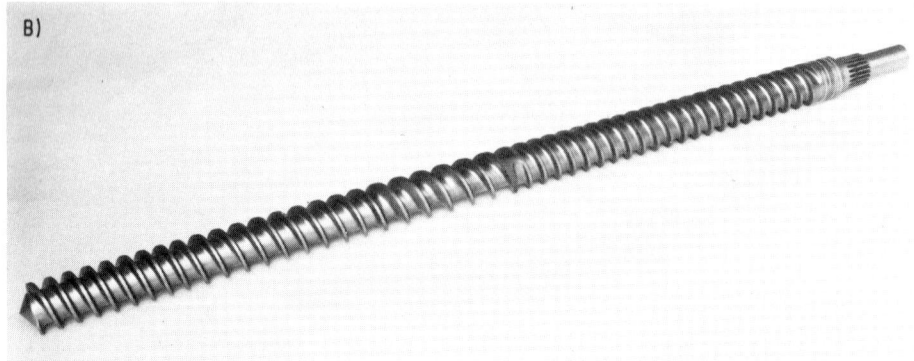

Figure 12 B). Vacuum screw. (Photo: Troester, Hannover, West Germany)

16.1.2 Extruder design

Rubber-processing extruders built for the jobs described in Section 1 are almost exclusively single-screw units (Figure 13). One distinguishes between the various machine types and series by the features of the extruder, and especially by the screws used — as detailed in 16.1.1.2.

Leading extruder manufacturers have developed modular systems for individual basic types and series, some of which permit different series to be equipped with the same basic elements. Such modular systems have many advantages: For the equipment manufacturer it results in more efficient manufacture. For the equipment user it means, for example, reduced down-time thanks to interchangeable replacement parts as well as greater flexibility in adjustment to changes in the production program (Figure 14).

16.1.2.1 Drive units

When considering the many possible uses, the question of the actual drive performance needed on extruders is most important, and should always be thought about thoroughly in relation to the likely working situations. Thyristor-controlled DC drives have been successful on modern extruders. Here and there three-phase shunt-wound- or variable frequency AC drives are used, as well as hydraulic motors and mechanically variable speed gearboxes.

16.1.2.2 Drive transmission

The extruder drive motors described in 16.1.2.1 can be coupled to the extruder through a variety of drive transfer elements.

In Europe, the most usual form of power transmission to the extruder gearbox is by high-performance v-belt or toothed belt. Exceptions to this are on big, cold-feed extruders

16.1 Extruders

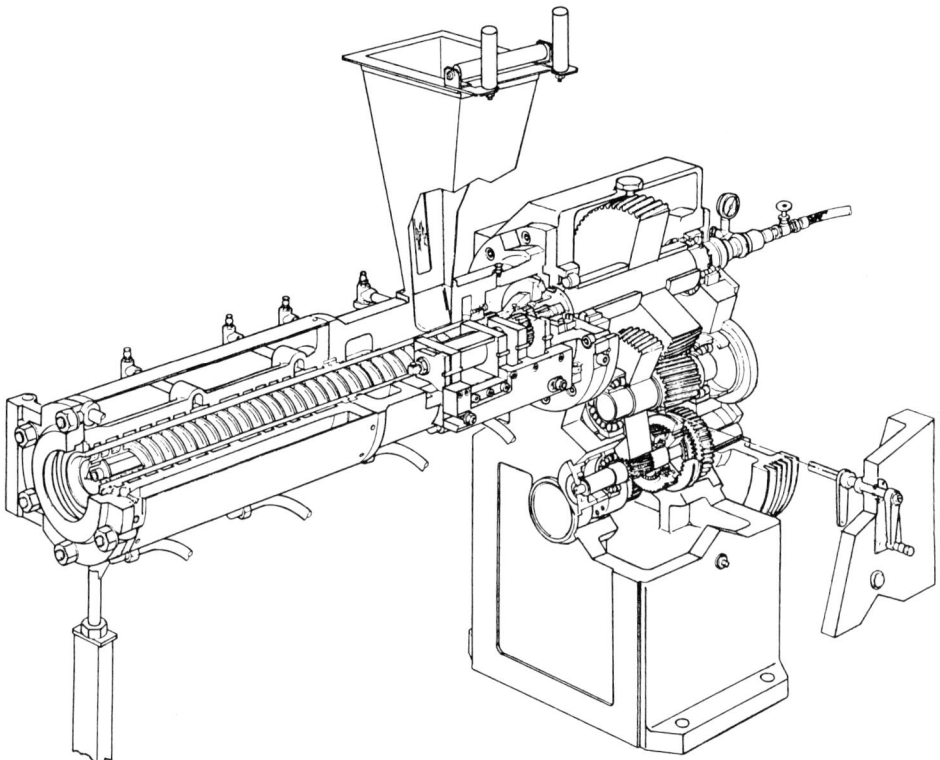

Figure 13. Cross-section of a rubber extruder. (Photo: Troester, Hannover, West Germany)

Figure 14. Cold-fed tandem extruders of Types QSM and GS. (Photo: Troester, Hannover, West Germany)

with screw diameters from about 200 mm upward and hydrostatic drives, to which the motor can be connected directly.
The above-mentioned belt drives are very flexible, quiet elements, which provide a variety of possibilities for positioning the motor and for adjustments or speed changes that may become necessary later. However, belt designs are not available to cover all of today's power ranges.
Alongside the recognized advantages of v-belts, there are some disadvantages to be noticed: in particular, belt sets of precisely the same length must always be used, and the belt tension must be correctly set and regularly readjusted.
Several extruder manufacturers favor a motor connected direct to the gearbox; this certainly results in compact construction, but allows little scope for possible speed adjustment later. Furthermore, the high speed of the gearbox input shaft produces higher noise levels.
All drives with a direct-coupled motor operate with an elastic coupling. In special cases clutches are also customary for short intermittent cycles with both direct and belt drives – for example in extrusion of short lengths on tube extrusion lines.

16.1.2.3 Gearboxes

All rubber extruders require a mechanical gearbox, unless they are direct-driven by a slow running hydraulic motor.
The normal form of construction that has been adopted is one that uses the machine frame as the gearbox housing. With this design the required reducing gears (usually ground and polished helical gears) and the oil storage tank are accommodated within the machine frame
Other extruder designs use gearboxes of proprietary construction, e.g. modular systems for spur gear-, bevel gear-, or worm gearboxes. Such constructions require the machine frame for each basic extruder to be specially designed, to accommodate the gearbox. Vertically mounted extruders are often fitted with such gearboxes. Hollow shafts are needed on gearboxes for feeding a heat-transfer fluid to the screw and the screw mountings.
Because of the much higher quality demands now made on gearboxes, especially for high specific performance and output, gearboxes of the highest quality are invariably necessary. Modern gearbox designs achieve exceptionally high efficiencies, and have considerable influence on the overall efficiency of the extruder. Extruder gearboxes are usually based on spur gears, in a two- or three-step design. Gearboxes for general purpose extruders are frequently units which allow changes in the speed and drive torque to be made in two or more steps.
On small machines with some 10 to 15 kW of drive power, gearbox lubrication results from immersion in the oil bath. Gearboxes on larger extruders are almost always provided with separately driven oil circulating pumps, oil filters and control devices. Automatic control of oil pressure is "state-of-the-art" today. Flow-control devices are also used.
The proper start-up procedure for extruders involves switching on the oil pump first, and then the main drive motor, after the "pump working" signal comes up. Conversely, if there is a failure of the lubrication system, the extruder is stopped, after a brief warning signal, to avoid gearbox damage.
The lifetime of a gearbox, which is taken as the basis for the design of the extruder drive, should be at least 35,000 to 40,000 operating hours, during which time short periods of overload, at up to 2.5 times the nominal torque, are permitted. Gearbox damage to today's precision units, given correct design and fault-free lubrication, is effectively unknown over decades of use.

16.1.2.4 Thrust bearings

The screw thrust bearing is an especially important extruder component. Some extruder designs are such that the thrust bearing, which supports the drive shaft axially, is built into the gearbox. This method of construction has the certain advantage of particularly compact bearings housed in the gearbox, but it does significantly reduce the flexibility that is really needed in an extruder manufacturer's program to cater for very different extrusion product requirements. Therefore it is advantageous to place the thrust bearing outside the gearbox.
Axial roller bearings are usually employed for the thrust bearing, with self-aligning taper rollers being preferred because of their self-centering behavior.
Design calculations for the thrust bearing are based on operating service life, and involve the maximum pressures to be expected on the extruder screw when running the envisaged production program.

16.1.2.5 Feed housing

Feed rolls and spiral undercuts in the feed housing are used to achieve trouble-free feeding of the compound into the extruder.
The feed roll, which forms a feed-roll pair with the conveying screw, is now the most usual material feeding aid on all cold-feed rubber extruders. The feed roll is designed so that its cylindrical surface projects to some extent into the feed opening, and forms a defined gap between itself and the rotating screw flights. For better feeding, the speed of the feed rolls is somewhat higher than the circumferential speed of the screw flights. The feed roll is usually driven by a pair of gears from the main shaft of the extruder. There are special demands on the feed rolls with regard to sealing against escape of the melt, on the bearing seals, on lubrication, and on ease and rapidity of cleaning. A hinged section that provides for sideways movement of the rolls is an advantage for cleaning, since it makes handling significantly easier; it also allows better access for replacing components.
With regard to bearings, there are various methods in use, and antifriction bearings have now completely replaced plain bearings. Extruders are largely fitted with coolable feed rolls for improved feeding action, since this arrangement increases the coefficient of friction between the feed-roll surfaces and the mix.
A rotating bank forms between the screw and the counter-rotating feed roll on extruders that are working well. A significant amount of plasticating action is provided in the feed zone, and conditions for uniform filling of the screw with little air entrainment are achieved.
Large hot-feed extruders, which are mostly located beneath the internal mixer, are fitted with feed rams to deal with lumpy compounds. If the mix is very sticky, a second, opposing ram is an advantage. The feed rams operate pneumatically or hydraulically, and — if there are two rams — alternately. In order to accept a larger feed hopper, such extruders are built with an extended feed zone. Very deep-cut screws or screws with larger than usual diameters, which are then tapered down towards the end, are also typical for such machines.

16.1.2.6 Barrel liners

Removable barrel liners provide a rapid and inexpensive means for repair when worn out after long periods of production. So-called wet liners, furthermore, provide a good basis for optimum temperature control of the extruder barrel.
Usually the liners have smooth bores. In the feed area (feed housing) they can be provided with auxiliary feed pockets behind the main one, or with spiral undercuts of large pitch (single or multiple start; 2 to 4D long).

All the measures mentioned serve, in the first place, to produce more wall contact, and so raise the conveying rate of the screw.

Liners are generally made from nitriding steel, and have nitrided surfaces. In exceptional cases exchangeable liners are not used in the feed zone. For the processing of particularly abrasive compounds, extremely wear-resistant, special alloy (centrifugally cast) bimetallic sleeves about 2 to 3 mm thick, are used. Their expected service lives vary enormously, and can only be estimated in relation to defined conditions of use. Producers of centrifugally cast barrels offer two essentially different groups of alloys:

a) alloys with a hardness of approx. 50 to 55 HRc, which usually exhibit improved corrosion-resistance, and

b) alloys of greater hardness − approx. 60 to 65 HRc, which can only be worked by grinding or electro-discharge erosion.

16.1.2.7 Extruder barrels

The length and the design detail of extruder barrels are arranged to suit specific processing requirements.

The first characteristic is the effective barrel length. This is 4 to 6 D with normal hot-feed extruders, 10 to 18 D − depending on diameter − for cold-feed extruders, and 14 to 24 D for degassing extruders.

From the design viewpoint one particularly distinguishes three different kinds of extruder barrel:

1. barrels with wet liners,
2. jacketed barrels,
3. peripherally drilled barrels.

The different kinds of extruder require very different designs in some aspects, but optimum design with regard to good heat transfer properties − that is to say, large contact surfaces and good flow characteristics − is involved in all cases.

Depending on the length of the extruder barrel, it is made in one piece or in several sections which are bolted together. The overall length is divided into thermal control zones, thus permitting individual control of the different sections.

The extruder barrel usually finishes at the end of the screw and carries the elements required for fixing the many different kinds of head to the end flange, which may be done by:

a) simply bolting on by means of hex head or socket screws,
b) hinged screws and nuts,
c) central screw thread,
d) two, three, or multi-element fixing clamp,
e) bayonet lock,
f) wedge clamping,
g) above tail clamps.

Fixing methods (a) to (f) can be readily operated manually on small extruders, but with (a), (b) and (f) there is a danger that uneven tensioning will occur around the circumference, and the possibility that poor sealing will result from improper operation. With (d) and (e), mechanical as well as hydraulic or pneumatic methods may be used satisfactorily, e.g. toggles or gear transmission.

Pin-barrel extruders have achieved special importance in the past few years. The pins in the barrel are outstanding aids to mixing or plastification. The number used depends on the application, as also does the number of planes (4 to 12) in which the pins are arranged, and the length of the barrel.

In each pin plane, depending on extruder size, there are 6 to 10 radial pins, which project across the barrel into the conveying section of the screw. The arrangement, depth of engagement, and the separation of the pins from each other are governed by the specific circumstances.

16.1.2.8 Extruder screws

The heart of every extruder is the screw. This has to be provided with widely differing screw geometries, mixing zones, shear zones, in a whole variety of ways for processing an extremely large number of compound variants. In addition to the geometrical make-up of extrusion screws for particular processing requirements, there are two key points: good wear-resistance and, increasingly on modern high-output cold-feed extruders (especially with smaller sizes), high torsional strength, to ensure that the greatly increased torque can be safely transferred. To achieve high output rates, rubber processing extruders have exceptionally deep conveying channels which, naturally enough, cause the torsional strength of the screw to be reduced. The result of this development is that special high-strength steels have to be used to meet the requirements. The screws require surface treatments like nitriding, stelliting, or surface hardening, to produce highly wear-resistant flights. The special properties of these steels need to be noted, and must also be taken into account when repair is carried out by surface coating or other methods.

High quality of surface on the screw flights in the conveying channel section is very important for achieving optimum operating behavior.

16.1.2.9 Temperature control of extruders

For temperature control, extruders are divided into separate zones. Screws should always be thermally controllable. The feed housing is normally only cooled and is generally set by a manually adjustable valve on the inflow. The feed rolls on large extruders should be coolable. The barrel is divided into zones, the number depends on the length of the extruder conveying section, and each one can be separately controlled to make it possible to arrive at the most favorable setting for the process.

Except in special cases, extrusion heads should always be properly controlled thermally, to impart optimum characteristics to this zone. For this purpose the die heads are provided with one or more (depending on sizes and use) control zones. In addition to temperature control by means of pressurized water or by oil, additional electrical-resistance heating is usually provided for the die ring. Electrically heated high-pressure water devices are usually used for thermal control of rubber extruders – oil-based units are less usual. In the cooling process one distinguishes between directly cooled and indirectly (heat exchanger) cooled units. The basic features of these control devices are as follows:

a) Simple continuous-flow water heaters with circulating pumps and direct mixing with cold water by means of thermostatically controlled solenoid valves.
b) Units like (a), which have closed-loop control based on automatic temperature regulators. Temperature sensors can be placed either close to the melt-flow region or in the circulating water stream. Generally the first arrangement is preferred, so as to approach the actual temperature as closely as possible, since a definite recording of the prevailing melt temperature is not really possible.
c) Units like (a), but with an additional heat exchanger for indirect cooling, using low-quality water.
d) Units like (b), but with the heat exchangers used in (c).

Units of types (a) to (d) are those most frequently employed; these are certainly distinct in design, and depend on the size of extruder and the task to be done, with regard to heating and cooling requirement, maximum temperature, and so on. Usually a temperature range of up to 140 °C is adequate for operation of rubber extruders; and the most precise temperature control with least possible control variation is sought.

As well as the foregoing equipment for use at ca. 140 °C, so-called high-temperature units for processing special kinds of rubber up to about 180 °C are occasionally used. In such cases oil-based devices are used more often than conventional pressurized-water controllers. With low-quality or hard cooling water, in addition to the indirect systems described, additional equipment is employed which comprises separate return coolers, circulating pumps and a storage tank. These units capture the hot water in the storage tank and cool it under thermostatic control to the cooling water temperature, by circulating it through heat exchangers. The cooled water can again be used in the system as required. Thus the used but clean water always remains in closed circuit, and losses are automatically monitored and compensated. Only clean, pretreated water should be used for make-up, to avoid gradually worsening the water quality in the closed circuit.

16.1.3 Extrusion heads

All head variants are governed by the basic requirements listed below:

a) shortest possible flow channels,
b) avoidance of dead spots,
c) highest surface finish quality (possible chrome plating) of the flow channel surfaces,
d) lowest possible deflections,
e) avoidance of larger cross-section changes,
f) uniform material output, particularly with flat-sheet dies and crossheads,
g) use of simply and easily exchangeable extrusion dies,
h) simple handling (with large heads, possibly using manipulators) and rapid cleaning capability when changing the formulation, to achieve short down times,
i) simple and reliable fixing (perfect seal against the pressurized melt),
j) good temperature control.

16.1.3.1 Pork-chop heads

Pork-chop heads serve to reduce the strand emerging from a preheating extruder into slices of moderate thickness. After cutting they must not stick together, and are guided to secondary processing machines like mills, calenders, or hot-feed extruders. Three essentially different designs are distinguished: those with –

a) fixed knives, which separate the strands against the rotating ends of the screw flights,
b) rotating knives, which also cut the extrudate against the screw flights or, alternatively, against a cross-shaped material guide running along the length of the melt,
c) a knife moved across the die exit.

16.1.3.2 Pelletizer heads

With pelletizers, rubber pellets are produced which, after cooling and drying, are used for subsequent processing on other, smaller extruders. Pelletizing- or granulating heads are the same in their basic design as strainer heads, and are simply supplemented by a variable-speed cutting device after the die plate. Straining is often combined with pelletizing, if this is necessary to achieve purity of the extrudate. The type of die plate, the cutting speed and the number of cutters can be adjusted to suit the output rate and the pellet size by exchange of die plate and altering the speed of the knives.

16.1.3.3 Strainer heads (Figure 15)

The simplest form of strainer head has a die plate and associated screen pack mounted in a simple flange immediately before the end of the extruder barrel. Normally, the screen surface area is made as large as possible to achieve long runs between screen pack changes. Increasing degree of contamination is monitored by means of a pressure measuring device before the screen pack, which can operate a safety switch. The front part of the strainer cover plate (flange for access to the die plate and screen pack) is normally provided with quick-fit bayonet or compression clamps to permit rapid exchange and cleaning of the screen. Strainer heads, as well as using an open form of construction at the outlet, can also be realized in numerous ways using dedicated head designs for shaping the extrudate. Depending on the way the strained compound is to be used, the head may be equipped as follows:

a) simple knives rotating at adjustable intervals can be employed for cutting slices or pieces,
b) adjustable-speed knife blades can be mounted on or off center to cut pellets directly: this is known as a pelletizer,
c) instead of cutting equipment, profile or tube forming attachments are also frequently installed.

Figure 15. Strainer with ram fed. (Photo: Troester, Hannover, West Germany)

16.1.3.4 Profile heads

Profile heads are the simplest kind of extrusion heads. They just guide the compound emerging from the screw to a die plate, which can easily be exchanged by unscrewing a clamping ring. The variety of profiles that can be used is limited only by the particular sizes of head and extruder.

16.1.3.5 Tube heads

16.1.3.5.1 Simple tube die heads (Figure 16)

Construction of simple tube heads is similar to that of profile heads. However, in addition, they have a spider or breaker-plate type of mandrel support, which carries the exchangeable mandrel. One attempts to minimize spider lines at points where the melt flow comes together by appropriate design of the spider arms or by using a breaker plate. In addition to carrying support air and a release agent (e.g. talc or silicone oil) into the tube, with large dies especially, the channels in the spider elements also carry the heat transfer medium for the mandrel.

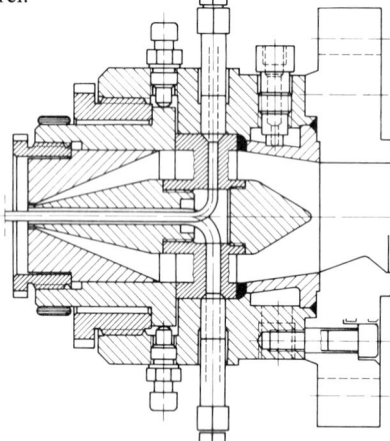

Figure 16. Tube die head
(Photo: Troester, Hannover, West Germany)

Tube wall-thickness correction is generally carried out by mechanical − or in exceptional situations hydraulic − radial adjustment of the die.

16.1.3.5.2 Multitube die heads

Multitube heads, with 2, 3, 5 or even 7 tubes, have a distribution housing that feeds into the individual heads arranged evenly around the circumference. Their construction is as described above.

16.1.3.5.3 Two-layer tube die heads

Two-layer tube die heads are used when tubes made from two different rubber compounds have to be produced (Figure 17). Such requirements can arise if, for example, an inner layer has to withstand the effects of a particular substance (gasoline, oil, acids, foodstuffs, etc.). A special external layer may be needed for good abrasion resistance, tolerance to certain substances, or even for special colors over the whole circumference or in segments (identification stripes, patterns, and so on).

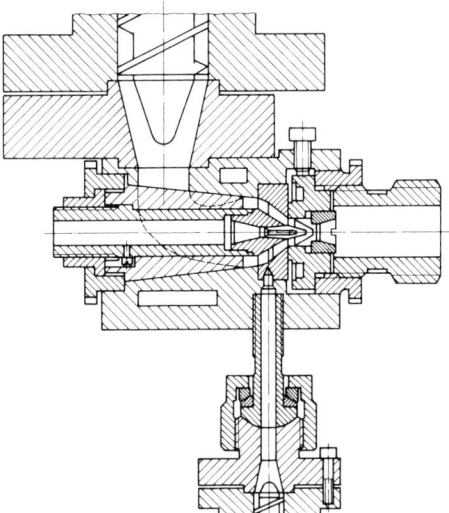

Figure 17. Two-component die head for steel-armored profiles
(Photo: Berstorff, Hannover, West Germany)

A two-layer tube head consists of a simple single-tube unit with an extended mandrel, and an attachment in the shape of a shortened crosshead which takes care of the distribution of the compound for the outer layer. The second layer can be fed vertically or horizontally.

16.1.3.6 Slit-tube heads

With such die heads, tubes are first of all formed in the usual fashion and are then slit lengthwise with one or more knives and laid flat on a haul-off belt as a single sheet, or as several strips.
This process is very efficient for making products for which the tolerances attainable by pure extrusion are adequate. Otherwise procedures described in 16.1.3.12 and 16.1.3.13 are recommended.
In addition to die designs similar to those described in 16.1.3.5, para. 1, there are others that instead of a spider employ an external, often hydraulically withdrawable, mandrel, thus avoiding disfiguring spider marks that frequently result when the divided flow comes together again.
Slit-tube heads are frequently used for making pre-products, for example sheet, liner strips, sheet for sole-plates and other molding preforms, but seldom for finished products.
With the help of lay-flat devices large, very thin-walled tubes can easily be laid flat after opening up, and then be hauled off.

16.1.3.7 Crossheads

Crossheads are used for coating strand-like products (e.g. cables, tubes, ropes). In this procedure the strand is drawn through the head, and the compound, diverted through a feed device, is applied in a pressure or tube-coating process. Most heads have adjustment mechanisms for thickness correction. The overall wall thickness can be influenced to a certain extent by changing the axial distance between mandrel and ring, whilst a uniform wall thickness is achieved either through radial adjustment of the die ring or, where that cannot be done, by displacement of a pivotable mandrel.

Most coating heads are arranged at 90° to the axis of the extruder. Heads at differing angles – 45, 30, or 60° – are called oblique heads. The axis of the crosshead or oblique head (according to the direction in which the strand runs) is not horizontal in some cases, but is arranged to suit individual requirements (e.g. of catenary cable lines).

16.1.3.7.1 Crosshead dies for sheathing

The simplest design of crosshead die is that for sheathing of round strands, and is used in large numbers for manufacture of cables, tubing, rope, pipes, and so on. In these cases the product is mostly provided with a round, tube-like sheath, which is generally uniform in thickness around the circumference. Tube dies or pressure-coating dies may be used, depending on the product.

Tube dies (outer ring and mandrel) have a short annular gap of 0.5 to 5 mm at the exit and produce a tube that is calibrated internally and externally. This is drawn onto the strand by shrinkage or with the help of vacuum.

Pressure dies, on the other hand, themselves create the high radial pressure required for coating the strand. They are profiled so that the melt (in the form of a tube) meets the strand at a predetermined angle, radially or obliquely.

16.1.3.7.2 Crosshead dies for two- and three-layer sheathing

Crossheads for two- and three-layer sheathing are used mainly for manufacture of cable and tubing. Two-layer sheathing dies can be made according to two different design principles:
a) Tandem construction – which means in practice that two single-layer crossheads are mounted one behind the other in a single housing,
   Advantage: simple construction. Disadvantage: long flow path for the inner layer.
b) Compact method of construction with two runners, one inside the other, for distributing the compounds. In this, one compound is led along the inner runner, through the compound in the outer runner.
   Advantages: short flow paths, good material distribution, general utility by exchange of runners, possibility of exchanging inner and outer layer – for example in the case of unmatched extruders. If necessary, sector or half-shell sheathing with different colored compounds is possible.
c) Disadvantages: Higher price, unfavorable temperature control of the individual melt channels.

Centering of the head is achieved in both variants by use of a radially adjustable die ring for the outer layer (as in 16.1.3.7.1), and for the inner layer, by angular adjustment around the long axis of the ball-socket mounted torpedo and mandrel.

Three-layer sheathing heads are normally a combination of two-layer heads 16.1.3.7.2 a) and b); thus the inner and middle layers are, in principle, produced with a head of type 16.1.3.7.2 b), and the outer layer by using a tandem head as described in 16.1.3.7.2 a).

The merit of such heads lies in the technical facility they provide for achieving two- or three-layer coating of a product in a single processing step; and for obtaining good adhesion between the individual layers by operating at high temperatures and pressures, and avoiding dirt inclusions.

The extruders can be arranged in many different ways. For example, to avoid forces, deformations and misalignments caused by thermal expansion, the extruders should be disposed around the head, regarded as a fixed point, and be freely expandable in the axial direction.

In addition to the heads described above, other well-known types are of modular construction, using disk units. This disk system gives a very compact head that can be completely dismantled. Each disk can be thermally controlled individually, and each extruded layer can be separately centered. Since the disks can be rotated, extruder connections can be made quite flexibly.

16.1.3.7.3 Sheathing of non-circular electrical conductors or of metal profiles

The principles contained in 16.1.3.7.1 and 16.1.3.7.2, in particular, are used:
Sector conductor sheathing can also be carried out with one or two layers. But an important additional element required is the guide for the continuous length to be coated. Generally, if precision guidance right up to the die entry is to be achieved, guide rollers are used. These are mounted either at the entrance to the crosshead, or within the torpedo — which consequently cannot be rotated. Another situation is in the production of metal-sheathed profiles that have to be partially or fully coated with one or more layers, which may be obtained from different compounds. Examples include armored profiles covered with extruded tubes, or having heavy foam rubber lips for trunk-lid seals for automobiles.

16.1.3.7.4 Covering of flat profiles

For covering of flat profiles up to about 400 mm wide, so-called flat-profile crossheads are used which operate, as do all crossheads, with a diverter or guide element for directing and distributing the stream of compound. In contrast to crossheads for sheathing circular cross-sections, the diverter is wedge-shaped. The construction and clamping of these heads are similar to those of sheet heads.
Multilayer sheathing of such profiles is not usual; on the other hand bands or stripes of other, perhaps colored, compounds can always be applied by using one or more extruders.

16.1.3.7.5 Double- or multiple-crosshead dies

Special arrangements that are possible include double- or multiple crossheads. These have common housings, similar to those described in 16.1.3.5, with a manifold contour to which two or more single-layer crossheads connect.

16.1.3.8 Special crosshead dies for filament and wire coating

Special heads for filament and wire coating (e.g. tire-bead wires).
These are easy-to-use crossheads in which the compound is simply fed under pressure from all sides onto the running strand, and is shaped at the die mouth.

16.1.3.9 Slot dies

In the broad sense, slot dies can also be envisaged for wide, thin profiles. However, they differ greatly in the way they are constructed, because of the special design requirements that must be satisfied to ensure uniform distribution of the melt stream across the width of the die. This is achieved by the use of deflector bars, baffles, guide channels, and so on. On the outlet side of the manifold, exchangeable pre-extrusion- and finished-extrusion profile strips are fixed by means of wedge cams. The same pre-extrusion strips are used for similar profiles. Such dies can be fitted with fully or partially automatic clamping mechanisms to speed up and simplify die- and compound changes.

16.1.3.10 Tread heads

For forming tread strips and side-wall strips for the tire industry, special subdivided dies — known as tread heads — are almost always fitted with hydraulic clamping and opening systems, so as to ensure the shortest possible downtime when changing heads or compounds. Other advantages of these hydraulically operated heads, aside from economic ones, are a reduction in hard physical work for the plant operators, and significantly better protection of the head against damage, especially of the sensitive sealing surfaces.

16.1.3.11 Piggyback extrusion heads

This type of head derives its name from the "piggyback" arrangement of extruders, one on top of the other. Such heads are used for manufacture of flat profiles from two, three, or four rubber compounds (Figure 18). In contrast to earlier layouts (e.g. Y-construction), the flow channels are placed symmetrically between the extruders and the die, and are kept relatively short (Figure 19). If there are more than two extruders working on a single head, it can be advantageous to regard one of them as the principal unit.

Figure 18. Two-component piggyback unit open for cleaning. (Photo: Troester, Hannover, West Germany)

The tire industry is the largest user of such systems, in which the different compounds are brought together in the head in the required shape, dimensionally accurate, under pressure and free from air inclusions.

In addition to the severe technical requirements met by piggyback heads, the facility for rapid clean-out and changeover, as well as ease of handling and pressure tightness, is of special importance. Modern piggyback heads are fitted exclusively with fully hydraulic opening and clamping systems, which also provide for total or partial opening of the head for cleaning. A change of head, including any pre- and post-extrusion profile strips and the overlap and flow-junction elements needed, is carried out with the help of hydraulic wedge clamps in the same simple way as with slot dies and tread heads.

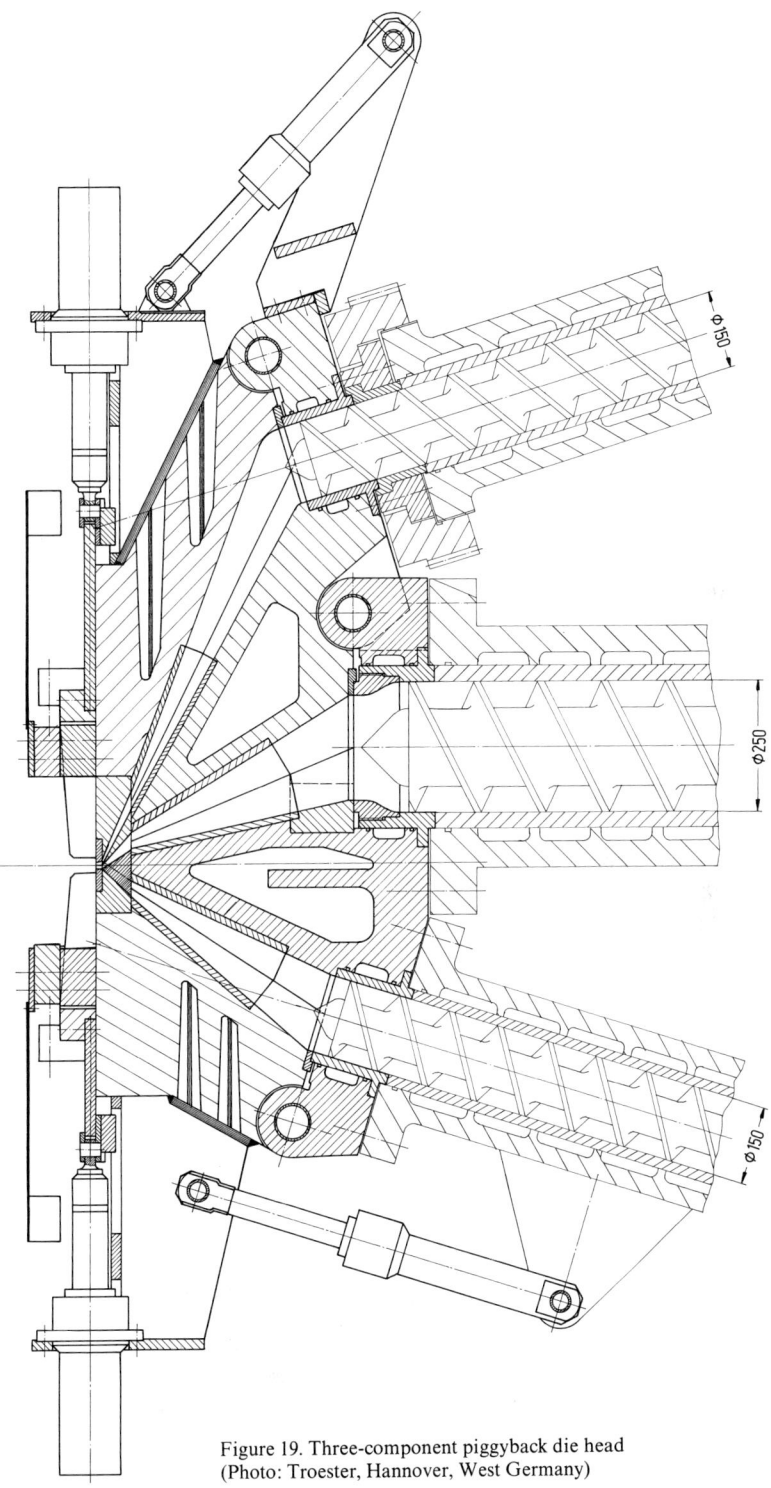

Figure 19. Three-component piggyback die head
(Photo: Troester, Hannover, West Germany)

In order to provide an even greater number of possible profile shapes, and, as far as possible, to achieve universal application of these units, universal piggyback heads have been developed. These allow the head to be changed over partially or fully to another product profile in a very short time with the help of exchangeable flow-channel inserts, cassettes, or loose parts.

16.1.3.12 Single-roll roller die

The single-roll roller die consists of an upper section with material guide channels, a preform die/final die assembly held in place by wedge fingers (Figure 20), and a lower section consisting of a rotating, cylindrical roll. Clamping together of the head as well as swinging the uppers parts of the head into the open position, is accomplished with the help of hydraulic cylinders. For fine adjustment or correction of product thickness, the distance between the roll and extrusion die can be changed, using a motor to shift the roll vertically. The main advantages of this head are that only part of the pressure necessary to overcome the resistance to melt flow has to be supplied by the extruder, because the roll has a hauling action, which makes it possible to work at significantly lower pressure and melt temperature. Also, products with large differences in wall thickness leave the die in a straight line without pulling, because they stick to the roll.

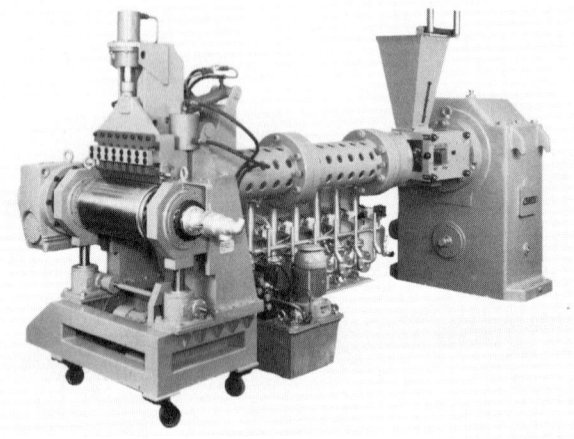

Figure 20. Single-roll roller die (EWK) with QSM extruder. (Photo: Troester, Hannover, West Germany)

16.1.3.13 Two-roll roller dies

16.1.3.13.1 Roller-head machines

Roller-head machines are the preferred choice for producing high quality rubber belts or strips to tight dimensional tolerances and with good surface quality. The preform head, in association with a hot or cold-feed extruder and its two-roll calender, make up a so-called roller-head unit (Figure 21 A). The task of the preform head is to feed the rubber compound coming from the extruder direct into the calender roll nip at a steady rate and a predetermined thickness.
This provides the possibility of producing sheets up to 20 mm thick in a single processing step with high precision and without air inclusions, since the usual rolling bank is avoided.

16.1 Extruders

It is also possible, by profiling a calender roll, to produce rubber sheets in a variety of contours. Instead of profiled rolls, modern calenders have quick-change roll shells which can be exchanged for other shells, pre-heated to working temperature, in less than five minutes, with the help of a special changeover device.

The preform head can be adjusted stepwise to individual product thickness and width by exchange of extrusion profile strips and head inserts.

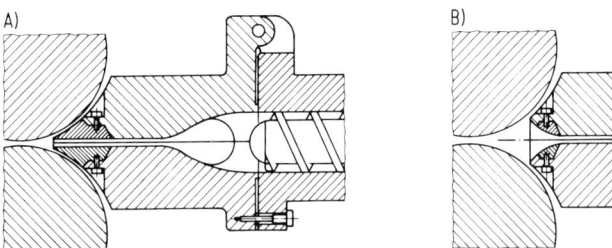

Figure 21. Schematic diagram of roller head and roller die. (Photo: Berstorff, Hannover, West Germany)

This type of head, like those already described, is equipped with fully hydraulic clamping and opening arrangements. The extruder, with the head, is mounted on a steel baseplate that can be moved lengthwise. Often, it can also be pivoted for easy screw removal; and for opening up and cleaning the die, it can be moved away from the calender. Roller-heads are used specifically for manufacturing roofing films, floor coverings, gasketing, sheet blanks, conveyor belts, and tire elements like inside liners and side strips.

16.1.4 Auxiliary equipment

16.1.4.1 Screen changers

Screen changers are devices that make possible rapid exchange of strainer screens like those of 16.1.3.3. One can divide the usual screen changers of this type into three variants, which are offered with manual or hydraulic operation:

a) single cassette screen changer,
b) double cassette screen changer,
c) two-fold segment screen changer.

With all three categories of screen changer the die plate is placed with the screen pack in a single- or multiple cassette. The procedure with single cassette screen changers is to push the new cassette with the clean breaker plate and screen in one direction through the cassette housing. This process is carried out hydraulically with a lead screw or a system of levers. At the same time another cassette, containing the contaminated screens, is pushed out the opposite side, and can be removed for cleaning.

In contrast, double cassette units operate with a double cassette that may be pushed in either direction – vertically or horizontally – for changeover.

Type c) units, segment screen changers, use a segment mounted outside the extruder barrel or die head as a cassette, which can carry two die plates and screen packs. The segment cassette is rotated about the mounting point by a suitable lever.

16.1.4.2 Die changers

Die changers are devices which permit the die on an extruder to be changed in a very short time, perhaps as little as two seconds; for example while the machine is running and producing any kind of profile. After the changeover there are only 3 to 4 meters of waste before a dimensionally accurate product is obtained, so long as the ratio of cross-sections of the two profiles is not too great.
The construction and hydraulic operation of the die changer is essentially the same as that of a hydraulic screen changer (16.1.4.1), with the die mounted in the double cassette in place of the die plate and filter.

16.1.4.3 Head changers

The complete head can be exchanged in a very short time with head change equipment. With such devices two or more heads are mounted on one turnable, which allows the heads to swivel into position in front of the extruder and be fixed to it by means of a hydraulic bayonet clamp. One or more additional heads may meanwhile be cleaned and prepared for later production runs.
When one run is finished the extruder is stopped, depressurized, and the head automatically uncoupled. The head is removed from the area of the extruder end flange by lateral displacement and exchanged inside the head magazine for the new head. When the new die head arrives at the extruder position it is pushed against the flange and coupled automatically to it.

16.1.4.4 Feeding devices

Hot-feed extruders are fed from rolling mills with strips or pigs. Cold-feed extruders process strips, sheets, or pellets, for which purpose a variety of feeding devices are used whose good'function is essential for continuous and uniform extrudate. Because of the very big differences in the feedstock form, these pieces of equipment are quite distinct from one another. Equipment for feeding strips or sheets, whose cross-section has not been matched to the extruder, can become relatively expensive.

16.1.4.4.1 Granule- or pellet-feed equipment

From the point of view of production reliability this feeding device is particularly suited for processing relatively "dry", non-sticky mixes. Certainly the production of granules or pellets requires extra work and is therefore expensive. In addition, storage and transport conditions are more difficult.
The compound is metered into the extruder from a storage bin, either direct or through a screw conveyor, a vibrator chute, or a conveyor belt. The feed hopper is equipped with a level indicator which controls the amount of material fed into it. A uniform level in the hopper guarantees good continuity of feed conditions for the extruder. Extruder sizes are limited to the small-to-medium size range (30 to 120 D), to which the granule or pellet sizes must be matched. A big advantage of granule feed is that one can easily blend different batches to obtain uniform mix quality.

16.1.4.4.2 Strip- or slab-feed equipment

Strips are either transported on pallets, wound on rolls, or prepared as lengthwise perforated sheets. The cross-section of the strip must be well matched to the particular extruder size to achieve uninterrupted operation.

Under favorable conditions, when the strips can run freely into the extruder hopper, only a few guide rollers are needed. In other situations a pair of stripper rolls or a conveyor belt may be necessary. The drive is controlled by an overhanging loop or dancer roll, and the strip delivered to the screw tension-free.

Large extruders can be fed direct with slabstock, so long as its cross-section matches the take-up ability of the screw. The design and operation of the strip feeders that have been described meets these requirements. One particular form of construction of strip feeder has a hydraulically driven roll pair that is mounted above the feed opening to deliver the strips of compound direct to the extruder. The drive torque is so adjusted that material held up on the rolls reduces the roll speed, and with it the strip feed rate. The result is that the quantity of compound fed to the extruder matches its demand. A disadvantage is that if the roll torque is exceeded, as when strips stick together, or if the accumulating material solidifies in the feed opening, the screw runs empty.

16.1.4.4.3 Slab cutting and feeding devices

To make it possible to feed wide slabs, combined cutting and feeding equipment has been developed. One distinguishes two types:

a) Those which cut strips from the end of one or more slabs and deliver them to the extruder hopper continuously on a conveyor belt. Control of the cutting and supply operations is done by a level-monitoring system in the feed hopper. This arrangement provides for a number of additional functions for automatic operation e.g. metal detectors in association with marking devices, automatic blanking cutters, and so on.

b) So-called zigzag cutters which, in contrast to a), do not cut right across the slab but leave a small uncut border on alternate sides. This results in a zigzag-shaped strip that is guided to the extruder by deflector and hopper rollers. Each successive cutting operation is initiated by a dancer roller.

The problem here is that variations in slab width, or defects, can on occasions lead to broken feed strips and thus to interrupted feed.

16.2 Continuous vulcanization lines

16.2.1 General

Historically, the rubber industry has switched over relatively rapidly to new processes where they are economic, and where products of the desired quality can be manufactured, despite the need to show improved long-term behavior for the products before the introduction of new technology.

Continuous vulcanization is almost always preferred to the discontinuous process, on the following grounds:

a) *Economy*

– utilization of extrudate heat so that only a small amount of energy is needed to raise the melt temperature to vulcanization temperature,

- better energy use on continuous lines (good insulation, no intermediate cooling, direct energy conversion with UHF or shear heads i.e. temperature increase without conduction of heat into the mix),
- high line speeds,
- no intermediate conveying of the extrudate,
- saving on manpower,
- smaller space requirement,
- smaller overall investment,
- energy saving,
- independence from steam generator.

b) *Quality*

- extruded profiles in desired lengths,
- constant profile geometry,
- clean surfaces without condensate water- or powder stains,
- regular and relatively simple quality control.

Over the past few years the continuous vulcanization lines described below were developed, with varying degree of success, and brought into production (Figure 22):

a) continuous-vulcanization pipe (steam, radiant heat, pressurized salt bath),
b) hot-air lines (extended, or three or more tiers),
c) fluid-bed lines,
d) salt-bath lines,
e) infrared lines,
f) Helicure lines,
g) UHF lines,
h) shear-head lines.

Except for type a), all these lines usually work at atmospheric pressure, or at only slightly raised pressure, and therefore normally require the use of vacuum extruders if bubble-free products, or products with controlled porosity have to be manufactured.
Low-pressure systems include:

a) the pressurized fluid bed – so far only in experimental versions for manufacture of round profiles like cables and tubes,
b) Helicure. Because of the flow resistance of the tube, a low initial pressure is produced near the extruder by the heat carrier or transport fluid, which falls to zero towards the end of the line.

As shown schematically below (Figure 22), all the continuous vulcanization methods have been introduced during the last thirty years. The two not dependent on thermal conduction were the last.

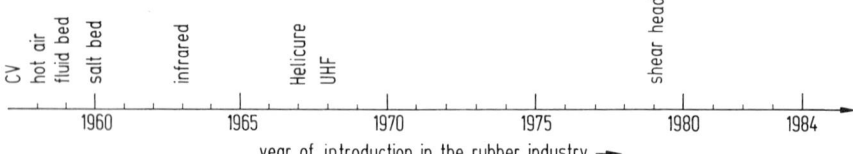

Figure 22. Continuous vulcanization lines

All continuous vulcanization lines are constructed in basically the same way, consisting of:

a) one or more (vacuum) extruders,

b) heating and temperature-maintaining equipment,
c) a cooling section,
d) haul-off, wind-up, or cutting equipment.

The principal differences are found in section b) of the lines, above all in the method of heating the compound, which can be achieved in various ways, using heat conduction or by energy conversion within the material itself.

16.2.2 CV pipe

These continuous vulcanization (CV) lines are of importance almost exclusively for cables, and are used in exceptional cases for other round products with a load carrier only: for example, tower lines for reinforced tubes with flexible mandrels.

16.2.3 Hot-air vulcanization

Hot-air lines are the oldest types of continuous vulcanization plant. Today, however, they are being used less and less as straightforward hot-air lines, but more in combination with UHF or shear-head lines, in which they perform the task of maintaining the temperature of the preheated compound for the required vulcanization time. Whilst pure hot-air lines are still often used for highly expanded foam-rubber tubes or sheeting, and can be as much as 100 to 150 m long (divided into sections with individual drives for each), hot-air sections integrated into UHF lines, for example, were only 6 to 8 m long up to a few years ago. More recently, however, greater lengths are being used on such lines so as to obtain an adequate degree of vulcanization at higher line speeds and to improve compression-set values. However, in order not to alter the overall line length greatly, the preference now is to use multi-tiered lines of modular construction with, at most, three sections 9 to 15 meters long. It is worth noticing that by the time the vulcanized product reaches the first deflector roll it is already sufficiently cured for the deformation to be within permitted limits. It is usual for the PTFE- or steel-mesh conveyor belts in multi-tiered hot-air lines to be powered separately, so that the line speed can be adjusted in each tier to compensate for the length changes caused by temperature differences or expansion of the foamed product. Steel-mesh belts are used almost exclusively for highly foamed products or for products with a large surface area, so as to achieve better heating of the underside.
Heating is carried out mostly by hot-air blowers that use electrical heaters or steam-heated heat exchangers.
The air temperature can be smoothly and independently adjusted in each section, normally within the temperature range 150 to 250 °C, but exceptionally up to 320 °C.
The flow of air can also be controlled. To reduce the energy consumption one does not operate with 100% fresh air, but uses one of the following methods:

a) closed-air circulation with fresh air addition in proportion to exhaust of plasticizer-vapor evolution,
b) preheating the fresh air by means of the waste air, usually in a counter-current heat exchanger.

In addition to the efficient use of energy that has been noted, the good insulation that is usual on modern lines, and avoidance of thermal bridges, also contribute to reducing energy costs. A result is that even tighter safety regulations are being met.
Advantages: a) clean, b) simple and observable, c) easier control of foamed-rubber profile production, because of slower heat-up.

Disadvantages: a) comparatively long lines because of poor heat transfer, b) poor efficiency, c) danger of surface oxidation, especially with NR- or peroxide-crosslinked compounds.

16.2.4 Fluid-bed vulcanization

These vulcanization methods use a pseudo fluid consisting of air-suspended ballotini – glass microspheres of 0.4 to 0.2 mm diameter – for heat transfer (Figure 23). The line consists of a tray with a steel wire-mesh base through which (alternatively) hot air or superheated steam can be blown, with the result that the spheres are lifted up and the whole mass is set in motion. The moving mass of ballotini and air or steam behaves practically like a fluid with a hydrostatic pressure that increases with depth, with the result that extrudates sink into the fluid to an extent determined by their specific gravity. Although some profiles, particularly hollow ones and tubes, cannot be totally covered with ballotini at the top of the bed, the preference is not to use hold down devices like steel belts or hold down rollers, so as to retain the advantages transporting of the extrudate along the line without distortion.

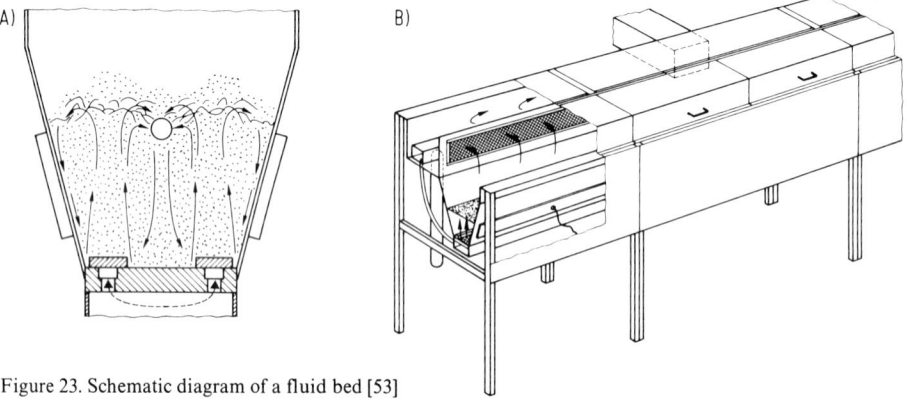

Figure 23. Schematic diagram of a fluid bed [53]

Fluid beds can be kept much shorter than hot-air channels. This is because the high heat capacity of the fluid mass of ballotini significantly improves heat transfer.
In addition to the heat supplied to the ballotini by the moving gas (or steam), the vulcanization bath is further heated at the external walls by electrical elements. To reduce heat loss and to allow as few ballotini or ballotini fragments (erosion) as possible to escape, a closed hood is located over the bath with pipe connections to:

a) simple evaporation pipes,
b) pipes for circulating hot gases,
c) heat exchangers.

Normally, fluid beds operate with air at temperatures up to 250 °C, but only in special circumstances with steam, and only then if materials have to be processed that are subject to oxidation (e.g. NR compounds or peroxide crosslinking compounds). Working with steam has the big disadvantage that in heating or cooling through a temperature range significantly above 100 °C, one must first of all use hot air, since otherwise the ballotini would stick together because of steam condensation and could no longer be kept in suspension. Heating of the walls alone is permissible only as long as the ballotini are kept in suspension, otherwise they could be melted by contact with the walls.

Positive characteristics of this process are:

a) no conveyor belt is needed,
b) no collapse of even delicate profiles,
c) free expansion of foamed products.

Disadvantages:

a) poorer efficiency, particularly when compared with LCM, due to large heat losses,
b) difficulties in curing many profiles since the ballotini cannot penetrate between narrowly separated ribs, or are not released after vulcanization (or only with difficulty),
c) adhesion of the microspheres to the extrudate, which damages the surface quality and makes a surface brushing operation necessary.

Pressurized fluid beds are being tested experimentally, but have not yet been used in production. Presumably the main reason for this is that complicated profiles can practically never be sealed, and although round ones (cables, tubes) can be sealed off for a certain time, because of a high degree of wear, loss of seal-tightness is to be expected after a relatively short period. With hollow profiles there is the difficulty that perfect internal pressure equalization must be achieved.

By far the simplest way of avoiding porosity is to use a vacuum extruder.

16.2.5 Salt-bath vulcanization

Salt-bath vulcanization is one of the so-called LCM methods (LCM = Liquid Curing Medium) and is by far the most common and universally used continuous curing method of profiles (Figure 24). The method was introduced by DuPont in 1960. The heat transfer agent is a eutectic mixture of salts, consisting of 53% KNO_3 (potassium nitrate), 40% $NaNO_2$ (sodium nitrite), and 7% $NaNO_3$ (sodium nitrate). The eutectic mixture has a melting point of about 140 °C and a density of ca. 2.15 g/cm^3 at room temperature, or of 1.9 g/cm^3 at 250 °C. It can be used at the vulcanization temperatures of all well-known compounds, and the usual operating range from about 180 to 250 °C can be covered without difficulty.

Figure 24. Salt bath. (Photo: Troester, Hannover, West Germany)

Other heat transfer fluids — mixtures of metals, glycols, glycerol, silicone oils, among others — have been tried, but have not proved to be so satisfactory as the salts mixture. There are two basically different designs of salt bath:

a) those using steel belts that have the task of pushing the low-specific-gravity product beneath the surface of the molten salt, and transporting it,
b) salt-bath/salt-spray lines in which the extrudate floats on the surface of the salt, and is irrigated from above with hot salt pumped up from the bath. These operate, if necessary, in combination with driven hold-down rolls in the back region.

At least in the case of solid profiles, the first procedure ensures that the product is reliably cured at constant speed and in straight form. However, since a certain amount of distortion of delicate profile sections cannot always be avoided, spray/salt-bath lines, which do not use conveyor belts, have some advantages in the manufacture of foam-rubber profiles (no problems in length adjustment with expanded products) and of finely ridged profiles.

Because, above all, of the complication of greater maintenance effort and the incomplete coverage of the profile with salt, these lines are used far less than those employing steel belts. Since a small amount of slippage of the profile beneath the belt is possible, most foam-rubber products can also be manufactured on this equipment. With highly expanded products length adjustment can be achieved by the use of two separately driven steel belts. The use of salt baths for highly expanded foam-rubber products is limited, however.

Modern salt-bath lines have to be fully enclosed and well insulated, except around the dimensionally adjustable entry and exit ports. The heating of the salt bath must be carried out, directly or indirectly, by electricity and the transport mechanism (conveyor belt or driven rolls) must be retractable at the end of production runs. In addition, to save costs, and above all to reduce waste-water contamination, the bath is fitted with a salt drip-back section at the outlet. This consists of hot-air jets and, if necessary, an intermediate mechanical stripper, preferably arranged in the region where the profile is running vertically upwards. All components on the line that come into contact with salt, like salt tanks, conveyor belts, deflector rolls, dip rolls, hold-down rolls, strippers, and so on, must be made from corrosion-proof materials.

Positive characteristics of salt-bath lines:

a) outstanding heat transfer, and therefore shorter length than equivalent hot-air or fluid-bed lines,
b) problem-free operation at high vulcanization temperatures within the working range of the compound, and at high production speeds,
c) easily insulated; low heat losses;
   (with fluid beds, because of the greater length required and the circulation of steam or air, there are greater heat losses),
d) very good efficiencies with established products,
e) no oxidation on the surface of the extrudate, since the salt is totally inert (important for NR and peroxide-crosslinkable compounds),
f) no swelling of the material (as can happen, for example, when curing in oil),
g) cleaner, spotless product surface.

Disadvantage of salt-bath lines:

a) with thick-walled hollow profiles or foam-rubber profiles the curing times are relatively long because of poor heat conduction, an effect that applies to all methods which depend on heat conduction; and the cure is uneven and related to the cross-section because of temperature gradients, although there is some compensation during cooling,
b) above all, with rough surfaces, highly convoluted profiles, and/or high speeds, there is a small amount of salt loss in spite of highly efficient stripping devices,
c) danger of collapse with products of low specific gravity (hollow profiles, tubes, foam-rubber products).

16.2.6 Infrared lines

Infrared lines are used almost exclusively for production of continuous lengths of silicone rubber. Because of the use of peroxide crosslinking for silicone rubber and of the related danger of surface oxidation, and also because of efficiency, these lines are preferred to those using hot sections continuously fed with hot air. Ceramic radiators are particularly suitable as heat sources, because of their high emissive power (ceramic dark emitter up to 0.96 µ), and of a favorable wavelength range for absorption by silicone rubbers. Normally, horizontal ovens are employed, although in favorable cases, when the profile is strong enough, vertical or combined vertical/horizontal lines are used, which have the advantage of uniform all-round energy input. Horizontal ovens have a highly reflective processing channel, which can expand freely in response to temperature changes.

The extrudate is moved through this channel on steel- or steel-mesh belts, and is heated direct by the ceramic radiator and by reflection from the channel walls to a maximum temperature of 250 to 300 °C.

Because of the groove low hot compression-set of silicone rubber, a special cooling zone can normally be dispensed with.

16.2.7 Helicure vulcanization

Helicure vulcanization is so called because of the coiled shape of the curing pipe (Figure 25). The most important section of the line is a helically coiled pipe that is filled with a liquid heat transfer medium, usually glycerol or glycol; this is continuously circulated, and simultaneously transports the extrudate and provides an initial excess pressure of 1 to 1.5 bar over a total pipe length of about 150 meters. The pressure depends on the length and diameter of the pipe, the dimensions of the extrudate, and the flow rate and viscosity of the liquid. As can be seen from the diagram of the line, the beginning of the pipe is connected to the extruder die head. The heat transfer fluid is pumped into the pipe a short distance from the beginning, so that the extrudate is carried with it to the end of the tube. The emerging fluid is returned from an overflow point into the storage container, where it is reheated to the vulcanization temperature by means of a heat exchanger, and then recirculated by the pump.

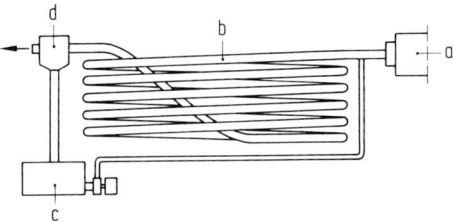

Figure 25. Helicure vulcanization [54]
a extruder die head, *b* vulcanization tube-coil, *c* reservoir with circulating pump, *d* flow-off reservoir

As already mentioned, use is made of the flow resistance of the pipe, so that at least at the beginning of the curing zone an excess pressure of up to 1.5 bar is produced; this is intended to reduce the porosity that would otherwise occur when not using a vacuum extruder, or to suppress it completely in harder compounds. At higher temperatures and/or with softer compounds, it is in any case absolutely necessary to use a vacuum extruder. The normal vulcanization temperature with this system is about 150 to 165 °C.

In order to reduce the length of the vulcanization tube and/or to attain higher pressures, some experiments with tower lines have been carried out. The Helicure process has not been put to practical use for the following reasons:

a) the development of many simpler and shorter lines,
b) vulcanization of straight profiles is not possible (curing to coiled form),
c) tension-free cure is impossible because of the transport mechanism (reduction of cross-section),
d) hollow profiles cannot be produced without the use of very elaborate pressure equalization systems,
e) flat profiles are not possible because of the circular cross-section of the pipe (pressure loss).

16.2.8 Ultrahigh frequency, or UHF, vulcanization

The principle of this method of continuous vulcanization differs in one essential from all other methods, namely that the heating does not come from the outside, i.e. by heat conduction, but is generated within the extrudate by UHF electromagnetic energy applied simultaneously across the whole cross-section.

A precondition for heating by electromagnetic field is adequate excitation of the compound. This means that the polarity of the polymer and of other components of the compound, including different carbon blacks, must be favorable. However, too high a polarity (combination of a polar polymer with highly polar additives) and above all uneven distribution of the polar components of the compound must be avoided, otherwise there is a danger of local over-heating. The extruded profile, which has to be heated to vulcanization temperature, is moved through the heating channel with the help of a conveyor belt. The channel is designed either as a waveguide or as a resonant chamber.

By correct dimensioning of the waveguide channels, the microwave energy is conducted practically loss-free. The strength of the electromagnetic field is a maximum at the center of the hollow conductor and decreases strongly towards the walls. Because of the small cross-section of the channel and of the uneven energy distribution, waveguide channels are used only for relatively small profiles (Figure 26).

Owing to this limitation, resonator chambers with much larger internal dimensions have been generally accepted. On these kinds of lines the microwave energy is not transmitted as it is in waveguides, but is reflected at the walls, so that in addition to standing waves, a large number of harmonics build up, leading to an equalization of the energy density over the full cross-section of the channel. Because of the decrease of energy density with increasing cross-section, the channel should certainly not be made greater than is absolutely necessary for the largest profile to be cured. The efficiency of the microwave unit and the uniformity of heating the product can be significantly improved by feeding in the energy at as many places as possible, by tuning the antenna radiation characteristic, by rapid movement of the conveyor belt and by series arrangement of UHF and hot-air units (temperature equalization).

As mentioned at the beginning, the extrudate temperature is raised to the vulcanization temperature in the UHF unit. To maintain the vulcanization temperature, hot-air channels are normally added, constructed in one length or in several tiers.

For slightly foamed rubber profiles it is advantageous, for length and speed matching, to provide two hot-air channels in series.

For highly blown products (insulation sheets or tubes which undergo a 10 to 15 volume change) special lines are required which permit completely free expansion to occur.

Advantages of the process:

a) very rapid heating across the whole cross-section of the profile (especially advantageous with large cross-sections, hollow profiles, or foamed rubber articles), providing high production speeds, uniform vulcanization overall, and rapid passage through the softening phase before vulcanization starts, giving very little deformation of even the most complicated profiles,
b) rapid change of energy input, as required (the best precondition for control functions),
c) non-polluting,
d) totally clean, spotless products,
e) good efficiency.

Disadvantages:

a) only applicable to polar compounds,
b) danger of surface oxidation with peroxide-crosslinked compounds (in contrast to salt baths).

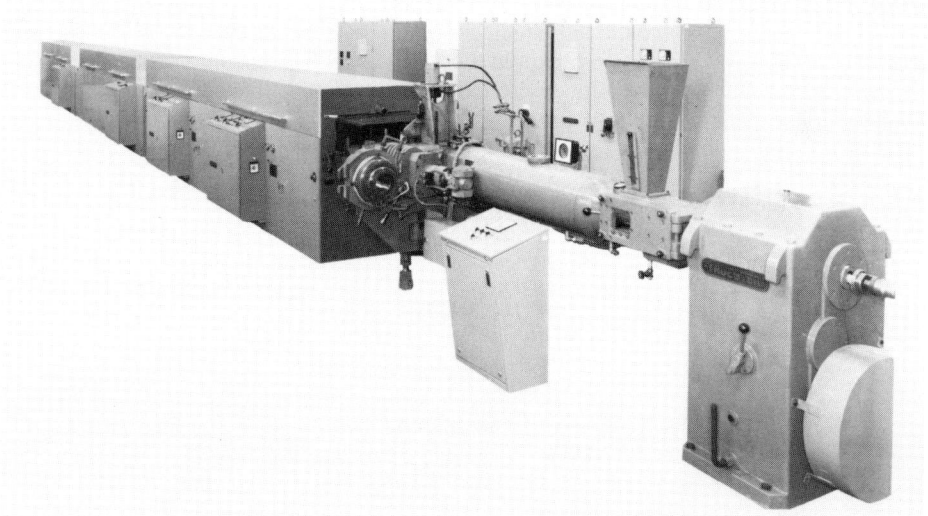

Figure 26. Microwave line. (Photo: Troester, Hannover, West Germany)

16.2.9 Shear head vulcanization

The principle of introducing energy at a shear gap was already established at the end of the 1950's (shear mandrel) and had been described by a number of authors; but is was not until 1979 that a production line acceptable to rubber processors was first introduced. Subsequent designs now exist which all use a shear mandrel and an enclosing jacket in the shear gap, but are otherwise distinguished predominantly by the configuration of the shear head, and by differences in the driven components of the extruder (Figure 27).

The compound fed from the extruder is sheared in a ring gap within the shear head, which is normally made up of a fixed cylinder and a rotating mandrel. The melt is raised to a high temperature by this device. The temperature increase depends on the geometry of the shear head, on the throughput rate, and on several characteristics of the compound. For a given shear head and a particular compound, it is primarily the output of the extruder and the mandrel rotation speed, and thus the dwell time of the melt, in the shear head, that are variable. The mandrel speed for optimum melt temperature is adjusted according to extruder output, the kind of compound, and the size and complexity of the profile. Normally the working temperature range is about 160 to 190 °C, which is much higher than usual for extrusion. To ensure that curing does not start prematurely, very precise, intensive thermal control of all parts of the shear head, and extremely well designed flow channels, including those in the head, are necessary. The better the design of extrusion head and die, the higher the melt temperature and production rate can be. Even though the average dwell time of the compound in the shear head at 3 to 12 s is extremely short, and there is therefore little danger of premature curing, even at elevated temperatures for most of the hot melt, account must be taken of the significantly lower rates of flow in the wall boundary layers by keeping the walls of the mandrel and its housing at a lower temperature (around 70 to 80 °C). Such lower mandrel and housing temperatures are also needed to ensure good adhesion at the walls to promote the introduction of shear energy.

Automatic melt temperature and mandrel-speed control systems have been developed, to maintain exit temperature from the head constant in spite of changes in throughput rate and unwanted viscosity or temperature variations in the compound. Even though the melt has already been raised to a high temperature level in the shear head, downstream heating sections are needed to maintain and equalize the temperature, or even to raise it further. While hot-air downstream sections are used most frequently because of their simplicity and total cleanliness, salt baths are often preferred for peroxide-crosslinked products, since by doing so surface oxidation and consequent stickiness of the product can be avoided.

There are various ways of mounting shear heads:

a) at 90 °C to the extruder (similar to a crosshead),
b) in piggyback arrangement, meaning obliquely above the extruder, but in such a way that extruder and shear head are aligned in the direction of product flow,
c) direct along the screw axis so that the mandrel becomes an extension of the screw (torpedo), and a part of the shear head housing rotates as well.

Advantages	Disadvantages
a) Simpler construction	More space required for the vacuum extruder, which is often very long
b) Very compact and space-efficient arrangement of extruder and shear head in the product-flow direction	Relatively heavy mass above the extruder
c) Very compact and space-efficient arrangement of extruder and shear head in the product-flow direction	Rotating sleeve difficult to seal against melts, no possibility of separate thermal control of mandrel, poor temperature control of the sleeve, additional torque on the screw and gearbox, no possibility of strainer and filter in front of the shear mandrel.

For coating tasks (steel-mesh, wedge profiles, cables), flow-optimized crossheads are attached to the shear heads. The photograph (opposite) shows a combination of two vacuum extruders, two shear heads, and a common crosshead, for producing sealing profiles consisting of steel-mesh core, solid rubber covering and extruded foam-rubber tubing.

Figure 27. Shear head. (Photo: Krupp)

16.2.10 Rotation vulcanization

For continuous vulcanization of many flat or profiled wide sheets, the Rotocure or Auma® machines have shown up well. The essential features of these machines are a large rotating heated drum and an endless steel- or rubber-covered steel-mesh belt running over two deflector rolls and a tensioning roll (Figure 28).

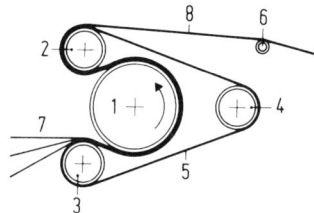

Figure 28. Rotation vulcanization line
1 heating drum, *2* front, upper deflector roll, *3* front, lower deflector roll, *4* back, deflector and tension roll, *5* pressure belt, *6* cooling roll, *7* product feed belt, *8* finished product

The sheet to be vulcanized is held against the heated drum by the pressure belt and transported along the line. The sheet thus receives a predetermined surface structure, and the development of bubbles and porosity is prevented.

The vulcanization energy is introduced by the heating drum and, if needed, additionally by infrared heating elements, which are mounted on the back of the steel belt. The continuous operating mode, unlike cyclical operation with hydraulic presses, provides the preconditions for uniform quality over the full length of the sheet.

16.3 Cooling

16.3.1 General

Extruded profiles or sheets leave the extrusion head at temperatures between 100 and 130 °C. In continuous vulcanization the temperature is raised to between 200 and 300 °C. The cooling sections must be able to cool the products to between 30 and 60 °C, depending on quality, before secondary operations (wind-up, stacking and so on), are carried out.

The length of cooling section required is dependent on the kind of cooling, the profile geometry, the profile temperature, the calorific value of the product, and the speed of the product line, and can nowadays be calculated extremely precisely.

If necessary, a simple heat exchanger using town water can also be provided, if additional cooling of the circulating water is needed.

The make-up of evaporated cooling water is done automatically by a ball valve or a level control. Fresh water or additional cooling can thus be quite simply turned on by a thermostat.

16.3.2 Cooling processes

16.3.2.1 Immersion cooling

Immersion cooling is the simplest of the cooling methods. The lines consist of one or several long water baths, through which the extruded product is pulled. The effectiveness of cooling can be improved if the water flow is turbulent. Often a special system of rollers is used to ensure that the product is fully immersed and cooled, and then lifted out, alternately, for evaporation to occur between sections.

The weight decrease due to buoyancy, that occurs on immersion in the cooling bath, eliminates the possibility of pressure marks being generated by the transport system – especially with thick unvulcanized products (truck tire tread, for example), and facilitates shrinkage.

16.3.2.2 Spray cooling

Spray-cooling lines are particularly efficient, since they permit evaporation to take place at the start of the cooling section, if the spray nozzles are properly arranged. The product should be sprayed simultaneously from all sides by clog-resistant radial whirl-nozzles which are fed either with fresh water or – to save water in the cooling circuit – from a reservoir. Such lines easily allow for cooling to be different zone by zone.

16.3.2.3 Contact cooling

Under this heading are included cooling lines (constructed of stainless steel or aluminum for good thermal conductivity) on which the cooling medium is separate from the hot product. Thin sheets and strips that should not come into contact with water can be very effectively cooled by the contact method.

There is, however, a risk that high atmospheric humidity combined with low cooling temperatures can give rise to condensation on the contact surface.

Contact cooling systems used industrially include:

Cooling rolls

These are extensively used in the manufacture of thin sheets and strips, as part of calendering lines for example. The cooling rolls are mounted in floating bearings, if the width allows, and are staggered vertically, an arrangement that simplifies thread-up and permits a favorably large contact angle.

A further feature of this roll arrangement is that the upper and lower sides of the product sheet contact the surface of the cooling roll alternately; thus it is cooled uniformly on both sides.

The driven cooling rolls are double-walled units provided with a special configuration of cooling channels, to which the cooling water is supplied through watertight rotary glands.

16.3 Cooling

Cooling belts

Cooling belts provide an alternative to cooling rolls. These are special steel conveyor belts which are sprayed intensively with coolant and act as a wall separating the underside of the product web from the coolant. With large cooling lines it might be appropriate to arrange several cooling belts above each other, to save space.
The advantage of such cooling belts is their extended layout. In comparison with cooling drums this simplifies thread-up, and operates in favor of good dimensional stability, because the product is provided with perfect support and is seldom, if ever, deflected. Lines like this are used for cooling webs and pellets.

Spiral cooling ramps

A further contact cooling possibility for granules is the use of cooling ramps, built either in straight lines (cascade- or multilevel arrangement) or in tower form as a helical channel. The cooling channel is double-walled and carries cooling water in the hollow space. (An arrangement without double walls can also be employed on air cooling lines, 16.3.2.4.) To move the pellets along, the cooling ramps are vibrated; this, and the effect of deflectors, results in the pellets being continually turned around so that different areas are brought into contact with the cooling surfaces.

Cooling troughs with conveying screws

These offer another possibility for cooling, wherein individual pellets undergo continuous rolling motion as they are transported along the cooled double-walled channel and hollow tubular drive shaft (also cooled) of the sheet metal helical screw. The benefit is that the material is kept in continuous motion, which prevents agglomeration even at low screw speeds.

16.3.2.4 Air-cooling lines

These lines are used to cool relatively thin-walled products. Despite the long dwell time this method is used for all products that must not come into contact with water.
The lines are laid out with several sections at different levels one above the other or, less frequently, in a looped arrangement.
The cooling blowers required are mounted either beneath the line or alongside it. They can also be placed outside the building. The air nozzle is installed close above the product web. In special situations with enclosed lines, cold air is used.
Cooling granules in air shafts by means of cold-air blowers is the basis of a special kind of air-cooling line. Using appropriate pipe conduit, the pellets can be transported to the filling, weighing or storage points while being cooled.

16.3.3 Cooling lines

Cooling lines are used to cool intermediate products in strip or pellet form, or for finished product as lace. Depending on the product, quite different transport systems are used, e.g. conveyor belts, mesh belts, wire-grid belts with rubber chains, roller tracks, steel belts, cooling drums, bucket elevators and so on.
Certain chemical and process-related aspects must be taken into account, as well as the shape of the product, when deciding on the cooling medium or system to be used. If, for

example, water is unsuitable because the product must not be brought into contact with it, air cooling or indirect cooling will be needed.

Other operations besides cooling are built into cooling lines — for example control (weighing, measuring), doubling, labeling (stamping, printing), cutting (lengthwise, across), drying (during wet cooling), wind-up, stacking, coating, pallet lay-down, and so on.

16.3.3.1 Slab-cooling lines

Cooling lines for sheeting (batch-off lines) are employed in the course of compounding and are coupled to the shaping equipment which follows the kneaders. Mills, extruders with slot dies, tubular dies, or flat-sheet die heads, or even roller-die lines are used as shaping devices (Figure 29).

Figure 29. "Batch-off" sheeting line. (Photo: VMI-Epe)

The best known form of construction uses a compact, revolving conveyor system with projecting rods that are continuously presented to the underside of the incoming sheet. The result is that the sheet hangs between the rods in loops and is carried through the water-spray cooling and drying lines, or through the air cooling line, and is laid on pallets in the normal wigwag fashion.

To prevent the web from sticking to itself, either immersion baths containing release agents (e.g. talc, zinc stearate) are installed before the cooling line, or powder applicators are used for this purpose after the cooling line.

16.3.3.2 Strip-cooling lines

A variety of different processes are used for cooling narrow strips, of which the following are examples:

16.3.3.2.1 Lines for the simultaneous cooling of several strips

They are built like cooling lines for sheeting, and use the same cooling methods.

16.3.3.2.2 Capstan lines for cooling strips of raw compound

On these very compact lines, the strips are placed on the rotating rods by a conveyor belt. Big loops are formed and hang down beneath the individual rods, which themselves turn slowly around the common perpendicular axis; the loops are first sprayed with parting agent and then cooled and dried in an airstream. Shortly before reaching the exit point the strips are taken off by a second conveyor belt and swung away or wound up, so that each rod can be freshly loaded immediately after reaching the exit position.

16.3.3.2.3 Mesh drums for cooling strips of raw compound

These units use a drum rotating about its horizontal axis, with the lower half dipping into a cooling bath, so that the helically wound strips find themselves alternately in the water or above it for evaporation. This rather simple, inexpensive design is adequate for many purposes.

16.3.3.3 Pellet-cooling lines

Cooling lines are needed as downstream equipment in pelletizing operations, to cool the pellets sufficiently to prevent agglomeration during storage. Many different variations of water-immersion or water-spray, and air- or contact-cooling lines are in use. The use of effective parting agents is even more important on pellet lines than on strip- and sheet-cooling lines, to prevent agglomeration and to guarantee trouble-free processing later.

16.3.3.4 Tread cooling lines

Such lines have the task of cooling the single or multicomponent extruded strip as quickly as possible to the maximum permissible temperature for storage, while holding the tightest possible tolerances.

To avoid distortions, these very long immersion or spray-cooling lines should be laid out in as extended a form as possible, and be equipped with several conveyor belts. Multi-tier construction is frequently necessary, for space-saving reasons. Deformations are kept to acceptable limits by using large radii at points of direction change. On transfer from one belt to another, auxiliary devices such as dancer rolls are needed. These can provide such good matching of one transport device to another that tension-free transport is possible. This is the precondition for controlled shrinkage.

Treadstrip-cooling lines incorporate additional operations besides cooling. The facilities required are listed below in the order of their arrangement on the line:

a) take-off rolls that can be swiveled or displaced sideways,
b) shrinkage roll train — speed difference of 5 to 10%,
c) throughput weight-tolerance scales (weight per meter),
d) dimension measurement devices,
e) cushion calender with doubling rolls,
f) connecting roll train with color printing or embossing attachments,
g) cementing unit,
h) exhaust conveyor,
i) upward and downward transporters, possibly with deflector devices,
j) immersion- or spray-cooling baths with special conveyor belts, deflector rolls, dancer rolls and so on,

k) water blow-off- and drying equipment,
l) tread cutter,
m) acceleration conveyor,
n) check weight scale,
o) booking station,
p) tread and cementer,
q) possibly automatic booking unit.

Depending on the design of treadstrip, some of these functions can be omitted, or replaced by others: for example, drive belting can be wound into cartons or on drums.

16.3.3.5 Side-strip-cooling lines

These lines are similar to those for treads. For reasons related to the nature of a particular product, some elements of the line, for example shrinkage conveyor, cushion calenders and the devices listed under 16.3.3.4 (m) to (q) are not necessary. Others are replaced by simpler ones − for example, like the cutting device and the wind-up.

16.3.3.6 Sheet-cooling lines

Cooling lines for rubber sheeting can be designed like those in sections 16.3.2.2 to 16.3.2.4. Normally some additional devices are needed for maintaining width and guiding the web, and providing for low tension transport; and also, if needed, for lengthwise and transverse cutting, doubling stations, and payout and wind-up equipment.
In connection with doubling and square-edged wind-up, treads may be wound into cassettes or onto reels. Tension controls as well as edge guiding or center guiding systems are being successfully employed.

16.3.3.7 Profile-cooling lines

These are usually needed for cooling the cured profile at the end of continuous vulcanization lines. On LCM lines the removal of residual salt contributes to the cooling effect. As a rule, multistage spray-cooling sections are brought into use. Flexurally stiff profiles can be cooled on extended spray-cooling lines. Air-cooling lines are used especially for flocked profiles.
All wet cooling lines for profiles require appropriate haul-off equipment and a drying section at the end which, in normal circumstances, is fitted with blowers and ring nozzle or a multitude of individually adjustable nozzles.

16.4 Measurement and control equipment

16.4.1 General

Electronic devices and control measures which lead to constant product properties are described in the following sections.
To ensure constant operational conditions, systems with feedback in the control circuit and dedicated set-value control are used. Because of the large number of individual process and product variants, and of the resulting multiplicity of configurations, only particular arrangements for special tasks are given as examples.

16.4.2 Extruder controls

Constancy of extruder and calender operating characteristics largely determines the uniformity of product properties. Especially direct influences are feeding, thermal control, and the screw-speed/pressure characteristics.

16.4.2.1 Feed devices

The arrangement of feed devices on cold-fed rubber extruders differs greatly depending on the operational circumstances. In feeding with special belts which move the slabstock from pallet systems to the extruder, a tachometer-roll is generally fitted on the output side of the belt, to record the feed rate. This quantity is used as the setpoint value for the feed belt.

A reliable feed method for all operational conditions (speed, slab quality) is the arrangement with preloaded dancers between extruder and feed belt. Greater control effort is needed here, however, in order to recognize and overcome problems in the start-up and changeover phases of slab feed.

16.4.2.2 Temperature control

Thermal control on rubber extruders and calenders is generally achieved with heating/cooling systems in which the heat exchange medium (usually water) is passed through the zones to be controlled. Normally the demand on cooling capacity during continuous extrusion is very large. Control of cooling is achieved by means of closed-loop feedback control of valve adjustments to modify the flow of coolant. In general the design in the control zones is physically such that there is a good deal of mass between the region of coolant influx and the temperature measuring point. The associated large thermal inertia makes itself felt in the frequency with which the supply of coolant to the controlled region is switched on and off, and results in a semi-continuous supply of coolant. Control points which do not have a large thermal inertia, because of design constraints, are increasingly being fitted with continuously operating systems.

16.4.2.3 Pressure/screw-speed control

One possibility for optimizing the continuous rubber extruder, using signals direct from the extruder, is to superimpose pressure control on DC drive machines that usually have speed control based on thyristor controllers. The actual value of melt pressure is measured before the die and transmitted to a reference point, where it is compared with the required pressure level. The drive speed is then controlled to keep the pressure conditions in the extruder constant.

The success of this arrangement is very dependent on the material and the die. There are three significant influences:

a) the pressure signal is generally not linear over the melt-throughput range involved,
b) the pressure signal exhibits time-dependent behavior when changes are made, for example when screw speed is altered,
c) the pressure signal alters with temperature/viscosity changes.

This kind of pressure/screw-speed control is frequently used with roller-head units and is effective in direct control of the size of the charge for the roller-head calender. A variant used in some situations is to have correction-signal feedback control for calender speed as well as for the extruder. The effectiveness of this kind of closed-loop control is strongly dependent on the geometrical conditions in the exit region of the die, and on the point at which pressure is measured.

16.4.2.4 Melt-temperature/screw-speed control

Another system of closed-loop control affecting extruder screw speed directly is used with shear heads. The temperature of the melt coming from the rotating shear mandrel and the speed of the mandrel drive are measured, and signals related to these quantities are used to keep the temperature constant despite differences in throughput, input temperature or melt viscosity. It should be noticed that the control system holds the temperature at the required value only at the measurement point. Temperature gradients or local overheating have to be limited or avoided by supplementary means.

16.4.3 Haul-off-speed control

Great importance attaches to the adjustment of the speed of conveyor belts (for cooling, vulcanization and so on) downstream of the extruder or calender. If the material output rate is non-uniform, the extruder take-off tension is ill defined, and unstable haul-off conditions can arise and bring about faulty stretched areas in the material. Direct measurement of tension is not generally possible in this region. Proposals to measure the speed of output of the compound at the die, with provision for some percentage reduction in the speed of the haul-off belt, or to match it to the extrudate, generally founder on the costs of such speed-measurement equipment. Alternatively, control of sag with the help of optical or distance-recording measurement systems and downstream closed-loop control of the haul-off can be successful; in such cases the material is not supported for a short distance after leaving the extrusion head. The drive of the haul-off roll train is then controlled to give a constant amount of sag. This drive also serves as the master drive for following motors. A change in the desired amount of sag is then synonymous with changes in other haul-off conditions (stretching ratios).

16.4.4 Measurement and control of product dimensions

The measurement of product geometry or weight is not possible until about 5 to 10 meters after it leaves the extruder or calender due to arrangement of the line. At low production speeds, above all, dead times are long, making the correction of short-term disturbances impossible. Basically these measurements are employed to keep the product within permissible tolerances over long production periods, and to regulate trends. Since the production speed on a given line can differ greatly from product to product, particular importance attaches to the design of this controller, whether the feedback is to the extruder or to the haul-off belt. The dead time can be regarded as advantageous when a controller is incorporated. With the help of a rotation pulse generator built into the conveying system, a new current value is fed back to the controller after a defined displacement from the extruder measurement point has occurred. The controller should generally be of the proportional (P) or proportional-integral (PI) type. If controllers are designed to be fully digital (DDC) the effect of the I-portion, in relation to the production speed, is simpler to define.

A variety of systems developed for specific products have been employed for geometrical and weight measurements.

16.4.4.1 Profile measurement

Two representative variants on measurement devices for profiles are:

- optical systems for measuring one or two principal dimensions. It is important for this purpose that the profile should always be presented in the correct attitude on the measuring plane, to avoid measuring errors due to tilting. It can, however, be disadvantageous that other geometrical features are not recorded.
- air-flow- or pressure measurement in a head which is provided with coarsely profiled cover plates and is fed with preheated air from a special compressor. This procedure gives an indirect measure of the cross-section of the profile. The set value for control is obtained by inserting a standard sample into the measuring head. Finely segmented profiles can be measured with this system, and it will record the loss of a lip, for example.

16.4.4.2 Thickness measurement

For thickness measurements on extended profiles, like intermediate products for tires, increased use is being made of traversing systems, replacing point-by-point measurement. These make it possible to obtain an almost continuous record of the thickness profile.
Single-point measurement can be carried out with:

- contact roll systems, that are placed on the product to provide a digital or analog measurement of displacement;
- systems having a measuring head which floats on an air-cushion a fixed distance above the product surface;
- laser systems that record the distance between the laser sensor and the surface. With a double-sided system the actual thickness of a test sample can be calculated and recorded;
- radiation absorption or back-scattering methods, which can be calibrated for thickness to establish a standard for each compound.

All contactless systems whose measuring heads are at an adequate distance from the surface are suitable for traversing operations. The traverse movement is recorded by a length-measurement system. Plotting of thickness values as a function of traverse displacement is done by microprocessor.
Each of the measurement processes listed can be used with good results only on specific problems.
Because of their mechanical form, contact rollers give an effective average measurement over a large patch. Depressions are not picked up if they are small in comparison to the roll radius. The weight of the measuring system itself can cause problems; weight compensation must be very carefully effected mechanically, otherwise the system will be liable to mechanical flutter.
Pneumatic systems have the same electrical elements as roller systems, but require a higher expenditure on auxiliary mechanical devices and energy supply units.
Laser systems for two-sided measurement must be operated within a very tight tolerance range against the upper and lower sides of the web. Sagging of the web can cause the working range to be exceeded and lead to inaccurate measurements. The surface structure is also important since the measurement principle depends on diffuse scattering of the laser beam by the surface. Laser measurement systems are extremely accurate.
Radiation methods are susceptible to long-term variations in the composition of compounds, and there is a specific operating range, according to the type of radiation, that must be adhered to. Radiation methods in general are incapable of picking up sharply defined contours, since the radiation covers a big area.

Thickness measurement on webs and films can be carried out with the same systems, which are frequently set up with a solidly mounted, temperature-controlled roll of large circumference as the measuring base.

16.4.4.3 Width measurement

Width recording can be regarded as another geometrical measurement technique. This is generally based on optical measuring procedures and is therefore not technically an extension of the systems so far described. In a production sense, width is often an interesting quantity, for example in connection with automatic insertion of lengthwise slitting systems, or as evidence about stretching ratios.

16.4.4.4 Weight measurement

Continuous weight measurement can be carried out with on-line scales. These have a measuring table with a self-powered roll train. The requirements are, among others, for vibration-free bearings (and if necessary, vibration-free foundations) and balanced drive elements.

16.4.4.5 Control of measured quantities

Every extrusion line and every multicomponent profile to be produced demands a special configuration of measurement system and feedback arrangements. With the greatly increased success of microprocessor controls, an appropriate tool is now available for building up these dedicated systems in a flexible fashion, specific to product if required (Figure 30).

Figure 30. Main console with microprocessor controller for a combined roller head/EWK line. (Photo: Troester, Hannover, West Germany)

16.4.5 Control of set values

A second essential measure for ensuring constant, reproducible operational conditions, in addition to the installation of control circuits, is the use of product-code controlled target values. A microprocessor system is needed for this. All target values and adjustable quantities (rotational speeds, slitting widths, cut lengths, temperatures and so on) which from experience are needed to ensure satisfactory manufacture of a particular product, can be set on the line before production start-up by the selection of a product code (number or name). Fine adjustments of individual units to specific tolerances, depending on design, can be made feasible.

References for Chapter 16

For the convenience of the reader the English titles of all publications in languages other than English are shown in parentheses.

1. Extruders

[1] *Lang, M.:* Mischungstechnische Grundlagen der Kunststoff- und Gummiindustrie (Mixing Fundamentals in the Plastics and Rubber Industries). C. Marhold Verlag, Halle, 1950.
[2] *Hanser, E. A.:* Handbuch der gesamten Kautschuktechnologie (Handbook of General Rubber Technology). Union dtsch. Verlagsges., Berlin, 1953.
[3] *Le Bras, J.:* Grundlagen der Wissenschaft und Technologie des Kautschuks (The Fundamentals of Rubber). Berliner Union, Stuttgart, 1956.
[4] *Schatz, O.:* Neuentwicklung Kaltgummiextrusion (A Recent Development in Cold Rubber Extrusion). Rubber Symposium, Gottwaldow, CSSR, 1963.
[5] *Schenkel, G.:* Kunststoff-Extrudertechnik (Plastics Extrusion Technology). Carl Hanser Verlag, München, 1963.
[6] *Baier, H., Romanowski, A.:* Kaltfütterung von Kautschuk-Spritzmaschinen (Cold Feeding of Rubber Injection Molding Machines). Gummi, Asbest, Kunstst. (1963) 7.
[7] *Romanowski, A.:* Stand der Extrudertechnik (The State of Extrusion Technology). Part I: Kautschuk-Extruder (Rubber Extruders). Kautsch. Gummi, Kunstst. (1964) 9, pp. 513/523.
[8] *Fellenberg, K.:* Spezialaufbereitung von Kautschuk-Mischungen im Vakuum (Special Compounding of Rubber Mixes in Vacuum). Kautsch. Gummi, Kunstst. (1965) 10, pp. 665/669.
[9] *Baumgarten, W.:* Kaltbeschickung von Kautschuk-Schneckenpressen (Cold Feeding of Rubber Extruders). Kautsch., Gummi, Kunstst. (1965) 10, pp. 670/674.
[10] *Baumgarten, W.:* Extrudieren von Kautschuk-Mischungen mit Entgasungs-Schneckenpressen (Extrusion of Rubber Mixes on Degassing Extruders). Kautsch., Gummi, Kunstst. (1966) 8.
[11] *Perlberg, S. E.:* Operation and Application of Transfermix. Rubber World (1967) 6.
[12] *Parshall, C. M., Salino, A. J.:* ... Shearmix/Transfermix. Rubber World (1967) 2.
[13] *Lehnen, J. P.:* Maschinenanlagen und Verfahrenstechnik der Gummiindustrie (The Machines and Processing Technology of the Rubber Industry). Verlag Berliner Union, Stuttgart, 1968.
[14] *Proksch, W.:* Tandem-Extruderanlage – eine Neuentwicklung für die Reifenindustrie (Tandem Extruder Lines – a New Development for the Tire Industry). Kunststoff u. Gummi (1968) 7.
[15] *Parshall, C. M.:* Continuous Mixing and Processing. IRI Manchester (1969) 4.
[16] *Menges, G., Lehnen, J. P.:* Gummiverarbeitung auf Einschnecken-Extrudern (Rubber Processing on Single-screw Extruders). Plastverarbeiter (1969).
[17] DT 1912459 (1969): Verarbeitung von Kautschukmischungen (Processing of Rubber Mixes).
[18] *Menges, G., Lehnen, J. P.:* Übertragbarkeit von Mischqualität und Ausstoß an Kaltgefütterten Einschnecken-Extrudern (Transferability of Mix Quality and Output on Cold-fed Single-screw Extruders). Plastverarbeiter (1970) 5.

[19] *Lehnen, J. P.:* Verarbeitung von Kautschuk-Mischungen auf kaltgefütterten Einschneckenextrudern (Processing of Rubber Mixes on Cold-fed Extruders). Kunststofftechnik (1970) 9.
[20] *Anders, D.:* Kaltgespeiste Extruder und ihre Einsatzgebiete (Cold-fed Extruders and their Areas of Use). Kautsch., Gummi, Kunstst. (1972) 3, pp. 103/108.
[21] *Schatz, O.:* Neuentwicklung auf dem Gebiet der Kaltgummiextrusion (Recent Developments in Cold Rubber Extrusion). Int. Rubber Symposium, Gottwaldov, CSSR, 8/1971.
[22] *Koch, H.:* Einsatzmöglichkeiten von kontinuierlichen Knetmaschinen im Mischsaal (Possible Uses for Continuous Kneaders in the Mixing Shop). Kautsch., Gummi, Kunstst. (1971) 8.
[23] *Dahlhoff, W.:* Konstruktion von Einschnecken-Extrudern (Design of Single-screw Extruders). Habilitation paper at the IKV, 12/1972.
[24] *Gohlisch, H.:* Transfermix für die Verarbeitung von Kautschukmischungen (The Transfermix for the Processing of Rubber Mixes). Gummi, Asbest, Kunstst. (1972), 9.
[25] *Menges, G., Harms, E.:* Neues Schnecken-/Zylinder-Konzept (A New Screw/Barrel Design). Kautsch., Gummi, Kunstst. (1972) 10.
[26] *Röthemeyer, F.:* Rheologische und thermodynamische Probleme bei der Verarbeitung von Kautschukmischungen (Rheological and Thermodynamic Problems in the Processing of Rubber Mixes). Kautsch., Gummi, Kunstst. (1974) 10, pp. 433/438, and (1975), 2, pp. 85/88.
[27] *Baumgarten, W.:* Beschickung kautschukverarbeitender Kaltbeschickungsextruder (Feeding Rubber Processing Cold-feed Extruders). Gummi, Asbest, Kunststoffe (1974) 4, pp. 238/252.
[28] *Harms, E.:* Pin-Type Cold Feed Extruder. Elastomerics (former Rubber Age) (1977) 6.
[29] *Harms, E.:* Einschnecken-Extruder mit Stiftzylinder (Single-screw Pin-barrel Extruders). Kautsch., Gummi, Kunstst. (1977) 10, pp. 735/739.
[30] *Capelle, G.:* Qualitätsverbesserung und Kostensenkung von kaltbeschickten Extrudern für Kalander-Beschickung (Quality Improvement and Cost Reduction by Use of Cold-fed Extruders for Calender Feeding). Gummi, Asbest, Kunststoffe (1978) 1.
[31] *Rüger, W.:* Spezial-Extruderspritzköpfe (Special Extruder Dies). Intern. Rubber Symposium, Gottwaldov, CSSR, 1977.
[32] *Harms, E.:* Kaltfütter-Extruder mit Querstrom-Mischzylinder (Cold-feed Extruder with Crossflow Mixing Barrel). Kunststoffe 69 (1979) 1.
[33] *Gohlisch, H. J.:* The QSM Extruder, a new cold feed Extruder. Intern. Rubber Conference, Venice, 1979.
[34] *Hofmann, W.:* Kautschuk-Technologie (Rubber Technology). Gentner-Verlag, Stuttgart, 1980.
[35] *Anders, D.:* Roller Head-Anlagen (Roller-head Lines). IKT, Nürnberg, 1980, pp. 1/12.
[36] *Gohlisch, H. J.:* Rationalisierungsmaßnahmen in der Extruder- und Kalandertechnik (Rationalization Measures in Extrusion and Calendering). Kautsch., Gummi, Kunstst. (1980) 12, pp. 1016/1021.
[37] *Harms, E.:* Ein neuer Extruder für die Kautschukverarbeitung ... modelltheoretische Überprüfung (A New Extruder for Rubber Processing ... Theoretical Model Study). IKV, Aachen, 1982.
[38] *Lehnen, J. P.:* Kautschukverarbeitung (Rubber Processing). Vogel Verlag, Würzburg, 1983.
[39] *Rüger, W.:* Anlage zur Herstellung von Zwei- und Drei-Komponenten-Laufstreifen und -Seitenstreifen (Production Lines for Two- and Three-component Tread- and Sidewall Strip). Kautsch., Gummi, Kunstst. (1983), 1, pp. 27/31.
[40] US-PS 2.744.287 (1956). *Parshall, C. M., Geyer, P.*
[41] CH-PS 420.581 (1965). *Maillefer, S. A.*
[42] US-PS 3.375.549 (1961). *Geyer, P.*
[43] DE 1.778.770 C3 (1968). *Koch, K.*
[45] DT 2.058.642 (1970), *Koch, K.*
[46] Europ.-PS 0.048.590 B1. *Gale, G. M.*
[47] DT 1.816.440 (1968). *Koch, H.*
[48] DT 1.912.459 (1969). *Menges, G., Lehnen, J. P., Harms, E.*
[49] DT 1.936.418 (1971). *Menges, G., Lehnen, J. P., Harms, E.*
[49a] *May, W.:* Das Einwalzenkopf-System EWk − ... (The Single Roll-head System EWk). Kautschuk u. Gummi, Kunststoffe (1984) 6, pp. 505/508.

2. Continuous Vulcanization

[50] *Baumgarten, W.:* Kontinuierliche Vulkanisation (Continuous Vulcanization). Kautsch., Gummi, Kunstst. (1966) 8, pp. 494/497.
[51] *Baumgarten, W.:* Salzbadanlagen (Salt-bath Lines). Gummi, Asbest, Kunststoffe (1968) 1/2, pp. 24/32, 102/110, and 126.
[52] *Bament, J. C.:* Kontinuierliche Vulkanisation nach dem LCM-Verfahren (Continuous Vulcanization by the LCM Process). Kautsch., Gummi, Kunstst. (1968) 2, pp. 69/73.
[53] *Davey, A. B.:* Continuous Vulcanization (Fluidized Bed). Rubber Age (1968) 4.
[54] *Schoenbeck, M. A.:* Continuous Curing of Extrusions (Helicure Process). Rubber Age (1969) 5.
[55] *Oettner, C. R.:* Kontinuierliche Vulkanisation im UHF-Feld (Continuous Vulcanization in a UHF Field).BAYER Information brochure (1969) 12, pp. 1/26.
[56] *Gohlisch, H. J.:* Continuous UHF Vulcanization. Rubber World (1970) 5.
[57] UHF-Anlagen für die Gummiindustrie (UHF Lines for the Rubber Industry). Company brochure Hertz Four 7/1970.
[58] *Ippen, J.:* Formulation for Continuous Vulcanization in Microwave Heating Systems. American Chemical Society, Chicago, 1970, pp. 1/17.
[59] *Wiseman, W. A.:* Continuous Curing with Hot Air. American Chemical Society, Chicago, 10/1970, pp. 1/8.
[60] *Boonstra, B. B.:* UHF Heating of Polymeric Materials. Akron Rubber Group (1970) 10, pp. 1/20.
[61] *Gohlisch, H. J.:* Salt Bath and UHF Methods. Rubber Age (1971) 4.
[62] *Humpidge, R. T., Newell, W. G., Morrell, S. H.:* Continuous Processing of Rubber (Fluid Bed). Journal of the IRI (1972) 4.
[63] *Shute, R. A.:* Microwave Heating. Rubber Age (1975) 2, pp. 31/38.
[64] *Focht, H.:* Mikrowellenenergie (Microwave Energy). Kautsch., Gummi, Kunstst. (1976) 4/5, pp. 187/191, and 272/275.
[65] Vulkanisation von Silikonkautschukprofilen (Infrarot) (Infrared Vulcanization of Silicone Rubber Profiles). Gummi, Asbest, Kunststoffe (1976) 7, p. 454.
[66] *van Amsterdam, C. J.:* Coming to Terms with Microwave Vulcanizing. European Rubber Journal (1977) 10, pp. 26/30.
[67] *Focht, H.:* UHF-Anlagen (UHF Lines). Gummi, Asbest, Kunststoffe (1979) 9, pp. 622/632.
[68] *Krieger, B.:* Improvement in Microwave Technology. American Chemical Society, Detroit, 10/1980.
[69] *Lue, Ven L.:* Continuous Vulcanization of Rubber. Rubber World (1980) 6, pp. 26/29.
[70] *Hofmann, W.:* Kautschuk-Technologie (Rubber Technology). Gentner Verlag, Stuttgart, 1980, pp. 472/487, and 516/529.
[71] *Ippen, J., Matenar, G.:* Kontinuierliche Vulkanisation (Continuous Vulcanization). BAYER Company information (1981) 10, pp. 1/23.
[72] *Kroksnes, F., Niehus, G.:* Scherkopf-Technologie (Smear-head Technology). DGF Conference, Nyborg, Denmark, 12/1982, pp. 1/11.
[73] Mikrowellen-Vulkanisations-Anlagen (Microwave Vulcanization Lines). Company brochure Troester (1983) 9.

3. Measurement and Control Equipment

[74] *Gohlisch, H. J.:* Kautsch., Gummi, Kunstst. 33 (1980) 2.
[75] *Bergweiler, E.:* Prozeßsteuerung bei der Flachfolien- und Tafelextrusion durch Mikrorechner (Process Control by Microprocessor in Flat Film- and Sheet-extrusion). Dissertation, RWTH Aachen, 1981.
[76] *Bergweiler, E.:* Meßverfahren zur Dimensionserfassung bei der Extrusion von Kunststoffen (Methods for Dimension Measurement in Extrusion of Plastics). Dissertation, RWTH Aachen, 1981.
[77] Kantenkontrolle bei Parkettelementen − Berührungsloses Messen mechanischer Größen (Edge Control of Parquet Elements − Contactless Measurement of Mechanical Quantities). Holz- und Kunststoffverarbeitung (1981) 2.
[78] *Rüger, W.:* Kautsch., Gummi, Kunstst. 36 (1983) 1, pp. 27/31.

[79] *Capelle, G.:* Kautsch., Gummi, Kunstst. 37 (1984).
[80] Radiometrische Meß- und Regelsysteme (Radiometric Measurement and Control Systems), Company brochure, FAG.
[81] Automation und Prozeßführung von Kautschuk- und Kunststoff-Verarbeitungsanlagen auf Mikro-Prozessor-Basis (Automation and Control of Rubber and Plastics Processing Lines by Microprocessor). Company brochure, Troester.
[82] Scherkopf-Anlagen (Shear-head Lines). Company brochure, Krupp Industrietechnik.
[83] DE-PS 3113747 A1 (1981) *Gerlach, J.*
[84] DE-OS 3036102 2 (1980). *Gohlisch, H. J.*

Acknowledgements

Use of leaflets, photographs, and drawings from the companies listed below is gratefully acknowledged:

Adamson Div., Wean United Inc., Pittsburgh/USA; Hermann Berstorff Maschinenbau GmbH, Hannover/West Germany; Stewart Bolling & Company, Cleveland/USA; Farrel Bridge Ltd, Rochdale/Great Britain; Guix S. A., Barcelona/Spanien; Krupp Industrietechnik GmbH, Hamburg/West Germany; Landshuter Werkzeugbau Alfred Steinl, Landshut/West Germamy; Ets. Lescuyer S. A., Bléré/France; NRM Corporation, Akron/USA; RAPRA, Shrewsbury/Great Britain; John Royle & Sons, Paterson/USA; Richard Tögel Maschinenbau, Hannover/West Germany; Paul Troester Maschinenfabrik, Hannover/West Germany; Uniroyal Ltd., Kitchener/Canada; VMI-Epe-Holland, Epe/Netherlands; Werner & Pfleiderer, Stuttgart/West Germany.

17 Fiber extrusion

M. Mayer

17.1 Introduction

The steady growth in world population carries with it an increasing need for textile fibers; and since production of traditional natural fibers like wool and cotton has remained virtually constant, the increased demand has to be met by man-made fibers.

By man-made fibers we understand fiber products made from polymers which resemble natural fibers in cross-section and properties and, like them, are either cut to fiber length or can be processed uncut as filament yarns.

Since the late seventies, the production of man-made fibers has exceeded that of natural fibers. The demands of the market are forcing the machine builders to develop ever more efficient equipment for the production of man-made fibers, to meet the basic demand for such fibers at reasonable cost.

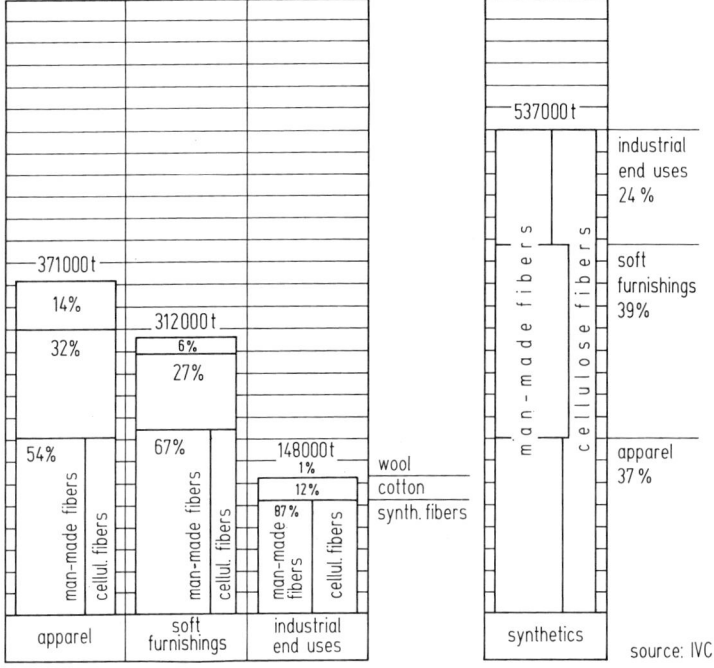

Figure 1. Fiber processing in the German Federal Republic

Man-made fibers are used not only for clothing, but for a host of other products as well. Polyester and nylon are the polymers most widely used to make man-made fibers, but in addition, the production of polypropylene fibers has increased considerably in recent years.

17.1.1 Areas of application for man-made fibers

The three main areas of application for man-made fibers are:
- apparel,
- soft furnishings,
- industrial products.

Figure 1 shows fiber processing statistics for the German Federal Republic for 1981 divided into the above categories. The total quantity of synthetic fibers processed was 537,000 t; and the total output of man-made fibers in the GFR amounted to 932,000 t.

When looking at fiber extrusion, it makes sense to divide spinning machines according to the three application areas.

17.1.2 The principle of fiber extrusion (Figure 2)

The polymer is melted in the extruder, passed through a filter to a manifold and distributed to individual spinning positions. Using precision gear pumps in the spinning heads, the molten material is drawn from spinnerets mounted in the spin packs, into filaments which are cooled by a quenching system. This area of activity will be called Section I. The filaments extruded and quenched in Section I are guided to Section II, the filament treatment section. In principle, Section I, with very slight differences, looks the same for the different fiber processes. It encompasses the following steps: melting, optional filtration, melt distribution, spinning, and quenching. In contrast, Section II is different for different products. Sections I and II are arranged either vertically, one on top of the other (Figure 2) or side-by-side. If the horizontal configuration is used, the extruded and quenched filaments have to be turned around and guided to Section II.

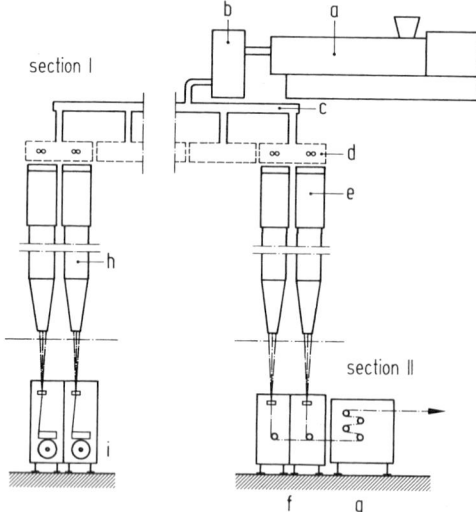

Figure 2. Principle of fiber extrusion
a extruder, *b* filter, *c* manifold, *d* spinning head with pump and spinneret, *e* quenching, *f* deflection, *g* haul-off unit, *h* interfloor tube, *i* take-up machine

It is also possible to arrange the spinning line directly after the melt reactor. In such direct spinning lines, the extruder for melting the chip is omitted and the polymer melt is routed directly from the poly-condensation system to the manifold by means of discharge pumps.

17.2 Extrusion section (Section I)

17.2.1 Extruders

17.2.1.1 History of development

In the early days of the production of melt-spun man-made fibers, the polymer chip was melted with the aid of a melt grid (Figure 3). In such systems, the chip is heated and melted in a vessel with a heating jacket. The so-called melt grid or grid spinner is mounted over the discharge funnel at the bottom of the vessel. By blanketing with inert gas, polymer degradation by oxidation is prevented. The booster pump at the bottom of the grid spinner conveys the melt to the individual metering pumps and the associated spin packs. This simple system, however, has some disadvantages:

- The melt capacity is low.
- In order to prevent thermal damage to the polymer, the temperatures of the melt system must be no higher than approximately 40 to 50 °C above the melting point of the polymer.
- Enlarging the melt surface means increased melt volume and consequently longer residence times.

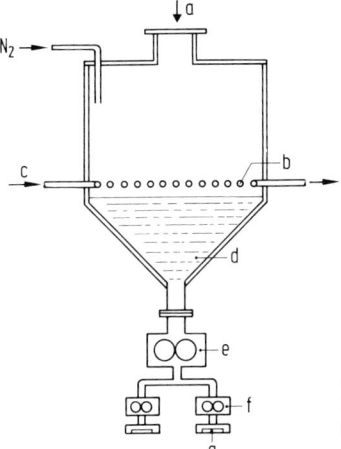

Figure 3. Melt grid
a chip hopper, *b* melt grid, *c* heating medium, *d* melt sump, *e* booster pump, *f* metering pump, *g* spinneret

Transport by gravity only and the virtually pressureless melting action by heat conduction result in extended and vastly differing residence times, especially at the walls of the system. This condition soon led to extruders, already known from rubber and thermoplastic processing, being introduced for melt spinning as well. In order to minimise the mechanical intricacy of the system, extruders with special melt distribution systems are used.
With the development of new fiber polymers, such as polypropylene, and with new spinning techniques such as high-speed spinning and spin-dyeing, the extruder has to carry out additional tasks. They include achieving specific molecular weight distributions or mixing additives into the melt.

17.2.1.2 Design and function of a spinning extruder

The distinctive feature of the extruder is the rotating screw which accepts the feed material, transports it, and discharges it again at increased pressure. Figure 4 shows the in-

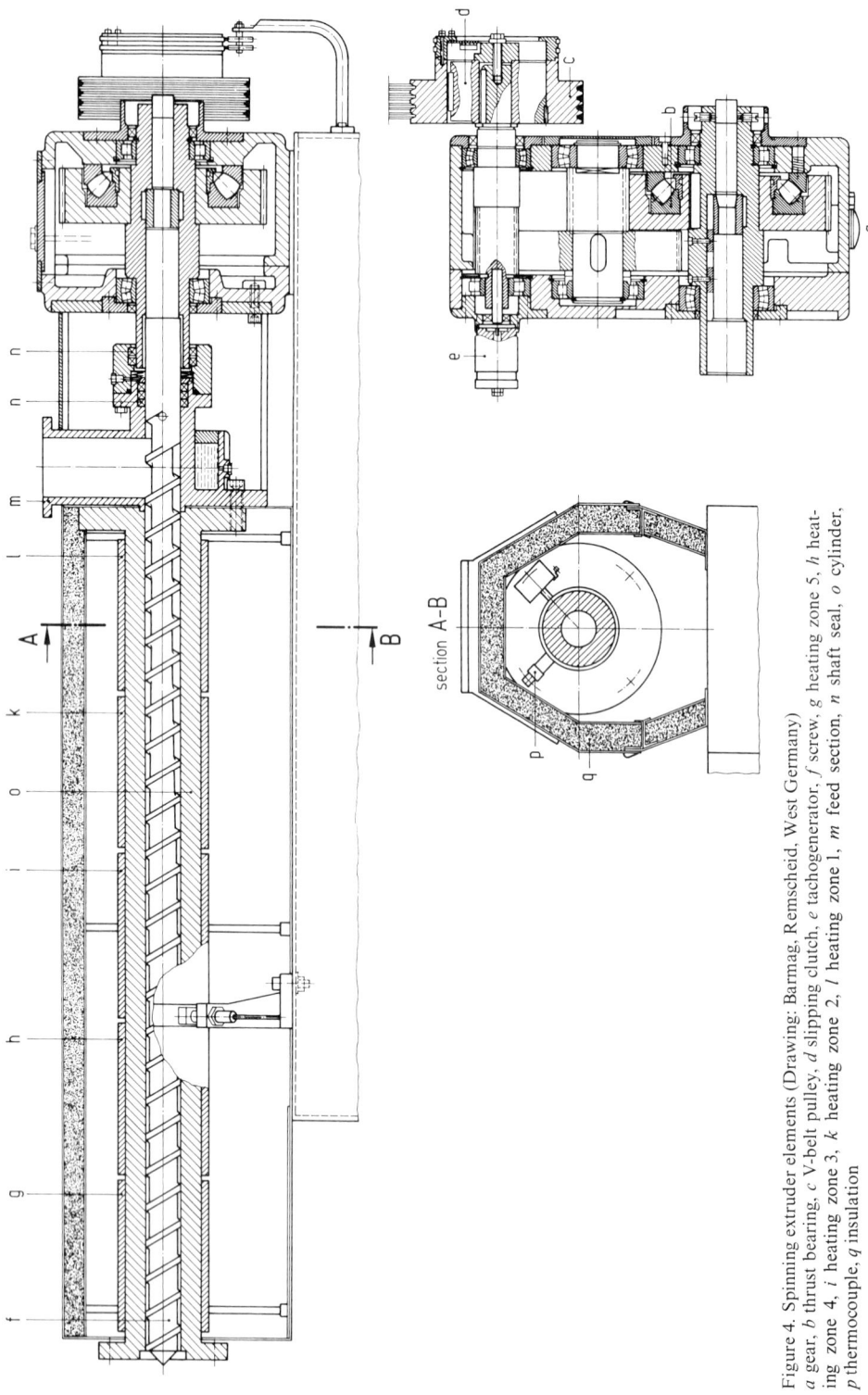

Figure 4. Spinning extruder elements (Drawing: Barmag, Remscheid, West Germany) a gear, b thrust bearing, c V-belt pulley, d slipping clutch, e tachogenerator, f screw, g heating zone 5, h heating zone 4, i heating zone 3, k heating zone 2, l heating zone 1, m feed section, n shaft seal, o cylinder, p thermocouple, q insulation

dividual elements of a single-screw extruder. The chip reaches the screw flights (*f*) through the feed section (*m*), which is cooled and temperature-controlled to prevent premature melting and bridging. The screw is driven by a drive unit consisting of motor, gearing (*a*), and thrust bearing (*b*), and feeds the polymer to the cylinder (*o*) where it is melted by friction and influx of thermal energy. For this purpose, cylinder heating zones (*g* to *l*) are needed, graduated in intensity and length according to heat requirements in order to eliminate temperature peaks. Because of the relatively high processing temperatures of polyester and nylon polymers (260 to 310 °C), the cylinder is insulated in its entirety (*q*) in order to keep heat losses and air conditioning costs to a minimum.

To prevent the chip from absorbing moisture through entry of air and to prevent it from degrading by oxidation during melting, dried nitrogen is usually added at the feed section. The extruder distributes the melt to several spinning positions, with each spinneret being fed by one spinning pump. Because the extruder feeds several spinning pumps, which at times have to be shut off individually or in groups in order to change spinneret or carry out maintenance work, special throughput control for the spinning extruder is necessary. In contrast to extruders which feed the die direct and in which the screw speed is kept constant, with spinning extruders the melt pressure is set at a preselected value at the extruder outlet.

17.2.1.3 Screw design

The melt quality at the extruder outlet is largely determined by the screw geometry. Figure 5 illustrates the most important parameters, with the designation of the individual zones reflecting their main function. In the transport zone, the chip or powder is transported, heated, and compressed. In the subsequent compression zone, it is further compressed and melted. In the metering zone some homogenization is also carried out, by the use of an appropriate mixing device.

There are several theoretical formulae for calculating the overall screw geometry.

For the extrusion of textile man-made fibers from nylon and polyester, a quasi-Newtonian behavior of the melt can be assumed [3]. Provided that the transport and compression zones are designed correctly, i.e. they supply the metering zone evenly with melt for the desired operating range, the simplified formula used for monofilament extrusion can be applied:

$$\dot{m} = 72 \cdot \varrho \left(D^2 \cdot h \cdot n - \frac{10.2 \cdot d \cdot h^3 \cdot p}{\eta \cdot L} \right).$$

The symbols used have the following meaning:

$\dot{m}$	throughput volume	kg/h
ϱ	melt density	kg/m^3
D	screw diameter	m
n	screw speed	min^{-1}
h	flight depth in the metering zone	m
p	pressure at the screw tip	bar
η	melt viscosity	Pa · s
L	length of the metering zone	m

It is further assumed that the pitch is $1D$, i.e. pitch $\varphi = 17°40'$, and the flight width $S = 0.1 D$.

Because of the high melting point and low melt viscosity (Table 1) of fiber-grade polymers, the screw clearance is of great importance. Melt capacity decreases with increasing clearance, which, at increasing screw speed, would be reflected in pressure fluctuations. This means that for fiber extrusion, the extruder screws must be designed to narrow tolerances. Too much clearance in the metering zone means that throughput is reduced by increased backflow over the screw flights.

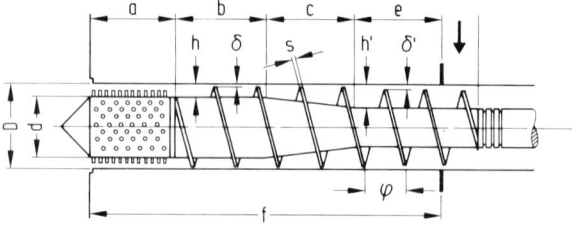

Figure 5. Screw geometry
a mixer, *b* metering zone, *c* compression zone, *D* cylinder diameter, *d* screw diameter, *e* transport zone, *f* effective length, *h* and *h'* flight depth, δ and δ' clearance, φ pitch, *s* flight width

Table 1. Physical properties of some fiber-grade polymers

		PA 6	PA 6.6	PET	PP
Melting point	°C	220	260	255	160
Enthalpy at process temperature	kJ/kg	650 to 700	730 to 780	480 to 520	700 to 750
Melt viscosity at process temperature	Pa · s	80 to 150	50 to 100	80 to 300	150 to 300
Process temperature	°C	255 to 280	280 to 200	280 to 300	230 to 270
Density	kg/m³	1140	1140	1390	906
Hygroscopic		yes	yes	yes	no

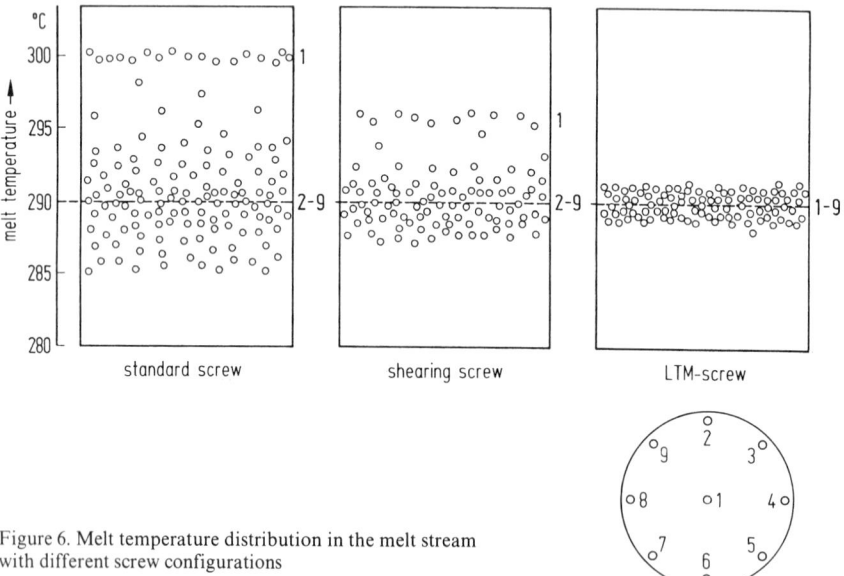

Figure 6. Melt temperature distribution in the melt stream with different screw configurations

The distribution of the melt into several partial melt streams after it leaves the extruder requires uniform temperature distribution at the extruder outlet in order to achieve uniform filament quality at the individual spinning positions. For this purpose, a mixing device at the end of the screw is necessary. Figure 6 indicates the melt temperature distribution measured at the screw outlet for three different screw configurations.

17.2 Extrusion section (Section I)

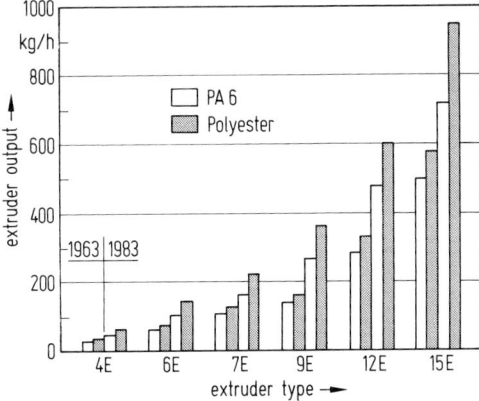

Figure 7. Extruder output

Figure 7 shows how by optimizing the extruder it has been possible to increase the extruder output over the past 20 years. Since polyester and nylon are hygroscopic, drying is necessary before processing can take place. If the chip heated in the dryer is conveyed direct to the extruder, an additional capacity increase of 10 to 20 per cent is possible.

17.2.1.4 Mixing of additives

In the production of carpet yarns, and less often with textile filament yarns for apparel use, spin dyeing is frequently used to obtain colored filaments. Because utmost color uniformity is required in the spun filaments, blending of pigments must be very homogeneous. This is achieved by using a special additive feeding and mixing system like that illustrated in Figure 8. The color master batch is melted in a small secondary extruder and fed into a dynamic mixer through a metering gear pump at the end of the extruder. Once the desired concentration has been set, it is kept constant by means of a control system even if individual spinning positions are shut off. The 3DD mixer has proved to be specially suitable for this purpose [4, 5].

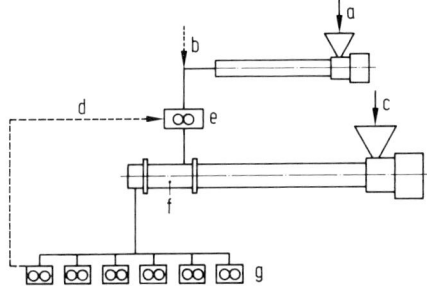

Figure 8. Spin dyeing with 3DD melt mixer
a color batch, *b* color dispersion, *c* chip hopper, *d* feed rate control, *e* metering pump, *f* 3DD melt-mixer extruder, *g* spinning beam

Function of the 3DD mixer (Figure 9): Bow-shaped grooves are milled into the surface of the rotating inner part of the mixer, and into the barrel, with the grooves distributed over the circumference. The melt enters the mixer axially, flows alternately into the outer and the inner grooves and, by the rotation of the inner part, is simultaneously sheared and cut into fine, disk-like partial streams. This explains the outstanding blending effect of the 3DD mixer.

568 Fiber extrusion [Ref. p. 613]

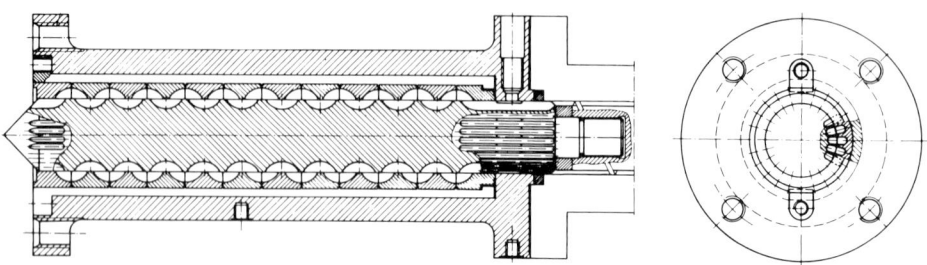

Figure 9. Principle of the melt mixer. (Drawing: Barmag, Remscheid/West Germany)

With the above-mentioned system, it is possible to use the spinning extruder not just as a melting and pump unit, but to blend different polymers, add stabilizers or anti-stats, or effect dye spinning as well.

17.2.2 Manifold

The melt from the extruder is distributed to the individual spinning positions by the manifold (Figure 10). It is the function of the manifold to route the melt to each individual spinning position under exactly the same conditions:

- The time required by the melt to flow from the extruder outlet to the appropriate spinning position must be the same for all positions.
- The melt temperature profile must be the same for each spinning position over the entire length and must be adjustable. Equal times of residence for the melt on the way from the extruder to the spinning position and constancy of melt temperature down to the individual spinning positions are essential for achieving identical fiber quality at each position, since the slightest difference in thermal history and residence time will be reflected in the filament characteristics.

Figure 11 shows a manifold for a small textile spinning line in which approximately 150 kg/h polyester or nylon 6 are spun. To ensure that residence times on the way to each spinning

Figure 10. Manifold between extruder and spinneret. (Photo: Barmag, Remscheid, West Germany)

17.2 Extrusion section (Section I)

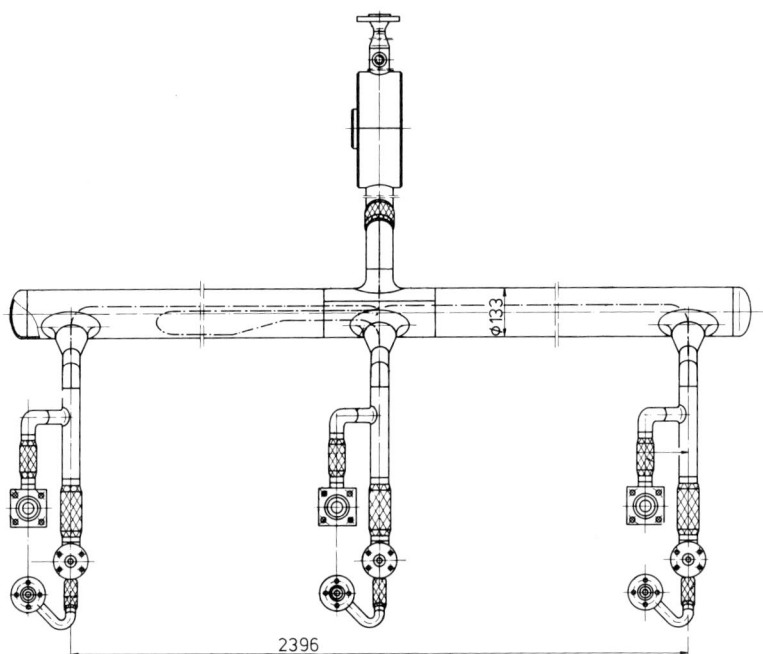

Figure 11. Example of a manifold

position are equal, the melt channels are designed with the same length and cross-section between the extruder and each spinning position. Figure 12 illustrates two versions:
The upper version shows the melt distributed in fork fashion between the extruder and the spinning position, whereas the lower version provides an individual melt pipe between extruder and spinning position which is of equal length to each position. The star-type distribution shown in the lower illustration has the advantage that even if one or several spinning positions are shut off, the residence time from the extruder to the operating spinning positions remains the same. This is not the case with the fork-type distribution, since the residence times in the individual channels vary if one or several spinning positions are shut off. For this reason, the star-type distribution is preferred.
The dot-dash lines in Figure 11 represent the melt pipes, which are of equal length to each of the three positions. The melt pipes are accommodated in a jacketing tube which is Dowtherm-heated. This design meets both requirements – equal residence time and the same thermal history at each position. In order to prevent thermal damage to the melt on its way from the extruder to the spinning position, the residence time must not be too long. Normally, residence times of approximately two to six minutes are established with proper dimensioning of the melt pipes. The resulting cross-sections and channel lengths cause a pressure drop between the beginning and the end of the manifold of 50 to 150 bar. Pressure ahead of the manifold at the extruder outlet is 20 to 50 bar higher. The melt channels have to be designed in such a way that no dead spots are created in which the melt can be retained for a longer period of time and thus suffer thermal damage. During flow through the pipe the melt in the center flows faster than that at the pipe walls. For this reason, some designs incorporate static mixers in the melt runner system. These are elements which subdivide the melt and direct the partial melt streams from outside to inside and vice versa and thus contribute to making the residence time more uniform. A possible drawback of static mixers is that they may cause too great a drop in pressure and create dead spots in which the melt can easily degrade.

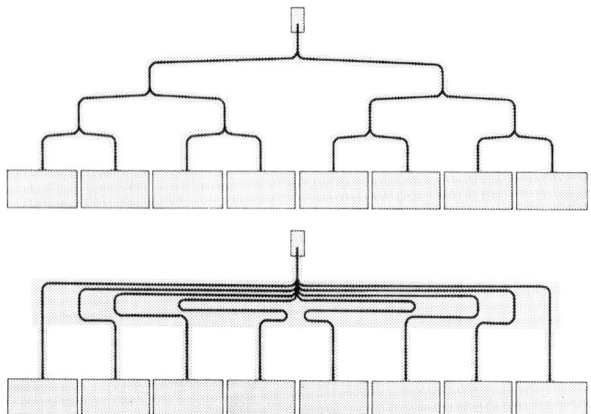

Figure 12. Different manifold layouts

The melt channel systems used in machines for the production of polyester, nylon 6, and polypropylene fibers are usually designed in such a way that they cannot be broken down into sections. For lines in which nylon 6.6 is processed, the system must be capable of being dismantled into sections for cleaning whenever necessary.

The manifold temperatures or the melt temperature depend on the materials processed, and are from 230 to 250 °C for polypropylene and 280 to 300 °C for nylon and polyester. Thus, during heat-up considerable thermal expansion of the individual manifold channels takes place. For this reason, elements which can compensate for the extended length must be installed at the appropriate locations. The manifold is considered to be a pressure vessel and consequently must conform to the regulations of individual countries.

17.2.3 Spinning head (Figure 13)

In the spinning head, the filaments are formed by pressing the melt through fine holes in the spinneret. From the manifold, the melt flows through channels to the spinning pump (a gear pump) where it is precisely metered and fed into the spin pack, from where the filaments emerge. It is the function of the spinning head to accommodate and heat the spinning pump and the spin packs.

The spinning head is designed as a hollow element which is usually Dowtherm-heated. The spinning pump, the spin packs, and the melt channels in between are laid out to achieve as even heating as possible.

On the right-hand side of Figure 13 is shown a spinning head for textile filaments. In this example, four threadlines per spinning position are produced, and thus four spin packs are used, each producing one threadline which, in turn, consists of several individual filaments. The melt is fed to the four spin packs by two twin pumps, i.e. each spin pack is supplied by one pump. The pumps are driven by shafts, c, which, in turn, are driven by bevel gears, b. Release couplings, a, are provided to shut off the individual pumps. The spin pack is inserted in the spinning head from the top and is clamped against the adaptor block to which the pump is fastened with a jack bolt. The melt channels then line up exactly, and unobstructed melt flow is ensured. A sealing ring at the contact pressure point keeps the melt from leaking out. The melt pressure at the pump inlet is approximately 20 to 50 bar; this is then increased by the melt pump to the pressure required for extruding the filaments from the spinneret holes. The melt pump is Dowtherm-heated all round. The adaptors between the manifold outlet and inlet to the pump, and between the pump outlet and the spin pack, can be dismantled for cleaning purposes. This is essential for nylon 6.6. Hollow spaces which occur above the spin pack or the adaptor block are filled with an aluminum-alloy heat sink to promote good heat distribution within the spinning head and to eliminate air currents, which might result in uneven temperatures. On the outside, the spinning head is carefully insulated, d, in order to keep heat losses to a minimum.

On the left-hand side of Figure 13 is another version of a spinning head. The four spin packs are supplied with melt by four twin pumps so that each threadline is produced by one pump. The usual production capacity of such a spinning head is around 10 to 50 kg/h, depending on the desired product.

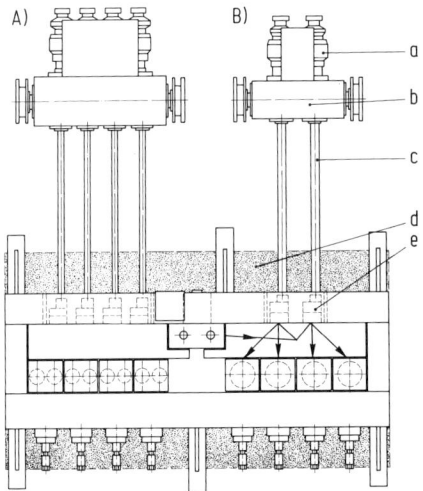
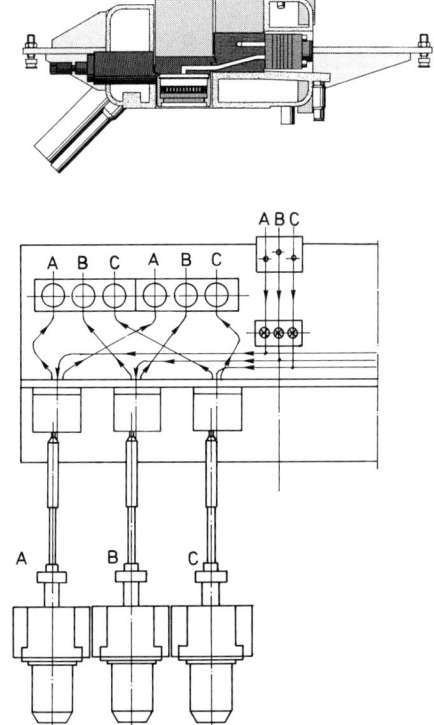

Figure 13. Spinning head
A) design with single pump,
B) design with double-pumps,
C) spinning head cross-section
a coupling, b pump drive, c drive shaft, d insulation, e gear pump

Figure 14. Spinning head for threecolor yarns

Figure 14 shows a spinning head in which three different melt streams, which could be of different colors, are fed by three pumps. These supply two groups of three spinnerets, each with the appropriate melt, to create two groups of three different colored filament bundles. The spinning head, like the manifold, is considered to be a pressure vessel and must therefore be approved in accordance with safety regulations.

17.2.3.1 Gear pumps for spinning

The melt coming from the manifold is conveyed to the spinning pump by channels in the spinning head. The pump feeds the melt at constant rate to the spin packs where the filaments are extruded.

Figure 15 shows how gear pumps work. One gear of the pump is driven by the drive shaft. A counter gear meshes with this gear. The melt enters the pump at manifold pressure and is carried in the gaps between the teeth to the outlet, from where it is routed through channels to the spin pack. The system and the conditions under which the spinning pump operates are described in Figure 13. The melt enters the two spinning pumps from the manifold at a pressure of approximately 50 bar. There are two twin pumps, each pair supplying two feed streams.

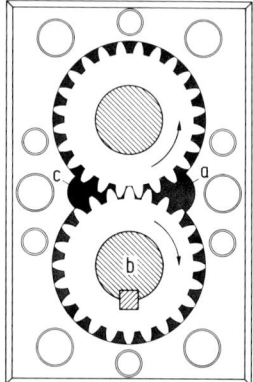

Figure 15. Gear pump function
a inlet, b drive, c discharge

Figure 16 is a schematic view of the twin pump in cross-section and rear elevation. The melt enters at the point a and is conveyed to the outlet b by the gears. The outlet of the left-hand pump in the cross-section is located on the mounting side, and the right-hand pump discharges through the hollow nonrotating axle, but also towards the mounting side.

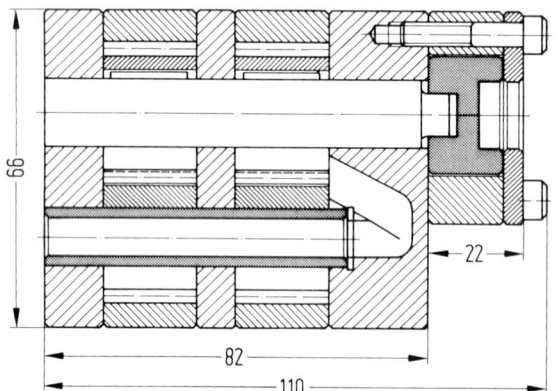
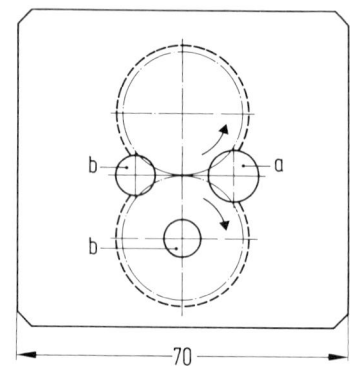

Figure 16. Double pump
a inlet, b discharge

Figure 17 shows the pump exploded view. The two gear pairs k/m and l/n are visible, as is the keyed drive shaft g that drives gears k and l, which, respectively, mesh with gears m and n to transport the melt. The melt fed by gears l/n passes through trunnion i to the discharge hole in the mounting plate f, melt from gears k/m goes to another hole in f.

Each pair of gears in this example has displacement/revolution of 1.2 cm³; and the twin pump output is 2×1.2 cm³ per revolution.

The speed of these precision gear pumps lies between 10 and 40 revolutions per minute, which, in this example, corresponds to an output of 720 cm³/h to 3600 cm³/h. The speed

17.2 Extrusion section (Section I)

should not drop much below 10 revolutions per minute, otherwise slight pulsations in melt flow might occur because of the tooth action.

The requirement that all melt-conveying openings, channels, recesses etc. have no stagnation points in which the melt might degrade, applies also to the spinning pump. Very high demands are made on the feed constancy of the pump. Metering precision from pump to pump must be kept within a tolerance range of approximately 0.5%.

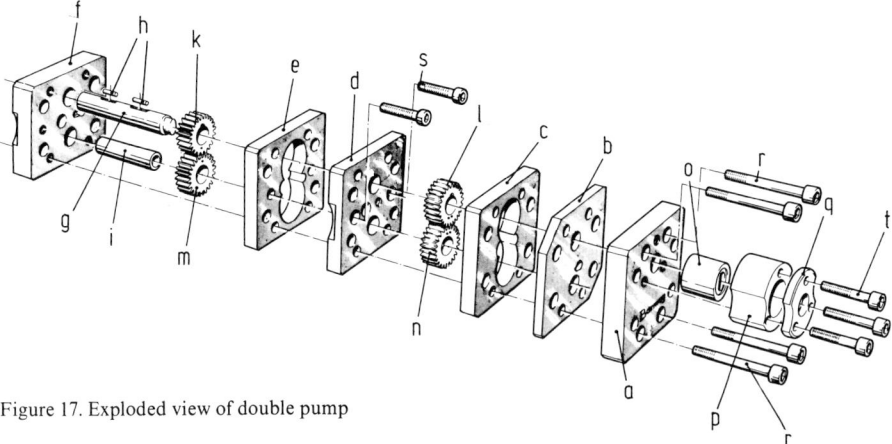

Figure 17. Exploded view of double pump

The pressures required to force the melt into the spin pack and through the spinneret holes are up to 450 bar. Pressure at the pump inlet is 50 bar. The pressure differential to be generated by the pump can amount to 400 bar. Even with these high pressures, feed constancy from pump to pump has to be within the 0.5% range.

The spinning head shown in Figure 13 and the gear pump described are used to produce apparel textile filaments. With a pump speed of 10 revolutions per minute, the output of such a unit is $4 \times 720 \text{ cm}^3/\text{h} = 2880 \text{ cm}^3$ for the four threadlines; at 50 revolutions per minute it is $4 \times 3600 \text{ cm}^3/\text{h} = 14,400 \text{ cm}^3/\text{h}$. If, for example, nylon were processed, this would mean an output of 2.8 to 14 kg/h. In addition to the twin two-gear pumps described, three-, four-, and five-gear pumps are being increasingly used – Figure 18. A three-gear pump has the gears arranged in one plane, with two melt outlets per plane. With a five-gear pump there would be four melt streams. Figure 18 shows a four-plane, three-gear pump providing, $4 \times 2 = 8$ melt streams. Spinning pumps for textile applications normally have output volumes of 0.3 to 3.5 cm^3/rev. per melt stream.

For the production of industrial yarns and carpet yarns, pumps with higher output capacity have to be used, to suit the heavier threadlines required. Such pumps usually have an output of 5 to 30 cm^3/rev.

To achieve the required high tensile strength, the melt viscosity of polyesters for industrial yarns is between 80 and 100 Pa · s at 290 °C. This means that the spinning pump has to generate a high discharge pressure in order to force the melt through the spin pack. Spinning pumps for the production of staple fibers have an output of 20 to 100 cm^3/rev. The temperatures of the melt in the spinning head and, consequently, also in the spinning pumps are between 250 °C and 300 °C, depending on the product being processed. These temperatures, together with the pressures and the resultant forces in the pump, and the required metering precision, make stringent demands on the gear pumps. The individual components of the pumps must be manufactured with the highest precision. This is done in air-conditioned manufacturing facilities. Great care is required during cleaning and assembly of the pumps. In order to attain the required high output precision, constant drive speed must also be ensured. In the spinning head shown (Figure 13), the pump drive shaft is powered by a bevel gearbox driven by a DC motor. Frequently, however, individual drives for each pump, driven by a synchronous motor and a static frequency changer, are used.

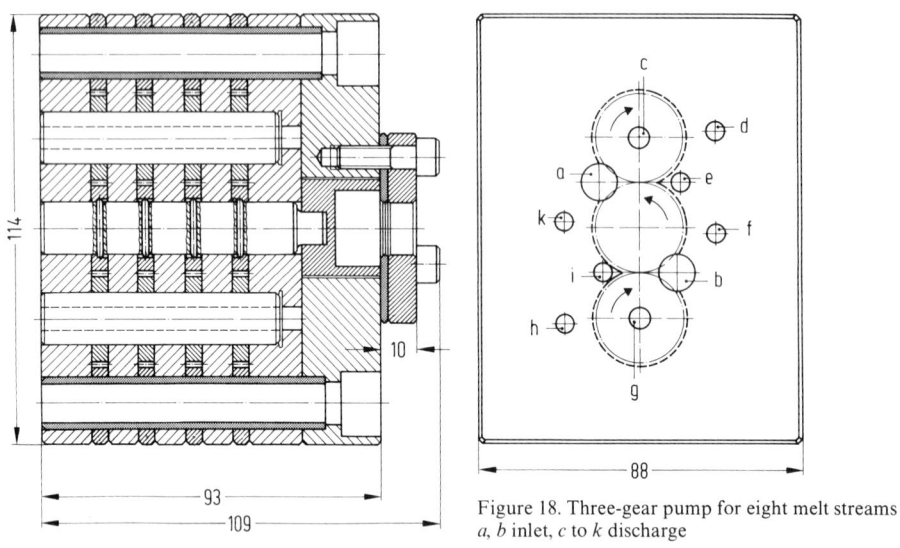

Figure 18. Three-gear pump for eight melt streams
a, b inlet, c to k discharge

17.2.3.2 Spin pack (Figure 19)

The melt reaches the spinning head from the extruder by way of the manifold, and is fed at high pressure to the spin pack (Figure 20). The melt stream is subdivided by the manifold to match the number of spinning heads used.

The melt stream is further divided in the spinning heads into even smaller melt streams in accordance with the number of spin packs. The filaments are formed as the melt emerges from the spinneret. This illustration shows how a spin pack is used on the spinning head (Figure 13).

Thus one textile threadline is produced per spin pack; in this example there are four threadlines per spinning head.

The spin pack is inserted in the spinning head from the top. Before this is done the insulation blocks must be removed from the spinning head. After the spin pack has been inserted it is pushed against the pump adaptor block by means of the jack bolt. The same rule applies to the spin pack as to all components — that the melt be conveyed carefully without stagnation points. By the same token, heating of the spin pack must be even, so that only minimal temperature deviations occur over the entire volume of the spin pack. This is especially true for the spinneret, from which the filaments emerge. The melt fed by the pump enters the spin pack, is forced through a filter bed, and from there flows through the perforated breaker plate into the space above the spinneret. The number of holes in the spinneret is determined by the number of filaments required.

Filter materials used are wire mesh, a (Figure 19) or sand, b (Figure 21). Filtration is also possible with sintered metal. Filtration largely determines the quality of the spun filaments and the reliability of the spinning process. The filter medium purifies, mixes, and distributes the melt evenly before it reaches the spinneret. For spinning of polyester and nylon, sand filtration is frequently used. It filters and effects mixing at the same time, and brings about final homogenization of the melt. Spin packs with sand filtration are usually designed for a pressure of 400 bar; in exceptional cases up to 1000 bar. Normal operating pressures are around 250 bar. When pressure has increased by approximately 100 bar, the spin pack is taken out, cleaned, and filled with freshly prepared sand.

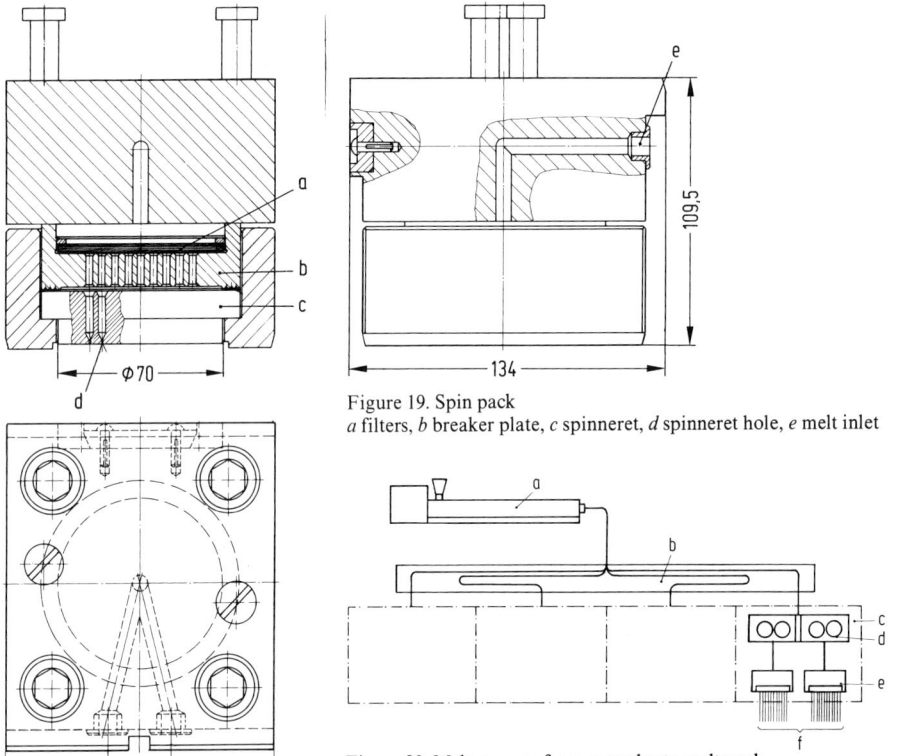

Figure 19. Spin pack
a filters, *b* breaker plate, *c* spinneret, *d* spinneret hole, *e* melt inlet

Figure 20. Melt streams from extruder to melt pack
a extruder, *b* manifold, *c* spinning head, *d* spinning pump, *e* spin pack, *f* filament output

Wire cloth filters are also used for spinning of polyester. In this case, a larger number of filters with 16,000 pores/cm^2 are used. Such spin packs are designed for pressures of 250 to 350 bar. Spinning is usually done at 100 to 150 bar. If the pressure increases because of dirt accumulation the spin pack is removed and cleaned. The purpose of filtration is not only to clean the melt, but also to ensure uniform pressure distribution over the entire surface of the spin pack.

The size of the filter surface is an important factor in determining what throughputs can be achieved with the spin pack. Frequently used values for polyester and nylon are 2.5 to 4 g/cm^2 per min. of filter surface. In rare cases, throughputs of up to 10 g/cm$^2 \cdot$ min are achieved. In the spin pack illustrated in Figure 19, the diameter of the filter surface is approximately 70 mm. Assuming a throughput of 3 g/cm$^2 \cdot$ min, the output would be approximately 114 g/min per spin pack. The melt flows out through fine holes in the spinneret; their diameter is between 0.15 and 0.8 mm. For spinning nylon and polyester filaments, the diameter is usually between 0.2 and 0.3 mm. Larger diameters of 0.8 mm max. are sometimes used for spinning high tex polypropylene filaments, for example for 20 dtex per filament (1 dtex corresponds to 1 g/10,000 m). The ratio between hole length and hole diameter L/d is often 2 for polyester and nylon.

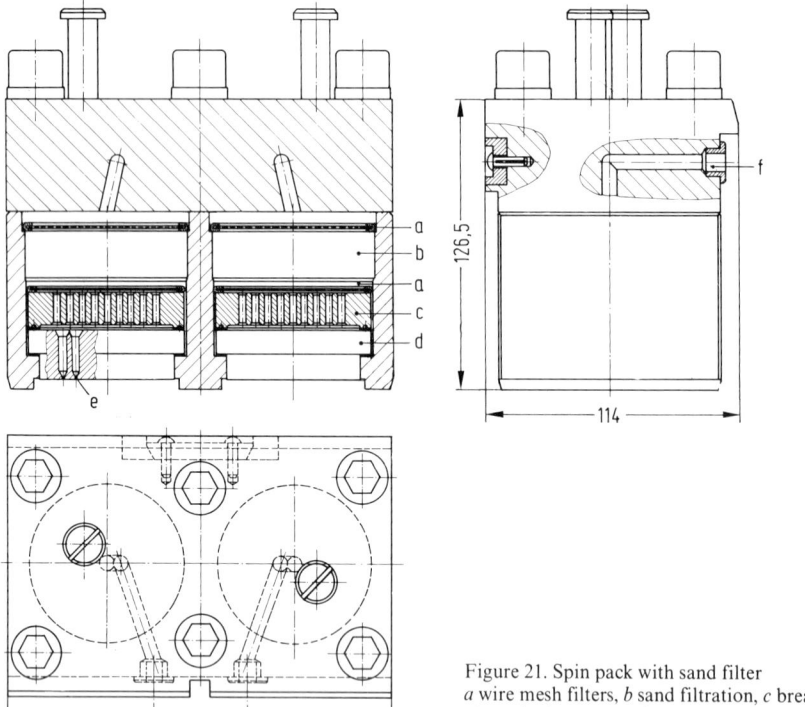

Figure 21. Spin pack with sand filter
a wire mesh filters, *b* sand filtration, *c* breaker plate, *d* spinneret, *e* spinneret hole, *f* melt inlet

For spinning polypropylene, even higher L/d ratios are used, up to approximately 8.
The spinneret holes must have excellent surface quality and, at the outlet, must be flush with the spinneret plate. If the spinneret plates have undergone frequent cleaning, wear shows at the spinneret holes. As a result, the individual filaments no longer emerge with the required uniformity, and this may lead to filament breaks. Therefore spinneret plates must be replaced after a certain length of time.

Figure 22 shows a dismantled spin pack. The spinneret plate, the breaker plate, and above it the filter, in this case wire mesh, can be seen.

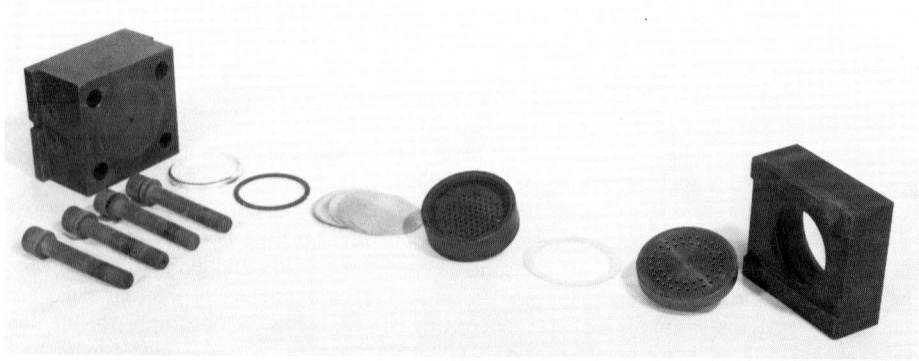

Figure 22. Dismantled spin pack. (Photo: Barmag, Remscheid/West Germany)

17.2.4 Quenching

The filaments emerge from the spinneret in the molten state. The function of the quenching system is to cool them to such a degree that the individual filaments are sufficiently solid to be hauled off downwards. Figure 23 shows the schematic design of the quenching system. The quench air is fed into a box, and flows through a filter pack into the quenching chamber itself, which is open in front or equipped with an air-permeable door. In the quenching chamber, the filaments are directed downward from the spinneret and cooled. In order to achieve completely uniform quenching of the individual filaments the quench air must not be turbulent. And to ensure this a filter pack (or similar arrangement) is provided to give uniform pressure distribution over the entire surface. The speed at which the liquid filament leaves the spinneret hole, is lower than the speed at which it is hauled off. This means that it is drawn down from the spinneret plate. However, cooling by quench air also begins at this point. If turbulence in the quench air were to occur here, cooling conditions would become uneven, and affect the drawing conditions for each individual filament, leading to unevenness of the filaments (fluctuations in thickness). The quench-air volume is set by means of a throttle valve. Turbulence occurring in this area must be carefully reduced, for example with perforated metal plates, in order to prevent pressure fluctuations and consequent differences in air speed, and uneven quenching conditions. The air speeds in quenching systems are around 0.3 to 1 m/sec., depending on the material being processed. For apparel textile filaments, the quenching length is usually around 1 to 1.5 m. For quenching of filaments for industrial applications (carpet yarns etc.) the quenching length is up to 2 m. An increased quenching length is required since the output rate for these end uses is higher, i.e. more material needs to be quenched per unit of time.

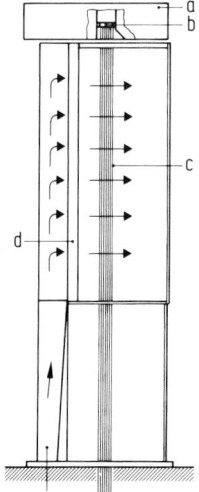

Figure 23. Principle of quenching system
a spinning head, *b* spinneret, *c* filaments, *d* filter, *e* air supply

The quenching of nylon and polyester is different in some respects from quenching of polypropylene filaments. Polypropylene filaments have a larger heat content than nylon and polyester, and thus, particularly for heavier filaments, the quench air is frequently cooled to about 13 to 18 °C. For nylon and polyester, quench air at approximately 20 to 22 °C is generally used.
Polypropylene is processed in the extruder and through the spin pack at a temperature of approximately 220 to 250 °C. This temperature is well above the melt temperature, and is

needed in order to achieve a homogeneous and uniform melt, free of crystal nuclei. In order to reach the solidification temperature of about 160 °C rapidly, the filaments must be quenched immediately after emerging from the spinneret. This produces the tenacity needed to draw out the filaments.

For polyester and nylon, the melt temperature is approximately 290 °C. The solidification temperature is only about 20 °C lower so this point is reached very quickly. Thus it is not necessary to quench immediately after the filaments leave the spinneret. The distance between the spinneret plate and the point where quenching begins is usually 50 to 150 mm.

The speed of the quench air over the entire quenching length is usually constant, i.e. a square speed profile is obtained. Towards the outside, the quenching chamber is closed by an air-permeable door. This serves to reduce the possibility of turbulence in the vicinity of the quenching system so that the freshly spun filaments cannot be affected.

17.2.4.1 Spinneret blanketing (Figure 24)

After repeated cleaning of the spinneret plates, the spinneret outlet no longer has sharply defined edges. As a result, the filaments no longer emerge from the spin pack at a right angle; this, in turn, leads to filament breaks and the need to replace the spinnerets.

Besides this wear-related phenomenon, deposits form around the edges of the spinneret holes during spinning and lead to the same problems. To remove the deposits, the spinneret plates are scraped with a spatula-type tool at intervals depending on the spinning conditions. It is essential that the spatula material is not so hard as to damage the surface of the spinneret plate. Especially when spinning nylon 6 and 6.6, scraping of the spinnerets can be necessary at short intervals. Since this entails interrupting production, efforts are made to lengthen these intervals as much as possible, and by blanketing with inert gas an extension of three or four times is possible. The gas is introduced below the spinneret plate, and instantly virtually eliminates oxidation at the plate surface, and thus reduces the deposits considerably.

Figure 24 shows how the protective gas is introduced from the channels below the spinneret. It penetrates between the individual filaments and blankets the surface of the spinneret. The protective gas is preheated to the temperature of the spinning head so that no cooling of the spinneret plate can occur which would lead to problems in spinning. As a rule, steam is used as protective gas, and less frequently, nitrogen.

Spinneret blanketing is not used in spinning of polyester and polypropylene.

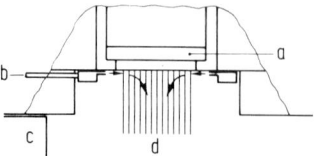

Figure 24. Spinneret blanketing
a spinneret, *b* blanketing gas inlet, *c* quenching, *d* filaments

17.2.4.2 Monomer exhaust

During spinning of nylon 6 and 6.6, substantial deposits of monomer form below the spinneret and in the upper portion of the quenching system. To prevent this, nylon spinning lines are provided with a monomer exhaust system. Figure 25 shows the fundamental scheme. A pipe connected to the exhaust blower sucks the monomers away from the area of the spinneret.

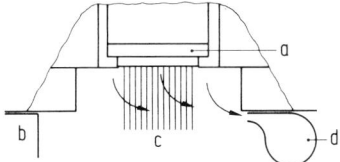

Figure 25. Monomer exhaust
a spinneret, *b* quenching, *c* filaments, *d* monomer exhaust

17.2.5 Heating system

To ensure temperature uniformity of all melt-carrying components between extruder outlet and spinneret, modern spinning lines are heated with Dowtherm vapor. Dowtherm is a thermally stable hydrocarbon. Its boiling point is in the range of the melt temperatures of spinning polymers. The melt temperatures of 230 to 290 °C used in spinning lines require only slight over-pressure in the Dowtherm vapor system. For a temperature of 290 °C, the required over-pressure is 1.1 bar. Spinning temperatures below the Dowtherm boiling point require operation under vacuum.

Figure 26 shows the heating system of a spinning line with two extruders, one manifold, and four spinning heads. From the Dowtherm boiler, which is heated by means of electric heating rods, the Dowtherm vapor rises upward into the spinning head. From there, it reaches the manifold which, as described under 17.2.2, is double-walled, i.e. the melt flows in an inner pipe surrounded by a second pipe which accommodates the Dowtherm. The Dowtherm vapor reaches the measuring head, which means it reaches the extruder outlet. The extruder itself is heated by temperature-controlled electric heater bands. The Dowtherm vapor system functions in such a way that the vapor immediately condenses at points which are below the vapor temperature. This releases the condensation energy of the vapor and results in outstanding temperature uniformity in the entire system. The Dowtherm condensate returns to the boiler through a return pipe. In determining the dimensions of the Dowtherm vapor system it is important to make the cross-section of the vapor feed and condensate return pipes large enough, since a considerable volume of vapor is needed for heat-up. The vapor system is vented at the highest point of the line. To counter thermal expansion in the vapor pipes, compensating elements are used. The Dowtherm boiler is equipped with safety valves which open if the pressure limit is exceeded. The pressure is then decreased by condensation of the Dowtherm vapor in condensation coils. The Dowtherm boiler, and all vapor and condensate pipes are insulated, in order to prevent heat losses.

17.2.6 Filters

Melt filters are increasingly being used between extruder and manifold (Figure 27).
The purpose of the filter is to remove gels and extraneous matter from the melt in order to achieve higher product quality and improved processability.
For spinning of polyester and nylon, filters with very fine pore sizes down to 20 to 30 µm are customary. The fine filters, the melt viscosity and the permissible pressure differentials, mean that large filter surfaces are required. In order to eliminate extended residence times of the polymer melt in the filter, the filter must be designed in such a way that a favorable ratio between filter surface and filter volume is achieved. With the filter described here, this is ensured by a concentric arrangement of the filter candles. The ratio between filter surface (m^2) and melt volume (m^3) is approximately 300:1.

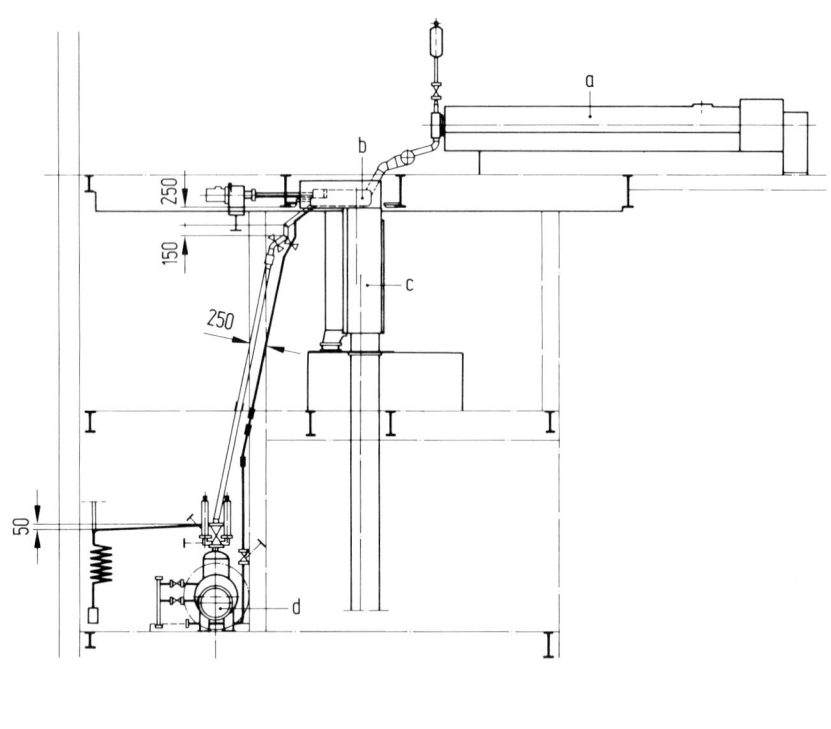
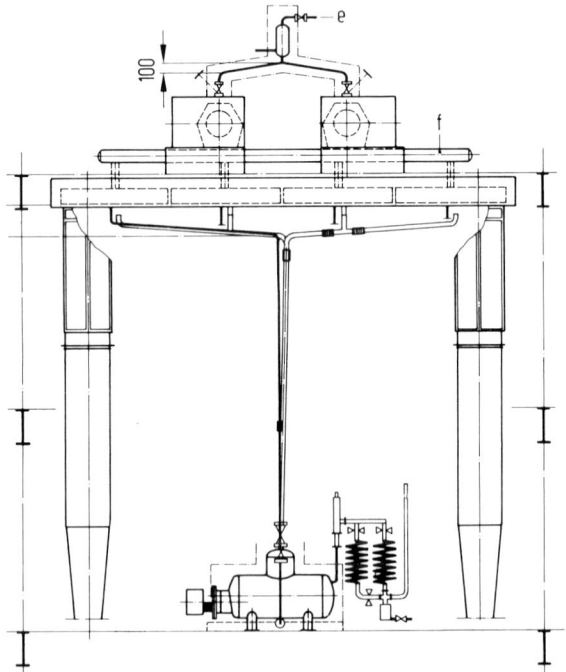

Figure 26. Heating system of a spinning line
a extruder, b spinneret, c quenching, d Dowtherm boiler, e connection for vacuum pump, f manifold

17.2 Extrusion section (Section I)

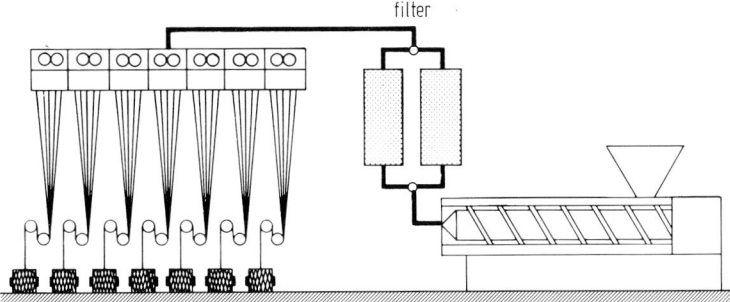

Figure 27. Spinning line with central melt filter

The filter candle consists of a support element surrounded by the filter medium. Figure 28 shows various filter media. Disks, arranged one behind the other, are also used, instead of filter candles (Figure 29).

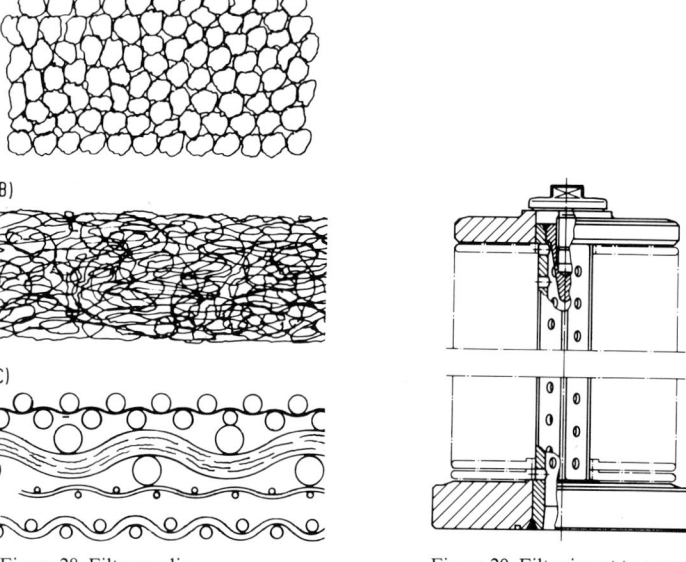

Figure 28. Filter media
A) sintered metal powder, B) metal fiber felt, C) wire cloth

Figure 29. Filter insert to accept disk filters

The filter is of non-stop design (Figure 30) with the filter insert set into the filter from the top. It consists of two identical filter chambers. The melt flows through the insert of one chamber while the insert of the second chamber is being cleaned. When the filter inserts become dirty, one switches over to the second chamber, in which a filter insert with clean filter candles has been placed. In order to prevent degradation of the melt, a nitrogen atmosphere is supplied to the second chamber during the time of the switch-over. Pressure indicators are incorporated at the inlet and outlet of the filter. As a rule, if a pressure increase of 100 bar occurs one switches over to the other filter chamber and the dirty filter is cleaned.

Depending on design, a filter candle may have a surface area of 0.1 to 0.5 m^2. The filters are Dowtherm-vapor heated and insulated.

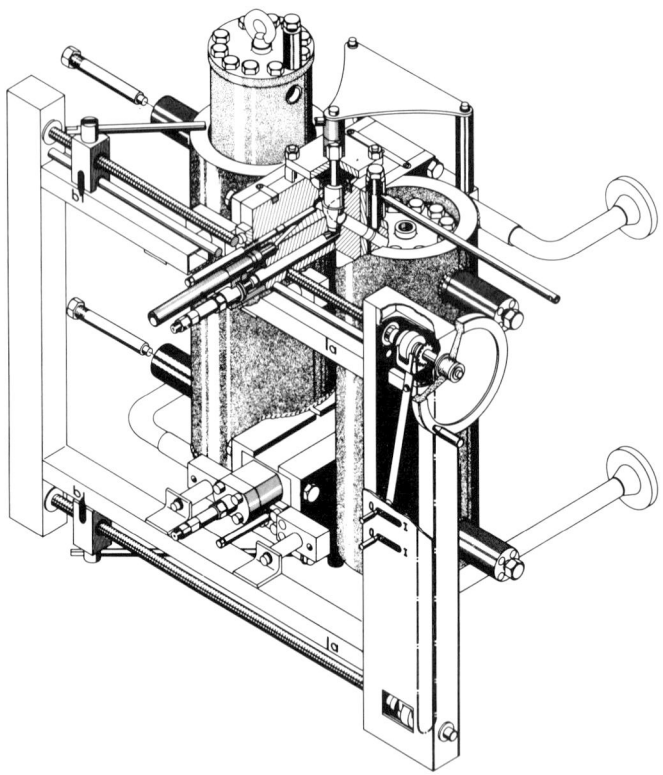

Figure 30. View of twin-chamber filter for filter change without interrupting production

For spinning of polyester and nylon, the throughput of the melt filter is approximately 200 kg/h per m^2 of filter surface. Quality and process improvements are achieved by using filters between extruder outlet and melt pipe. This becomes particularly evident during spinning at high filament haul-off speeds, since in this case especially severe demands are made on polymer melt purity and homogeneity. In production, it has been proved that the lifetime of the spin packs (Figure 31) can be extended three to five times by use of filters. And thanks to fewer yarn breaks, it has been possible to increase the efficiency of spinning lines.

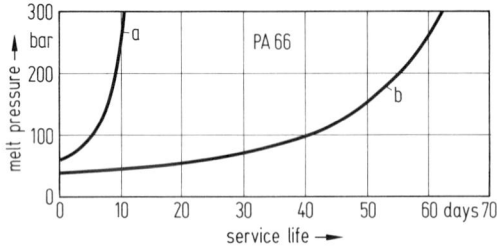

Figure 31. Spin pack service life; a without filter, b with filter

Use of filters in the filament spinning line is particularly advisable if the quality of the polymer chip varies.

17.2 Extrusion section (Section I)

17.2.7 Examples from the extrusion section (Section I) (Figure 2)

Table 2 gives a list of spinning line configurations.

Table 2. Examples of spinning line layout

Area of application		Textile yarns (apparel) Fig. 32	Industrial yarns (tire cord) Fig. 33	Carpet yarns 3 color Fig. 34	Polypropylene staple fiber Fig. 35
Number of threadlines		96	16	48 Combined into 16 3-color threadlines)	8 (each threadline has 2600 filaments)
Material		PET	PET	PP	PP
Viscosity	Pa·s	20 to 30	80 to 100	–	–
Output	kg/h	250 to 450	150 to 420	180 to 540 (for 3 extruders)	625
Extruder					
screw D	mm	120	150	90	170
L	mm	$24 \times D$	$24 \times D$	$30 \times D$	$33 \times D$
LTM mixer		yes	yes	no	no
drive capacity	kW	70	97	51	238
number of heating zones		6	6	5	6
heating capacity	kW	58	92	42	128
Melt filter		twin-chamber filter			single-chamber filter
filter area	m^2	1	no	no	3.5
filter material		pleated wire cloth			pleated metal fiber felt
pore size	μm	25			80 μ
Spinning head					
number of spinnings heads		6	8	4	4
number of threadlines per spinning head		16	2	12 (combined into four 3-color threadlines)	2
spinneret diameter	mm	70	220	100	
filter material		sand	screen cloth	screen cloth	screen cloth
Spinning pumps					
number of pumps per spinning head		8 twin pumps	2	6 twin pumps	2
throughput	cm^3/rev.	2×2.4	1×10	2×10	1×70
Quenching system					
number of threadlines per quenching unit		8	2	6	1
length	mm	1200	1400	1700	1750
width	mm	650	475	660	670
quenching air volume	m^3/h	1417	1200	4000	–
Heating system		Dowtherm		Dowtherm	Dowtherm
Heating capacity	kW	69		42	64

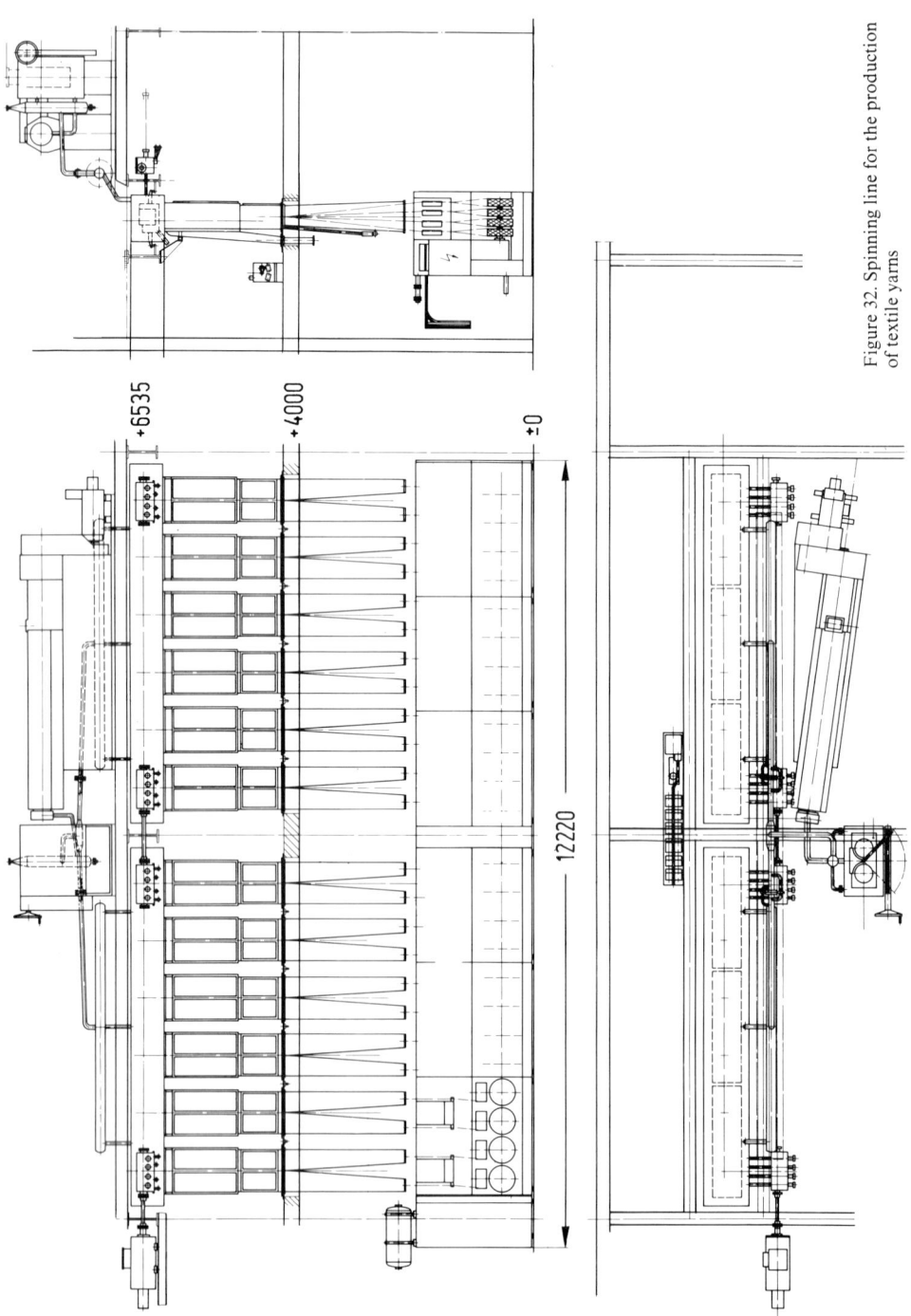

Figure 32. Spinning line for the production of textile yarns

17.2 Extrusion section (Section I)

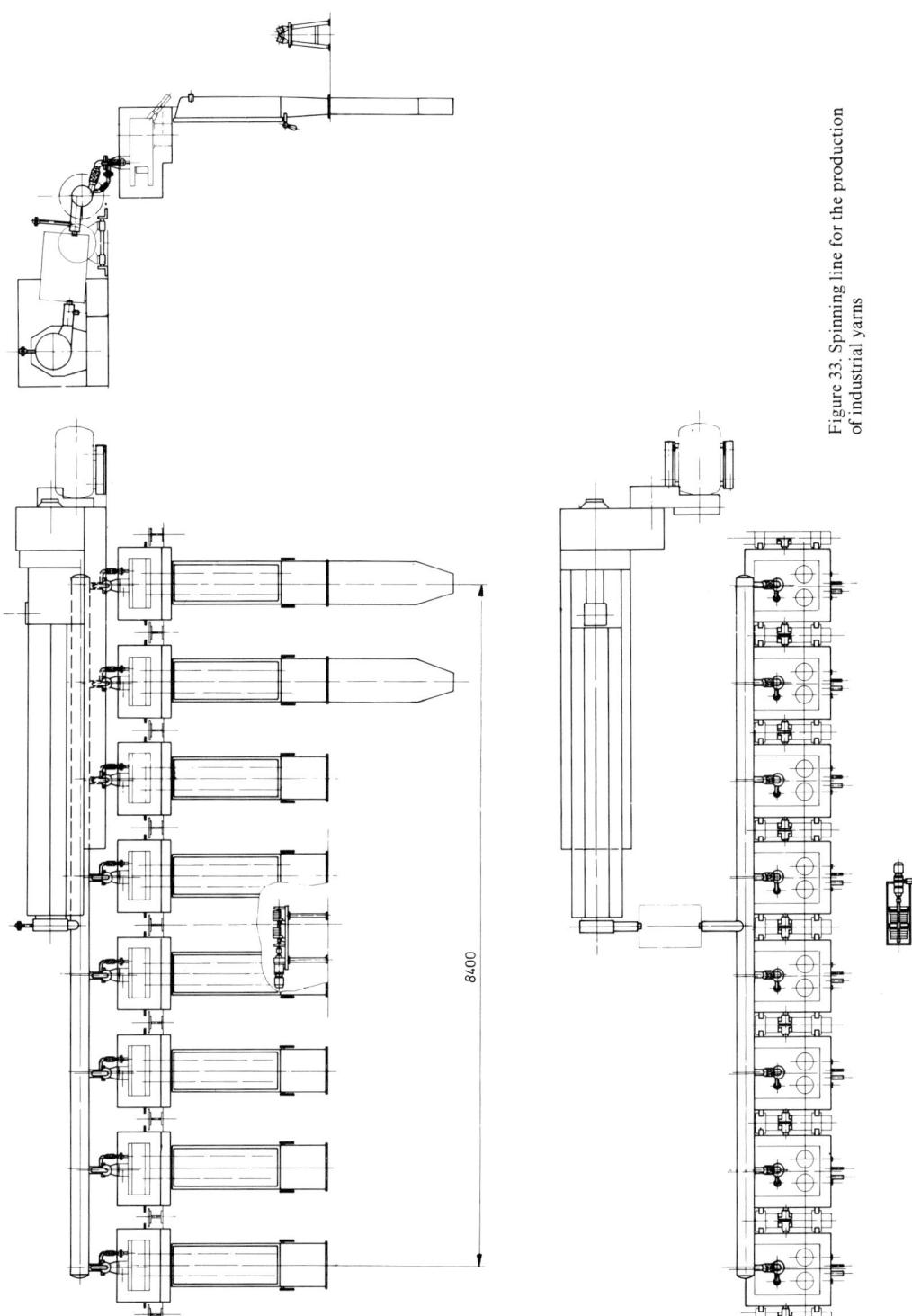

Figure 33. Spinning line for the production of industrial yarns

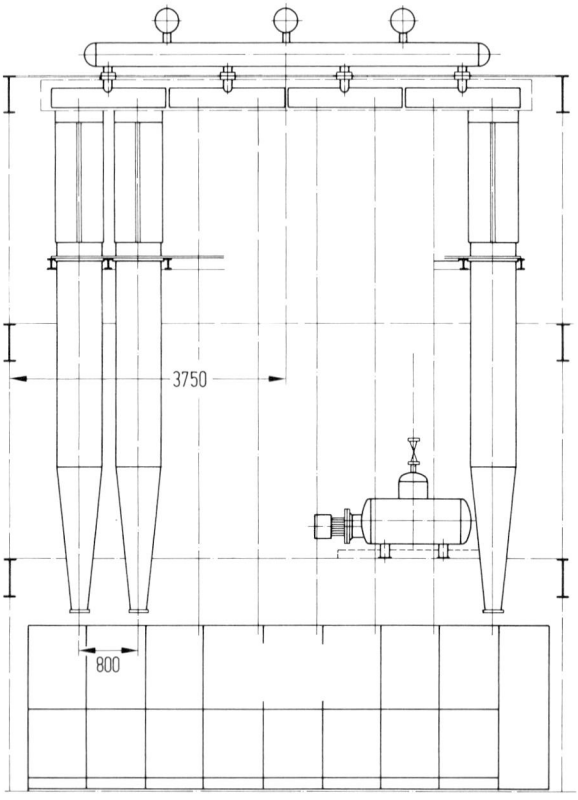

to Figure 34

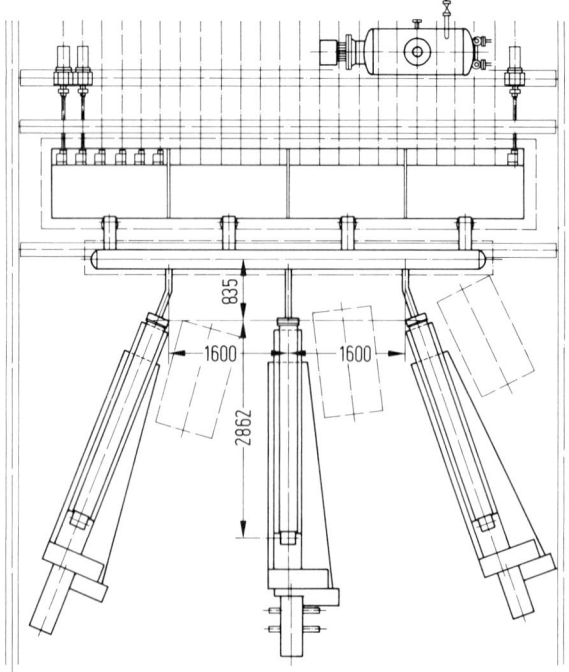

to Figure 34

17.3 Filament treatment section (Section II)

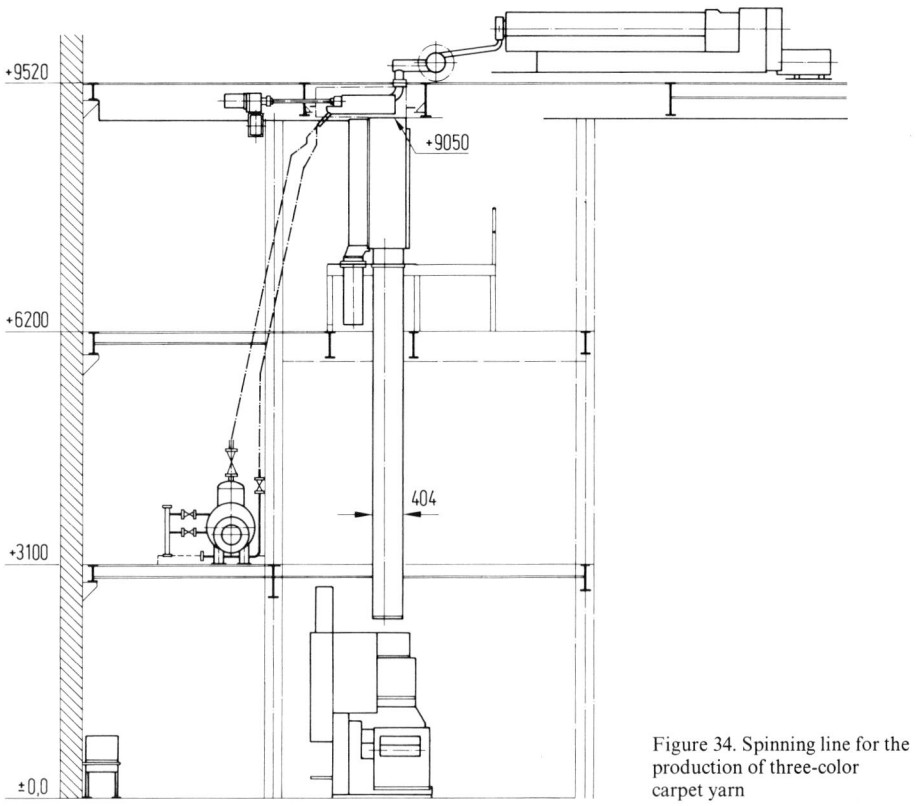

Figure 34. Spinning line for the production of three-color carpet yarn

17.3 Filament treatment section (Section II)

Section II consists of three different machine types:

1. Machines in which the freshly spun filaments are either wound into packages or are deposited in cans. The filaments are not subjected to any further treatment beyond a pre-orientation related to the haul-off speed of the filaments. But, depending on the end uses of the fiber, it has to undergo additional process steps.
2. Machines in which the freshly spun filaments are immediately drawn and taken up.
3. Machines in which the freshly spun filaments are drawn and subjected at once to further treatment.

In the extrusion section, the filaments emerge from the holes in the spinneret plates in a liquid state and, with quenching applied at the same time, are accelerated to the speed at which the yarn is taken off in the filament treatment section with the aid of the appropriate devices. The speed of the filaments emerging from the spinneret holes is low compared with the take-off speed in the filament treatment section. This means that with simultaneous quenching, the filament draw-down is high.

Example: Let us assume the diameter, D, of the spinneret hole to be 250 µm, and the individual filament weight to be 6 dtex. According to the approximate formula $d = 10 \times \sqrt{\text{dtex}}$,

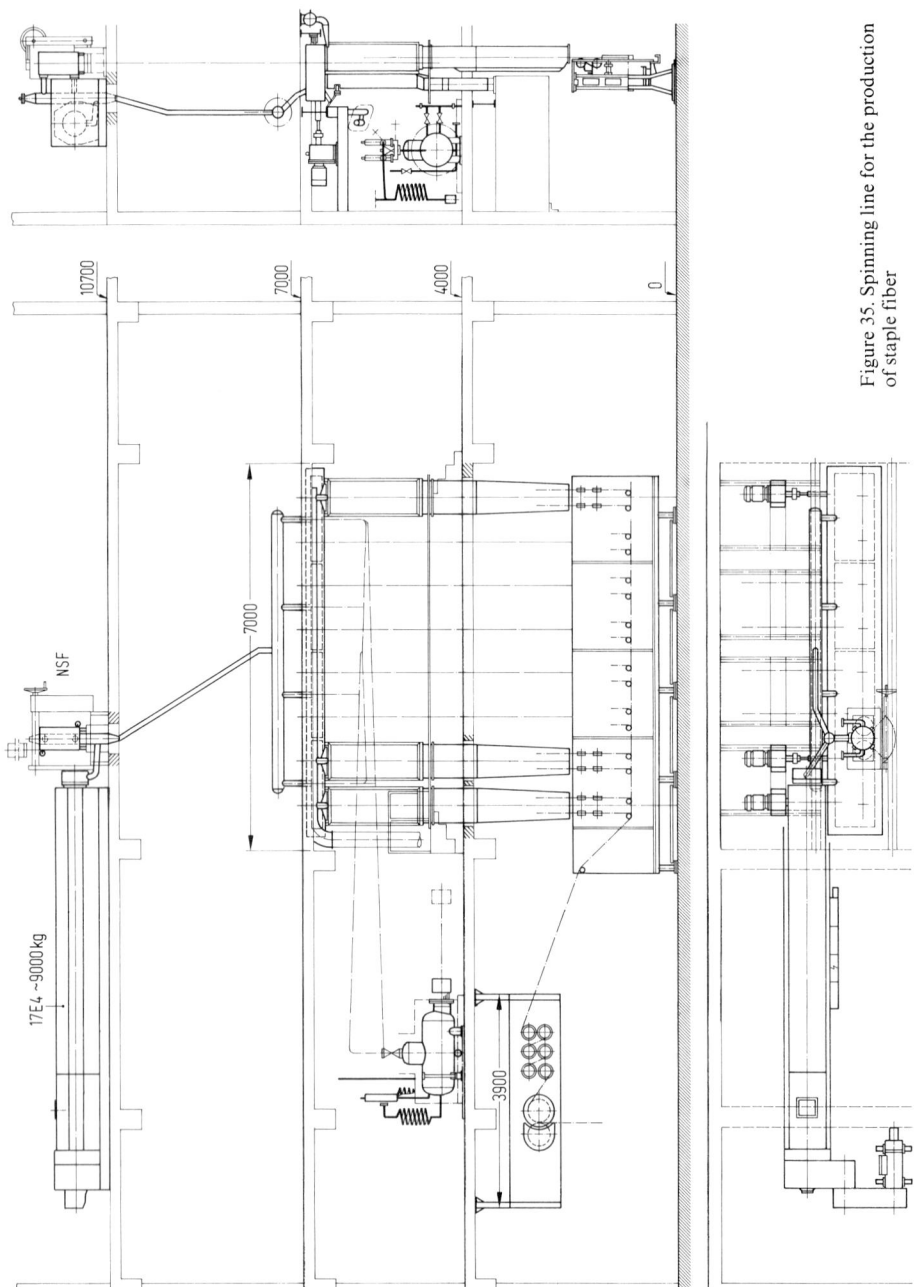

Figure 35. Spinning line for the production of staple fiber

the diameter of the filament is 25 µm. This means that the ratio of spinneret hole cross-section to filament cross-section is 100:1 i.e. the spin draw ratio is 100:1.

The take-off speed of the filaments and the consequent orientation of the molecules are essential factors in determining the quality and usability of the filaments produced.

17.3.1 Machines for take-up and depositing of freshly spun fibers

17.3.1.1 Take-up machines

The extruded threadlines are wound onto bobbins. If yarns for textile applications are produced, for example for apparel end use, the speeds of conventional machines are within the range of 1000 to approximately 1800 m/min. The yarns produced by this method are drawn on a draw twister and then wound onto bobbins. Figure 36 shows the supply package (A) and the threadline, which is drawn between two godets (B) and (C) to be subsequently taken up by the spindle (D).

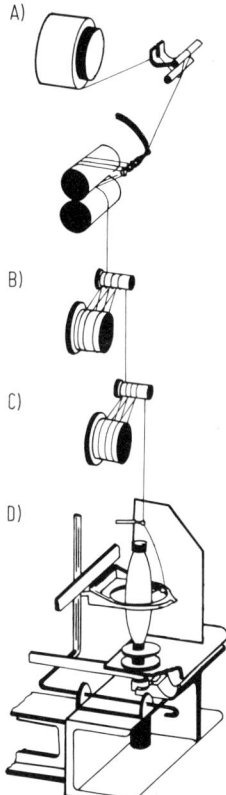

Figure 36. Yarn path in a draw-twisting machine
A) spinning bobbin, B) godet, C) godet, D) twisting spindle

Yarns which are produced within the above-mentioned speed range are very sensitive because the preorientation is relatively low, owing to very slight spin drawing. The packages have a very short shelf life and must therefore be processed further within a few days. In addition they have a relatively unstable package build and are unsuitable for transport. This means that packages produced at conventional speeds must be further processed at the producer's plant.

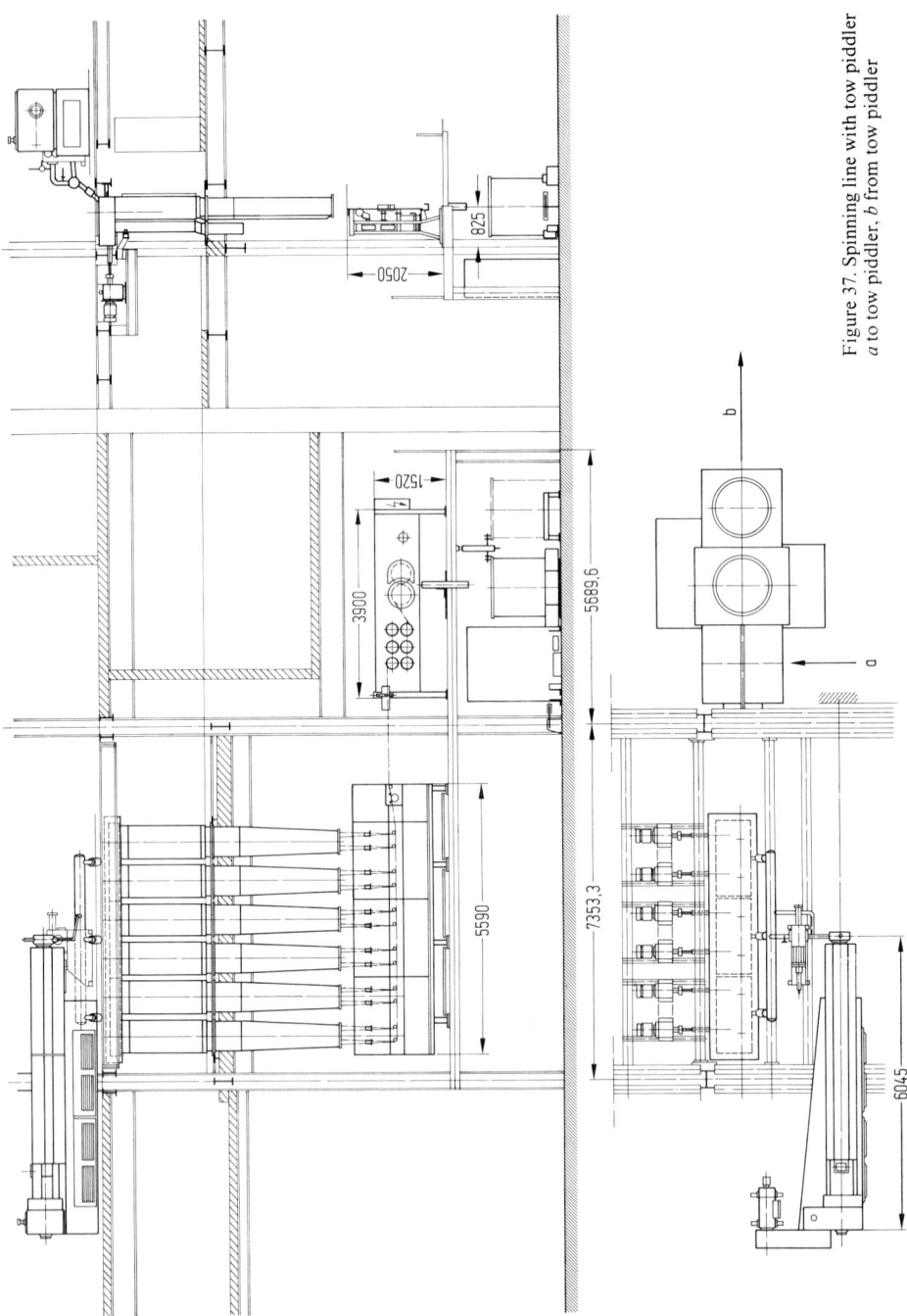

Figure 37. Spinning line with tow piddler
a to tow piddler, *b* from tow piddler

17.3 Filament treatment section (Section II)

Because of the limited durability of the material, its poor transportability, and unsatisfactory processability on texturing machines, spinning machines with conventional take-up are becoming less and less important for the production of textile yarns for apparel end uses. Such lines are no longer built.

17.3.1.2 Spinning line with tow piddler

Figure 37 shows a spinning machine or, more precisely, the extrusion component and the adjacent filament treatment section. The threadlines at the end of the extrusion section are deflected horizontally towards the filament treatment section and are combined as they leave the individual spinning positions. The take-off machine in the filament treatment section hauls off the tow using rotating rolls and conveys it to the sunflower wheels which feed it into a can underneath. The can is oscillated back and forth so that it fills evenly. Once it has been filled, it is automatically replaced by an empty can.

The take-off speed for staple fiber lines is the same as for the conventional take-up machine, in the range of approximately 600 to 1800 m/min. Therefore, what has been said before also applies to this product. However, since the large cans do not have to be shipped to other companies and since further processing takes place immediately, there are no problems in treating the material.

In a draw unit (Figure 38), the tow is taken off from several cans and drawn with simultaneous heat treatment. After drawing, the tow is conveyed to a crimper where the filaments are finely crimped to give them properties that resemble those of natural fibers.

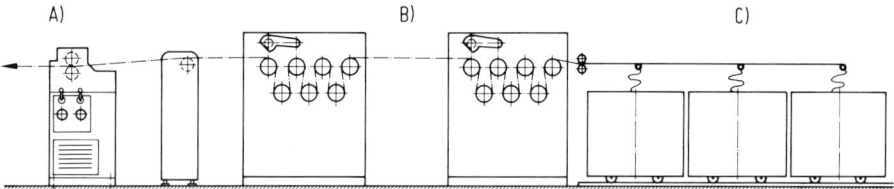

Figure 38. Draw unit and crimper for the production of staple fiber
A) crimper, B) draw units, C) can creel

The crimped tow is packed in cartons and cut into staple fiber at another location. Cutting can also be done directly following the crimper. The schematic drawing shows various intermediate treatment steps in the draw line. The cut fibers are spun to threadlines in subsequent processing steps. Spinning, in this context, means that the cut fibers, with lengths of approximately 30 to 80 mm, are twisted into a yarn. (In principle the same effect as on the spinning wheel.) The cut PET, PA or PP fibers can also be blended with natural fibers such as wool and cotton, and the blend can subsequently be spun into yarn.

17.3.1.3 Production of POY (preoriented yarn)

If the take-off speed of the filaments is increased beyond the range described under 17.3.1.1, at take-off speeds of approximately 3000 m/min and higher, the preorientation of the filaments is so high that one talks about preoriented yarn. POY packages keep well, i.e. they can be stored over a longer period of time and, in addition, possess excellent mechanical stability so that transport is no problem.

Production of POY yarns became possible after take-up machines capable of being operated at the required high speeds became available. Figure 39 shows a machine for the production of POY packages. The threadlines coming from the spinning machine are taken off by two S-wrapped godets and are then guided to the take-up head. On the machine shown, four threadlines per winder are taken up. The take-off speed of the threadlines is a) in the range of 3000 to 4000 m/min for polyester and b) 4200 to 5500 m/min for nylon.

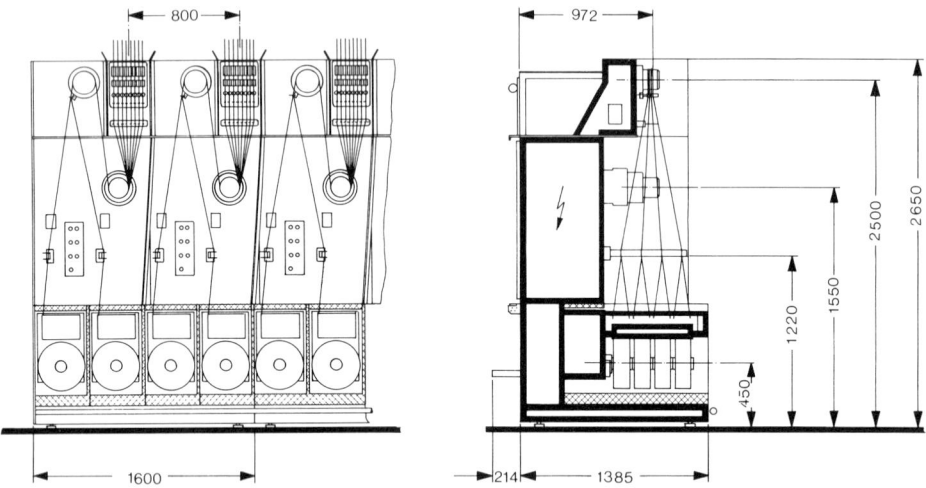

Figure 39. POY-take-up machine

On the machine shown in Figure 40, the threadlines coming from the spinning machine are taken off directly by the take-up head. The advantage of a godetless machine lies in its simplified design and easier handling. The take-up head has to fulfil special requirements in order to ensure good packages. (These requirements will be described in more detail under 17.4.1.)

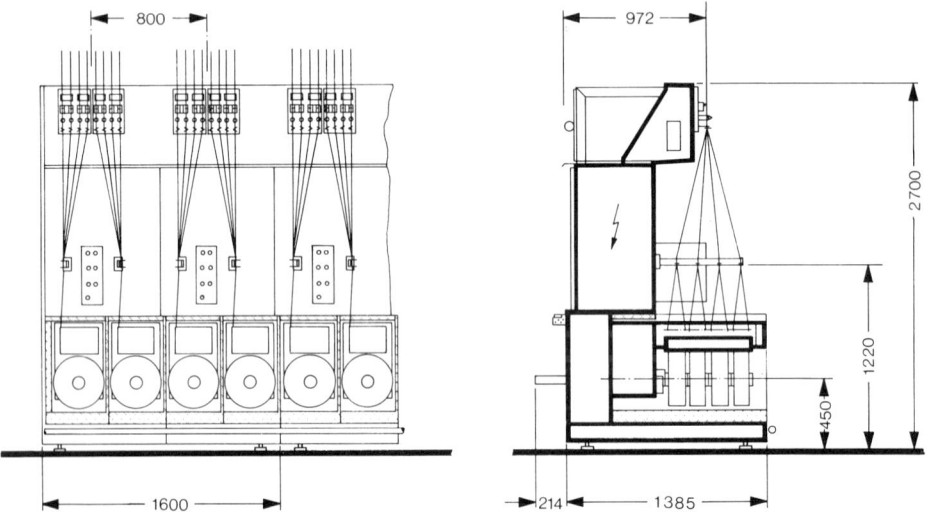

Figure 40. Godetless POY take-up machine

17.3 Filament treatment section (Section II)

Although preorientated, POY packages are not yet fully drawn. Figures 41 and 42 show schematically the dependence of residual draw ratio on the take-off speed for polyester, nylon 6, and nylon 6.6.

The take-off speed is shown on the horizontal axis, and the product of take-off speed and residual draw ratio on the vertical axis. This value provides a measure of the production capacity of the spinning machine.

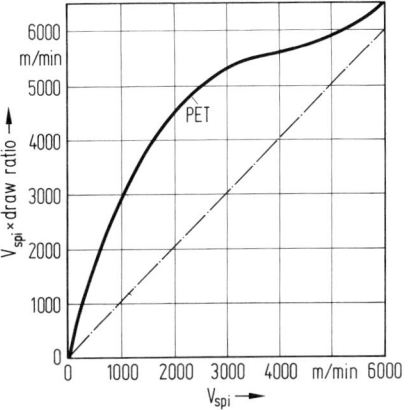

Figure 41. Dependence of production on spinning speed for polyester

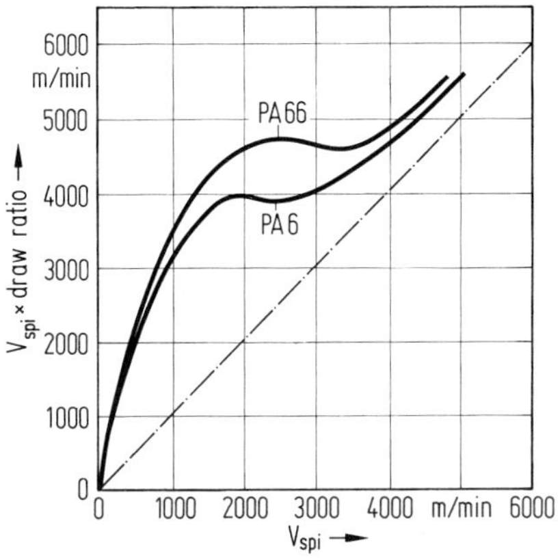

Figure 42. Dependence of production on spinning speed for nylon 6 and nylon 6.6

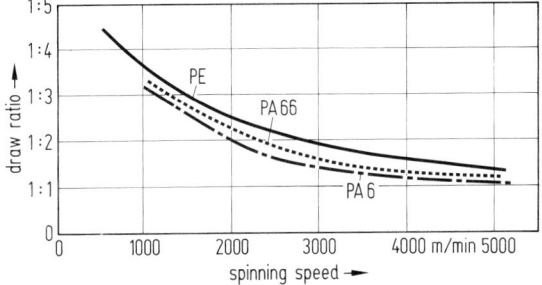

Figure 43. Dependence of draw ratio on spinning speed

Example: A POY polyester package is produced at a take-off speed of 3500 m/min. It is supplied to a second machine and drawn. After drawing, the yarn should be 100 dtex. Figure 41 shows that with a spinning speed or take-off speed of 3500 m/min, the product $V_{spi} \times$ draw ratio is 5500 m/min. Based on this, the residual draw ratio which must be set at the machine for further treatment is 1.57 (5500/3500). This means that the production capacity of the POY machine is higher by factor $P = 1.57$, since a threadline is taken up with a yarn count of 1.57×100 dtex and is then drawn to the required 100 dtex on the machine. If the take-off speed were increased further, up to the range beyond 6000 m/min, the curve shown would approximate a straight line, i.e. draw ratio is approaching 1, and the spun material is again fully drawn. Thus the production capacity of the take-up machine becomes proportional to the take-off speed. For polyester, the take-up speeds are usually 3000 to 4000 m/min. Speeds in the range up to 6000 m/min and higher are very problematic for polyester from a spin-technology standpoint and are therefore not used on a production scale. Nylon is a different matter. Here production speeds of 5000 m/min and higher can be applied.

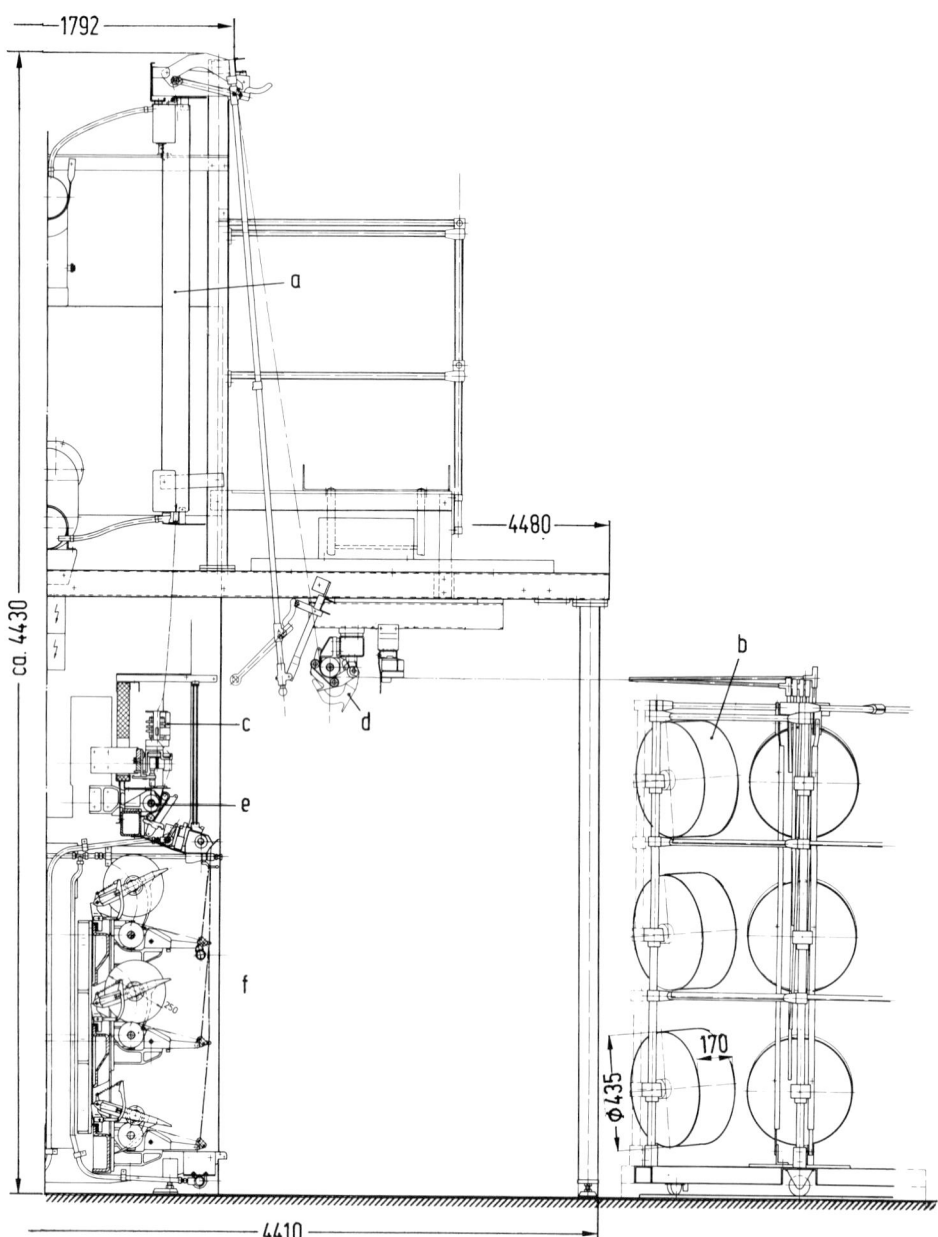

Figure 44. Texturing machine for supply of POY packages
a heating tube, *b* POY spools, *c* false twister, *d* feed unit 1, *e* feed unit 2, *f* wind-up

POY packages are frequently processed on texturing machines to impart bulk and stretch to the yarn (Figure 44). This is accomplished by incorporating a false-twist device and a heater within the drawing zone. As the yarn is fed into the drawing zone, it is simultaneously twisted by the false-twist unit and passes over the heater plate in the twisted condition. The yarn is deformed plastically by the combination of drawing, twisting, and heating and is heat-set in this state by cooling between the heater and the twisting unit. After the false-twist unit, the twist is removed, but the yarn tries to return to the twisted condition because of the heatsetting "memory". A yarn produced in this way has a high degree of stretch. A process to produce a yarn with less stretch and more bulk employs a second heating zone in which the stretch yarn is allowed to partially relax.

17.3.2 Machines for drawing freshly spun filaments

17.3.2.1 Spin drawing of textile yarns

Under 17.3.1.1 through 17.3.1.3 machines have been described in which the spun filaments are wound into packages or deposited in cans. Further treatment of the filaments did not take place. The filaments then had to be drawn into threadlines in subsequent process steps and, if necessary, undergo further treatment. A special case was when the filaments were taken off so fast that they were fully drawn or almost fully drawn.

If fully drawn yarns are needed, the filament treatment section must consist of a spin-draw machine. Drawn yarns are used for a variety of textile applications, for example for woven cloth or for the manufacture of curtains.

The production capacity of the spin-draw machine is proportionate to the take-up speed. This means that the take-up speed of the machine should be as high as possible, in order to achieve good production capacity. Modern spin-draw machines for polyester and nylon yarn permit speeds of up to approximately 6000 m/min.

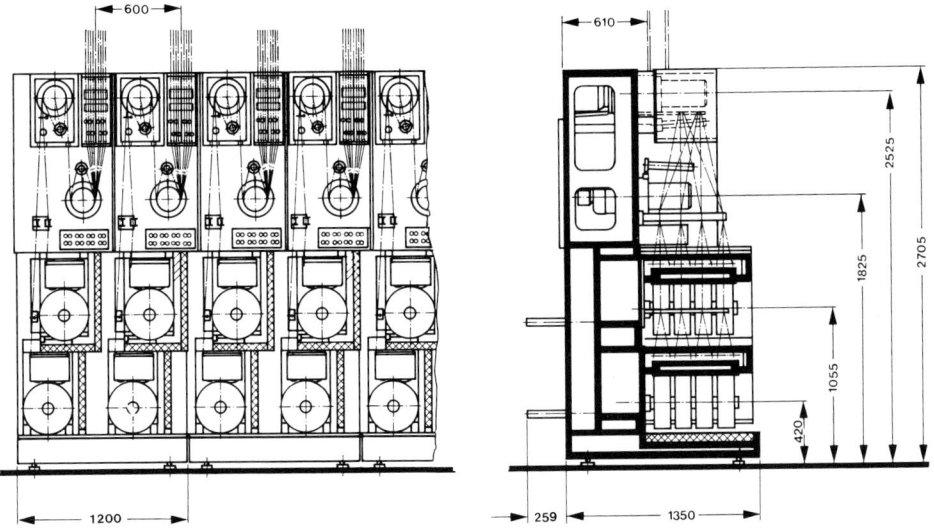

Figure 45. Spin-draw machine for industrial yarns

Figure 45 shows a textile spin-draw machine on which eight threadlines per position are drawn and taken up. The eight threadlines coming direct from the spinning machine are wrapped several times around the first godet; from there they proceed to the second godet, around which they are again wrapped several times (Figure 46). The speed of the second godet is higher in proportion to the draw ratio that the threadlines undergo. Both godets can be heated. For the first godet, the temperature for textile yarns lies in the range 60 to 80 °C approximately and for the second, around 120 to 180 °C. From the second godet the threadlines run to the take-up heads with four threadlines being taken up on each. The technical demands made on spin-draw machines are very stringent because of the high speeds and the heating of the threadlines by godets, which must be extremely precise. Temperature differences of only a few degrees from godet to godet result in differences in the drawn yarn which are reflected in lack of dyeing uniformity. Because of the high temperatures of the second godet and the simultaneous high circumferential speeds in production machines, these godets are mounted in an insulated housing. In this way, larger heat losses are avoided. Because of the technical intricacies, spin-draw machines for textile yarns are relatively expensive in the configuration described. For this reason, attempts have been made to develop spin-draw processes which are less complicated technically. An example is the H4S high-speed spin-draw process developed by Ems-Chemie/Inventa, Figure 47. On this machine, the wrap on the first, cold, godet is approximately 180°. Next there is a godet pair, and again several wraps are made. The duo, too, has cold rolls. The duo then conveys the four threadlines per position to the take-up head.

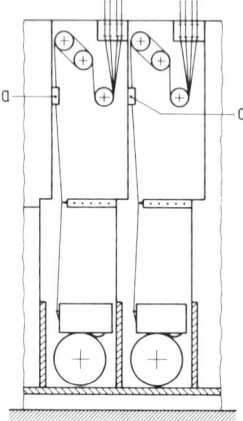

Figure 46. Yarn path in a spin-texturing machine for textile yarns. (Photo: Barmag, Remscheid/West Germany)
a godet, *b* air-bearing roll

Figure 47. H4S high-speed spin-draw process
a heat treatment

In order to reduce the high degree of shrinkage resulting from cold drawing to the required level, a steam injector incorporated ahead of the take-up provides the required heat treatment [7].

17.3.2.2 Spin drawing of industrial yarns

Severe demands are made on industrial yarns. They are used for tire cord, conveyor belts, canvas, safety belts, sewing thread, and so on. Usually, high tenacity coupled with relatively low elongation is required.

17.3 Filament treatment section (Section II)

Figure 48 shows a spin-draw machine for industrial yarns. Two yarn ends per position are drawn and taken up. Coming from the spinning machine, the yarn ends are first wrapped around a small diameter godet. From there, the yarn ends go to the first heated godet. A slight tensioning of the yarn takes place between the two godets which ensures that the yarn ends run onto the heated godet with a certain pre-tension, so that no slippage can occur. Pre-drawing takes place between the first and second godet, which receives several wraps. Final drawing takes place between this godet and a godet pair. The temperature of the duo is 250 °C. From the duo the yarn is conveyed to another godet which, as a rule, runs more slowly in order to impart relaxation to the yarn. From here, the yarn proceeds to the take-up head by way of a final godet. The take-up head in this machine is designed for fully automatic, wasteless transfer.

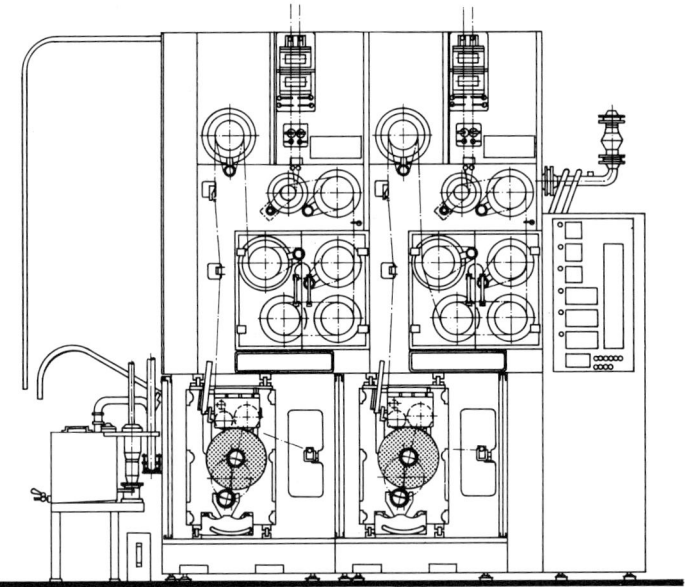

Figure 48. Spin-draw machine for industrial yarns

Spin-draw machines, in which extruding, drawing and take-up occur in one step, are increasingly being used because their high operating speeds and compact design offer an attractive ratio of investment cost to production capacity. In cases where the shrinkage of an industrial yarn has to be very low (e.g. < 3% in hot-air measured at 190 °C), two-stage processes are used, i.e. the yarn ends are taken up conventionally at speeds of approximately 400 to 1500 m/min and subsequently are fed to a draw line where they are drawn to their final tenacity by means of heated draw units, and then wound into packages. The winding speed is between 200 and 400 m/min. Because of the low speed and the associated higher residence time under heat, shrinkage is greatly reduced. With this slow process, hot-air shrinkage values of 3% and less are achieved.

Under 17.3.1 machines have been described in which the spun yarns are wound onto bobbins or deposited in cans. Under 17.3.2 machines have been mentioned for haul-off and drawing of the spun yarns, thus supplying a finished product. In the following section, machines will be described for direct further treatment of the filaments subsequent to drawing.

598 Fiber extrusion [Ref. p. 613]

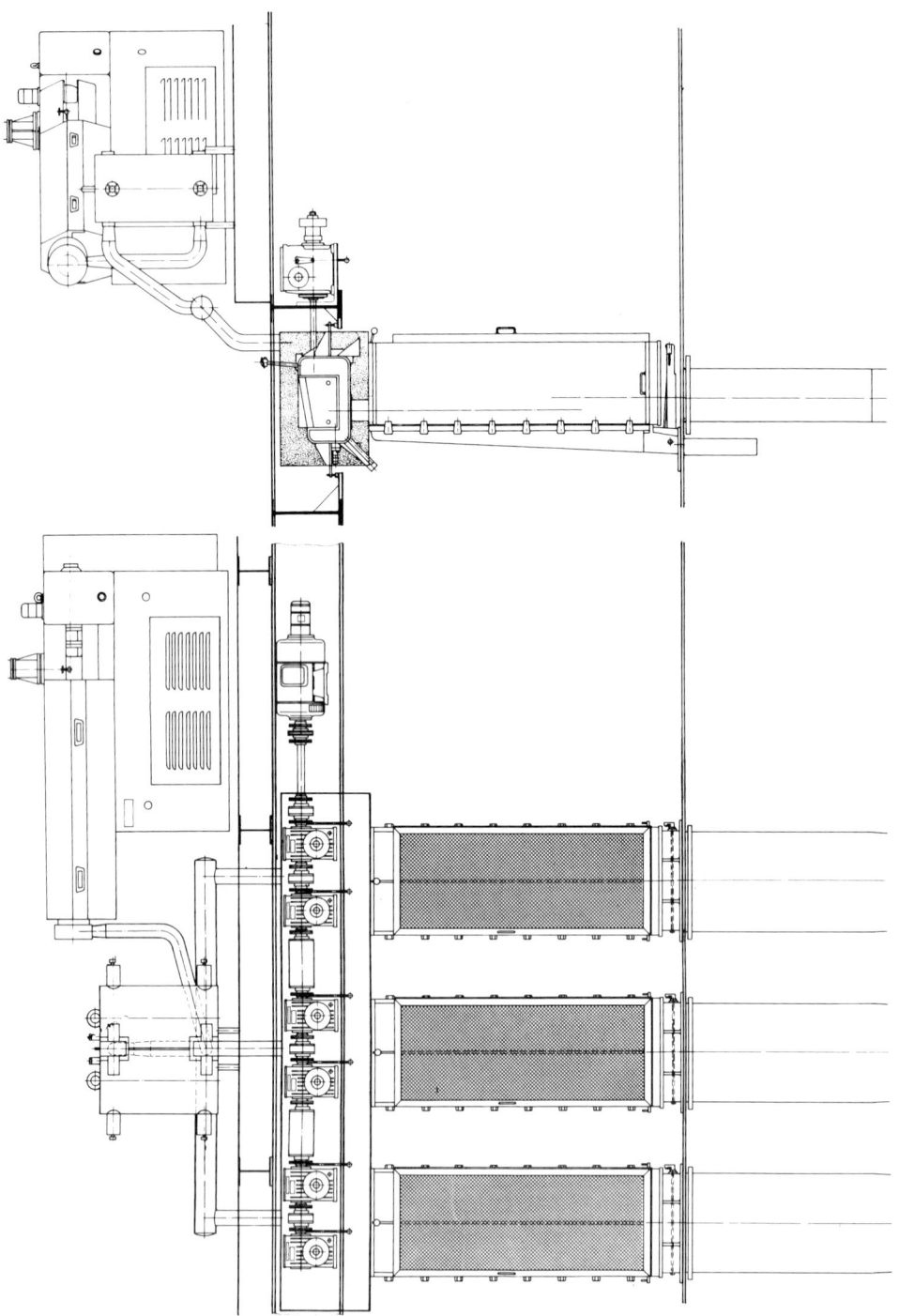

17.3 Filament treatment section (Section II)

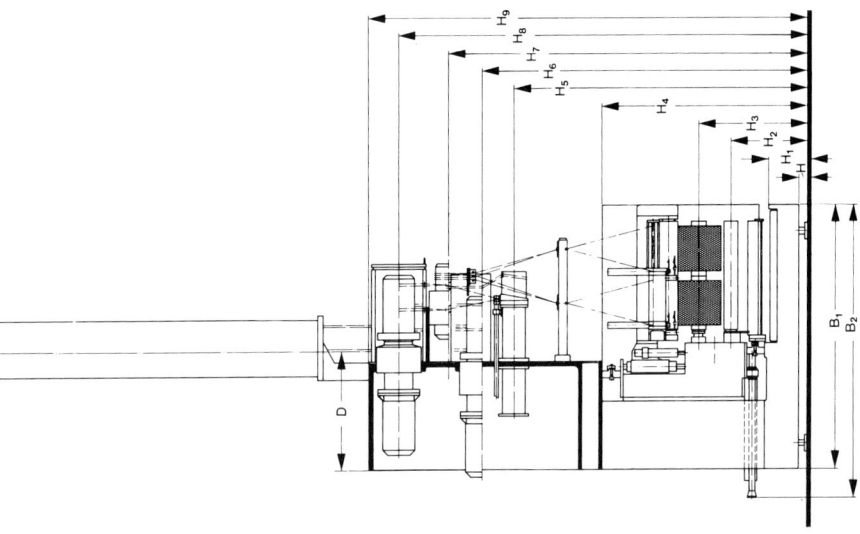

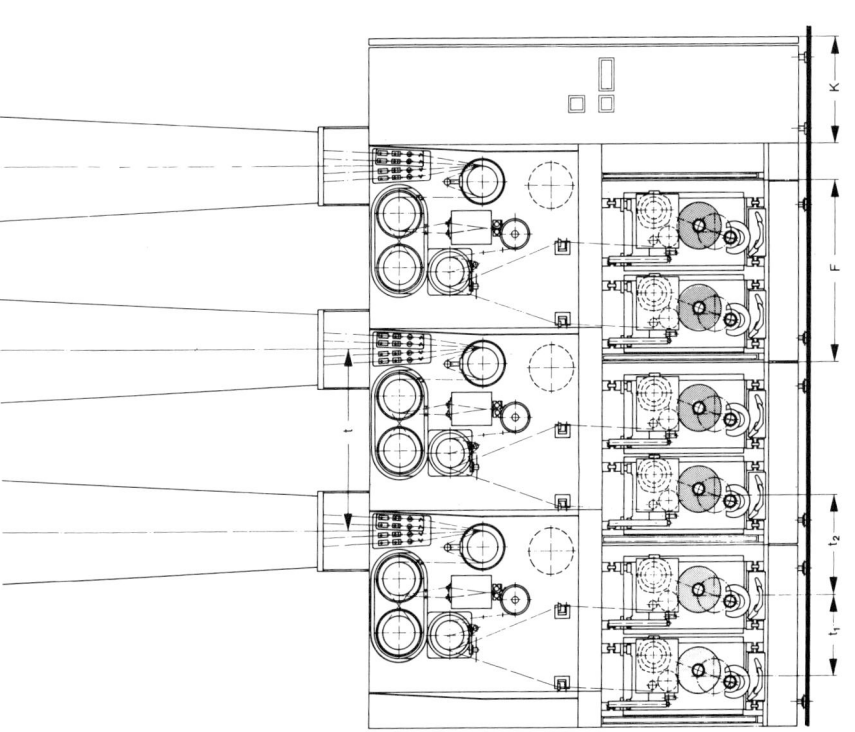

Figure 49. Spin-draw texturing machine for carpet yarns

17.3.3 Machines for drawing and further treatment of freshly spun filaments

17.3.3.1 Spin-draw texturing machine for the production of carpet yarn (Figure 49)

The spin-draw texturing machine is capable of drawing, texturing, and winding four yarn ends per position. The yarn ends are drawn between the first heated godet and the duo, which is also heated. After leaving the godet duo, which is insulated to prevent heat losses, the yarn ends are crimped with steam or air in a texturing jet (to be described in more detail under 17.4.8); after cooling they are conveyed – by means of a take-off godet – to two take-up heads provided with fully automatic transfer. The line speeds on these machines are in the range of 2000 to 3000 m/min, with a trend towards higher speeds.

17.3.3.2 Production of staple fibers for carpet end use

Figure 50 shows a machine which to a large extent corresponds to the spin-draw texturing machine described under 17.3.3.1. Instead of going to take-up heads, textured yarn ends are combined into a tow and guided horizontally to a haul-off unit. From there, the tow proceeds to a cutter which cuts it into staple fibers. A blower then conveys these to a baling press. As already mentioned, the speeds of these machines are between 2000 and 3000 m/min.

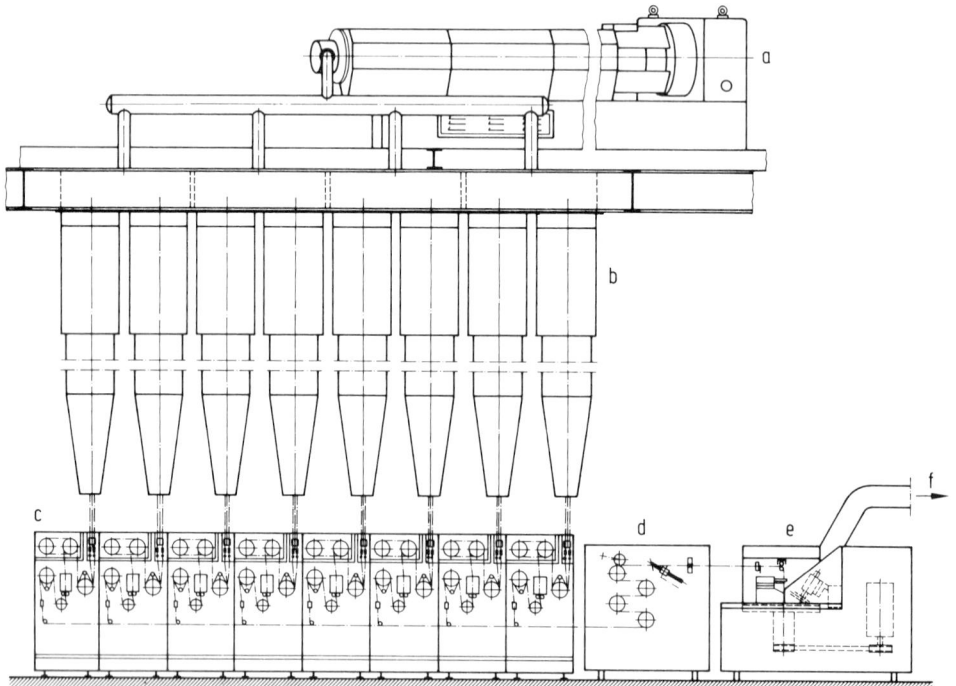

Figure 50. Spinning machine for the production of carpet staple fiber
a extruder, b spinning line, c draw-texturing, d haul-off, e high-speed cutter, f baling press

In the examples described under 17.3.3.1 and 17.3.3.2, individual process steps – spinning and haul-off, drawing, texturing, winding, or cutting – are combined in one machine. In both cases, the speed of the technically intricate machine is set as high as technologically possible, in order to achieve good overall cost efficiency on the line.

In the following section, a machine that combines the functions of spinning, haul-off, drawing, crimping, and cutting of fibers to staple is described. In order to achieve an efficient line, the design speed was set low enough for the individual components to be laid out in a relatively simple and cost-effective way.

17.3.3.3 Fiber extrusion line for in-line operation

Figure 51 shows a machine in horizontal configuration, with a length of approximately 20 m for the production of polypropylene staple fibers. The line has up to 16 spinning positions, which can either be equipped with a small extruder each or be fed from a central extruder which conveys the melt direct to the spinning head, with or without intermediate spinning pump.

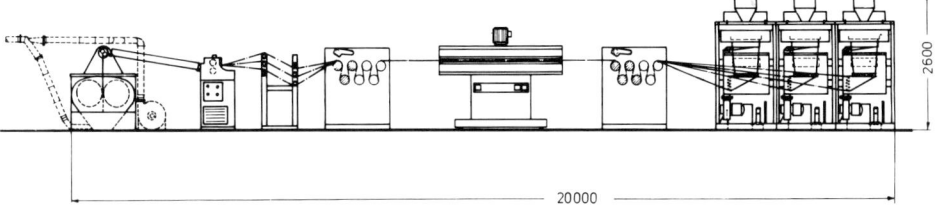

Figure 51. Fiber extrusion line for continuous production of polypropylene staple fiber

The filaments emerging from the spin packs are cooled in a short quenching system, guided to the first draw unit, and from there to the second draw unit. Drawing takes place in between. Between the draw units the filament sheet runs through a heater. Following the second draw unit, the tow is crimped in a stuffer box from where it is taken to a cutter, cut into staple fibers, and transported to a baling press by means of a blower. The drawing speed of this line is 120 m/min. (For the lines described under 17.3.3.1 and 17.3.3.2 it was approximately 3000 m/min.) Production capacity per position is 70 kg/h, i.e. for the line shown this would be $3 \times 70 = 210$ kg/h. One spinneret per spinning position is used with a maximum number of 40,000 holes. This means that on the 16-position line, up to 640,000 filaments can be spun, drawn, crimped, and cut. Instead of the small individual extruders, a central extruder can be used. In that case, each spinning head is provided with a pump.

17.4 Important components of the filament treatment section (Section II)

17.4.1 Take-up heads

The "end product" of the spinning machines described so far is the package, except on the spinning line for the production of staple fibers, where the tows are usually deposited in a can.

Exceptionally severe demands are made on the winder, such as speeds up to 6000 m/min, yarn counts from 15 dtex all the way up to several 1000 dtex, package diameters of over 500 mm, package weights of more than 150 kg per spindle, up to eight yarn ends per spindle, manual or fully automatic operation, low noise levels despite high speeds, and so on.

The economy of the spinning lines is determined by the availability of appropriate take-up units. Figure 52 shows a winder which can wind up to eight yarn ends at speeds of 6000 m/min max.

Figure 52. High-speed winder for up to eight yarn ends and speeds up to 6000 m/min
(Photo: Barmag, Remscheid, West Germany)

The package must have a good cylindrical shape and uniform Shore hardness. These two requirements make severe demands on the traverse system of the winder. It must be designed in such a way that the yarn reverses very quickly and is nevertheless treated gently over the entire stroke and at the reversal points. If the turn is too slow at the reversal points (which means that more material is deposited there), the edge build-up is too high. In the course of package build this is "flattened" down to a fairly cylindrical shape. It leads, however, to hard spots at the reversal points resulting in yarn damage and, above all, increasing the danger of overthrown ends, i.e. individual yarn ends protruding from the sides of the packages, which can cause yarn breaks during take-up in subsequent processes. The take-off speeds of subsequent treatment machines are sometimes in excess of 1000 m/min. Good package build further requires that the take-up tension is not too high and furthermore is as constant as possible over the entire traverse width. In cases where the take-up tension can be influenced only within certain limits, for instance in godetless high-speed spinning (described under 17.3.1) it may be too high for good package build. The winder therefore must be capable of reducing excessively high yarn tension.
During the traverse of the yarn, it is inevitable that tension peaks will occur in the yarn, because of unequal lengths in the traverse triangle. This must be compensated by the design features of the winder. In this way the danger of overthrown ends and the formation of slippage-prone yarn layers is reduced. The individual layers have to be securely deposited on the package so that during package build or subsequent take-off no slippage of yarn layers and consequently no yarn breaks occur. In addition, the yarn tension fluctuations during package take-off should be as low as possible.
An example will show how stringent the demands are for secure yarn deposit on the package. Figure 53 illustrates the package shapes that are produced on winders currently in use all over the world. Let us look at winder type SDE-900 with a package diameter of 435 mm, on which four packages per spindle are wound. Each package has a volume of 26.6 l. Assuming a yarn count of 40 dtex, the length of the yarn wound would be 6650 km. A single incident of serious yarn slippage would lead to a yarn break during subsequent take-off of the package.
The winder shown in Figure 52 is equipped with a yarn traverse system which consists of a combination of camshaft and grooved roll (Figure 54).

17.4 Important components of the filament treatment section

The yarn end running to the winder is first guided through the traverse guide of the camshaft and then conveyed to the grooved roll. With this combination it is possible to provide the camshaft with a gentle turn-around. In conjunction with the lightweight thread guide – it only weighs about 1 g – this results in low mass forces at reversal and a low wear rate and long service life for the thread guide. The grooved roll has a very small turn radius, since there are no reciprocating masses to be moved. This, again, results in the required fast yet gentle reversal of the yarn end. Also, the grooved roll must guide the yarn with great precision right up to deposit on the package. By overfeeding the grooved roll, i.e. with the circumferential speed of the grooved roll faster than the yarn, a certain tension decrease is possible for building good packages at high speeds, even in godetless processes.

Another requisite for good package build is the uniformity of the yarn tension over the entire traverse stroke. Thanks to the varying depth of the grooves, the grooved roll makes it possible to compensate for the different lengths in the traverse triangle, thus ensuring even tension distribution over the stroke.

The time needed to wind the yarn ends can vary considerably. Example: Let us again assume a winder with a package diameter of 435 mm and four packages per spindle, with each package having a volume of 26.6 l. Polyester is wound at a speed of 5000 m/min in a spin-draw machine as described under 17.3.2. The denier of the wound yarn is 40 dtex. This results in a production capacity of the machine of 20 g/min per yarn end. The package volume is 26.6 l. For polyester, this corresponds to approximately 26.6 kg (winding factor 1 kg/l). The resulting doff cycle is 26,600/20 = 1330 min, i.e. 22 hours approximately.

In a second example, the doff cycle of a spin-draw texturing machine will be calculated (17.3.3). Package dimensions are: 250 mm stroke length with a diameter of 260 mm. This corresponds to an "R" type winder. This package size is used worldwide for textured carpet yarn.

The take-up speed of the spin-draw texturing machine is 2500 m/min. Let us assume a yarn count of 1300 dtex. This results in a production capacity of 325 g/min. The package volume is 11.9 l. For textured material this corresponds to 5 kg (winding factor 0.4 kg/l), representing a doff cycle of 5000/325 = 15 minutes.

These two examples clearly demonstrate that for short doff cycles a take-up head is needed for fully automatic and wastefree yarn transfer. If in the example of the spin-draw texturing machine for carpet yarn the packages were to be transferred manually and the yarn ends to be re-strung, which would require approximately 2 minutes, a yarn waste of 11 per cent would occur, which would make the entire line uneconomic.

For this reason, for short package doff cycles (predominantly for industrial yarns and carpet yarns) take-up heads with fully automatic and wastefree transfer are used. Figure 55 shows a take-up head for fully automatic transfer at speeds up to 6000 m/min. Fully automatic take-up heads are being increasingly used for textile take-up machines as well, although for these packages doff cycles may be several hours. The object of this move is to minimize labor costs.

The main components of staple-fiber lines (17.3.3.2 and 17.3.3.3), where the spun filaments are not wound into packages, are: spinning line, tow piddler, draw unit, crimper, cutter.

Spulköpfe der Baureihe SW4S, SW46S und SW4R haben die gleichen Spulenformate
Take-up heads of the series SW4S, SW46S and SW4R have the same package shapes
Las cabezas bobinadoras de la Serie SW4S, SW46S y SW4R tienen los formatos de bobina iguales

Baureihe SW4S, SW46S	Durchmesser	Ges.-Hülsenlänge/⌀	Volumen/Hub	Volumen/Stroke	Volumen/Carrera
Series SW4S, SW46S	Diameter	Total tube length/⌀			
Serie SW4S, SW46S	Diámetro	Longitud total del tubo/⌀			
S / RS	260 mm	300/75 mm	11,9 dm³ / 250 mm	24 dm³ / 250 mm	
SD / RD	360 mm	300/75 mm			
SLD / RLD	360 mm	410/75 mm	33,6 dm³ / 350 mm	2 × 16,3 dm³ / 2 × 170 mm	3 × 9,6 dm³ / 3 × 100 mm	4 × 6,7 dm³ / 4 × 70 mm
SSL / RSL / R	260 mm	600/75 mm		2 × 11,9 dm³ / 2 × 250 mm	3 × 8,1 dm³ / 3 × 170 mm	4 × 5,7 dm³ / 4 × 120 mm
SSD / RSD	360 mm	600/75 mm	52,9 dm³ / 550 mm	2 × 24 dm³ / 2 × 250 mm	3 × 16,3 dm³ / 3 × 170 mm	4 × 11,5 dm³ / 4 × 120 mm
SSD	435 mm	600/75 mm	78,6 dm³ / 550 mm	2 × 35,7 dm³ / 2 × 250 mm	3 × 24,3 dm³ / 3 × 170 mm	4 × 17,15 dm³ / 4 × 120 mm

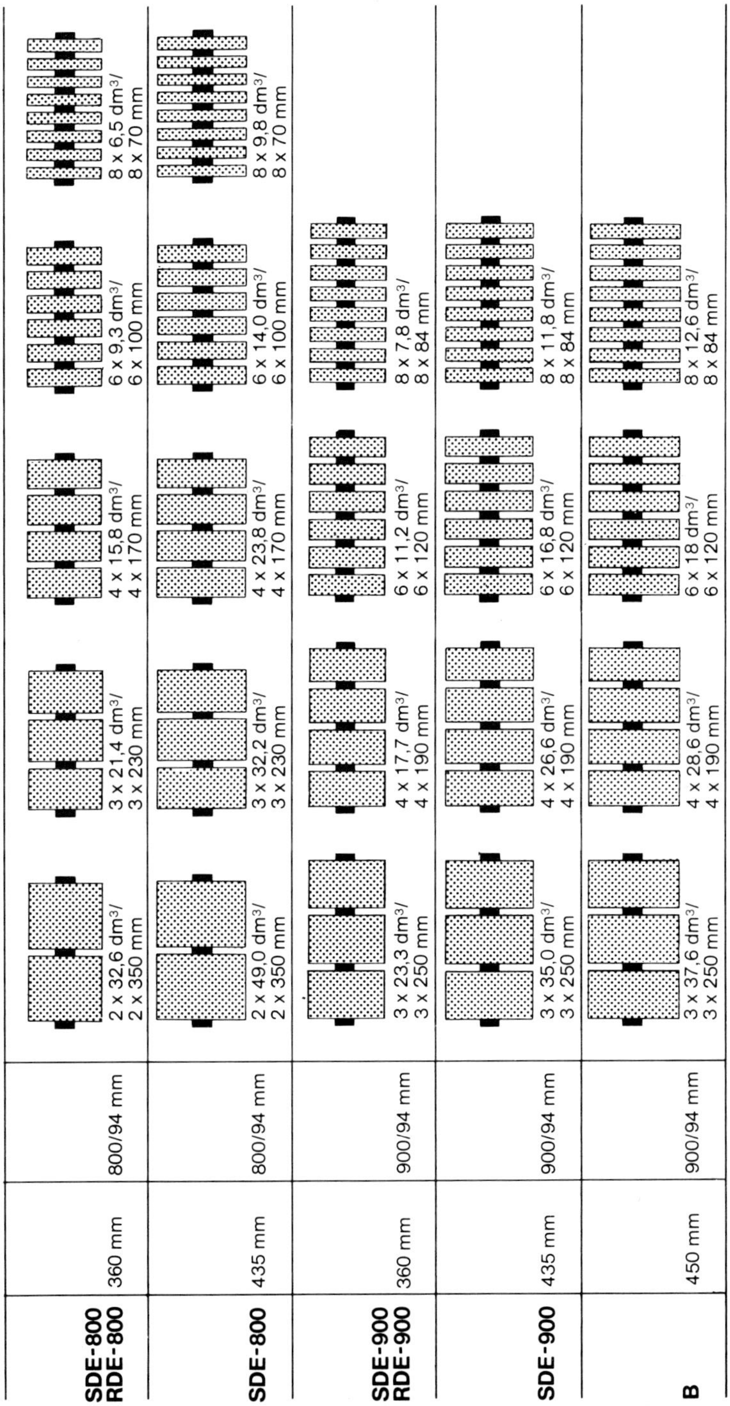

17.4 Important components of the filament treatment section

SDE-800 RDE-800	360 mm	800/94 mm	2 × 32,6 dm³/ 2 × 350 mm	3 × 21,4 dm³/ 3 × 230 mm	4 × 15,8 dm³/ 4 × 170 mm	6 × 9,3 dm³/ 6 × 100 mm	8 × 6,5 dm³/ 8 × 70 mm
SDE-800	435 mm	800/94 mm	2 × 49,0 dm³/ 2 × 350 mm	3 × 32,2 dm³/ 3 × 230 mm	4 × 23,8 dm³/ 4 × 170 mm	6 × 14,0 dm³/ 6 × 100 mm	8 × 9,8 dm³/ 8 × 70 mm
SDE-900 RDE-900	360 mm	900/94 mm	3 × 23,3 dm³/ 3 × 250 mm	4 × 17,7 dm³/ 4 × 190 mm	6 × 11,2 dm³/ 6 × 120 mm	8 × 7,8 dm³/ 8 × 84 mm	
SDE-900	435 mm	900/94 mm	3 × 35,0 dm³/ 3 × 250 mm	4 × 26,6 dm³/ 4 × 190 mm	6 × 16,8 dm³/ 6 × 120 mm	8 × 11,8 dm³/ 8 × 84 mm	
B	450 mm	900/94 mm	3 × 37,6 dm³/ 3 × 250 mm	4 × 28,6 dm³/ 4 × 190 mm	6 × 18 dm³/ 6 × 120 mm	8 × 12,6 dm³/ 8 × 84 mm	

Figure 53. Package shapes

Figure 54. Traverse-motion system with camshaft and grooved roll. (Photo: Barmag, Remscheid, West Germany)

Figure 55. Take-up head for fully automatic transfer at speeds up to 6000 m/min (Photo: Barmag, Remscheid, West Germany)

17.4.2 Tow piddler

Figure 56 shows the spinning line schematically, with the fiber tow being deposited into cans by means of a tow piddler. Underneath, in this case, there is a round can which rotates and moves back and forth, and is filled evenly with the deposited fiber tow. When the can is full, it is automatically replaced by an empty can. The fiber tow is hauled away by the sunflower wheels. These sunflower wheels are arranged in such a way that the fiber tow is pulled into the narrow gap between the teeth of the two wheels and from there transported vertically downward into the can.

Maximum speeds for sunflower wheel deposit are around 1800 m/min. In order to realize the advantages of higher preorientation and, consequently, improved shelf life, tow piddlers have been developed for speeds of 3500 m/min. In actual production, such speeds have hardly been introduced as yet, since pulling the tow out of the can still poses problems, impairing the efficiency of the line.

17.4.3 Draw unit

The filled cans are passed to a draw unit where the tow is pulled out of the cans and drawn between two or more draw units. The godet rolls of the draw unit can be heated to suit the process conditions. Depending on the physical properties desired, further heat treatment between the draw units may be required.

A machine was described in 17.3.3.3 in which the tow is drawn directly after spinning, i.e. it is not deposited in a can. This is possible because the speed of the line is 120 m/min.

The speeds of the draw units in staple-fiber processing are between 100 and 400 m/min. The forces required for drawing the fiber tows may be several tons.

17.4.4 Crimper

After passing the draw unit, the tow is conveyed to the crimper, with the partial tows running parallel to each other combined by doubling, and pulled by air cylinder-

17.4 Important components of the filament treatment section

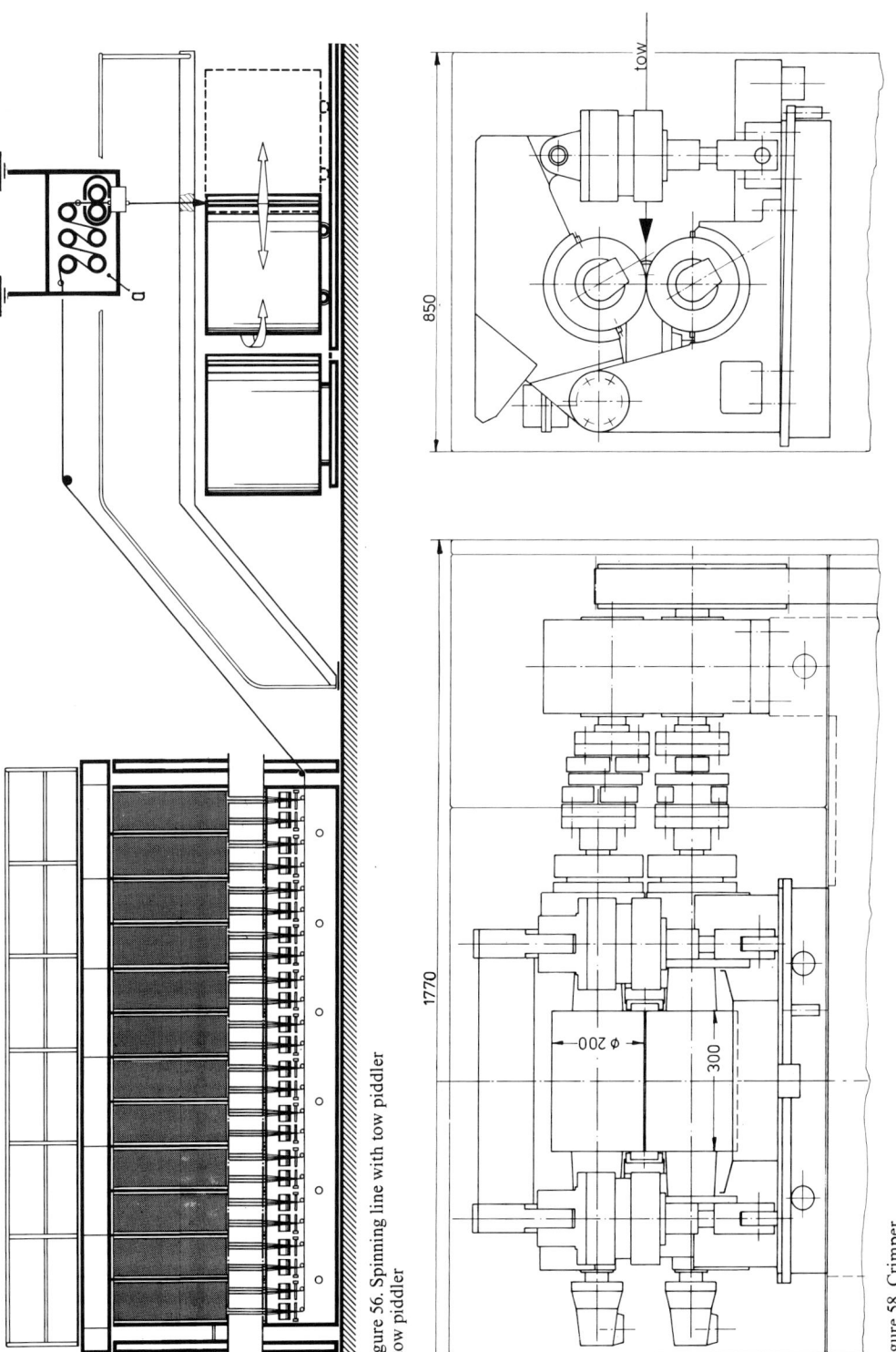

Figure 56. Spinning line with tow piddler
a tow piddler

Figure 58. Crimper

pressurized nip rolls into the crimper (Figure 57). The combined tow then passes into the stuffer box immediately in back of the nip rolls. This is closed at the outlet end with an adjustable flap. Thus, as the fibers are fed through the stuffer box against the resistance of the flap and the friction at the walls, a high pressure is developed in the box. The axial forces on the individual filaments cause them to become crumpled (or 'crimped') into small alternating bends. Usually there are about 10 bends per 10 mm filament length.

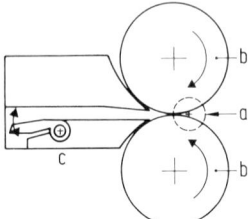

Figure 57. Principle of crimper
a yarn-in, *b* conveying rolls, *c* stuffer box

The crimper shown in Figure 58 has a maximum take-off speed of 350 m/min. The maximum tow size which can be processed in the crimper is 2 million dtex. Drive capacity is 36 kW. The width of the stuffer box is 300 mm, and the diameter of the nip roll 200 mm. For heat treatment of the crimped filaments steam is frequently supplied to the stuffer box. The crimped fiber tow emerging from the crimper is either deposited in cartons and further processed at another location, or conveyed direct to a fiber cutter and cut into staple fiber. The entry speed of the crimped tow into the cutter matches the exit speed of the crimper.

17.4.5 Cutter

Figure 59 shows a cutter that can be used in an integrated staple-fiber process as described in 17.3.3.2. Integrated means that spinning, drawing, texturing, and cutting take place in one process. Consequently, the speed of the fiber bundle is high. The cutter shown here is suitable for speeds up to 6000 m/min.
The tow is taken off by a closely spaced godet pair, and deposited on a horizontally mounted cutting wheel. As the deposit increases in height, the filaments are pressed through the blades by a rotating wheel, and carried to the outside by centrifugal force. An aspirator sucks the cut fibers away and transports them to a baling press. The blades on the cutting wheel can be set to give the desired length of staple fiber.
In another process for producing staple fiber, the drawn but uncut fiber tow is deposited in cans and taken to a converter. The converter pulls the fiber tows out of the cans, and draws them, under appropriate heat treatment, to an extent that causes every filament of the fiber tow to break. The break points, however, are distributed in such a way that the entire tow does not rupture. This results in tows of staple fibers which can be subjected to subsequent processes as desired.

17.4.6 Godets

All spinning lines, with the exception of godetless POY lines, need godets to take off the spun filaments. If the filaments are to be wound into packages or deposited in cans (17.3.1.2), cold godets are used. Their only function is to haul off the filament and transport it further. Heat treatment does not take place.
In those cases in which the yarn is subjected to further treatment (17.3.2 and 17.3.3), the filaments have to be heated by means of heated godets (with the exception of the H4S process described under 17.3.2.1). This process demands good temperature uniformity.

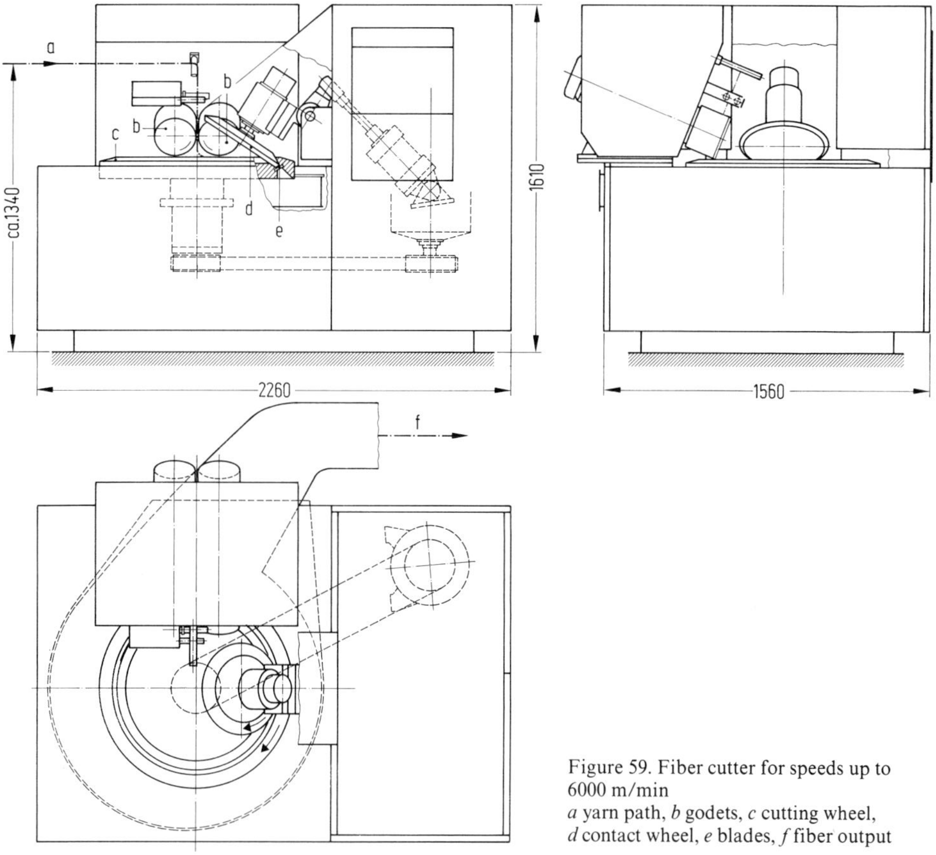

17.4 Important components of the filament treatment section

Figure 59. Fiber cutter for speeds up to 6000 m/min
a yarn path, *b* godets, *c* cutting wheel, *d* contact wheel, *e* blades, *f* fiber output

During spin drawing of textile, industrial, and carpet yarns, as a rule only temperature differences of a few degrees C are allowed. Figure 60 shows temperature profiles of godet rolls that are used for this type of machinery. The main elements of a godet unit are:

- the rotating godet shell
- the shell bearing
- the synchronous electric motor, flanged onto the bearing
- the heater around which the godet shell rotates, and
- the sensor which transmits the temperature signal to the temperature controller.

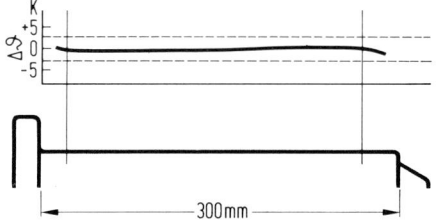
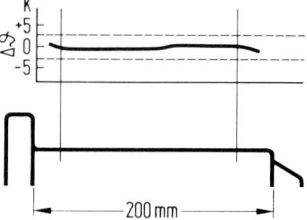

Figure 60. Temperature profile of heated godets

The Vapotherm godet shown has a temperature spread within $\pm 1\,°C$ over the entire godet length. This is achieved by the following design (Figure 61):
The rotating godet sleeve contains a closed, annular chamber filled with water. This space is evacuated. Within the godet shell, there is a stationary induction coil around which the shell rotates. Alternating voltage is connected to the induction coil and the alternating magnetic field passing through the rotating godet shell induces a high current, resulting in heating of the inner jacket of the shell (same principle as for transformer with primary and secondary coil). This heats and evaporates the water. The resulting steam condenses on the inside wall of the outer jacket. This means that all the heat removed from the godet surface comes from the inner jacket with the help of the steam. The large heat of vaporization of steam, which is immediately released on condensation at the cooler spots, ensures temperature uniformity over the entire roll surface.

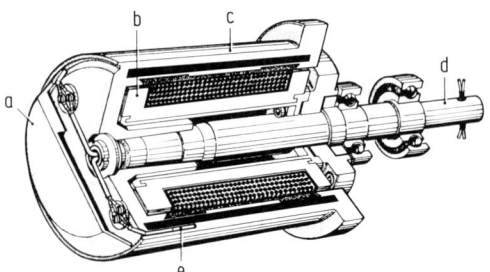

Figure 61. Cross-sections of a godet with extremely uniform temperature profile. (Photo: Barmag, Remscheid/West Germany) *a* cover, *b* coil carrier, *c* godet, *d* shaft, *e* temperature sensor

Sensors arranged close to the surface are connected to the temperature control system through the godet shaft. Transfer from the rotating shaft to the control system is effected by a contactless signal transmitter. The heating capacity of the godets described is 14 kW for a length of 300 mm and 11 kW for the 200 mm version. The maximum surface temperature is 250 °C, the maximum speed 6000 m/min. Good product quality depends on having the proper roll surface, as well as good temperature uniformity. On the one hand, the surface must be wear-resistant, and on the other it must provide the required friction conditions for the yarn. As a rule, hard-chromed surfaces, either polished or matt, or ceramic coated (plasma-coated) surfaces are used. The plasma coating consists of a mixture of aluminum oxide and chromium- or titanium oxide. An advantage of plasma coating compared with a chromed surface is the higher wear resistance. Disadvantages are the higher price and increased sensitivity to damage by hard objects.
In many spinning processes, the godet roll is operated in combination with an idler roll, in which case the threadlines are wrapped around both the godet and the idler roll. Because of the small diameter of the idler roll, the speed may be over 50,000 revolutions per minute at 6000 m/min. For this reason, idler rolls for high-speed spin-draw processes have air bearings. In order to facilitate yarn string-up, the idler rolls are often driven by an air turbine which is switched off once all threadlines have been strung up. Godets that run at high temperatures and high speeds are protected against excessive energy losses by an insulation hood.
For optimal adaptation of the godet to the yarn path, it is pivotable on a ball-and-socket joint over $\pm 10\,°$ and can be fixed in the desired position.

17.4.7 Spin finish application

Freshly spun filaments must have spin finish applied before they are wound up or deposited in cans, drawn, and subjected to further treatment. These spin finishes are either emulsions of water and oil, or pure oils. The spin finish imparts a certain cohesion to the filaments, eliminates electrostatic charge and provides good sliding properties, so that friction at the thread guide is not excessive and subsequent processing can take place. The spin finish is frequently applied with rollers, Figure 62. The roller rotates slowly in a trough filled with spin finish so that its surface is wetted with a thin film. The threadline sliding along the roller accepts the spin finish. By setting the number of revolutions, the quantity of spin finish to be applied can be controlled. Increasingly nowadays spin finish is applied by jets. They offer the advantage of being able to meter the quantity of spin finish exactly, with appropriate displacement pumps.

Figure 62. Spin finish rollers.
(Photo: Barmag, Remscheid, West Germany)

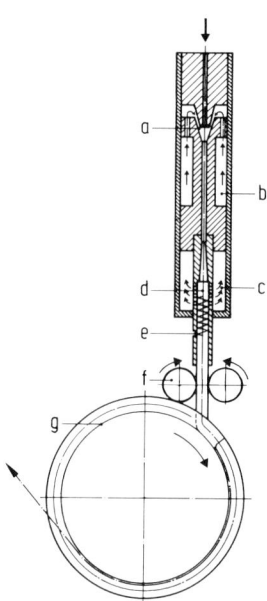

Figure 63. Texturing jet principle. (System Barmag)
a injector, *b* air inlet, *c* air outlet, *d* stuffer box,
e plug formation, *f* conveying wheels, *g* cooling drum

17.4.8 Texturing jet

In processes for the production of crimped filaments, a texturing jet is used after drawing. Crimpers or stuffer-box type crimping devices cannot be used at speeds as high as 2000 to 3000 m/min. In these cases, texturing or crimping units that work pneumatically are used. Figure 63 shows the fundamental design of such a texturing jet. The yarn end entering the texturing chamber is fed to the texturing jet by means of hot air or steam and forms the texturing plug in the surface box. The hot air or steam leaves the stuffer box radially and is then removed by an exhaust line. The plug formed in the stuffer box is fed by means of conveying wheels at constant, adjustable speed. With this system, it is possible to set the desired degree of crimping precisely with the three parameters: air or steam temperature, air or steam pressure, and the speed of the conveying wheels. Controlled feeding of the textured plug through the conveying wheels makes the system largely insensitive to spin-finish types with different frictional behavior, different filament cross-sections, etc. The conveying wheels take the plug to a rotating cooling drum which is perforated and is connected to a vacuum blower system. This system sucks the textured plug against the cooling drum and the air is removed through the plug, which is thus cooled. Cooling the textured plug is necessary in order to prevent the crimp being drawn out while dispersing the plug and taking off the yarn.

A cold take-off godet disperses the plug at the cooling drum and guides the yarn to take-up or to the fiber cutter. The speed of the take-off godet is approximately 25 per cent lower than the speed of entry of the yarn into the texturing chamber, because of the contraction caused by crimping.

17.4.9 Automation

Data logging and automation are being used increasingly in spinning lines. Of special interest is the use of automation for take-up machines, to remove the heavy packages and deposit them on creels. Figure 64 shows the function of such an automated system. In spinning lines, the automated device is generally called a "doffer". It consists of:

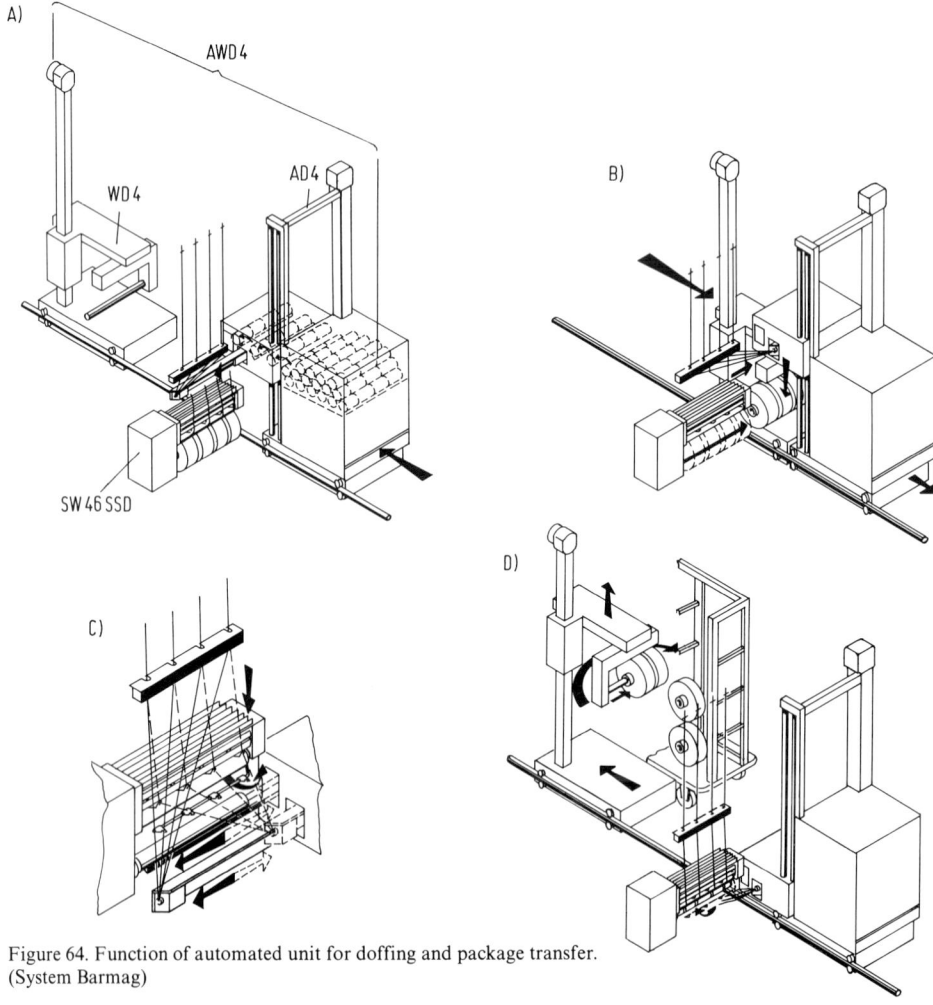

Figure 64. Function of automated unit for doffing and package transfer. (System Barmag)

– a component for yarn handling and supply of empty tubes,
– a component to remove full packages and deposit them on creels.

The yarn handling component, AD4, moves to a position which is ready for doffing and, in this case, accepts the four yarn ends in an aspirator jet (B). Simultaneously the package handling component, WD4, accepts the full packages (B), takes them to a stand-by creel and deposits the packages individually on the creel (D). At the same time, the AD4 component places empty tubes on the bobbin chuck, starts the take-up head, and strings up the four yarn ends (C).

The entire spinning line is monitored by a computer which tells the doffer the positions at which packages are ready for transfer. A weighing device can be integrated in the doffer to weigh the individual packages, and calculate the denier in the computer on the basis of the package doff time and the weight measured, and thus effect denier control, eliminating the need for this type of control at a later point. At the same time, the computer records the number of complete and short packages and prints labels, which are affixed to the packages for identification purposes. This is important for detecting possible faults in the spinning line at an early stage, and separating the packages involved.

Transport systems also lend themselves to automation. These systems take the packages from spinning to inspection and sorting stations, to the shipping department, the stock room, or to the next processing machine.

Process control, and hence direct intervention in the overall process, is now being introduced step by step. It requires the development of new sensors to monitor a variety of parameters and transmit them to the computer, which then makes the appropriate corrections in the overall system. Examples of such parameters are extruder temperatures, spinning-pump speeds, quantity of spin finish applied, godet temperatures and speeds, take-up speeds, and so on. This development will bring many changes on spinning lines in the next few years.

References for Chapter 17

For the convenience of the reader the English titles of all publications in languages other than English are shown in parentheses.

[1] *Gathmann, E.:* Internal report, Barmag, 1984.
[2] Chemiefasern (Chemical Fibers). Company brochure, Glanzstoff 1960.
[3] VDMA Verband Deutscher Maschinen- und Anlagenbau e.V., Fachgemeinschaft Gummi- und Kunststoffmaschinen (Ed.): Kenndaten für die Verarbeitung thermoplastischer Kunststoffe (Characteristic Data for Use in Processing Thermoplastic Materials), Vol I: Thermodynamik (Thermodynamics). Carl Hanser Verlag, München, Wien, 1979, pp. 209, 519, 523, 526.
[4] *Jacobi, H. R., Cegla, J.:* Auslegung von Extrudern für quasi-newtonsche Kunststoffschmelzen (Design of Extruders for Quasi-Newtonian Polymer Melts). Kunststofftechnik 12 (1973) 8, pp. 213/217.
[5] *v. Dahl, W.:* Verteilungsproblematik bei der Zugabe von 8 Additiven in Spinnsysteme (Distribution Problems in Feeding Eight Additives into Spinning Systems). Company brochure, Zimmer, Frankfurt/Main, 1980.
[6] *Pohl, N.:* Dispersives Mischen mit dynamischen Mischern (Dispersive Mixing with Dynamic Mixers). In: Praktische Rheologie der Kunststoffe (Practical Rheology of Plastics). VDI-Verlag, Düsseldorf, 1978, pp. 177/196.
[7] *Lückert, H., Busch, M.:* Stand und Entwicklung des PET-Anlagenbaus (Status and Development of PET Line Design). Chemiefasern (1983) Jan., pp. 29/38.

18 Ram extrusion of PTFE and UHMWPE

P. Stamprech

18.1 Description of the process

Ram extrusion is a pressure-sintering process for the continuous production of profiles from high-molecular-weight polymers. It is used in particular in the processing of PTFE and UHMWPE.

The important parameters which underlie the process will be discussed in the following pages. Since ram extrusion is predominantly used for processing PTFE, consideration of this material will be very much to the fore.

Although PTFE and UHMWPE are classified as thermoplastics, because of their high melt viscosities — in the region of 10^{10} Pa · s — it is not possible to employ the normal processing techniques used for thermoplastics. Over the years, however, ways have been found with UHMWPE, by adapting machinery and modifying the polymer, to allow normal processing methods to be used.

But with PTFE one must make use of a pressure-sintering technique which is very similar to that used in the production of sintered metals. Ram extrusion (sinter extrusion) is a process for continuous production of profiles from discontinuous material feed.

Precompressed powder is transferred by a ram into a heated die of the required cross-section. The individual charges sinter together under the effects of heat and pressure to generate a continuous extrudate. Whilst, because of the great shear sensitivity of PTFE, tapering or spreading in the die has to be completely avoided, in the processing of UHMWPE shaping of the extrudate can be carried out. Very high pressures are built up which require the die and the machine to have correspondingly heavy dimensions. In fact it is possible to process UHMWPE in PTFE extruders, while the converse would lead to poor results.

The present state of the art is such that PTFE can be processed into simple profiles, tubes and rods. Because of its lower tendency to over-shear, UHMWPE can even be turned into complicated profiles. The processing areas of the products are listed in Table 1.

Table 1. Dimensions of ram-extruded profiles

		PTFE	UHMWPE
Pipe:			
Diameter	mm	7 to 300	25 to 300
Min. wall thickness	mm	1.5	0.8
Rod:			
Diameter	mm	4 to 125	10 to 250

Figure 1 shows the principle of the process in schematic form. A PTFE pipe, for example, is manufactured as follows: PTFE powder is charged into the extrusion pipe by means of a loading device and compressed into a ring by a ram. The ram is then pulled away, fresh powder is added, and the compression process repeated. In addition to moving the powder, the compression has to move the whole succession of ring preforms into the heated zone of the extrusion pipe, where they sinter together to form a tube and leave the die as a continuous extrudate.

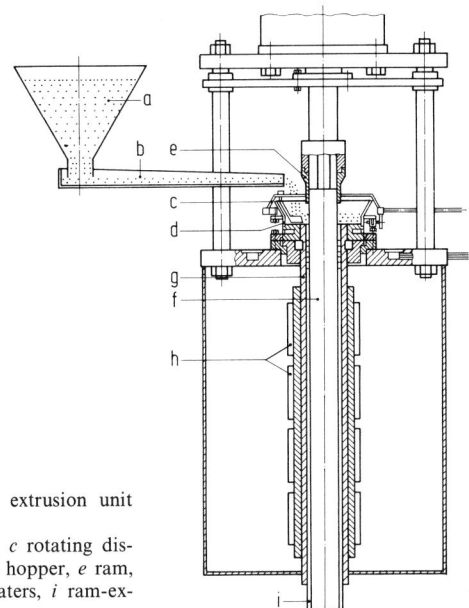

Figure 1. Schematic layout of an extrusion unit for producing tubing
a dosing hopper, *b* vibrating tray, *c* rotating distributor pan with spreader, *d* feed hopper, *e* ram, *f* mandrel, *g* sintering pipe, *h* heaters, *i* ram-extruded tube

Because of the large thermal expansion coefficient of PTFE, high friction forces arise at the walls of the extrusion die as the extrudate moves forward. This produces the pressure required for densification of the powder.

Horizontal and vertical arrangements of the extruder can, in principle, be used. However, dosing, which is critical for the quality of the extrudate, is simpler to carry out with a vertical extruder, and this form of construction has been adopted in recent years. With UHMWPE, on the other hand, the outstanding free-flowing nature of the powder, and the behavior of the melt, make it possible for thin-wall, complicated profiles to be produced even with horizontal extruders.

18.2 Processing machines

In the past, most PTFE processors made their own machines, especially ram extruders. By so doing it was possible to adapt machines to each individual requirement, and to take know-how into consideration. This kind of operation was changed by a new development in the construction of ram extruders (Figure 2). FAG Kugelfischer put forward the idea of two vertically arranged extruders, constructed according to the methods of Hoechst AG. The proposal extended to two machine sizes, from which extrudate up to a maximum diameter of 300 mm can be produced. And even multiple extrusion, which makes it possible to work very efficiently, is envisaged. By addition of auxiliary equipment, the machines can be equipped so that they are capable, essentially, of automatic operation.

In extrusion of UHMWPE one can dispense with vertical machines. Ram extrusion of UHMWPE was developed by Ruhrchemie AG in Oberhausen. Plastikmaschinenbau GmbH offers horizontal extruders with press closing forces of some 40 to 1250 kN, which are used largely for extruding UHMWPE (Figure 3). The die is flange-bolted onto the horizontal machine, thus creating a compact unit.

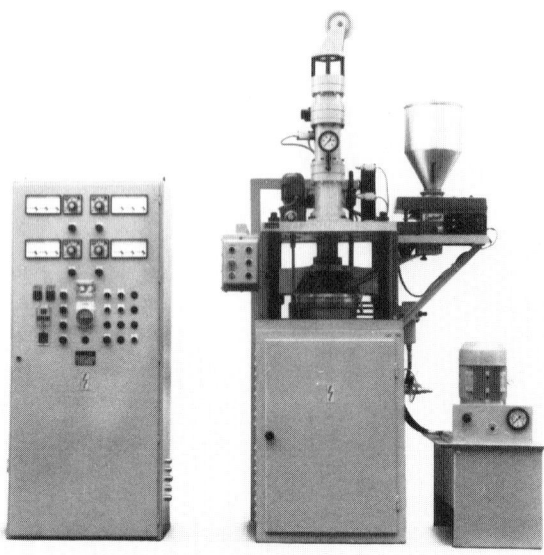

Figure 2. Ram extruder.
(Photo: FAG Kugelfischer,
Schweinfurt, West Germany)

Figure 3. Ram extruder with
flange-mounted rod profile die.
(Photo: Plastikmaschinenbau GmbH)

18.3 The extrusion die

Whilst the machine is seen simply as a power source, one can describe the die as a reactor. Not only does it shape the product, but it also determines the quality of the product by the combined effects of pressure and temperature. But naturally the quality of the product can only be as good as the ingoing material allows.

18.3 The extrusion die

There is a basic distinction in extrusion between rod- and pipe dies. Because of the larger cross-section, that must be properly sintered, rod dies are usually made longer than pipe dies. Since dies represent a large cost element, it is necessary in practice to depart from the ideal of making the optimal tool for each product dimension, and to seek compromises which allow the existing tools to be combined with each other. With this procedure greater flexibility in production is achieved, and one can have recourse to a larger stock of tools. Generally, extrusion dies consist of the following elements:

- extrusion pipe
- mandrel (if extruding hollow profiles)
- ram (Figure 4).

Figure 4. Parts of a sixfold tube die.
(Photo: FAG Kugelfischer, Schweinfurt, West Germany)

18.3.1 The extrusion pipe

The extrusion pipe is the most important, and at the same time the hardest die component. In Figure 5 the layout of the extrusion pipe is explained and a definition proposed at the same time.
To obtain the best possible extrudate the inner surface of the extrusion pipe must be accurately parallel and as smooth as possible, being honed to ensure this. If chroming or nickel plating is carried out on the surface, additional honing is needed to attain the required dimensions. The outer surface of the die is unimportant for extrusion, and turned or planed surfaces are adequate. Corrosion protection of the outer surface has been shown to be advantageous.
Because of the high pressures developed at the high working temperatures (around 400 °C for PTFE) the pipe wall should be made as thick as possible. In no circumstances should it be less than 6 mm, and it should increase with increasing diameter. The length of the heating zone (HL) is a significant factor in controlling the pressure. Thus in designing a tool it is particularly important to choose the length of the heating zone correctly.
The heated length affects the extrusion speed as well as the extrusion pressure, and thus influences the efficiency of the process. To reach the required processing temperature, heating bands are used, usually of 1 kW capacity. Depending on the size of the extruder, a number of bands are concentrated together and controlled as a group. In general, ram extruders are provided with more than four heating zones, which can be controlled in-

dependently of each other. Even temperature of the surface of the tool is ensured by attaching well fitting sleeves made of aluminum. The overall length of the heated part of the pipe matches the length of the aluminum sleeves. In order to be an effective heat store the thickness of the sleeves should not be less than 20 mm. The temperature control sensors are placed in the sleeves between the heater bands. Particular attention must be paid to ensuring that there is no air gap between the sensor and the surface, since this would lead to false readings.

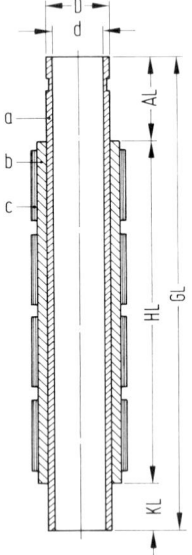

Figure 5. Schematic layout of an extrusion pipe
a sintering pipe,
b aluminum sleeve,
c heater bands,
d inner diameter of extrusion pipe,
e outer diameter
AL warm-up zone length
HL sintering zone length
KL cooling zone length
GL overall length

In the design of multiple tools the use of aluminum blocks has been abandoned. In order to ensure uniform heat transfer into several extrusion pipes, molten salts are now used as the heat transfer medium. This guarantees good contact between the transfer medium and the extrusion pipe and prevents the large temperature variations that can occur with air gaps in the system. The disadvantage of this system is that temperature grading is difficult to achieve.

18.3.1.1 Choice of material

As already indicated, the extrusion pipe is the most important and most costly die component. Thus the proper choice of material and its preparation are extremely critical.
Before we discuss the experience gained with various steels it should be stated that a smooth, honed inner surface to the tool is a basic requirement for obtaining good-quality extrudate.
Since the working temperature of the tool is set at 400 °C, one must use non-scaling steels. There are four suitable types available, each with its advantages and disadvantages. Chrome-nickel steel No. 1.4571 is most frequently used for extrusion pipes. This V4A steel has the advantage that it does not scale at high temperatures. Its disadvantages are the softness of the surface and poor machinability. V4A steels also promote PTFE skin build-up rather easily on the contact surfaces, bringing with it the need for frequent cleaning of the tool. Because of the soft surface, scratches or grooves can be caused when the honed

bore is cleaned. This mechanical damage to the smooth surface favors build-up of surface deposits.

For all these reasons, dies made from this steel have a much shorter life than those with a hard inner surface.

Another recognized material for dies is steel No. 1.4550. This material is not used for extrusion pipes in Europe since it is a specification of American origin and has a composition corresponding to U.S. standard St 347. Both these materials are nickel steels, but in contrast to V4A steel are Niobium-stabilized.

The experience in America with this steel is that surface deposits occur to a smaller extent than with normal Cr–Ni steel. These results have been confirmed experimentally in the laboratory. The disadvantage of this superior material is that lengthy and, above all, costly special treatments are usually needed, sharply raising the cost of the tools and limiting the flexibility of the processor.

When a tool made from this material is new and little used, the surface build-up is not so rapid as with V4A steel. With older tools with scratches on the internal surface, the difference from V4A steel is hardly noticeable

The most frequently specified type of die is an extrusion pipe made from St 52 steel honed and surface-treated. Protective chroming of only a few hundredths of a millimeter provides the die with properties not available with the two other types of material mentioned.

The lower rate of deposit build-up, easier cleaning and harder surface of chromed dies result in longer service life. Further advantages of the higher strength are better machinability and the fact that thick-walled pipes are easier and cheaper to work than those made from Cr–Ni steel. The extra cost of chroming that must be met is usually less than the difference between normal and Cr–Ni steels.

The disadvantage of this material occurs with diameters below 20 mm, when problems arise in chrome-plating the internal surfaces of long pipes, causing porosity in the protective layer. At such small dimensions, one should fall back on the material types previously mentioned, or use the latest plating method.

The method that has been used successfully in recent years is the Kanigen process of chemical nickel plating. Not only has it the advantage that uniform non-porous layers can be applied, which avoids an additional honing operation, but it can even provide bores of less than 20 mm with corrosion protection. Chemical nickel plating is being increasingly used in the manufacture of ram-extrusion tools, since nickel offers the best corrosion protection against HF (hydrofluoric acid), which can be released as a decomposition product during extrusion.

Extensive experiments have demonstrated that chemically nickeled tools disfavor film build-up in the heated zone, and that extrusion pressures are lower than with all other methods.

The problems with extrusion pipes for UHMWPE are less serious, because of the lower processing temperatures – between 180 and 200 °C – and the much less corrosive melts. A nitrided steel is normally adequate. One must certainly pay attention to the stability of the tool, since the working pressures, which can be as much as 10 times greater than with PTFE, can only be controlled with difficulty.

Surface treatments, like those described for PTFE operation, also have beneficial effects with UHMWPE, and bring about a reduction in working pressure.

18.3.2 The mandrel

For manufacturing hollow structures, the mandrel constitutes the second friction-generating surface, the first being the extrusion pipe. To keep the level of shear in the die gap as low as possible, the quality and trueness of surface of the mandrel must match that of the extrusion pipe. The surface of the mandrel must also be smooth and of constant diameter. For diameters less than 50 mm the pipe is usually machined from solid rod; above 50 mm diameter, pipe stock is used.

Since in tube extrusion one must work with tight tolerances between mandrel and ram, and between ram and extrusion pipe, with diameters above 80 mm it is necessary to cool the mandrel in the region of the ram stroke. If the diameter is greater than 120 mm it is recommended that the mandrel also be heated in the heated zone of the pipe, and cooled over the range of ram motion by employing an insulating barrier.

This rather expensive procedure is required in most situations, since the heating capacity may not be sufficient to deliver the heat required by conduction through the mandrel.

Generally, mandrels are cylindrical and extend some 100 to 200 mm into the cooling zone. It is advantageous to provide a device on the extruder with whose help an adjustment to the height of the mandrel can be made. This can provide a means of influencing the shrinkage behavior of the extrudate.

In exactly the same way as with extrusion pipes, chrome- or nickel-plated mandrels can be produced from Cr–Ni steels and from normal St 52 steel (chrome- or nickel-plated). The same criteria apply to mandrels as to extrusion pipes, and chrome- and nickel-plated mandrels are preferred for this purpose to those made from Cr–Ni steels.

18.3.3 The ram

As has already been explained, tight tolerances are needed in the manufacture of tubing. Consequently the diametral play between the ram and the mandrel, and the ram and the extrusion pipe is only 0.1 to a maximum of 0.3 mm. With extrusion of rod this can be widened to 0.5 mm for greater diameters.

In order to compensate for thermal expansion the mandrel must be cooled. Similarly, the sealing zone of the extrusion pipe must be mounted in a cooled plate.

The choice of ram material depends on the dimensions of the extrudate. With solid rods larger than 10 mm in diameter, and with tubes of wall thickness greater than 15 mm, brass, gun metal or bronze are fully satisfactory. But if the tube walls are thinner or the diameter of the rods smaller, one must change to harder materials. Good results have been obtained with hardened steel rams. However, in this case the extrusion tube must be fitted with a hardened sleeve. With hardened rams, wear problems, in particular, can be avoided.

18.4 Processing problems

In the processing of ultra-high-molecular-weight polymers, exemplified by UHMWPE and PTFE, welding together of the individual ring charges by sintering is the major problem. Localized faults and poor quality finished product can usually be traced back to this area.

The most important quantity is ram pressure, and this has a very strong influence on welding quality: this is called extrusion pressure in what follows. If extrusion pressure is too high the surfaces between the individual charges will be pressed so smooth that no mechanical keying between the individual charge elements comes about. The consequence is the occurrence of so-called 'tablets', meaning that individual charges have not welded together (Figure 6).

On the other hand, if the extrusion pressure is too low an unfavorable degree of densification of the powder will be achieved. And in this case, too, poor mechanical properties will result.

18.4 Processing problems

The magnitude of extrusion pressure required depends first of all on the powder. The operating window of a material is determined by the hardness of the powder particles, and therefore by their load acceptance limits; this is important because the extrusion pressure, which is dependent on the die dimensions, can only be influenced to a small extent. Listed below are the factors which make corrections to extrusion pressure possible:

- die dimensions,
- surface quality of the die,
- extrusion speed,
- temperature profile along the die,
- densification rate of the powder,
- depth of insertion of the ram,
- use of a movable mandrel
  (in extrusion of hollow profiles).

Figure 6. Faulty tube (left) caused by tablet formation and (right) good-quality, well welded tube.
(Photo: Hoechst AG, Frankfurt a. Main, West Germany)

Because of its shear sensitivity, the pressure limits with PTFE are narrower than with UHMWPE. Hence, special attention has to be given to extrusion pressure. Whilst the first four factors relate to ram extrusion in general, the last three have to be considered principally in the manufacture of hollow profiles or rods of very small diameter.

18.4.1 Dimensioning the die

Looking across the full range of ram-extrudate sizes, i.e. from thin-wall tubes to solid, large-diameter rods, the extrusion pressure can lie between 25 and 800 bar. The magnitude of the extrusion pressure depends in the first place on the dimensions of the extrudate itself.
If one considers a rod tool, there is a large pressurized cross-section available in relation to a small friction-generating outer surface. Since friction at the outer surface is responsible for the pressure build-up, very low pressures are generally generated in the extrusion of rod. With large dimensions, rods larger than 50 mm in diameter for example, the pressures are so low that the use of brakes is necessary. By 'brakes' is meant devices which cause a pressure increase during extrusion. These can take the form of removable shoe brakes or calibration dies at the end of the extrusion pipe. The horizontal method of construction of ram extruders offers advantages in this case, since the weight of the extrudate cannot contribute to lowering the extrusion pressure. In contrast, with vertical machines the extrudate mass must be slowed down by 'brakes' before a pressure increase becomes necessary, in order to prevent abrasion of the extrudate in the extrusion pipe.
However, if a mandrel is mounted inside this rod die, so creating a pipe die, the pressurizing surface becomes progressively smaller as the extrudate wall becomes thinner. But, in contrast, the frictional surface area, in the extreme, can increase to nearly double that of the rod die. This is the reason that with ram extrusion the pressures vary over a very wide range. The lower range of pressure is developed in the manufacture of rods, or of pipes with wall thicknesses above 15 mm. And the upper range of pressure occurs if one wants to use ram extrusion to manufacture thin-walled tubes, i.e. those with walls from two to three millimeters thick, or very thin rods.

18.4.2 Surface quality of the die

As with extrusion speed, the surface quality of the die also influences the extrusion pressure. In principle, one can assume that back pressure will be higher the rougher the surface is.

It should be noticed that when extrusion starts with a clean tool, a PTFE film builds up on the inner surface, particularly where the temperature is highest. This film becomes thicker as extrusion proceeds. At the beginning of extrusion, slippage of PTFE on the thin PTFE film is an advantage, since, as a result, the extrusion pressure decreases after running for some time. It is especially easy with thin-walled tubes to achieve the optimal pressure region because of this. However, as extrusion continues the thickness of this film increases more and more, with the result that a reduction of the cross-sectional area occurs, accompanied by shearing of the individual densified charges in this zone of the die. This shearing becomes externally evident in erosion of the individual charges. And tablet formation, as it is called, can also occur. If this happens, extrusion must be stopped, the tool must be cleaned, and a new start made.

Despite the view that is sometimes expressed that the inner surface of a die should be matt finished in order to promote a more rapid build-up of the film, and consequently a reduction of friction forces inside the die as quickly as possible, in practice smooth bores are preferred, because one obtains a longer service life for the same die dimensions.

Film build-up is strongly dependent both on extrusion pressure and on the PTFE powder used; and at high extrusion pressure it progresses more rapidly than at low pressure. If a PTFE powder with low granule hardness is available, as is the case, for example, with powders used for extrusion of rod, the film is created more rapidly than with hard powder granules.

The negative effect of film build-up that occurs with PTFE is not so evident with UHMWPE. Because of its lower shear sensitivity, taper may even be introduced into the die without its leading to disturbances to the extrudate. Such procedures cannot be considered in PTFE processing.

18.4.3 Extrusion speed

The extrusion speed has a quite significant effect on the pressure, particularly when pipes are being extruded. However, the speed is limited by the length of the sintering zone. It is generally the case that the pressure is higher, the higher the extrusion speed. However, in order to achieve optimal machine efficiency, one has to find the best compromise between pressure build-up and degree of sintering. It is pointless to make a die with a very long heating zone to achieve high output speeds: the result would be that the extrusion pressure would become very high because of the large surface area, and in the extreme case the pressure limit of the powder would be exceeded, resulting in tablet formation. If the sintering length is made too short, only low extrusion speeds are achieved. This means, with small dimensions, that operations are often uneconomic. During the extrusion of heavy-section rods the speed has little influence on pressure. In this case the extrusion speed only affects the sintering of the extrudate. The larger the rod diameter becomes, the closer one approaches the lower pressure region.

Under unfavorable conditions, the pressure required to densify the powder may not be reached; in this case, as already indicated, the extrudate is slowed down by application of mechanical, pneumatic, or hydraulic 'brakes', so that the necessary pressure for densification can be built up.

In the processing of UHMWPE, the length of the sintering zone can be adjusted to an extrusion speed determined by economic/efficiency considerations. But care must be taken to design the tool to handle the high extrusion pressures generated.

18.4.4 Temperature control

Whereas the processing temperatures for UHMWPE lie between 180 and 200 °C, for sintering PTFE temperatures above the crystal melting point, 327 °C, are necessary. In practice temperatures in the range from 360 to 400 °C have proved to be suitable for ensuring rapid sintering. At temperatures above 380 °C and with the relevant length of dwell time in the sintering zone, one must expect some thermal damage to the PTFE.

Heating the material in the extrusion cylinder to ca. 380 °C produces a volume increase of some 30%, which causes the PTFE to be forced against the inner walls of the pipe, thus generating counter pressure in the tool.

If, however, one tries to densify PTFE in an unheated pipe, the pressure increases to a very high level after quite a short time. This shows that heating the tool causes a pressure drop, in spite of the thermal expansion. Control of temperature between the pressurizing zone and the first heating zone is therefore particularly important. The greater the unheated length at the beginning of the pipe, the greater is the pressure increase. This is of special importance in the extrusion of thin-walled tubes. On the majority of machines this length is determined by how the die is supported, and should not exceed 200 mm. Experiments have shown that reducing the unheated length by about 50 mm, when making 50 mm diameter tube with a wall thickness of 3 mm, produced a pressure reduction of 100 bar. For this reason it is very important in the extrusion of thin-walled tubing and of thin rods to mount the heated section of the die as close as possible to the compression zone, in order to achieve optimum processing conditions for the PTFE powder.

The temperature profile is of prime significance for the control of pressure, and the last heating zone also provides some control of the pressure. In this respect, however, there is a difference in extrusion behavior between rod and tubing.

For processing UHMWPE some two-thirds to four-fifths of the heated length is run at temperatures of 180 to 200 °C, and often at as high as 230 °C. In the last one-third to one-fifth of the heated portion of the tool the temperature is reduced, so that the extrudate leaves the die at about 130 °C. This stepwise, slow cooling prevents the occurrence of stresses which might give rise to cracking.

18.5 Materials for ram extrusion

Next to deciding on the processing machinery and the die, the choice of raw material is one of the most important factors affecting ram extrusion. In the same way as with HMW polyethylene, with PTFE there are certain requirements laid on the raw material which must be fulfilled to guarantee trouble-free extrusion; these are:

- good free-flowability,
- high packing density,
- good granule stability,
- granule hardness to suit the application, and
- cleanliness.

Since, basically, only powders can be processed by ram extrusion, in order to be able to fill up the profile cross-section prescribed in the best fashion they must have good flowability and a high bulk density. These two properties determine whether or not the powder can be used for ram extrusion.

Since extrusion pressures in the ram extrusion of PTFE range from 25 to 800 bar, many grades of material are available for the different levels of pressure (Table 2). Whilst Hostaflon® TF 1502 and 1645 are suspension-polymerized powders which have not undergone heat conditioning, Hostaflon® TF 1101 has been heat treated. Heat treat-

Table 2. Powder properties for Hostaflon grades

		Hostaflon TF 1645	Hostaflon TF 1502	Hostaflon TF 1101
Mean particle size	μm	650	500 to 600	600 to 700
Packing density	g/l	800	700 to 800	600 to 700
Recommended processing pressure	bar	25 to 125	25 to 150	60 to 800
Optimum processing pressure	bar	50 to 80	50 to 100	200 to 450
Kind of suspension-polymerization powder		agglomerate particle: not heat-treated	solid particle: not heat-treated	heat-treated

ment results in significantly greater granule hardness and an associated higher pressure stability.

Hostaflon® TF 1502 is a solid granule, whilst Hostaflon® TF 1645 consists of agglomerated granules.

Grain stability is only important for granular polymer grades, Hostaflon® TF 1645 and other Hostaflon® compounds produced in a similar way. (These are PTFE/filler mixtures produced by the same process.) Since granular polymers are manufactured by a special process, the individual particles, which consist of many primary nuclei, are pressure- and shear sensitive. If nucleus stability is low, it can happen that with low pressure on the powder the granular particle disintegrates, and flowability is lost. Pressure can arise even during transport or when dosing the powder for processing. Therefore special importance attaches to the dosing devices used for processing PTFE powders during ram extrusion.

PTFE grades that have not been heat treated are suitable for the pressure range from 25 to 150 bar, while heat-treated grades can be used for pressures from 60 to 800 bar.

The allowable processing pressure defines the range in which the powder can be processed without harming the mechanical properties.

One should always operate above the low pressure limit, otherwise full densification of the powder will not be achieved. If the upper pressure limit is exceeded, a fall in output, and tablet formation can result. The optimum pressure lies somewhere in the middle of the recommended range, and one should always take care to operate within this range.

In Figure 6 the formation of tablets in a tube of 3 mm wall thickness can be clearly recognized. In this case the tablets have been formed by using too high a pressure on unsintered powder. The same tube size produced from a sintered powder with hard particles provides an example of fault-free processing. To achieve this, attention must be paid to precise dosing during individual working strokes.

Product quality standards to be applied during tests on PTFE products have been laid down by the Fluorinated Plastics Group of the Association of the Plastics Processing Industry (GKV) in Frankfurt. Standards for heat-treated and untreated powders are summarized in Table 3.

Table 3. Strength and extension of ram-extruded product (GKV minimum values)

		Sintered PTFE powder Hostaflon TF 1101	Unsintered PTFE powder Hostaflon TF 1645/1502
Tensile strength	N/mm^2	19.6	24.5
Extension at break	%	200	250

As the table shows, higher strength is required for untreated powders with soft particles. Because of the better compressibility of the powder in the pressure range from 25 to 150 bar, this is easy to achieve.

It has been shown in practice that these grades can be processed particularly well into rods greater than 20 mm in diameter and pipes with wall thicknesses above 10 mm.

Presintered powders, because of their hard, pressure-resistant particles, are more suitable for use with product dimensions at which higher extrusion pressures occur. Therefore, these materials are used for rods less than 20 mm in diameter and for thin-walled pipes with wall thicknesses less than 7 mm. The minimum usable pipe wall thickness is around 1.6 mm. Attempts to reduce wall thicknesses still further are, in fact, in progress; but, because of the high pressures, they encounter such serious processing problems that the efficient operating limit is almost reached. Additionally, with pipe sheathing, for permeability reasons the sheathing layer should not be made too thin.

Recently developed, presintered PTFE grades with an extremely high packing density offer the processor the possibility of achieving higher throughputs on his machines. This applies certainly only if the sintering and compression behavior in the tool permits it. If, for example, in the processing of a powder with a packing density of 650 g/l the ram-stroke rate is already at its limit, the use of a material with a packing density of 1,000 g/l permits the output to be increased by some 50%. This is of special importance for the small values of wall thickness, which require only a short sintering time.

Furthermore, the recently developed Hostaflon® grade TF 1502 exhibits somewhat lower shear sensitivities than conventional products. This is based on a lower melt viscosity and provides the products with a larger processing window.

As a general observation, powders that have not been subjected to heat treatment are more sensitive to surface build-up in the die and to temperature control than presintered products.

Because of the reduction in melt viscosity, it has also been possible to improve the mechanical properties of the thermally conditioned products. This is shown in Table 4.

Choice of UHMWPE grades proves to be far simpler. Within the product range, which is not especially large, Hostalen® GUR 415 and the standard grade 412 are satisfactory as general-use powders for ram extrusion. Hostalen® GUR 415, the powder with the highest F 150/10 yield value, 0.5 to 0.65 N/mm², and a packing density of around 400 g/l, has also been generally accepted for extrusion of large-diameter rods and for complicated thin-wall profiles. Table 4 gives a comparison of the mechanical properties of PTFE and UHMWPE grades.

Table 4. Comparison of the mechanical properties of PTFE and UHMWPE grades

Rod 23 mm ∅	Unit	Hostaflon TF 1645	Hostaflon TF 1502	Hostaflon TF 1101	Hostalen GUR 412	Hostalen GUR 415
Tensile strength	N/mm²	24 to 26	26 to 29	21 to 23	41	44
Extension at break	%	280 to 350	350 to 450	300 to 350	450	450
Density	g/cm³	2.15	2.17	2.16	0.94	0.93
Shore hardness, D DIN 53 505	–	56	55	56	64 to 67	61 to 65
Ball indentation hardness DIN 53 456	N/mm²	20	22	22	40	38

18.6 Extrusion of solid profiles

The extrusion of solid profiles is the simplest form of ram extrusion processing. It is simple because the demands placed on the powder and the die, and all the boundary conditions, are not so severe as with pipe extrusion.

Since effective sintering of the extrudate is directly related to the wall thickness, longer dies are generally used for ram extrusion of solid profiles, especially rods, in order to raise the output rate to economically interesting levels.

Extensive investigations on PTFE have been carried out so as to be able to determine extrusion speed as a function of the heated length by computer calculation. The result was that a working basis could be provided for the processor, which made design calculations on tools possible.

The experiments were carried out with rod tools having diameters from 9 to 100 mm. The maximum achievable output was determined empirically. The smallest values called for by the GKV (Association of the Plastics Processing Industry) were taken as lower limits. The experimental conditions are summarized in Table 5.

Table 5. Experimental conditions for ram extrusion of PTFE rods

Die diameter D	Length of heating zone HL	Number of heating bands	Temperature of control zone, in extrusion direction				Output speed
			1	2	3	4	
mm	m		°C	°C	°C	°C	m/h
100	1.6	12	360	370	370	360	0.35 to 0.65
45	1.6	12	370	380	380	360	1.0 to 2.5
22	0.8	6	380	390	370	–	2.0 to 4.0
9	0.6	4	390	390	–	–	6.0 to 12.0
9	0.33	2	390	–	–	–	4.0 to 7.0

Following the scheme in Table 5, the extrusion speed was varied over a given range until the product quality reached the minimum acceptable according to GKV standards. This speed was adopted as the limit value. The results are summarized in Table 6.

Table 6. Maximum speeds for ram extrusion of PTFE rods

Die diameter D mm	Length of heating zone HL m	Maximum extrusion speed v_{max} m/h	Minimum sintering time t_{min} min	v_{max} (for $L = 1$ m) m/h	Extrusion pressure bar	Hostaflon grade used
100	1.6	0.55	175	0.34	39*	TF 1502/1645
45	1.6	2.0	48	1.25	60	TF 1502/1645
22	0.8	3.5	14	4.4	140	TF 1502/1645/1101
9	0.6	11.0	3.3	18	700	TF 1101
9	0.33	6.0			250	TF 1101

* Pressure obtained by brakes

To better illustrate the comparison of dies, it is advantageous to calculate the minimum dwell time t_{min} (min) in the sintering zone from the heated length, HL (m), and the maximum speed v_{max} (m/h):

$$t_{min} = 60 \cdot HL/v_{max} \text{ (min)}. \qquad (1)$$

The values for t_{min} are also shown in Table 6 and relate to extrusion of 22 mm rods from a number of untreated and thermally pretreated (sintered) powders, chosen to produce a range of extrusion pressures.

A double logarithmic plot of the minimum dwell times against the corresponding die diameters in Table 6 gives a straight line.

18.6 Extrusion of solid profiles

Following Eq. (1), one can also plot the maximum extrusion speed v_{max} for a reduced heated length (HL = 1 m) against the die diameter. This, too, gives a straight line. Both straight lines appear on Figure 7.

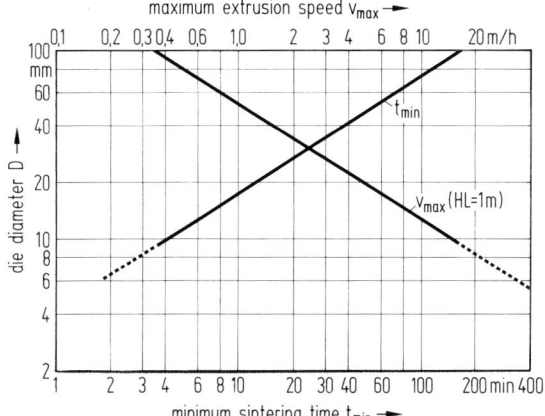

Figure 7. Minimum sintering time and maximum extrusion speed for ram extrusion of pipes

The t_{min} straight line is described by the equation:

$$\log t_{min} = x \cdot \log D + \log C \tag{2}$$

or

$$t_{min} = C \cdot D^x. \tag{3}$$

The two unknowns, C and x, are determined from Eq. (2) by inserting pairs of values from the corresponding straight lines in Figure 7.

Using values of t_{min} and D in mm from Eq. (3) we obtain:

$$t_{min} = 1/12 \cdot D^{5/3}. \tag{4}$$

Finally the combination of Eqs. (1) and (4) gives:

$$t_{min} = 60 \text{ HL}/v_{max} = 1/12 \cdot D^{5/3}$$

and

$$v_{max} = 720 \text{ HL}/D^{5/3}. \tag{5}$$

The factor 720 applies only when values of v_{max} in m/h, HL in min, and D in mm are used.

The maximum allowable extrusion speed can be obtained from Eq. (5) as a function of heated length and die diameter for usual ram extrusion heating conditions.

While the surface area available for heat transfer and heat transmission increases linearly with die diameter, the extrudate mass and the heat capacity of the extrudate increase as the square of the diameter. This is why the exponent 5/3, with which the diameter appears in Eq. (4), must lie between 1 and 2.

Similar relationships can be set up for UHMWPE. On the input side of the die the temperature over the first two-thirds to four-fifths of the length is kept within the range 180 to 230 °C, and in the remaining one-third to one-fifth at 140 to 160 °C. Then, for given temperature conditions the dwell time, t (min), of PE extrudate in a die length of 1 m, is calculated as a function of diameter D (mm) from the relationship:

$$t = \text{const.} \ D^n,$$

where
$$\text{const.} = 5 \cdot 10^{-3} \text{ to } 5 \cdot 10^{-2},$$
$$n = 1.5 \text{ to } 3,$$
$$D = 4 A/U.$$

A is the cross-sectional area of the extrusion die, and
U is the contour length of cross-section.

It turns out that throughput is directly proportional to the length of the die, and so the preceding formula can be expressed in terms of the die length required. Just as with PTFE, this relationship is independent of the extrusion pressure. Since large quantities of profiles are produced from ultra-high-molecular polyethylene, it is the 'output' rate calculated at the thickest point in the profile that has to be obtained. The same is, in fact, true for PTFE.

Whilst, because of machine design, UHMWPE rod extrusion can only be carried out horizontally, with PTFE it can be done both vertically and horizontally. Horizontal extrusion has the advantage that to reach the lower pressure limit the weight of the rod does not have to be supported by brakes, as it does in vertical extrusion.

Because of the homogeneity of the molten material, it is normal to produce UHMWPE solid profiles up to 250 mm diameter by ram extrusion. Solid PTFE rods are produced up to a maximum diameter of 125 mm.

Because of the high processing temperatures of PTFE and the relatively rapid cooling rates, stresses originating in crystallization differences build up, and can influence the quality of the extrudate. The length of the cooling section (CL) is also important. It should be between 300 and 500 mm long in order to guarantee as uniform cooling of the extrudate as possible. In pipe extrusion, especially with thin walls, one can keep the CL short.

However, with rod extrusion it is necessary, particularly with large diameters, to provide longer cooling lengths, so as to avoid stresses in the extrudate as far as possible.

With vertical extrusion the scope for doing this is limited by working-space height, whilst in horizontal extrusion, cooling behavior can be better idealized. The horizontal method is used for extrusion of UHMWPE, but it does, certainly, lead to very long dies.

In practice, for this reason only solid profiles up to some 50 mm in diameter have proved successful. Larger diameters are only used for special situations, where manufacture in one piece is an absolute requirement.

18.7 Extrusion of pipes and hollow profiles

In the extrusion of hollow profiles the densified powder must slide over two surfaces — those of the extrusion pipe and the mandrel — producing a counter pressure, because of friction; naturally, therefore, extrusion pressures are higher than in rod extrusion. In the extrusion of UHMWPE, this can be countered with appropriate machine drive power, but with PTFE one runs into the danger of exceeding the upper pressure limit of the powder and getting tablet formation. It is therefore obvious that all measures contributing to a reduction of extrusion pressure should be employed. These parameters will be discussed separately.

First of all, however, continuing from rod extrusion, we shall consider the calculation of output speed.

In the extrusion of pipes, similar conditions to those used for the extrusion of rods can be found. One then proceeds on the assumption that the minimum sintering time in pipe dies can be obtained by deducting from the time needed to fully sinter a solid rod, the time that would be required for sintering the core (diameter that of the mandrel) if this consisted of PTFE. This procedure is illustrated in Figure 8. If, for example, a 45 mm rod needed a

18.7 Extrusion of pipes and hollow profiles

sintering time of 48 min. and one of 26 mm diameter 19 min., the result for a pipe die with 45 and 26 mm inner and outer dimensions would be 48 minus 19 = 29 min. The practical result for these dimensions was 28 min.

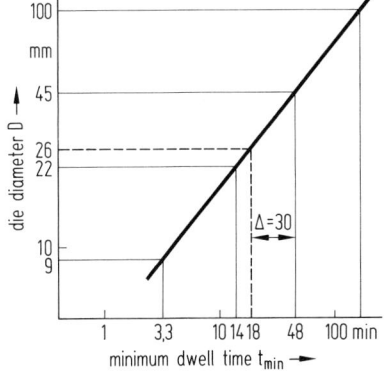

Figure 8. Die diameter D vs. minimum dwell time for ram extrusion

With a 45/36 mm die the calculated value was 15 min. and the practically determined one 12 min. (see Table 7)

Table 7. Maximum speed for ram extrusion of PTFE pipes

Die dimensions D mm	Length of heating zone HL m	Maximum extrusion speed $v_{max.}$ m/h	Minimum sintering time $t_{min.}$ min	$v_{max.}$ (for $L = 1$ m) m/h	Extrusion pressure kp/cm^2	Hostaflon grade used
45 × 26	1.6	3.5	28	2.20	550	TF 1101
45 × 36	0.8	4.0	12	5.0	750	TF 1101

The error in calculated values is, in fact, larger the larger the pipe wall thickness. This means that the maximum extrusion speeds calculated in this way for thin-wall pipes are always somewhat lower than can be achieved in practice, which constitutes an additional safety factor.

For calculating the maximum extrusion speed (v_{max}) for pipes as a function of heated length and die diameter, the equation:

$$v_{max} = \frac{720 \, HL}{D^{5/3} - d^{5/3}} \tag{7}$$

applies for usual conditions of ram extrusions, with

- D the diameter of the extrusion pipe in mm,
- d the diameter of the mandrel in mm,
- HL the length of the heating zone in m.

For pipes with wall thickness greater than 10 mm, the same temperature control rules apply as to rod extrusion. Thin-wall pipes with walls about 2 to 5 mm thick are insensitive to temperature impact because of the short dwell time in the sintering section (the heat-up time is short). Temperatures in the 380 to 410 °C range can therefore be used without the risk of thermal damage.

Because during cooling the extrudate shrinks onto the mandrel, the condition in pipe extrusion is the reverse of that in rod extrusion. A pressure increase occurs. If the temperature in the cooling zone is raised, the extrudate is kept away from the mandrel and a fall in pressure comes about. Pressure regulation by means of temperature has a strong influence in pipe extrusion on the final dimensions of the extrudate; this is not so in rod extrusion.

The powder heating section should be kept as short as possible. In most cases this is dependent on the dimensions of the extruder. However, it should never exceed 200 mm in length. The shorter one makes the heating section, the lower is the resulting extrusion pressure. This is of great importance, especially when extruding thin-wall tubing.

As has already been mentioned, in ram extrusion of hollow profiles the factors listed below are important for reducing the extrusion pressure:

- the powder densification speed,
- the depth of the ram stroke,
- the movable mandrel.

The way these factors operate will be described below. These considerations apply equally to PTFE and UHMWPE.

18.7.1 Powder densification rate

Powder densification rate influences the extrusion pressure, just as do all the factors so far discussed. The faster one compresses the powder, the greater is the pressure increase. Since in ram extrusion of PTFE one always tries to operate in the optimum range of pressure, it is recommended that low ram insertion speeds be employed. The densification speed should not exceed 10 mm/s. This limitation is not necessary with UHMWPE.

One achieves two things by slow densification:

a) The powder charge contains air which can escape through the narrow annular gap between ram and die during slow densification.
b) During the compression process the extrudate cakes onto the tool. If now the densification is rapid, it is suddenly torn free from the die walls, causing the pressure to rise abruptly. With low rates of densification the pressure is applied slowly to the adherent extrudate, and rises until gradual freeing of the extrudate from the die walls occurs, which normally generates a low pressure.

18.7.2 Ram stroke depth

The size of the individual powder charges is determined by the depth of the ram stroke. Smaller charge sizes require a higher extruder stroke rate and involve higher densification rates. Therefore proper reconciliation of tablet thickness with the powder densification rate has a great influence on the end result.

On the other hand, if one operates with a very large tablet thickness, there is the risk that densification differences between charges will occur, which can give rise to quality variations. A tablet thickness of 10 to 15 mm has proved successful in practice.

18.7.3 The movable mandrel

The use of a movable mandrel has proved to be the most important pressure-reducing element in the extrusion of thin-wall pipes, and it is due to this one element that the range of uses of ram extrusion has broadened greatly.

18.7 Extrusion of pipes and hollow profiles

The operating principle is described below

During densification of the powder the pressure builds up, and operates on the extrusion pipe and on the mandrel. In order to be able to move the extrudate along in the tool, the pressure-dependent friction forces on the pipe and the mandrel must be overcome by the ram.

Since the mandrel is mounted on an individual hydraulic cylinder, with the valve open it can be moved along by the ram through components of frictional force acting on the mandrel. The result is that the ram has only to overcome frictional forces on the pipe surface, and with just one surface available as a pressure-generator the pressure goes down.

When the ram, with the mandrel, reaches the low pressure point – at the end of the compression period, after which the extrudate slowly expands – the mandrel is withdrawn to its starting position. This causes renewed pressure loading on the powder, arising from friction against the mandrel.

By using the moving mandrel, the overall extrusion pressure is divided into two partial pressures, some 30 to 50% smaller than the original value. The moving mandrel also makes it possible to extrude pipes with extremely thin walls – 2 to 3 mm – under optimum pressure conditions for the powder.

Only by use of this technique has the production of thin-wall pipe at relatively low pressures become possible, and the problem of tablet formation been largely overcome. By combining all effective parameters optimally, minimum wall thicknesses of as little as 1.6 mm can be achieved.

The pressure generated in the production of UHMWPE piping is many times that needed in the manufacture of profiles with solid cross-sections. Since the pressures arising in ram extrusion of UHMWPE are considerably higher than those needed for ram extrusion of PTFE, one might expect damage to the material properties when using the above-mentioned process for extruding UHMWPE pipes. Also, the extruder design might be economically deterring in view of the high design pressures.

To be able to overcome this problem one makes use of a process similar to that used for PTFE, which is described in European Patent Specification 0.008.434 B1.

The invention lies in a ram extruder for producing plastics tubing, consisting of a die heated with heating elements, a reciprocating ram within the die, and a mandrel which defines the inner diameter of the tube and is rigidly fixed to the ram. It appears that the transport of the powder and its compression during plastication into the desired pipe shape can be realized by simultaneous forward and backward movement of mandrel and ram, without interrupting the process of ram extrusion. Thus it is not required that the mandrel should always be stationary, nor that the ram be held in the compression position while the mandrel is withdrawn to the starting position, as is the case with PTFE.

It has been shown that in ram extrusion of UHMWPE pipes by an extruder fitted with this system, considerably lower pressures are developed than in ram extrusion of a corresponding solid profile. The rigid connection between ram and mandrel results in a considerable simplification of the extruder drive, as well as reducing extrusion pressure, since ram and mandrel can be operated by a single hydraulic unit.

The length of the mandrel is normally the same as that of the die from the feed zone to the exit. The mandrel diameter corresponds to the inner diameter of the extruded pipe, allowance being made for shrinkage. The mandrel is solidly fixed at one end to the ram, but is not normally supported on the discharge side of the die, since this would interfere with the production of a uniform tube. Thus, during the start-up process a guide-sleeve must be slipped into the die over the mandrel. This is carried through the die with the extrudate and removed at the end. The horizontal method of construction makes this procedure necessary. By use of this start-up aid the mandrel remains centrally located until it is completely surrounded by extrudate, and in this way the uniformity of the wall thickness is ensured.

Pipes with normal wall-thickness tolerances can be produced by this technique. The condition for its success is that the powder should flow regularly, thereby guaranteeing uniform charge dosing.

References for Chapter 18

For the convenience of the reader the English titles of all publications in languages other than English are shown in parentheses.

[1] *Steininger, A., Stamprech, P.:* RAM-Extrusion von Polytetrafluorethylen (RAM Extrusion of Polytetrafluorethylene). Kunststoffe 60 (1970), pp. 290/294.
[2] *Steininger, A., Stamprech, P.:* Extrusionsverarbeitung von Polytetrafluorethylen-Füllstoff-Mischungen (Extrusion Processing of Polytetrafluoroethylene/Filler Mixtures). Kunststoffe 63 (1973), pp. 558/563.
[3] *Stamprech, P.:* RAM-Extrusion von Polytetrafluorethylen (RAM Extrusion of Polytetrafluorethylene). Plastverarbeiter 31 (1980), pp. 61/68.
[4] DE-PS 2829232 C2 (1976). Ruhrchemie AG.
[5] EU-PS 0008434 B1 (1982). Ruhrchemie AG.

19 Extrusion welding

Peter John

19.1 Introduction

Plastics welding is a process employed for joining thermoplastics. The heat-softened surfaces of the materials to be welded are joined by applying pressure, with or without additional material. On the basis of scientific findings, it is nowadays assumed that at the boundary zone of the fusion face, the materials flow into and diffuse with one another. In the process, the macromolecules in the high-viscosity melt become entangled [1, 2].

All the welding processes used at present are listed in DIN 1910, part 3. Among the twenty or so welding techniques mentioned in this standard, there are, in addition to extrusion welding, two other processes with which a large proportion of the welding tasks covered by extrusion welding can also be performed. These are the heated-tool butt-welding method and the hot-gas welding process. Schematic diagrams of these three techniques are shown in Figure 1.

High-quality, economic processing of semi-finished plastics product is nowadays only possible if the optimum technique for the particular application in question is selected from these three welding methods.

19.2 Extrusion welding

Extrusion welding is a welding technique invented during the early sixties. A number of different variations were subsequently developed, and particularly in recent years the process has gained increasing acceptance. Its main characteristic is that the welding (or filler) rod consists of a homogeneous, completely plasticated strand (extrudate), which is produced with an extruder or, as a variation, also by means of an extruder-like mixing chamber, and is presented direct into the welding gap without being cooled. Prior to this the groove faces in the base material are in the majority of cases heated to welding temperature by hot air and in some cases also by radiant heat from halogen lamps. Down to a certain depth, the weld faces are transformed into the thermoplastic state. The pressure exerted on the melt causes the plastic parts of the material to flow into one another and join together. A firm joint is formed on cooling [3].

Today, there are five variations of the extrusion welding method. Although they differ with regard to technique, equipment and application, they possess the following common characteristics:

– the filler rod emerges from a plasticating unit (e.g. extruder) in the form of a strand.
– the filler rod is homogeneous and completely plasticated.
– the fusion faces of the parts being welded are in a thermoplastic state from the surface down to a certain depth.
– the welding procedure is performed under pressure, thus permitting a certain degree of flow to take place.

This flow process in the first two methods is indicated in Figure 1 by arrows. The quality of the weld is largely dependent on this flow process, since it gives rise to a vortex formation at the interfaces, resulting in thorough mixing and interpenetration of the material in the welding zone.

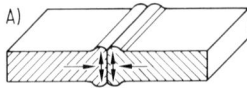
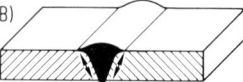

Figure 1. Welding process
A) hot-plate butt welding,
B) extrusion welding,
C) hot-gas welding

19.3 Fields of application

Extrusion welding is employed in the following areas of semi-finished product processing:

- In tank construction, where the main task is joining thick-walled parts. This is the case if a tank wall has to be welded to a tank floor, or if flat sheets — which for production reasons cannot be manufactured in the required size — have to be welded together from small sections. Other examples are the welding-on of pipe connections or loose flanges and the attachment of struts or reinforcements.
- In effluent engineering and civil engineering, when heated-tool butt welding is not possible and the fusion faces at the weld areas are not flat. This is the case, for example, in the construction of shafts and collecting pits for effluent, the welding in of additional junctions in existing pipelines and in the joining of pipe sections when the final weld between two fixed points can no longer be performed by the heated-tool butt-welding method. With large-sized pipes with diameters exceeding 1.5 m, extruder welding is the only technique possible, since there are no heated-tool butt-welding devices suitable for the purpose. Extrusion welding has also been used successfully for the lining of timber and concrete pits, the welding of brickwork collars onto pipelines, and in the repair of damaged pipelines.
- In apparatus engineering, in cases where the geometric shape of the parts in question prevents the use of the heated-tool butt-welding method, or hot-gas welding cannot be used for quality reasons, e.g. in the production of filter plates, bases, trays and diverse other types of construction.
- In pipeline engineering, when heated-tool butt welding is not possible or possible only under very difficult conditions, e.g. when producing pipe bends from segments or from T-pieces with junctions of various diameters, which have to be welded in at different angles. Extrusion welding is also indispensable for the production of reducers from pipe segments or sheet material. Because of their size, they have to be produced in the workshop.
- For special tasks mainly involving the welding of films, webs or thin sheet, e.g. the lining of earthworks or protection of buildings against groundwater from outside or leakage of dangerous media from inside. A further important field of application is the waterproofing of large-area refuse dumps against surface water [4].

19.4 Extrusion welding equipment

19.4.1 Extruder unit with movable welding head

The extruder unit with a movable welding head is used for continuous welding (Figure 2). The apparatus consists of a pellet-fed single-screw extruder with an adjustable barrel heating system and controllable screw speed. The welding speed is determined by the melt

19.4 Extrusion welding equipment

delivery and the latter in turn by the screw speed, which is smoothly variable. At the orifice of the extruder is connected an approx. 2 m long metal-braided PTFE tube through which the weld material is conveyed to the welding head. It possesses high thermal stability and low frictional resistance. A voltage harmless to the welder is applied to the steel braiding of the tube, and the adjustable current heats the braiding in a similar way to a resistance heater, and on start-up melts the solidified strand in the tube. During operation, the continuous heat supply prevents the melt from cooling.

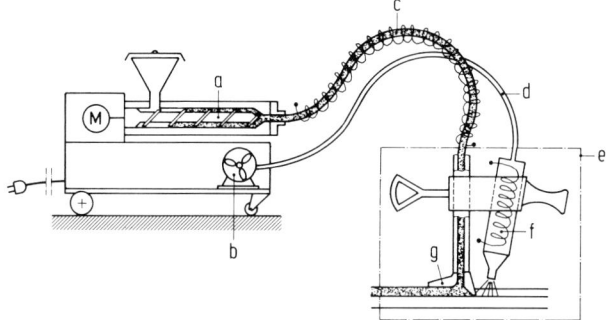

Figure 2. Extrusion welding apparatus with movable welding head
a extruder, *b* air supply unit, *c* heated tube, *d* air tube, *e* welding head, *f* hot-gas torch, *g* welding shoe

The tube leads into the welding head consisting of a mounting device with a handle, a welding shoe and an air heater. During the welding operation, the extrudate is pressed into the welding groove via a welding shoe made from PTFE. The air heater installed in front of the welding shoe, and usually consisting of a conventional hot-gas welding appliance, obtains the air needed to heat the base material via a tube leading from a blower. The welding parameters − melt temperature, melt throughput, hot air quantity and hot air temperature − are adjustable, thus enabling the welding operation to be optimized and reproducible weld seam quality to be obtained.

19.4.2 Mobile welding apparatus

The mobile welding apparatus is likewise used for a continuous welding method which has been largely automated. It was developed specifically for the welding of HDPE sheets, both for prefabrication in the workshop and for use on-site, main importance attaching to the latter applications. The apparatus consists of a motor-driven trolley on which is mounted a smoothly adjustable single-screw extruder for melt delivery. It is equipped with open-and closed-loop control units, and all the ancillary units necessary for welding purposes, Figure 3.
The sheets of film are welded to each other by lap joint, the joint faces having first been cleaned by wire brush. The nozzle of an air heater, pushed into the overlap, heats the joint surfaces to welding temperature. A downstream die, which is fed from the extruder, extrudes the filler rod between the plastified zones where the sheets are to be joined. Preheating of the sheets extends beyond the width of the extrudate, in order thereby to obtain a perfect bond with the base material at the edges of the filler rod. The quantity of heat introduced into the sheet material must be kept to a minimum so as to avoid edge waviness during welding. This is done by exposing the material for a short period to a relatively high air temperature, which results in high welding speeds. The necessary welding pressure is applied by a roller system. It oscillates in pendulum fashion, in order to adjust to

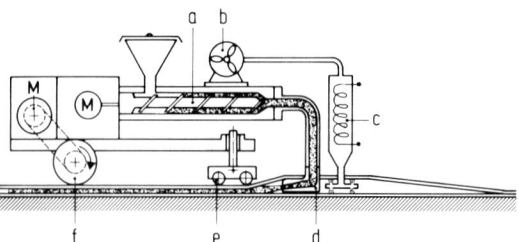

Figure 3. Mobile extrusion welding apparatus
a extruder, *b* air supply unit, *c* hot-gas torch, *d* die, *e* pressure rolls, *f* drive

any irregularities in the surface of the substrate. Preheating of the base material, and the quantity and temperature of the filler rod can be adjusted to obtain an optimum welding speed. This is necessary to offset the various environmental influences prevailing on site.

19.4.3 Welding apparatus for manual transfer of the filler rod from the extruder to the welding gap

An extruder with a very high output is used with this method. Welding accessories used include PTFE tubes, conventional hot-air blowers and pressure rollers and pressure applicators (Figure 4). The welding operation itself is carried out intermittently as follows:
The surface of the prepared gap is heated by hot air over a length of approx. 500 mm. The joint faces are thus in a plastic state. Heating of the material is carried out by moving the hot-air blower to and fro with a fanning motion. Shortly before the heating operation is finished, a strand of plasticated filler rod is conveyed from the extruder into a flexible PTFE tube, which prevents the filler rod from cooling before welding commences. The different thicknesses of filler rod which are required depend on the shape of weld, and are obtained by interchanging dies. During welding, the length of rod is allowed to slide from the tube into the weld gap, where slight pressure is applied to it by means of a roller mounted on the tube or by a pressure applicator. Using the pressure applicator, and starting from one end, the filler rod is then pressed in firmly in the direction of welding and the welded seam smoothed and shaped until the filler rod has cooled. This rolling motion on the filler rod ensures that the necessary pressure is applied and that any entrained air is expelled. This pressure also prevents the formation of shrinkage cavities. The end of the rod is angled to ensure good overlapping with the next length.

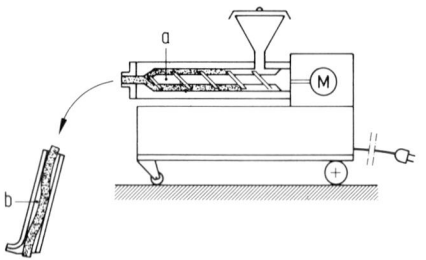
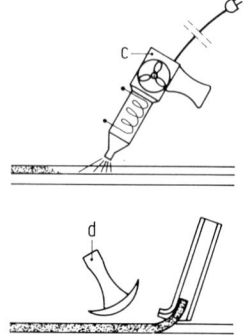

Figure 4. Extrusion welding unit using manual transfer of welding material
a extruder, *b* PTFE tube for conveying filler material, *c* hot-gas torch, *d* pressure applicator

19.4.4 Welding apparatus with extruder-like plasticating chamber

The apparatus incorporates a plasticating system possessing neither barrel nor screw. It consists of a mobile supply unit and a welding head, which are joined to each other by a cable and a conveying tube for the welding rod (Figure 5). A length of rod approx. 4 mm thick is unwound from a reel, fed to the welding head and is pushed into a heated mixing chamber by means of counter-rotating drive rollers, the speed of which can be adjusted. Here, the filler material becomes molten and is mixed thoroughly in the chamber. At the end of the mixing chamber is located a welding shoe, through which the extrudate emerges. In the mixing chamber, the welding wire acts as a piston which melts. A halogen lamp, which radiates its heat onto the weld gap via a gold-plated parabolic reflector, is used for preheating the base material. The unit mounted on the welding head to control the lamp voltage enables the preheating of the base material to be varied as required during the welding operation. This welding apparatus can also be equipped with a conventional hot-air blower. The welding speed can be varied by adjusting the rate at which the rod is conveyed into the mixing chamber.

The technique described has also been partly automated, the welding head being mounted on a trolley. By this means it is possible to weld flat sheets, rectangular tanks, and sockets on insulating sleeving for district heating lines.

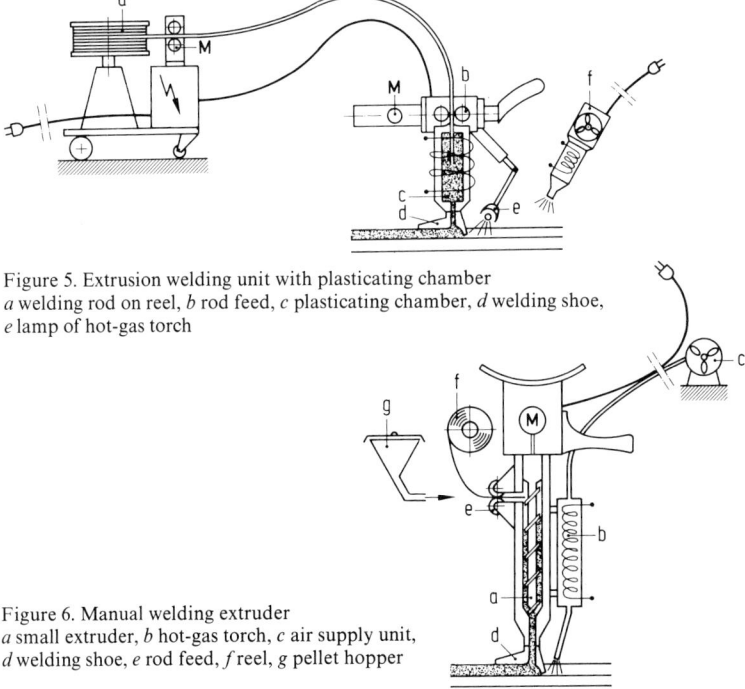

Figure 5. Extrusion welding unit with plasticating chamber
a welding rod on reel, *b* rod feed, *c* plasticating chamber, *d* welding shoe, *e* lamp of hot-gas torch

Figure 6. Manual welding extruder
a small extruder, *b* hot-gas torch, *c* air supply unit, *d* welding shoe, *e* rod feed, *f* reel, *g* pellet hopper

19.4.5 Manual welding extruders

The manual welding extruder is designed in portable form (Figure 6). It consists of a mini-extruder driven by a motor similar to those used in small machine tools. The speed of the motor is not adjustable. The base material is preheated by a conventional hot-air apparatus, the air for which is obtained from a compressed air system or other external air supply. The welding extruder is fed with pellets or coiled welding rod. The advantage of the latter is that, irrespective of the position of the apparatus, the material can at any time be fed in by the screw, pelletized and plasticated. The welding shoe is mounted direct on the orifice of the extruder. Since the drive speed cannot be adjusted, the welding speed is predetermined. Preheating of the base material must be adapted accordingly. The extent of use of this technique far exceeds that of other systems. Since all the individual parts are combined into a single unit, the apparatus is relatively heavy. As with all other methods, it is possible to use different welding shoes.

19.5 Welding shoes

Welding shoes are made by hand from PTFE sheet material. Their design is adapted to the individual weld form and the different material thicknesses to be welded. They are easily interchangeable and possess two lateral guides which prevent the melt from escaping at the sides [5]. With the welding shoe, the pressure necessary for welding is applied to the joint faces via the filler material. Welding shoes, which must be longer than some minimum length, ensure that the welding pressure is applied for a minimum time, so that the joint can be formed at the weld faces and shrinkage of the filler material be compensated for by a certain holding pressure. The "nose" at the front of the welding shoe prevents filler material from flowing forward; instead, it presses against the nose and in so doing pushes the welding head automatically in the welding direction. Before welding commences, the welding shoe must be heated, since a rough, uneven weld surface is obtained if it is carried out with a cold shoe. The hot air around the welding head, or the hot extrudate led past the welding shoe on a wooden or plastics sheet, is a suitable means of heating the shoe. The shoe can, of course, also be heated separately on a hot plate or with a conventional hot-air blower.

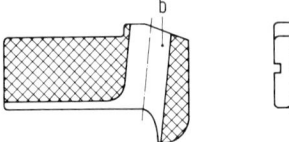

Figure 7. Welding shoe
a support groove, *b* melt feed, *c* smoothing track, *d* nose

19.6 Preparation of the seam

The weld gap employed in tank construction and apparatus and pipeline engineering normally has a groove angle of between 45° and 60°, depending on material thickness and type of weld. A very important requirement for the quality of the weld is that the material at the joint faces be clean and unoxidized. The faces and adjacent areas must, therefore, be machined directly before welding is carried out. Electric hand planes, scrapers, milling cutters or other machining tools can be used for this purpose. This measure is particularly important in the case of materials which at high temperatures have a tendency to oxidize at the surface, or when repair work is carried out on parts in which the surface has been attacked by chemicals. Here the damaged areas must be machined off.

19.7 Start-up of welding apparatus

When a welding apparatus is made ready for use, all the heating units must first of all be switched on in order to melt the plastic. To avoid damage to the equipment during this phase, particular care should be taken to ensure that no places (e.g. at the nozzle, at the welding head, at the extrudate outlet) remain unheated. If necessary, these must be heated with ancillary devices or auxiliary blowers. For safety reasons, some apparatuses are equipped with an interlock device which prevents the drive motor from being switched on too early.

During operation of the extruder, frictional heat produced within the filler material adds to that from the electrical heating. Since thermal equilibrium is necessary during welding, this operation can start only when the melt has stabilized at the required temperature. The extrudate temperature is checked by inserting the probe of a high-speed temperature indicator immediately at the exit from the welding shoe. Fixed thermometers do not supply exact temperature measurements. However, they can serve to identify slow changes in temperature not immediately discerned by the welder (e.g. temperature drop due to a breakdown in the heating unit during welding).

Material which may possibly have been overheated and, as a result, thermally degraded, cannot be used for welding purposes and is regarded as waste. This risk is present in particular at the start-up stage.

19.8 Welding (or filler) material

The quality of the weld also depends on the quality of the filler material. Depending on the welding process, material in pellet or rod form may be used. The melting behavior and density range of the filler material must match the base material. Filler material of unknown origin may not be used, nor may material that has already been processed once (regranulated scrap).

If shrinkage cavities are present in the filler rod, they will also be found in the welded seam. It is, therefore, important to use only completely homogeneous rod.

Shrinkage cavities also form on the inside of the weld when the pellets are damp from condensed moisture. Even the slightest precipitation can give rise to shrinkage cavities in the pellets. Only predried material, warm to the touch, should be used. For this purpose, it suffices to preheat the pellets in batches in a drying oven.

Because of their electrostatic behavior, pellets and rod in containers already opened attract dirt and should be protected against dusting.

19.9 Preheating of the base material

Depending on the welding process and the application, preheating of the base material is carried out with hot air from a conventional blower or by heat radiation from a halogen lamp. In tank construction and apparatus and pipeline engineering, hot air at a temperature of approx. 300 °C is generally used. Preheating causes relevant areas of base material to become plastic down to a depth of approx. 0.5 mm. This can be checked by inserting a thin metal wire directly in front of the welding shoe or by scratching with a drawing pin or small screwdriver. Preheating with air has the advantage that heating of the base material is relatively gentle, with a slow fall in temperature towards the edge zones. Preheating by halogen lamp supplies parallel heat rays via the parabolic reflector. These heat the base material, with an abrupt fall in temperature towards the edge zones. This method of preheating is generally employed when welding films, since hot air causes shrinkage of the films as well as waviness.

The amount of hot air required depends on the shape of the nozzle. All-purpose nozzles with round or oval cross-section require larger quantities of air than special nozzles adapted to the individual weld shape. In some cases they are designed so that they project into the weld cross-section [5].

19.10 Welded seams

Extrusion-welded seams are normally produced in a single operation. Double-V butt welds, or other forms of weld with back-up, require two operations. The fact that only one welding operation is necessary can be regarded as the reason for the economic efficiency of extrusion welding and the quality of the welded joint.

If the welding gap cannot be filled in one operation because the material is too thick, the output of the welding extruder too low or the necessary welding pressure too high, a multi-run weld is needed. After each run, the flash is stripped off with a scraper; this enables any notches present to be levelled out.

19.11 Finishing the weld

Subsequent machining of welds is not permitted and is not necessary as a rule, since individual welding techniques — performed by proficient welders — usually result in uniformly smooth surfaces with notch-free edge zones.

When the welding shoe has become misshapen through age, or the geometry of the parts to be welded is complex, extrudate escapes at the sides of the shoe. This causes flash to form which has no link with the base material and must be removed. It is stripped with a scraper down to the welded areas of the edge zones whilst in a moderately warm state (Figure 8).

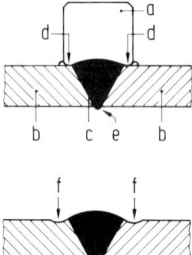

Figure 8. Finishing the weld edges
a welding shoe, *b* base material, *c* weld melt, *d* unwelded area, *e* root, *f* worked edge zone

19.12 Practical examples

Guidelines DVS 2205 part 3 and DVS 2205 part 5, supplement, contain practical examples of extrusion-welded seams. In these guidelines, the designer will find a suggested solution to almost every welding problem. The design examples listed have been compiled by experts in the field and have proved themselves in practical usage. Figure 9 shows an example of an optimum weld design for joining two flat sheets. A single-V butt weld is first of all formed on one side and the sheet turned through 180°. The V-groove is milled out to ensure a good join with the backing layer and the second butt weld made.

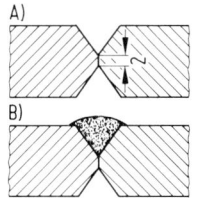
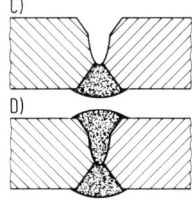

Figure 9. Formation of an X-weld
A) prepared edges, B) welded single-V butt weld, C) specimen turned over and root of opposite side milled out, D) finished weld

19.13 Economic efficiency

In Table 1 are shown the times required to produce a 1 m long single-V butt weld. In this comparative study of the welding of 25 mm thick HDPE sheets, it can be seen that production of the extrusion-welded seam only takes about 1/6 of the time needed with the high-speed hot gas welding technique. Since high personnel costs can only be absorbed by efficient production techniques, extrusion welding is shown to be an effective method of improving economic efficiency [6].

Table 1. Comparison of the production times for a 1 m long and 25 mm thick single-V butt weld on HDPE sheet

	Extrusion welding min	High speed hot gas welding min
Specimen preparation (cutting the groove)	12	12
Preparation for welding (clamping, tacking)	15	10
Welding	15 (1 seam filling, 2 men)	264 (32 round seams, 3 mm thick, 1 man)
Trimming the top layer	10	5
Total production time	52	291

19.14 Welding factor

Temperature and time are both factors which influence the strength of plastics. This naturally also applies to welds produced by extrusion welding. A distinction is thus made between "short-term welding factor" and "long-term welding factor", which is determined by the tensile creep test.

19.14.1 Short-term welding factor

The short-term welding factor (fz) is determined from the ratio of results obtained in the short-term tensile test on welded specimens to those obtained with unwelded specimens. The procedure is described in guideline DVS 2203 part 2. Guideline DVS 2203 part 1 specifies a minimum short-term welding factor value of 0.8. This factor, established from accelerated tensile tests, applies to loading times up to one hour and provides an initial guide to the strength of the weld. A visual examination of the fracture pattern can also supply information on whether the weld quality is good or bad (e.g. brittle fractures, fractures with ductile portions, stretching in the base material etc.).

19.14.2 Long-term welding factor

When estimating the service life of a welded component, it is always necessary to take account of the long-term welding factor. The creep curves of welded and unwelded specimens are first of all established. As Figure 10 shows, the long-term welding factor (fs) is then determined from these two curves, fs being the quotient of the stress values which in the welded specimen and the reference specimen of the base material result in the same time-to-failure values. Further details on the determination of long-term welding factors are given in guideline DVS 2203 part 4. For extrusion-welded joints, guideline DVS 2203 part 1 specifies a minimum long-term welding factor of 0.6.

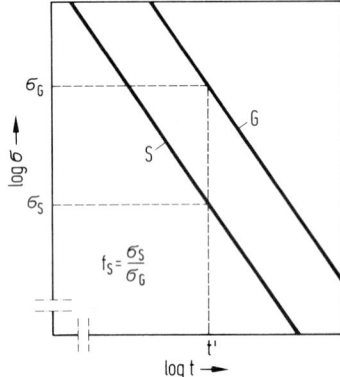

Figure 10. Determining the long-term welding factor. Relationship between static stress and time to break – double logarithmic plot
S creep curve of welded joint, G creep curve of base material, f_s long-term welding factor, σ_s stress on weld at time to failure t', σ_G stress on base material at time to failure t'

Since extrusion welding is a manual technique, a high long-term welding factor value is largely achieved by proper temperature control, optimum preheating of the base material and the skill of the welder. The plasticating and homogenizing of the filler material in the extruder also has a considerable influence on the quality of the weld.
A plastics technician, in whose training welding is an important element, requires additional training to achieve high weld quality. A qualified plastics welder who has for a number of years been working chiefly in the extrusion-welding field, can produce joints with a long-term welding factor of 0.9 [7]. This value has up to now only been achieved in the partly automated heated-tool butt-welding process.

19.15 Chemical resistance

Although plastics are resistant to a large number of chemical substances, there are certain media which lower creep strength. Two different methods are used to test chemical resistance. In the first method, stress-free specimens are immersed in the test media and the change in strength and elongation behavior subsequently checked. In the other method, internal-pressure creep tests are performed on pipes filled with the test media, and the so-called chemical resistance factor is determined from the results. This factor quantifies the influence of stress when plastics are subject to the action of chemicals [8].

Extrusion-welded pipes were tested by the second method in order to ascertain whether the welded seams possess the same resistance factor as the base material. These tests have shown that if the base material is chemically resistant, this is also true of the weld. If the creep strength of the base material is lowered by chemical attack, this applies in like measure to the welds [9].

19.16 Quality assurance

Quality assurance embraces the following prerequisites:
- The design must be specifically adapted to the operating sequence of the extrusion welding process. This means, among other things, that the welded seams must be arranged so that working with a particular welding apparatus is obstructed to the least possible extent.
- Both the semi-finished product to be welded and the filler material must be suitable for the extrusion welding process, if the weldability of the components is to be ensured. Submission of works certificates is advisable.
- The welding apparatus must be in faultless condition and fully perform all the functions incorporated by the manufacturer.
- The welders must be experienced in extrusion welding, familiar with the welding apparatus and, where possible, have been trained on it.
- The given welding parameters must be adhered to during welding. If a new material is to be used, weldability in accordance with DVS 2201 must first be established, in order then to determine the optimum welding parameters for the material.
- Welding should, if possible, be carried out in the workshop. If circumstances do not permit this, care must be taken to ensure that the working area is protected against adverse weather conditions.

19.17 Testing

If unfamiliar materials are being used, or the welding jobs concerned have not previously been carried out in this form, a working specimen must first of all be prepared and tested by a destructive method in accordance with the DVS guidelines. These comprise the accelerated tensile test (DVS 2203, part 2), the tensile creep test (DVS 2203, part 4) and the technological bending test (DVS 2203, part 5). The requirements to be met by these test methods are stated in DVS 2203, part 1. Although the accelerated tensile test is quick and easy to perform, the test results do not differentiate sufficiently between various weld qualities, unless the latter exhibit extreme differences. The technological bending test is better suited to provide information on the quality of the welded seam. Apart from the size of bending angle attained, it is also necessary to evaluate the appearance of the fracture pattern. Only these two together can give information on the deformability of the joint and thus on the quality of the weld.
The tensile creep test gives the most accurate measure of the strength of the weld, since it provides results which can be employed as weld efficiency values in strength calculations. However, this test requires a lot of expenditure in time and money, since, for example, testing periods of several years are necessary for the HDPE grades on the market [10]. Not until recently has an attempt been made to shorten these times by testing in a solution of wetting agent.
The peel test has proved a useful method of testing extrusion-welded, overlapped films, Figures 11 and 12. In this method, which is not described in literature concerned with weld

testing, the two ends of the film are pulled apart, as can be seen in the diagrams. This causes stresses to arise in the weld in a form and size not encountered under practical conditions. However, the appearance of the fracture pattern (ductile or brittle fracture) gives reliable information on weld quality.

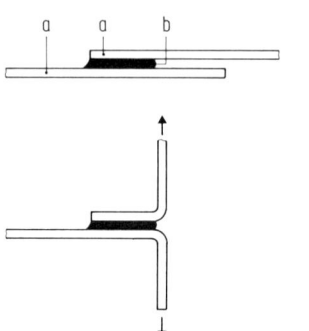

Figure 11. Overlap welding
Top: overlapping film webs
with extruded weld seam, bottom: peel test
a film, *b* weld

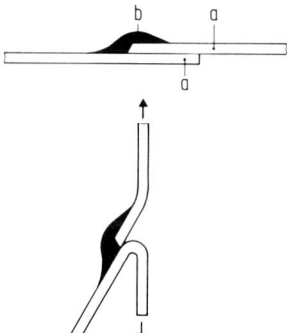

Figure 12. Melt application welding
Top: film web with surface welding
at the overlap, bottom: peel test
a film, *b* weld

Only if the working specimen conforms to the quality requirements, can the component be welded, using the parameters determined for it.

The finished weld can be tested by a non-destructive method, visual testing being of primary importance. In this regard, attention should be paid to freedom from notches at the weld edge zones, to penetration beads, and to a uniform surface. Using ultrasonic and X-ray tests it is possible to detect a lack of adhesion at the groove faces, and shrinkage cavities or vacuoles on the interior of the weld.

19.18 Prospects

Whereas in the past extensive scientific studies have been made of other welding techniques – in particular of the heated-tool butt-welding method – no investigations of this kind have been carried out into the extrusion welding process. Thus the high quality standards achieved with this technology can be attributed to the practical experience acquired and to the performance tests carried out. Nevertheless, there is a need for scientific studies aimed at developing clearcut relationships between welding parameters and weld quality, and making better use of the advantages of the method.

It is expected that extrusion welding will increase in importance, and also that it will be used increasingly for lining FRP structures.

References for Chapter 19

For the convenience of the reader the English titles of all publications in languages other than English are shown in parentheses.

[1] *Potente, H.:* Zur Theorie des Heizelement-Stumpfschweißens (On the Theory of Hot-plate Butt Welding). Kunststoffe 67 (1977), pp. 98/102.
[2] *Gaube, E.:* Der Kunststoff im Apparate- und Anlagenbau (Plastics for Apparatus and Plant Construction). Z. Werkstofftechnik 12 (1981), pp. 2/10.

[3] Deutscher Verband für Schweißtechnik e.V.: Richtlinie DVS 2209, Part 1: Schweißen von thermoplastischen Kunststoffen, Extrusionsschweißen, Verfahren – Merkmale (Welding of Thermoplastic Materials, Extrusion Welding, Processes – Characteristics).

[4] *Knippschild, F. W., Taprogge, R., Tronow, K.:* Großflächen-Dichtungselement aus Niederdruck-Polyethylen (Large-area Watertight Membranes from Low-density Polyethylene). Kunststoffe im Bau 4 (1977), pp. 154/160.

[5] *Bemelmann, K.:* Optimierung der Extrusionsschweißnaht durch gezielte Vorwärmung der Fügeflächen und geeignete Ausbildung des Schweißschuhs (Optimization of Extrusion Seam Welding by Selective Prewarming of the Joint Surfaces and Proper Construction of the Welding Shoe). DVS-Berichte, Vol. 84.

[6] *Gumm, P., Hausdörfer, D., Muth, W.:* Extrusionsschweißen, ein neues Verfahren zum Verbinden dickwandiger Teile aus Hart-Polyethylen (Extrusion Welding, a New Process for Joining Thick-walled Rigid Polyethylene Parts). Kunststoffe 61 (1971), pp. 108/114.

[7] *John, P., Hessel, J., Gaube, E.:* Eine Neuentwicklung auf dem Gebiet des Extrusionsschweißens (A New Development in Extrusion Welding). Kunststoffe 75 (1985), pp. 11/13.

[8] *Kempe, B.:* Prüfmethoden zur Ermittlung des Verhaltens von Polyolefinen bei der Einwirkung von Chemikalien (Test Methods for Determining the Behavior of Polyolefins in the Presence of Chemicals). Werkstofftechnik 5 (1984), pp. 157/172.

[9] *Hesse, J., Hausdörfer, D., Kempe, B.:* The Influence of Oxidizing, Surface Active and Swelling Fluids on Welding HDPE Joints. IIW-Doc. XVI-453-84.

[10] *Diedrich, G., Gaube, E.:* Zeitstandfestigkeit und Langzeitschweißfaktor von geschweißten Rohren und Platten aus Hart-Polyethylen und Polypropylen (Static-load Strength and Longterm Weld Factors of Welded Pipes and Plates made from PP and Rigid PE). Kunststoffe 63 (1973), pp. 793/797.

20 Feeding of extruders

W. Mücke

20.1 Introduction

Scarcely 20 years ago, what we know as extruder feeding was an activity usually carried out manually: granules were poured into a machine hopper out of bins or sacks that had to be brought in from elsewhere with much effort. But in plastics processing too, automation has moved in, relieving operating personnel of much tiring and dusty work. What used to be done by muscle power is now taken care of automatically by conveyors.

It was not just the demand for better working conditions that prompted such progress; equally there were the increasing technical requirements of customers, and the introduction of new plastics and applications that required new techniques to be applied to the pellet immediately before it reaches the extruder. This often entails blending-in additives like pigments, or drying the material. Today it has become a matter of course that this also is done automatically. Thus, feeding today covers a broader spectrum of activities than it used to do.

The basic task of transporting the raw materials, though, has remained the same; it must be accomplished dust-free, as far as possible, and with equipment that is adaptable to the various types of products to be conveyed. Even today, an old means of conveying is still in use: the screw conveyor – but more on this and other techniques in the following sections.

20.2 Screw conveyors

There are two basic types of screw conveyors: one with a rigid screw working in a rigid housing; the other with a flexible screw-helix working in a flexible metal or plastics tube. Both types are driven by electric motors and can be used to convey either pellet or powder in bulk. Usually there are several discharge openings to feed several extruders simultaneously (Figure 1).

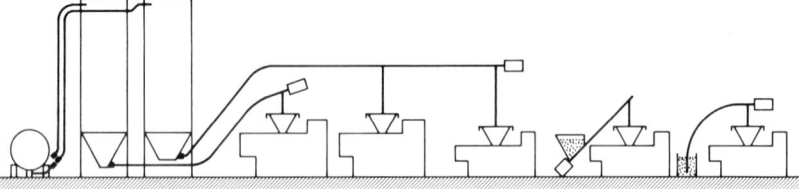

Figure 1. Screw feeders of various types

The length of a screw conveyor is limited by its mechanical stability (Table 1). Therefore, whenever the distances to be covered are long, several conveyors have to be operated in series. If their lengths have to be changed, it usually involves a great deal of rearranging: replacing the screw and shortening or lengthening the housing [1, 2].

Table 1. Screw conveyors for feeding extruders

Screw	Flexible ribbon	Flexible helical ribbon	Flexible circular-section helix
Housing	rigid with bends	flexible	flexible
Material conveyed	powder, slow-flowing	powder, slow-flowing; pellet	powder, free-flowing; pellet
Distance conveyed m	60	15	15
Flow rate m³/hr	4	8	25

20.3 Pneumatic conveying systems

Pneumatic conveying systems can be adapted to changes in operating conditions much more easily. The lines, consisting of pipes or tubes, can be shortened or lengthened easily and adapted to space requirements optimally regardless of whether they have to go over or under an obstruction. And should it become necessary to increase the flow rate, it usually suffices to add a larger blower or compressor.

Pneumatic conveying equipment can operate with suction or pressure, or a combination of both in alternating sections; and can serve one extruder, or many.

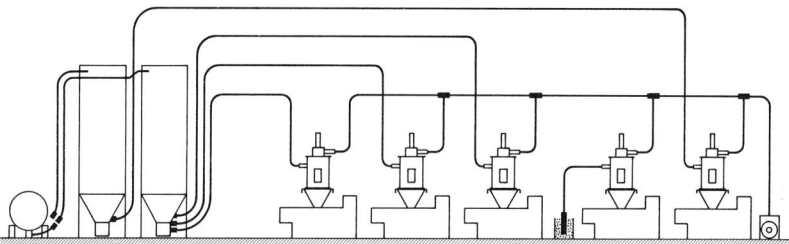

Figure 2. Separate conveying units with central blower

Figure 3. Distribution panel.
(Photo: Filterwerk Mann & Hummel GmbH, Ludwigsburg, West Germany)

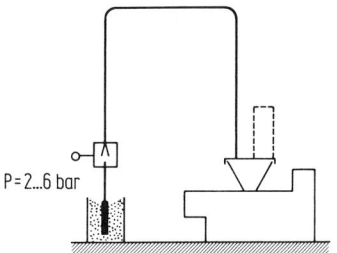

Figure 4. Suction/pressure conveyor

20.3.1 Single-machine conveying systems

The simplest form of material supply consists of placing a conveying unit on the hopper of each extruder and connecting it to the material supply bin by means of a conveying line (usually a flexible hose whenever the distances to be covered are short).
Conveying air can be supplied to each unit by individual blowers, or centrally for a larger number of units (Figure 2).
One feature that all single-machine conveying systems have in common is that each unit has its own conveying line, and more than one unit can convey at the same time, even with a single source of air. Whenever a different product is to be conveyed, the intake end of the conveying line has to be connected with a different storage bin. Recoupling stations with quick-change couplings are used for this (Figure 3).

20.3.1.1 Suction/pressure conveying units

This system requires compressed air at 2 to 6 bar (Figure 4). Low pressure is produced in the first section of the conveying line, and high pressure in the second one, by applying the injector principle. The conveying line can empty direct into the machine hopper, which must, however, be sealed off and equipped with a filter whenever the products to be conveyed are dusty. This type of system may place a considerable load on the plant compressed-air supply, especially if more than one device is being used. The compressed air must be free of oil and water.

20.3.1.2 Suction conveying units

The suction transfer unit, also known as a mini-conveyor, usually requires no compressed air to convey pellet, and differs from suction/pressure conveying, in that it is equipped with a blower, usually placed on top of the bin of the product to be conveyed (Figure 5).

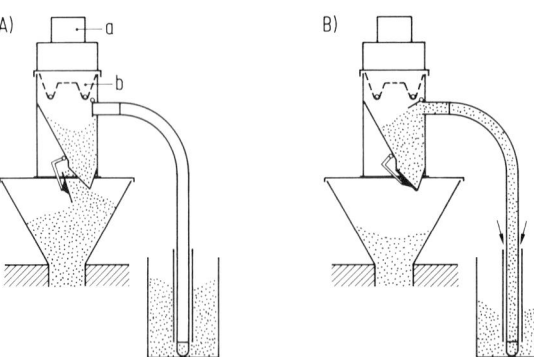

Figure 5. Suction conveyor
A) rest position, B) conveying position

This operating principle is quite simple. After the discharge valve in the product bin closes automatically, the blower produces a reduced pressure for a specified length of time, which draws the pellet through the conveying line to the bin on the extruder hopper. A filter there separates the pellet from the conveying air. As soon as the blower is shut off, the partial vacuum is filled and the pellet itself pushes the discharge valve open and flows out. Then a new conveying cycle begins, unless the valve cannot close because the machine hopper is full [3].

20.3 Pneumatic conveying systems

During breaks in conveying cycles, the filter is cleaned of adhering particles. Various techniques for doing this are described in Section 21.2.3.

Being fully automatic, suction conveying units are usually equipped with malfunction indicators, which respond whenever the material supply runs out. This can be done by monitoring valve opening when conveying is finished, for instance. However, provision for signalling malfunctions that only occur as isolated events should not be made [4].

The flow rate obtainable with suction conveying units depends on many parameters, such as the properties of the product to be conveyed, the length of the line and its configurations, including the number of pipe bends and vertical sections. Figure 6 shows a typical curve for PE pellet being drawn through lines of various diameters.

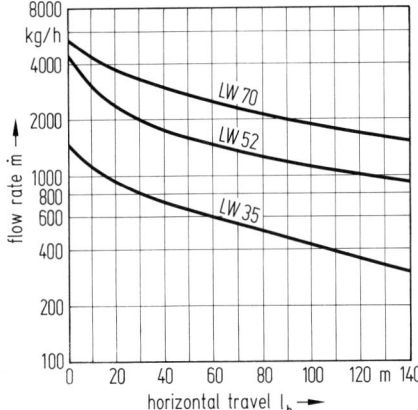

Figure 6. Throughput of suction conveyors for PE pellet

20.3.2 Multiple-machine conveying systems

These systems are used whenever more than one extruder is to be loaded with the same material. Features common to all multiple-machine conveying systems are that one supply line per material type is mounted above the processing machines, all machines are connected to a common source of air, and only one unit can convey material at any given time (Figure 7).

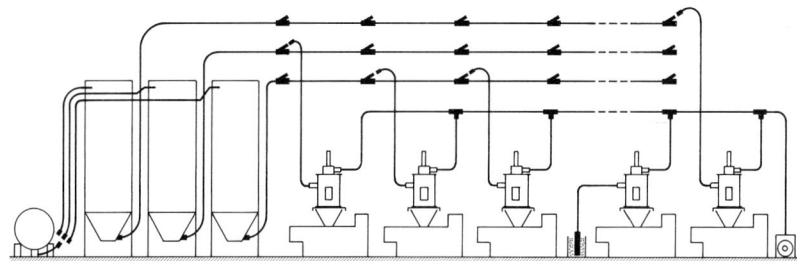

Figure 7. Multipoint conveying system

20.3.2.1 Pressure conveying

This is the classic form of pneumatic conveying system, which is suitable for high flow rates over long distances; it is available in a number of versions. All types keep the conveying line under pressure right up to the extruder hoppers, making it necessary to seal off all connections carefully, especially in systems carrying dusty pellets, to avoid an undesirable discharge of dust. Various publications are available on designing such systems: references [5] to [7].

20.3.2.1.1 Pressure conveying systems incorporating pipe-switching junctions

These systems have pellet bins on the extruders, with two indicators for minimum and maximum level, plus filters (Figure 8). Whenever an indicator registers its minimum level, the appropriate pipe-switching junctions move into place to create the corresponding conveying route, while the blower, and in turn the rotary discharge valve, come on to begin conveying. The maximum indicator stops conveying when it registers full; then the filter is cleaned.

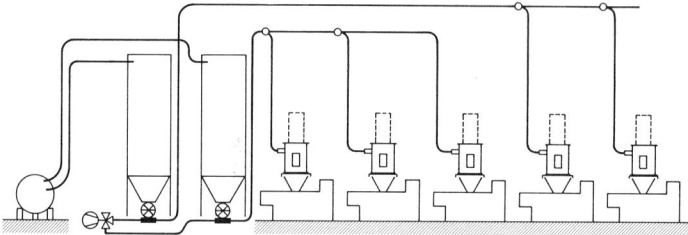

Figure 8. Multipoint pressure-conveying line

The two-way pipe-switching junctions shown in Figure 8 can be replaced by a multidirectional pipe-switching junction for each machine group.

20.3.2.1.2 Systems with ring lines

Systems featuring a ring line (Figure 9) work without pipe-switching junctions. Then the junctions shown in Figure 8 are replaced by pipe loops (Figure 10 A). During conveying, air flows through all pipe loops, while the pellet is separated out wherever a material bin over a machine hopper is not full. In order to avoid unnecessary recirculation, a probe can monitor the flow of material downstream from the final pipe loop and shut down the system after a specified period of time.

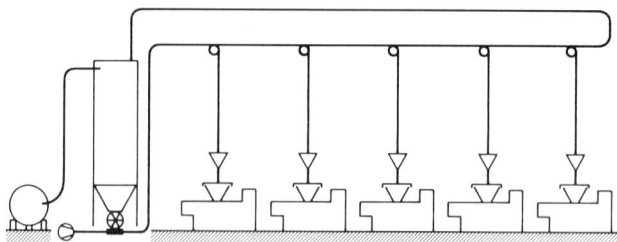

Figure 9. Pressurized-loop conveyor line

Since the pipe loops cause a rather large pressure drop as a result of the 360° circulation, use of pipe loop systems is limited to smaller installations.
Pressure losses can be reduced by using material containers shaped like separation bins (Figure 10 B), making pipe loops superfluous [8].

20.3 Pneumatic conveying systems

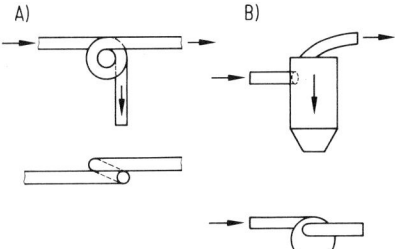

Figure 10. Separation devices
A) pipe loop, B) separator tank

20.3.2.2 Suction conveying

The storage bins for multimachine suction conveying systems are similar in design to those described in Section 20.3.1.2. The principles of operation of both types are comparable, the only difference being that multi-units have a common source of air and are usually connected by one or more common conveying lines. Conveying special products stored in containers beside the extruder creates no problems (Figure 11).

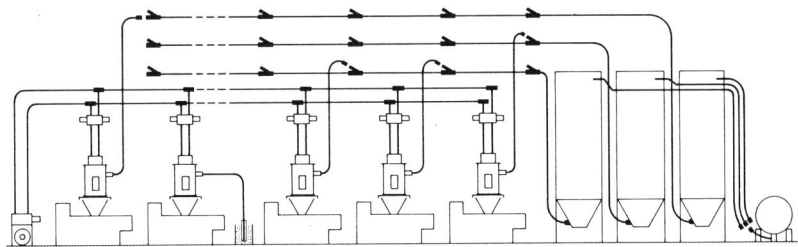

Figure 11. Multipoint suction conveying system for powders

Multimachine suction conveying systems are suitable for conveying either pellet or powder, depending on the filter system used (Figures 12 and 13). Since these systems work on the suction principle, small leaks do reduce their flow rate, but do not cause dust discharge. By selection of an appropriate filter system, dust discharge during filter cleaning can be eliminated. All vibration methods will accomplish this, as well as the suction and

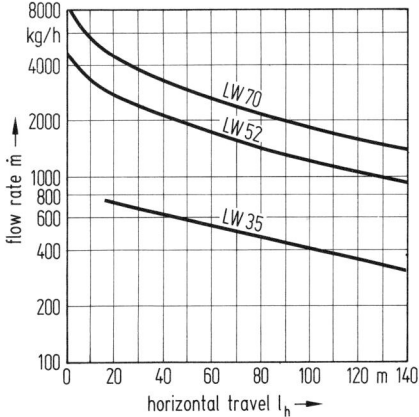

Figure 12. Throughput of KFG multipoint system for PE pellet

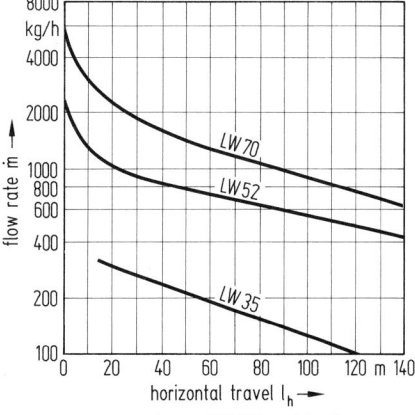

Figure 13. Throughput of KFG multipoint system for dry blend

recirculating-air scavenging methods described in Section 20.3.3.3. Although with the last method mentioned, the filter is cleaned under pressure, neither the pellet chamber of the conveying station nor the extruder hopper is subjected to pressure [9, 10]. Figure 14 shows the same type of conveying station with coupling connections as is shown schematically in Figure 11.

Figure 14.
Conveying station KFG 645.
(Photo: Filterwerk Mann & Hummel GmbH, Ludwigsburg, West Germany)

Multimachine conveying systems are usually equipped with central controls, with which the individual conveying stations can be turned on and off, and their operating functions monitored; an illuminated panel displays the state of the conveying process and facilitates operation (Figure 15). Information on the number of operations carried out (to determine material consumption) or any indications of malfunctions can also be added, by connecting the system to a printer or data-processing equipment for each extruder.

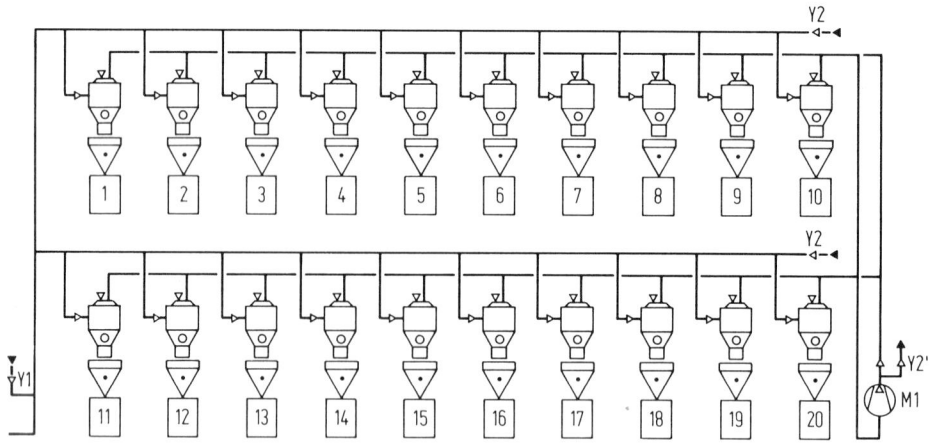

Figure 15. Illuminated diagram of the KFG system

20.3.3 Filter systems

In the interest of dust-free processing, various filters matched to the characteristics of the material to be conveyed are used in pneumatic conveying systems. Depending on whether the product to be conveyed is in pellet form or consists of very fine clinging grains, dirt

20.3 Pneumatic conveying systems

will build up differently on the filters. Whereas one can dispense with filter cleaning if only pellet is conveyed, with powder conveying it is absolutely necessary. With the systems described here, it is usually carried out in the breaks between operations, using one of the following three methods:

- mechanically by means of vibrators (run pneumatically or by electric motors),
- pneumatically using air from an outside source (pulses of compressed air) or
- pneumatically using air from the source of conveying air.

The author does not intend to go into detail on the first two cleaning methods, but will deal with the different ways of applying the third, which is used during suction conveying.

20.3.3.1 Cleaning filters by pressure scavenging

Pressure scavenging entails blowing air through the filter in the direction opposite to that of conveying, either by reversing the conveyor-blower or by exchanging suction and pressure connections by means of a shuttle valve (Figure 16). The scavenging air must escape to atmosphere through a second filter, the resistance of which creates an undesirable pressure below the conveying unit (all bin connections must be sealed off effectively!). Since the filter for scavenging air is not cleaned automatically, one must avoid conveying materials with very small particle sizes; otherwise they would have to be cleaned off manually.

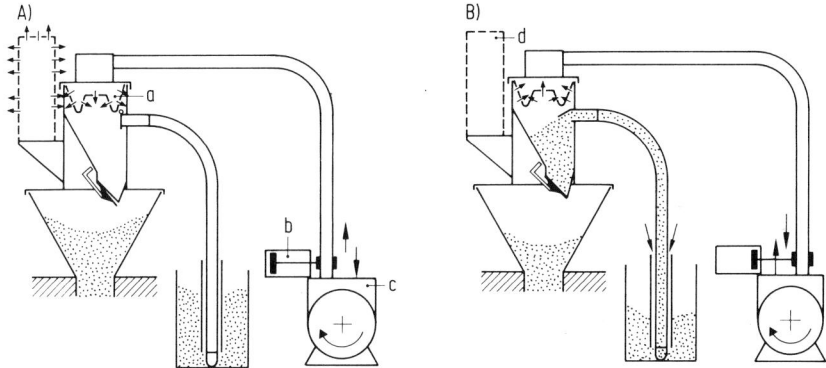

Figure 16. Conveying unit with filter cleaning by air-stream reversal
A) flushing, B) conveying; *a* conveying-air filter, *b* convey/flush valve, *c* blower, *d* flushing-air filter

20.3.3.2 Cleaning filters by suction scavenging

When scavenging is done by this method, air is drawn through two separate filter chambers alternately (Figure 17). Thus a low pressure area is created below the conveying unit, caused by the filter resistance in the chamber being cleaned [11].

20.3.3.3 Cleaning filters by scavenging with recirculated air

The space below the conveying unit remains depressurized with this method since the same quantity of air blown into the filter by the blower for cleaning purposes is drawn off again (Figure 18). Air flows out of a jet rotor located in a filter cartridge, blowing off the particles adhering to the filter medium with great force. As the scavenging air is discharged, it sets

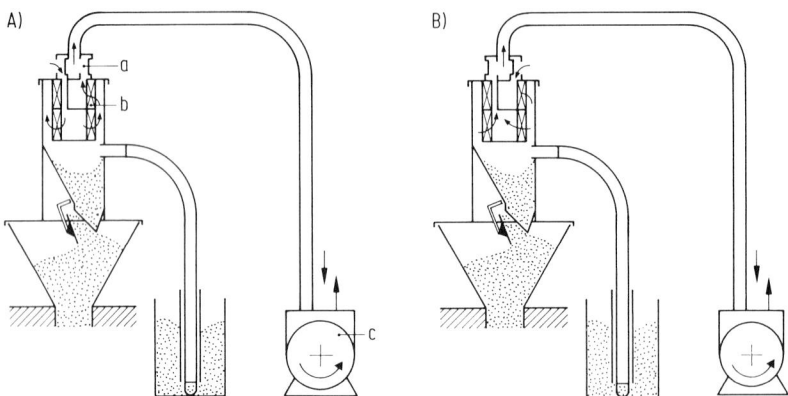

Figure 17. Conveying unit with filter cleaning by suction purging
A) flushing 1, B) flushing 2; *a* convey/flush valve, *b* conveying-air filter, *c* blower

the jet rotor spinning so that it reaches all parts of the cartridge. Unlike the method described in the preceding section, in this method the filter is cleaned in a single step, that is to say whenever a valve on the blower is moved from suction operation (= conveying), to operation with recirculated air (= scavenging) [8].

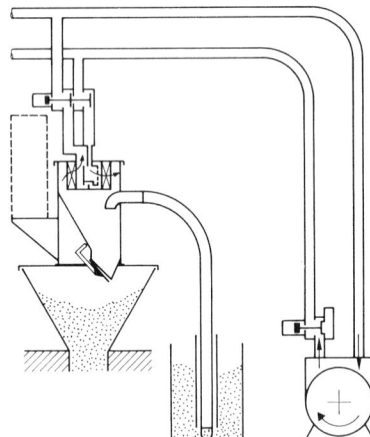

Figure 18. Conveying unit with filter cleaning by flushing with circulating air during the cleaning phase

20.4 Auxiliary equipment

Many processing methods require that plastics be treated in certain ways just before extruding: adding pigments, lubricants, foaming agents or other additives by means of metering systems. Auxiliary equipment may also be required to extract moisture by means of heated, sometimes dried air. Such units may be located either directly on top of the extruder or centrally. Whenever material is pretreated centrally, special attention must be paid to the transport of the product to the extruder if the material is not to suffer adverse effects during movement.

20.4.1 Metering systems

Depending on the degree of accuracy required, either gravimetric or volumetric metering systems can be used. Gravimetric units are expensive, but accurate; volumetric units are cheaper and quite robust, but do not give so high a degree of accuracy with bulk materials as gravimetric ones. Both types of system are available for continuous operation with either liquid or solid raw materials [9].

20.4.1.1 Metering by redirecting granulate flow

Certain materials do not require a particularly high degree of metering accuracy, and this is the case with the addition of ground reclaim material, for instance, where a low-cost method can be used. A metering flap valve is placed directly upstream of the conveying unit (sometimes integrated into the unit itself), and swings back and forth between two intake ports during conveying (Figure 19). In this way, the two products are drawn in and discharged alternately as layers in the conveying unit. As they flow out, they are mixed together adequately.

Waste material that accumulates during production is often reground at once, adjacent to the extruder. Since the quantity of reclaimed material is usually subject to fluctuations, the proportion of ground to new material can be increased when an indicator on the regrind bin signals a surplus.

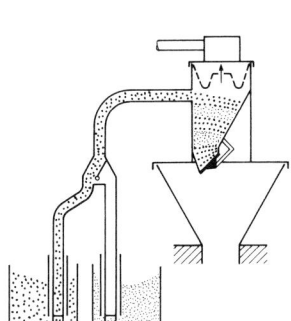
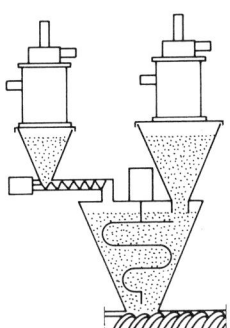
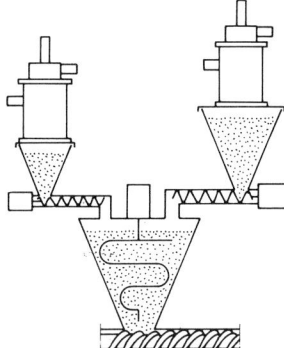

Figure 19. Dispensing flap-valve in the conveying pipe

Figure 20. Partial metering with a quasi-continuous dispensing process

Figure 21. Full metering with a quasi-continuous dispensing process

20.4.1.2 Metering without a main component

This method is a form of semicontinuous metering, in which the main component is fed unmetered into the mixing bin. This bin is substituted for the extruder hopper and is used to mix the main material with one or more additives (Figure 20). Although it appears as if the main material were unmetered, in reality it is flowing towards the mixer at a constant rate because of the action of the extruder. The dosing units, therefore, must add their components in proportion to the extruder's output. This can be accomplished by changing the speed of a metering screw or by changing the number of chambers used in a chamber dosing unit.

20.4.1.3 Metering of all components

When the main component is also fed through a dosing unit, the metering process becomes independent of the processing machine (Figure 21). Then the metering ratio is determined by the speeds of the screw dosing units or the ratio of the numbers of chambers used in a chamber dosing unit.

For metering reclaimed material ground at the extruder, chamber units of the EFG type (Figure 22) can be equipped with a special automatic control to vary the quantity metered in relation to the quantity of reclaimed material accumulated. The tolerance range can be set as desired. Maximum and minimum level sensors are fitted to the bin for ground material on the reclaim mill. As long as the level remains between the two indicators,

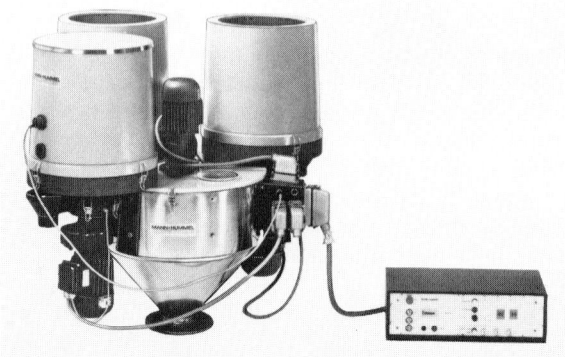

Figure 22. Coloring unit EFG 200 with full metering of three components.
(Photo: Filterwerk Mann & Hummel, Ludwigsburg, West Germany)

materials are metered according to the setting on the unit as soon as it is switched on. When the amount of reclaim being produced exceeds that metered, the level in the ground material bin is allowed to rise to the maximum level indicator. This causes the quantity metered to be increased in small steps until the level drops below the maximum, or the preset maximum quantity of reclaimed material is reached (Figure 23). Whenever the minimum indicator registers too little reclaimed material, the principle of operation is simply reversed.

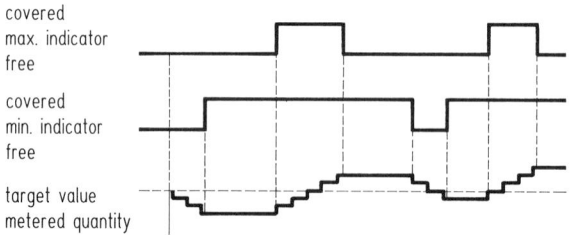

Figure 23. Principle of rework control by use of an EFG metering unit

20.4.2 Drying systems

Plastics absorb water at a rate depending on the relative humidity of the air in which they are stored (Table 2). If the storage temperature is constant, this relationship can be represented by sorption isotherms (Figure 24). If a material is stored in air having a high relative humidity, it will adsorb moisture; conversely, it desorbs moisture whenever the relative humidity is low. Since the water-adsorption capacity is dependent primarily on the prevailing relative air humidity, this characteristic can be utilized in drying.

20.4 Auxiliary equipment

Table 2. Equilibrium moisture content – various plastics stored in air at 23 °C, relative humidity 50% [15]

	Unit	CA	CAB	ABS	PA 6	PA 66	PBT	PC	PMMA	PPO	PET
Equilibrium moisture	%	2.2	1.3	1.5	3	2.8	0.2	0.19	0.8	(0.1)	0.15
Water content acceptable for injection molding	%	0.2	0.2	0.2	0.15	0.03	0.02	0.08	0.05	0.02	0.02

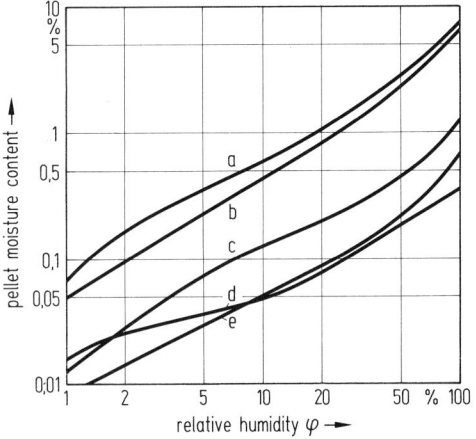

Figure 24. Sorption isotherms for various resins. a PA, b PA + 30% glass fiber, c ABS, d PET + 30% glass fiber

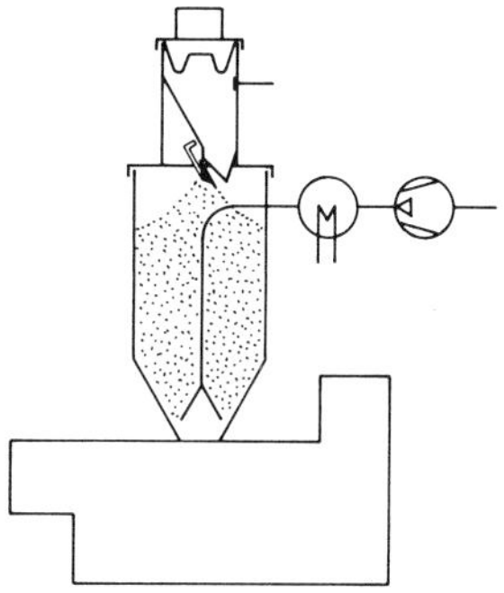

Figure 25. Fresh-air drier

20.4.2.1 Drying with ambient air

The simplest method is to heat the plastic in an oven. When the temperature of air is raised, its moisture content drops and the plastic gives off water until equilibrium is established again. This process takes quite a long time, and will not produce satisfactory results even after many hours if the ambient air contains too much water (on moist summer days, for instance). This is because the equilibrium moisture content will then be too high to allow the required residual moisture content in the granule to be reached. A warm-air dryer positioned directly above the extruder hopper and automatically filled by conveying equipment is more effective under these conditions (Figure 25). The drying is carried out by means of ambient air heated in an air heater and forced through the material in the drying bin by a blower. Since ambient air is used here as well, results are only as good as those attained with a drying chamber.

In order to save energy, some dryers operate on an air blend, with much of the air being recycled, the rest being drawn in as ambient air.

20.4.2.2 Drying with dried air

If the results obtained with the systems described above are not satisfactory, or if good results are to be ensured regardless of adverse weather conditions, then the moisture must be extracted from the air before it is heated. This is done by routing the air through adsorption agents, usually silica gel or a molecular sieve. In this way, the relative humidity of the drying air can be reduced to less than 0.1% ($\cong -35\,°C$ dew point*) at 20 °C. Compared with an original ambient temperature of 35 °C, and a relative humidity of 80% ($\cong 31\,°C$ dew point), this amounts to a reduction in the moisture content of the air by a factor of 200 and a reduction in the equilibrium residual moisture in the plastic by the same factor. Though such a low value is useful, in practice it can only be attained after a very long drying time. Figure 26 shows these interrelationships, taking PET as an example.

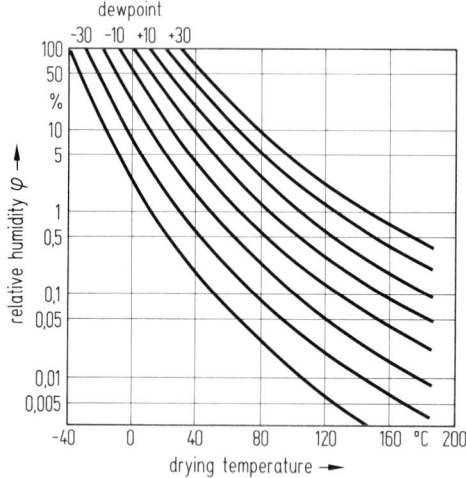

Figure 26. Equilibrium moisture content of PET as a function of dew point and drying temperature

Using air with such a low moisture content as this reduces drying time by as much as 50%, at the same drying temperature and residual moisture level in the pellets. This reduces the time that they must remain in the dryer at drying temperature; it may also reduce the drying temperature itself, thus reducing energy costs as well.

Such units work in a closed loop, with the result that only the moisture taken up by the air as it passes over the pellets has to be removed (Figure 27). This means high efficiency despite the energy required to regenerate the adsorption agent.

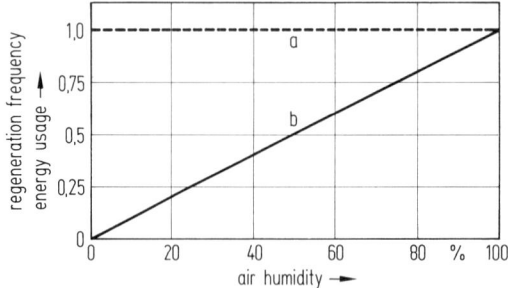

Figure 27. Energy usage in dew-point-dependent regeneration
a time-controlled, *b* dew-point-controlled

Regeneration takes place in two stages. First the water stored in the bed is desorbed using ambient air heated to 250 °C; the adsorber bed is then cooled using cold air. Most dryers use ambient air for this, but unfortunately, under humid/warm climatic conditions, this causes adsorption of a lot of moisture, and poor cooling because heat of adsorption is released during the adsorption. The result is a poorly regenerated adsorber bed.

---

*) Temperature at which the air is saturated with its water content (fog formation).

20.4 Auxiliary equipment

Another type of air dryer is available that treats air in such a way that its dew point is not adversely affected by weather. Dryers of this type use either dried air or the air discharged from the drying bin for cooling purposes. Unlike the ambient air, this discharged air is very dry, with a dew point well below 0 °C.

Figure 28 shows a dryer with a double bed adsorber and Figure 29 the system display panel of such a dryer. While the air discharged from the drying bin flows through blower M1 and the left adsorber bed, a partial flow is bled off through a bypass line for cooling purposes, routed through the right bed and fed back into the main flow of air again, through a second bypass. Thus, the heat energy in the flow of cooling air coming from the right bed is added to the drying air instead of being wasted.

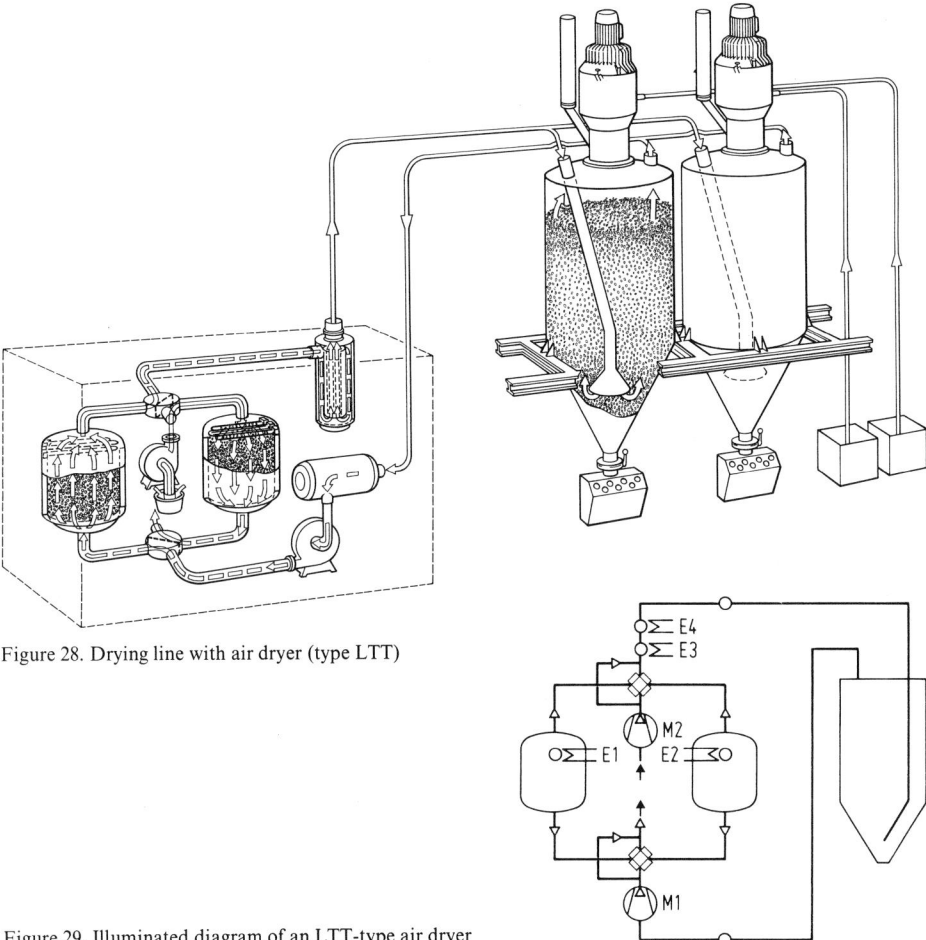

Figure 28. Drying line with air dryer (type LTT)

Figure 29. Illuminated diagram of an LTT-type air dryer

In this way, this regeneration method has a double energy-saving effect: firstly, since no moisture is adsorbed from the ambient air during cooling, no energy is required to desorb it; and secondly, adding air warmed up while being used to cool down the drying air reduces the amount of energy needed to heat up the left bed to drying temperature.

Still another energy-saving provision is that regeneration is not initiated at fixed time intervals, as is common with other systems, but instead only when indicated by the dew point of the drying air. This means that the adsorber bed is always optimally loaded with moisture regardless of the humidity of the air to be dried. Therefore, the energy required for regeneration is proportional to the moisture loading of the air (Figure 28).

The sensors used to measure the dew point must have good long-term stability because only then can uniform drying be ensured. Instruments with digital displays have proved acceptable in monitoring the prevailing dew point and drying-air temperature. Analog outputs make it possible to register temperatures continuously.

20.4.2.2.1 Pellet dryers using air drying

Smaller units, suitable for mounting directly over an extruder hopper, have only one chamber for the adsorption agent. This means that drying must be interrupted for the duration of regeneration.

Dryers employing two adsorber beds make it possible to dry continuously: while air is being dried through one bed, the other bed can be regenerated (Figures 28 and 29). Very low dew points can be obtained by dimensioning the adsorber beds generously.

Dryers with more than one adsorber cartridge also regenerate only one cartridge at a time (Figure 30). Regeneration takes place sequentially by moving valves or turning a carousel.

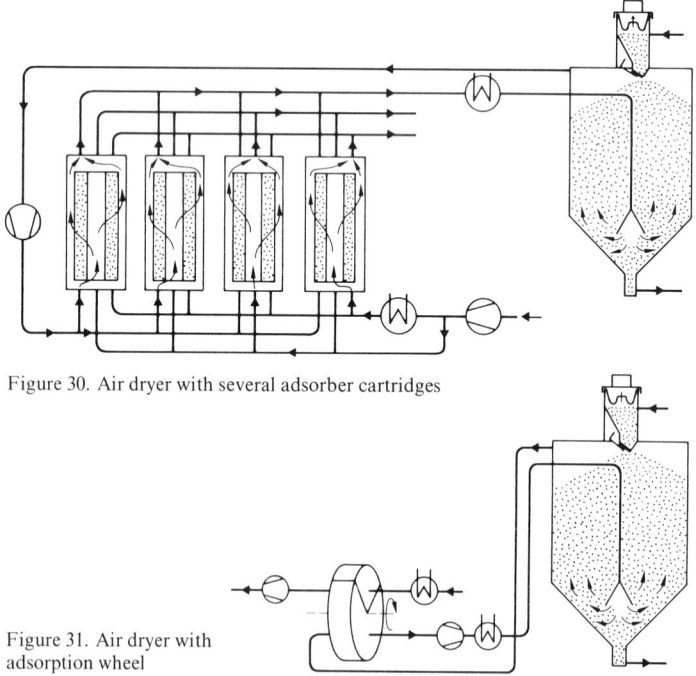

Figure 30. Air dryer with several adsorber cartridges

Figure 31. Air dryer with adsorption wheel

The air dryer described in Figure 31 works continuously: a wheel coated with lithium chloride and porous in an axial direction rotates while the air to be dried flows through it and moisture is adsorbed by the hygroscopic salt. Regeneration takes place on the opposite side. The dew point obtainable is about $-30\,°C$.

Other types such as turbulent-bed dryers or vacuum dryers are not so interesting for use in extruder feeding, and thus are mentioned here only for the sake of completeness.

References for Chapter 20

For the convenience of the reader the English titles of all publications in languages other than English are shown in parentheses.

[1] DBP 2 516 625.
[2] *Weber, M.:* Strömungsfördertechnik (Flow Conveying Technology). Krauskopf Verlag, Mainz, 1974.
[3] *Siegel, W.:* VDI-Forschungsheft No. 538 (1970).
[4] *Flatow, J.:* VDI-Forschungsheft No. 555 (1973).
[5] DBP 2 721 899 C2.
[6] DBP 2 755 671 C2.
[7] DBP 2 540 672.
[8] DBP 2 755 671 C2.
[9] Dosieren in der Kunststofftechnik (Metering in Plastics Technology). VDI-Verlag, Düsseldorf, 1978.
[10] *Meier, B., Tonne, F. J.:* Grundlagen zur Trocknung von Kunststoffgranulat (Fundamentals of Drying Polymer Pellet). Company brochure, Du Pont de Nemours GmbH, Düsseldorf, West Germany.

21 Extrusion recycling of plastics waste

H. Tenner

21.1 Introduction

In the mid-fifties the first attempts were made to reclaim plastics waste and to make it useful for reprocessing. For the most part primitive reclaim techniques were used, and unsatisfactory results obtained. However, forward-thinking processors and machinery builders at that time recognized the future importance of recycling. They worked together to develop concepts and equipment which are still in use today.

Quite early on, reclaiming by melt extrusion played an important role, especially when good quality was required.

Since then plastics recycling has developed into a high-level technology comparable with other outstanding technologies in plastics processing.

Many circumstances contribute to the problematic aspects of recycling. Among these is the fact that when primary products are manufactured, thought is seldom given to scrap recovery methods.

Problems in re-use arise with colored scrap, coatings on metal, textiles or paper, and with coextrusions.

Additional influences include UV- and thermo-oxidative ones, dirt, foreign bodies and moisture.

The various technologies required for reclaiming are also determined by the form of the materials, such as fibers, films, pipes, profiles, and many others.

In practice today recovery methods are of three basic types:

Chemical processes	Mechanical processes	Thermal processes
− pyrolysis	− granulating	− extrusion
− hydrolysis	− densification	
	− agglomeration	

A deeper study of the different recycling processes shows the following:

- pyrolysis requires great technological effort and big investment. Thus, the economics of this approach remain in doubt.
- hydrolysis is only useful for particular kinds of plastics; end products are monomers, the basic components of the plastic.
- granulating, densification, and agglomeration are reclaiming processes which do not affect quality, since they are merely physical processing steps. They can be used independently, but are usually part of complete extrusion systems.

Extrusion of waste materials reveals a totally different situation:

- specific reclaim costs are low,
- investment costs are low; production capacities range from 50 kg/h to well above 1000 kg/h,
- high-quality product is obtainable thanks to the well-defined and controllable nature of the process,
- the waste materials can be modified during the process,
- there is a wide variety of applications, and practically all thermoplastics can be processed,
- in special cases a final product is manufactured directly from the waste, without intermediate pelletizing.

21.2 Recycling via the melt – Stages of the process

With well-designed plants, the overall process should be carried out in precisely defined, individually controllable steps.
The complete plant is only as good as its individual components, which must be properly interfaced to create a smoothly operating processing unit.
Figure 1 shows the typical flow sheet and a schematic layout of a recycling plant for film and fiber waste. Incoming material is fed (a) past a metal detector (b), to a granulator (c); pneumatic conveyor (d) then moves the chopped material to a stock silo (e). Conveying air is released through a cyclone or exhaust filter (f). The crammer feeder (k) receives clean material direct from the stock silo, or contaminated material via a washing plant (g). A metal separator (i) is placed between crammer feeder and stock silo. The fluffy cut-film scrap is densified and force-fed to the extruder (l) by the crammer feeder. The extruder plastifies, homogenizes, degasses (if required), and conveys the melt into the screen changer (o) which filters contaminating materials from the melt. The melt-pressure measuring unit (n) indicates when internal screen changing occurs. The melt-temperature measuring unit (m) measures the melt temperature. Upon passing the screen changer the melt streams through the die head (p) and through lace dies with linear or circular hole patterns, to be cut by the die face pelletizer (q).
The lens-shaped pellets are cooled by water for an appropriate distance and then move to the pellet dryer (s). The dryer removes surface water from the pellets, but leaves some residual moisture, which is necessary for further processing.
The cooled pellets are then run over a classifying screen (t) to separate over- and undersized particles. Depending on the conveying distance involved, either a blower (u), or a vacuum system with a star valve is used to convey the pellets into the storage silo (v). A station for weighing and filling bags (w) or other containers can also be provided.
The individual processing steps are:

- granulation
- conveying
- storage
- densification
- washing

- metal separation*
- plastication and homogenizing*
- degassing*
- melt filtration*
- pelletizing*

The starred * steps are quality-defining and quality-influencing factors – the others are regarded more or less to be quantity- or efficiency-influencing elements.

21.2.1 The individual steps in the process and their purposes

21.2.1.1 Granulating

The economy of the entire plant is affected if the wrong granulator is chosen. The following are the most important criteria.

21.2.1.1.1 Effect of the screen hole size

The throughput rate increases with increasing screen hole size, and the specific energy consumption of the mill drops (Figure 2). To achieve the highest levels of output from the mill at any flake size one must also provide good suction to draw the material through the screen holes (see Figure 3). This may affect the size distribution of the flakes, however,

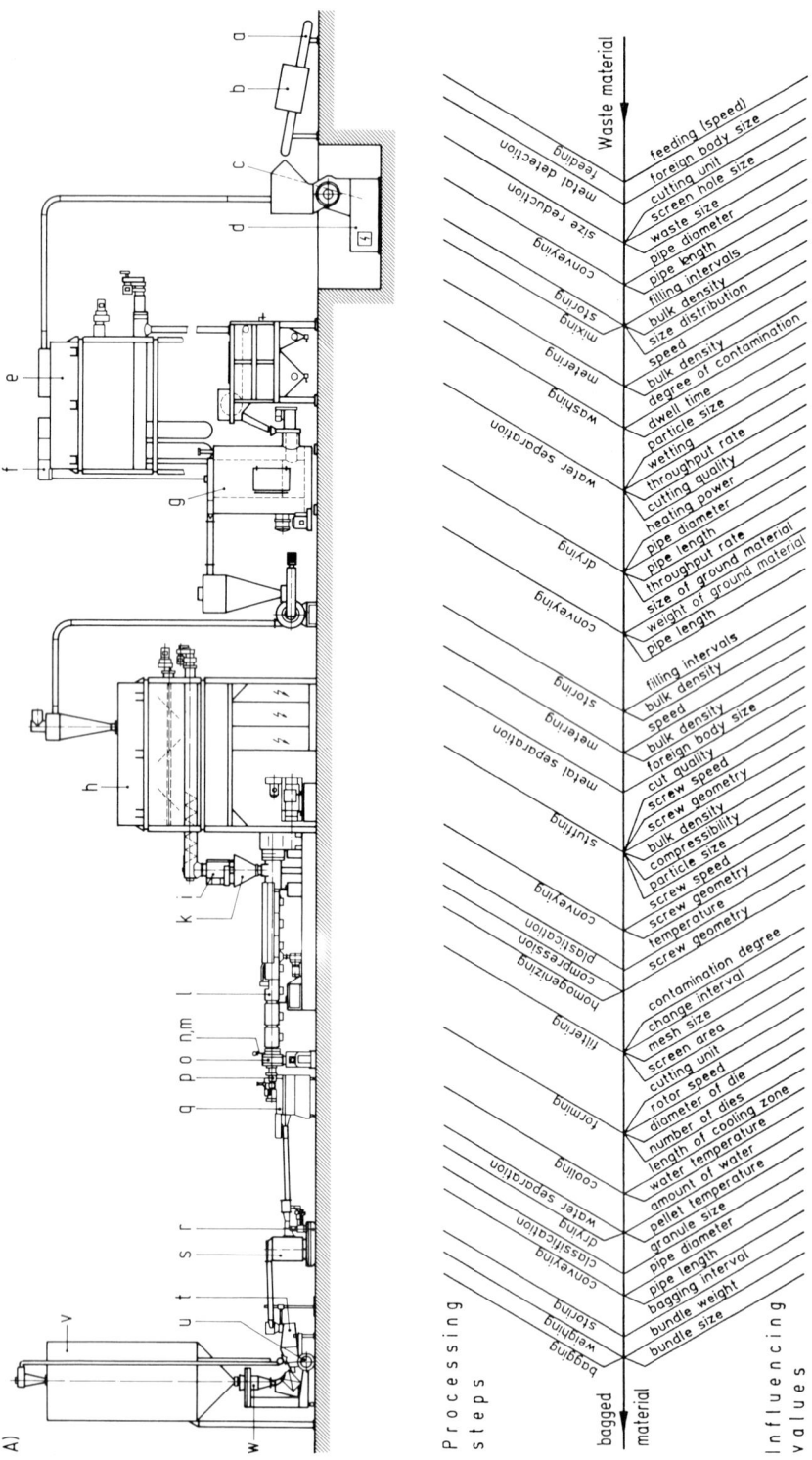

Figure 1. Principle of reclaiming waste plastic (film, fibers, etc.) by extrusion
a conveyor belt, *b* metal detector, *c* granulator, *d* conveyor blower, *e* stock silo, *f* exhaust filter, *g* washing plant, *h* stock silo, *i* metal separator, *k* crammer feeder, *l* extruder, *m* melt-temperature sensor, *n* melt-pressure sensor, *o* screen changer, *p* pelletizing die head, *q* water-cooled die-face pelletizer, *r* water container with circulating pump, *s* spin-dryer, *t* classifying screen, *u* pellet-conveyor blower, *v* pellet silo, *w* bagging scale

which in turn can affect the output of the extruder (see Figure 4). Additionally, bigger flakes lead to difficulties during conveying, storage and discharge of the silo. However, if the characteristics of the feedstock are known, the optimum processing conditions can be sought. Lower throughput with bigger flakes is only partly due to lower bulk density, as was shown during other tests. The main causes are poor conveying and compression behavior, and problems arising in the area between crammer feeder and the extruder.

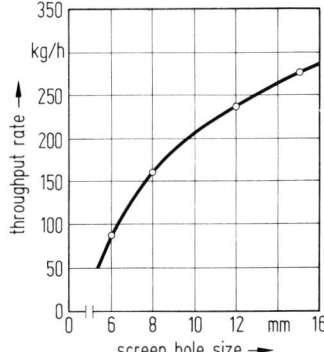

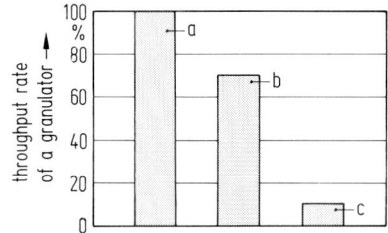

Figure 3. Effect of suction on the throughput of a granulator when grinding PE film [1]
a very good, *b* insufficient, *c* no suction

Figure 2. Effect of screen hole size on throughput rate of the granulator
Rotor diameter 240 mm, width 400 mm, drive power 11 kW, double oblique cut, material LDPE film 120 µm

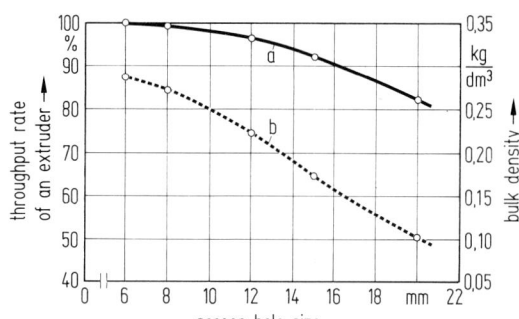

Figure 4. Effect of screen hole size *a* on throughput of the extruder, *b* on bulk density

21.2.1.1.2 Effect of knife wear

When flakes are made using a screen of the right size, but with worn knives, problems similar to those experienced with large flakes are observed. The cut surfaces are crumpled and ragged-edged, and can hook into each other; this results in a lower bulk density and poor conveying behavior and finally in a reduction of the granulator throughput rate. This then also reduces the performance of the extruder (Figure 5).

Figure 5. Effect of knife wear
a on throughput of granulator,
b on throughput of extruder,
c on energy consumption of granulator.
Granulator: rotor diameter 400 mm, rotor width 600 mm; Waste material: LDPE 120 µm film

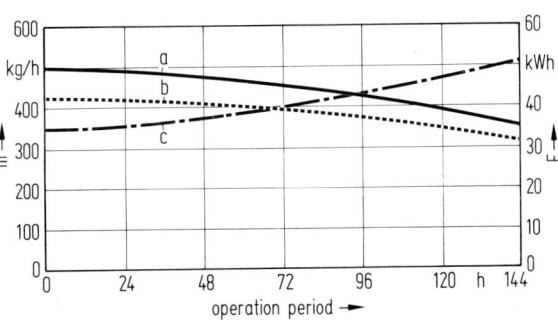

21.2.1.2 Plastication, homogenizing, degassing

These are processing steps which, in practice, are often done simultaneously. The definition of "plastication" adopted here is "the transformation of thermoplastics under pressure and temperature from the solid state to a melt condition".

"Homogenizing" is the internal and external mixing of one or more materials to achieve a continuous phase and uniform temperature distribution.

The individual processing steps must take into account the fact that materials to be recycled have already been modified by thermal history and oxidation, and that they consist of mixes of materials of different melt index, lubricant content, fillers or polymer type.

If it is necessary to modify material properties — by filling, reinforcing or alloying, for example — then special screw-plasticating systems, called kneading extruders, can be employed.

For simple, standard purposes and moderate thoughput capacities, single screws are dominant. Higher output rates and more sophisticated applications often require the advantages of the twin-screw extruder.

The essential differences between the individual systems with regard to:

feeding and conveying behavior,
shear work,
mixing and homogenizing efficiency, and
dwell-time spectrum

are considered below.

21.2.1.2.1 Feeding and conveying behavior

Feed and conveying behavior affects output and with it the product quality. Fluctuations in output with fixed operational settings may result in quality variation.

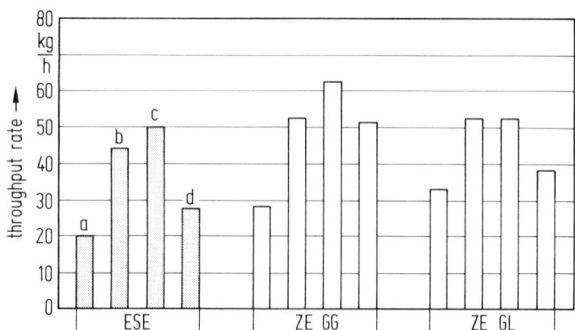

Figure 6. Intake behavior of different screw systems with various types of material [3].
a ground PUR 0.646 kg/dm^3, b HDPE powder 0.550 kg/dm^3, c PA pellets 0.737 kg/dm^3, d HDPE powder 0.815 kg/dm^3.
ESE = single screw, 40 mm diameter, ZE GG = twin screw, 34 mm counter-rotating, ZE GL = twin screw, 34 mm co-rotating

A practical trial was made to compare the three primary extruder screw systems using various materials and screw speeds. To eliminate the effect of die resistance, trials were made without a die attached. The results are shown in Figure 6. Clearly, the worst of the three systems in feed behavior with powder materials was the single screw. It is also clear that material "d", which had the worst fluidizing characteristic, ran best on counter-rotating screws and that with this system we achieve on average the highest feed rate and least scatter in rate between materials.

The intake and feed rates of all three systems are largely determined by bulk density; but material form and flow characteristics also have an effect.

Results can be improved by applying rational measures or auxiliary devices, like a crammer feeder.

Within a certain range of bulk density the crammer feeder, in association with special extruder feed zones, makes it possible, without forced feeding, to achieve the same high output rates with poorly flowing materials like film flakes, as with free-flowing agglomerate (Figure 7).

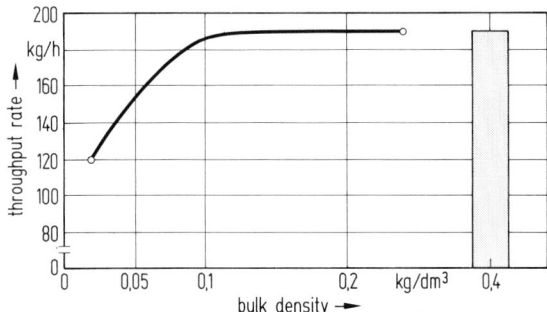

Figure 7. Extruder throughput rate versus bulk density of flakes
a flakes fed by crammer feeder, *b* agglomerate, normal feed
Extruder 70 mm diameter with enlarged feed opening

21.2.1.2.2 Shear work

Shear work is a relative measure of plastication and homogenizing. It is not just the total amount of shear work that counts; the way it is done is important.

For shear rate in the single-screw extruder, the following simplified expression applies:

$$\frac{dv}{dx} = \frac{V_0}{h} \qquad (1)$$

V_0 peripheral speed at screw root,
h flight depth

and for the counter-rotating twin-screw extruder the following:

$$\frac{dv}{dx} = \frac{V_{02} - V_{01}}{s_k} \qquad (2)$$

V_{01} peripheral speed at screw root
V_{02} peripheral speed at screw tip
s_k gap between screw root and tip at intermesh.

An intensive degree of homogenization and plastication is achieved by means of this 'difference function' shearing. Plastication is gentle, because shear forces are generated by internal kneading instead of external friction.

The size of the shear gap, as well as peripheral speed, influences the melt pressure in the shear gap and therefore the energy put into the melt.

The results of studies on the effect of the size of the (intermesh) gap on melt pressure are shown in Figure 8 [3].

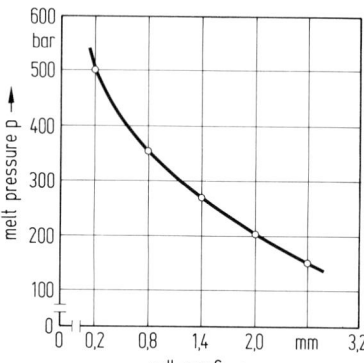

Figure 8. Melt pressure p in the roll gap of a counter-rotating twin-screw extruder as a function of the size of the roll gap s_K

The co-rotating twin-screw extruder

With co-rotating screws conveying is not done on the displacement principle as it is with the counter-rotating twin-screw extruder. Its effects are similar to those of the single-screw extruder. In contrast to the counter-rotating extruder, the maximum shear rate in the gap for the co-rotating mode is generated by the sum rather than the difference of surface speeds. The following simplified expression applies.

$$\frac{dv}{dx} = \frac{V_{01} + V_{02}}{s_k}. \tag{3}$$

Using this arrangement in conjunction with high screw speeds, shear rates are achieved many times higher than with the counter-rotating twin-screw extruder. In addition, the higher operational speeds of the co-rotating twin-screw extruder give higher output rates with the same screw diameter. Up to 100% more output at a given speed than with the counter-rotating twin-screw extruders can be obtained if material properties and quality requirements will permit this.

21.2.1.2.3 Mixing efficiency

The single-screw extruder exhibits good outer — meaning longitudinal — mixing. However, the diagonal and internal mixing capability is worse (marble effect with different colours). Counter-rotating screws have good inner, but bad outer mixing efficiency, which is caused by higher back flow. Good longitudinal mixing effects can be achieved by suitable design and optimization of the gaps between flights, as well as by incorporating mixing elements. Co-rotating screws have the best longitudinal and diagonal mixing properties of the three systems, because of their open, axial pumping action.

21.2.1.2.4 Dwell time

The dwell-time spectrum gives a measure of the time taken by each particle of the melt to move a defined distance. In relation to thermal homogeneity a narrow dwell-time spectrum is more beneficial, and thermal overload or uncontrolled streams of "cold particles" are probably less likely. Practical experiments were carried out to investigate the behavior of the three different systems under different operational conditions with the aid of contrast agents [5].

As a result of these trials it was found that the dwell-time spectrum depends strongly upon the mode of operation and the material. Low screw speed and a high degree of filling result in better values than with conditions reversed.

The co-rotator shows the broadest spectrum, and the single-screw unit varies from poor to good. Best continuity with a narrow spectrum is obtained with the counter-rotator.

21.2.1.2.5 Degassing

Degassing is unnecessary for most reclaim operations on polyolefins, but is absolutely essential with hygroscopic polymers, like PA or PET.
The degassing process involves:

- a brief boiling process and
- a lengthy diffusion process.

The lengthy diffusion process is usually not completed, because the residence time of the melt in the vent region is relatively short.
The four most relevant targets in designing for effective degassing are:

- long dwell time,
- high temperature,
- increased shear deformation,
- large melt surface-area, i.e. low degree of screw filling in the venting zone.

With extrusion processes carried out efficiently, these factors are only variable within a certain range. Opportunities to vary degassing through temperature, shear deformation and dwell-time changes are limited in most practical situations.

The filling degree is a real variable, however, and can be optimized. Figure 9 shows the dependence of the effectiveness of the degassing on the degree of filling. In practice melt filling in the vent zone is decreased by increasing either the pitch or the screw diameter. The use of multistart screw elements (up to four starts) in the vent zone also provides a bigger effective screw surface. Placing a few vent zones one behind the other is a further option for improving the effectiveness of venting.

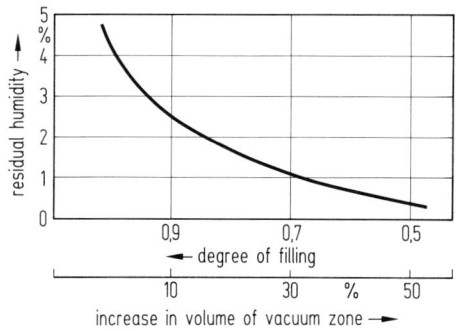

Figure 9. Influence of the degree of screw filling on the degassing efficiency of a counter-rotating twin-screw extruder

The effectiveness of such measures differs among the three screw systems. The best practical basis for good venting is found with the counter-rotating twin screw, because of the following features:

- long dwell time associated with relatively low screw speed, and well balanced conditions before and after the vent zone provided by the special conveying characteristics;
- thin melt lamina at the top of the screw flights directly below the vent opening.

The degassing capabilities of both single-screw and co-rotating-screw extruders have been greatly improved by long-term technical development efforts.

Each of the screw systems described has its special advantages. Choice of the best system for a given task is made with respect to:

- process requirements,
- quality standards,
- output rate,
- cost and available funds.

21.2.1.3 Melt filtration

This processing step is important in two ways, with respect to:

- quality of the product and
- efficiency of the plant.

The melt filter separates contamination from the melt, typically:

- dust
- paper, rubber, adhesive tapes
- additives, such as printing ink, pigments, fillers etc. on or in the plastics
- foreign bodies like metal particles, wood, stones.

The degree of contamination can range from less than 0.1% for in-house production scrap to 0.1% to 0.5% for film scrap or other used material gathered over a period of time. Material having aluminum, paper or cloth mixed with it can often have up to 10% foreign matter.

Foreign matter leads to production problems, e.g. bursting of a blown-film bubble, or blocking of nozzles of injection molding machines, and results in a reduced or totally unsatisfactory final product quality.

The degree of contamination permitted depends upon the nature and quality of the intended final product. In the case of films the permitted particle size should be lower than

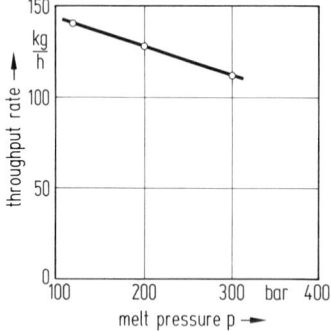

Figure 10. Reduction of throughput rate and rise of melt pressure with increased screen fouling

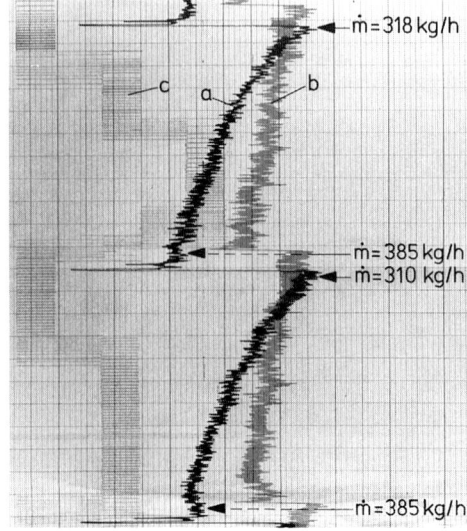

Figure 11. Relationships between melt pressure, throughput rate and energy consumption
a melt pressure (bar), b drive power (kW), c heating power (kW)

20 μm to allow production of 30 μm film without problems. However, for injection molding applications, particle sizes even greater than 100 μm are acceptable. Thus the fineness of the filter has to be chosen to suit the material quality required or the use to be made of the secondary feedstock.

Quality disadvantages may appear if filtering is too coarse, while very fine filtering can bring economic disadvantages. Fine screens reduce throughput rates and have to be changed frequently (Figures 10 and 11). Figure 10 shows how extruder output depends on melt pressure.

Melt pressure, however, is a function of the screen fineness, the surface area of screen and, of course, the degree of contamination. Measurements have shown not only how much the throughput is reduced, but also that there is an increase of the specific energy consumption (see Figures 11 and 62). The diagram shows that a pressure difference of 115 bar caused a throughput reduction of 75 kg, or 19%. Screen changing intervals of less than 30 minutes are uneconomical.

Too large a filter area results in entry and exit volumes so large that there is risk of thermal degradation, material stagnation, and poor self cleaning (as in the case of a color change).

Figure 12 shows which screen sizes have to be used for different operating conditions.

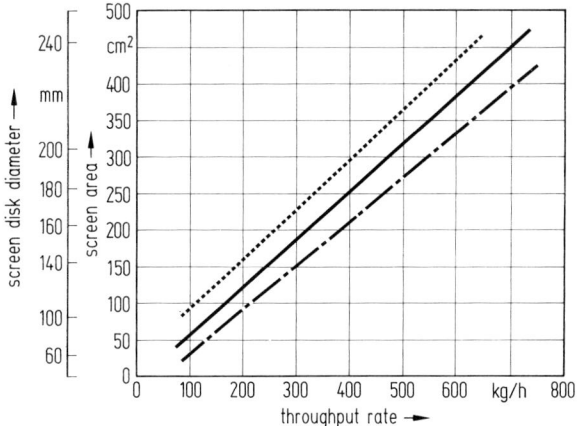

Figure 12. Screen sizes required for different throughput rates in recycling of film scrap

Table 1. Dimensions and designation of filter screens

	Units	Coarse	Medium	Fine
Mesh width	μm	500	250	70
No. of mesh	per cm^2	150	600	6000
Wire dia.	mm	0.37	0.15	0.05
Open surface	%	35	37	35

Screens may be classified as coarse, medium, and fine, for simplification purposes (see Table 1). One coarse plus one or two fine screens are needed for film manufacture. Injection molding, and pipe or profile extrusion require only one or two rough screens. As already mentioned, the most suitable system can only be chosen by taking the kind and amount of contamination, and the required screen size, into account.

21.2.1.4 Pelletizing

The goal of the pelletizing process is to produce a final product with the narrowest possible pellet size range and best possible pellet form.
All other feedstocks used in plastics processing, especially those cut from strands, chunks and other forms, as well as irregular shapes, will not be considered as pellets, but as ground material.
The form, size and uniformity of the pellets determine the ease of subsequent processing.
Strand pellets are normally 2.5 to 4 mm in diameter and 3 to 5 mm in length. Lenticular pellets are 3 to 5 mm in diameter and 1.5 to 3 mm long. Table 2 shows the influence of bulk density, pellet form and size on output rate. It is shown that, in spite of bulk-density differences of some 23% with lenticular pellets, the throughput variation is only 3.7% at most. For cylindrical pellets, a bulk-density difference of 5.5% results in an output difference of 7.2%. The output rate was decreased by 25% when using ground material, and the film thickness was not constant. Considering these values, it appears that bulk-density differences of up to 25% have no great influence on the throughput rate. Often secondary (or regenerated) feedstock is processed with a certain proportion of virgin material. In such a situation segregation may occur with different pellet sizes and pellet forms. Thus a lenticular pellet, typical of virgin material, should be aimed for in reclaim operations. Comparison trials with different pellet forms have confirmed that similar forms have the least tendency to segregation [5].

Table 2. Throughput rate versus pellet form and bulk density

	Throughput rate – kg/h					
	Lens pellets			Strand pellets		Ground material
Screw speed [rpm] \ Bulk density [kg/dm^3]	0.415	0.473	0.542	0.495	0.513	0.348
80	12.3	12.2	12.5	12.1	12.7	9.25
150	25.5	26.3	26.7	25.0	26.2	18.8
200	33.7	34.5	35.0	32.5	35.0	25.5

21.3 Applications, products, special properties

Thermoplastic waste is divided into the following categories for practical purposes:

- bulk plastics: LDPE, HDPE, PS, PP, ABS, SAN, PVC,
- engineering plastics: PA, PET, PC, and similar resins,

and into the following types:

- in-line factory waste (unused),
- used waste.

Unfortunately the waste is not often available in a segregated condition, and mixtures of widely differing materials cause great problems in reclaiming and subsequent processing.
Table 3 shows the compatibility of different thermoplastics [6]. The reclaiming processes differ according to the quality and type of the thermoplastic waste material; these may be classified as follows:

21.3 Applications, products, special properties

- standard process for in-line waste,
- washing and reclaiming process for contaminated waste,
- reclaiming plus compounding in a single process for filling and alloying at the same time,
- special process for engineering plastics,
- direct reclaiming from waste to the final product, with filling if necessary.

Table 3. Miscibility of different thermoplastics. The miscibility decreases from 1 to 6, i.e. 1 means very good miscibility, 6 incompatibility

	Standard PS	HIPS	SAN-copolymer	ABS	PA	PC	PMMA	POM	PVC	PP	LDPE	HDPE	PBT
Standard PS													
HIPS	1												
SAN-copolymer	6	6											
ABS	6	6	1										
PA	5	4	6	6									
PC	6	5	2	2	6								
PMMA	4	4	1	1	6	1							
POM	6	6	6	5	6	6	5						
PVC	6	6	2	3	6	5	1	6					
PP	6	6	6	6	6	6	6	6	6				
LDPE	6	6	6	6	6	6	6	6	6	6			
HDPE	6	6	6	6	6	6	6	6	6	6	1		
PBT	6	6	6	5	5	1	6	6	6	6	6	6	
PET	5	5	6	5	5	1	6	6	6	6	6	6	6

21.3.1 Standard recycling of production waste

In this category LD- and HDPE film, HDPE- and PP tapes and fabrics, PS punching grids, as well as pipes, profiles, and injection molded parts are important. The waste is normally segregated and largely free of dirt or other impurities.

Figure 13 shows the flow diagram of a typical plant. For small to medium output rates, the single-screw extruder is most frequently used; a tapered-feed-zone screw is an advantage.

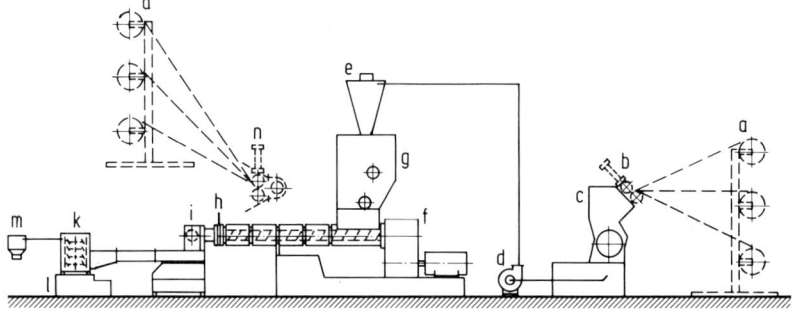

Figure 13. Standard reclaiming plant for small and medium output rates
a film roll support, *b* feeding device, *c* cutting mill, *d* conveyor blower, *e* cyclone, *f* single-screw extruder, *g* storage and feed container, *h* screen changer, *i* pelletizing die head (die face pelletizer), *k* spin-dryer, *l* water circulation unit, *m* cyclone, *n* film roll support for direct feeding of extruder

The short screw and the two-stage tandem screw (Figures 32 D and 32 E) have been specially designed for such applications. Waste in roll form needs no pregranulation, and can be fed direct.

For storage of film flakes, a combined horizontal crammer feeder with integrated silo is suitable. An extra film flake silo is then unnecessary (Figure 43).

Final products are lens-shaped pellets, which can be converted, alone or mixed with virgin material, into new products like film, tapes, sheets etc. The quality that can be achieved depends upon the degradation suffered during initial processing and on the material type.

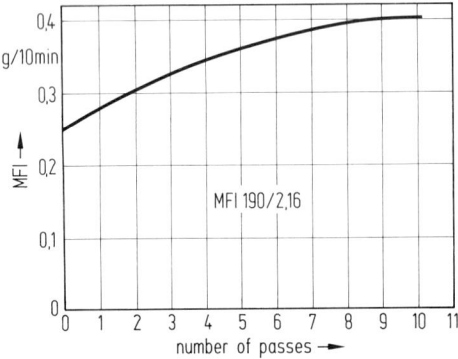

Figure 14. Change of LPDE melt flow index on repeated processing

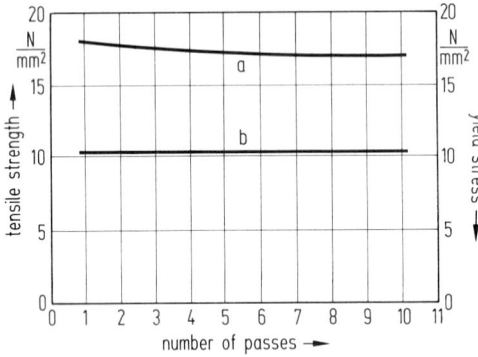

Figure 15. Change of tensile strength and yield stress of LDPE with repeated processing
a tensile strength, b yield stress

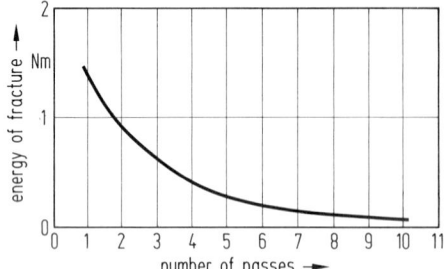

Figure 16. Change of mechanical properties of high impact PS with repeated processing [7]

Changes that occur in polyolefins are relatively small. Molecular weight reduction, caused by processing, especially multiple processing, is compensated by crosslinking reactions, so that the processing properties can remain more or less constant. Such effects are shown in Figures 14 and 15. The situation is different for styrene copolymers. After each processing cycle the tensile properties deteriorate (Figure 16).

After about four processing cycles the toughness has severely deteriorated, and the effectiveness of the rubber impact modifier has been reduced through crosslinking. Finally the properties of the material are no better than those of a standard polystyrene.

21.3.2   Washing and recycling of contaminated waste

Increased recycling of in-line waste has led to a shortage of clean waste material and therefore made it increasingly necessary to reclaim used and contaminated scrap. Scrap material of this kind is mostly bulk plastics from sacks, shopping bags, shrink-wrap, bottles and canisters, bottle crates and so on. Suitable secondary feedstock is only obtained from such scrap by using a washing process. A range of suitable plant designs (Figure 17) is commercially available.

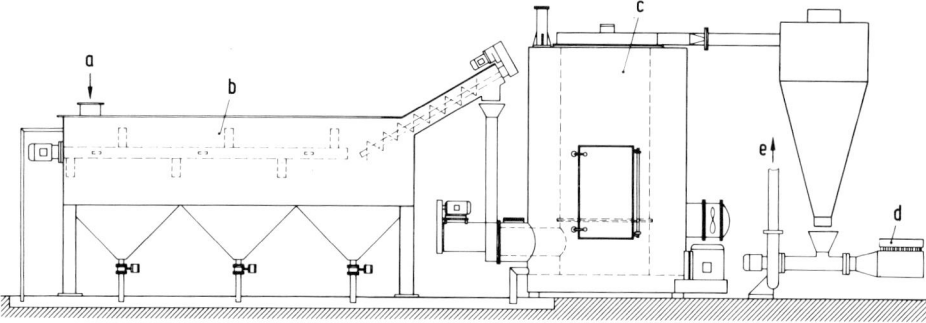

Figure 17. Washing plant for plastics waste (schematic)
a material supply (flakes), *b* washing track, *c* spin-dryer, *d* thermal dryer, *e* material discharge (washed and dried)

The process starts with size reduction in either a dry or a wet system. After dry grinding, ground material is blown into a stock silo and then continuously discharged (*a*) into a washing trough (*b*), by metering screws.
Two counter-rotating paddle shafts slowly transport the material through the washing trough. The turbulence generated rinses the dirt from the plastic particles; the dirt sinks to the lowest point of the trough, from where it is cleaned out at regular invervals. The cleaned waste material floats and is discharged by a metering screw, and most of the water is removed. The metering screw conveys the ground flakes into a drying system (*c*), consisting of spin-dryer and a hot-air dryer (*d*). After leaving the drying system the flakes (*e*), with a residual moisture content of 1–2%, are guided into the second stock silo and from there to the extruder where they are converted into lenticular pellets. The washing process does not affect the quality of the secondary feedstock. However, damage to the plastic during its "first life" does affect quality. Agricultural film and bottle crates will have been subject to significant changes of physical or optical properties, such as:

– worsening of mechanical properties, including tensile strength and impact resistance,
– alteration of the surface structure through cracking and crazing,
– color changes like yellowing, or reduction of transparency.

Such damage can in practice be partly or wholly eliminated by blending the scrap with virgin material and/or special stabilizers and additives.

21.3.3 Integrated recycling and compounding, including filling and alloying

Typical application: PP film scrap has to be reclaimed, and at the same time 10 to 35% of fillers, 3 to 6% lubricants and 2 to 4% color concentrate must be incorporated.

Full compounding recycling doesn't mean recovery of materials as they were. Rather, this art involves both recycling and upgrading the materials to a specific standard in a single processing step. This saves costs and creates a more valuable secondary raw material.

Figure 18 shows the design concept of such a plant. After the usual size reduction to a flake size of approx. 8 mm, the PP is conveyed into a silo (a, b). After passing the metal separator (e) the material comes to a volumetric feeder (f). Color concentrate (c) is added by another feeder (d). The feeder (f) guides a precisely defined amount of material into the crammer feeder (g).

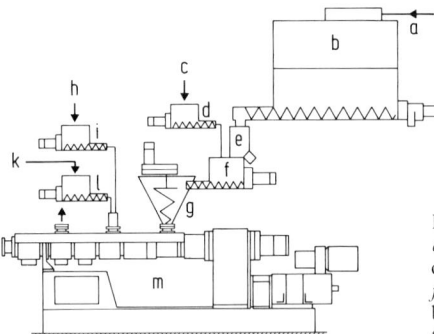

Figure 18. Reclaiming and compounding plant a polymer feed (PP), b flake silo, c color concentrate feed, d metering, e metal separator, f metering, g crammer feeder, h calcium carbonate feed, i metering, k additive feed, l metering, m twin-screw extruder

Here the flakes are densified and force-fed continually into the extruder (m). In this case a twin-screw extruder is required for plastication because good dispersion capability is needed for filling. Calcium carbonate ($CaCO_3$) (h, i) and PE-wax (k, l) are fed through a second feed opening by two volumetric feeders. The entire line, including the interlocked feeders, is monitored and controlled by a microprocessor. The secondary raw material produced is used for injection molding. By filling with calcium carbonate, shorter injection cycles, improved rigidity, higher heat-distortion temperature and lower shrinkage are achieved. Lubricants improve the flow characteristics of the melt. The special problems of this reclaim operation are the dosing accuracy of the feeder screws and the intensive dispersion needed for incorporating fillers and, additionally, the requirement for an excellent standard of degassing and removal of moisture from the filler. The effect of the filler on the secondary feedstock is shown in Figures 19–21. Figure 22 shows the density increase with increasing filler content, and the resultant cost reduction. In this case the filling is especially cost-effective, because the compounding and reclaiming operations were combined at a single cost not very different from that of reclaiming alone.

21.3.4 Reclaiming of engineering plastics waste

These are high-grade materials; in reclaiming them, possibly better, although rather different properties, can be obtained to make the material suitable for a wider range of uses than the original applications. Waste from high-grade thermoplastics, like PA or PET, is available in relatively large amounts as textile fabrics, fibers, film from x-ray and video-tapes, and bottles. The waste comes from products for which the optical end use specifications are extremely high. Therefore only relatively small amounts of that waste can normally be re-used for its original purpose. Thus the basically high-class waste ought, in principle, to be regenerated for other critical applications.

21.3 Applications, products, special properties

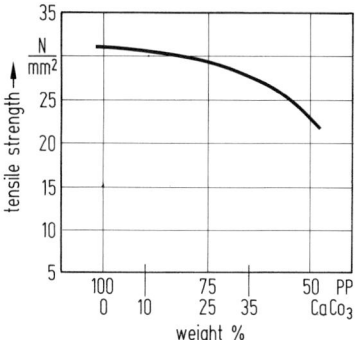

Figure 19. Tensile strength as a function of $CaCO_3$ content

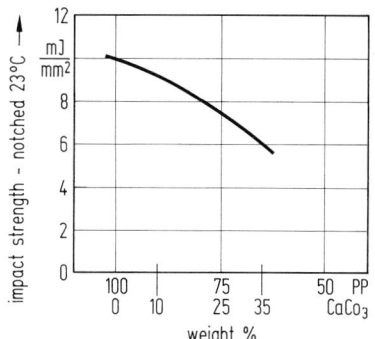

Figure 20. Notched impact strength versus $CaCO_3$ content

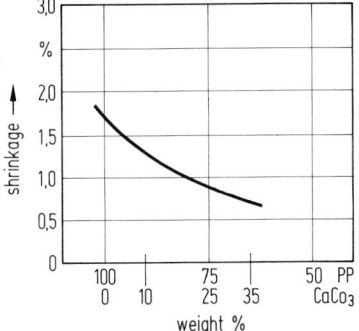

Figure 21. Shrinkage versus $CaCO_3$ content

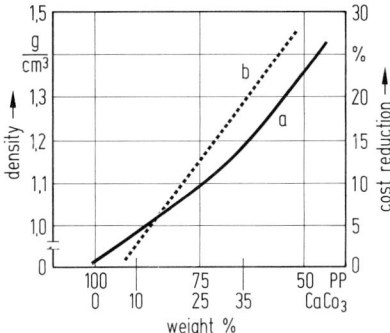

Figure 22. Change of density and price of filled PP with $CaCO_3$ content
a change of density with filler content
b change of price with filler content

This can be done by incorporating mineral fillers, or by reinforcing with glass beads or glass fibers. Single-screw and twin-screw extruders are used for the purpose as follows:

– for reclaiming without filling or reinforcing: the single-screw extruder
– for reclaiming and filling and/or reinforcing in a single process: the twin-screw extruder.

Because of their hygroscopic properties, plastics like PA and PET tend to degrade by hydrolysis during processing, with decrease of molecular weight and melt viscosity, and an associated worsening of physical properties. The humidity in the waste must be removed before processing to guarantee the best possible quality of the secondary raw material. This is achieved by intensive drying in special silos. Recommended drying temperature is between 100 and 150°C and the drying period two to four hours. To eliminate atmospheric oxygen locally, dry nitrogen can be used to blanket the feed zone of the extruder.

In fiber and film reclaiming, besides dealing with the hydrolysis problem, it is essential to achieve the following:

– efficient silo storage and continuous discharge,
– trouble-free extruder feed,
– gentle plastication with intensive degassing.

Horizontal silos with large-volume metering screws are used for storage. Trouble-free feed is achieved by having a tapered feed zone on the single-screw and by free-meshing large-volume feed screws on the twin-screw unit.

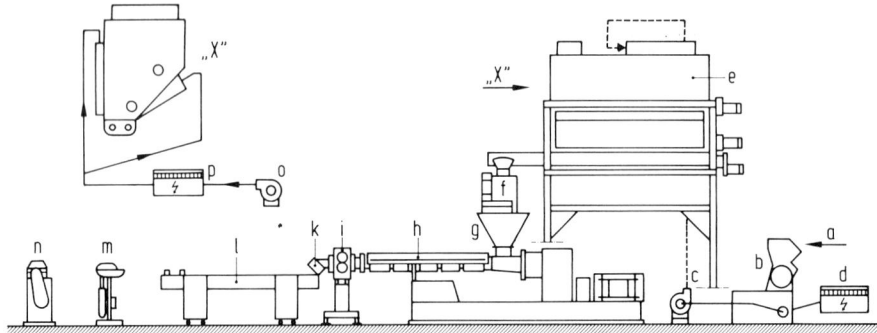

Figure 23. Plant for the recovery of PA or PET fiber and film scrap
a PA or PET fiber or film scrap, *b* cutting mill, *c* conveyor blower, *d* el. heat exchanger, *e* preheating silo, *f* metal separator, *g* crammer feeder, *h* single- or twin-screw extruder, *i* screen changer, *k* strand die head, *l* cooling bath, *m* blow dryer, *n* strand pelletizer, *o* auxiliary hot-air conveyor blower, *p* auxiliary heat exchanger

Special screw geometries, if necessary in combination with a heated feed zone, ensure gentle plastication. Figure 23 shows a plant for use with such materials. There is a direct connection between the solution viscosity and the molecular weight of polymers. Intrinsic viscosity (100 ml/g) is relatively easy to determine and so can be used as a quality control parameter (Table 4).

Table 4. Physical properties of PA 6 before and after processing by single- and twin-screw extruders

		Starting material	Run on single screw	Run on twin screw
Relative viscosity	η rel.	2.5	2.31	2.38
Tensile strength	N/mm^2	71	63	65
Impact strength	mJ/mm^2	69	64	68

To compare the effectiveness of simple single-screw- and the more advanced twin-screw technologies, the single-screw extruder was fed with glass fibers and PET flakes together (cold-feed technique), and on the twin-screw extruder glass fibers were fed into the melt separately via a second feed opening (hot-feed technique similar to the scheme in Figure 18).
Figures 24, 25 show results obtained on single- and twin-screw extruders.

21.3.5 Direct recycling from film scrap to blown film

Naturally, every plastics processor would like to reprocess in-plant waste straight into final product. This is often possible in cases like injection molding or extrusion of thick-walled commodity materials. In film manufacturing it is practical with very low quality film. Such a method was developed by one leading blown film manufacturer who uses this now patented process [9] with very good results in his own factory.

21.3 Applications, products, special properties

Figure 24. PET waste mixed with 30% glass fiber on a single-screw extruder: hopper cold-feed

Figure 25. PET waste mixed with 30% glass fiber on a co-rotating twin screw extruder: feed into melt

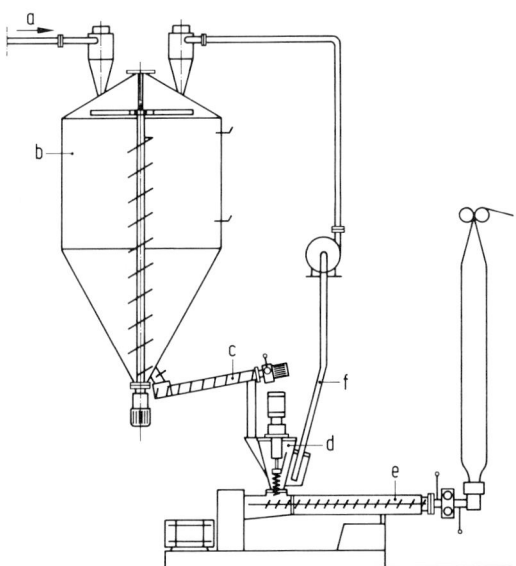

Figure 26. Direct reclaiming, from film scrap to blown film
a material feed (flakes), *b* silo mixer, *c* conveyor screw, *d* crammer feeder, *e* blown film extruder, *f* pneumatic return conveyor

Figure 26 shows a schematic diagram of this plant. Ground film scrap (*a*) is pneumatically conveyed to a mixing silo (*b*). A screw conveyor (*c*) guides the mixed flakes into a crammer feeder (*d*). Predensification and forced feeding of the blown film extruder (*e*), on which film is produced in the usual way, follows. In this process, a certain amount of material is pneumatically removed from a pocket (21.4.2) located beside the crammer feeder when it reaches a defined, adjustable, level. Material removed is fed back to the mixing silo.

By keeping the filling level constant, a constant rate of extruder feed and thus a constant melt discharge rate at the film blowing head are achieved.

In this case only slightly contaminated production waste is used, and the film made is dimensionally accurate and of good quality.

21.4 Typical line design concepts and line components

There is a wide variety of different systems available to suit the differing types and forms of waste material. Suitably designed lines and machines are available for almost every purpose.

To understand how the many different systems have arisen, one needs to know in what quantities and converted forms the various waste materials are found. Five of the biggest industrialized countries, USA, Japan, FRG, France, and the UK produced approx. 48 million tons of bulk plastics in 1984. This divides into 28 million tons of PE and PP, 10 million tons of PVC, and 10 million tons of other plastics. From the 28 million tons of PE and PP some 19 million tons has probably been used to produce films and similar products. Reprocessing film scrap into compact, solid products therefore affords a solution to another problem – disposal.

Voluminous and poorly flowing materials must be fed continuously to the plastifying screw; bridging and agglomerations must be avoided. The first leads to output fluctuations, the second to machine overload and stoppage, which, if persistent, may cause machine damage.

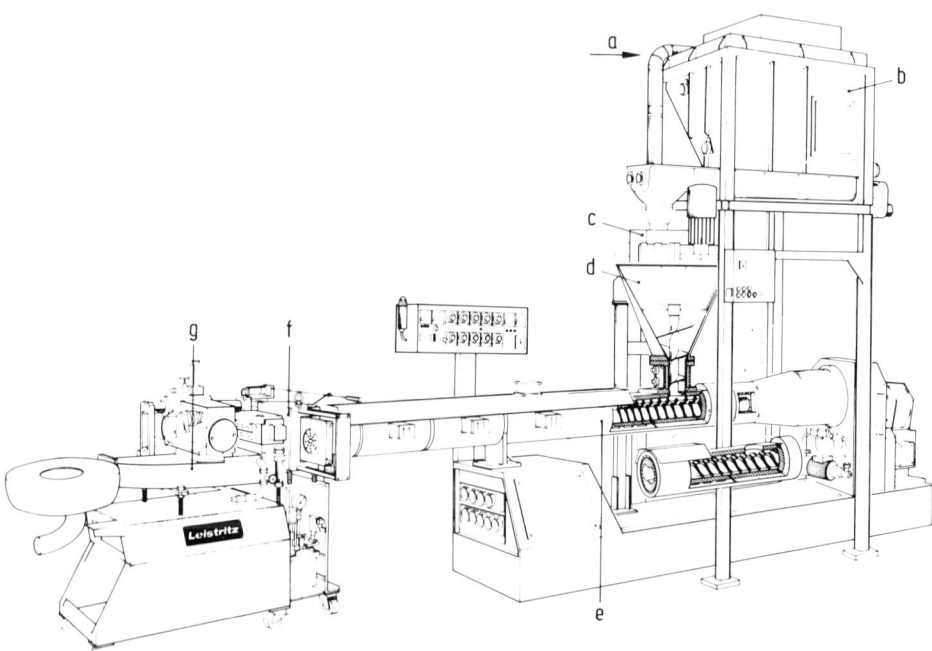

Figure 27. Large-capacity plant for film and fiber waste recovery
a material feed (flakes), *b* flake silo, *c* metal separator, *d* crammer-feeder, *e* extruder, *f* screen changer, *g* die-face water-cooled pelletizing system

21.4 Typical line design concepts and line components

21.4.1 Classification of reclaim lines according to their uses

21.4.1.1 Processing of film and fiber waste, with preliminary size reduction: output greater than 200 kg/h

The flake silo is a typical feature of the line shown in Figure 27. It is often used as a buffer between chopping mill and extruder, in conjunction with the vertical crammer feeder. The flake silo must be suitable for throughput rates above 200 kg/h, otherwise an economical and continuous operation of the line is not possible. The extruder may be either a single-screw of standard or special design as described below, or a twin-screw extruder of either co- or counter-rotating type.

21.4.1.2 Processing of film and fiber waste, with preliminary size reduction: output below 200 kg/h

The difference between this line (Figure 28) and the one shown in Figure 27 is that silo (a) and crammer feeder are one unit, and the crammer feeder screw is of horizontal design. With this small throughput rate the silo volume is suitable for continuous operation. A single-screw extruder is used for plastication.

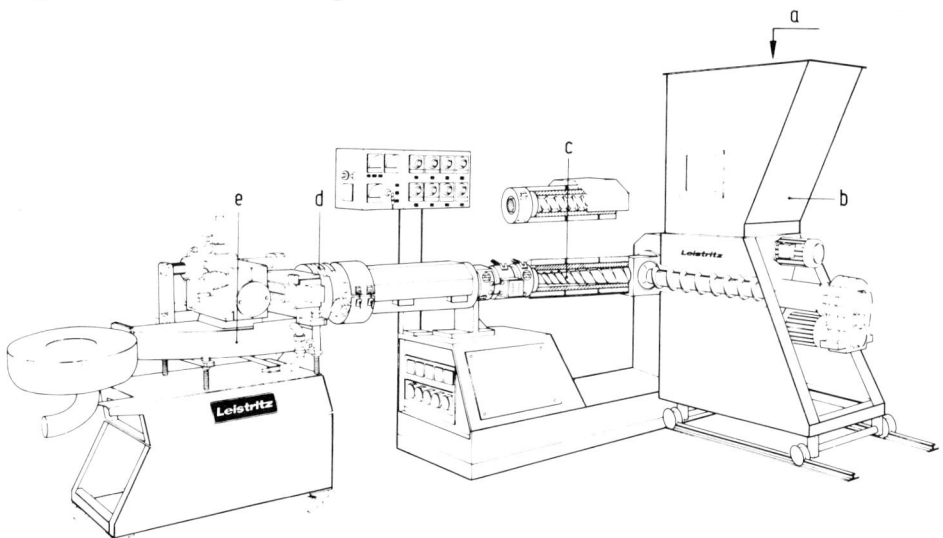

Figure 28. Small-capacity plant for film and fiber waste recovery
a material feed (flakes), b crammer-feeder container, c extruder, d screen changer, e water-cooled die-face pelletizing system

21.4.1.3 Processing of film and tape scrap (direct feeding) without size-reduction: output below 200 kg/h

The design of line shown in Figure 29 is typical of those used for reclaiming roll material. Only in this way can a continuous operation be guaranteed. If irregular scrap is fed manually to the extruder, substantial process variations cannot easily be avoided, and these can reduce the quality of the product recovered. Another disadvantage is that the operator must remain near the extruder. This design is economic only for outputs below 200 kg/h.

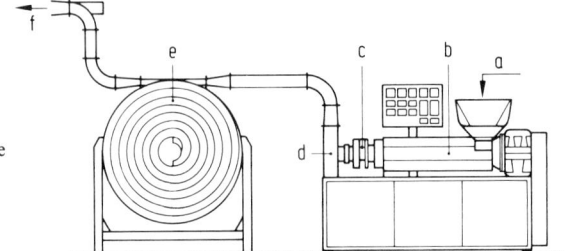

Figure 29. Plant for recovering film and tape scrap using direct feed of unchopped waste
a feed of unchopped material, *b* extruder,
c screen changer (manually operated),
d die-face pelletizer (air-cooled),
e spiral air cooler, *f* reclaim pellet

21.4.1.4 Processing of all kinds of plastics waste, including films, tapes, sheets, hollow bodies etc. – with size reduction but without external crammer feeder

The outstanding advantages of this design (Figure 30) are universal applicability, compact construction, and very low installed cost. The arrangement involves an extruder with an independently driven cramming screw concentric with the plasticating screw, and provides for relatively constant feeding of irregular wastes. The cramming screw is also suitable for use as a feeding screw for materials with high bulk density.

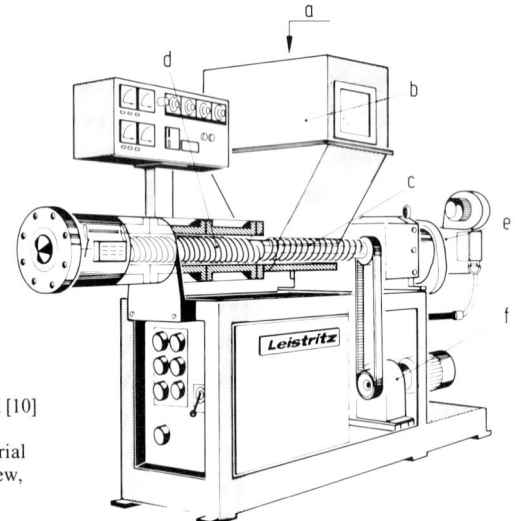

Figure 30. Two-stage waste recovery plant RCA-X [10] with integrated stuffing screw
a material feed (flakes or milled material), *b* material stock container, *c* stuffing screw, *d* plastication screw,
e main drive, *f* crammer-feeder drive

Because few, if any, cold compression problems occur in the transition zone, and because of the general utility of this system, the energy consumption is relatively low.
In general this design integrates the stuffing or feeding screw, the stock silo (up to 2 m^3 volume) and the plastication unit into one machine.

21.4.1.5 Integral processing and compounding of all kinds of plastics waste, possibly with lace extrusion of end product

The design shown in Figure 31 is intended for use without direct feeding (21.4.1.1/21.4.1.2 and 21.4.1.4), and where high quality product is required. Co-rotating- or counter-rotating twin-screw extruders are used. To suit special applications auxiliary devices may be incorporated.

21.4 Typical line design concepts and line components

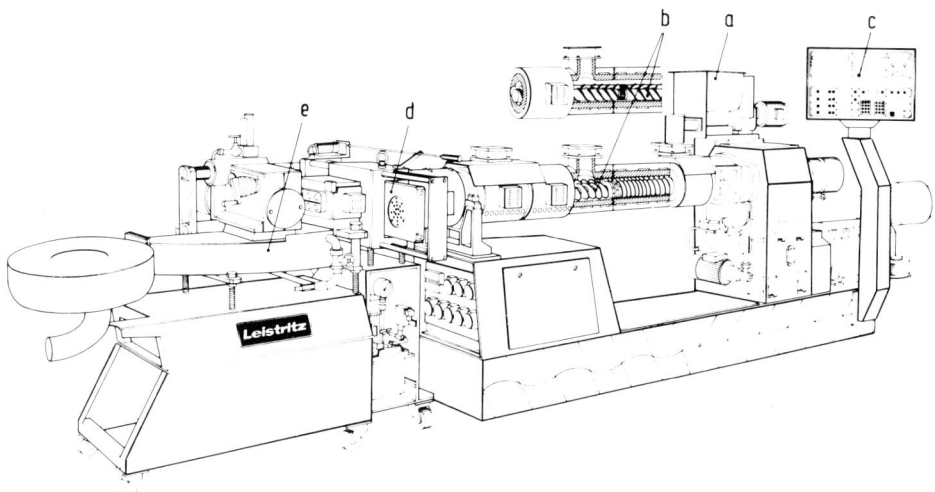

Figure 31. Reclaim compounding and extrusion plant for all kinds of thermoplastic waste, optional end product manufacture
a feeder, or crammer feeder, *b* plastication unit, single or twin screw, *c* microprocessor control, *d* screen changer, *e* pelletizing system (die-face or strand) or other downstream equipment

This method is used for high-value reclaiming operations integrating simultaneous modification, such as filling, alloying, and reinforcing, and is essential if end products like pipes, sheets etc. are to be produced in-line. For other kinds of operation, eg. film blowing, a single-screw extruder can be used.

21.4.2 Classification of reclaim extruders with respect to design criteria

Many different types of extruder are used within the range of lines described under 21.4.1. Their principal design features are described below:

21.4.2.1 Standard single-screw extruder with modified processing unit

Every manufacturer makes a number of variations of the basic design, drive system and processing elements. The standard single-screw extruder with slight modifications is used successfully for reclaim operations, for example with the feed section of the screw enlarged for processing voluminous and light waste. A grooved barrel in the feed zone may also be an advantage.

21.4.2.2 Adiabatic operation, single-screw extruder (Figure 32 A)

This design of extruder is still sometimes used for reclaiming. It typically features a short screw and high screw speed, which means that it is suitable for only a limited processing range.

21.4.2.3 Single-screw extruder with increased screw diameter at feed zone

This special type of single-screw extruder is the machine typically used for reclaiming film-, fiber-, tape-, and other scrap having a low bulk density.

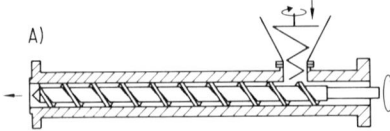
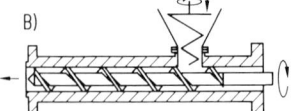

Figure 32 A). Single screw — standard

Figure 32 B). Single screw — adiabatic

Two different designs are available:

- cylindrical plastication and metering screw section with tapered feed-screw sections (Figure 32 C)
- screw with cylindrical feed section, tapered transition and cylindrical metering section (Figure 32 D).

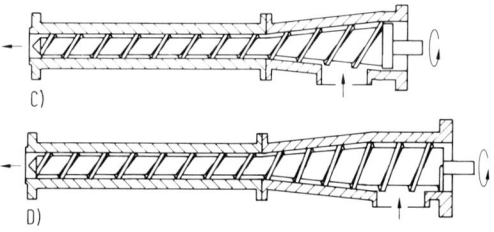

Figure 32 C), D). Single screw with enlarged feed zone

Because of the large size of the feed zone, even materials having an extremely low bulk density can be fed reliably. At a bulk density of below 200 g/l a stuffing aid is necessary. Above 200 g/l it is normally not needed. With such a screw design, materials with poor dry-flow properties, like PP, PA and PET fiber- and fabric waste can also be reclaimed satisfactorily. Heating of the screw-feed section is an advantage except with PP, for which the feed section is grooved and cooled intensively to improve feed and conveying behavior. Machines operating on this principle are also used for reclaiming unground film- and tape scrap. These applications also require an enlarged feed opening.

21.4.2.4 The two-stage extruder

In the two-stage extruder (Figure 32 E) the feed and plastication screws are driven co-axially. Thus the material stream goes from feed- to metering section without deviation. The feed screw can either feed or stuff. A large feed opening about 5 D long aids flake intake.

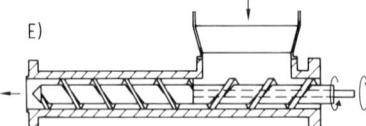

Figure 32 E). Two-stage extruder

The feed screw (a) is driven via geared V-belts by a steplessly variable DC-drive. The plastication screw (b) is directly driven by an axially located gearbox (Figure 30). Either a DC-motor or a normal three-phase AC motor serves as the main drive. The speed settings of the feed and plasticating screws depend on the bulk density and output rate required, and are, in practice, electronically controlled. The problem with this design is the small space available for building-in the bearing of the feed screw on the shaft of the main screw. The experience so far regarding the life expectancy of the bearings is very encouraging, however.

21.4.2.5 The short-screw extruder

The short-screw extruder (Figure 32 F) has no feed screw; its screw carries out feeding and plastication duties simultaneously. A short screw with a big diameter and low pitch angle has the same circumferential path as a long screw of small diameter and high pitch angle.
A screw with a big diameter in the feed section accepts flakes of low bulk density well. Because manufacturing costs are low and feeders are not used, this kind of machine can be manufactured inexpensively. Further advantages are its simple operation, and the fact that it is extremely robust for handling badly contaminated waste materials.
A disadvantage of the short-screw system is that it cannot handle a wide range of bulk densities and that all bulk densities cannot be processed at the maximum output rate.
A pair of additional feed rolls enable unground film scrap to be fed direct from a roll of film.
Both systems are intended for small and medium output rates (max. 230 kg/h). Because of their simple and compact construction their installed cost is relatively low.

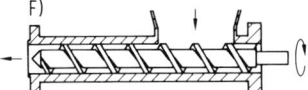

Figure 32 F). Short-screw extruder

21.4.2.6 The short-screw plasticator (Figure 32 B)

This is another version of the short-screw extruder; it is used especially for scrap which is mixed, contaminated or filled with foreign bodies (general waste).
The screw is 5 D long and has three starts. A plane front face rotates against the stationary surface of the barrel. The space between these two surfaces is the shear zone. Its action is that of a disk plasticator. The screw has a pure conveying and feeding duty. The material is densified by counter pressure and friction between barrel and screw. The material is sheared in the space in front of the screw and heat generated by this action plasticates the material.

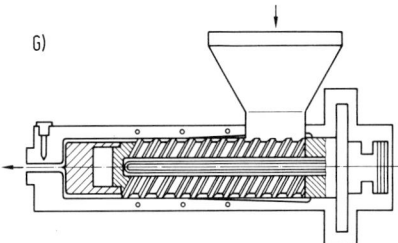

Figure 32 G). Short-screw plasticator [10]

21.4.2.7 The retruder (Figure 32 H)

This system is intended to accomplish size reduction, densifying and plasticating in one unit. It is designed for films up to 3 mm thickness, and for tapes and hollow bodies made of LDPE, HDPE, PP, PS, and soft-PVC.

The unit consists of a cylindrical material-delivery container which has a rotary cutting unit on its base. The barrel is arranged radially to the container, and the screw, which is driven on the outlet side, as in the vertical extruder, draws material from the container. The waste material is chopped up and heat-dried in the delivery container by the effects of mechanical shear and friction. The extruder screw receives the prepared material, plasticates and homogenizes it, and discharges it as melt on the drive side of the screw.

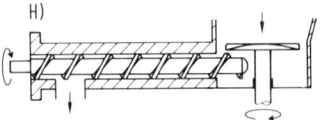

Figure 32 H). Retruder [11]

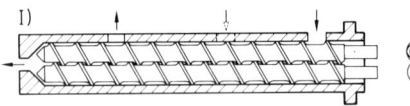

Figure 32 I). Twin-screw co-rotating

21.4.2.8 The co-rotating twin-screw extruder (Figure 32 I)

This type of machine has been well accepted for many years for the preparation of plastics feedstocks.
For reclaiming, the following specific design criteria are important:

- large-volume single- or bilobal screw systems
- screw volume enlarged by open meshing of feed zones and increased root-diameter clearance
- special screw geometries to provide the possibility of shorter processing length.

21.4.2.9 The counter-rotating twin-screw extruder (Figure 32 K)

Specific design criteria for reclaiming are:

- higher screw speed,
- lower specific screw torque,
- greater flight clearance and bigger roll gap,
- increased length of 20 to 27 D,
- higher heating power,
- large-volume feed section.

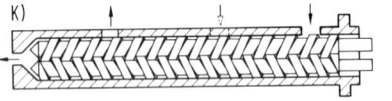

Figure 32 K). Twin-screw counter-rotating

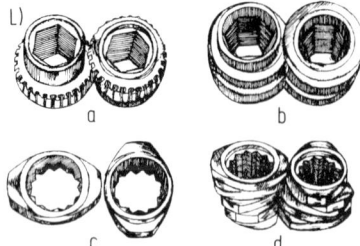

Figure 32 L). Mixing, homogenizing and kneading elements for twin-screws
a mixing element, *b* homogenizing elements for counter-rotating screws, *c* kneading disks, *d* kneading blocks for co-rotating screws

To improve the effectiveness of mixing and homogenizing, the counter-rotating system has mixing and homogenizing elements. With the co-rotating system the plastication capacity must be improved by kneading elements (Figure 32 L).

21.4 Typical line design concepts and line components

Table 5. Design of single-screw recovery systems: machinery data and throughput rate

Machine type	Screw diameter mm	Screw length L/D	Drive power kW	Throughput rate kg/h	Preferred application	Special design features
Single-screw, standard	50 to 200	20 to 35	20 to 250	50 to 1000	granular materials, practically all thermoplastics (pellets, ground materials, agglomerate); also with feeder, for film and fiber waste	classical construction, drive and barrel variants according to manufacturer
Single-screw, adiabatic operation	40 to 100	10 to 15	12 to 55	40 to 180	film scrap of low bulk density (with feeder)	short plastication screw, high screw speeds
Single-screw with enlarged feed zone	60/90 200/280	25 to 35	30 to 250	80 to 1200	direct feed films tapes, fiber, fabrics; in modified design also PA and PET-fiber and film waste, esp. at very low bulk densities	tapered feed zone or enlarged cylindrical feed zone with tapered transition, partial screw heating, extremely large feed opening
Two-stage extruder RCA X	110 130*	10 + 6 10 + 6	20 60	90 to 120 150 to 230	film and tape scrap with widely varying bulk density, low-cost design	big screw dia. and short screw length, separately driven feed screw coaxial with the plastication screw, extremely large feed opening
Short-screw extruder	70 to 125	10 to 12	15 to 100	50 to 150	all kinds of film and tape scrap; also direct feed for simple reclaiming applications, low-cost design	big screw dia., low screw pitch, short screw, extremely large feed opening
Short-screw plasticator system FNP	60 100 120	5	32.5 53 107	25 to 50 75 to 150 150 to 300	processing of various thermoplastic mixes, including very dirty scrap containing impurities	short screw, combined with disk torpedo, special screw heating
Retruder: combination of granulator, preplasticator and extruder	100	13.5	45	200 to 500	all kinds of film scrap	combination of granulator and pre-plasticator, with single screw, driven on the discharge side. Melt discharge laterally

* depends on the processing material

Table 6. Design of twin-screw recovery systems: machinery data and throughput rate

Machine type	Screw diameter mm	Screw length L/D	Drive power kW	Throughput rate kg/h *	Screw speed $\min^{-1}$	Preferred application	Special design features
Twin-screw co-rotating	34 to 135	20 to 35	10 to 700	30 to 3500	30 to 300	for continuous recovery and compounding jobs possibly with extrusion of a final product. Incorporation of glass fibers into PA, PET, PP and similar	co-rotating screw system, high screw speeds and segmented barrel (with some manuf.)
Twin-screw counter-rotating	34 to 170	20 to 35	10 to 250	20 to 1300	20 to 150	for combined recovery and compounding applications, poss. at same time extrusion of a final product, incorporation of high pigment loads, extrusion of thin-walled final products against high counter-pressure	counter-rotating screw system, low screw speeds; screws and barrel either in cheap one-piece version or as segments
Twin-screw with extended discharge screw	95 125	14/20	50 75	280 450	10 to 90 10 to 80	general recovery applications with nearly all thermoplastics, processable without venting	counter-rotating twin-screw system at feed and plastication section, single screw at discharge section

* depends on material to be processed

21.4.2.10 Twin-screw with extended metering screw (Figure 32 M)

This is a special design which has been proved over many years in reclaiming. It has the good feeding behavior and the low specific-energy consumption of a twin-screw unit, while retaining the operating safety factors and long-life advantages of a single screw.
A comparison of all the single- and twin-screw extruders described above, with the relevant technical data, is shown in Tables 5 and 6.

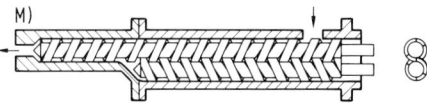

Figure 32 M). Twin screw with extended discharge screw [12]

21.4.3 Wear and wear prevention in reclaim extruders

The main factors concerning wear on screws and barrel are:

21.4.3.1 Kind and degree of contamination of the material

The user should sort his waste material, and clean it appropriately. Washing is often needed in preparation for extrusion. Unfortunately these two actions are often neglected in practice. Besides cleaning it is also essential to remove metals before size reduction and extruder feeding.

21.4.3.2 Fillers, reinforcements or pigments

If the application and economic considerations allow, the use of pigments or fillers that cause less wear can be considered.

21.4.3.3 Technique of adding fillers and reinforcement

Feeding these components into the melt is preferable to feeding into unplasticated material.

21.4.3.4 Operational conditions

High screw speeds, low temperatures, high melt pressures and high melt viscosities all tend to increase wear on the extruder.

21.4.3.5 Machine designs

Along with work to increase the specific performance of extruders, more attention has been paid to improving the wear properties of screws and barrels.
Today, the flights of single screws are stellited and the barrel bore is lined with special alloys based on cobalt, tungsten, nickel and chrome.
Twin screws, also, are stellited, or the flights are covered with a sintered laminate.
One of the latest procedures is to manufacture screw elements of fully hardened tool steel. Twin-screw extruder barrels are made from two bi-metallic tubes slit longitudinally and

pressed or welded together along the slit line. Barrel liners made from stellite or similar material, and types made of fully hardened tool steel are also on the market. These liners are clamped together as barrel sections or shrunk into the barrel.

Depending on the application and type of wear-resistant system adopted, increases in the service lives of such barrels and screws in the ratio of 1:2 to 1:4 can be achieved, with corresponding reductions in specific costs.

21.4.4 Size reduction: cutting mills

The choice of mill type (also called a 'granulator') depends upon:

- capacity requirements
- type and form of the material to be treated.

The extruder output rate determines the size of the machine, and in general:

$$\dot{m}_c \approx 2\dot{m}_E \qquad (4)$$

Where:

$\dot{m}_c$ capacity of mill
$\dot{m}_E$ capacity of extruder.

The design of the mill, and especially the design of the cutting rotor, is defined by the type and shape of the scrap material. An enclosed rotor is used for solid parts, and an open rotor for hollow bodies, film waste and so on (Figure 33).

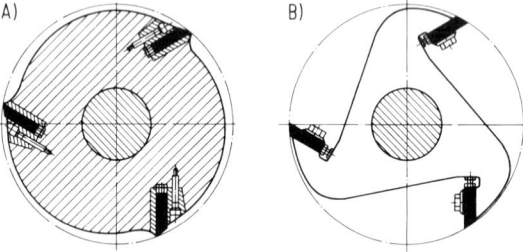

Figure 33. Rotor design for granulators
A) open rotor, B) enclosed rotor

Rate of throughput for a given material can be completely different for different scrap forms (e.g. PE pipes, hollow bodies, film). Different polymer types require different mill sizes for the same throughput rate.

Performance tests have shown differences of up to 300% in energy requirement (see Figure 34).

Until recently, the granulators on the market nearly all used the same design principle. A rotor with a few blades rotated against one or more stator blades to chop up the material. A screen placed below the rotor governed the particle size according to hole diameters (Figure 35). This basic principle still applies; however, there have been improvements in recent years including the so-called double helical cut between rotor and stator blades (Figure 36). This has resulted especially in:

- lower specific-energy consumption and higher throughput at the same drive power,
- lower noise levels because noise peaks generated by conveying air pulsating in the space between the blades do not occur with helical setting.

21.4 Typical line design concepts and line components

There have been other successful developments on noise level reduction in granulators; the most significant ones are:

- isolation of the material to be ground from frame and ancillaries,
- charging hopper designed as silencer,
- charging opening designed as noise lock.

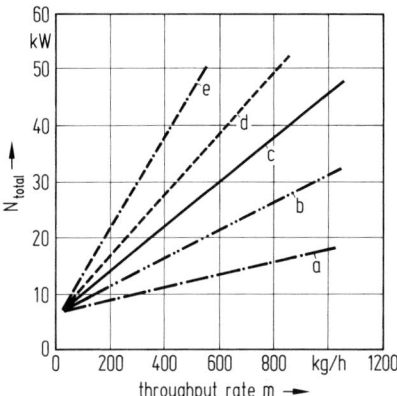

Figure 34. Throughput rate and energy consumption for grinding different types of plastics
a PS sheet, *b* PVC thermoformable,
c PE pipes, *d* PE film 100 µm, *e* PP film, 50 µm biaxially oriented

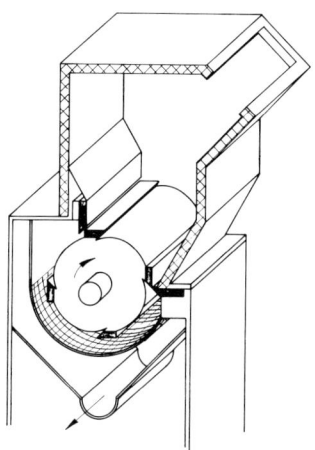

Figure 35. Schematic view of a granulator

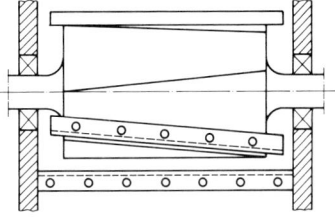

Figure 36. Principle of double offset cut

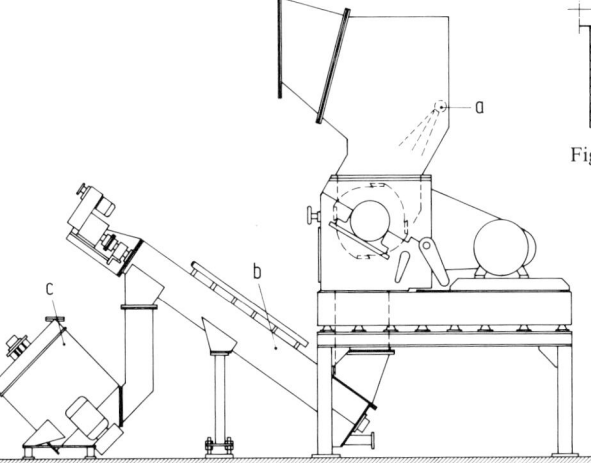

Figure 37. Washing granulator
a water inlet
b screen conveyor screws,
c dryer

These measures, together with the double helical cut, have brought noise levels down from above 100 dBA to less than 90 dBA.

Wash-grinding is important, especially for the treatment of contaminated scrap. The material is ground and washed in a continuous process by the intense vortex created by jetting water (*a*) into the granulating chamber (Figure 37).

The flake-water mix is discharged from the bottom of the granulator by screen conveying screws (*b*) into the dryer (*c*).

21.4.5 Metal detection and metal separation

It is essential to remove metal to avoid machine damage, and to improve product quality. The following three systems are used:

1. inductive metal detectors,
2. inductive metal separators,
3. solenoid grids.

Inductive metal detectors and metal separators use the same principle:
A highly selective resonant circuit ($Q \sim 500$) is fed with its resonant frequency from an HF generator. The scrap material is guided through a suitably located detecting coil which forms part of the circuit. If metal enters the coil it removes energy from the magnetic field by induction of eddy currents and upsets the resonance of the circuit. The resonance amplitude changes to an extent dependent on the size of the metal part.

21.4.5.1 Metal detector

The detector provides a visual or acoustic signal, and reverses the conveyor belt transporting the plastic scrap to the granulator, unless the offending metal is removed (Figure 38).

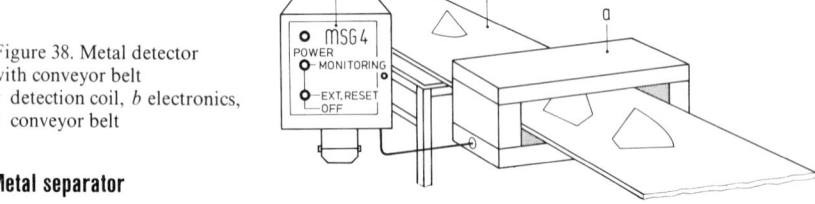

Figure 38. Metal detector with conveyor belt
a detection coil, *b* electronics, *c* conveyor belt

21.4.5.2 Metal separator

The operating principle of the metal separator is shown in Figure 39. Metal-free material runs straight through the separator, but if the detection coil reacts to metal, a solenoid valve operates a double cylinder to deflect and remove the contaminated material through a flap.

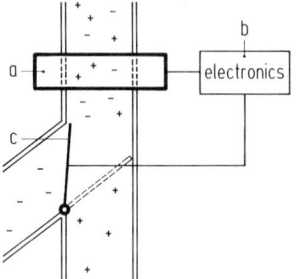

Figure 39. Inductive metal separator
a detection coil, *b* electronics, *c* pneumatically or hydraulically operated separation valve

After an adjustable "separation period" passes, the flap swings back. The period is automatically extended if more metal parts follow within the set period.

21.4.5.3 Solenoid grid

The solenoid grid is used merely as a supplement to the inductive metal separator.

21.4.6 Bulk storage and homogenization

The storage of chopped film, fiber, and fabric scrap is full of difficulties. Discharge of the silo is the key problem, with the bulk density and the shape of the ground particles playing an important role.

As a result of many years' experience with many different materials, a special silo (Figure 40) was evolved. The horizontal arrangement of the silo body has the advantage that it can be built relatively low and long. The great length has a favorable effect on the discharge behavior of the large metering screw, which is placed below the inclined plate in the lower half of the silo; and the stirring and mixing elements, located above the metering screws, therefore have a better mixing and dispersion efficiency. The material is discharged by a variable-speed screw. Silos as described are built in capacities from 5 to 30 m^3. Hot-air heating is an option available on this type of silo.

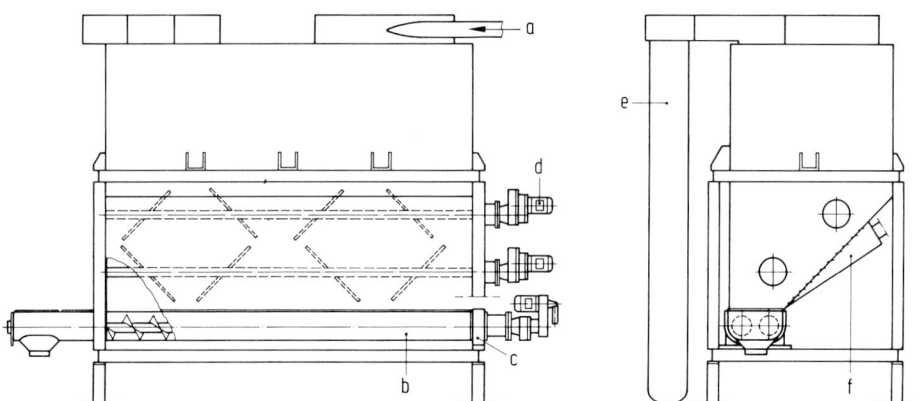

Figure 40. Special silo for poorly flowing materials
a material feed, b discharge screws, c distributor gearing with drive, d mixing and stirring screw with drive unit, e filter element, f hot air supply

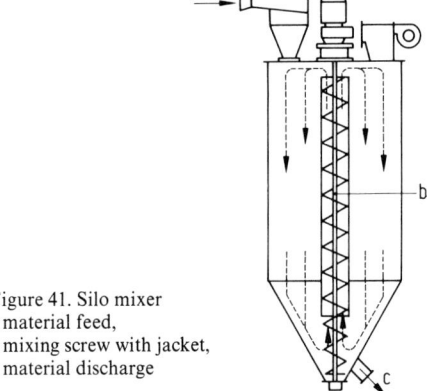

Figure 41. Silo mixer
a material feed,
b mixing screw with jacket,
c material discharge

A mixing silo (Figure 41) is used for storing and mixing granular (easy-flow) materials. The mixing element is a vertical screw in a jacket. This kind of mixing silo is made in sizes from 1 to 50 m^3.

There are many other storage and mixing systems for use with difficult bulk materials in reclaiming processes.

21.4.7 Extruder feeding – the crammer feeder

In many cases, easy processing of specific material forms, such as film flakes, fiber and fabric scrap, is only possible by use of a crammer feeder. This device is used for precompression and forced feeding of low-bulk-density material, typically below 0.3 kg/dm^3.

Two types of design are distinguished:

– the vertical crammer feeder (Figure 42),
– the horizontal crammer feeder (Figure 43).

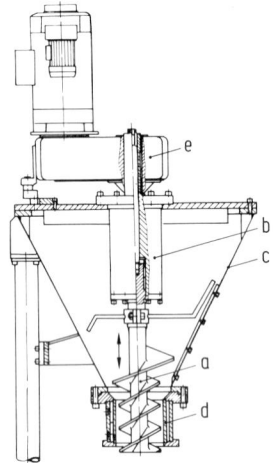
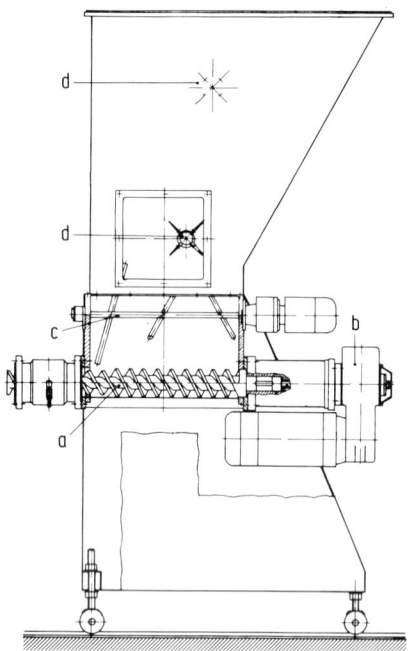

Figure 42. Vertical crammer feeder
a axially adjustable stuffing screw, *b* screw bearing, *c* material hopper, *d* intake pipe, *e* drive

Figure 43. Horizontal or side-stuffer
a stuffing screw, *b* variable drive, *c* stirrer, *d* filling-level indicator

Both units operate on the same principle. However, the vertical crammer feeder is intended for higher throughput rates. The horizontal crammer feeder is used for smaller throughput rates and serves additionally as a mixing and stock silo. The horizontal crammer can normally only be mounted on single-screw extruders.

Usually both devices are driven by armature-current-limited DC-drives, which largely avoids overfeeding of the extruder. The vertical stuffing screw is tapered in the intake zone. Adjustment to very different bulk densities is done by axial gap variation. With the horizontal crammer feeder this is unnecessary. To achieve a constant feed rate while the bulk density of the feedstock varies, the crammer feeder and extruder can be interfaced with an electrical control, which governs the speed of the crammer-feeder screw in relation to current consumption of the extruder main drive.

21.4.8 Screen changer – melt filter

Generally these devices are divided into three categories:

– discontinuous screen changers,
– continuous screen changers,
– melt filtration devices.

21.4 Typical line design concepts and line components

With the first two categories screens have to be changed, while with the third either there is no screen, or if there is one it need not be changed, because it is cleaned mechanically, or by the melt stream (back rinsing). 'Discontinuous' as compared with 'continuous' refers to the continuity of the production process. Discontinuous units interrupt the melt stream during the changeover. In some cases they even stop the production process. With continuous devices, by diverting the melt stream during the screen change, interruption is avoided. The choice of system depends upon the polymer to be processed and the operating parameters. Basic options are as follows:

- low degree of contamination, small to medium throughput rate: discontinuous device.
- standard or high degree of contamination, medium to high throughput rate: automatic or semi-automatic discontinuous device, provided that strands are allowed to break during change; if strand break is not allowed, a continuously operating unit must be used.
- very high degree of contamination, containing a large proportion of solid foreign bodies: melt filter.

To be sure of dependable and trouble-free operation the following design criteria should be adhered to:

- pressure drop in the screen carrier should be as small as possible,
- flow channels must be streamlined, short and without dead spots,
- sealing of moving parts must be guaranteed, whether with frequent changes or in continuous operation,
- screens must be simple and quickly exchangeable,
- devices must be accident-proof.

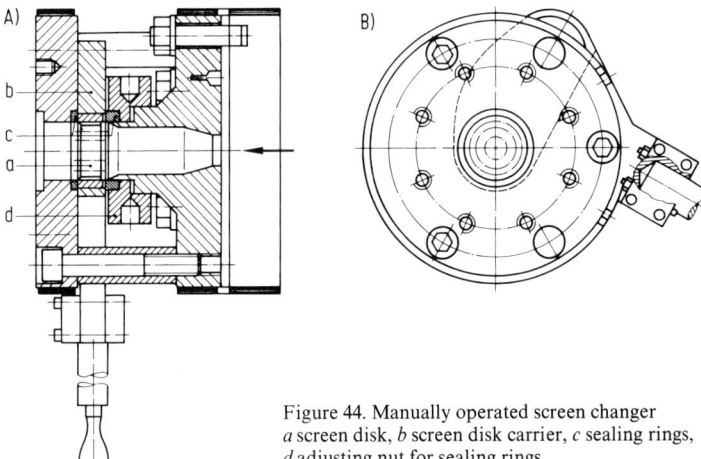

Figure 44. Manually operated screen changer
a screen disk, *b* screen disk carrier, *c* sealing rings, *d* adjusting nut for sealing rings

21.4.8.1 Discontinuous screen changers

There is a wide variety of this type of device. The manually operated screen changer (Figure 44) is the simplest typical example of this category. Such appliances are available with screen diameters of up to about 80 mm. During the changing process the extrusion line must be stopped. This type is only economic with long intervals, 4 to 6 hours, between changes, or if the degree of contamination is low, or if throughput rates are small.

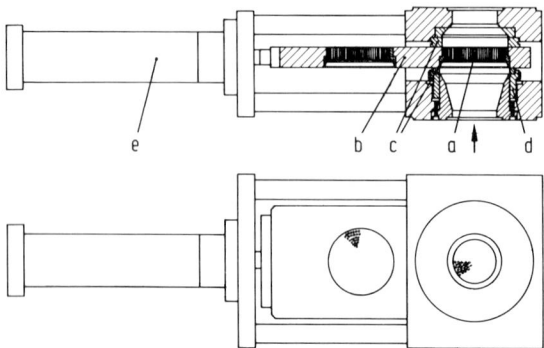

Figure 45. Hydraulically operated, flat-disk screen changer
a screen disk, *b* screen disk carrier, *c* sealing rings, *d* adjustment screw for sealing rings, *e* hydraulic cylinder

Figure 45 shows the discontinuous, hydraulically operated, flat type of screen changer, which is normally used industrially.
Changing screens with this kind of unit is done in less than one second. Nevertheless, the melt stream is momentarily interrupted and, with strand pelletizing operations in particular, the strands break. When a hot-face pelletizing unit is in use, this slight melt-stream interruption is not significant.
The design of the sealing mechanism for the slide plate is the main difference between units of this type. There are hydraulic, or plate-spring braced, or PTFE-ring melt-pressure sealed systems available.
The systems used in reclaiming usually have aluminum-bronze rings on their inlet and outlet sides, which are pressed against the hardened sliding plate by set-screws.
Often the inlet seal is constructed with a deformable sealing lip which is further tightened by the melt pressure itself. Flat screen changers can have screen disk diameters up to 600 mm.

21.4.8.2 Continuous screen changers

The sliding-bolt screen changer (Figure 46), proven in many years of practical operation, is a device in this category. The device may optionally have one or two screen-changing bolts. Each of these would have two screen-carrier plates. The unit is of simple, robust construction, and has short, direct melt channels. Screen changeover requires only a matter of seconds. The system suits practically all polyolefins, polystyrene and similar materials. Contaminated screen exchange takes about two minutes. Strand breakage does not occur.
Operational sequence: When the contamination of the screen in position A) reaches the allowed maximum, the screen-carrier bolt is slowly moved hydraulically across the direction of melt flow. Position B) shows the change from screen *1* to screen *2*. At this position screen *2* becomes flooded with melt and displacement of the airlock takes place. In operational position C), screen *2* is in the new working position. The melt streams straight through the bolt housing, screen and screen carrier. The fouled screen *1* may now be removed. To suit higher output rates, changers with two parallel screen-carrier bolts can be supplied.
A new development which, as a system, can be categorized as a screen changer and filter unit, is the continuously operating screen changer shown in Figure 47. The essentials of the unit are front and rear housing walls (*a*) and (*b*), filter disk (*c*), and drive unit (*d*). The screen plates are positioned in nests separated by webs in the filter disk.
Operational sequence: The filter disk and the inner surface of the housing walls are clamped against each other to obtain a melt-proof seal. The filter disk is rotated by the ratchet drive, (*d*), by means of the teeth on the outer ring. Rotation speed depends on the degree of

contamination in the melt. The melt passes axially, (*e*), through the rear housing wall, filter disk and front housing wall into the extrusion die. When access to the next screen space is opened up by the rotating motion of the filter disk, the melt enters and covers the filter surface, the filter chamber is filled and air escapes through a vent channel. This process is continuously repeated.

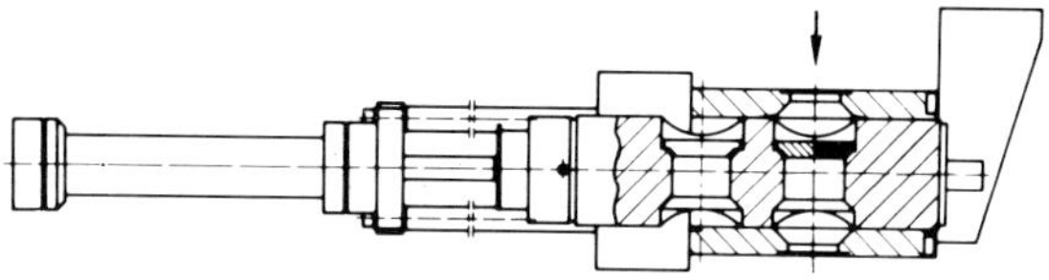

Figure 46. Continuous screen changer type LSWE with sliding screen carrier

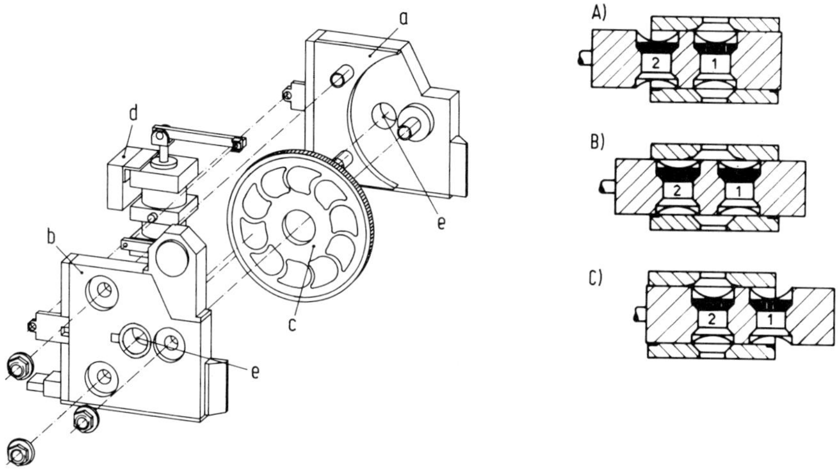

Figure 47. Construction of a continuous screen changer [13]
a and *b*, front and rear housing walls resp., *c* filter disk, *d* drive, *e* melt flow

21.4.8.3 Melt filtration devices

During the past few years some new developments have been commercialized, which is evidence of continuing demand for better automatic melt-filtration units. Three typical examples are described below, starting with that of Figure 48.
Operational sequence: The melt (*a*) flows into the intake chamber (*b*) of the filter, through two perforated disks (*c*) containing a screen pack (*d*). Filtered impurities are removed at regular intervals by a so-called scavenging finger (*e*) and evacuated through separate channels (*f*) and (*g*). Between scavenging events the finger is in "parking position", which means in an area in which the perforated disk has no holes.
As soon as the melt pressure exceeds a certain level, the finger is set into rotation by a separate drive. The number of rotations is directly related to the degree of contamination. Back rinsing of the screen and simultaneous removal of foreign bodies is brought about by the difference in pressure between outlet chamber (*k*) and back-rinsing channel (*l*).
A unit that combines the principles of the separation device and the continuous filter is shown in Figure 49.
Operational sequence: The melt flows axially through a filter disk (*a*). On the feed side of the filter disk is a rotating flat-spiral scraper (*b*) which moves the impurities towards the center of the disk, where they are discharged axially by a screw (*c*). The discharge screw and cleaning rotor are variable-speed-motor driven (*d*). The filtered melt flows either into a

screen changer for fine filtering, or direct into the pelletizing head. Another variation, for special applications, works without its own drive and without a discharge screw. In this case the extruder screw drives the scraper. Discharge of impurities direct to the outside is effected by a pressure-controlled bypass.

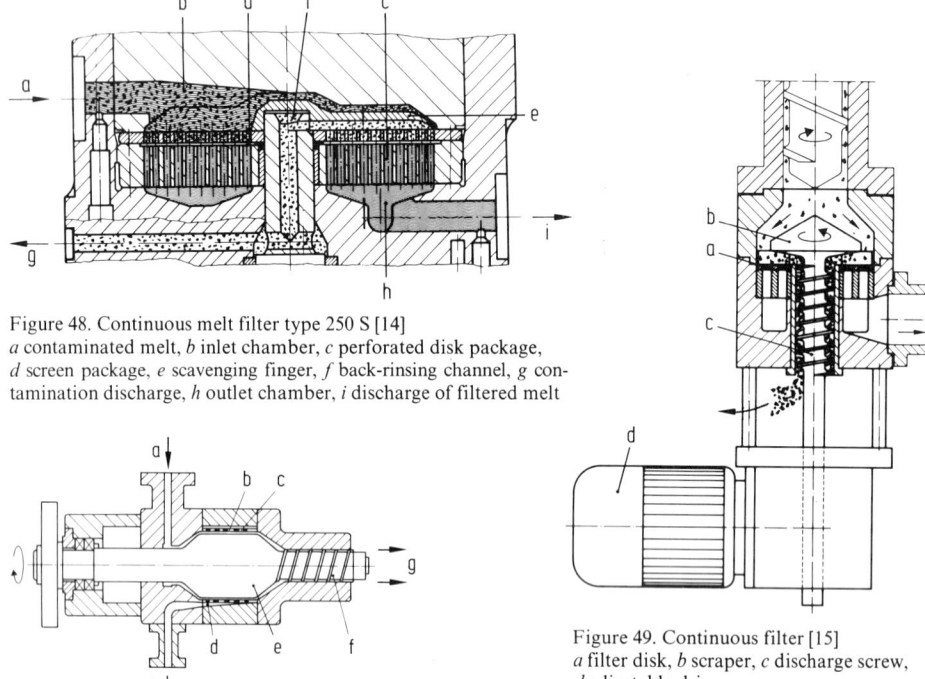

Figure 48. Continuous melt filter type 250 S [14]
a contaminated melt, b inlet chamber, c perforated disk package, d screen package, e scavenging finger, f back-rinsing channel, g contamination discharge, h outlet chamber, i discharge of filtered melt

Figure 49. Continuous filter [15]
a filter disk, b scraper, c discharge screw, d adjustable drive

Figure 50. Continuous separation head [16]
a material to be filtered, b filter body, c filter bores, d channels, e scraper shaft, f restrictor device, g filter residue, h filtered material

Table 7. Comparison of screen changer and filter systems

Category	Screen diameter	Throughput rate	Filter fineness	Max. degree of contamination	Purging loss	Price index[3]
	mm	kg/h[2])	per mm[1])	%	%[1])	
Discontinuous screen changer	70 to 80	30 to 300	20 to 100	0.1	–	30
Manual automatic	60 to 600	1000 to 5000	20 to 100	0.5	–	100
Continuous screen changer, automatic	60 to 2 × 250	100 to 2000	20 to 100	0.5	–	100
Melt filter devices:						
Melt filter	250	300 to 500	50 to 150	5	5 to 15	300
Pre-filter	150	350	400	10	7 to 15	200
Separation head	200					
	600	350 to 700	–	10	7 to 15	400

[1]) Filter fineness depends on the filter screen used
[2]) Throughput rate depends on degree of contamination and polymer type
[3]) The price index is related to a comparable throughput of 300 kg/h

21.4 Typical line design concepts and line components

Another device developed for impurity separation is shown in Figure 50.
Operational sequence: Material to be filtered (*a*) is passed into the inner chamber of a hollow cylindrical filter body (*b*). The melt leaves through the many filter holes (*c*) and flows through channels (*d*) away from the outside of the filter body. The filter deposits are freed by scraper elements, mounted on a rotating shaft (*e*), and are conveyed to an outlet having a restrictor device (*f*). The scraper elements are pressed against the inner surface of the filter body pneumatically, hydraulically or mechanically. The filter body is made in one piece with a smooth inner jacket and an outer jacket ribbed in the circumferential direction. The filter bores have a diameter of 80–200 µm. Table 7 compares different screen changing and filter systems.

21.4.9 Pelletizing

Reclaiming operations require universal, variable and easy-to-handle pelletizing systems with outputs ranging from 100 to 500 kg/h and for special cases up to 1,000 kg/h. The following systems have, on the whole, been generally accepted:

21.4.9.1 Strand pelletizing

In spite of its relatively large space requirements and its high noise level, which make it very intrusive, this system is still of great importance. Most probably this is because it is simple to operate and therefore suitable for use by unskilled personnel. Figure 51 A shows the process schematically and Figure 51 B the principle of a strand cutter. Key features of this system are the following:

- die head with holes in line or on a circle (flat or round die),
- thermostatically controlled cooling-water temperature,
- powerful blower (water removal),
- strand pelletizer with collector intake and large number of blades (low rotor speed even at high throughput),
- sound proofing.

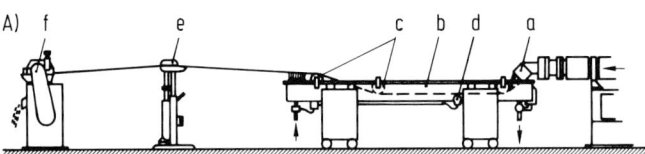

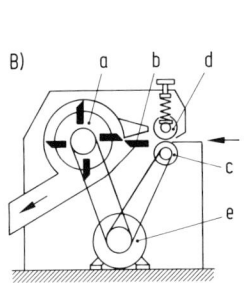

Figure 51 A). Strand pelletizing – schematic
a die head, *b* cooling bath, *c* strand guides, *d* thermostat,
e blow dryer, *f* strand pelletizer

Figure 51 B). Strand pelletizer – cutting unit
a cutting rotor, *b* stator knife, *c* intake roll, *d* pressure roll, *e* drive

The strand take-off speed should not exceed 60 to 70 m/min and the number of strands should not exceed 40. Pellets from the strand pelletizer are cylindrical. The pellet diameter is determined by the take-off speed. Pellet length can be altered by changing the speed (change wheels or adjustable drives) of the intake rolls.

21.4.9.2 Die-face pelletizing

There are two basic types of cooling and conveying system:
− air-cooled systems
− water-cooled systems.

21.4.9.2.1 The air-cooled die-face pelletizer

Normally, this is a simple unit with a limited range of applications (HD and LDPE) and low throughput rates (approx. 100 kg/h). Operation is similar to that of the central cutter used in PVC pelletizing.

The water-cooled die-face pelletizer is used more in reclaiming than for any other purpose except for PVC pelletizing, and is the most popular face-pelletizing system. The principle of operation of all systems is similar. The rotor cuts the melt strand directly at the die and throws the pellets into a water trough or water ring. With some designs a water vortex is generated in the rotor chamber to prevent the molten pellets from sticking together. On demand, cooling water conveys the pellets through special cooling channels into the dryer. Afterwards a cyclone separates the pellets and the residual heat of the material provides final drying. Such a system is shown schematically in Figure 52.

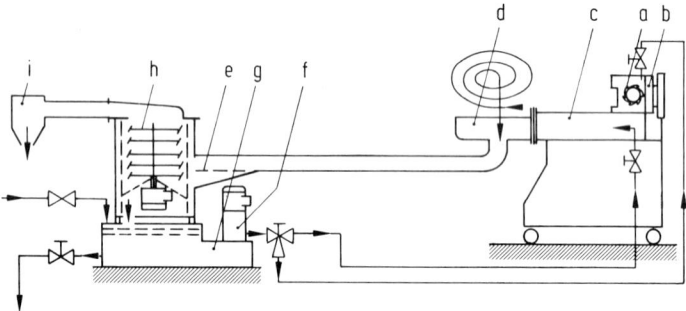

Figure 52. Principle of a die-face pelletizing system with water circulation and drying
a cutting unit, *b* die head, *c* water trough, *d* spiral cooler, *e* water separator, *f* circulating pump, *g* water tank (heatable), *h* dryer, *i* cyclone

The following die-face pelletizers are distinguished by the design of cutting device and water supply:

21.4.9.2.2 The water-cooled die-face pelletizing system with central cutting rotor

A water mist is generated in the rotor chamber to prevent the pellets from sticking together. Afterwards they are thrown into a water trough and cooled (Figure 53).

21.4.9.2.3 The water-cooled die-face pelletizing system with off-axis rotor cutting

This is similar to the preceding system but die- and rotor axis are displaced parallel to one another. This strongly affects the designs of die head and knife (Figure 54).
With the central method of cutting, strand holes are located in one or more concentric circles. This calls for bigger discharge surfaces and a larger die head. However, knife guiding and knife manufacture are simpler, because spring knives directly touching the die face can be used.

21.4 Typical line design concepts and line components

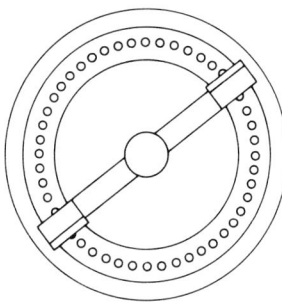
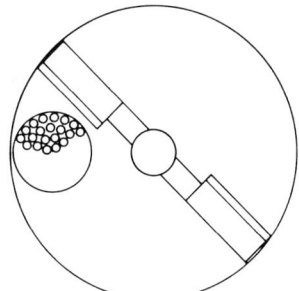

Figure 53. Central die-face pelletizer – schematic

Figure 54. Parallel-axis die-face pelletizer – schematic

On the other hand the parallel eccentric design has a simpler die plate (comparable to a strainer plate). Therefore it is possible to locate many strand holes on a small cross-section and to have smaller discharge surfaces and a smaller die head. With this principle of operation the problem is the cutting rotor. It has to be relatively big and must not touch the die plate (precise gap setting). The principal use of both systems is to pelletize LD and HDPE.

21.4.9.2.4 The water-cooled die-face pelletizing system with milling cutter rotor, type LHWG (Figure 55)

A helical milling cutter or a rotor with built-in blades is used as the cutting rotor. The die head strand bores are placed in line. The die head consists of the die body (*b*) and the heat-insulated die plate (*c*). The cutter unit (*d*) is positioned in line with the die head on two slide bars. By taking a light milling cut on the face of the unhardened die plate, simple

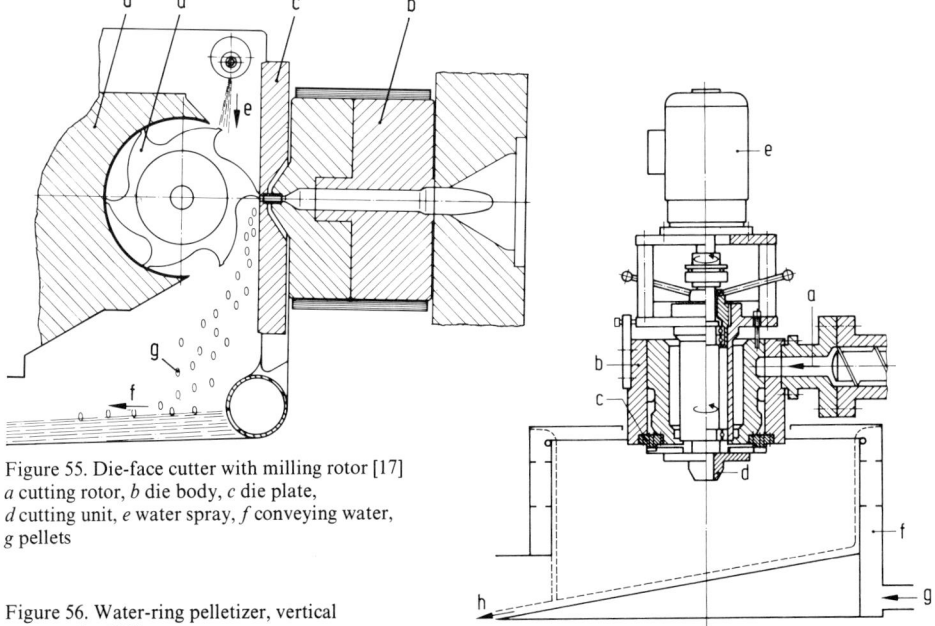

Figure 55. Die-face cutter with milling rotor [17]
a cutting rotor, *b* die body, *c* die plate,
d cutting unit, *e* water spray, *f* conveying water,
g pellets

Figure 56. Water-ring pelletizer, vertical construction [18]
a melt supply, *b* die head, *c* die plate, *d* knife head, *e* drive, *f* water-ring housing, *g* water inlet, *h* pellet/water-mix outlet

and reliable adjustment of the cutting unit is guaranteed. This type of unit is suitable for pelletizing almost all thermoplastics, excluding some PA and PET types and rigid PVC. Throughput rates are from 10–1,000 kg/h depending on type of material.

21.4.9.2.5 The water ring pelletizer

Figure 56 shows the schematic layout in section. Melt outlet may be either vertical or horizontal. Cutting is carried out centrally in both cases by rotating knives that are either springloaded against the die face or adjusted to run with a small clearance. This system usually requires no water mist in the rotor area. After cutting, the pellets are thrown tangentially into the water ring, which is circulating in the water housing at high speed. From this stage, as in all other systems, cooling and transport to the dryer takes pace. This system is widely applicable and, excluding PVC, suitable for nearly all thermoplastics. For economic reasons it is used only for output rates of 500 kg/h upwards. The pellets produced on die-face pelletizers may be lenticular, spherical or cylindrical, depending on polymer type. Pellet length can be changed by altering cutter speed, and diameter by changing the die plate. For a more complete survey of criteria relating to individual systems see Table 8.

Table 8. Survey of throughput rates, polymers and pellet dimensions for pelletizing systems

System	Throughput rate kg/h	Polymers	Pellet diameter mm	Pellet length mm	Suitable for brittle materials	Suitable for soft materials	Pellet form
Strand pelletizer	5 to 1000	○	1 to 6	1 to 6	–	–	C
Hot face air-cooled pelletizer	20 to 100	○	4 to 6	1 to 2.5	+	–	L
Hot face water-cooled pelletizer with central cutting unit	20 to 600	◒	2.5 to 6	1 to 3	+	–	L/C
Parallel-axis cutting unit	20 to 3000	◑	2 to 6	1 to 3	+	–	L/C
Milling cutter unit	10 to 1000	◔	1.5 to 6	1 to 3	++	++	L/C
Water-ring pelletizer	500 to 5000	◔	1.5 to 6	1 to 3	++	+	L/C

The figures in the table are approximate.
○   practically all thermoplastics
◒   LD and HDPE
◑   LD and HDPE, poss. PS, ABS and similar resins
◔   almost all thermoplastics except PVC and special PA-PET types
C   cylindrical pellets
L/C   acc. to type of polymer, lens-, spherical- or cylindrical pellets
++   very suitable
+   suitable
–   not or only partly suitable

21.5   Energy balance – specific energy consumption

Extrusion plants, especially reclaiming lines, are great energy consumers. By choosing the wrong equipment, or unsuitable screw geometries, or poor operational conditions e.g. wrong temperature profile for a certain polymer or over-long intervals between screen changes, the energy consumption may be unnecessarily high. The energy flow diagram in Figure 57 shows where energy consumption and energy losses arise during the operation of a reclaim plant (excluding washing). The biggest consumers in the plant are the extruder (approx. 28%) and the granulator (approx. 12%).

21.5 Energy balance — specific energy consumption

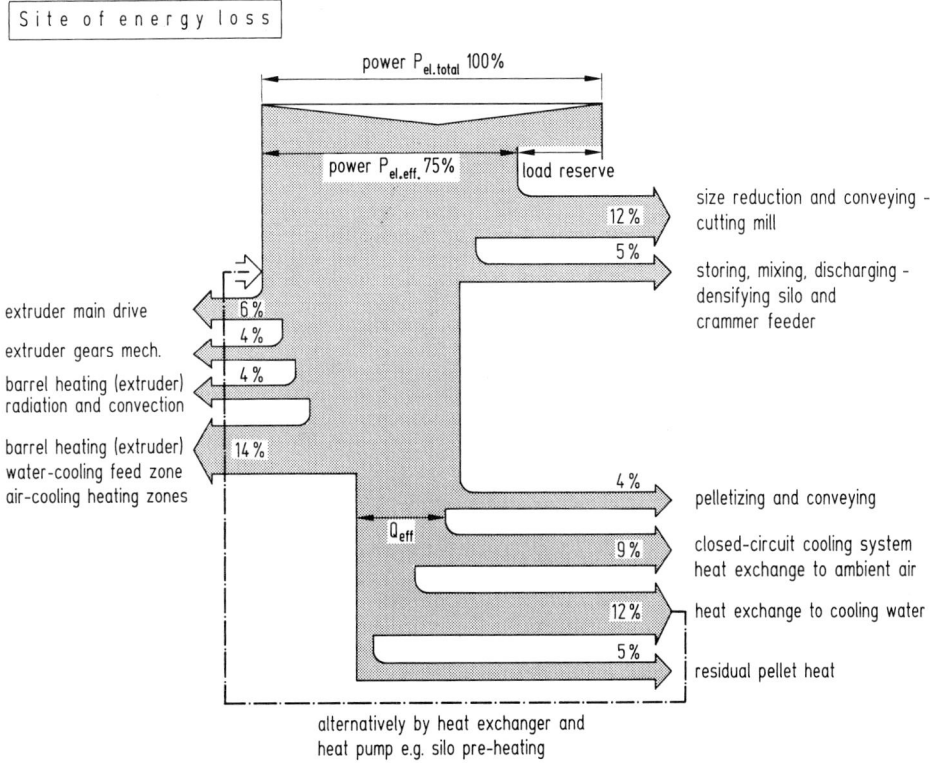

Figure 57. Energy flow diagram of a reclaim extrusion line

21.5.1 Energy consumption in size reduction

Figure 58 shows the specific energy consumption of different waste materials. Material "a" needs approx. 0.03 kWh/kg as against material "e" with 0.1 kWh/kg, a difference of more than 300%.
Possibilities for energy saving:

- use of a granulator with double helical cut,
- high-rate removal of granulate,
- correct choice of screen hole size,
- monitoring knife wear and re-sharpening.

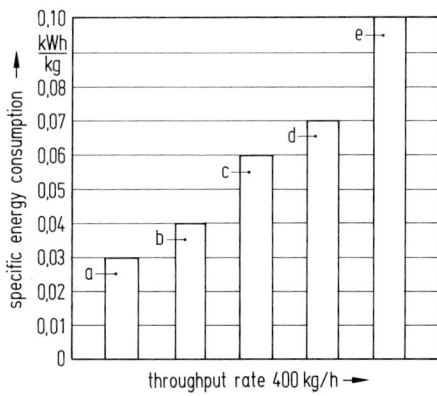

Figure 58. Throughput rate and energy consumption for grinding various types of plastics
a PS sheet, b PVC thermoform sheet, c PE pipes, d PE film 100 µm, e PP film 50 µm biaxially oriented

21.5.2 Energy consumption in extrusion

The material dictates the general level of energy consumption, which can be affected positively or negatively in a number of ways. The relevant factors have been investigated and quantified in a series of practical trials [19].

21.5.2.1 Specific energy consumption versus screw speed with different polymers and extruder systems

Figures 59 and 60, respectively, show the throughput rate and specific energy consumption in relation to screw speed for single- and twin-screw extruders.

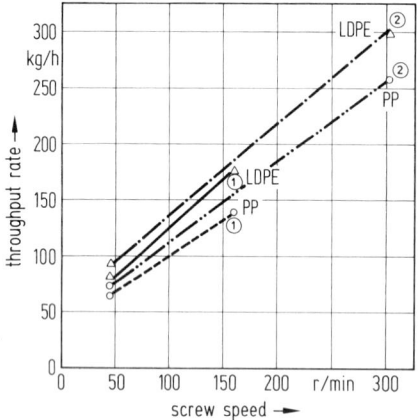

Figure 59. Throughput rate versus screw speed for a single screw (1) with tapered feed zone 70/120 mm diameter and co-rotating twin screws (2) 67 mm diameter
○ = PP film 80 µm, △ = LDPE film 60 µm, flake size approx. 8 mm

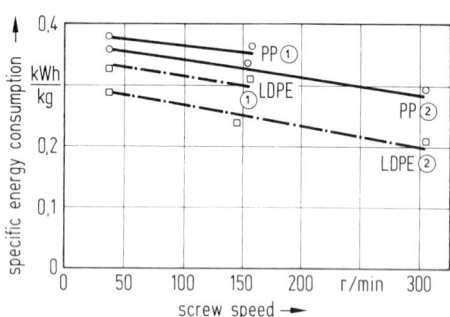

Figure 60. Specific energy consumption for drive and heating of twin screws 67 mm diameter, (2), and a single screw with enlarged feed zone, (1), in processing of: LDPE-(□) and PP flakes (○)

Figure 60 may be interpreted as follows:

- specific energy consumption with PP film scrap is approx. 23% higher than with LDPE,
- specific energy consumption of the single screw is approx. 23% higher than that of the twin screws,
- specific energy consumption is reduced by up to 25% at higher screw speeds (throughput rates).

21.5.2.2 Specific energy consumption versus flake size

Flakes of different sizes were processed under identical extruder load conditions (approx. 90%) and the energy consumed by the drive and by barrel heating was recorded. The results, shown in Figure 61, indicate that the specific energy consumption is almost unchanged for flake sizes from 6 to 12 mm. Above 12 mm it increases significantly. The twin screw reacts more than the single screw to bigger flakes.

21.5 Energy balance – specific energy consumption

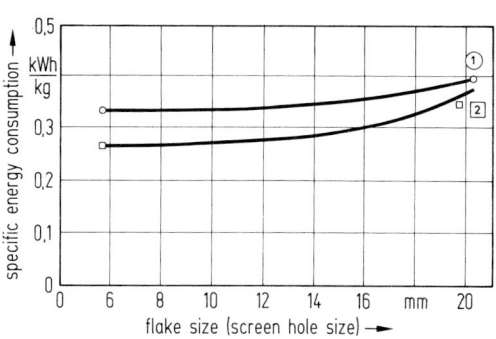
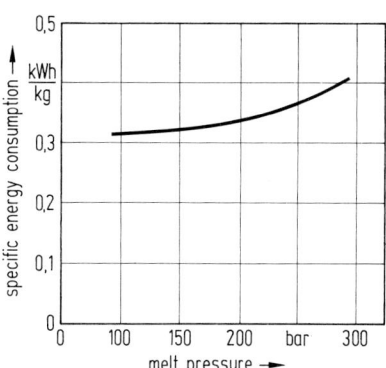

Figure 61. Specific energy consumption versus flake size (*1*) Single screw with enlarged feed zone 70 mm diameter, (*2*) Twin screws, co-rotating, 67 mm diameter

Figure 62. Specific energy consumption versus melt pressure. Single-screw extruder 70 mm diameter, 80% load LDPE agglomerate, MFI 190/2, 16 = 0.6

21.5.2.3 Specific energy consumption in relation to melt pressure

The results of the trials (21.2.1.3) on the decrease of the throughput rate at higher melt pressures suggested that this effect would be accompanied by an increase in specific energy usage. This was confirmed by relevant trials (Figure 62) whose results show some 25% increase in specific energy consumption (0.32 to 0.4 kWh/kg) between 100 bar and 300 bar melt pressure.

21.5.2.4 Other influences on the specific energy consumption

The easiest way to reduce the specific energy consumption, and at the same time the most neglected one, is to have the right barrel-temperature program.
Normally the range of temperatures is higher on reclaim extruders than on film blowing extruders, and higher melt temperatures result in a lower specific energy consumption. Also, attention to screw geometry is important.

Final evaluation of the energy balance in the reclaiming process

Approximately 30% of reclaim production costs are energy costs. The investigations described show that these costs can be favorably influenced by the right choice of plant and machine design, and operational method. This means:

- using the right temperature program
- operating in the optimum load range
- using the most suitable screw configurations
- having the right particle size
- operating in the right melt pressure range.

Further energy-saving measures, like reducing radiation losses and recovering heat from cooling water, will need to receive more attention in future.
If suitable heat-recovery systems were used, energy recovery of up to 15% on the extrusion plant could be realized, and used for heating the polymer before it is fed to the extruder.

References for Chapter 21

For the convenience of the reader the English titles of all publications in languages other than English are shown in parentheses.

[1] *Leschauski, K., Gorzitzke, W., Röthele, S.:* Forschungsprogramm Wiederverwertung von Kunststoffabfällen (Research Program, Recycling of Plastics Waste). Teilproject 3. VKE, Frankfurt (1981), p. 39.
[2] *Hensen, F., Knappe, W., Potente, H.:* Handbuch der Extrusionstechnik, Teil 1 (Plastics Extrusion Technology, Part 1). Carl Hanser Verlag, München, Wien, 1986.
[3] Unpublished Experiment Reports, Leistritz GmbH Maschinenfabrik, Nürnberg, West Germany.
[4] *Rauwendaal, J.:* Analysis and Experimental Evaluation of Twin-screw Extruders. Polym. Engin. and Science 21 (1981), pp. 1092/1100.
[5] *Tenner, H.:* Verwerten von thermoplastischen Kunststoffabfällen (Recycling of Thermoplastic Waste). VDI-Verlag, Düsseldorf, 1979, pp. 119/121.
[6] *Jenne, H.:* Verwerten von thermoplastischen Kunststoffabfällen (Recycling of Thermoplastic Waste). VDI-Verlag, Düsseldorf, 1979, p. 193.
[7] *Jenne, H.:* Verwerten von thermoplastischen Kunststoffabfällen (Recycling of Thermoplastic Waste). VDI-Verlag, Düsseldorf, p. 184.
[8] DE-PS 2837135 C2. (1984) Bischof & Klein GmbH & Co., Lengerich, West Germany.
[9] DE-PS 3334394 2 (1983) Leistritz GmbH Maschinenfabrik, Nürnberg, West Germany.
[10] Company brochure, FN, Herstal, Belgium.
[11] Company brochure, Öswag, Linz, Austria.
[12] Company brochure, Leistritz GmbH Maschinenfabrik, Nürnberg, West Germany.
[13] Information from Gneuß Kunststofftechnik GmbH, Bad Oeynhausen, West Germany.
[14] DE-PS AS 2502669 (1977) Werner E. Siemens, Espelkamp, West Germany.
[15] DE-PS 2324581 C2 (1983) Leistritz GmbH Maschinenfabrik, Nürnberg, West Germany.
[16] DE-PS 3239030 A1 (1983) Josef Gail, Aichach, West Germany.
[17] DE-PS 2809680 C2 (1983) Leistritz GmbH Maschinenfabrik, Nürnberg, West Germany.
[18] DE-PS 3116153 C2 (1983)
DE-PS 3116117 C2 (1983)
DE-PS 2825639 C3 (1981)
DE-PS 2825288 C3 (1983)
Hermann Berstorff Maschinenbau GmbH, Hannover, West Germany.
[19] Unpublished Experiment Reports, Leistritz GmbH Maschinenfabrik, Nürnberg, West Germany.

Index

abrasion 473
Accuflow coextrusion feedblock 245f
accumulator head blow molding machine 427f
accumulator head with ring (annular) piston 365
acoustic insulation with foams 443
acrylonitrile, residual content 42
adaptive control 192
adaptor blocks 570f, 574
adaptor dies 246, 272f
adaptor, pressure measurement in 320
adaptor process – see feed-block process
additive feeding and mixing systems 567f
additives 26, 29, 431f, 445ff
adhesion
– in coextrusion 412
– – – coating 332
– – – – active factors 335ff
– promoters 237, 333, 336, 412f, 433
– promoting groups 333
– promotion by gas 139
adhesive bands 261
adhesive coating 348
adhesive layers 125
adhesive sleeves 75
adhesive strength, improvement 333
adjustment drives, hydraulic servo- 420
adsorption wheel for air drying 660
after-shrinkage 108, 206
air cooling ducts 88
– – in elastomer extrusion 549
– – ring 463
– – zones 71
air drying with adsorption media 659
– – with adsorption wheel 660
– –, pellet dryer 48, 660
air jet 273
– – for drying monofilaments 322
air knife 149, 166, 181f, 186f, 191, 267
air showers 149, 267
air-blower nozzles 86
air-cooled die-face cutter 39
– pelletizer 39
air/water pelletizing equipment 48
alarm procedure 328
aliphatic hydrocarbons as blowing agents 434
Alkor process 139
alloying 26, 41
allyl compounds, crosslink promoting 492
Alquist drive on winders 468
Anaconda process (MDCV process) 506
angled extrusion heads 530
anisotropy of strength 286
annealing rolls 182

annular-piston accumulator heads 408, 415, 416ff
– – with bolt mandrel support 416
– –, capacities 419
– – with overlapped heart curves 419
– –, ring channel fed 417
– –, side-fed, with guide ring 418
– –, two-stage heart curve fed 417
anti-blocking agents 172
antioxidants 472
antistatic treatments 187
antistats 172
anti-stick agents 550
apparent viscosity 146
applications
–, coextruded films and sheets 223
–, crosslinked PE 490
–, extruded films and sheets 221f
–, extrusion welding 634
–, film tapes 287f
–, foamed materials 443
Armocel process 450, 482
aroma barriers 161f
atactic component in PP 260
audio film 261
Auma machines 547
Autoflex die 230f
autogenous working of extruders 513
automation 52f, 74, 141, 191
– in fiber extrusion 612f
– of film blowing 119f
– of film-tape lines 305ff, 317
– of sheet extrusion 249
– of start-up and shut-down 52f
auxiliary extruder 237
auxiliary machine 31
availability of compounding lines 50
awning cloth 309
awning fabrics 285
axial winder 107f (see also central winder)
axis crossing 211
azo-compounds as chemical blowing agents 474

back flow on extruder screws 668
backflow-check flights 381
backward degassing 29f
bagging film 113
baler twine, 285, 286ff, 295
ballotini 540
banana rolls 183
Barfilex process 299
barrel
– clamps for die heads 524

barrel
- construction materials 100, 374
- cooling 484
- - by individual blowers 375, 462
- - by liquid 375
- degassing 225
- design for elastomer extruders 524
- design for recycling extruders 689 f
- grooved in feed zone for recycling 683
- insulation 565
- liners 523 f
- temperature control 525
- - - by liquids 461
- - - with oil 484
- with wet liner for elastomer extrusion 524
barrier films 129 ff
- layer 138, 140, 160
- materials 127 ff, 130
- -, properties in coextrusion 130
- properties 96 f, 262, 331, 412
- - of stretched films 262
barrier screw 387
BASF, foam extrusion process 450 f
basket height control in film blowing 120
batch-off lines 550 (see also sheet cooling lines)
bead at roll gap entry 205, 207 f
beaded socket 76
belt drives, adjustable 373
belt drives for elastomer extruders 522
belt take-off 83, 93, 466
bending-compensated lamination equipment 359
bending-compensated roll-pair 359
Bexphane process 280
biaxial drawing lines 263 f
- - of PP 261 f
biaxial stretching of sheet 248 f
biaxially oriented film 257, 261
- -, precursor casting lines 198
bimetallic cylinder 317
bimetallic feed-zone bushing 524
Binder-Schmidt process 72
Biot number 69
Black Box process 246 (see also feed-block process)
blade slitting 108, 239
blades, heated 239
blind-grooved section 386
blocking of blown films 98, 106
block calibrators 86, 88
blocking tendency 335
blow molding
- - extruder, grooved feed design 380 f
- - machines 423, 425 ff
- - -, extrusion dies 388 ff
- - -, mold-closing speeds 424
- - -, parison stretching and pre-welding 424 f
- - -, pinch-off bars 424
blow moldings, crushing resistance of 420

blow moldings, dimensions of 370
- -, multi-layer for hot-filling 413
- -, VC content 375
blow-up ratio 112, 127, 440 f
blower cooling of barrels 462
blowing agents 433 f, 444 ff
- -, aliphatic hydrocarbons as 434
- -, azo-compounds as chemical 474
- -, carbon dioxide as 434
- -, chemical 431, 447 ff, 473 f
- -, chemical, properties 435
- -, chlorinated hydrocarbons as 434, 474 f
- -, fluorinated hydrocarbons as 434
- -, hydrocarbons as 474
- -, injection of 437, 461
- -, nitrogen as 434
- -, physical, 431 f, 434, 446, 451, 465, 473 f, 483
- -, -, properties 474 f
- -, -, pumps for 465
- -, -, storage containers for 465 f
- -, tumbling 475
blowing heads 82
- - for coextrusion 139
- - for HDPE films 112
blowing mandrel 424
-, working pressures 111
blown film
- -, assessment criteria 97 f
- -, blocking 98, 106
- -, coextruded, properties 130 f, 135 f, 140
- - coextrusion 125 f
- - coextrusion, high spot dispersion 138
- - -, pressure measurement 141
- - -, processing temperatures 128
- - -, thickness control 119 f, 141
- -, convertibility 98
- -, dies 101 f, 293
- -, dimensional precision 98
- -, extrusion of 95 ff
- - lines
- - -, thick LDPE film 115
- - -, thin LDPE film 114 ff
- -, manufacture from recycled scrap 678 f
- -, optical quality 98
- -, pretreatment 109, 115
- -, recycling 678 ff
- -, stretching precursor 274
- - stretching process 280
- - tape lines 303
board 203 (see also sheet)
board, roll-stack process for manufacture
  of 203, 214 f
boil-shrinkage 311
- of HDPE monofils 316
- of PA6 monofils 313
- of PET monofils 312
- of PP monofils 315

Index

bombage 211
bonding agents 128f, 336, 412f, 433
bonding layers – see tie-layers
bowing of calender rolls 211
brakes in ram extrusion 622ff
breaker-plate 175, 229, 463
–, annular 62f
– mandrel support 62ff, 389, 413f, 528
bristles 309
bubble collapsing 105
– – devices 141
– diameter control 105, 119f
– profile 119f
– – with HDPE 112
– stability 96, 103, 113, 118, 127
bulk materials conveying, pellet and powder forms 646

cable
–, catenary form 500
– covering 93
– –, with elastomers 528f
– – lines 90ff
– –, pressure processes 529f
– –, tube processes 529f
–, heavy 93
–, insulation with crosslinked PE 489f, 497ff, 507
–, light 93
– strands, PP 295
calender 203, 207ff, 210, 211, 231, 247
– feed unit 27, 38
– feeding 38
– lines, arrangements 212
– for PVC/ABS processing 217
– roll 209f
– – adjustment 213
– – bending 210
– – bending, compensation for 210f
– – construction methods 210
– – gap loading 207
– – individual drives 211, 233
– – thermal control 213
calendering materials 217
calendering without roll bead 534
calibration 478ff
– baskets 105, 112, 115, 117, 120
– in blow molding 102
– – – – of HDPE 112
– – – – of LDPE 104f
– and blowing unit 424f
– dies, dimensions and positioning of 68
– equipment for foam extrusion 463f
–, external 65
– in extrusion of PVC foam 486
– forces 423
–, friction forces 486

calibration inserts 66
– mandrel 453, 455, 463
– of pipes 65ff
– with precooling 66
– of profiles 83ff
– techniques for precise container openings 365
calibrator separation 486
calibrators in foam extrusion 449f
calorific equation of state 409
can offtake (drawn yarn) 590f, 606
capacity measuring device 93
capstan haul-off 91f
– lines for cooling raw-compound strip 551
– take-off units 91, 498
carbon black in crosslinked PE 492
carbon dioxide as a blowing agent 434
carpet 309
carpet backing 285, 287f, 289
carpet-backing tapes 261
carpet staple fibers, production 600f
– – fibers, spinning machine 600
– warp yarns 289
– yarn 567
– – manufacture 600
– –, PP 583f
– –, spin-draw machine 598ff
– –, spinning line for three-color 587
Carreau equations 409
carrier-bag films 96, 113, 116, 132, 136
– – with PE outer layer 342
carrier material 128
carton board 332
– – profiles 85
cascade extruder 30, 145, 188, 227f, 267f
cascade filter 267
cascade temperature control 145
casting roll 148f, 168, 263, 267, 273
– – drive 182
– –, emboss design 194
– – surface finish 273
– – temperature 273
– – –, effect of 164f
catenary form of hanging cable 500
caterpillar take-off 72f, 83, 466
cavity (transfer) mixer 515
CCV (catenary-type continuous vulcanization) 499
cell growth 437
cell nuclei 477
cell size 433
cell structure 443, 473, 474f, 477f
Celuka foam extrusion process 449, 474, 479f
center-fed die 465
centering of extrusion dies 530
– screws 59f
– of spider-mandrel-holder blowing heads 101f
– of spiral-mandrel blowing heads 101f
central drive 190

central winder 106, 107f, 114, 153, 267
– –, computer-controlled drives 108
– –, principle of 107f
centricity measuring device 92
centrifugal drier 675
chalk as filler for waste 676f
chalk, coated, as filler for waste 473
chamber metering unit 659
chamber profile 80
cheese winding 312
chelators 472f
chemical resistance, extrusion welds 642f
chill-roll process 143ff, 203, 273
– –, for film tapes 307
– –, haul-off units 181ff
– –, roll arrangements 168
chlorinated hydrocarbons as blowing agents 434, 474
chromium plating of slot dies 229
circular knife 84
– –, rotating 239
– –, slitting 108
circular saws with hardened blades 469
– – with swarf removal by vacuum 239
circumference-driven winder 106 (see contact winder)
civil engineering, use of extrusion welding in 634
clamp
– closing speed in blow molding 424
– closing-speed control system 424
– force, specific, blow molding m/cs 424
– systems, for intermittent parison extrusion 424
– units 423ff
clamping systems for elastomer slot die heads 531f
clamps for continuous pipe-blowing units 423
– for extrusion blow molding 370f
cleaning of coextrusion dieheads 140
cleaning rolls 182
clips for stretching 274f, 278
clutch coupling 522
coater, for primers, lacquers, glues, dispersions 355f
coathanger dies 85, 229f, 464
coathanger version of spider dies 390
coating 143, 341ff
–, calender 247
–, extrusion 250f, 331
–, extrusion, lines 339ff
–, extrusion, typical products 340, 347
– lines, extruder throughput 338
– unit 343
coextruded blown-film dies, cleaning 140
coextruded blown film lines 137
coextrusion 82, 93, 95, 118, 143, 148, 229, 268f
– 271f, 449
– blow molding 369, 412
– of blown films 125ff
– of blown films, combinations 126

coextrusion coating 331
– of corrugated pipes 78f
– dies, blow molding 412ff
– dies, slot 178f
– dies, tubular film 139
– of flat films and sheets 245f
– of foamed films 458
– of foamed profiles 450
– of foamed profiles (2-stream) 479ff
–, operties of barrier materials for 130
–, quality improvements by 134
–, reduction of material costs by 131, 139
– of solid-skin/foam-cored profiles (2-stream) 479
– systems for blow molding 369, 412, 427
– of thin layers 127
cold-feed extruder 513f
cold-feed technique 678
cold feeding of elastomer extruders 513f
cold parison shearing 424
color pigments 433
coloring 140
compatibility of plastics in waste compounding 672f
composite corrugated pipe 78f (s.a. corrugated pipe)
– films 95
– profiles 80f
compounding 27 (see also waste compounding)
– extruders, single-screw 41, 50
–, filling and alloying when recycling 676ff
–, for film and sheet extrusion 223f
– lines 26ff
– –, feeding of 43f
– –, on-stream factor 50f
– PVC for foam extrusion 473
–, resin modification during recycling 682f
compression zone 565f
computer-controlled drives for central winders 108
computer monitoring on spinning lines 613
concentration, melt solution 26, 30f
condenser films 261
conductor ends, joining of 92
conical bore, impedance 404f
conical ring gap, impedance 404f
contact calibration device 247
contact cooling in elastomer extrusion 548ff
contact-cooling rolls 338f, 548
contact point of melt in extrusion coating 338
contact ring, thermostatted 463
contact roll for thickness measurement 555
contact time on cooling roll 238
contact winder 106, 112, 114, 153, 240f
– – with pre-hauloff 117
– –, principle (sketch) 107
– – with two winding stations 107
container, openings in 365
container, wall thickness distribution 420, 422
container, woven tape construction 288
contaminated waste 675f, 685

contaminated waste, washing mills 691
contamination, degree of in recycling 670
continuous parison extrusion in blow molding 370
continuous self-cleaning filter units 697f
continuous vulcanization (CV) 537ff
– – in pipes 539
contraction of melt between die and roll-gap 339
control,
–, bubble diameter 105, 119f
–, screw speed 176
–, stored program 250
– techniques for elastomer extrusion 552f
–, thickness, in flat-film extrusion 192
–, winder drive 155
conveying
– coefficient n of metering pumps 436
– control, metering flap-valve 655f
– granular and powdery bulk product 646
conveying-and-kneading sections 387
conveying systems, filter cleaning 655ff
– –, individual 651
– –, multipoint 649
– –, pneumatic 649ff
– zone 565
conveyor belts 309, 535f
conveyor units, single-machine 647f
cooled die-head process for solid rods 89
cooling
– air ring 102f, 112, 267
– bands 549
– blowers 103, 549
–, blown film 102
–, blown film, HDPE 112
– channels 92f, 461, 486
– – in pipe extrusion 68f
– – with telescopic section 92f
–, cold gas method in blow molding 425
–, effect on morphology 152
–, elastomer extrudate 547f
–, external in film blowing 102, 118, 141
–, extruded flat film 148
–, extruded PVC foam 486
–, film and sheet 237f
– grids 105f
–, hopper unit 525
– line
– – for intermediate elastomeric products 549ff
– – for profiles 83, 552
– – for rubber sheeting 551f
– – for tire treadstrip 552
– monofilaments 308
– ramp, vibratory 549
– rate 236
– roll 149f, 167, 182ff, 237f, 290, 338, 340, 548
– –, contact temperature 149f
– –, contact time 238
– –, contact type 339f, 548

cooling roll, dimensioning of 150f
– –, materials and surface finishes 237
– –, temperature for extrusion coating 336
– roll stack 236
– roller conveyor 237f
– spiral 99
– trough with conveying screws 549
–, water-bath method 274
copolyamides, coextrusion of 130
core plate 84
core profile 81
corona treatment 185, 333, 341
– – – of aluminum foils 335
co-rotating twin screw 29, 31, 37, 41f, 46ff, 50
– – – extruders, ZSK 44f
– – – kneaders 32f, 44
corrugated composite pipe, coextrusion of 78f
corrugated pipe 78ff
co-stabilizers 472
counter-bending, 210f
– force 210f
counter-rotating twin screw kneader 32
– –, meshing 32, 37, 48f, 50
– –, non-meshing 30, 49f
covering layer, coextruded films 160
Cr-Ni steels for extrusion pipes 621f
crammer
– or metering screw, coaxial with plasticator 684
– screws 40, 43, 46, 227 (see also stuffer screws)
– –, coaxial with plasticating screw 682
– –, horizontal 681
– units 663, 667, 676, 678f
– units for recycling 694
crank drives, electromechanical 423
crank sheet-cutter 239
crimper 591, 601, 606f
crimping of yarns 595
cross-cutting 239f
–, crank-type sheet cutter 239
– devices 108 (see also guillotine)
– saw 469
cross-flow heat exchanger 357
crossflow/mixing (CFM) extruder 518f
(see also pin-barrel exruder)
crosshead die, tandem set-up for elastomers 530f
crosslink indicator 497
crosslink-promoting allyl-compounds 492
crosslinkable PE (XPE), compounding 41
crosslinked PE, applications 490
– –, properties 489f
crosslinker 431, 445ff
crosslinking
– control 496f
– control by water supply in silane process 496
– degree 509f
– effectiveness, comparison 504
– in extruder barrel 447

crosslinking
- after extrusion, thermoplastics 489 ff
- before foaming 431
- and foaming in microwave or hot-air tunnels 445 ff
- -, free-radical/thermal 493 f
- -, heat of reaction in 494 f
- -, liquid bath 505
- in nitrogen pressurized pipes 498
- pipe
- - -, catenary 500
- - -, inclined straight 499 f
- - -, pressurized 497
- - -, vertical 499 f
- processes for PE 490
- - for PE, choice criteria 510
- -, radiant heat 501 f
- -, temperature, dwell-time, thickness effects 499
- -, UHF radiation 504
- -, UV light 507
cross-mixing 668
cross winding 326
crush resistance of blow moldings 420
crystalline structure, fine 167
crystalline superstructure 258 f
crystallization 65, 206
- -, degree of 260
- -, effect of cooling on 186
curl of film 131, 139, 162, 166
cutter in feed section for recycling 686
cutter with parting tools 74
cutting devices 73 f, 292 f, 296, 463
- - for flat films and sheets 239
- - for opening up film bubbles 453
- - for profiles 84
- - for staple fibers 608 f
cutting and edge-trimming equipment 347
- equipment for foams 469
- to length, foamed pipe 469
- to length, foamed PVC extrusions 487
cutting mills 663, 673, 690 ff
- -, double helical (or offset) cut 690 f
- -, energy usage 691, 703
- -, knife wear 665
- -, rotor designs 690
- -, sound insulation 690 f
- -, suction, effect of 663 f
- -, throughput 663 f
- of staple fibers 600 f
CV pipes 538 f
CV process for crosslinking of PE 499

dancer accumulator, horizontal 498
- arm 301
- roll control 353
dancer rolls 107 f, 117, 190, 239, 351, 537 f, 551
data gathering 53, 120 f

data gathering on film and tape lines 304
data gathering on monofilament lines 328
DC drives 174
- - with torque limit and smooth speed adjustment 484
- motors 372
- shunt-wound motors 360, 461
- winder drives 154 f
DDC (direct digital control) 192, 249
decomposition half-times for peroxides 493
decorative tapes 288
deflector bar - see turner-bar
deflector-mandrel die 85
deflector rolls 183
degassing 26 f, 29 ff, 37, 41 ff, 44 ff, 47
  145 f, 175, 245, 663, 669
- -, effect of degree of screw filling 669
- extruders 167 (see also vacuum extruders)
- extruders for elastomers 519
- screws for SB 167
delamination processes in weld- and pinch-off
  areas 412
densification rate of powders in ram extrusion 630
design, rheological 400 f
DFDR (elastically DeFormable Die Ring) 421
diameter control unit 91 ff
diamond-faceted mixing section 379, 385 f
die
die change in elastomer extrusion 536
die design
- -, blown film 101
- -, channel sections, parallel arrangement 403 f
- -, channel sections, series arrangement 403 f
- -, dimensioning, most important parameters 405
- -, dimensions 621
- -, equation for series connected channels 403
- -, foam extrusion 463 f
- -, foam extrusion, PVC 485 f
- -, pipe extrusion 65
- -, ram extrusion, surface quality 622
- exit gap, deckling of 177
die-face cutter 48
- -, air cooled 682, 700
- pelletizer 459
- pelletizing 700 ff
die-face underwater pelletizer
- - - with central cutter 701
- - - with milling cutter 701 f
- - - with parallel eccentric cut 701
- underwater pelletizing 663, 680 f, 700 ff
die gap adjustment 399
- - -, automatic 177
die gap variation 418
die-land zones 409, 414, 418, 463, 486
die lips 125
- -, flexibly adjustable 145 f
die/mandrel profiling in exit gap 420 f

die-mouth disk 84f
die plates 307, 318, 319
– –, for producing monofilaments 319
– ring, elastically deformable (DFDR) 421f
die wall, layer formation in ram extrusion 625
diffusion of gases from small to larger bubbles 477
digital controller (DDC) 554
dimensional control 119f
– – in elastomer extrusion 556
dimensioning of center-fed dies 390
dimensioning of cooling rolls 151
dip blow molding 363
dip cooling 72
dip cooling in elastomer extrusion 547
diphenyl, safety regulations 318
–, vapor as heat transfer medium 318, 570, 579ff
direct gassing 433, 446f, 451, 457f 471, 473
– –, metering physical blowing agent for 436
– – –, units 466
– – processes 431ff
– –, single-screw extruders for foamed film 454
– –, single-screw extruders for foamed LDPE pipe 453
– –, with twin-screw extruders 483
direct spinning lines 563
direct winder 107 (see also central winder)
discharge extruder 29
– screw 37f, 46
– screw, extended 689
disk calibrator 65
disk-element build-up of coating die-heads 530f
disk filter 582f
disk plasticator 685
dispersing 26f, 37, 41ff
dispersions, applicator roll-stack for 356f
displacement volume, gear pump 176
doctor blade 356
doff cycle, calculation of 603
doffer 614
doffing, automatic 613
dosing units – see metering screws
double-circuit air circulation system 326
double- or multiple-crosshead dies for elastomers 531
double- or multiple spinning pumps 319, 571ff
double-offset cut, in cutting mills 690f
double overhead take-off 92
double-roll haul-offs 458
double turret winder 467
double-wall rolls for polishing stacks 209
double winder 92f, 107
doubling 237
Dowtherm heating 570, 579ff
drainage pipes 78
draw
draw bath 311, 312, 316
draw-down calibration 87f

draw-down of melts 29, 96
draw ratio 274, 286f, 288
draw ratio, adjustment 327f
draw rolls 276
draw-twist machines 590f
draw twisting 589f
draw units for filaments 591, 600f, 606
drawability 259f
drawing – see also stretching
– baths 325f
–, biaxial 143, 248, 261
–, clips for 274f, 278
– and crimping units (crimpers) for staple fibers 591
– of fibers 595ff
– of filaments 577, 587, 595ff, 600f
– of film tapes 285f, 292f, 297f
– force 297f, 308, 323
– lines 274
– –, dimensions 278ff
– –, for neck-in-limited-drawing 281
– –, throughputs 278ff
– of monofilaments 308
–, neck-in restricted 281, 286
– process 260
– –, raw materials for 261
– of sheets 248
–, uniaxial 143, 273f, 285f
– units 297f, 310, 312, 323
– –, drives for 323, 324f
– – with seven godets 323f
– zone 276
drawing-office films 261
drawn films 156, 162, 260f (see also orientation)
– –, heat conditioning 267, 274f
– –, winding 282
drive 100
–, elastomer extruders 520
–, extruder 372f
–, foam extruders 461
–, output end of screw 686
– power for extruders 376f, 379
– –, extrusion blow molding of PO 379
– –, extrusion blow molding of PVC 376f
– –, grooved extruders 377ff
– –, tension controlled 360f
– torque with grooved feed zones 384
drum off-wind, traversing 498
drum wind-up unit 498
dry blends, for foam extrusion 475
dry laminating 347ff
– – equipment 359
dryer, forced convection type 356f
dryer-hoppers 173
dryers 356f
drying 341ff, 345f, 348
– with adsorption media 658f
– agent, silica gel as 657

drying cabinets 317, 657
- conditions 173
- with dry air 657 ff
-, effect on extruder throughput 567
- equipment for extruder feeding 656 ff
- media, molecular sieves as 657 f
- of monofils 322
- of pellets 172 f
- of pellets in monofil manufacture 317
- tunnels 347
dwell time
- - per unit of length 404, 409
- - in sintering zones (minima.) 626
- - in slot dies 147
- - spectrum 388, 398, 668 f

E modulus of film tapes 289 f
EATP test methods 301
edge guidance 105
edge pinning devices 182
- - air jet 181
edge strip 144 f, 263, 469
- -, recycling 172 ff, 290, 301
- - and roll cutting equipment 347
- - vacuum removal of 108 f, 183, 290
edge trim 167, 187, 239, 268, 340 f, 346 f, 466
- - from coextruded films 180
- - compounding 109
- - recycling 301
- - recycling, automatic 279
- - regeneration with coextruded flat films 163
elastomer
elastomer compounding 47
elastomer extruder 513 f
- -, barrel 524
- -, coating equipment 553
- -, cold feeding 513
- -, DC drives, thyristor controlled 519
- -, drives 520, 522
- -, fixings for die heads 524
- -, hopper unit 523
- -, hot feeding 513
- -, hydraulic motors 519
- -, temperature control 553
- -, thrust bearing 523
elastomer extrusion 513 ff
- -, clamping the die head 531 f
- -, control of product parameters 556
- -, control techniques 552 f
- -, cooling 547 ff
- -, feeding devices 536 f
- -, granule/pellet feed 536
elastomer extrusion head 525 ff
- - -, change 536
- - -, closure systems 532
- - -, crossheads, double or multiple 531
- - -, multitube 528

elastomer extrusion head, pork-chop 526
- - -, profile 528
- - -, slit-tube 529
- - -, tube 528
- - -, tube, two-layer 528
- -, measurement and control equipment 552 f
- -, pellet feed devices 536
- -, pin barrel 524 f
- -, pressure/speed control 553
- -, screen changer 535 f
- -, screws 524
- -, shear screws 515
- -, single screw extruder 520 ff
- -, slab-cutting equipment 537
- -, strip feed devices 537
- foamed products 541 f
- slot dies 531
- vacuum extruder 519
- web-thickness measurement 555
electrical insulating film 261
electrical insulation conduit, corrugated 78
electromechanical crank drive 423
electron accelerator 359, 508
electron-beam crosslinking 491, 497, 507
- dryer 358
- unit, capital costs 509
electrostatic charge, removal from films and sheets 248
- pinning devices 149, 182, 198
embossing 237
- of flat films 166
- of foamed PVC profiles 487
energy
- balance in recycling process 705
- flow diagrams for recycling 704
- input, specific 52
- requirements in drying with adsorption media 658 f
- usage
- - in extrusion of waste 704 f
- - in size reduction for recycling 704
Engel process 506
engineering plastics
- -, compounding 42
- -, waste compounding 676
environmental aspects 357
environmental problems in extrusion coating operations 336
EPDM cable mixes, compounding 50
equation of state, thermal 409
equilibrium moisture content in air 657
equilibrium swelling, check on crosslinking 497
ethylene copolymers, crosslinking after extrusion 489
ethylene, residual content 29 f
EVA blown film 97
- - -, coextrusion 126, 128
- - -, properties of coextruded 133 ff

EVK kneader, single screw 49
EVK screw 514
EVOH
–, coextrusion 127f, 130
–, properties in coextruded blown film 133f
exhaust gas combustion units 357
expanding mandrel 425
expanding and pre-welding device, blow molding 425
expansion ratio, linear 440
explosion safeguard 357
extension at break
–, of HDPE monofils 316
–, of PA6 monofils 313
–, of PET monofils 311f
–, of PP monofils 315
–, PP tapes 289
extrudate oxidation 333, 336f
extrudate swelling 60 (see also strand swelling)
extruder
–, autogenous working 513
–, cascade arrangement 30
–, for coextrusion blown-film lines 137
–, crosslinking in 447
– drive 372
–, drive capacity 377f, 379
–, drive for foam extrusion 459
–, for extrusion blow molding 371f
–, for extrusion welding 637f
–, feed for waste compounding 666f, 677f
– feeding
– –, drying equipment 656ff
– –, metering devices 654f
– –, with roll film for recycling 673f
– –, suction (vacuum) conveying systems 651
–, for flat film 163, 174
–, for flat film, throughput 164
–, with heat insulation 175
–, metered feeding of 654f
–, mixing mechanisms in 668f
–, for orientation lines 265
–, with pin barrel 517f
– platform, height adjustment 373
–, pressure profile 388
–, for profiles 83
– throughput, blow molding PO 379
– –, blow molding PVC 376f
– –, effect of packing density on 663f
–, water cooling 225
extrusion blow molding 363ff
– – –, clamp force, specific 424
– – –, clamp units 370f
– – –, coextrusion dies for 412ff
– – –, continuous extrusion 369f
– – –, economic importance 366ff
– – –, history of development 363ff
– – –, insert positioning 418

extrusion blow molding
– – –, intermittent extrusion 369ff, 414ff
– – –, lines, economic value mc/s (FRG) 366
– – –, lines, modular screws for 385f
– – –, part dimensions 370
– – –, PC 368
– – –, PE 368
– – –, POM 368f
– – –, PP 368
– – –, PPO, 369
– – –, PS and S-copolymers 368
– – –, PUR 369
– – –, raw materials 367f
– – –, surfboards 419
– – –, tube removal devices 424
– – –, twin-screw extruders 371f
– – –, wall-thickness control 419f
– – –, waste removal 427
extrusion calendering 216, 218, 227
extrusion coating 250, 331, 458
– –, environmental impact 336
– –, impingement point of melt 338
– – and laminating lines 349
– –, materials 332
– –, pretreatment 332f
– –, substrate 342f
– –, superficial weight 337
– –, web speeds, throughputs 337
– –, web width 337
extrusion dies, centering 530f
extrusion of elastomers 513ff
(see also elastomer extrusion)
– of flat films 259
– of foamed profiles 479
extrusion heads (die heads, dies) 84f
(see also slot dies, dies, extrusion dies)
– – for blow molding 388ff
– –, change devices in elastomer extrusion 535f
– –, disk construction type 530f
– – for elastomers 525ff
– – for multilayer parisons 413
– – for sheathing fibers and wires 531f
– – for tire profiles 532ff
– –, two-component 529ff
extrusion lamination 331, 339ff
extrusion laminator, single 349f
– –, with two rotation directions 349
extrusion lines
– –, feeding of 646ff
– – for stretched films 263f
– melt spinning process 307
extrusion pipes
– –, choice of material 618f
– –, chroming 619
– –, Cr-Ni steels for 618f
– –, heating 617f
– –, V4A steels for 618f

extrusion
- of pipes with foamed walls 482 f
- of plastics waste 662 ff
- of profiles, foamed 482
extrusion screws 27
- of sheet, materials processed 203
- speed, ram extrusion, max. with PTFE tubes 629
- speed, ram extrusion, effect on pressure 622
extrusion welding 633 ff
- -, applications 634
- -, economics 641
- - equipment, 639 ff
- - -, with extruder/melter chamber 637
- - -, hand transfer of extrudate 636 f
- - -, mobile 635 f
- - -, with mobile welding head 634 f
- -, examples of products 640 f
- -, melt overlay welding 644
- -, overlap welds 644
- -, preheating 639 f
- -, preheating with halogen lamp 637, 639
- -, quality assurance 643 f
- -, seams 640
- -, shrinkage cavities 639
- -, variants 633
- -, weld factors 641 f
- -, welding shoes 638
- welds, finishing of edges 640
- -, preparation 640 f
- of woodlike sheets from one melt stream 481 f

false twister 594 f
false-twist machines 594 f
fault analysis 53
feed
- angle 382 f
- bushes 62
- grooves (helical) 523
feed-metering screws 43 f
feed opening for direct gassing 461
- pockets 523
- ram 523
- rolls 297 f, 523
- screws, conical 673 f
- tension 297 f
- zone, wear 524
- zones, drive torque for 384
- zones, grooved 62, 99, 111, 137, 144 f, 174, 225, 293, 365, 372, 380 ff, 413 f, 461
- -, -, operating behavior 384 f
- -, throughput and torque requirements 62
feedblock 148, 178
- process 148, 178 f, 229
- process, combined with multichannel die 179 ff
- process, compared with multichannel die 180
- process, Dow Chemical 246

feeder extruder 413
feeding
feeding behavior, back-pressure independent 383
- compounding lines 43 f
- - - with twin-screw extruders 46 f
- devices 536
- devices on cold-fed rubber extruders 553
- of extruders 646 ff
- - -, drying systems 656 ff
- - -, metering equipment 654 f
- - -, suction (vacuum) conveying systems 651
- fillers and reinforcements to melt zone 689
- film and tape waste without size-reduction 681 f
- of pellet in elastomer extrusion 536
- screws, two-start 46 f
- of strips or slabs 537
felts 285, 288, 309
FEM for calculations on multi-dimensional flow 403
fiber extrusion 561 ff
- -, automation 612 ff
- -, filtration media 574 ff, 581
- -, heating systems 579 ff
- -, in-line operation 601 f
- -, melt filtration 579 ff
- -, melt temperature distribution 566
- -, mixing in of additives 567 f
- -, treatment section 562, 587 ff
fiber raw materials, properties 566
- waste recycling 683 ff
- - -, after size-reduction 681
fibrillation 311
FIFO principle 414 ff
filament 562
- comb 322
- crimping 611 f
- density, monofil lines 330
- draw units 593, 601, 608
- drawing 577, 587 ff, 594 ff, 600
- drawing and secondary treatments 601 f
- -, drawn, treatment section 562, 585 ff
- extrusion (see also fiber extrusion)
- -, deposits on spinneret plates 575
- -, monomer exhaust blower 578
- -, quenching 577 f
- -, spin finish applicators 611
- -, spinneret hole diameters 577
fiber orienting 589
- preorientation 587, 591, 591 ff
- quenching 589
- spin finish application 610 f
- take-up or spooling 589 ff
- yarns 567
fill-level constancy control 679
- control systems 656 f
- indicator signal 656 f
filled thermoplastics for films and sheet 218
filler 43, 224, 432 f

Index

filler
- addition 43 f
- addition to melt 689
- -, chalk as 676
- for foamed PVC materials 473
- grain structure 43
- metering 42 ff, 436
- -, mineral 677
- mixing 44, 245
- pourability 43
filler rod 636, 639
filling 41
- degree and effect on degassing 669
- degree and properties of regranulate 676 f
- plastics waste with mineral fillers 677
film 331
- -, biaxially oriented 257, 261
film blowing 95 ff
- -, automation 119 f
- -, bubble-basket height control 120 f
- -, calibration 102 f
- -, cooling 102, 112
- -, -, contact 280
- -, -, external 102 f, 118, 141
- -, -, internal 139 f
- -, -, water 119, 280
- -, creasing during 106, 111
- -, for film tapes 295
- - heads 101 f, 267
- - - for foamed material 453
- -, horizontal, with foamed material 452, 463, 483
- -, thickness equalization 109 f, 112
- -, thickness tolerance, control of 120
film, calendered 216
- -, circular knives for trimming 187
film, coextruded
- -, -, applications 223
- -, -, edge trim 179 f
- -, -, five-layer 131, 136, 246
- -, -, layer arrangements 160 f
- -, -, layer thickness measurement 250
film converting 331 ff
- drawing calender 207, 208 f
film, extruded
- - applications 221 f
- -, -, compounding of raw materials for 223 f
film extrusion, temperature profile 219 f
- -, thickness control 220
- -, foamed 442 ff
- -, foamed, extrusion of 463
- -, heavy-duty 96 f, 106 f, 113, 115
- -, -, sack, 113, 115
- -, -, shrink, 113, 116
- for jacketing glass jars 458
- -, melt-laminated 96, 113 f, 161 f
film, multilayer 125, 135, 143 f, 160 f, 246, 272 f
- neck-in 276

film, oriented 156, 161 f, 257 ff, 278, 278 f
- -, -, production line 264
- -, -, properties improvement 261
- -, packaging 96
- -, roll-stack process for manufacture of 203, 214 f
- -, roofing 535
- -, single-layer 218
- -, stretch, 161
- stretching line, schematic 266
- - - -, transverse drawing 276
- - - -, waste rework 267
- strips, removal by suction 185
- surface, pretreatment 184 f
film tape 260 f, 273
- - applications 287 f
- - cross-sections 287
- - drawing 285 f
- - E modulus 289 f
- - EATP shrinkage test methods 303
- - extrusion 284 ff
- - heating equipment 292 f, 298
film tape lines
- - -, automation 305
- - -, data recording 304
- - -, measurement and control equipment 305 f
- - -, process control 305
- - -, process monitoring 305 f
- - patents 284
- - production lines 299
- - raw materials 284
- - slitting 296
- - splitting 285 f
- - stretching 292 f, 296 ff
film thickness, control in flat film extrusion 193
- - distribution 220
- -, optimization 194
film waste
- -, mesh sizes used in processing 671
- - production line throughputs 302 f
- - recycling 678 ff
- - - with single-screw extruders 683 f
- - - with size reduction 680 f
- - - without size reduction 682
- -, textile, secondary processing of 285 f
- - winding equipment 292 f, 300
- -, world production of 284
- -, welding 634
- -, wide 113, 116
filmtruder 31, 41
filter 229
filter basket 35
filter-basket die, see breaker plate etc.
filter candles 581
- change 35 ff
- changing systems, data 698
- cleaning on conveying lines 653 ff
- cloth 175, 318

filter
- disk changer 685 f
- fabric 288
- inserts, cleaning 271 f
- media in fiber extrusion 574 f, 581
- media, sand 318, 574 f
- mesh, classification 671
- –, nonstop 581 f
- pack 307, 463
- pack, lifetime 582
- –, pressure build-up by contamination 581
- throughputs 698
filtration 26 f, 31 f, 37, 267, 269 f, 295, 318
- equipment 175
- equipment in pneumatic conveying 652 ff
- of melts, 31 f, 37, 574 f, 579 ff, 670 f, 694 ff
- unit, continuous 698
fine films 96 f, 107, 143, 203
finishing freshly spun filaments 611 f
finite element method (FEM) 403
fishing line 309
fishtail die 464
five-layer blowing head 139 f
five-layer coextrusion 127
five-layer films 131, 136, 246
flame retardant 433
flame treatment 333
flange spools, parallel winding 327
flash kettle 41
flash trimming 424 f
flat-belt drives for godets 324
flat disk screen changer 696
flat film 143 ff, 203 ff
- – applications 159
- –, coextruded 160 ff
- –, –, edge trim 163
- – coextrusion 245 f
- – cutting equipment 239
- – embossing 166, 194
- –, extruded 216
- – extrusion 143 ff, 192, 272
- – –, cooling 148
- – –, flow diagram 167 ff
- – –, lines 163
- – –, neck-in 182 f
- – –, process control systems 191 f
- – –, process stages 165
- – –, roll surface 166 f
- – –, roll temperatures 164, 167 f
- – –, temperature control units 182 f
- – –, temperature measurement points 191
- – –, throughputs 243
- – –, water-bath cooling 198
- – haul-off equipment 239
- – haul-off speeds 167
- – properties 158
- – quality, influences on 245

flat film, raw materials 156 f
- – – –, melt index 157
- – – thickness control, 96
- – – – equalization 168, 183, 240
- – – – measurement 193
- – – – tolerance 145
- – – transverse stretching 278
- – – waste 173 f
- – – winding 153
- – – winding equipment 188 f, 240 f
flat profile crosshead dies 530
flat profiles from rubber compounds 531
fleeces – see felts
flex dies 230 f
flexibly adjustable die lips 145 f, 192 f
flight surface hardening on screws 174
floor coverings
- – – in sanitation and agriculture 444
- – – in sports arenas 443
floor heating 490, 507
floorings 534
flow
- anomalies in HDPE 402
- channels
- –, annular cross-section, calculation 407 f
- –, circular cross-section, calculation 406
- directors, pressure conveying lines 650
- marks 101 f
- –, multidimensional, calculation 403
- pattern in a roll gap 207 f
- restrictor bushing 230
fluid-bed vulcanization 540 f
- – lines 538, 540 f
fluorinated hydrocarbons as blowing agents 434
fluoro-chlorohydrocarbons as blowing agents 474 f
flushing air 425
flying-knife 108
FM-3 computer 249
foam degassing 41
foam extrusion
- –, BASF process 450
- –, bubble slit into two webs 453 f
- –, calibration equipment 463 f
- –, calibrator 451
- –, Celuka process 449
- –, choice of pipe material and recipes 431 ff
- –, contra-rotating twin-screw extruders 484 ff
- –, cutting to length in 487
- – dies 463 f, 485
- –, extruder drives 461
- –, film blowing heads for 453 f
- –, foaming agents and additives 433
- –, haul-off units 466, 486
- –, hot-spots 433
- –, laminating equipment 468 f
- –, metering equipment 460
- –, mixing devices 460

foam extrusion
– –, mixing the raw materials 460
– –, multistage processes 447 f
– –, outer skin 474, 481 ff
– –, PE for 431 f
– –, PP for 431 f
– –, processing steps 430
– –, PS for 431 f
– –, PVC for 431, 471 ff
– –, PVC, calibration 486
– –, PVC compounding for 473
– –, PVC, cooling 486
– –, PVC, cutting to length 487
– –, PVC, dies 485
– –, PVC, haul-off equipment 486
– –, PVC, recipes 472
– –, PVC, selection criteria 487
– –, raw material feed 435
– –, Reifenhäuser process 449, 479
– –, screw design for twin-screw extruders 484
– –, screw lengths 461
– –, secondary operations 431
– –, single-screw extruders 430 ff, 449
– –, single-screw tandem extrusion lines 455
– –, single-stage process 444 ff
– –, temperature, pressure, gas evolution in 477 f
– –, twin-pipe die 464
– –, twin-screw, cylindrical 485
– –, two-stage processes 447 f
– –, waste, compounding 441 f
foam modifier 475
foam process, theory 477
foam process for thermoplastics 437
foam waste, re-extruded 432, 435
foam webs, windable 452
foamable PVC formulations 476
foamed films
– –, blowing process 452
– –, coextrusion 458
– –, direct gassing – single screw-extruder 454
– –, extrusion of 455
– –, laminating 441 f
– –, from PS 118, 444
– –, winding equipment 467 f
foamed hollow profiles 450, 480
– materials, for acoustic insulation 443
– – from crosslinked PE 447, 490, 507 f
– – from crosslinked polyolefins 447
– –, cutting equipment 469
– – for packaging applications 443
– –, stacking equipment 469
– –, thermal insulators 443
– –, waste compounding lines 441, 459
foamed pipe, cutting to length 469
foamed rubber 544
– – sheet 539
– – tubes 539

foamed semifinished products 443 f
– solid profiles, coextrusion 481 f
– thin films 453 f, 458
foaming
– aids 475 ff
–, crosslinking during 431
–, degree of 474, 486
–, free 478
–, inward 478 f
foil, aluminum, corona treatment 335
folding carton board 342
foodstuff packaging 413
force feeding, degree of 383
force transfer 100
form-and-fill films 96, 113
forward degassing 29
four-component coextrusion blowing head 138
fourfold off-wind unit 468
four-layer coextrusion 413
four-quadrant drives 360
– drives on polishing stacks 235
four-roll S-haul-off 466
four-wheel gear-pumps 574
Fourier number 69, 151, 388
– –, increase in conveying direction 388
free-running roll 183
freeze line 105, 112, 120
fresh-air dryer 657 f
friction 328
friction coefficient
– – between monofil and godet 323
– – between solids and feed-zone surface 382
– forces during calibration 486
– heat, from grooved bush 384
– wind-up 300
frost line 105 (see also freeze line)
fuel canisters with low surface resistance 412
Fujikura process (FZCV process) 505
furniture industry 449
Furukawa process 447

gap load 205 ff
gap load in calenders 207
gap load between cooling and pressure rolls 336
gap widths 125
gas crosslinking process 501 f
gas permeability 134
gas supersaturated solution in melt 478
gas tightness 118
gasketing 535
gear pump – see also metering pump
– – 32, 34 f, 40, 175 f, 263, 267, 265, 271
– – 296 f, 320, 455, 562, 571 ff
– –, metering 567
gel particles 263, 579
gel/sol test 497
gels 111

geo-textiles 288
glass fiber, metering 45
– – reinforcement 677f
– – rovings, mixing 47
– spheres, reinforcement with 677
gloss 472
glue laminating 346, 348
glues, applicator roll-stack 356f
godet duo 597
godet rolls 323, 324, 592, 595ff, 609, 610
– –, flat belt drives for 324
– –, heating of 608f
– –, surface of 610
– –, temperature profile on heated 609
– –, thermally controllable 292f,
– –, Vapotherm type 609f
– –, wear resistance of 613
grain structure of fillers 43
granulation 26f, 663, (see also pelletizing)
granulation in recycling 699ff
granulation of waste 671f
granulators, performance data 702
gravimetric dispenser 654f
gravure roll coating 341, 355f
grooved bush 99, 111, 293, 461
– –, cooled 380
– –, designs for blow molding extruders 381
– –, temperature control of 414
grooved extruder 372, 374f, 378, 380ff
– – drive power 385, 386f
– –, PE screws for mixed feeds 379
– – pressure profile 383
– – screw performance characteristics 385
– – screws for HM-HDPE 379
– – series, design data 386
– – throughputs 385f
grooved feed zones, operation of 384
grooved flight barrier 519
grooved rolls 325
grooving and longitudinal cutting equipment 455
guillotine 455
gusseting device 116

H4S rapid spin-draw process 596, 608
hand welding extruder 637f
haul-off
– belts 72
– caterpillar, speed control 500
– devices for
– – – blown HDPE film 112f
– – – blown LDPE film 105f
– – – foam extrudates 466
– – – foam PVC extrudates 486
– – – pipes 72f
– – – profiles 83f
– – – roll-cast films and sheets 181, 239
– – – sheets, drive power 241

haul-off ratio 440f
–, reversing or rotating 109f
–, reversing with turn-bar 110, 114
– rolls 105f, 340, 536
– speed
– –, cast flat films 167, 181
– –, control 90, 554
– –, filament wind up 591f
– – in pipe extrusion 68
–, for thermoforming films 186
– –, upper limits with LDPE blown film 104
– units, biaxially stretched film 198
hawsers 285 (see also ropes, cables)
HCV (horizontal continuous vulcanization) 499
HDPE
– blown film 96, 97
– – –, bubble haul-off equipment 112
– – – – shape 112
– – – coextruded, properties of 134f
– – – die (head) 111f
– – – lines 110f
– coextrusion 125, 128f
– flow anomalies 402
– monofilament, extension at break 316
– –, tensile strength of 316
– slip/stick flow effects 402
heart curve 417f
– – feed system 417
heat conditioning oriented films 267, 274
heat-initiated radical crosslinking kinetics 493
heat recovery 357
– removal 99, 381
– retention 487
heat setting 280, 324
– – of monofils 308
heating, diphenyl vapor (Dowtherm) 570, 579ff
heating, of slot dies 177
heating systems in fiber extrusion 579ff
heat transfer coefficient in cooling of pipes 72
heat-up program, DDC controlled on film line 191
heavy-duty bags 136
– film 96f, 106f, 110
– sack film 113, 115f
– shrink film 113
height adjustment, extruder platform 373
helical grooves 381f
Helicure vulcanization 538, 543ff
Henry's law 438
high-modulus tapes 288f, 295
– yarn 302
high-performance blown film lines, performance data 99f
– extruders 62 (see also grooved feed zones)
high-speed spinning 563f
– yarn take-up (winder) 602
high-voltage cable 497f
– tester 92f

Hitachi cable process 504
HM (high MW) films 110f
HM HDPE, staggered-leg spider dies for processing of 399
hollow articles, continuous blow molding of 78
hollow bodies, see blow moldings, containers
hollow profiles 80f
– –, foamed 450f, 480
– –, ram extruded 628ff
homogenization 26f, 29, 437
– section of twin-screw extruders 686
– zones 375, 565f
homogenizing extruder 29f
hopper
– coloring devices 656
– cooling 525
– degassing 29, 226
hopper-dryer systems 656ff
hopper metering devices 172, 654ff
– – and mixing devices 460
– units on elastomer extruders 523
hot-air dryer 657f
– ovens, monofil drawing 311ff, 315, 326
– shrinkage 597
– vulcanization 539f
– – lines 538f
– – –, air circulation with fresh air make-up 539
– – –, prewarming fresh with waste air 539
hot body with ceramic insulation 462
hot and cold mixer combinations 224, 475
hot cutting, profiles 469
hot-draw ratio (melt draw-down) 319f
hot drum 547
hot-feed extruder 511
– technique 678
hot feeding of elastomer extruders 511, 536
hot filling 412
hot-melt coating 331
hot-melt-fed extruder 265, 271
hot-sealability 331, 336f
hot-set test for checking crosslinking index 497
hot-spots in foam extrusion 433
hot wedge cutter 239
hot-wire cutting 424
hydraulic motors for elastomer extruders 520
hydraulic servo-adjustment drives 420
hydrocarbons, aliphatic, as blowing agents 474
– halogenated, as blowing agents 474f
hydrolysis in plastics recycling 662
hyperbolic winder 241

ICI process, bioriented films 280
idler rolls 610
immersion baths filled with anti-stick agents 550
immersion saws 74, 84
impact drier 356
– knife 74, 84

impact knife for flash removal 424
impact strength 96
impedance 404
individual conveyor units with central blower 647
individual drive, on calenders and polishing stacks 219, 233
induction measurement process 77f
industrial cloth 285
industrial fibers 583f
– –, H4S fast spin-draw process for 596
– –, spin drawing 596ff
– –, spinning lines 585
– –, spin-draw machines 597
inert gas blanketing, monofil extruder screws 318
– – –, spin die/melt grid 563, 578
injection molding process, structural foam parts 447
in-line operation in fiber extrusion 601
in-mold labeling 427
insert sensor for temperature measurement 639
inserts, positioning in blow molding 409
intensive melting section 389
intermediate products – see semifinished products
intermittent parison extrusion 369f, 414ff
internal air exchange 104, 115, 116, 125, 141
internal calibration 65
internal cooling in film blowing 102, 104, 139f
internal stress distribution in pipes 65, 72
internal stresses 206
– – in PTFE rods 628
ionizers 248
ionomer
–, blown film 97
–, extrusion 126, 129
–, properties of coextruded blown film 134f
IR (Infrared)
– equipment for continuous vulcanization 538, 543
– heating elements 547
– radiation for crosslinking PE 498
– radiation for equalization of stresses 206, 248
– thickness measurement 105
– thickness measurement units 250
iris diaphragm 103, 112f
island, slot die 230

jets, spin finish application 611
–, texturizing 611

K value of PVC for foaming 432, 472
Kanigen process 619
kicker (for chemical blowing agent) 435, 474
kneader 225
kneading blades 32
kneading disks 29, 32, 227
kneading elements in twin-screw extruders 686
kneading section, conveying 387
knead-stock 207f, 236

knife
- bars 296
-, chopping 108
-, circular 84
-, circular, rotating 239
- cylinder 299
knife-cylinder cutters 191
knife, saw-toothed 108
- wear in cutting mills 665
knitting tapes 288 f, 295
knot strength 307
- - of HDPE monofils 316
- - of PA6 monofils 313
- - of PET monofils 312
- - of PP monofils 315
Ko-Kneader 37 f, 42, 47 f, 48 ff, 227
Kombiplast 38 f

labeling in the blow mold 427
lacquer circulation with viscosity control 356
lacquer-laminating equipment 348 f
lacquering 331, 344, 346, 348
lacquering/laminating machines and products 350 ff
lacquers, applicator roll systems 356 f
lamellar packets 259
laminar distributor 229
laminating films 96, 113, 161 f
laminating media 331
laminating rolls, bending of 359
laminating stack 455
lamination 143, 237, 342 f, 345 f, 348 f
- equipment in foam extrusion 468 f
- by extrusion 342
- extrusion lines 339 f
- of foam films 441 f
- off-wind equipment 468
- with paper, moistening equipment 469
- with preheating 469
laminator roll-pair 338
laminators 247
large area filters 269 f
large blow molding machines 424
large blow moldings 366
laser systems for thickness measurement 555
layer thickness ratios 125
layflat width 115
LCM (liquid curing medium) 541 (see also salt-bath)
LDPE 29
- blown film lines 99, 113 ff
- - -, applications 95
- - -, coextruded 134 f
- - -, properties 96 f
- coextrusion 126 ff
- film blowing 95
- - -, haul-off equipment 105
- - -, haul-off speeds 104
- foamed semifinished product 442 ff

LDPE pipes, foamed 453
-, residual ethylene content 29
- two-layer films 125
leakage flow in spinning pumps 320
leakproofing in over-pressure calibration 67
lens pellet 672
lift, hydraulic 248
light stability, synergy 472
limiting surface tension of bubbles 438
line-contact pressure on polishing stacks 191
line pressure 336 (see also gap load)
linings, automotive 443
linings for wood and concrete pits 634
lip adjustment 146
liquid cooling of barrel 375
LLDPE
- blown film 97
- - -, coextruded, properties 134 f
- - - lines 117
- coextrusion 127 f
- compounding 30
long-life filter 294
longitudinal cutting equipment 108 f, 183, 185 f
longitudinal mixing 668
- - effect 388
longitudinal shrinkage 440
longitudinal stretching 277 f
- - of coextruded PP films 278
- - lines, films 263
- - -, monofilaments 314 f
low-temperature resistance of oriented films 261
LTM mixing zones 295 (see also Transfermix)
LTM screws 294
lubricants 131, 172, 432, 451, 676
-, external 473
-, heat transfer 506
-, internal 473
Luvitherm process 236

machinability 331
- of blown films 98
- of coextruded blown films 130 f, 135, 140
Maillefer screw geometry 29
Maillefer screws 388, 515, 518
man-made fibers 561
- -, applications 562
mandrel holder with offset spiders 398 f
-, movable, in ram extrusion 630 f
- movement, axial 463
- in ram extrusion 619 f
- retainer plate 84
- retainer, with perforated ring 62 f
- temperature control 390
manifold
- channel contour for center-fed dies 392
-, dimensioning 145 f
- system for slot die, design 230 f

manifolds for melt spinning 376 ff
marking devices 84
master computer station 121
masterbatches 447
materials properties – see LDPE, HDPE, PA, PP
mats 309
MDCV process (Anaconda) 506
MDPE, coextrusion 128 f
measurement and control
– – – equipment in elastomer extrusion 552 f
– – – – on film-tape lines 304 f
– – – in pipe extrusion 76 ff
– – – in profile extrusion 90
– – – technology 119, 191
– precision for blown films 98
measuring device for monofil diameter 328
melt accumulator 414
– –, separate 414 f
melt-application extrusion welding 643 f
melt contraction 339
– distribution manifold 568 f
– drawability 29, 96
melt-fed extruder 265, 271
melt filter 100, 263, 290, 295, 463,
　　663, 694 ff, 697 ff
– –, scraping devices 697 ff
– –, throughput 582
– filtration 32, 37, 574 f, 670 f
– –, in recycling 794 f
melt flow, components of 409
– fracture 111, 118
– grid 563
– index 96, 431
– manifold in coathanger dies 413
– mixer 269
– pipes 569
– preorientation 295
– pressure 319
– – control 90
– – measurement point 191 (see also pressure)
melt-pressure speed control 327 f
melt, quasi-Newtonian behavior 565
– roll-laminating calender 208, 246
– spinning 563
– –, distribution manifolds 568 ff
– strength 96
– temperature 119
– – control by smear head speed 546
– –, measurement point 191
– throughput on monofilament lines 330 f
– web 204
melting extruder 271
memory effect 204
metal detectors 537, 663
metal fiber felt as filter medium 581
metal foils 332
metal powder, sintered 581

metal separation 689, 692 f
metal separator 224, 663, 678, 680
metering 42 f, 435
– belt weigh-feeder 46 f
– devices for foam extrusion 460
– – in extruder feeding 654 f
– – on hoppers 460
– of fillers 245
– flap in conveying line 655 f
– in foam extrusion 484
– of ground waste 655 ff
– physical blowing agent for direct gassing 436
– pump 567 f
– pump with adjustable stroke 436
– screws 44, 174, 227, 646, 655 f, 676
– systems 654 ff
– unit for direct gassing 465
– –, chamber type 659
– weigh-feeder 46 f
– zone 565 f
MFI 95 f (see also melt index)
microcomputers 53, 177, 191, 193, 249, 327
microprocessor control for roller-head line 556 f
– – in sheet extrusion 249
microprocessors 92, 120, 147, 184, 249, 361, 366
microwave 501, 544
– equipment 545
– resonator chambers 544 f
milk films 136
mill goods (irregular granules from cutting mills) 672 f
milling cutter pelletizing 48
miniature extruder 637 f
miscibility, plastics in waste compounding 673
mixed wastes, compounding 681
mixer, dynamic 567
mixer, static 569
mixing 26 f, 31 f, 37 f, 41, 44, 47
–, of chopped waste 693
– devices for foam extrusion 460
– effects in extruders 668
mixing elements 27, 29, 111, 137, 144 f, 167, 174
– – 293, 317, 365, 377 f, 385 f, 389, 462, 566
– – on twin-screw extruders 686
– of fillers 245
–, hopper units 460
– rings 385
– screws 295
– technique for foamable PVC dry-blends 475
– tip 385 f
– woodflour into PP 45
MM-HDPE, blown films 97
model studies 383
modifier 432
modular construction 350
– – principle 47, 52
– screw for grooved extruder 385 f
– systems 520

moistening devices for paper laminating 469
moisture content, maximum with PET and PA 317
molecular chains, orientation 307
molecular sieves as drying media 657f
molecular-weight distribution 96, 260, 563
monofilament
- cooling 308
- cross-sectional shapes 307
- diameter control 328
- dieplates 318f
- dies 318ff
- -, heating of 318
- drawing 308
- drying 322
- extrusion 307ff, 565
- - process steps 310
- heat setting 308
- hot shrinkage 312ff
- knot strength 312f
monofilament lines
- -, data recording 328
- -, design by nomogram 329f
- -, hot-air ducts 327
- -, melt throughput 328f
- -, production speed 307
- -, suction cones on 327
- -, threadline density 329
- -, working widths 328
-, low pressure PE 316
-, polypropylene 315
- production processing parameters 312
- quench baths 322f
- single-stage drawing 314ff
- - - in hot air oven 315
- - - in water bath 314
-, two-component 309
- two-stage drawing 310
- winding equipment 326
- woven fabric 309
- wrap-angle on godet 323
monomer exhaust in filament extrusion 578
motor, AC shunt-wound 372
multichannel die process 178
multichannel dies 148, 229, 272f
multicomponent profiles 82
multicut Transfermix 517
multigap process to limit film-draw neck-in 281
multilayer adaptor 271
- blow moldings 412
- - -, continuous parison coextrusion for 365
- dies 245f
- film blowing heads 397f
- films 125, 143, 160, 246, 273
- films, polyolefin 135
- sheets 246
- stretched films 271
multilayer tube extrusion 412ff

multipacks 340
multiple dies in ram extrusion 617f
multiple extrusion dies 365
multiple processing, change of MFI and strength 674f
multiple-strand dies 90
multipoint conveying equipment 649ff
- vacuum conveying systems for powder and pellet 651
multipurpose extrusion coating and laminating line 347
multipurpose haul-off 347
multiroll application 341
multiroll winder 267
multistage foam extrusion process 447f
multitube die heads for elastomers 528
MVX kneader 49

neck-in 192, 285f
- during film stretching 280
- during flat film extrusion 182, 193
needle blowing process 424f
needle rolls 299
needles 292
nets 309
nickel plating, chemical 619
nip-roll haul-off units 84, 466
- - for double-wall ribbed sheet 261
nitrogen
- blanketing 172, 677
- - spin-extruder feed section 565
- as blowing agent 434
- pressurized pipe for IR crosslinking of PE 501f
Nokia process 503
nomogram for designing monofilament lines 329f
non-contact thickness measurement 120, 191
nonstop filter 581
nucleating agents 431, 438, 445, 473f, 477
- - and additives for foam extrusion 433
nucleation of PP 261f
nucleators 289, 474

offset spider heads 399f
- - - for HM-HDPE and PP 399
- - mandrel support
- - - -, rheological design 400f
- - - - in tube heads 399f
offsetting roll axes 211
off-wind unit for extrusion lamination 468
off-wind web tension, guide values 325
off-winding 340ff
oil, temperature control with 462, 525
online weigh-table 556
optical cables, extrusion lines for 90f
optical fiber 309
optical properties of blown films 98

Index

optimization of film thickness 194
– of process control in flat film extrusion 192
orientation 206, 258f, 263, 440
– in filaments 589
– of molecular chains 307
outdoor carpeting 287
outer layer, foamed 132
output 278, 328f (see also throughput, melt throughput)
overlap welding 643f
over-pressure calibration 65, 67
oxidation on extrudate surfaces 333, 336
oxygen 334
– permeability 163
ozone 333
– shower 333ff

PA
– coextrudate in blown film 119
– coextrusion 126
– copolymers, coextruded blown film properties 134f
–, extrusion blow molding resins 369
– fiber and film waste, compounding 677f
– flat films, water uptake 166
– moisture content, maximum 317
– pipes 57
– predrying of waste 677
– waste compounding 676f
PA6
– coextrusion 130
– extension at break 313
– fiber spinning grades 566
– monofils, tensile strength 313
– properties in coextruded blown films 134f
– recycling, effect of properties on 678
– spinning speeds 593
PA6.6
fiber spinning grades 566
– spinning speeds 593
package doffer, automatic 612
package forms 603f
packaging
– applications for foam materials 443
– fabrics 289
– films 96
– materials, manufacture by extrusion coating 332
packing density 382
– –, effect on extruder throughput 665
pad-chain haul-off 83 (see also caterpillar take-off)
paper 332
–, laminating with 469
–, upgrading 331, 339f
paperlike films 110
parallel connection of die zones 404
parallel winding 326
– – on flange spools 326

parison die – see tube die
– lengthening (sag) 414
– manufacture 389f
–, multilayer 414
– removal devices in extrusion blow molding 424
parting agent 551
pattern breaker 300
PC coextrusion 128f
– – for blow molding 368
– – control 361
PE
– blow molding extruder design 378
– for carrier bags 342
– coated aluminum tapes 342
– coated paper 342
–, crosslinkable, compounding 41
–, crosslinked, applications 490
–, crosslinked, properties 489
–, crosslinked shrink films and tubes 490
– crosslinking, carbon black in 492f
– –, CV process 499
– –, post-extrusion 489ff
– –, steam processs 498f
– – by UHF field 504
– – by UV light 507
– for extrusion blow molding 368
– for foam extrusion 431f
– foamed crosslinked products 447, 490, 507f
– foamed pipe, single-screw direct-gassing line 452
– pipes, crosslinked 506f
– shear-head die for crosslinking 506f
peel test 332f
pellet
– containing blowing agent, manufacture of 465
– cooling lines 551
– driers, hot air type 660
– feed equipment for elastomers 536
– preheating 172f 249
–, PS foamable 458
–, PVC extrusion foamable 475
– weighing as thickness control of blown film 120
pelletizer head for elastomers 526f
pelletizing equipment 35, 45, 48
– underwater 30, 31f, 35ff, 40
permeability to oxygen 132f
peroxide
– crosslinker 448f
– crosslinking parameters 495
– crosslinking of silicone rubber 543
– decay, rate constants 495
– radical polymerization of PE 491, 497
peroxides for crosslinking of PE 492f
PET
– bottles, stretch blown 367
– drying with low-humidity air 658f
– moisture content, maximum 317
– monofilaments, extension at break 311f

PET monofilaments
- -, heat shrinkage 313
- -, tensile strength 311 f
- pre-drying of waste 677
- spinning speeds 593
- as spun fiber raw material 566
- waste compounding 677 f
photo-crosslinking of PE by UV light 507
photographic film 261
photoinitiators 507
physical properties
- -, coextruded tubular film 136
- -, stretched films and film tapes 261
piggyback extrusion heads 532 ff
pigments for PVC foams 473
pin barrel for elastomer extrusion 524 f
pin-barrel extruder 516 ff
- - with cold feed 521
- - single-screw 49
- - throughput 519
pinch-off edges of blow mold 424
pinning aids 182
pinning devices, electrostatic 149, 182, 198
pipe
- applications 56 f
- calibration 65 ff
- cooling 71 f
- coverings 444
pipe dies 59 ff, 486, 616 f
- -, automatic centering 78
- -, design calculations 410
- -, inserts for 65
- -, operating ranges 60
- -, process parameters 411
- - PVC, rigid, DIN specifications 60 f
- -, sixfold, for ram extrusion 617
- -, throughputs 61
pipe extrusion 35 ff, 56 ff
- -, cooling zones 68 ff
- -, foamed-wall pipes (Armosic) 482 f
- -, haul-off speed 68
- -, line construction 634
- -, measurement and control techniques 76 ff
- -, ram method for PTFE, HDPE 628 ff
pipe, foamed 442 ff
- haul-off units 72 f
-, internal stress distribution in 65, 72
pipe loops in pressure conveying 654
-, ram extruded PTFE, UHMWPE 628 ff
pipe socketing, equipment for 75
pipe sockets, shapes 75
-, twin-screw extruder for 59
-, wound 57
Pirot process 280
planet-gear/single-screw extruder tandem units 448 f
planet-roll extruders 37, 48 ff, 208, 226, 227 f
- mixing section 377

planetary saw 74
plasma coating 610
plasticating screw for waste 682, 684
plastication 27, 32, 37, 41,
- chamber, extruder-like for welding 637 f
- zones with phase separation 388
- -, designs for grooved extruders 386 f
plasticators 37, 50
plastics, compounding of 26
-, - - engineering 41 f, 676
-, - - filled 41
-, - - reinforced 41
-, - - thermoplastic engineering 41 f
plastics, equilibrium moisture content in 657
plastics, grinding of, rates and energy needs 690 f
plastics melts, filtration of 26 f
plastics, recycling of 662 ff
plastics, sorption isotherms of 657
plastics waste, engineering plastics, compounding 676 ff
- -, extrusion 662 ff
- -, filling with mineral fillers 677
- -, recycling unit 682
- -, reinforced with glass fibers/spheres 677
plastification, plastifying - see plastication
Plastiscrew 515
PLCV (pressurized liquid salt cont. vulcaniz'n) 506
PMMA, coextruded facing layers 82 f
pneumatic conveying equipment 647 ff
- - -, filter units 652 ff
pneumatic thickness measuring system 556
polishing rolls 186 f
polishing stack 203 f, 205, 207 f, 210 f, 211, 219 231 f, 447
- - central drive 232 f
- -, column guidance of rolls 234
- - cooling 236
- - deformations 231 f, 233 f
- -, double-wall side supports 234
- - drive 235
- - drive, four quadrant 235
- - drive, individual 211, 231 ff
- - drive power 241
- -, floating roll bearings 235
- - gap loading 244
- - heating capacity 241 f
- -, lamination and coating attachments 237
- - line compression 191
- - roll arrangements 211, 235 f
- - roll, double-wall 208 f
- - roll setting 213
- - roll temperature control 213
- - safety devices 248
- - wrap angle 235 f
polyester
-, coextrusion 128, 130
- films, oriented 278

polyethylene – see LDPE, HDPE, HM-HDPE
polyethylene, high-molecular (HM-HDPE) 386
polymer modification during recycling 682f
polyolefin (see also LDPE, HDPE, HM-HDPE, PE, PP)
–, calibration dies for extruded 66f
–, compounding lines 27ff
–, foamable materials 447f
–, foaming 434
–, multilayer films 135
–, pipe extrusion lines 58, 61
–, pipes 57f
–, powder, compounding 32
–, range of pipe dies for 64
–, screws for 379
–, three-zone screws for blow molding 378
polystyrene and S-copolymers, compounding 41f
POM for extrusion blow molding 368f
Pope winder 354
pore-size regulators 446, 473, 474
pork-chop extrusion head for elastomers 526
post-cooling equipment 427
– haul-off 183
– rolls 148, 182
– sections 88
post-foaming 441
– duct 454
post-homogenizing extruder after kneader 514, 515f
post-lacquering unit 343ff
post-shrinkage 111
post-stretching 278, 309, 312
pourability of fillers 43
powder applicators 550
power usage, specific 83
POY (PreOriented Yarn) 591f
–, spool production, texturing machine 594
PP
– carpet yarn 583f
– coextrusion 126
– extension at break of monofils 315
– extrusion blow molding polymers 367
– fiber spinning grade 566
– film, biaxial stretching 261f
– film waste, compounding 675f
– flat films, transparency 164f
– foam extrusion resins 431f
– foamed semifinished product 443f
– morphology 258f
– morphology, effect of cooling rate 152
– nucleation 261
– offset spider mandrel dies for 398f
– screws for flat film extrusion 167
– staple fibers 583f, 588
– tensile strength of monofils 315
– two-stage drawing lines 279
– wood flour mixtures 44f
PPO (modified) extrusion blow molding grade 369

precooling in calibration 67
predrawing 597
predrying of PA and PET waste 677
pregassing 446
pre-haul-off unit 106f, 112, 116, 189
preheat extruder 526
preheating
– in extrusion welding 639
–, halogen lamp, extrusion welding 637, 639
– in lamination 469
– of pellet 173, 249
premixing 37f, 435
– raw materials, foam extrusion 460
preorientation in filaments 587, 590f
preorientation of melts 283
preplastication 225
pressure 119, 320
pressure anisotropy coefficients, feed bushes 383f
pressure-coating process for cable sheathing 529f
–, control of 320
pressure conveying 649f
– – systems, level valves 650
– – –, pipe junctions 650
– – –, pipe loops 650
– – –, ring-main system 650
– – –, separation devices 650f
–, increase by contamination of filter 581f
–, levels in tubular film dies 111
–, measurement in adaptor 320
–, –, tubular film coextrusion 141
pressure pipe, socket on 75f
pressure profile
– –, extruder barrel (incl. grooved) 389
– –, grooved extruder 383
– –, roll gap 205, 207
–, role in ram extrusion 620ff
pressure rolls 194, 336, 338ff
– rolls, coatings on 338f
–, speed control by, elastomer extrusion 553
Pressurized Liquid Salt Cont. Vulcanizing (PLCV) 506
pressurized water equipment 187
– water temperature control 462f, 525
prestretching 309, 312
pretreatment of surfaces 168, 184, 342f
–, blown films 109 115
– in extrusion coating 333, 340
preweld tongs 425
prewelding 425
prewelding of parisons in blow molding 425
primer 334
–, application systems, extrusion coating 334
– coating 333, 342f, 345f
– – units 341ff, 356f
printability 331
printing of foamed PVC profiles 487
probe roll 121

process computer 192, 249
- control, film-tape lines 304f
- - systems 121, 327, 361
- - -, flat film extrusion 191
- - -, monofil lines 327f
- data gathering 362
- models 192
- monitoring 327f
- - on film-tape lines 304f
processing aids for PVC 472f
- range in foaming of thermoplastics 431f
- temperatures in coextrusion of blown films 128
product-code controlled target values 557
production costs, reduction by coextrusion 131
- speeds on monofil lines 307
- waste, clean, recycling of 673f
profile dies 83f, 397f
- -, components of 85
- extrusion heads for elastomers 528
- haul-off 83f
- quotient 394f, 396
- relaxation, heat radiators for 88
profiled tubes 80
profiles
-, applications 81ff
-, calibration 83, 85
-, coextrusion of foamed 450, 481ff
-, cooling lines 83, 552
-, cross-section, measurement procedure 555
-, cutting equipment 84
-, dimensions, control 90
-, dimensions, measuring methods 555
-, extrusion of 80ff, 90, 482
-, foamed 442ff, 479, 482
-, -, with higher density outer layer 480
-, -, of low density 451
-, -, with solid outer layer 479f
-, haul-off equipment 83
-, metal-sheathed, rubber covering 531
-, PTFE and UHMWPE, dimensions of 614
-, steel armored 529
-, twin-screw extruders for 59
-, winders for 83
Profitmaster system 249
programmable logic controller (PLC) 361
protective films for sheet 240
protector dies – see tread heads 532
PS
- and copolymers, extrusion blow molding types 368
- foam extrusion types 431f
- foamed-film blowing 118
- foamed semifinished product 442f
- pellet for foaming 458
PTFE
- powder metering 615f
- ram extrusion 614ff

PTFE sintering 614f
- solid rods, internal stresses 628
- tapes 539
pumps, gear, five-wheel 573f
-, -, three-wheel 573 (see also gear pumps)
- for physical blowing agents 465
punch grid 167
punching foamed semifinished products 487
PUR for extrusion blow molding 369
PVC
- calendering compounds 217
- compounding 37, 473
- compounding lines 40ff
- dry-blend for foam extrusion 472
-, extrusion blow molding compounds 368
- film blowing 119
- films, foamed, extrusion of 483
- foam extrusion 431f, 471f
- - -, compounding for 473
- - -, cutting to length 487
- - - formulations 472
- - -, haul-off equipment 486
- foamed semifinished products 442ff, 475f, 486
- K values for foaming 432, 472
-, pigments for 473
- processing aids 473
- profiles, foamed, printing of 487
PVC, rigid
-, -, dies 61, 66
-, -, flat film extrusion 187
- - foam 475
-, -, foamed semi-product, blanking and stacking 487
-, - - pipe, guide formulation 476
-, - - profiles, guide formulation 476
-, - - sheet (free-foamed), guide formulation 476
-, -, lines for pipe manufacture 58
-, -, profiles, applications 81
-, -, screw design for extrusion blow molding 376
-, -, woodlike foamed semifinished product 471, 473
-, -, - - profiles 487
PVC, soft
-, -, foam 475
-, -, profiles for automobile industry 81
-, spider-mandrel dies for blow molding 398f
-, twin-screw extruders for 57
-, VC content of 368
PVC/ABS blends for calendering 217
PWDS system 421f
pyrolysis 665

quality
- control in extrusion welding 643f
- improvement by coextrusion 134
- of sheet and flat films, effects on 245
quench-air speed 577f
quench-air temperature 577

quenching 562, 601
- baths for monofils 322 f
- in filament extrusion 577 f

radial play between screw and barrel 376
radiation crosslinking 448 f, 491, 501 ff
radiation dose 508
radiation heating 502 ff
radiation-crosslinking tubes, water-cooled 502
radical crosslinking, reaction steps 491
radical transfer reactions 492
radiometric thickness measurement 555
ram extruder 513, 615
- - with shear gap 506
ram extrusion
- - brakes 621 f
- - dies 616 ff, 625 f
- -, -, matt surface 622
- -, hollow profiles 628 ff
- -, mandrel heating and cooling 619 f
- - mandrel, movable 633 f
- - materials 623 f
- -, metering of PTFE powder 614 f
- -, multiple dies 617
- - of pipes 628 ff
- -, powder densification, effect of pressure 630
- - pressure, influences on 620 ff
- -, processing temperatures 623
- -, PTFE, film formation during 622
- -, -, pressure for 624
- -, -, rod tools, design 626
- -, -, sintering time, minimum 627, 629
- -, PTFE types 623 ff
- -, PTFE and UHMWPE 614 ff
- -, ram 617, 620 f
- -, ram insertion depth, effect on pressure 630
- -, sintering 620 f
- - six-fold pipe die 617
- -, size of die, effect on pressure 621
- -, skin formation 618
- -, solid profiles 625 ff
- - speed, effect on pressure 622
- - speed, vs. pressure for PTFE pipe 628 f
- - stroke rate 630
- -, tablet formation 620 f, 624, 628
- -, temperature control, effect on pressure 623, 630
- -, UHMWPE types for 614, 625
ram feed 527
raw compound, strip cooling 551
raw material costs in coextrusion 131, 140
- - feed in foam extrusion 435
raw materials
- - for calendering 217
- - for extrusion coating 332
- - for film tapes 284
- - for ram extrusion 623 ff
- - for sheet extrusion 217

reaction heat in crosslinking 494 f
rectangular spinning dies 319
recycling 127
-, compounding and extrusion line for 683
- and compounding processes 676 ff
- contaminated wastes 675
-, crammers for 694 f
-, cutters in extruder feed section for 685 f
-, degree of contamination 670
-, effect on properties of PA6 678
-, energy balance 705
-, energy flow diagram 703
-, energy usage 702 f
- extruder 174
- -, grooves for barrel feed zone 683
- -, wear and wear prevention 689 f
- film scrap to blown film 678 ff
-, filtration of the melt 694 f
-, melt process 663 ff
-, pelletizing for 699 ff
- plastics waste 172 f, 662 ff, 682 f
- - -, preceded by size reduction 680 f
- - -, with size reduction during 690
- - -, without size reduction 681 f
- - -, sorted, production 673 ff
-, screws with conical feed zone for 683 f
-, screws with cylindrical feed zone for 683 f
-, short-screw extruder for 685
-, single-screw extruder for 684
-, single-screw extruder, large end-dia. for 683 f
-, - -, m/c data and throughputs 687
-, - - with modified screw sections for 683 f
-, solution viscosity, change on 678
-, twin-screw extruder for 688
-, - -, co-rotating for 682, 686
-, - -, counter-rotating for 682, 686
-, - -, m/c data and throughputs 688
-, two-stage extruder for 684
reduction gear-boxes 373, 461
reel change 188 f, 340
reel, filament wind-up 326
refuse dumps, waterproofing of 634
refuse-sack films 96, 116, 117
regrind 173
-, effect of filling on properties 676 f
- extruders with degassing 454
- layer 160
regulating screw 39
Reifenhäuser foam extrusion process 443, 475
reinforcement 27, 41, 677
reinforcing cloths 288
release agent, feed through spider (tube head) 528
repelletizing 109
repelletizing of foam waste 441 f, 459
residual draw-ratio 593
residual moisture after degassing, vs. filler content 669 f

residual monomer content
– – –, acrylonitrile 41
– – –, ethylene 29 f
– – –, styrene 41 f
resistance heating elements 375
resonator chambers for microwaves 544
restrictor bar 229 ff
– gap length in side-fed dies 396 f
– gap width in side-fed dies 396 f
– zone 272
Retruder 685 f
rheological design 400
ribbed hollow sheet 84
– – –, extrusion 247
– – –, profile cross-section 247
– – –, roll haul-off 247 f
rigid films 143
ring-gap dies 257
ring-main feed on pressure conveying lines 654
rod dies for PTFE extrusion, design 626
roll
roll adjustment, polishing stacks and calenders 219 f
roll-application of spin finishes to filaments 610 f
roll arrangements on polishing stacks 211
– – on polishing stacks for cooling 168
– axes, offsetting 211
– bearings 211
– bending 210
– – on laminators 359
roll change
– –, flying 340, 350 f, 352 f
– –, sequence on an off-wind unit 352
– – system, automatic 188 f
roll-coating process 356
roll-contact cooling 143
roll embossing process 311
– feed to extruders 673
roll gap setting 213
– – flow pattern 207 f
– – pressure profile 205, 207
– goods, waste compounding 681
– haul-off units 466
– – –, power usage 244
– stack 513, 536
roll stretching 257
roll-surface deposits 112
roll surface, flat film extrusion 166
– temperatures, flat film extrusion 164, 167
roll-train processes for making films and sheets 203
– –, film and sheet production parameters 214 f
roll wrap-round 205 f
roller bubble-collapse system 115
– conveyors for cooling 237 f
– die 535 (see also two-roll die)
– head 532 (see also two-roll die)
rolls, take-off 182
roofing films 535

rope strands 261, 273, 287, 289, 303
ropes 285, 287, 309 (see also hawsers, cables)
ropes, sheathing with elastomers 530
rotary cutter 469
– device for four-component coextr. blowing head 138
rotary-wheel heat exchanger 357
rotating cross-cutter 469
– die 110, 112
– extruder and die 110, 112
– haul-off/winder combination 109, 112
– or reversing haul-off equipment 109 f
rotation vulcanization 547
rotational speed control 268
rotocure machines 547
rotor arrangements on cutting mills 690
rubber – see also elastomer
rubber, mixes (compounds) 513 ff
rubber sheet, cooling lines 552
rubber, thermoplastic, compounding of 40 f
rubber, UHF crosslinking of 504

sack films 96 f, 103
sacking tapes 261
sacks 285, 287
safety
– devices on polishing stacks 248
– precautions with tunnel dryers 357
– regulations for diphenyl heating 318
sag, control of 554
salt-bath crosslinking 498, 506
– lines for continuous vulcanizaton 538, 541 ff
sand as filter medium 318, 574 f
sandwich films 443
– sheet 443
saw-toothed knife 108
saws with digital setting of cut length 248
SB copolymers, degassing screws for 167
scavenging finger 703 f
Schmidt clutch 235
scraper elements in melt filters 698
scratch resistance 473
screen changer 100, 229, 463, 678, 680 f, 694 ff
– –, automatic 228
– –, continuous 696 f
– –, in elastomer extrusion 535
screen sizes in processing film waste 671
screen-change devices 30, 31, 35 f, 175, 227 f, 663
– –, data for 698 f
– –, discontinuous 685 f
screw characteristics on grooved extruders 383 f
screw conveyor 536, 646
screw cooling with compressed air 375
screw displacement for ring gap variation 376
screw elements in fully hardened tool steel 689 f
screw evaporator 26, 31

Index

screw kneader 26, 31, 42, 49, 227
– –, twin screw ZSK 46f, 47
screw length for foam extrusion 461
screw play 565
screw ram accumulator 415
screw sections, double-flight 385f
screws
– with deep metering zone for PO 380
– for elastomer extrusion 525
– with flow-dividing elements 516ff
– with flow-dividing mixing zones 514
–, land-surface hardening of 174
–, materials for 100, 317, 374
– for PP flat film 167
– for recycling 683f
– with shear sections 514f
– with slotted-disk homogenization zone for PO 380
–, surface quality of 525
–, temperature control of 484, 525
– for twin-screw extruders 484
– for twin-screw spinning extruders 563f
–, wear resistance of 525
–, – – –, in recycling 689
seal 138
–, large area, against surface water 634
sealability 161f
secondary treatment in extrusion of foams 430
segment calibration 87
self-cleaning time 398, 414
semifinished products 430ff
(see also intermediate products)
– – with compact skin 474
– –, foamed 430f, 442ff, 476
– –, foamed, applications 443f
– –, foamed, thickness and density of 442
– –, quality needs in ram extrusions 624
– –, tandem single-screw extruder for 456
separation fabrics 288
separation force 191 (see also calender gap loading)
separator units on pressure conveying lines 650
series arrangement of die sections 403
setting 289, 290
setting zone 263
sewing thread 309
SFDR (static flexible die ring) 422
SFDR, center-fed die 420
shear elements 27, 111, 137, 144, 167, 174, 295, 317, 365, 383, 385f, 387ff
shear groove principle 269
shear rates in homogenising extruders 667
shear rings 385
shear screws for elastomer extrusion 514
shear torpedo 29
shear work in compounding of waste 667f
sheathing
– dies with fixed-centered die inserts 92
– –, two-layer 93

sheathing with extruded foams 465
– of flat profiles with elastomers 531
– of metal profiles with elastomers 531
– of non-circular conductors with rubbers 531
– of sector conductors 530
sheet 203
sheet calibrator 464
sheet, coextruded, applications of 223
–, –, thickness control of 220
sheet coextrusion 245f, 479
sheet cutting equipment 239
–, extruded, applications of 221f
– extrusion 203, 223f, 249
– – automation of 249
– –, haul-off power for 241f
– –, haul-off units for 239
– –, raw materials for 217f
– –, temperature profile in 218f
– –, thickness control in 220
– –, throughput rates in 242
–, foamed 443ff, 454
–, foamed PVC 471
–, line offloading of 240
–, lines for extruded 216
–, quality, effects on 245
–, single-layer 218
–, stretching of 248
–, thickness distribution in 220
–, thin, welding of 634
–, woodlike 481
sheeting, cooling lines in elastomer extrusion 550f
shell calibration 87
short-screw extruder 111
– – for recycling 685
– plasticator 685
shrink films 96, 507
– –, fine 96, 113f
– – and tubes from crosslinked PE 490
shrinkage 85, 99, 219, 220, 260, 289f, 440, 597
– cavities in weld seam 639
– factor 486
– force 99
–, hot air 597
– measurement 301
– measurement (EATP) methods for film-tapes 303
side extruder 29, 41 (see also auxiliary extr.)
side-fed dies 101f, 390f, 414
– –, coathanger type 390
– –, design of manifold for 391f
– –, knit lines in 391
– –, manifold contours in 390, 392
– –, restrictor gap lengths in 396
– –, restrictor gap widths in 396
side-strip cooling lines 552
side stuffer 694f
silane crosslinking 497, 509f
– –, active parameters 496

silane crosslinking of PE 490
– – of PE, steam treatment 509f
– –, reaction mechanism 495f
silica gel as drying agent 657
silicone oil as heat transfer agent 505
silicone rubber, peroxide cross-linking of 543
silicone rubber strands, continuous crosslinking 543
silo
– mixer 693
– for poorly flowing materials 693
– storage of chopped waste 693
similarity theory 383
simulation devices 121
simultaneous biaxial stretching 261, 274ff
single-layer crosshead die 530
single-layer films and sheets 218
single-roll roller die 534, 556
(see also single-roll dies)
single-screw
– direct gassing for foamed PE pipes 452
– – – for windable foamed webs 452
single-screw extruders
– –, adiabatic, for recycling 683
– – for blow molding 378
– – for blow molding polyolefins 378
– – for compounding 41, 50
– –, conical feed, for waste compounding 678
– – for direct gassing 451ff
– – for elastomers 520ff
– – for film tapes 293ff
– – for flat films 225
– – for HDPE tubular film 111
– – for LDPE tubular film 99
– –, operating domain of 375
– – for orientation lines, throughput 265ff
– – for polyolefins 57
– –, pressure and temperature profile of 374
– – for recycling 683f
– – for sheet extrusion 225
– –, throughputs of 265ff
– – for waste compounding 673f, 678
single-screw extrusion lines
– – – for paper-coated foam sheet 454
– – – for repelletizing foam waste 459
single-screw tandem extruder 448f
– – – for foam films 455
– – – for foam sheets 457
– – – for one-side melt coating 457
– – – for PS pellet with blowing agent 458
single-spider mandrel-holder die 398f
single-stage foam extrusion process 444ff
singles, extrusion lines for insulating wire- 91f
sintered metal filters 318, 574
sintering in ram extrusion 620
– time, minimum in ram extrusion of PTFE 627, 628f
– zone, minimum dwell time in 626

six-layer structures 413
size-reduction of wastes 663, 675, 680ff
– – – for recycling 690
– – –, throughput and energy usage in 690f
skin build-up in ram extrusion 618f
skin, on extruded foam 474, 478ff
slab, elastomer, cutting and feeding devices 537
sleeve calibrator 65
sleeve-disk calibration 67
sleeved (quick-change) cooling rolls 209
slide-bolt screen changer 696f
slip agent 30f, 41
slip processes in flow of HDPE 403
slit-tube die heads for elastomers 529
slitting blades 141
slitting equipment 141
slitting of film-tapes 278, 296
slot dies 143, 146, 176ff, 204, 208, 216, 227, 229f 257, 267, 273, 293, 295, 338, 340, 447, 458
– –, automatic lip adjustment on 192, 230, 249
– –, calculation of pressure drop in 230
– –, chrome-plating of 229
– – for coextrusion 178f
– – design of melt distribution in 230
– – for elastomers 531f
– – with flexible die-lips 272
– –, heating of 177 231
– –, materials for 229ff
– –, residence time in 147
– – for sheets 230f
slotted-disk mixing sections 387f
slotted disks 379, 386
smear device 101
smear head vulcanization 545f
smear heads 545f, 554
– – on continuous vulcanization lines 538
– – for crosslinking PE 507
– –, mandrel, for melt temperature control 546, 554f
smooth-roll applicators 356
smoothing bars or rolls 356
solenoid grid for metal removal 692
solid outer skin on semifinished product 474
solid-phase pressure forming 160
solid profiles 80
– –, extrusion of 625ff
solid rods 82, 89
– –, cooled die process for 89
solvent recovery lines 357
sorption isotherms for various plastics 657
sound insulation 443
– – on cutting mills 690f
specific clamp force in extrusion blow molding 424
specific power requirements 83
speed limits on twin-screw kneaders 51
spherulites 258f
spherulitic structure 164

Index

spider 528
spider lines 63, 64
spider mandrel blown-film die 59, 101 f
– – support, deformations in 400 f
– – tube die 389 f, 398 ff
– – – – for PVC 398
– – – dies, single and offset 398 f
spider-type mandrel support 413 f
spin-draw-texturing machine for carpet yarns 598 ff
spin-drawing of industrial fibers 596 ff
spin-drawing of textile fibers 595 f
spin dyes 565, 567
spin extruder 317
– – screw designs 565 f
– – modules 564
– – pressure control 565
– – throughputs 567
spin-finish jets 611 f
spin head throughputs 572
spin packs 318, 562, 574 ff
– – with sand filtration 576
– – with wire mesh 574 ff
spinneret 562 f
– blanketing with inert gas 578
spinneret holes
– –, deposits on 578
– –, diameters for filament extrusion 575
– –, surface quality of 576
– –, wear indications in 576
spinning
– beam 567
– grid 563
– heads 562, 571 f, 576
spinning line for carpet staple fibers 600
– – for industrial fibers 585
– – for PP staple fibers 588
– – for textile fibers 584
– – for three-color carpet yarns 587 ff
spinning lines, can take-up on 591
– –, computer monitoring of 613
– –, data for the extrusion section 583
– –, mass throughput rates of 583
– –, process control on 613
– –, threadline speeds on 593 f
spinning pump 318, 319, 562, 570 ff
  (see also gear pump)
– – drive 320, 573 f
– – metering accuracy 573
– – pressures 570, 573
– – throughputs 573
spiral grooves in extruder feed zones 381 f
spiral mandrel 389, 414
– – dies 63 f, 101 f, 111
split (network) fiber 261, 273
split fiber manufacture 299
splitting of film-tapes 286 f
splitting tendency 287, 289

spool – see package (of yarn)
SPPF (solid-phase pressure forming) process 160
spray cooling 66, 72
– – in elastomer extrusion 548
– ring 86
– tank 88
spreader rolls 183
stabilizers 432
stacker 455
– for foamed materials 469
stacking of foamed semifinished product 469
– platform with hydraulic lift 248
stagnation zones 409
stamping device 84
staple fiber
– – cutting 601 f
– – cutting equipment 608 f
– – drawing and crimping equipment 591
– –, PP 583
– –, PP, production of 601
– – production 591
– – texturing 600 f
start-up and shut-down process automation 53
start-up waste 167
static mixer 100, 405
steam crosslinking process 498 f
– crosslinking tubes 499 f
– curing process for crosslinked PE 498 f
– vs. radiation crosslinking, economics of 505
– treatment in silane crosslinking 509 f
steels for screws and barrels 100, 317, 374
steel-mesh belts 539
storage capacity of a winder 153 f
storage system 372
strainer 227, 527
strainer heads 527
strainer screens, exchange devices 535
strand
– haul-off procedure 46
– off-wind 326
– pellet 676
– pelletizer 46 ff, 458, 678, 699 f
– pelletizing 41, 699 f
– swelling 69, 65, 68
strapping tapes 261 287 f
strength, anisotropy of 286
stress-crack resistance 420
stress relaxation 236
stress relaxation by IR radiation 248
stretch blow molding 363
stretch-blown PET bottles, market growth rate 367
stretch films 114, 117, 136, 161
stretching – see drawing (uniaxial, biaxial)
strip and slab feed to elastomer extruders 537
strip cooling lines, elastomer extrusion 550
stroke depth, extrusion ram 630
stroke rate in ram extrusion 630

structural foam parts made by injection molding 447
stuffer box 608
– screw, horizontal, vertical 694
styrene copolymers, compounding 41 f
styrene, residual content 41 f
substrate for extrusion coating 342, 343
suction cones on monofilament lines 328
– conveying devices 648
– conveying lines, throughput with PE pellet 649
– conveying systems in extruder feeding 651
– on cutting mills 663 f
– knife 149, 181 f
– nozzles for drying monofilaments 322
– removal of film edge trim 185
suction-cup lifter 469
suction-cup stacker 239 f
suction/pressure conveying lines 648 f
Sumitomo process 504
superheated steam crosslinking 503 f
supersaturation of dissolved gases in melts 478
support air, feed 390, 399
support air, for parisons 528
support roll 340
support-roll drive 105 (see also contact winder)
surface decoration for wood-like PVC profiles 487
surface quality of dies, effect on ram pressure 622
surface weight in surface coating 337
surface winder 106 (see also contact winder)
surfboards, manufacture by blow molding 419
swelling of extrudates 60 (see also strand swelling)
swing system 423
synergy in light stabilization 472

tablet forming in ram extrusion 620 f, 622 f
tachogenerators on casting and other rolls 191
tacho-roll 553
take-off of filaments 589 ff
take-off of tubes and pipes 75, 83 f
take-up heads 601 ff
– –, fully automatic changing 597, 603
– –, high-speed 602
– –, yarn-traversing devices for 602 f
take-up machines 562, 589
take-up, POY (preoriented yarn) 591 f
take-up speeds 593 f
tandem arrangement of crosshead dies 530
tandem extruders 521
– – for foamed semiproducts, tech. data on 456
– – using single screws 448, 455 f
– extrusion coating/laminating equipment 343
– lacquering/laminating machine 348
– process for cable sheathing 93
tangential coating procedure 356
tank construction, welded 634
tank storage system for physical blowing agents 465
tape extrusion 97
tape types 286 (see also film tapes)

tapes, recycling on single-screw extruders 681 ff
target values, product-code controlled 557
tarpaulins 287
tearing elements 387
tearing sections 385, 389
Technoform profile drawing 87
  (see also vacuum calibration.)
telephone wires, extrusion lines for producing 92 f
temperature control devices
– –, calender and polishing-stack rolls 213
– –, effect on pressure in ram extrusion 623
– –, flat film extrusion 183
temperature controller 525
temperature distribution in melt channel 294
temperature measurement in flat film extruson 191
– –, contactless 120
temperature of melts 119
temperature profile in film and sheet extruson 218 f
temperature resistance 134
temperature rise during flow through a die 409
tennis racquet strings 309
tensile strength of HDPE monofils 316
tensile strength of PA6 monofils 313
– – of PET monofils 311 f
– – of PP monofils 315
– –, PP tapes 289
tension bolts 177, 192 f
tension control 340
tension-controlled drives 360
tension force controllers 340, 354
tension measurement between haul-off and calibration 90
tension measurement rolls 351, 355 f
tension reduction control 354
tenter frame 257
testing methods, coating and laminating materials 332
– –, extrusion welded seams 643
– –, leaks in containers 425 f
textile fiber 583 f
– – spin-drawing 595 f
– – spin-drawing machine 595 f
– – spinning machines 584
texturing 602
– jets 600 f
– machines 594 f
– machines for spooled POY 594
thermal equation of state, Spencer and Gilmore 409
thermal insulation from foamed materials 443
thermal shrinkage
– – of PA6 monofils 313
– – of PET monofils 313
thermocouple-comb 405
thermoformability 161 f
thermoforming films 143 f, 156, 162, 167, 172
– –, applications of 159
– –, coextruded 160
– –, haul-off equipment for 186

thermoplastic engineering polymers, compounding 41f
thermoplastic rubber 41
thermoplastics
-, compatibility of various 672
-, filled, compounding of 42f
-, foaming mechanism in 437
-, miscibility of 673
-, post-extrusion crosslinking of 487ff
-, processing range for foamed 431f
- reinforced, compounding of 45f
thermosets 41
-, compounding of 48f
-, pelletization of 49
thickness control
- - in coextrusion of blown films 142
- - in extrusion of sheets and films 220
thickness distribution in films and sheets 220
thickness equalization
- - in blown film 109, 112
- - in coextrusion of blown film 138
- - in extrusion of flat film 168, 183, 240
thickness measurement 193, 339ff, 343f, 348
- - on coextruded film 264
- - by contact roll 555
- - devices 168, 177, 183, 343, 347
- - on elastomer webs 555f
- - on flat films 193
- - by laser 555
- - by non-contact methods 120, 199
- - by pneumatic systems 555
- - by radiometric systems 555
thickness tolerance
- - control in blown film manufacture 120
- - on flat films 145
thin films, 447
- -, foamed 453, 458
thin layers by coextrusion 127
three-component piggyback extrusion heads 533
Three-DD mixer 567f
three-layer coextrusion 127, 413
- - of films 131, 136
- sheathing die 528
- tubular film dies 139f
three-stage foaming process for polyolefins 434
three-zone screws
- - for blow molding polyolefins 378
- - for polyolefins 379
- - for PVC 376
throttle valve 230
throttling device 39
throttling quotient 383
throughput-rates
-, BOPP lines 278f
-, coating extruders 338
-, granulators 668
-, grooved extruders 385

-, and haul-off speed in flat film extrusion 243
-, HDPE in grooved extruders 380
-, pellet form and packing effects on 666f, 672
-, pelletizing systems 702
-, PO film blowing extruders 379
-, PVC film blowing extruders 376f
-, recycling extruders with filters 582, 698f
- in sheet extrusion 242
-, single-screw extruders (oriented films) 265ff
-, single-screw recycling extruders 687
- in size reduction (grinding) of plastics 691
-, spinning extruders 567
-, spinning heads 572
-, spinning lines 582
-, spinning pumps 572ff
-, suction conveyor systems 648
-, suction conveyors (PE pellet) 649
-, twin-screw recycling extruders 688
thrust bearings, on elastomer extruders 523
thyristor control 360
- - of DC drives 372, 553
- - of shunt-wound DC motors 235
thyristor-fed DC drives for elastomer extruders 520
tie-layers 126f, 135ff, 160
tie molecules 260
tire industry, extrusion dies 532ff
tire treadstrip cooling lines 551
torpedo 84f, 417f, 463
torque on twin-screw kneaders 52
torsion modulus measurement 497
tow, haul-off unit 600
Transfermix 516f
transparency 134
- of PP flat films 166
transport zone 565 (see also conveying zone)
transverse shrinkage 440
transverse stretching of films 276
- - of flat films 278
traverse mechanisms 326f
tread heads 532
Troester mixing zones 514, 516
Troester shearing sections 515
tube cross-section changes 401
tube cutting 424f
tube dies (die heads) 405, 528
- - for continuous production of preforms 389f
- -, for elastomers 528
- -, offset-spider mandrel holder for 398f
tube process in cable sheathing 530
tubular film, foamed 474 (see also blown film)
tubular-ram accumulator - see annular-piston
tufting tapes 287
tumble mixing of blowing agent 475
turner-bar unit 106, 343ff, 347, 359f
turret winder 156, 183f, 267
- - with automatic roll change 240
- - with contact rolls 352f

twin blowing head 116
twin dies for foam pipe extrusion 464
twin screws
– –, conical 485
– –, co-rotating (co-rot.), meshed 30 f
– –, –, –, for foam extrusion 484 f
– –, counter-rotating (cr-rot.), non-meshed 30 f
– –, cylindrical, for foam extrusion 485
twin-screw extruders 145, 225
– – with barrel degassing 226
– – with conical screws 484 f
– –, co-rot, meshed 29 ff, 37 f, 41 f, 46 f, 50
– –, –, for recycling 682, 686
– –, –, for shearing 667 f
– –, –, for waste compounding 678 f, 681
– –, cr-rot., meshed 32, 37, 47, 50, 57, 59, 483 ff
– –, –, non-meshed 30, 49
– –, –, for recycling 682, 686, 691
– –, –, for waste compounding 681
– – with direct gassing 483
– – for extrusion blow molding 371
– –, feeding with 46 f
– –, kneading elements for 686
– – with long discharge screws 689
– –, mixing/homogenization sections of 686 ff
– – for pipes 59
– – for profiles 59
– – for PVC 57
– –, recycling 676
– –, –, sizes and throughputs of 688
twin-screw kneaders, co-rotating 44 ff, 47
– –, co-rotating, meshed 32 ff, 48 f, 227
– –, counter-rotating, meshed 34, 48
– –, –, nonmeshed 32 ff
– –, speed limits of 51
– –, torques of 52
twin-screw/single-screw tandem unit 449
twin-spool haul-off 93
two-chamber filter 267, 271, 582
two-chamber system in quench baths 321 f
two-component extrusion heads 529 f
two-component monofilaments 309
two-flight screw sections 385 f
two-layer blowing heads 139
– dies for sheathing 93
– films 136
– films with inter-layer bonding agent 136
– process 412 f
– tube extrusion heads 528 f
two-roll extrusion dies 534 ff, 553 f
two-roll haul-off units 241
two-stage cooling rings 118
two-stage drawing – see also drawing
– – lines for PP films 279
– – of monofils 310 ff
– – process 261, 261 ff, 274 f, 276 f
– – process, factors affecting 264

two-stage drawing, process steps for 264
– – process, temperature profile with PP 265
two-stage extruder for recycling 684
two-stage process for extrusion of foam 445 ff
two-station blow molding machines 427

UHF cross-linking of PE 504
– cross-linking of rubber 504
– field, crosslinking in 504
– units for continuous vulcanization 538
– vulcanization 544 ff
UHMWPE
–, ram extrusion of 614
–, ram extrusion, types for 625
ultrasonic thickness measurement 76
underwater pelletizing 30, 31, 33, 37, 36, 41,
– – equipment 36
uniaxial drawing 274, 285 ff
– – line, multilayer film 281
– – lines 267
– drawn films 261, 280
– stretching – see uniaxial drawing
upholstery industry 443
UV absorber 472 f
UV light crosslinking 507 f

V4A steel for extrusion pipes 618 f
vacuum
vacuum block calibration 88
vacuum chamber 104
– conveying – see also suction conveying
– – equipment 460
vacuum degassing 29, 44, 46 f, 538
– – with double-hopper 225
– –, extruder barrel for 225
– extruder 538, 541 (see also degassing extruder)
– hopper 173
– short calibration 87 f, 89
– spray bath 486
– – tank 88, 486
– tank 88
– tank calibration 65, 84 f, 89, 486
– tumble dryer 317
– water tank 486
– zone 519 f
Vapotherm godets 609 f
VC content in PVC and containers 368 f
VCV (vertical type continuous vulcanization) 499
vee-belt drive 372, 461
vertical crammer 694 ff
vertical CV unit for cable manufacture 497 f
vertical extruder 372
vibrating tray 615
vibrator-chute hopper with fill-height control 536
vibratory channel 46
virgin/rework HM-PE mixtures 386

viscosity
–, apparent 146
–, control of 356
–, solution, change on recycling 678
volume metering 435f, 654f
– – device 654
volume shrinkage on crystallization 65
vulcanization, continuous
–, –, foamed products 541
–, –, foamed rubber profiles 542
–, –, lines 537ff
–, –, with microwaves 544
–, –, on rotating drum 547
–, –, salt-bath lines for 538, 541f
–, –, with smear heads 546f
–, –, UHF equipment for 538, 544f

wallpapers 288
wall-thickness control
– – by elastically deformable die ring 421f
– – in extrusion blow molding 419f
– – for a large (25 l) container 420
– – systems 365, 371f
– distribution in containers 420
– profile, axial and perimeter 422
– programming device 421f
warp-beam films 261, 282
washing equipment 663f
– – for contaminated waste 675f
washing mills for dirty waste 691
waste
– chopped, silo storage and mixing of 693
waste compounding 662ff
– –, compatibility of plastics in 672f
– –, engineering plastics 676f
– –, extruder feed for 666f, 677f
– – in foam product manufacture 441
– – lines for foamed materials 441f
– –, PA, PET, fiber and film 677f
– –, PP film 676
– –, roll goods 681f
– –, shear work in 667
– –, single-screw extruders for 673f, 677f
– –, tangled mixed 681
– –, twin-screw extruders for 677f, 681
– contaminated 675f
–, –, recycling processes for 675f
–, –, washing lines 675
–, energy usage in extrusion of 702ff
–, film 671
–, pelletizing of 671f
–, recycling for secondary uses 673f
–, refeed on oriented film lines 267
–, reground 655ff
– removal in extrusion blow molding 427
– – station 427

waste, as roll goods 674
–, size reduction of 663, 665, 675
–, uses for 665f
waste-water pipe-socket 75
water bath process 295
– – – quenching in flat film extrusion 198
water-cooling in film-blowing 119, 280
– on extruders 225
–, heat transfer calculations 68ff
water removal sponge 322
water-ring pelletizer 37, 48, 701f
– pelletizing 44, 46
water treeing in cables 497
water uptake of PA flat films 166
water-vapor permeability 134, 163
waxes 332
wear 38, 43, 45, 47, 51, 53, 111, 174, 225, 374, 384, 385, 388
– in feed zone 523
– and prevention in recycling extruders 689f
– in spinneret holes 576
weaving tapes 287, 289, 295, 302
web-cooling lines 552
web division 466
– edge control 108f, 112, 115, 139, 343f, 359f
– edge thickening 338
– path control lines 359
– speeds in extrusion coating 337
– tension control 106f, 354f
– tension measuring equipment 183
– width in extrusion coating 337
wedge clamp, hydraulic 532
weight measurement, continuous 556
– metering 268, 435f, 654f
weld lines 418
– –, parisons 417
– –, with side-fed dies 391
– –, with spider-mandrel dies 398
weld seams
– –, extrusion-welded 640
– –, –, chemical resistance 642f
– –, –, finishing 640f
– –, preparation 638
– –, quality 423f
– –, testing 643f
weldability 134f, 140
–, improvement by coextrusion 130
welding
– device for connecting conductor ends 92
– factors in extrusion welding 641f
– of films, webs, and thin sheets 634
– head, movable 634
– machines, mobile 635f
– materials, for extrusion welding 639
– rod – see filler rod
– shoe for extrusion welding 637f
welding together of divided melt streams 409

wet laminating 356
wide films 113, 115
width measurement, optical methods of 556
winder 91 f, 106, 467
–, Alquist drive 468
–, combination contact/central 107 f, 114
–, double 91 f, 107
–, turret 468
winding 290
– characteristics 154
– characteristics, programming of 190
– cores 190, 352
– equipment
– – for film tapes 292 f, 300
– – for flat films 188, 240
– – for foamed films 462
– – for monofils 326
– factor 154
– of filaments 589 ff
– of flat films 153
– forces, control of 327
– of oriented films 282
– of profiles 84
–, speed of fiber take-off during 591 ff
– tension characteristic 354
window profiles 82 f
– –, extrusion lines for 58
– –, foamed 471
wind-up units 340 ff, 352 f
wire cloth for melt filtration 581
– covering dies 397
– drawing/annealing machine 93

– guiding device 92
– mesh as melt filters 574 f
– preheating unit 91 f
wire-mesh tumbler cooling 551
wires, drawn plastics 309
wires, insulation extrusion lines for 91 f
wooden slat layflat system 112
woodlike PVC semi-finished products 471, 739
– foamed PVC profiles 487
– sheets, extrusion of 481
Woodlite process 451, 481
working range, monofil lines 329
– –, single-screw extruders 375
worn parts, exchangeable 520
wound film, reel capacities 153 f
wrap angle of monofils on godets 323
wrap angle on polishing rolls 235 f

XPE cable, extrusion lines for 92
–, foamable, compounding for 41
X-ray film 260

yarn traverse 300, 602 ff
yarns, crimping of 595, 606 f
–, drawing of 595 ff
–, high-modulus 302
–, industrial 583, 585, 596 ff
–, textile 583, 595 f

zigzag cutter 537
zip-fasteners 309

Hanser
AUTHORITATIVE
TECHNICAL BOOKS
BY EXPERT
AUTHORS

Plastics Technology and Polymer Science

Rauwendaal, C.
Polymer Extrusion
By Dr. C. Rauwendaal, Raychem Corporation, Menlo Park, USA.
568 pages, 391 figures, 31 tables. 1986. SPE-sponsored. Hardcover.
ISBN 2-446-14196-0

The extruder is indisputably the most important piece of machinery in the polymer processing industry. This explains why there are some comprehensive books on polymer extrusion. In order not to duplicate other texts on extrusion, this book emphasizes new trends and developments in extrusion machinery.

The extrusion theory is covered as completely as possible; however, the mathematical complexity has been kept to a minimum. This has been done to enhance the ease of applying theory to practical cases and to make the book accessible to a larger number of people. The emphasis of this book is on demonstrating how the extrusion theory can be applied to actual extrusion problems, such as screw design, die design, troubleshooting, etc.

An important reason for this book is to report on the many new developments that have occurred in the field of extrusion since about 1970. Several of these have impacted and will impact the state of the art in extrusion quite dramatically. This is the most up-to-date and comprehensive text on polymer extrusion. It covers machinery, theory, and practice. This book is intended primarily for practicing polymer process engineers and chemists, but it also serves as an excellent text for those who have no previous experience in extrusion.

Michaeli, W.
Extrusion Dies
Design and Engineering Computations. By Dr. W. Michaeli, Rubber Flooring Materials, Foams, and Synthetic Leather Department of Carl Freudenberg, Weinheim FRG.
462 pages, 205 figures, 12 tables, 1984. SPE-sponsored. Hardcover.
ISBN 3-446-13667-3

This book sets guidelines for the design and the computational engineering analysis of extrusion dies for polymers. The engineer as well as the student learns how to overcome the unsatisfactory empirical engineering methods for designing extrusion dies by using computational approaches which have been tested in practice.
Various types of extrusion dies, guidelines for their design, approaches to computational engineering analysis, as well as their limitations are presented. This book introduces todays computational approaches developed by the industry and institutes, and presents the resulting mathematical model of the transport phenomena (flow and heat transfer) in the extrusion dies.

Carl Hanser Verlag
P.O.Box 86 04 20
D-8000 Munich 86
Fed. Rep. of Germany